19th International Union of Materials Research Societies - International Conference in Asia (IUMRS-ICA 2018)

Innovation of Advanced Material for Green Technology and Sustainable Development

IOP Conference Series: Materials Science and Engineering
Volume 553

Bali, Indonesia
30 October - 2 November 2018

Editor:

Evvy Kartini

ISBN: 978-1-7138-0895-4
ISSN: 1757-8981

Printed from e-media with permission by:

Curran Associates, Inc.
57 Morehouse Lane
Red Hook, NY 12571

Some format issues inherent in the e-media version may also appear in this print version.

Additional copies of this publication are available from:

Curran Associates, Inc.
57 Morehouse Lane
Red Hook, NY 12571 USA
Phone: 845-758-0400
Fax: 845-758-2633
Email: curran@proceedings.com
Web: www.proceedings.com

TABLE OF CONTENTS

ADVANCED FUNCTIONAL AND STRUCTURAL MATERIALS

COMPUTATIONAL MATERIALS, MODELING AND SIMULATION

MATERIALS PROCESSING AND CHARACTERIZATION

NANO SCIENCE AND TECHNOLOGY, NUCLEAR SCIENCE AND TECHNOLOGY APPLICATIONS

POLYMER-, BIO- AND SOFT MATERIALS

ENERGY AND ENVIRONMENT MATERIALS

OTHER RELATED TOPICS

IOP Conf. Series: Materials Science and Engineering **553** (2019) 011001 doi:10.1088/1757-899X/553/1/011001

19th International Union of Materials Research Societies – International Conference in Asia (IUMRS-ICA) 2018

IUMRS-ICA 2018

19th International Union of Materials Research Societies International Conference in Asia

ANVAYA HOTEL AND RESORT,
BALI-INDONESIA
30 OCTOBER – 2 NOVEMBER 2018

ORGANIZED BY

MRS-INA
MATERIAL RESEARCH SOCIETY OF INDONESIA

Accepted, peer reviewed papers from the IUMRS-ICA 2018

Edited by
Evvy Kartini *et.al*
Indonesia

2018

IOP Conf. Series: Materials Science and Engineering **553** (2019) 011001 doi:10.1088/1757-899X/553/1/011001

PREFACE

The 19th **International Union of Materials Research Societies – International Conference in Asia (IUMRS-ICA) 2018** was held in Bali on **30 October – 2 November 2018** . The main theme of IUMRS-ICA 2018 is "***Innovation of Advanced Material for Green Technology and Sustainable Development".*** The conference is organized by Materials Research Society-Indonesia (MRS-INA) with the aim to provide international forum for presenting scientific results in the field of materials science by scientists, academicians, industries and government from national, as well as regional and international.

The pupose of the IUMRS-ICA 2018 is to gather Indonesian Materials Researchers, and to promote international collaboration and cooperation among scientists and academia from regional and developed countries. This should be accomplished by presence of invited world-class speakers for the scientific program. In this event, the fundamentals, innovations and industrial applications of Materials Sciences and Technology will be discussed.

The Committee received over **200 scientific articles** showing a diverse and stimulating program, from more than 40 national and international institutions, covering 20 countries (UK, Turkey, Taiwan, Czechia, Pakistan, India, Austria, Germany, Ethiopia, Egypt, USA, Malaysia, Singapore, Thailand, Japan, Bangladesh, China, Australia, Korea and Indonesia). The publication committees were responsible for arranging approximately **150 full peer review** of the manuscripts that will appear in several media of international publications, such as Progress in Natural Science: Materials International, IONICS, Journal of Atom Indonesia, Indonesian Journal of Materials Sciences as well as the IOP Conference Proceedings. After selection and peer review, there are 66 articles to be published in IOP Materials and Engineering.

There are several International Symposia covering subjects in material research, namely (1) Energy and Environment Materials; (2) Advanced Functional and Structural Materials; (3) Polymer-, Bio- and Soft Materials; (4) Nano Science and Technology; (5) Computational Materials, Modeling and Simulation; (6) Materials Processing and Characterization; (7) Nuclear Science and Technology Applications; and (8) Other Related Topics. I truly acknowledged all the publication team and reviewers who have been working very hard on reviewing all those submitted articles.

The conference was remarked by the Presidential Advisory Council of the Republik of Indonesia, Professor Sri Adiningsih, and a Plenary Speech from the President of International Union of Materials Research Societies (IUMRS), Professor Soo Wohn Lee, and other distinguish international speakers.

IUMRS-ICA 2018
Prof. Dr. Evvy Kartini
President of MRS-INA

Prof. Dr. Anne Zulfia
Chair of IUMRS-ICA 2018

ORGANIZATION

IOP Conf. Series: Materials Science and Engineering **553** (2019) 011001 doi:10.1088/1757-899X/553/1/011001

IOP Conf. Series: Materials Science and Engineering **553** (2019) 011001 doi:10.1088/1757-899X/553/1/011001

29. Prof. Dr. Shahidan Radiman, *UKM, Malaysia*
30. Prof. Dr. Stefan Adams, *NUS, Singapore*
31. Dr. Jamie Schultz, *Director of ANSC ANSTO, Australia*

INDONESIAN ADVISORY BOARD

1. Prof. Dr. Ridwan, *Director of CSTAM BATAN*
2. Prof. Dr. Ir. Herman Yuwono, *Head of Metallurgy Departement, University of Indonesia (UI)*
3. Dr. Azwar Manaf, *University of Indonesia (UI)*
4. Dr. Bambang Sugiono, *University of Indonesia (UI)*
5. Dr. Bondan Tiara Sofyan, *Kemenhankam*
6. Dr. Agus Purwanto, *Sebelas Maret University (UNS)*
7. Prof. Dr. Khairurrijal, *Bandung Insitute of Technology (ITB)*
8. Dr. Agung Nugroho, *Bandung Insitute of Technology (ITB)*
9. Prof. Dr. Darminto, *Sepuluh Nopember Institute of Technology (ITS)*
10. Prof. Dr. Ratno Nuryadi, *The Agency For the Assessment and Application Technology (BPPT)*
11. Dr. Andika Pramono, *The Indonesian Institute of Sciences (LIPI)*
12. Dr. Muhammad Taufik, *State University of Malang (UNM)*

ORGANIZING COMMITTEE

1. Chair : Prof. Dr. Anne Zulfia (Universitas Indonesia – UI)
2. Co Chairs : Dr. Teguh Yulius Surya Panca Putra
3. Secretary : Dr.Eng. Jaka Fajar Fatriansyah, M.Sc., Wahyuaji Natottama Putra, ST., MT, Dr. Sudaryanto, Yualina Riastuti Partiwi, Rida Nurul Shelni Rofika
4. Treasury : Dr. Eni Sugiyarti, Noer'aida
5. Coordinator of Conference : Dr. Maykel Manawan
6. Coordinator of Congress : Dr. Sunit Hendrana
7. Coordinator of Workshop : Prof. Dr. Darminto
8. Coordinator of Exhibition and Poster : Yustinus Purwamargapratala, ST., M.Si.
9. Finance : Dr. Mas Ayu Elita Hafizah
10. Coordinator for Publication : Dr.Eng. Jaka Fajar Fatriansyah, M.Sc.,, Dr. Alfian
11. Coordinator for Accommodation, Transportation and Tour : Enggay Sugati, S.E., Agus Sujatno
12. Coordinator of Press Conference : Purnomo
13. Webmaster : Dr. Chairul Hudaya, Dr. Sudaryanto
14. Easychair : Dr. Nofrijon Sofyan

SCIENTIFIC COMMITTEE

IOP Conf. Series: Materials Science and Engineering **553** (2019) 011001 doi:10.1088/1757-899X/553/1/011001

IOP Conf. Series: Materials Science and Engineering **553** (2019) 011002 doi:10.1088/1757-899X/553/1/011002

Peer review statement

All papers published in this volume of *IOP Conference Series: Materials Science and Engineering* have been peer reviewed through processes administered by proceedings editors. Reviews were conducted by expert referees to the professional and scientific standards expected of a proceedings journal published by IOP Publishing.

Magnetic and Electromagnetic Wave Absorption Characteristics of $BaFe_{11}Ti_{0.5}Zn_{0.5}O_{19}$ in X Band

Dedi[1*], Novrita Idayanti[1], Tony Kristiantoro[1], Nanang Sudrajat[1], Arighi Gustinova[2], Indah Primadona[3]

[1]Research Center for Electronics and Telecommunication, Indonesian Institute of Sciences (LIPI).
[2]Departement of Metallurgical Engineering, Jenderal Achmad Yani University, Bandung, Indonesia.
[3]Research Unit for Clean Technology, Indonesian Institute of Sciences, Bandung, Indonesia.

dedi030@lipi.go.id

Abstract. Research has been made on $BaFe_{11}Ti_{0.5}Zn_{0.5}O_{19}$ compound by high energy milling (HEM) with variation of milling t = 2, 8, and 12 hours. This research was conducted to find out the influence of milling time on magnetic and electromagnetic wave absorption properties of $BaFe_{11}Ti_{0.5}Zn_{0.5}O_{19}$ compound. Materials were characterized by X-Ray Diffraction (XRD), Permagraph C, Scanning Electron Microscope (SEM), and Vector Network Analyzer (VNA). XRD patterns of bulk sample had shown the material forms a single phase $BaFe_{12}O_{19}$ (Barium hexaferrite). A SEM image of bulk sample shows hexagonal structure. A VNA spectrum obtained from 12 hours milling exhibit the highest electromagnetic wave absorption compared with 2 and 8 hours milling, with 99,19% absorption at frequency 11,7 GHz in X-band. Agglomeration occurs on the sample powder may due to electrostatic force among particles and a SEM image confirmed it. Based on that results, the addition of milling time can improve the electromagnetic wave absorption of $BaFe_{11}Ti_{0,5}Zn_{0,5}O_{19}$ powder.

Keywords :Barium hexaferrite, High energy milling, Electromagnetic wave absorption.

1. Introduction

The ferrite-based magnets are proven to be effective for electromagnetic wave absorbing materials (EWAMs) applications that work at frequencies from hundreds MHz to GHz. EWAMs can convert Radar-emitted electromagnetic waves into heat energy. As a result, the wave reflected and received by the receiver becomes very weak. The hexaferrites, notably $BaFe_{12}O_{19}$ (BaM) is able to reduce power and radiation on the Radar due to their high values of saturation magnetization (Ms) and coercivity (Hc) along with low cost, excellent chemical stability and corrosion resistance [1] .

A wide range of possible compositions of these ferrites were synthesized by various preparation techniques and different substitutions. As cation substitutions is one of the ways to modify the

IOP Conf. Series: Materials Science and Engineering **553** (2019) 012001 doi:10.1088/1757-899X/553/1/012001

properties of BaM in order to meet the requirements, especially for EWAMs. Previous researchers have modified and improved the properties of M-ferrites by replacing Ba^{2+} by Sr^{2+}, La^{2+}, Pb^{2+} ions or substituting Fe^{3+} ions by trivalent, divalent, and tetravalent cations [2–5].

Here, we reported the milling time effect of high energy milling (HEM) on magnetic properties and electromagnetic wave absorption of Barium hexaferrite doped with Ti^{4+} and Zn^{2+} element of $BaFe_{11}Ti_{0,5}Zn_{0,5}O_{19}$ compound in X-band.

2. Experimental method

$BaFe_{11}Ti_{0,5}Zn_{0,5}O_{19}$ samples were synthesized by mechanical alloying method via high energy milling (HEM). Iron oxides (Fe_2O_3), Barium carbonate ($BaCO_3$), Magnesium oxide (MgO) and Titanium oxide (TiO_2). These oxides in proper stoichiometric were commercially procured from Merck. All of oxides were mixed with dry milling in HEM machine for 2 hours and calcinated at 900°C for 3 hours. Calcinated powder were sieved up to 425 mesh and pelletized using hydraulic press machine in stainless steel dies of 12 mm diameter and further sintered at 1200°C for 3 hours. Magnetic properties of sintered tablets were measured by Permagraph C (Magnet-Physik Dr. Steingroever GmbH). Afterthat sintered tablets were pulverized with HEM machine in various milling time for 2, 8, and 12 hours, respectively. X-Ray diffraction (XRD) pattern of pulverized powder was performed with Rigaku Ultima IV diffractometer with Cu-Kα radiation. Grain morphology characterized using Scanning Electron Microscope (SEM, Hitachi SU3500]. X-Ray Energy Dispersive Spectroscopy (EDS) was also carried out. The microwave properties conducted with Vector Network Analyzer (VNA, ADVANTEST R3770). Microwave absorption properties were evaluated by reflection loss (*RL*) which can be calculated by the equation:

$$\text{Return Loss (dB)} = -20 \log|\Gamma| \tag{1}$$

$$\Gamma = 10^{\left(-\frac{\text{ReturnLoss}}{20}\right)} \tag{2}$$

$$\text{Absorption (\%)} = (1 - \Gamma) \times 100\ \% \tag{3}$$

3. Results and discussion

The XRD pattern of pulverized powder sample in Figure 1 shows a typical structure of BaM with P63/mmc space group (ICCD: 98-004-7018). All peaks of substituted BaM with milling time variation appear in 2, 8 and 12 milling.

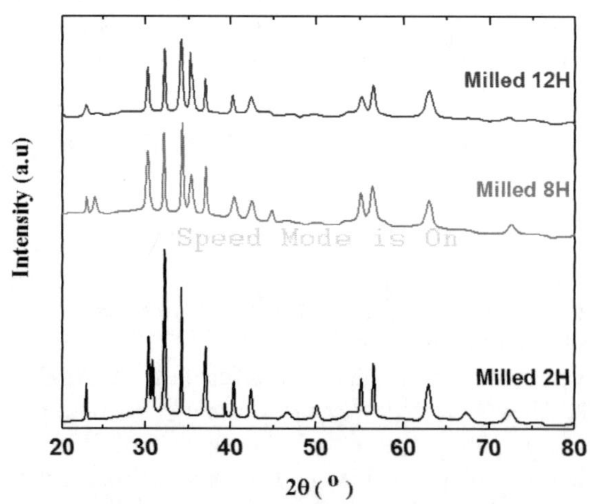

Figure 1. XRD pattern of $BaFe_{11}Ti_{0,5}Zn_{0.5}O_{19}$ milled powder.

The magnetic properties for bulk sample are shown in Figure 2. The results show that the undoped BaM has largest intrinsic coercivity (H_c) of 230,434 kA/m, and drastically H_c decreased as Fe^{3+} ions substituted by Ti and Zn ions.

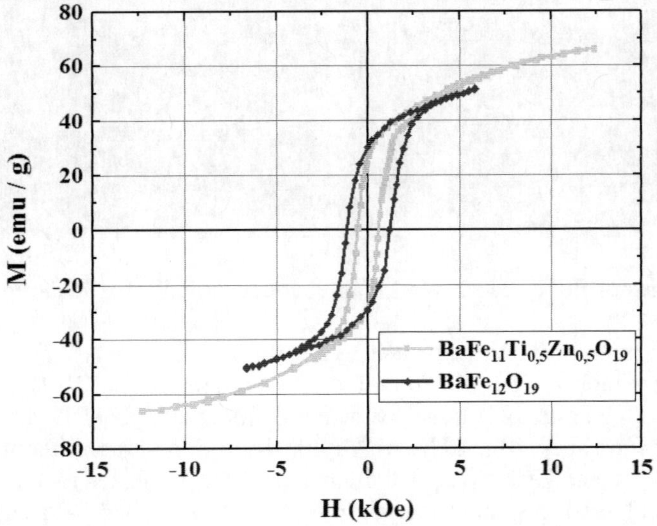

Figure 2. Hysteresis curves of $BaFe_{12}O_{19}$ and $BaFe_{11}Ti_{0,5}Zn_{0.5}O_{19}$.

Table 1 represent the magnetic properties of $BaFe_{12}O_{19}$ and $BaFe_{11}Ti_{0,5}Zn_{0.5}O_{19}$ with their bulk density. The saturated magnetization (Ms) of $BaFe_{11}Ti_{0,5}Zn_{0.5}O_{19}$ increased to 66,01 emu/g from 51,46 emu/g of $BaFe_{12}O_{19}$, and vice versa the coercivity decreased due to Fe^{3+} ion substituted by Ti-Zn ions.

Table 1. Remanence, magnetization saturation, coercivity, and bulk density.

Compound	Mr (emu/g)	Ms (emu/g)	Hc (kOe)	Bulk density (g/cm^3)
$BaFe_{12}O_{19}$	34,29	51,46	2,869	4,849
$BaFe_{11}Ti_{0,5}Zn_{0.5}O_{19}$	27,71	66,01	0,623	4,941

A SEM image of $BaFe_{11}Ti_{0,5}Zn_{0.5}O_{19}$ pellet shown in Figure 3a, some pores were exhibited on the surface of sample due to lack of holding time (3 hours) during the sintering process, so then bulk density (BD) value of sample didn't achieve to 5,20 g/cm^3 as theoretical density of Ferrite. Hexagonal structural exhibited in the sample as shown in the Figure 3a, and has random orientation.

It has been proven that $BaFe_{11}Ti_{0,5}Zn_{0,5}O_{19}$ synthesis through mechanical alloying method is able to form a hexagonal crystal structure with an incomplete angle due to the addition of Ti^{4+} and Zn^{2+} element as doping. Hexagonal crystals with perfect angles may be achieved with appropriate stoichiometric calculations as well as changes in process parameters during sintering.

Figure 3b shown the morphology of $BaFe_{11}Ti_{0,5}Zn_{0.5}O_{19}$ powder on 10.000x magnification of a SEM image, it shows the presence of agglomeration in the sample which has been pulverized for 12 hours, according to Leroch et al. [6]; agglomeration occurs due to Van der Waals bonding, coulomb force, and electrostatic forces between particles.

IOP Conf. Series: Materials Science and Engineering **553** (2019) 012001 doi:10.1088/1757-899X/553/1/012001

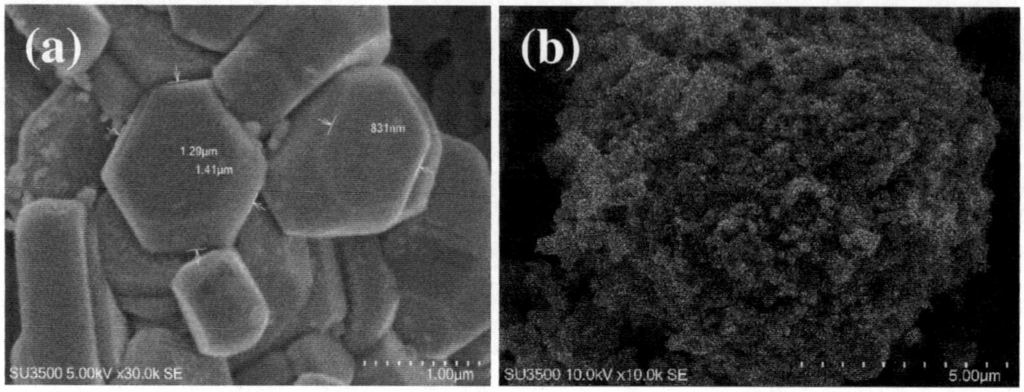

Figure 3. SEM images of $BaFe_{11}Ti_{0,5}Zn_{0.5}O_{19}$ tablet and as powder for 12 hours of milling time.

The microwave absorption properties of $BaFe_{12}O_{19}$ and $BaFe_{11}Ti_{0,5}Zn_{0.5}O_{19}$ have been measured by VNA, compared to reflection loss (RL) of $BaFe_{10}TiZnO_{19}$ from reference [7]. The RL results shown in Figure 4 and Table 2, with 2 hours pulverizing time shown fluctuating in RL values, with the highest RL was -13,9247 dB at frequency 10.8 GHz, which the absorption value of the wave was reached to 79.87%. The absorption percentage is much better than that of the pre-tested metal plate which is shown in Table 2. In 8 hours pulverizing sample, the highest RL value was -24.0395 dB at frequency11.92 GHz and calculated wave absorption achieved to 93.72% and 6.28% reflect back. For 12 hours pulverizing results shows the highest RL value upto -41.851 dB at frequency 11.7 GHz and calculated of electromagnetic wave absorption was 99,19%. This sample has a better wave absorption value than previous results that only reached -25 dB [7].

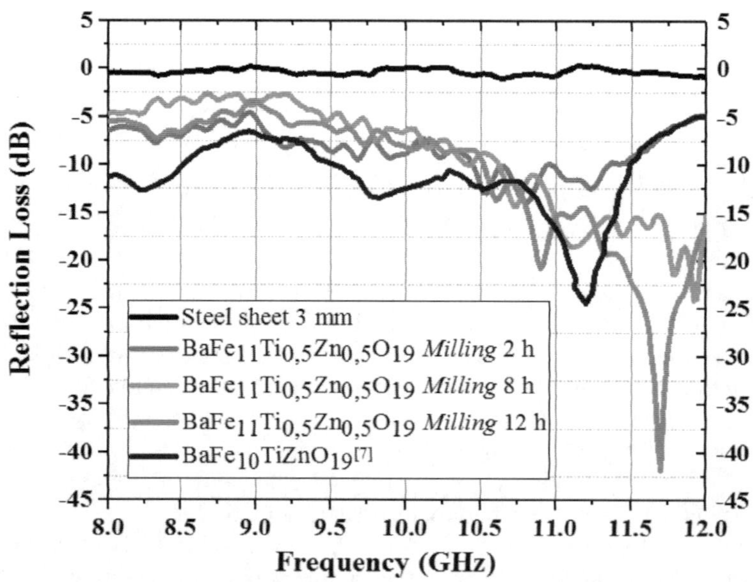

Figure 4. Reflection loss (RL) of $BaFe_{11}Ti_{0,5}Zn_{0.5}O_{19}$ and other samples as comparison.

These results increase considerably compared to $BaFe_{12}O_{19}$ sample due to doping effect of Ti and Zn elements and pulverizing $BaFe_{11}Ti_{0,5}Zn_{0.5}O_{19}$ particle by longer milling time. Also, these results suggest that the mechanical alloying process can produce a good EWAM. The EWAM is a material that not only has a large electromagnetic wave absorption value, but also has a wide band of absorption due to the EWAM can absorb in many frequency ranges [8].The more absorption values that are formed means the uptake of the EWAM is getting better. Other factors that may affect the results of VNA testing are sample density, sample surface condition, and preparation process prior to testing. Solid and flat surface samples without air gaps will provide more accurate test results. The absorption mechanism of electromagnetic waves can be explained through two overviews; the magnetic and the conductive aspects. Both of these aspects cause the penetrated electromagnetic waves into the EWAM will lose its energy. When electromagnetic waves reach the conductive material, the electric field of the wave will be absorbed and cause a Coulomb force. The existing Coulomb force on the conductive material will interact and subsequently produce an electric current and convert to heat and kinetic induction energy. The kinetic induction occurs due to the emergence of a magnetic field. The magnetic field arising from the conductive material will be reinforced by the magnetic properties of the EWAM.

Table 2. Wave absorption values of $BaFe_{11}Ti_{0,5}Zn_{0.5}O_{19}$ powder.

Sample	Reflection Loss(dB)	Frequency (GHz)	Γ	Wave absorption (%)
Steel plate 3 mm	-1,0310	10,6	0,8881	11,19
$BaFe_{11}Ti_{0,5}Zn_{0.5}O_{19}$ 2h milled powder	-13,9247	10,8	0,2013	79,87
$BaFe_{11}Ti_{0,5}Zn_{0.5}O_{19}$ 8h milled powder	-24,0395	11,92	0,0628	93,72
$BaFe_{11}Ti_{0,5}Zn_{0.5}O_{19}$ 12hmilled powder	-41,8510	11,7	0,0081	99,19

4. Conclusion

Based on research results, it has been concluded that mechanical alloying process succeeds in forming a hexagonal crystal structure as $BaFe_{12}O_{19}$. The electromagnetic wave absorption values of $BaFe_{11}Ti_{0,5}Zn_{0.5}O_{19}$ powder keep increasing along with the length of milling time, with the highest reflection loss (RL) value is 12 hours milling time which is equal to -41,851 dB at frequency 11,7 GHz with wave absorption achieved to 99.19 %.

Acknowledgment

The financial supports of this work by the DIPA project 2017 of Research Center for Electronics and Telecommunications (P2ET), Indonesian Institute of Sciences (LIPI), and INSINAS Project 2017-2018 of The Ministry of Research, Technology, & Higher Education the Republic of Indonesia, are appreciatively acknowledged.

References

[1] S. Singhal, T. Namgyal, J. Singh, K. Chandra, S. Bansal, "Acomparative study on the magnetic properties of $MFe_{12}O_{19}$ and $MAlFe_{11}O_{19}$ (M = Sr, Ba and Pb) hexaferrites with different morphologies", Ceramics International 37 (2011) 1833–1837.

[2] P.A. Marino-Castellanos, A.C. Moreno-Borges, G. Orozco-Melgar, J.A. Garcia, E. Govea-Alcaide, "Structural and magnetic study of the $Ti4^+$-doped barium hexaferrite ceramic samples: theoretical and experimental results", Physica B 406 (2011) 3130–3136.

[3] A. Gonzalez-Angeles, J. Lipka, A. Gruskova, J. Slama, V. Jancarik, V. Slugen, "Magnetic comparison of BaCa and BaSr substituted Hexaferrite powders", Journal of Physics: Conference Series 217 (2010) 1–5.

[4] T.M. Meaz, C.B. Koch, "X-ray diffraction and Mossbauer spectroscopic study of $BaCo_{0.5x}Zn_{0.5x}Ti_xFe_{12-2x}O_{19}$ (M- type hexagonal ferrite)", Egypt Journal of Sol–Gel Science and Technology 26 (2003) 197–203.

[5] A.Septiani, T.Kristianto, Dedi, N.Idayanti, Y.Taryana, "Magnetic and microwave absorbing properties of Al and Zn substituted barium hexaferrites in X-band region", 2016 International Conference on Radar, Antenna, Microwave, Electronics and Telecommunications, (ICRAMET) 149-151.

[6] S.Leroch and M. Wendland, "Influence of Capillary Bridge Formation onto the Silica Nanoparticle Interaction Studied by Grand Canonical Monte Carlo Simulations", Langmuir (2013) 29 (40), 12410-12420.

[7] Nenni, M. Delina, W. A. Adi, Y. Taryana, "Pengaruh Substitusi Ion Ti-Zn Terhadap Sifat Kemagnetan dan Sifat Penyerapan Gelombang Elektromagnetik Material Sistem $BaFe_{12-x}Ti_{x/2}Zn_{x/2}O_{19}$", Spektra: Jurnal Fisika dan Aplikasinya, Universitas Negeri Jakarta, (2017) (In Indonesia)

[8] N. Awalin, "Studi Penyerapan Gelombang Elektromagnetik Rentang X-Band Dengan Menggunakan Penyerap PANi Konduktif dan Barium M-Heksaferit Terdoping Ion Zn (0,3<X<0,9)", Undergraduate Theses, Surabaya: FMIPA – Institut Teknologi Sepuluh November, (2017) (In Indonesia)

Selected Application of Linear Composites Containing Side Emitting Optical Fibres

Dana Křemenáková[1], Jiří Militký[1], Rajesh Mishra[1*], Mohanapriya Venkataraman[1], Michal Petru[2]
[1]Technical University of Liberec, Faculty of Textile Engineering, Department of Material Engineering, Liberec 46117, Czech Republic
[2]Technical University of Liberec, Faculty of Mechanical Engineering, Department of the Design of Machine Elements and Mechanisms, Czech Republic.

rajesh.mishra@tul.cz

Abstract. The main aim of this contribution is description of side emitting plastic optical fibres basic properties and their efficient embedding into fibrous structures for creation of textile structures with active visibility in shadows. For preparation of textile structures containing side emitting plastic optical fibres the special weaving technology is used. Illumination system with light emitting diode (LED) is used as light source. Light intensity of textile structures is compared with light intensity of fibres. Incorporation of side emitting plastic optical fibres into fibrous tube provides sufficient side emission especially for plastic optical fibres with larger diameters. The corresponding technology has been developed and the possible applications were proved in real conditions. The use of side emitting POF in textile tube are still in the first stage of industrial realization and it will be necessary to optimize their composition and functionality for various targeted applications including surface effects, doping and use of special dyestuffs.

Keywords: side emitting optical fibres, safety textiles, active illumination, light intensity model, embedded optical fibres.

1. Introduction

Standard polymer optical fibre (POF) is a dielectric waveguide transferring light or infrared radiation across its axis by the mechanism of total internal reflection on the interface of two materials with different refractive indices. The main requirement for classical POF is to prevent side emission causing loss of transferred light. Description of classical POF and their application especially as sensors are described in [1]. In the side emitting plastic optical fibres (SEPOF) the light leaks out from their surface. Side emission occurs if the light incidence angle is smaller than critical angle. This effect can be obtained by the increasing of cladding refractive index, decreasing of core refractive index or by the change of incident light angle. It is also possible to use multiple micro-bending of core or cladding; additives causing reflection or fluorescence into core/cladding or to create geometric asymmetry in the core/cladding system. There are various commercial types of patented SEPOF including methods of their preparation [2, 3].

The SEPOF can be used for creation of optically active textile structures providing opportunities to highlight people and objects without the need for external light exposure. Due to the transmission loss, the intensity of radiation emitted in any direction decays exponentially along the straight fibre axis with increasing distance from the light source [4, 5].

IOP Conf. Series: Materials Science and Engineering **553** (2019) 012002 doi:10.1088/1757-899X/553/1/012002

An optical fibre integrated in weaving patterns can be described by a sequence of locally bended and straight sections. For this case the transmission loss is based on the distance between threads, thickness of threads and weaves. Local bends should be suppressed in the cases of using SEPOF for achieving active visibility at higher distances from light source. It is therefore necessary to use special embedding of SEPOF into textile structures e.g. in the form of tubes wrapped by textile yarns.

The aim of this paper is to describe optical properties of SEPOF and effective covering by fibrous structures to create safety textiles with active visibility under dark conditions. Braiding technology was used to create textile structures containing SEPOF [6]. The optical fiber end was connected with light energy source after cutting with heated wire and then by polishing with diamond powder. Light emitting diode (LED) was used as light source having illumination intensity 43.9 Wm^{-2}. A device was developed for measurement of light intensity on surface and cross section at various distances from light source. The dependence of intensity on surface and cross section with regard to the distance from light source is expressed by the exponential type model with the rate parameter attenuation factor. Second parameter defining quality of illuminating system is light intensity at the input to the fibre. The influence of the optical fibre type and diameter on the attenuation factor at the surface and cross section is quantified. Light intensity is compared between textile structures and optical fibres.

2. Experimental methods

Light rays are transmitted through optical fiber causing attenuation [7-8]. It depends on wavelength, type of fiber, it's structure (i.e. crystallinity and orientation), impurities and accompanying substances (dopants), the distance from the source, and also on outer geometry (micro-bends, macro-bends, surface damage). When bending the fibers, some part of the emitted light rays is dependent on the ratio between fiber diameters and bending radius [9-11]. Bends in end emitting optical fibers inside textile structures are actually needed to achieve their surface illumination [11]. The bending of side-emitting optical fiber leads to loss of uniformity in illumination .

Attenuation is characterized by decrease of illumination intensity in optical fibers and it is undesirable when used in telecommunications technologies, due to loss of information. Due to the huge differences in the illumination intensity the attenuation is usually expressed in decibels. Attenuation coefficient α in decibel [dB] is defined as logarithm ratio between illumination power at input P_1 and at output P_2

$$\alpha = 10\log(P_1 / P_2) \quad (1)$$

Eq. (1) implies that change in power ratio by one order is shown by change of attenuation by 10 dB. The attenuation rate α_L is defined as ratio of attenuation coefficient and distance between measuring powers P_1 a P_2.

$$\alpha_L = \alpha / L = \frac{10}{L}\log(P_1 / P_2) \quad (2)$$

The unit of the attenuation rate is dB/m. The attenuation rate is ideally constant [14], but generally it is a nonlinear function of the length L. In the case when α_L = constant, it follows equations (1) and (2), see [11]

$$P_2 = P_1\, 10^{-\alpha_L L/10} \quad (3)$$

Working fiber length Lp is the length of the side-emitting optical fiber, which can be used realistically At the end of this length illuminated power P_{Lp} is still sufficient. For this work attenuation α_{Lp} = 20 and 30 dB were selected. The working length of the optical fiber can be calculated from Eq. (4).

$$L_p = \frac{10}{\alpha_L}\log(P_1 / P_{Lp}) \quad (4)$$

IOP Conf. Series: Materials Science and Engineering **553** (2019) 012002 doi:10.1088/1757-899X/553/1/012002

Special device was developed for measurement of illumination intensity in straight condition. This on-line computer controlled device is composed of a light sensor, step driver, control unit, measuring channel and input/ output rollers. A schematic of this device is shown in Fig. 1.

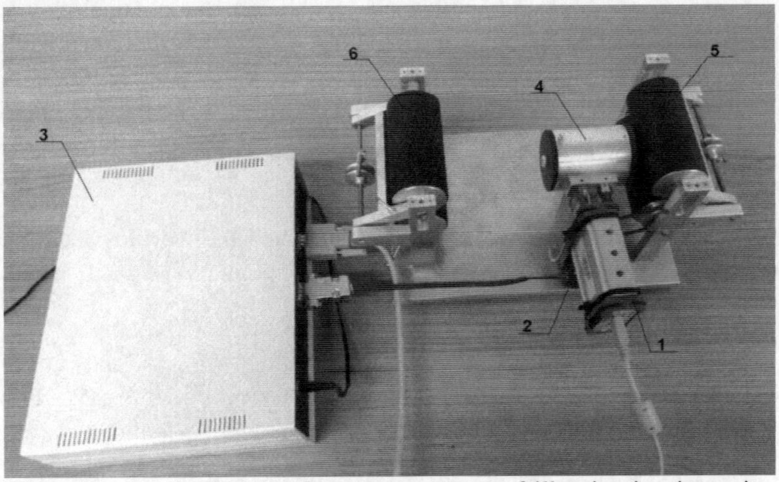

Figure 1. On-line device for measurement of illumination intensity in straight state, 1- light sensor, 2– step driver, 3– control unit, 4– measuring channel, 5– output rolls, 6– input rolls

An illumination system with light emitting diode (LED) was developed and used as light source. The fiber end was connected with light energy source after cutting with heated wire and polishing with diamond powder. Optical fibers of type “Grace standard” and “Grace flexi” with different diameters were used (table 1).

Table 1. Optical fibres specification.

Grace fibres specification	**standard**	**flexi**
Core/ Cladding material	PMMA/ Polyvinylidene fluoride	PMMA Polytetrafluorethylene
Diameter	from 0,2 - 3 mm	2 - 14 mm
Core refractive index	1.49	1.475
Cladding refractive index	1.41	1.34
numeric aperture	0.48	0.48
maximal input angle	57.4	75
mass density	1190 kg m^{-3}	1190 kg m^{-3}
wavelength	400 - 900 nm	380 – 780 nm
temperature of use	20 – 70 °C	15 – 80 °C

3. Results and Discussion

Illumination intensity is measured as function of distance from source and is shown in Figure 2 and Figure 3. Experimental illumination intensity can be used for creation of regression model (see eq. (5)) and corresponding parameters can be obtained by nonlinear or linearized regression using least squares criterion. Eq. (3) can be written as

$$P(L) = P(0)\,10^{-\alpha_L L/10} \tag{5}$$

where P(L) is illumination intensity at the distance from source L, P(0) is illumination intensity at fiber input and α_L is attenuation rate. By logarithmic transformation of eq. (5) the straight line $\log P(L) = -\alpha_L L/10 + \log P(0)$ is obtained. Slope of this straight line k can be used for calculation of mean attenuation rate α_L = -10 k and intercept q can be used for calculation of illumination intensity on the fiber input P(0) = 10^q. By use of regression, straight line parameters k and q, coefficient of determination R^2 and parameters P(0) and α_L were obtained, as shown in table 1.
For expressing working length of optical fiber $L_{p,\,an}$ illuminating power P_{Lp} must be obtained on the end of this length with sufficient value of attenuation coefficient α_{Lp}

$$P_{Lp} = P(0)\,10^{-\alpha_{Lp}/10} \tag{6}$$

Illuminating power P_{Lp} can replace $P(L)$ in Eq. (5) and the working length of optical fiber Lp can be calculated as

$$L_p = \frac{10}{\alpha_L}\log(P(0)/P_{Lp}) \tag{7}$$

Working length calculated for attenuation coefficient α_{Lp} from 10 to 20 dB is given in table 2.

Table 2. Parameters of regression model (eq. (5)).

Parameters		„Grace-standard" 0,25 mm	„Grace-flexi" 14 mm
Illumination intensity on the fibre input P(0)	[Wm^{-2}]	0.000007	0.01367
Attenuation rate α_L	[dB mm^{-1}]	0.0071	0.00254
Working length of optical fibre L_p for attenuation α_{Lp}=10 – 20 dB	[mm]	1409 and 2817	3937 and 7874

Illumination intensity model curve as standard power function (Eq.(5)) derived from assumption of constant rate of attenuation is shown in Figure 2 and Figure 3 as grey curve [1]. It was found, that at short distances from light source illumination intensity is sharply decreasing especially for higher diameter optical fibres (higher than 1 mm). Estimation of parameters P(0) and α_L is not accurate and standard power function (Eq.(5)) is not suitable for these purposes. Black piecewise solid line in Figure 2 and Figure 3 is so called LLF2 model, it is linear piecewise function consisting of two different linear sections. This model is based on the assumption that in small distances from light source there is no uniformity in side emission due to aperture and critical angle. In second part the illumination intensity is slowly decreasing with distance from source L (system is adjusted). Local slopes of LLF2 are sensitivity coefficients a_1, a_2. Corrected illumination intensity at the fibre input is $P_{cor}(0)$. LLF2 model is described by equation (it is linear regression spline with one knot)

$$LLF2 = P_{cor}(0) + a_1 L + a_2 (L - L_c)_+ \tag{8}$$

where function $(x)_+ = 0$ when x is negative and function $(x)_+ = x$ if x is positive. L_c is distance of transition between first and second phase. Parameters of LLF2 for optical fibre were found by use of modified linear regression [18] as given in table 3. Working length of optical fiber L_{pcor} for attenuation $\alpha_{Lp\,cor}$ from 10 to 20 dB was calculated by using Eq.(7) as given in table 3.

IOP Conf. Series: Materials Science and Engineering **553** (2019) 012002 doi:10.1088/1757-899X/553/1/012002

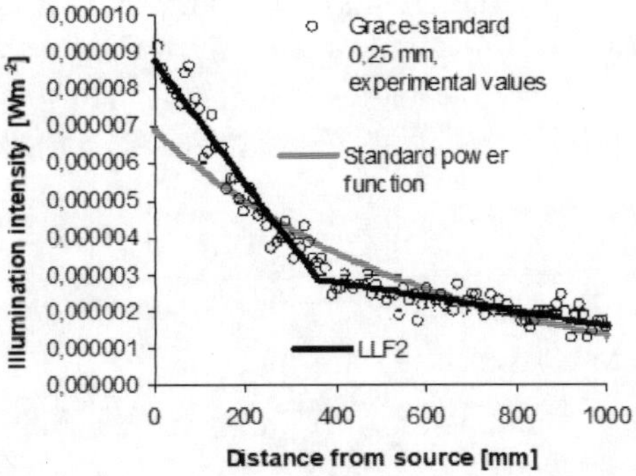

Figure 2. Illumination intensity of fiber "Grace-standard" – fiber diameter 0,25 mm.

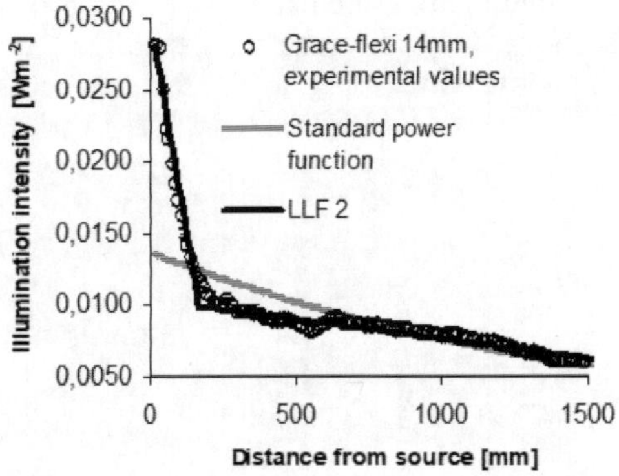

Figure 3. Illumination intensity of fiber "Grace-flexi" – fiber diameter 14 mm.

Table 3. Parameters of smoothing curves of illumination intensity calculated by using of LLF2

Parameters	**„Grace-standard" 0,25 mm**	**„Grace-flexi" 14 mm**
Minimal residual sum of squares S $[Wm^{-2}]^2$	1.372 E-11	0.00037
Corrected illumination intensity on the fibre input $P_{cor}(0)$ $[Wm^{-2}]$	0.000009	0.02993
Slope of first straight line a_1 $[Wm^{-2}\ mm^{-1}]$	-1.64 E-08	-0,000121
Slope of second straight line a_2 $[Wm^{-2}\ mm^{-1}]$	-1.96 E-09	-2.72 E-06
Distance of transition between first and second phase L_c [mm]	359.9	165.0
Working length of optical fibre L_{pcor} for attenuation $\alpha_{Lp\ cor}$ =10 and 20dB [mm]	1378 and 1791	2709 and 3699

3.1 Illumination Intensity of Woven Structures

Weaving technology was used for preparation of textile structures containing side emitting POF[6, 11]. The differences between side illumination of POF with or without textile cover are shown in Figure 4.

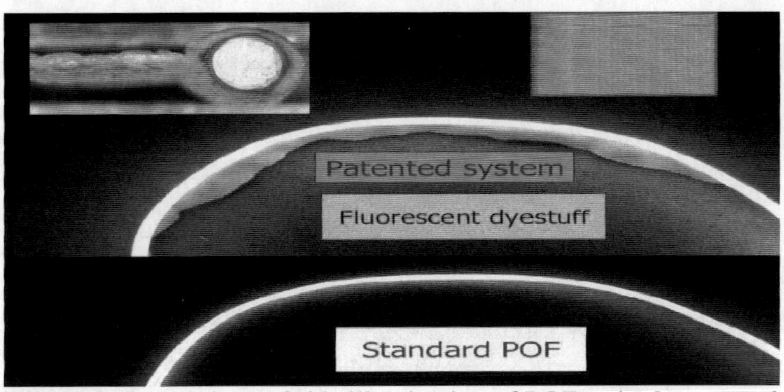

Figure 4. Comparison of side illumination of POF and POF covered by textile.

Illumination intensities as function of distance from source are shown in Figure 5 for POF and the same POF covered with textile. It is visible that POF in textile cover shows higher side emmision intensity.

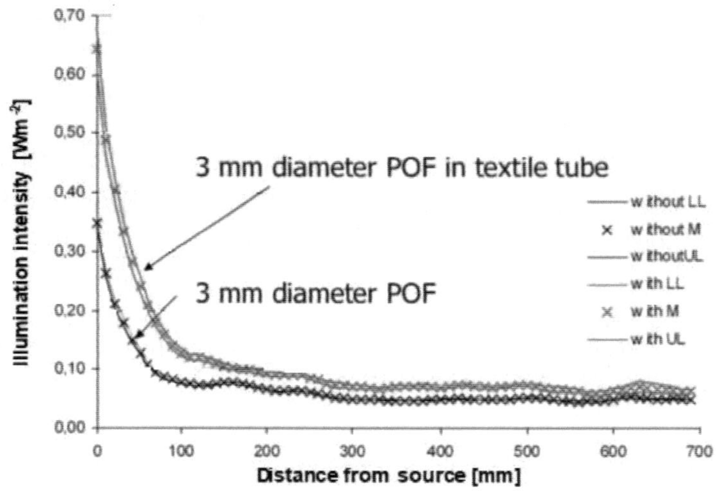

Figure 5. Side illumination intensity of POF and POF with textile cover.

The parameters of LLF2 for POF and POF in textile cover is calculated by modified regression and given in the tab. 4.

Table 4. Parameters of smoothing curves of illumination intensity calculated by using of LLF2.

Type	3mm-diameter POF	3mm-diameter POF in fabric	2mm-diameter POF	2mm-diameter POF in fabric
Critical point [Wm^{-2}]	0.123	0.203	0.115	0.202
slope1 [$Wm^{-2}mm^{-1}$]	-0.0043	-0.0071	-0.0012	-0.0033
slope2 [$Wm^{-2}mm^{-1}$]	-5.40 E-05	-0.00011	-9.96 E-05	-0.00022
intercept1 [Wm^{-2}]	0.826	1.439	0.340	0.793
intercept2 [Wm^{-2}]	0.097	0.152	0.120	0.200
Lc [mm]	172.9	182.9	192.8	192.77

3.2 Applications of Line Light System

Side emitting POF in textile cover is useful for active safety applications which provide sufficient side emission especially with larger diameters. These structures can be used for many types of applications as silhouette visualization in the dark conditions: useful for active visibility of road users, persons (jackets, school bags, handbags, backpacks), objects (cars, bikes, baby carriage, wheelchairs), animals (straps for dog, bridle for horses), security lighting systems (cars outline, identifying open doors of cars in the dark, restrictions on roads), setting of limits (parking barriers, end of carpets, stairs, etc.), emergency lighting (hospitals) - lighting in corridors, lifts, edge visualization, etc., highlight information, aesthetically complement - separate application area may be a different color effects related to the design, creation of so-called emotional fabrics characterize the mental state respectively feelings - information system. Some applications of POF with textile cover are given in Fig. 6.

(a)protective textiles (b) hospital irradiation room (c) hospital bed

Figure 6. Applications of POF with textile cover.

Optimized POF in textile tube can be also used in the emergency lighting facilities (LIHS) marking of escape routes, etc. Illumination lighting with application of POF in textile tube (SEPOF Grace, diameter 6 mm, polyester textile cover of white color) in Na Plesi hospital (CZ) for lighting of irradiation room (Figure 6 (b)) and bed (Figure 6 (c)) were successfully realized.

The energy consumption and total cost of lighting were evaluated for hospital corridor of 16 m length. For sufficient illumination the LIHS (totally 6 W) or 4 light bulbs (each 40 W, i.e. totally 160 W) or alternatively 3 fluorescent tubes (each 15 W, i.e. totally 45 W) were installed and compared. The luminosity was measured by luxmeter (see. Figure 7).

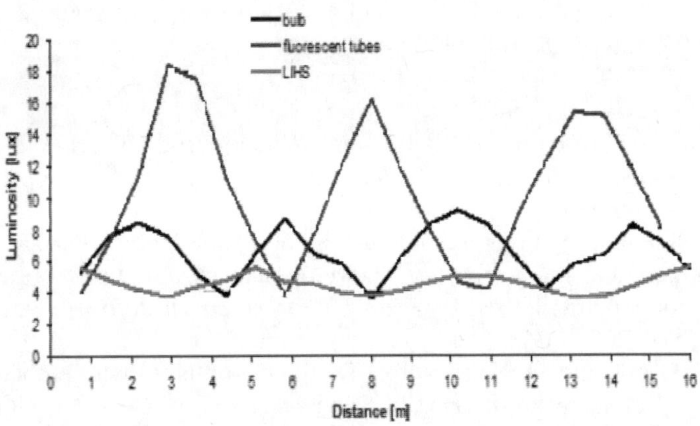

Figure 7. Comparing of luminosity of different lighting systems.

Line illumination by LIHS is evenly covered in the space (ratio between maximum and minimum luminosity is 1.5) and has minimal consumption of energy. Illumination by light bulbs and fluorescent tubes is uneven with over glare and low shine regions. Ratio between maximum and minimum luminosity is for light 2.3 and for fluorescent tubes even 4.6.

Table 5. Different illumination systems comparison.

Illumination system	Daily voltage	Daily consumption	Year consumption
Light bulb	12,8 kWh	33 CZK	12 147 CZK
Fluorescent tube	3,6 kWh	9 CZK	3 416 CZK
LIHS	1,4 kWh	4 CZK	1 329 CZK

For light illumination of 100 m hospital corridor of 8 hours daily during one year (price level 2,6 CZK per kWh) is price for different system given in table 5. It is visible that the lowest total cost is for LIHS illumination system i.e. 1329 CZK.

4. Conclusions

Lighting system [11] with (LED) was developed. Prototypes for measurement of illumination intensity in side emitting optical fibre was developed. Data treatment and evaluation was carried out. Quality of illumination system was determined as illumination intensity at fibre output. Parameters characterizing quality of optical fibres are attenuation rate and working length. Illumination intensity and mean attenuation rate are described as function of distance from the light source. There is considerable variation in the mean attenuation rate (upto 240 mm from light source for optical fibre. It was found that the inclusion of optical fibres into active safety systems will provide adequate visibility with larger fibre diameters. The adequate working length of fibres was obtained as about 2 m. Optical fibers can be used in straight form or with macro-bends for active visibility textiles. Textile covers improve the durability of the fibers.

Acknowledgement

This work was supported by the Ministry of Education, Youth and Sports of the Czech Republic and the European Union – European Structural and Investment Funds in the frames of Operational Programme Research, Development and Education – project Hybrid Materials for Hierarchical Structures (HyHi, Reg. No. CZ.02.1.01/0.0/0.0/16_019/0000843) and by project "Modular platform for autonomous chassis of specialized electric vehicles for freight and equipment transportation", Reg. No. CZ.02.1.01/0.0/0.0/16_025/0007293.

References

[1] J. Zubia, and J. Arrue, " Plastic optical fibers-an introduction to their technological processes and applications," *Optical Fiber Technology*, vol. 7, pp.101-140, 2001, doi.org/10.1006/ofte.2000.0355.

[2] J.R. Zarian et. al., Side lighting optical conduit, US Patent 5987199 (1999).

[3] A. Harlin, M. Makinen, and A.Vuorivirta, " Development of polymeric optical fiber fabrics as illumination elements and textile displays, " *AUTEX Research Journal*, vol. 3 pp. 1-8, 2003, doi.org/autexrj.org/No1-2003/0048.

[4] A. Endruweit, A. C. Long , and M.S. Johnson, " Textile Composites with embedded optical fibers, " in *8th International Conference on Textile Composites (TEXCOMP-8)*, Nottingham, UK, pp. 16-18, Oct 2006.

[5] B. CaiI, J. Xiao-li, and C. Zhang, " The characteristics and Application of Polymer Optical Fiber," *Journal of Wuhan University of Technology – Material Science*, vol. 18, No. 4. pp. 25-34, 2003, doi.org/10.1007/BF02838388.

[6] D. Křemenáková, V. Lédl, J. Militký, B. Bůbelová, and B. Meryová, Active emitting safety means, Utility model No. 24997, Industrial property office, March 2013.

[7] D. Marcuse, *Theory of Dielectric Waveguides*, Academic Press, San Diego, pp. 75-80, 1974.

[8] A. W. Snyder, and J. D. Love, *Optical Waveguide Theory*, 2nd ed., Chapman and Hall, London, 1983.

[9] V. Koncar, " Light Emitting Fabrics for Photodynamic Therapy of Skin and Nail," The International Istanbul Textile Congress, Istanbul, Turkey, May 2013.

[10] J. Wang, B. Huang, abd B. Yang, "Effect of weave structure on the side-emitting properties of polymer optical fiber jacquard fabrics," *Textile Research Journal*, vol. 83, pp. 1170-1180, 2013, doi.org/10.1177/0040517512471751.

[11] D. Křemenáková, R. Mishra, J. Militký, J. Mareš, and J., Šesták, in *Selected properties of functional materials, Part-II: Optical Properties of Textiles*. OPS Kanina 2013.

3D-Modelling and Casting Simulation for The Fabrication of Megaprostheses Implant by Vacuum Centrifugal Casting

B. Suharno[1*], S. Supriadi[2], Sutiyoko[3], D. Setyawan[1], W.M.F. Rahmatulloh[1], and G. Ramahdita[1]

[1] Department of Metallurgical and Materials Engineering, Faculty of Engineering, Universitas Indonesia, Depok, Indonesia

[2] Department of Mechanical Engineering, Faculty of Engineering, Universitas Indonesia, Depok, Indonesia

[3] Ceper Polytechnic for Manufacturing, Klaten, Indonesia

suharno@metal.ui.ac.id

Abstract. Recently, a high number of bone fracture incidents per year urges researchers in developing megaprostheses as artificial body parts. However, the complicated design and requirements for this custom-made prostheses are the key challenge. In accordance, this study focused on the preparation steps of 3D-modelling and casting simulation for the fabrication of megaprostheses implant by vacuum centrifugal casting. The reconstruction steps of 3D-modelling were started by observing the medical image data (i.e. CT-scan images). As the 3D-modelwas ready, the gating system was designed by placing two mandible bone models as the casting part with sprue and ingates where the design was respected to the gating system design for centrifugal casting. Moreover, the variation of casting part position,numbers, shapes and dimensions of the ingate, as well as the total surface area of ingates (Si) was applied while creating the gating system designsfor casting simulation. Chemically Pure Titanium (CP-Ti) with the pouring temperature and time of 1700°C and 4s, respectively, were used as the parameters for casting simulation. The solidification test results showed that the shrinkage cavities tended to decrease in volumetric scale with increasing the total surface area of the ingates, while the casting part position, number, shape, and dimension of ingate showed the difference in effectivity of the solidification of molten metal on the shrinkage cavities formation.

Keywords: 3D-modelling, Biomaterials, Bone fracture, Casting simulation, Implant, Megaprostheses

1. Introduction

Recent interest in the development technologies on the artificial body part or better known as prostheses has been motivated by the search of materials and fabrication methods with good mechanical properties as well as its biocompatibility that supports the human body. The special prostheses that being used inside the human body, called endoprostheses, has been proven to help the patient for improving the prognosis and the survival rates to amputation [1]. Moreover, the replacement of the low-quality bone with custom-made endoprostheses implant or better known as megaprostheses implant is capable of being used on the case of musculoskeletal pains, malignant tumors, and large skeletal defects [2]. This megaprostheses implant has been widely used in oncologic orthopedic surgery and is one of the leading ways for medical treatment of sarcoma patients [3].

Titanium and its alloys are considered as metallic materials for megaprostheses implant due to its good corrosion resistance and biocompatibility, low density and low modulus elasticity with good mechanical properties that suited to the human bone structure [4]. Commercially Pure Titanium (CP-Ti) and Ti-6Al-4V alloy are the most common materials used for biomedical applications, yet the Ti-6Al-4V has a possible toxic effect for permanent implant applications by releasing its aluminum and vanadium contents [5]. In the case of biocompatibility, titanium can create the thin passivation layer of titanium oxide (TiO2) on its surface that protects the materials from being corroded. Due to the high chemical activity of titanium and its alloys at high temperature, casting of the molten of titanium or its alloys is one of the precarious manufacturing processes for maintaining its composition purity from crucible and mold contaminations as well as atmospheric gases. Centrifugal casting has been developed mainly on the manufacturing of circular pattern parts on aircraft and automobile industries that provides high density and mechanical properties.

High dimensional accuracy is the main concern for implant development [5-6]. Thus we divide our process into two steps. Firstly, 3D-modelling focused on implant dimension accuracy, extracting model, and design for the casting process. Secondly, casting simulation was aimed to lower production cost by eliminating trial and error process, predicting the shrinkage cavity, optimizing the design, and maximizing cost efficiency.

This study was carried out to evaluate the 3D-model reconstruction and casting simulation of the lower jaw bone (or the mandible bone) as the megaprostheses implant model using 3D-modelling and casting simulation software. The 3-D model reconstruction steps, gating system design, and solidification simulation as well as the shrinkage cavities formation was studied for bone replacement using casted titanium with customized design. The casting simulation, the effects of the variation of casting part position, number, shape, and dimension of the ingate were also studied for solidification test on casting simulation of the megaprostheses implant model.

2. 3D-Modelling

A 3D model was prepared by modifying and converting the digital image data (DICOM files) from CT-scan images to 3D anatomical models using Osirix MD. Figure 1 showed the reconstruction steps started from the isometric view of CT-scan images where the bone fractures were on the lower jaw bone (or the mandible bone) and split into three parts such as the right mandible bone, chin bone (i.e., including the mental protuberance and mental tubercle of mandible bone), and the left mandible bone. Based on the structure of skeletal defects observation, the right mandible bone was chosen as a reference model for generating the megaprostheses implant that further modification could be able to replace the left mandible bone and chin bone. However, the actual right mandible bone showed the coarse surface contour on its surface topography. Smoothing and rendering were needed in order to make an easier modification for generating the megaprostheses implant model.

A 3D model of the megaprostheses was prepared by modifying and converting the digital image data (DICOM files) from CT-scan images to 3D anatomical models using Solid work CAD software, as shown in Figure 2. Figure 2a shows reconstructed jaws consist of the jaw bone and the jaw titanium. Figure 2.b shows the implant model of titanium which was reconstructed. The implant model is lower jaw bone from condyle to the mid of lower jawbone. The implant will be connected to the original jawbone using a connection plate embedded in the implant.

3. Casting Simulation

Casting simulation was conducted using Solidcast 7 to evaluate cooling behavior. Figure 3 shows the configuration of the gating system for the casting process. It consists of sprue as pouring target of the molten titanium and ingate to supply molten titanium from the sprue to the casting parts. Different orientation of the casting part is shown in Figure 3.b-c. To obtain casting product with low porosity, a variation of the ingate size and shape were carried out as shown in Figure 3 d-g.

IOP Conf. Series: Materials Science and Engineering **553** (2019) 012003 doi:10.1088/1757-899X/553/1/012003

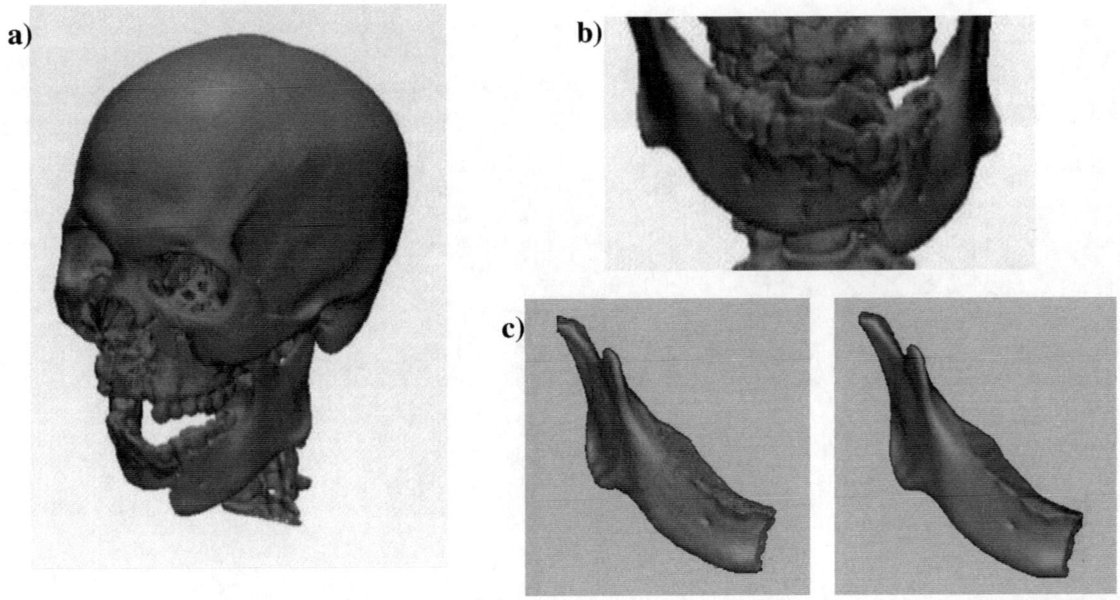

Figure 1. The reconstruction step images from the patient with the fracture through the mandible bone: a) isometric view, and b) fracture view, and c) from left to right: before (as an actual) and after smoothing and rendering the surface using 3D-modelling software.

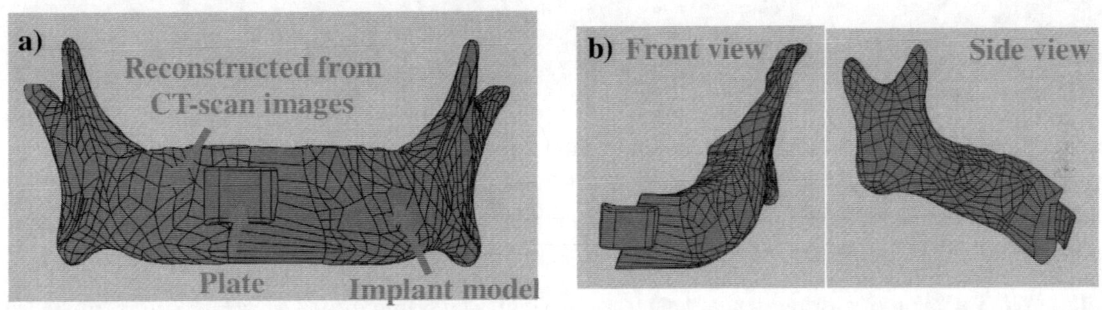

Figure 2. 3D-modelling images of the mandible bones: a) after reconstruction from medical image data with additional part of the plate for joining purpose; and b) the mandible bone model which was selected for megaprostheses implant.

Systematic of simulation variation of gating systems, includes gating dimension and geometry are shown in table 1. Design A1, A2, A3 and B1 were using three ingates under different gating size, while B2 and B3 were using four ingates; i.e. B4 , B5, C1, C2, and C3 using rectangle and circle. Besides, a various number of ingates were also studied with a maintained cross section of the ingate at 207.54 mm^2.

IOP Conf. Series: Materials Science and Engineering **553** (2019) 012003 doi:10.1088/1757-899X/553/1/012003

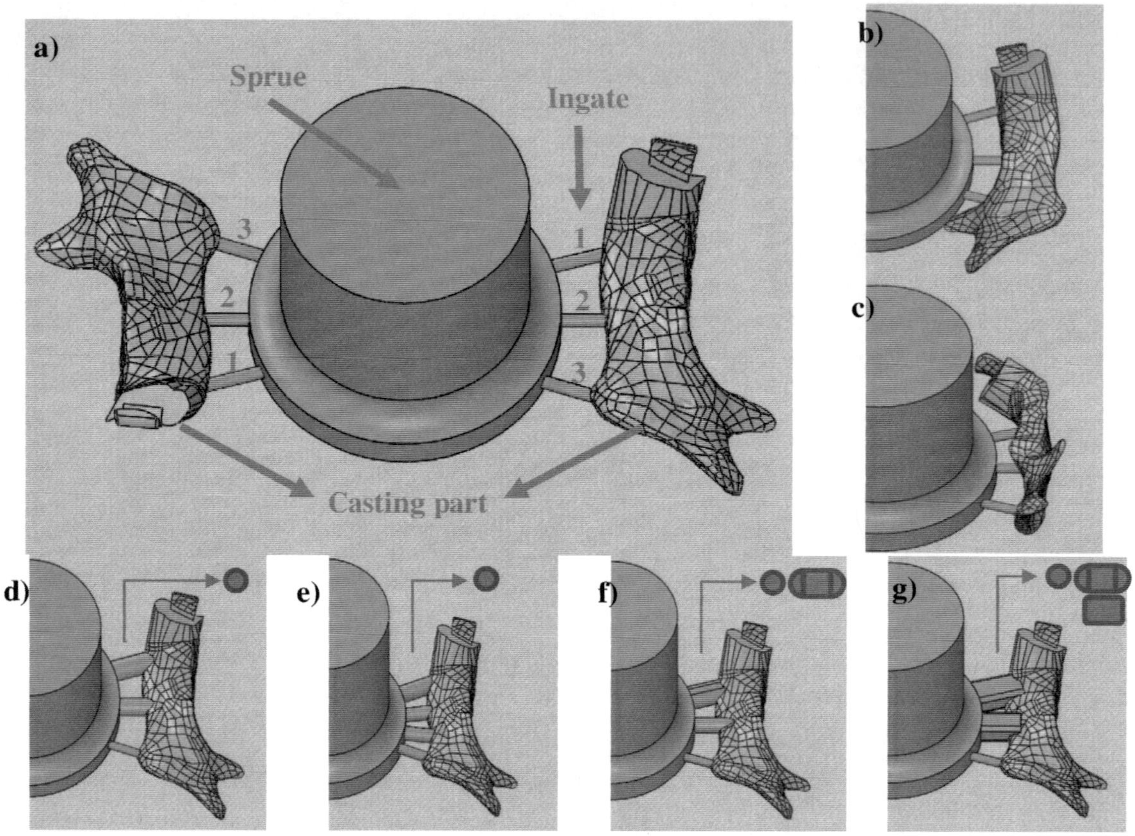

Figure 3. The variation of gating system for casting simulation of the mandible bone model: a) – c) Design A series: casting part positions with total surface area of ingates (Si) < theoretical surface area of ingates (Asm); d) – f) Design B series: number and shape of ingates with Si ≈ Asm; and g) Design C series: with Si >Asm.

Table 1. The variation of casting design based on the shapes and dimensions of ingate with their surface areas.

Design series	**Dimension of ingate (mm)**				**Surface area of ingate (mm^2)**	
	Ø1	Ø2	Ø3	Ø4	Total (Si)	Theoretical (Asm)
Design A1	4.65	4.65	4.65	-	50.97	207.54
Design A2	4.65	4.65	4.65	-	50.97	207.54
Design A3	4.65	4.65	4.65	-	50.97	207.54
Design B1	12.00	10.00	4.50	-	207.63	207.54
Design B2	10.00	8.00	8.00	6.00	207.43	207.54
Design B3	12.00	7.00	6.00	6.00	208.21	207.54
Design B4	Ø = 10, l = 10, w = 5	10.40	5.00	-	208.20	207.54
Design B5	Ø = 12, l = 12, w = 5	8.00	5.00	-	208.07	207.54
Design C1	Ø = 12, l = 12, w = 7	Ø = 12, l = 12, w = 3	5.00	-	303.93	207.54
Design C2	Ø = 12, l = 12, w = 8	Ø = 12, l = 12, w = 5	5.00	-	334.93	207.54
Design C3	l = 12, w = 16, fillet Ø = 5	l = 12, w = 14, fillet Ø = 5	5.00	-	368.93	207.54

IOP Conf. Series: Materials Science and Engineering **553** (2019) 012003 doi:10.1088/1757-899X/553/1/012003

Simulation steps using solid cast 7 are shown in Figure 4. Sprue, ingate and casting part were arranged and meshed as shown in in Figure 4a. Then it was covered with mold material (Figure 4b). The filling simulation was carried out in order to observe hot molten metal entering the cavity of the casting part. Solidification of the casting part was conducted to evaluate shrinkage that promotes porosity. Chemically Pure Titanium (CP-Ti) with the pouring temperature and time of 1700°C and 4s, respectively, was used as the parameter for casting simulation.

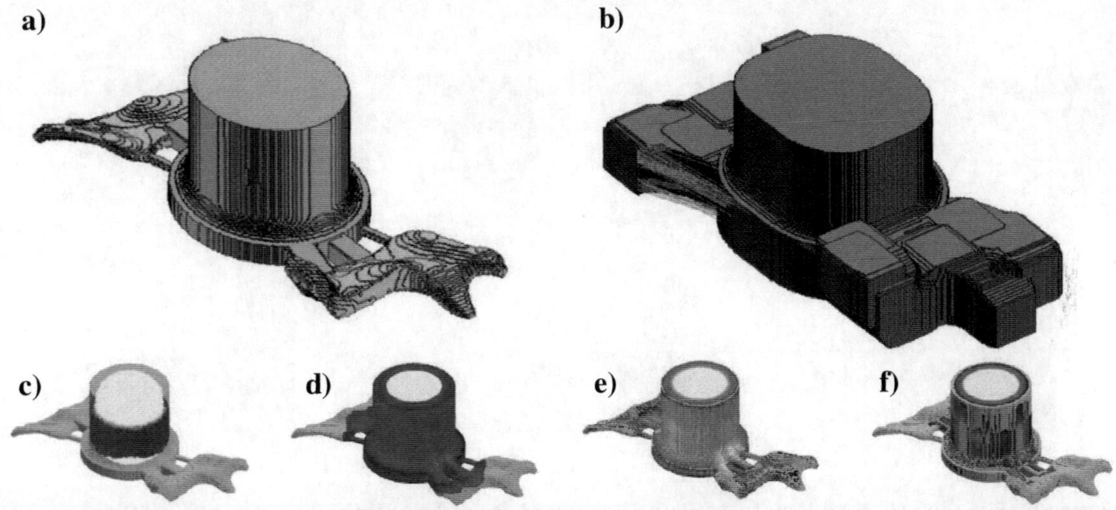

Figure 4. The isometric views of the solidification test using SolidCast 7 casting simulation software: a) meshing and b) molding of the casting parts and its gating system; c) – d) the filling time of 60 and 90 %, respectively; and e) – f) the solidification time with the solid part of 1 and 5 %, respectively.

4. Simulation Results

Evaluation of the casting simulation of the titanium casting was focused on the porosity which leads to mechanical failure on the implant. The porosity of the titanium casting is one of the main issues in titanium casting due to the low density of the titanium which tends to make a reduction of titanium during solidification. The void during solidification can be reduced with appropriate gating system to provide enough supply of molten titanium [6]. Shrinkage of titanium was reported at 1-2 % which means the higher tendency for shrinkage porosity [7].

In the design for the implant, additional riser was limited to maintain dimensional accuracy. Therefore, minimization of the gating system without riser was applied in this simulation. The result has shown a variation of gating systems provide various shrinkage phenomena. Shrinkage was found in the thick section of the part as shown in Figure 5 in which the gas void is high at smaller ingate. Gas void volume decreased with an increasing cross-section of the ingate. The optimum gas void is achieved at C type gating system especially C3 gating system.

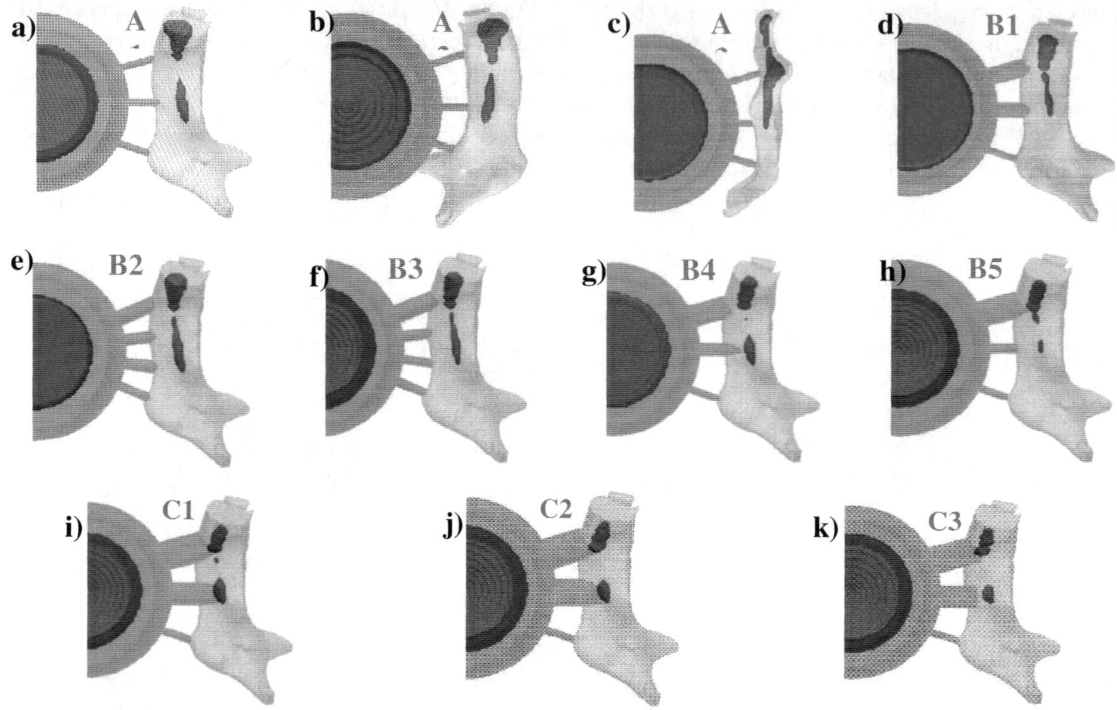

Figure 5. The top views of shrinkage cavities resulted from the solidification test using SolidCast 7 casting simulation software: a) – c) Design A series with Si <Asm; d) – h) Design B series with Si = Asm; and i) – j) Design C series with Si >Asm.

5. Conclusion

This study has successfully performed initial steps for megaprostheses fabrication, both 3-D modeling and casting simulation, using chemically pure titanium as the material. The investigation has shown that increasing total surface area and details of the ingates could lead to the minimization of shrinkage cavities as well as enhancement of effectivity. In this work, C3 design shows the best result that will further proceed for the real fabrication process.

Acknowledgment

This study was conducted under the research funding of Hibah Publikasi Internasional untuk Tugas Akhir Mahasiswa (PITTA) – DRPM Universitas Indonesia, contract number 2394/UN2.R3.1/HKP.05.00/2018.

References

[1] Calori M.G., Colombo, M., Malagoli, Mazzola, M. Bucci, Megaprosthesis in post-traumatic and periprosthetic large bone defects: Issues to consider, Reparative Orthopaedic Surgery Department, Orthopaedic Institute G. Pini, University of Milan, Italy, 105–110 (2014), DOI: 10.1016/j.injury.2014.10.032

[2] Capannaa R , Francesco Muratoria F, Campob, Modular megaprosthesis reconstruction for oncological and non-oncological resection of the elbow joint, Department of Orthopaedic Oncology, AziendaOspedaliero-UniversitariaCareggi, Florence, Italy, 78–83 (2016), DOI: 10.1016/j.injury.2016.07.041

[3] Majóa J., R. Cubedo R. and Pardoc N., Treatment of Osteosarcoma. A Review Group, Study Tumours AparatoLocomotor, Sociedad Española of Surgery Ortopédica and Traumatología (SECOT), Barcelona, Spain, 54(5), 329-336 (2010), DOI: 10.1016/S1988-8856(10)70255-8

[4] Zivic, F., Babic, M., Grujovic, N., Mitrovic, S., and Adamovic, D., Influence of loose PMMA bone cement particles on the corrosion assisted wear of the orthopedic AISI 316LVM stainless steel during reciprocating sliding, Wear, 300, 65–77 (2013), DOI: 10.1016/j.wear.2013.01.109
[5] Fu, T., Zhan, Z., Zhang, L., Yang, Y., Liu, Z., Liu, J., Li, L., and Yu, X., Effect of surface mechanical attrition treatment on corrosion resistance of commercial pure titanium, Surface and Coatings Technology, 280, 129–135 (2015), DOI: 10.1016/j.surfcoat.2015.08.041
[6] J. D. Cotton, L. P. Clark, H. R. Phelps, Titanium investment casting defects: A metallographic overview JOM: the journal of the Minerals, Metals & Materials Society 58(6):13-16 (2006), DOI: 10.1007/s11837-006-0172-z
[7] Nishimura F, Watari F, Nakamura H, Fukumoto R, Morita N., Casting accuracy and shrinkage of titanium castings made with zirconia investments. ShikaZairyo Kikai. 9(6), 850-7 (1990) [Article in Japanese]

The Study of the Effect of Final Tin Slag as Partial Substitution in Portland based Concrete towards Corrosion Resistance of Reinforcement Steel against Chloride Environment with Linear Polarization Method

Irma T. Aryani[1], Rini Riastuti[1*], Muhammad Ibnu[1], Afif Shidqi[1]
[1] Department of Metallurgical and Materials Engineering, Faculty of Engineering Universitas Indonesia, Kampus Baru UI Depok 16424, Indonesia

irmatriaryni@gmail.com

Abstract. The final tin slag is the final product of the tin smelting process which contains oxide compounds such as SiO_2, CaO, Al_2O_3, and Fe_2O_3 which have similarities to the content of the oxide compounds found in Portland cement (OPC). The existence of the final tin slag has not been widely used. For that reason, the final tin slag will be very useful if it is used as a cement substitute as a raw material for making concrete. This study aims to analyze the resistance of cement and steel reinforcement mixtures on concrete to a corrosive environment using linear polarization method. In this study, the composition of the final tin slag used was 10%, 20%, 30%, and 40%. The final tin slag used came from PT. Timah Tbk on Bangka Island Indonesia. Then, the 0.5 w/c ratio concrete will be cured for 28 days and will be immersed with a 3.5% NaCl solution for 6 days. The results presented that the reinforcement steel embedded in concrete OPC60-FTS40 was the sufferest towards corrosion since it had the highest corrosion rate in each immersion day among all the concrete samples and followed by OPC80-FTS20, OPC70-FTS30, OPC100, and OPC90-FTS10 in six days of immersion.

Keywords: final tin slag, Portland cement (OPC), concrete, corrosion resistance, linear polarization

1. Introduction

Major development of infrastructures raises the demand for concrete and its aggregates significantly [1]. Concrete is a mixture of cement, aggregate, and water which is commonly used in construction of a building. A common issue in building is corrosion in reinforcing steel of concrete due to chemical reaction between the steel and outer environment through porosity within the concrete. The porosity in concrete allows reaction between environment and systems within the concrete and eventually leads to corrosion. Furthermore, Indonesia is a maritime nation with a huge risk for corrosion occurring in construction concretes resulting from high salinity of its seawater. Aggressive ions, including chloride, in such environment accelerate the corrosion process. The corrosion is detrimental for any construction as it reduces quality and lifetime of a building [2]. The corrosion in reinforcement steel occur due to the oxide compounds form hydroxides that are very alkaline (pH 12-13) when there is additional water in the hydration reaction [3]. Those alkaline condition trigger the form of passive layer which will be damaged cause the penetration of aggressive ion such as chloride ion [4]. Replacing component is a challenge for negative effects resulting from material substitution associated with strength improvement in concretes [5]. Therefore, it is critical to use proper mixture of concrete aggregates which reduces porosity in order to impede corrosion process and mitigate its impact on a building construction.

Application of waste materials including slag, glass, calcium carbide residues, and so on as material substitution is trending recently and studies had been conducted on its application for years [6]. Processing result of tin ore or slag is abundant in Indonesia; however, its utilization is still not widely accepted and this study was expected to improve the prudent use of existing resources. Final tin slag is a byproduct of smelting which contains various oxide compounds including SiO_2, CaO, Al_2O_3, and Fe_2O_3 all of which are similar to materials found in Portland cement (OPC) that is commonly used in mixing a concrete [7]. Final tin slag possesses a potential to be used as raw material in preparing a concrete due to its common compounds.

This study aimed to identify characteristics of OPC and mixture of cement and final tin slag on corrosiveness of reinforcement steel in concrete by using Linear Polarization that served to analyze corrosion resistance property of reinforcement steel evaluated by corrosion rate of each concrete.

2. Experimental method

2.1 Materials

Materials for concrete included a mixture between Ordinary Portland Cement (OPC) produced by Semen Gresik Manufacturer, Final Tin Slag (FTS) the byproduct of tin processing in PT. Timah Tbk in Bangka, coarse aggregates such as gravel sized 20-40 mm, and fine aggregates such as sand. Table 1 described observation result by X-Ray Fluorescent (XRF) which indicated oxide compounds of both OPC and Final Tin Slag (FTS) cement:

Table 1. OPC and Final Tin Slag (FTS) Chemical Compositions [8].

	CaO	SiO2	Al2O3	Fe2O3	SO3	TiO2	P2O5	MgO	SnO2	Na2O
OPC	66.00	21.50	5.50	3.90	2.75	-	-	5.00	-	-
Slag	5.65	39.66	13.18	8.77	-	6.84	5.26	1.91	4.00	3.75

Material of reinforcing steel was AISI 1045 medium carbon steel with diameter 10 mm and length 140 mm. Chemical compounds of AISI 1045 are outlined in table 2 below:

Table 2. AISI 1045 steel bar composition.

C	Fe	Mn	P	S
0.420 – 0.50 %	98.51 - 98.98 %	0.60 – 0.90 %	≤ 0.040 %	≤ 0.050 %

2.2 Sample Preparation

First, AISI 1045 reinforcing steel was sliced into 5 parts with size 140 mm and diameter 10 mm for each sample. The steels were then polished by angle grinder to remove dirt from its surface. Both ends of steel were then mounted by hardener-mixed-resin with mounting height 20 mm respectively thus left 100 mm of steel unmounted. One end of mounted steel was affixed with copper wire that served to conduct the current during polarization test.

Concrete containing final tin slag was destroyed by ball mill till it reduced to particle size of 100 Mesh. Final tin slag was then mixed with OPC, coarse aggregates sized 20-40 mm, and fine aggregates i.e. sand. All of materials were mixed with water by 0.5 water/binder ration and stirred afterward. The mixture was then poured in cylinder-shaped concrete mold with diameter 100 mm and length 200 mm. Mounted steel was implanted in the middle of concrete mold as illustrated by Figure 1 below.

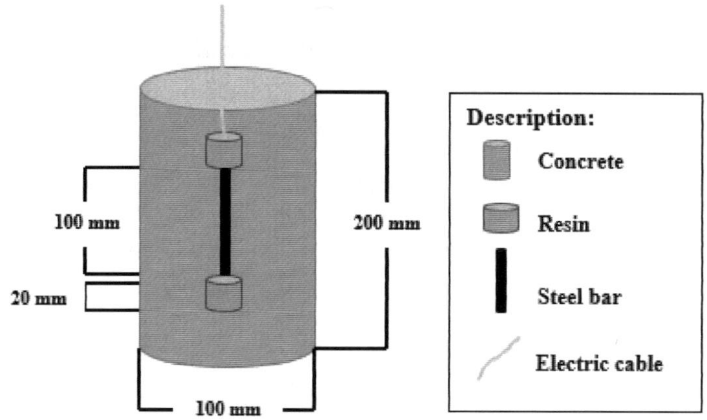

Figure 1. Concrete specimen's dimension.

Figure 2. Concrete Specimen.

The concrete was immersed with water for 28 days in order to achieve its maximum strength. Afterwards, it was soaked with 3.5 % NaCl solution with interval of 0, 3, and 6 days and corrosion behavior were evaluated by Linear Polarization within each interval.

2.3 Concrete Composition

The composition of OPC and Final Tin Slag (FTS) in the concrete is shown by Table 3 below:

Table 3. Formula of tin slag and OPC as raw materials in concrete composition

Concrete Sample	OPC (wt %)	Final Tin Slag/FTS (wt %)	w/b ratio
OPC100	100	0	0.5
OPC90-FTS10	90	10	0.5
OPC80-FTS20	80	20	0.5
OPC70-FTS30	70	30	0.5
OPC60-FTS40	60	40	0.5

2.4 Linear Polarization Test
Gamry Instrument Potentiostat PCI4G750 with Global software was used to analyze corrosion rate in reinforcing steel by Linear Polarization method. Linear Polarization test was conducted regularly based on immersion period of concrete in 3.5% NaCl i.e. in day 0, 3, and 6. The initial and final potentials used to determine the curve of Linear Polarization test result were -0.5V and 0.1V. The test result used in this study was corrosion rate which derived from analysis by comparing potential curve (V) and Current (A). Corrosion rate of each sample was then compared to identify corrosion resistance of each concrete.

(a)

(b)
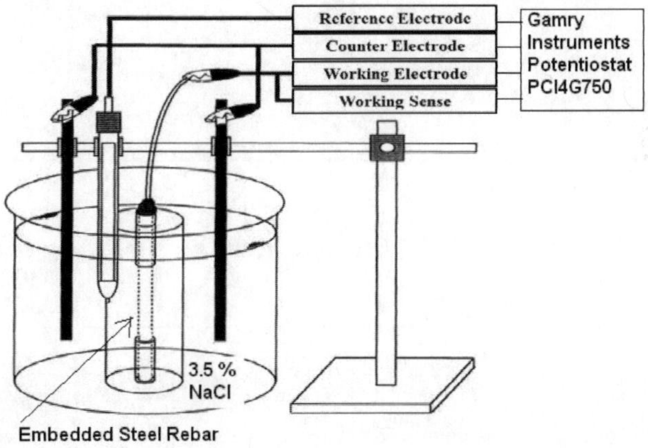

Figure 3. a) Linear Polarization Test Circuit b) Scheme of Linear Polarization Test Circuit [4].

2.5 Compressive Strength Test
The compressive strength test was conducted at Structures and Materials Laboratory, Department of Civil Engineering, Universitas Indonesia. The machine press was used to determine the compressive strength subjected to the given load of each concrete after 28 days of curing.

3. Results and discussion

3.1 Linear Polarization Test
In the 0 day of immersion, Linear Polarization curve suggested concrete with the highest to the lowest current density were in following order: and OPC60-FTS40, OPC80-FTS20, OPC100, OPC90, and OPC70-FTS30. OPC60-FTS40 concrete had the highest current density i.e. 0.95255 μA/cm^2 and OPC70-FTS30 concrete had the lowest current density i.e. 0.48551 μA/cm^2. The graphic also indicated that OPC60-FTS40 concrete was located at the rightmost of the curve and OPC70-FTS30 concrete was at the leftmost.

The third day of immersion revealed that the graphic moved closer to the right end than in previous day for all samples excluding the OPC 80-FTS20 concrete which was located at the leftmost. Graphic movement to the right side indicated greater penetration by chloride ion into the concrete that resulted in greater rates of electron transfer from catodic to anodic. Linear Polarization graphic in Figure 4b) suggested that OPC60-FTS40 concrete had the highest current density than the rest of concretes (2.65830 μA/cm^2) while OPC80-FTS20 concrete had the lowest current density (0.00008 μA/cm$^{2)}$.

IOP Conf. Series: Materials Science and Engineering **553** (2019) 012004 doi:10.1088/1757-899X/553/1/012004

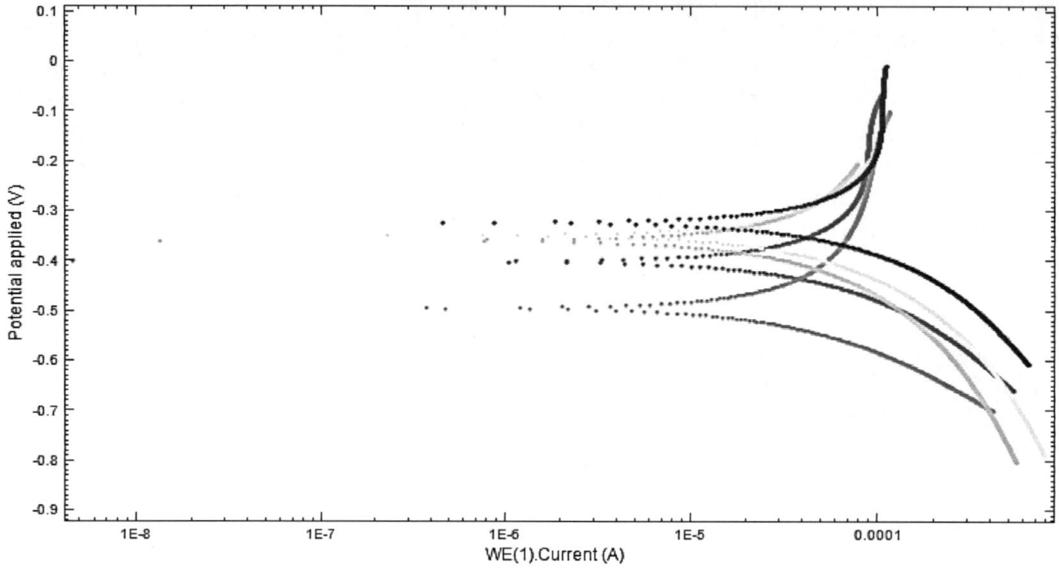

a)

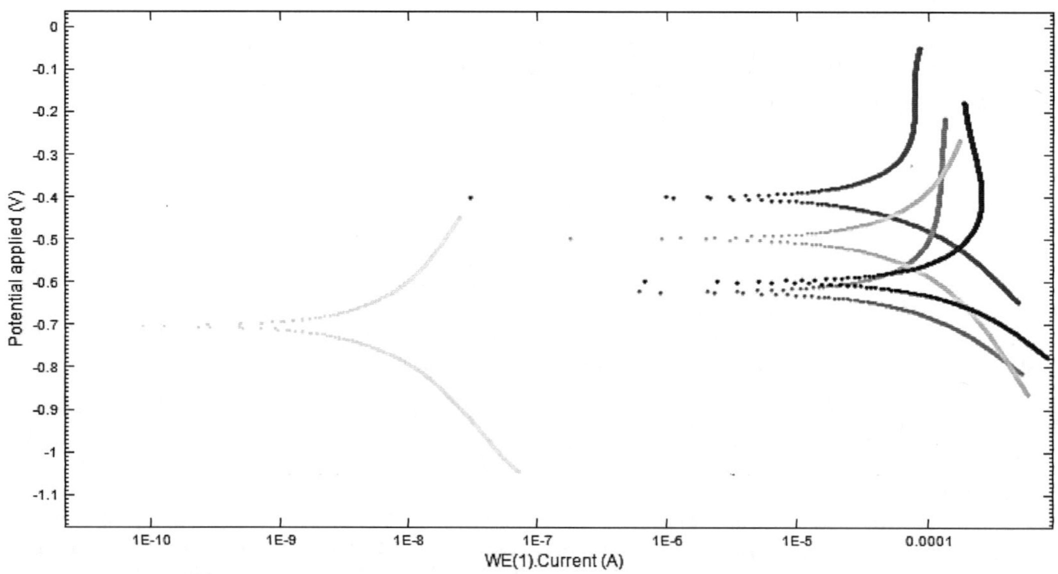

b)

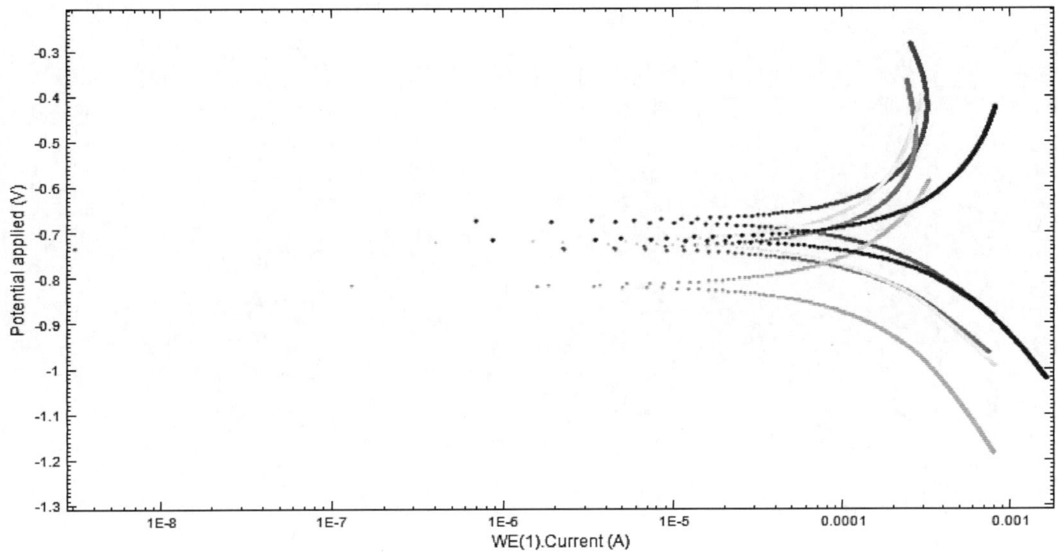

c)

Figure 4. Comparison of Linear Polarization Curves of Reinforcement Steel in Portland mixed Final Tin Slag (FTS) Concrete Day 0 immersed, b) Day 3 immersed, c) Day 6 immersed.

The sixth day of immersion revealed Linear Polarization graphic at the right side as illustrated in Figure 4c. Concrete with the highest to lowest current density were in following order: OPC60-FTS40, OPC80-FTS20, OPC70-FTS30, OPC100, and OPC90-FTS10. OPC60-FTS40 concrete had the highest current density i.e. 4.32580 μA/cm^2 and OPC90-FTS10 concrete had the lowest current density i.e. 1.74650 μA/cm^2. The current density of all samples increased in comparison with the previous result which suggested greater rates of chloride ion transfer penetrating the concrete which resulted in faster corrosion rate.

Table 4. Quantitative parameter of each steel bar in concrete composition measured in 3.5% of NaCl solution for 6 days from Linear Polarization curve.

Concrete Sample	Immersion time (day)	βa (V/decade)	βc (V/decade)	Ecorr (mV)	Icorr (μA/cm^2)	Corrosion rate (mpy)
OPC100	0	0.11648	0.08565	-400.020	0.49914	0.22852
	3	0.13614	0.09506	-397.430	0.53602	0.24540
	6	0.16538	0.12249	-675.480	1.75270	0.80242
OPC90-FTS10	0	0.16741	0.09527	-496.980	0.49490	0.22658
	3	0.32268	0.14330	-625.010	1.35360	0.61972
	6	0.18858	0.14616	-733.230	1.74650	0.79962
OPC80-FTS20	0	0.22239	0.17259	-339.520	0.85118	0.38968
	3	0.17347	0.16974	-697.650	0.00008	0.00004
	6	0.26096	0.17632	-721.210	1.84120	0.84296
OPC70-FTS30	0	0.14971	0.11546	-358.390	0.48551	0.22227
	3	0.20575	0.16826	-495.650	0.93303	0.42717
	6	0.24767	0.17864	-825.110	1.76700	0.80900
OPC60-FTS40	0	0.18644	0.12729	-319.890	0.95255	0.43611
	3	0.25791	0.16416	-592.760	2.65830	1.21702
	6	0.25226	0.19494	-716.970	4.32580	1.98048

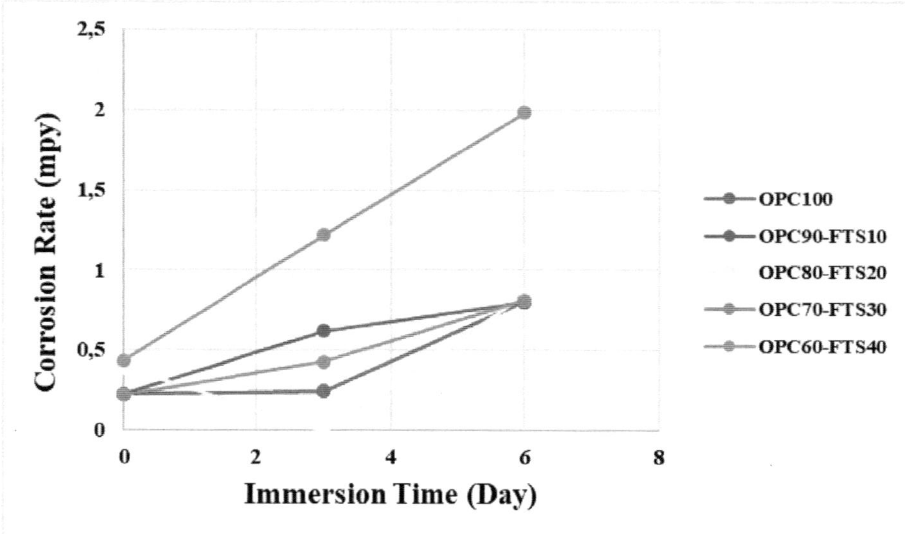

Figure 5. Comparison graphic of reinforcement steel corrosion rate in OPC 100, OPC90-FTS10, OPC80-FTS20, OPC70-FTS30, and OPC60-FTS40 concrete.

Comparison of corrosion rate of all five samples were demonstrated in graphic in Figure 5. Corrosion rate was directly proportional with current density. The higher its current density, the faster its corrosion rate woud be. It is because current density indicates rates of chloride ion transfer. Corrosion rate in the 0 day revealed OPC60-FTS40 concrete had higher corrosion rate than other samples i.e. 0.43611 mpy, while OPC70-FTS30 concrete had the lowest corrosion rate i.e. 0.22227 mpy.

The third day of immersion indicated an increase in corrosion rate with the exception of OPC80-FTS20 concrete which had a decrease in corrosion rate instead up to 0.38964 mpy. The decrease in corrosion rate was resulted from hydration reaction within NaCl solution. The reaction induces chemical bonding between tricalcium aluminate (C_3A) and chloride ion in pore solution which produced Friedel's salt and reduced porosity in concrete as well as rate of electron transfer [9,10]. OPC60-FTS40 concrete had the highest corrosion rate i.e. 1.21702 mpy while OPC80-FTS20 concrete had lowest corrosion rate i.e. 0.00004 mpy.

Corrosion rate in the sixth day of immersion suggested a significant improvement in each sample. It was due to greater penetration by chloride ions than in previous days. OPC60-FTS40 concrete had the highest corrosion rate i.e. 1.98048 mpy while OPC90-FTS10 concrete had the lowest corrosion rate i.e. 0.79962 mpy.

The analysis of Linear Polarization curve on corrosion rate revealed that OPC60-FTS40 concrete which was added with 40% of Final Tin Slag (FTS) always had the highest corrosion rate than the other concretes. Meanwhile, OPC90-FTS10 concrete which was added with 10% of Final Tin Slag (FTS) had the lowest corrosion rate after 6 days of immersion.

3.2 Compressive Strength Test

The compressive strength of each concrete sample are shown by Table 5 below:

Table 5. Compressive strength of each concrete.

Concrete Sample	Compressive Strength (MPa)
OPC100	19.220
OPC90-FTS10	4.6988
OPC80-FTS20	12.634
OPC70-FTS30	14.658
OPC60-FTS40	12.346

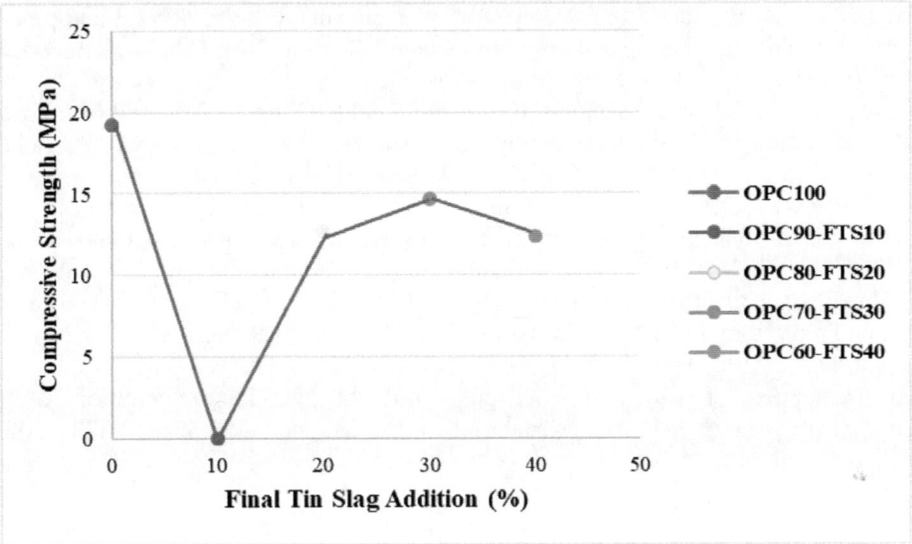

Figure 6. Compressive strength comparison of each concrete.

The results of the compressive strength test are given by Figure 6 above. It shows that concrete OPC100 had the highest strength with amount of 19.220 MPa and sequentially followed by OPC70-FTS30, OPC80-FTS20, OPC60-FTS40, and OPC90-FTS10 with the amount of 14.658, 12.634, 12.346, and 4.6988 MPa in approximately.

4. Conclusions

The conclusions of the test result are first the reinforcement steel embedded in concrete OPC60-FTS40 was the sufferest towards corrosion since it had the highest corrosion rate in each immersion day among all the concrete samples and followed by reinforcement steel in concrete OPC80-FTS20, OPC70-FTS30, OPC100, and OPC90-FTS10 in six days of immersion. After six days of immersion, OPC90-FTS10 provided the lowest value of corrosion rate. The second is the concrete sample which provided the highest compressive strength was OPC100 while the weakest was OPC90-FTS10. It shows that OPC90-FTS10 gave the highest corrosion resistance, but it was not strong as the other concrete.

Acknowledgment

The authors would like to thank to the Directorate for Research and Community Engagement Universitas Indonesia (DRPM-UI) for the financial support.

IOP Conf. Series: Materials Science and Engineering **553** (2019) 012004 doi:10.1088/1757-899X/553/1/012004

References

[1] N. Kisku, et al, A critical review and assessment for usage of recycled aggregate as sustainable construction material, Constr. Build. Mater., 131 pp. 721-740, 2017. DOI: 10.1016/j.conbuildmat.2016.11.029

[2] D. Jones, Principles and prevention of corrosion, Upper Saddle River, N.J.: Prentice-Hall, 1996.

[3] J. Broomfield, Corrosion of steel in concrete, London: Spon Press, 2003.

[4] A. Sagüés, E. Moreno and C. Andrade, Evolution of pH during in-situ leaching in small concrete cavities, Cement and Concrete Research, vol. 27, no. 11, pp. 1747-1759, 1997.

[5] A. Neramitkornburi, et al, Durability against wetting-drying cycles of sustainable lightweight cellular cemented construction material comprising clay and fly ash wastes, Constr. Build. Mater., 77, pp. 41-49, 2015. DOI: 10.1016/j.conbuildmat.2014.12.025

[6] A. Arulrajah, et al, Recycled waste foundry sand as a sustainable subgrade fill and pipe-bedding construction material: engineering and environmental evaluation, Sust. Citi. Soc., 28, pp. 343-349, 2017. DOI: 10.1016/j.scs.2016.10.009

[7] Agung Cahyadi, et al, Corrosion Behavior of Carbon Steel in Concrete Material Composed of Tin Slag Waste in Aqueous Chloride Solution, IOP Conference Series: Materials Science and Engineering. Volume 299, 2018. DOI: 10.1088/1757-899X/299/1/012047

[8] Boy Ahmad, Effect of Steel Slag Substitution in Geopolymer Concrete on Compressive Strength and Corrosion Rate of Steel Reinforcement in Seawater and Acid Rain, Universitas Indonesia, 2013.

[9] Shamsad Ahmad, et al, Chloride-Induced Corrosion of Steel in Concrete: An Overview on Chloride Diffusion and Prediction of Corrosion Initiation Time, International Journal of Corrosion, Volume 2017, Article ID 5819202, 9 pages, 2017. DOI: 10.1155/2017/5819202

[10] S. Han, Influence of diffusion coefficient on chloride ion penetration of concrete structure. Construction and Building Materials. vol. 21, no. 2, pp. 370-378, 2007. DOI: 10.1016/j.conbuildmat.2005.08.011

The Effect of Zn in as-Cast Mg-Zn Alloys for Biodegradable Materials

Asep Ridwan Setiawan[1], Al Azhary Putera Satria[1], Arif Basuki[1], Untung Ari Wibowo[1]

[1]Materials Science and Engineering Research Groups, Faculty of Mechanical and Aerospace Engineering, Institut Teknologi Bandung, Bandung 40132, Indonesia

asep.ridwans@material.itb.ac.id

Abstract. This research presents the analysis of microstructure, mechanical, and corrosion behavior of three Mg-Zn binary alloys, Mg-1Zn, Mg-3Zn, and Mg-5Zn. The result of metallography for as-Cast Mg-Zn alloys with different Zn content shows that the presence of Zn will cause grain refinement effect to the microstructure of Mg-Zn alloys and allows the formation of γ phase of Mg-Zn precipitation along the grain boundaries. The addition of Zn also significantly increases the mechanical properties of magnesium alloys corresponding to fine grain strengthening and second phase strengthening effect. Immersion test in Ringer's solution show that the addition of Zn in magnesium alloys reduce its corrosion rate by improving magnesium alloys corrosion potential.

Keywords: corrosion properties, grain refinement, magnesium alloy, mechanical properties

1. Introduction

The application of Magnesium alloys as implant material is the most developed alternative due to its close mechanical properties to human's bone so it will not cause stress shielding effect during bone healing process [1]. In addition, the usage of magnesium alloy as implant materials does not require second surgical method to carry out the implant plate as it is dissolved inside human body. The problem found in magnesium alloy is that the degradation rate is too fast, so it will be dissolved faster than the human's bone recovery time. Refining the alloy composition is one of the keys to overcome that problem.

Previous research found that Mg, Ca, and Zn did not have cytotoxicity [2], it is also found that Zn is important nutrition with one of the most abundant amounts in human body [3], so it is naturally safety for biodegradable implant applications. Finding the optimum Zn content in Mg alloys paid a lot of attention in developing biomedical implant materials. Several researchers were also discussed the mechanical properties of Mg-Zn alloys and shown that the presence of Zn may increase the mechanical properties of Mg alloys while also increase the corrosion resistance of Mg [4]. This work tries to systematically study the effect of Zn towards microstructure, mechanical properties, and corrosion behavior of Mg-Zn alloys fabricated from Mg sacrificial anode and pure Zinc in as-cast condition.

2. Experimental Methods

Magnesium alloys were prepared from Magnesium ingots (99.3%) and Zinc ingots (99.975%) through casting process. Magnesium and Zinc were prepared to get final zinc content 1 wt%, 3 wt%, and 5 wt% alloys. All the raw materials surface was grinded to remove the oxide layer. Casting operations were conducted in a graphite crucible, using induction furnace protected by CO_2 + HFC-134a mixed gas to eliminate the presence of oxygen inside the furnace and thus suppress loss of magnesium. The casting temperature is between 700°C and 730°C with 20-30 minutes holding time. The molten magnesium alloys then poured to a mold and formed the ingots when it solidifies.

Table 1. Chemical composition of Mg-Zn alloys.

Mg Alloy with Zn	Detected elements (wt.%)									
	Al	**Mn**	**Si**	**Cu**	**Ni**	**Fe**	**Cd**	**Pb**	**Zn**	**Mg**
1%wt	0,1	0,495	0,0495	0,01985	0,00099	0,02975	0,00025	0,0000	0,999	98,307
3%wt	0,1	0,485	0,0485	0,01955	0,00097	0,02925	0,00075	0,00018	2,99277	96,321
5%wt	0,1	0,475	0,0475	0,01925	0,000095	0,02875	0,00125	0,0003	4,98795	94,335

2.1 Microstructure

The Microstructure were examined by standard metallography process. Samples for microstructure observation are cut from ingots and grinded on sandpapers up to 5000 grit then polished with alumina paste and etched in a solution of chromic acid. Microstructure observations of the alloys were conducted using optical microscope.

2.2 Mechanical Properties

Mechanical properties were tested by tension test based on ASTM B557M-02A l. The tension test was carried out with Tarno Grocki testing machine. Testing specimen are cut from the cast ingots and machined to form the tension test specimen. Specimen with a gauge length of 45 mm and diameter 9 mm with screwed grip was selected for tensile test.

2.3 Immersion Test and Hydrogen Evolution Measurement

Immersion tests were carried out in simulated body fluid (SBF) using Ringer's solution (compositions are listed in Table 2) in accordance with ASTM G31-72. The temperature of the solution was adjusted at 37 ± 1°C to simulate the average temperature of human body. The immersion was held for 4 days. After immersion process done, the samples were taken out from the solution and cleaned with chromic acid (200 g/l CrO_3) to remove surface corrosion products without removing any amount of metallic Mg. Then the samples were rinsed with distilled water, dried in air, then measure the mass loss of sample to calculate the corrosion rate

The hydrogen evolution volume was measured along with immersion test in SBF. The evolved hydrogen was collected by installing a funnel and a burette above the specimen (shown in Figure 1). The hydrogen formed by the reaction between Mg alloys with SBF will be caught by the burette and push the SBF solution down so the volume can be measured. The hydrogen evolution rate, VH (ml/cm^2 day) was calculated from the total hydrogen released for the whole immersion test duration. The hydrogen evolution rate, VH (ml/cm^2 day) have relation to corrosion rate, RH (mm/year), and can be calculated using following equation;

$$R_H = 2.279V_H \quad (1)$$

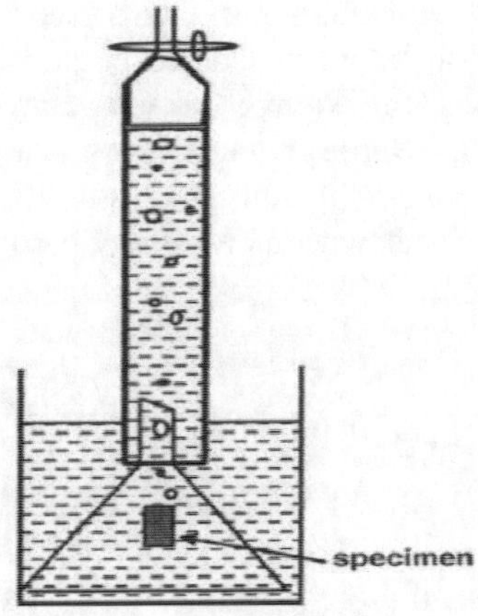

Figure 1. Sketch of immersion testing installation.

2.4 Electrochemical Measurement

Electrochemical tests were carried out at 37 ± 1°C in a chamber with Ringer's solution on a VersaStat® automatic laboratory corrosion measurement system using a standard three-electrode configuration, with a saturated calomel as a reference, a graphite electrode as the counter and the sample as the working electrode. The working surface is ground on sandpapers up to 1000 grit, and then is cleaned by the alcohol and dried in air with only one side of 0.05 cm^2 exposed to the Ringer's solution for the test. All the measurements are measured at a scanning rate of 1 mVs^{-1}.

Table 2. Chemical composition of simulated body fluid used in the study.

Compound	Amount(mg/l)
NaCl	6000
KCl	400
$CaCl_2.2H_2O$	270
$C_3H_5NaO_3$	3120

3. Results and discussion

3.1 Microstructure and Chemical Composition

Figure 2 illustrates the microstructure of Mg-1Zn, Mg-3Zn, and Mg-5Zn. It can be seen from. the microstructure shown in Figure 2b and 2c that the grain size of Mg-5Zn is finer than Mg-3Zn. The pure Mg have average grain size of 350 μm [5], while the grain size measured in the microstructure of Mg-3Zn and Mg-5Zn are 86.7μm and 56.94μm. It shows that the presence of Zn cause grain refining effect in casting of Mg alloys. It was widely accepted that the presence of Zn limits the grain growth and drive the nucleation of α- Mg and thus the grain size will be refined [6]. The refinement efficiency of a solution element can be determined by the calculation of a growth restriction factor (Q). It has been documented [7] that Zn had a higher Q value than Al (4.32) and Y (1.70). Zn has 5.31 growth restriction factor value, meaning that Zn has more effective grain refinement effect.

Figure 2a shown different typical microstructure than the others. This is because the presence of Cu that contaminate the molten in casting process. The contamination of Cu indicated when undissolved Zn ingots discovered after casting so electron probe micro-analyzer (EPMA) characterization was conducted (Table 3) to detect the content of Zn. The result shows that there was Cu content inside the alloys was relatively high. As can be seen in the Figure 2d, there are 16 wt% of Cu but that is not the exact fraction of Cu since EPMA is semi-quantitative characterization. As the casting process carried out at 700-750°C, a eutectic reaction between Cu stirrer and molten magnesium probably occurs because the stirrer is the only possible source of Cu. Based on a previous research [8], the microstructure shown in Figure 2a have same characteristics as Mg-Zn-xCu alloys which is show thick grain boundary. The eutectic intermetallic (MgZnCu) are solidified along the branches of the dendritic structure. Therefore, it is also stated that with increasing Cu content cause the degree of forming of dendritic structures will increase.

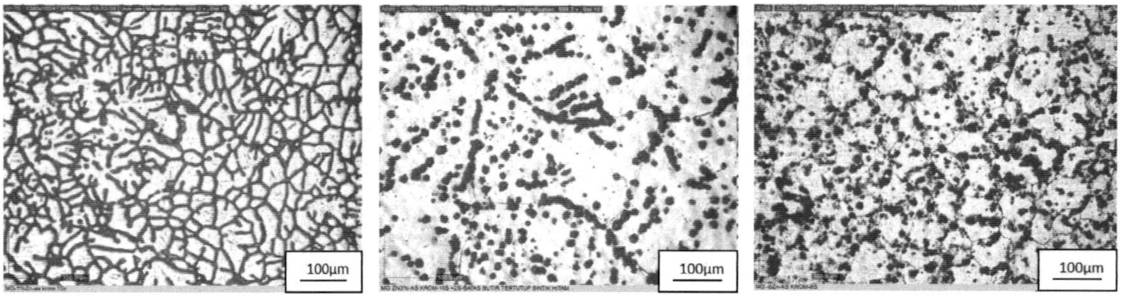

a) b) c)

Figure 2. Microstructure of magnesium alloy with a) 1wt% Zn, b) 3wt% Zn, and c) 5wt% Zn.

Table 3. EPMA Analysis results of Mg with 1wt% Zn.

No	Element	Crystal	Peak WL(nm)	K-ratio	Mass%
1	Mg	RAO/CHI	0.99006	0.76696	77.75
2	Mn	LiF/CH4	0.21026	0.00714	0.64
3	Cu	LiF/CH4	0.15363	0.16777	16.03
4	Zn	LiF/CH4	0.14309	0.05813	5.59

3.2 Mechanical Properties

Table 4 shown the mechanical properties of Mg-xZn alloys as the result of tension test. The presence of Zn increases the tensile strength and yield strength of Mg-Zn alloys. The increase of strength can be explained by Hall-Petch relationship, about grain boundary and strength relation. On the other hand, based on the Mg-Zn binary phase diagram, the maximum solubility of Zn in Mg at room temperature is 1.6%wt. The excess of Zn content may generate the solid solution strengthening. In addition, when Zn content is 5 wt%, several MgZn phase will precipitate along grain boundaries which improving the strength of Mg-Zn by dispersion strengthening [8]. So, these grain refinement, solid solution strengthening, and second phase strengthening contribute to improve the mechanical properties of Mg-Zn alloys. In the other hand, the presence of second phase can decrease the elongation of the alloys because it is dispersed along grain boundary could be new crack source. As shown in table 2, Mg-3Zn have higher elongation than Mg-5Zn. While Mg-1Zn failed at much lower elongation because the contaminant of Cu that makes this alloy become too brittle as MgZnCu particles may become the nucleation sites of microcracks [9].

Table 4. Mechanical properties of Mg-Zn

Material	Yield Strength (MPa)	UTS(MPa)	Maximum Elongation (%)
Mg-1%Zn	56	111.6	1.2
Mg-3%Zn	76.5	166.62	9.3
Mg-5%Zn	77.6	188.6	7.77

IOP Conf. Series: Materials Science and Engineering **553** (2019) 012005 doi:10.1088/1757-899X/553/1/012005

3.3 Immersion Test and Hydrogen Evolution Measurement

Figure 3 show the hydrogen evolution rate of Mg-Zn alloys as function of immersion time. According to hydrogen evolution measurement, the corrosion rate of Mg-1Zn, Mg-3Zn, and Mg-5Zn respectively are 487.17 mmpy, 1.488 mmpy, and 5.081 mmpy. For Mg-1Zn, the result only listed until 4 hours of testing because the hydrogen evolution occurs too fast as the hydrogen formed excess the burette volume capacity in 4 hour and 20 minutes. Based on an earlier research [10], it is proven that the presence of Cu in Mg based alloys, allow the formation of Mg_2Cu precipitation in grain boundary which is an electrochemically noble phase. The phase formed exhibits hydrogen generation around 300 times more than pure magnesium and cause the test results for Mg-1Zn relatively much higher than Mg-3Zn and Mg-5Zn. Whereas for immersion test, the result for the samples respectively are 162.55 mmpy, 2.03 mmpy. And 8.041 mmpy. Based on immersion and hydrogen evolution test, it is shown that alloying with Zn reduce the corrosion rate of magnesium [11] except for Mg-1Zn which is caused by the contaminant of copper in the alloy. It also can be seen that Mg-3Zn has higher corrosion resistance than Mg-5Zn. This happen because with higher content of Zn exhibits more presence of secondary phase in the alloy. The potential difference between the phases causes a micro-galvanic effect in the alloy which finally increase the corrosion rate as the volume fraction of the second phase increase [11].

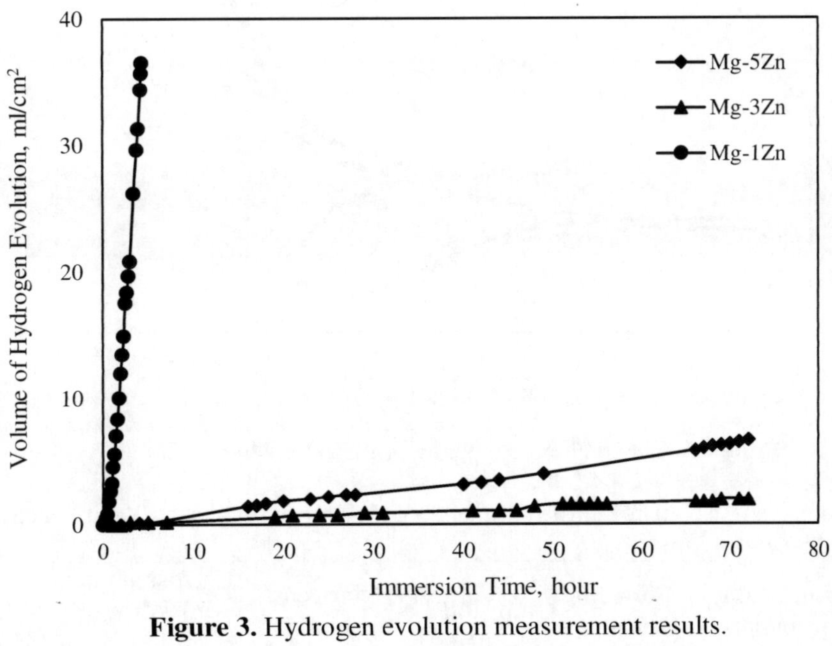

Figure 3. Hydrogen evolution measurement results.

Previous research [12] mentioned that there exist a very fine particles, Mg_xZn_y secondary phase, locate on the grain boundary of Mg-Zn alloys. The phase called by Mg_xZn_y because the size of this particles is several hundred nanometers. Thus, it is so difficult to precisely determine the chemical composition of that phase. Although this white particle exist in Mg-xZn alloys has a relatively low volume fraction, it will contribute to the corrosion resistance of Mg alloys since it is nobler than Mg matrix. The volume fraction of the white particles increases as the content of Zn increase. This result fits the corrosion rate obtained in this research where Mg-3Zn have better corrosion resistance than Mg-5Zn.

3.4 Electrochemical Measurement

The electrochemical polarization curves are shown in Figure 4 and the electrochemical data based on the curves shown in Table 5. The variation of data obtained caused by the difference of chemical content of each alloy and the phase formed in casting process. This result confirms the corrosion rate obtained in immersion and hydrogen evolution measurement which is also indicate that Mg-3Zn and Mg-5Zn have the significantly lower corrosion rate than Mg-1Zn. The polarization curves for both Mg-3Zn and Mg-5Zn show similar characteristics, the anodic sides are visible with passivation tendency below the breakdown potential point, the point where the polarization curve's slope become more linear. The existence of passivation tendency at the anodic side indicate the presence of oxide films formed on the surface of the Mg alloys [13,14]. However, the fact that the anodic current densities slowly rise with increasing anodic potential below the breakdown potential means the oxide film formed has porous structure [15].

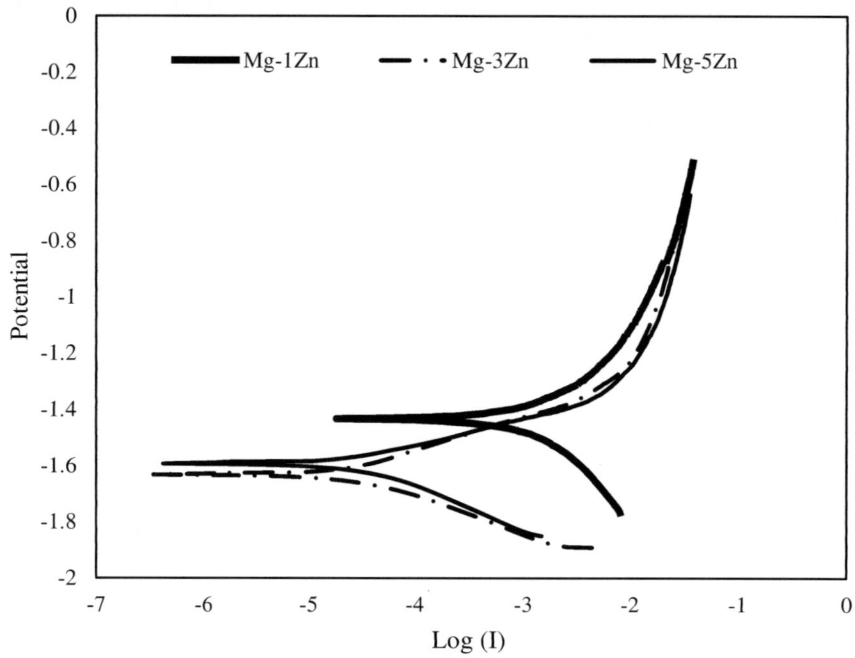

Figure 4. Electrochemical behavior measurement results.

Table 5. Corrosion rate calculated from immersion, hydrogen evolution, and electrochemical measurement of Mg-Zn alloys in Ringer's solution.

Materials	Corrosion potential, Ecorr (mV vs SCE)	Current density, Icorr (μA)	Corrosion rate, Ri (mmpy)	Hydrogen evolution rate, V_H (ml/cm^2day)	Corrosion rate for Hydrogen evolution R_H (mmpy)	Corrosion rate for immersion test, C_R (mmpy)
Mg-1Zn	-1.432	300.181	54.305	213.745	487.17	162.55
Mg-3Zn	-1.621	3.135	0.5744	0.6504	1.4882	2.03
Mg-5Zn	-1.591	3.035	0.5492	2.2032	5.021	8.041

The corrosion rate obtained by electrochemical measurement are contrast with immersion and hydrogen evolutions result where Mg-3Zn were better than Mg-5Zn. This could be happened because current density/I_{coor} calculated from electrochemical curve is a function of surface area. While the passive layer of Mg alloys formed and exposed to SBF, the porous structure attacked and forming a pitting corrosion. As pitting corrosion happen, only several areas will be attacked extremely by corrosion so the result of corrosion rate calculated by Tafel curve and immersion test might have different results.

4. Conclusion

The influence of Zn in as-cast Mg-xZn alloys were studied in this paper. The presence of Cu contaminant in Mg-xZn alloys improve strength, reduce ductility, refine the grain size, and increase the corrosion rate. Thus, increase the hydrogen gas generation significantly. The average grain size of Mg-xZn alloys reduce as the Zn content increase which also cause the strength increase meanwhile the ductility of alloys decreases. The corrosion rate measured by immersion test shows that Mg-3Zn have better corrosion resistance. While for electrochemical test, Mg-5Zn has higher corrosion resistance.

Acknowledgment

This research was financially supported by the 2018 PTUPT Research Grand, Ministry of Research, Technology and Higher Education of the Republic of Indonesia.

References

[1] M.P. Staiger, A.M. Pietak, J. Huadmai, G. Dias, Magnesium and its alloys as orthopedic biomaterials: A review, Biomaterials. (2006). doi:10.1016/j.biomaterials.2005.10.003.
[2] L. Yang, E. Zhang, Biocorrosion behavior of magnesium alloy in different simulated fluids for biomedical application, Mater. Sci. Eng. C. (2009). doi:10.1016/j.msec.2009.01.014.
[3] H. Tapiero, K.D. Tew, Trace elements in human physiology and pathology: Zinc and metallothioneins, Biomed. Pharmacother. (2003). doi:10.1016/S0753-3322(03)00081-7.
[4] N. Sezer, Z. Evis, S.M. Kayhan, A. Tahmasebifar, M. Koç, Review of magnesium-based biomaterials and their applications, J. Magnes. Alloy. (2018). doi:10.1016/j.jma.2018.02.003.
[5] S. Cai, T. Lei, N. Li, F. Feng, Effects of Zn on microstructure, mechanical properties and corrosion behavior of Mg-Zn alloys, Mater. Sci. Eng. C. (2012). doi:10.1016/j.msec.2012.07.042.
[6] E. Zhang, D. Yin, L. Xu, L. Yang, K. Yang, Microstructure, mechanical and corrosion properties and biocompatibility of Mg-Zn-Mn alloys for biomedical application, Mater. Sci. Eng. C. (2009). doi:10.1016/j.msec.2008.08.024.
[7] Y.C. Lee, A.K. Dahle, D.H. Stjohn, The role of solute in grain refinement of magnesium, Metall. Mater. Trans. A Phys. Metall. Mater. Sci. (2000). doi:10.1007/BF02830349.
[8] C.J. Boehlert, K. Knittel, The microstructure, tensile properties, and creep behavior of Mg-Zn alloys containing 0-4.4 wt.% Zn, Mater. Sci. Eng. A. (2006). doi:10.1016/j.msea.2005.11.006.
[9] Y. Zhang, X. feng Huang, Y. Ma, T. jun Chen, Y. dong Li, Y. Hao, Effects of Cu addition on microstructure and mechanical properties of as-cast Mg-6Zn magnesium alloy, China Foundry. 14 (2017) 251–257. doi:10.1007/s41230-017-6094-2.
[10] S.K. Oh, H.W. Kim, M.J. Kim, K.S. Eom, J.S. Kyung, D.H. Kim, E.A. Cho, H.S. Kwon, Design of Mg-Cu alloys for fast hydrogen production, and its application to PEM fuel cell, J. Alloys Compd. (2018). doi:10.1016/j.jallcom.2017.12.257.
[11] Y. Ding, C. Wen, P. Hodgson, Y. Li, Effects of alloying elements on the corrosion behavior and biocompatibility of biodegradable magnesium alloys: A review, J. Mater. Chem. B. (2014). doi:10.1039/c3tb21746a.
[12] Y. Song, E.H. Han, D. Shan, C.D. Yim, B.S. You, The effect of Zn concentration on the corrosion behavior of Mg-xZn alloys, Corros. Sci. (2012). doi:10.1016/j.corsci.2012.08.037.
[13] Y. Song, D. Shan, R. Chen, E.H. Han, Investigation of surface oxide film on magnesium lithium alloy, J. Alloys Compd. (2009). doi:10.1016/j.jallcom.2009.04.137.
[14] Y. Song, D. Shan, R. Chen, E.H. Han, Corrosion characterization of Mg-8Li alloy in NaCl solution, Corros. Sci. (2009). doi:10.1016/j.corsci.2009.03.011.
[15] N. Hara, Y. Kobayashi, D. Kagaya, N. Akao, Formation and breakdown of surface films on magnesium and its alloys in aqueous solutions, Corros. Sci. (2007). doi:10.1016/j.corsci.2006.05.033.

IOP Conf. Series: Materials Science and Engineering **553** (2019) 012006 doi:10.1088/1757-899X/553/1/012006

The Influence of Yttrium Addition on The Corrosion Resistance and Tensile Strength of as-Cast Mg-Y Alloys

Asep Ridwan Setiawan[1], Herdi Ardiyana[1], Arif Basuki[1], Untung Ari Wibowo[1]
[1]Materials Science and Engineering Research Groups, Faculty of Mechanical and Aerospace Engineering, Institut Teknologi Bandung, Bandung 40132, Indonesia

asep.ridwans@material.itb.ac.id

Abstract. This study shows the effect of Yttrium addition (2-6 wt%) on the mechanical properties and the corrosion resistance of Mg-Y binary fabricated from sacrificial magnesium anode. The beneficial of Mg anode source was tremendous for magnesium alloy especially for corrosion resistance. Improvement of mechanical properties due to the addition of Yttrium was caused by grain refinement solid solution strengthening. Tensile strength increased with the addition of Yttrium. Mg-Y binary is immersed in the SBF solution to see the corrosion resistance. The hydrogen evolution was increased from 0,32 mL/cm^2 day, 1 mL/cm^2 day, and 34,42 mL/cm^2 day for Mg-2%wt Y, Mg-4%wt Y, and Mg-6%wt Y respectively.

Keywords: Corrosion potential, corrosion rate, grain refinement, magnesium anode source, Mg-Y binary, mechanical properties, solid solution strengthening

1. Introduction

Currently, the biodegradable implant material which are widely used for bone recovery was magnesium-based alloys. This type of implant can reduce stress shielding due to its modulus of elasticity similar to the bone. As its degraded by time, no additional surgical treatment was required after the bone healed [1]. As we know the body needed magnesium in some amount during bone formation [2]. Mostly the constraint of the magnesium was the fast corrosion rate and low mechanical strength which will affect the healing to fail [3]. In this work, Mg-Y alloys was fabricated through casting process. Base material that was used is magnesium anode source. The presence of trace element in magnesium anode source such as Mn give an advantage to increase corrosion resistance [4]. The increase of corrosion resistance is significant with alloying element. The alloy that was used in this research is yttrium. Yttrium can form intermetallic compounds with the impurities that was reduced their bad influences [5]. Yttrium also has a good biocompatibility based on survey on the high rare earth element background regions in China that was the tolerant daily intake of Y around 4.2 mg.day^{-1} [5]. The mechanical properties increased due to solid solution strengthening mechanism and grain refinement. The microstructure was determining of corrosion behavior.

2. Experimental Methods

In this work, Mg anode source and Mg-Y master alloys was melted in graphite crucible by heat induction furnace until reach liquid phase. The temperature of the liquid phase is between 700°C-750°C. The mixture of HFC-134a, N_2, and CO_2 gas were used for shielded gas during the casting process the targeted composition of the alloys was Mg-2Y, Mg-4Y, and Mg-6Y (wt.%). The chemical compositions of the as-cast Mg-Y alloys were determined by AAS methods, and the results are shown in Table 1. After casting, the ingot was cut into test specimen. The specimen was characterized by SEM, EDS, XRD, metallography, immersion test, and electrochemical test.

Table 1. Chemical composition of alloys in wt.%.

	Composition (wt.%)								
Alloy	Mg	Al	Zn	Mn	Fe	Ni	Cu	Si	Y
Alloy 1	balanced	0.092	0	0.46	0.028	0.00092	0.018	0.046	2
Alloy 2	balanced	0.084	0	0.42	0.025	0.00084	0.017	0.042	4
Alloy 3	balanced	0.076	0	0.38	0.023	0.00076	0.015	0.038	6

2.1 Microstructure and Mechanical Properties

The microstructure of the as cast alloy was observed using optical metallography. The specimen was cut to 1 cm x 1 cm, etched with 2% oxalic acid before microscope observation. The mechanical testing was undertaken by following ASTM B557 M procedure. The dimension of the dog bone specimen is 45 mm in gauge length and 9 mm in diameter with screw rounded on grip.

2.2 Immersion Test

The immersion test was conducted to measure the hydrogen evolution and the corrosion rates of the alloy. The specimen size was 2 cm x 2 cm in square. Total surface area of the specimen compared to the solution immersion test was not less than 1 cm^2:20 mL according to ASTM G31-72. Before test, the surface of the specimen was grinded up to 2000 grit sandpaper. Ringer's solution was used for test medium with the composition showed in Table 2. The test temperature was set to 37°C±1°C. The hydrogen evolution rate or VH (ml/cm^2/day) and the corrosion rate or RH (mmpy) was calculated using the following equation.

$$RH = 2.279VH \tag{1}$$

Time of immersion was prolonged to 4 days. After immersed, the specimen was removed from the hydrogen evolution installation and immersed on chromic acid with composition about 200 gram/l and hold until 30 min to removed corrosion product without removing any amount of metallic Mg. Finally, the specimen was rinsed on distilled water and cleaned ultrasonically in acetone or alcohol after that the specimen dried in air. Before and after immersion test, the specimen was weighed, and the data was recorded. The corrosion rate was calculated using the following equation.

$$Ri = \frac{k \times mass\ loss(gram)}{time(hours) \times area\ of\ exposure(cm^2) \times density\ of\ magnesium\ alloy(\frac{g}{cm^3})} \tag{2}$$

K that used for mmpy is 87600 according to ASTM G31-72.

Table 2. Chemical composition of simulated body fluid used in immersion test.

Compound	Amount (mg/l)
NaCl	6000
KCl	400
$CaCl_2.2H_2O$	270
$C_3H_5NaO_3$	3120

IOP Conf. Series: Materials Science and Engineering **553** (2019) 012006 doi:10.1088/1757-899X/553/1/012006

2.3 Electrochemical test

The electrochemical test was used for measuring the corrosion potential and current of the alloy in Ringer's solution medium. The specimen size is plate with the exposed area closed to 0.05 cm2 approximate to O ring seal. Before test, the specimen surface was ground up to 2000 grit. The temperature of the test was set to 37oC±1°C which is simulated normal human temperature. The data was recorded and analyzed by VersaStat® automatic laboratory corrosion measurement system using standard three-electrode configuration, with a saturated calomel as a reference, a graphite electrode as the counter and the sample as the working electrode.

3. Results and discussion

3.1 Microstructure and mechanical properties

The microstructure revealed that grain refinement increases due to the addition of Yttrium as shown in Figure 1. As the yttrium content increased the grain size was degreased. Grain refinement could make the mechanical properties increase related to Hall-Petch equation. The result was confirmed by tensile test result (showed in Table 3) which show increasing tensile strength as the Yttrium content increased. Beside of that the hardness data also confirmed that was increased mechanical properties due to increasing of Yttrium content. This could indicate the strengthening mechanisms to this alloy was due to Yttrium addition. Alloy 1 was not tested because when machined the specimen failed to form because alloy 1 was most brittle than the others. The force that applied for each alloy to form dog bone specimen was same and indicate that Alloy 1 was most brittle and weaker than the others. There was need strengthening mechanism to increase mechanical properties especially for toughness. Mechanism of non-basal slip was studied for increasement of mechanical properties especially for its properties. The non-basal slip system was pyramidal slip that accommodate Mg-Y binary to increase the ductility [6]. The tensile properties were increased as the Y content was increased but there is a reduction in ductility from the addition of 4% weight Yttrium to 6% weight Yttrium due to reduction of grain size [5].

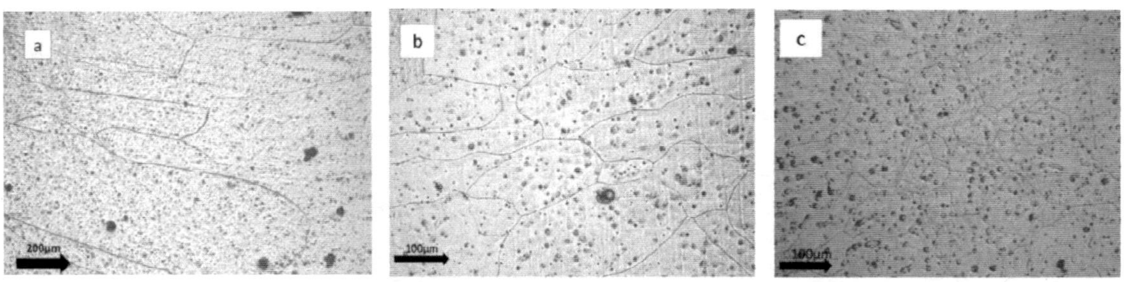

a. Mg-2Y b. Mg-4Y c. Mg-6Y

Figure 1. Microstructure of Mg-Y alloy.

Table 3. Mechanical properties of Mg-Y alloys.

Alloy	Hardness (HV)	Yield Strength (MPa)	Ultimate Tensile Strength (MPa)	Elongation at Break (%)
Mg2Y	38	-	-	-
Mg4Y	48	125,7	146,19	1,57
Mg6Y	60	133,6	164,26	1,55

3.2 Immersion Test

The immersion test results were depicted in Figure 2. Calculated corrosion rates as the function of hydrogen evolution measurements were 0.74 mmpy, 2.29 mmpy, and 78.44 mmpy for Mg-2%wtY, Mg-4%wtY, and Mg-6%wtY respectively. The higher corrosion rates in Y addition above 2%, was likely caused by higher secondary phase in the alloy which lead to micro galvanic corrosion especially for Mg-6%wtY [7]. As additional of yttrium, the amount of secondary phase will be increase [8]. Based on cast processing the phase distribution of primarily phase and secondary phase should be random and not uniform. This type of microstructure can accelerate of corrosion rate of the magnesium [7].

The corrosion rate according to mass loss measurement were 1.2 mmpy, 5.5 mmpy, and 104.7 mmpy for Mg-2%wtY, Mg-4%wtY, and Mg-6%wtY respectively. Both of two calculated, the optimum composition for corrosion rate of Mg-Y binary according to this research is Mg-2%wtY and this study related to common research [5].

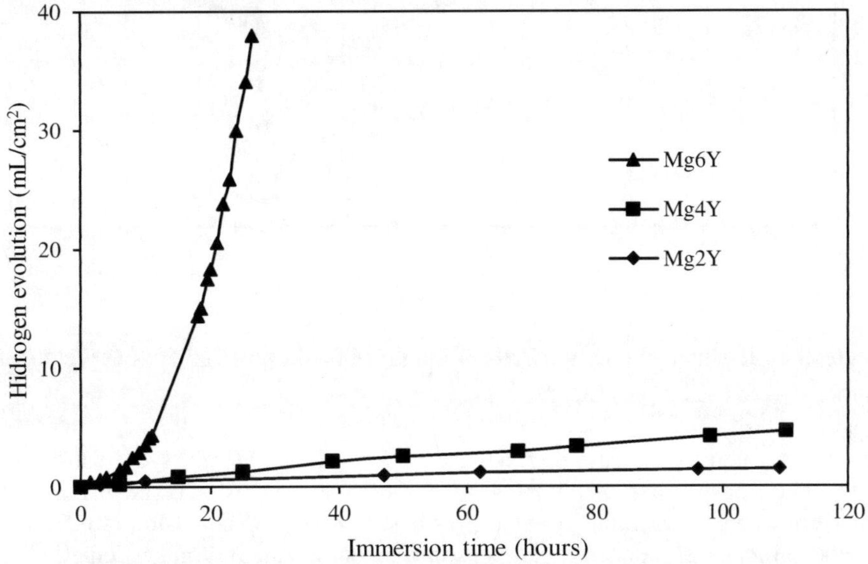

Figure 2. Hydrogen evolution rates measurement on Mg-Y alloys.

3.3 Electrochemical Test

Based on electrochemical test, the data were current density and potential corrosion for each alloy. Tafel extrapolation were used from origin. Current density that could calculated from this origin were 0.96 A/m^2, 0.61 A/m^2, and 0.26 A/m^2 for Mg-2%wt Y, Mg-4%wt Y, and Mg-6%wt Y. I corrosion equals to corrosion rate [3]. So, the corrosion rate decreased gradually with increasing Y content. The corrosion data that resulted based on Electrochemical test and immersion test was contradiction. The differences data caused by differences of modus corrosion between Immersion test and electrochemical test. Time of exposure in medium corrosion from each test was different significant. Formation passive layer depends on time of exposure in medium corrosion. Passive layer gives a tremendous effect on corrosion rate [9]. Passive layer that formed based on electrochemical test and immersion test was different. Passive layer that formed on electrochemical test was uniform and solid while passive layer for immersion test was porous.

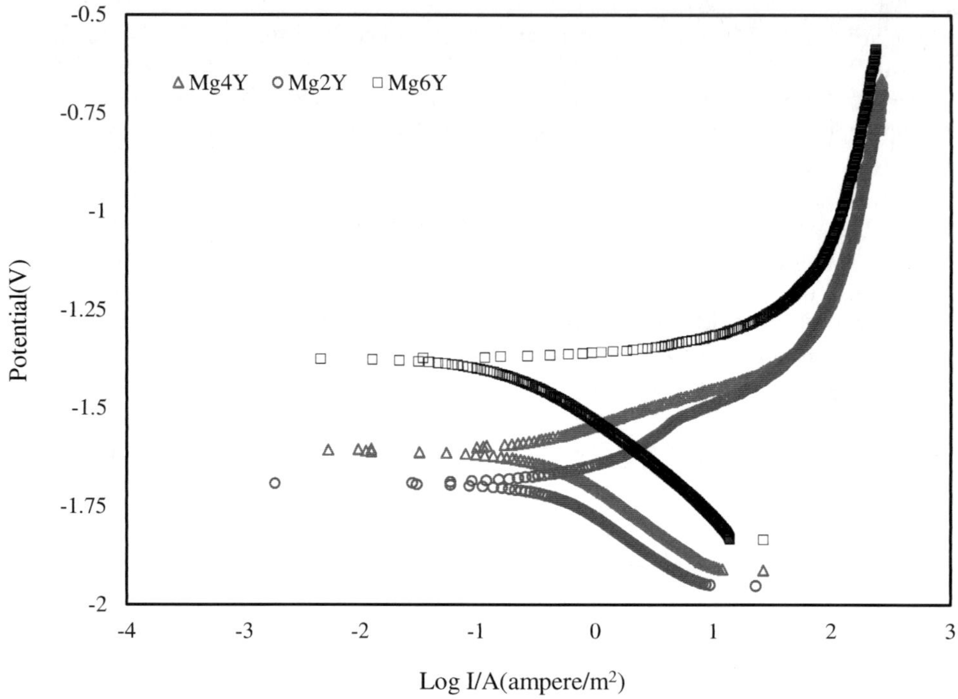

Figure 3. Electrochemical measurement results of magnesium alloy with different yttrium addition.

4. Conclusion

The tensile strength and ductility increase due to increasement of Y content. Mg2Y was best corrosion resistance from the others based on immersion test. Mg6Y was best corrosion resistance from the others based on electrochemical test. The immersion test was best to revealing corrosion rate for those alloys because in electrochemical test, there would be some error in calculation current density due to transitive area that was exposed in electrolyte

Acknowledgment

This research was financially supported by the 2018 PTUPT Research Grant, Ministry of Research, Technology and Higher Education of the Republic of Indonesia.

References

[1] A.C. Hänzi, P. Gunde, M. Schinhammer, P.J. Uggowitzer, On the biodegradation performance of an Mg-Y-RE alloy with various surface conditions in simulated body fluid, Acta Biomater. (2009). doi:10.1016/j.actbio.2008.07.034.
[2] W. Frank, The history of biodegradable magnesium implants: A review, Acta Biomater. (2010). doi:10.1016/j.actbio.2010.02.028.
[3] A. Groysman, Corrosion for everybody, 2010. doi:10.1007/978-90-481-3477-9.
[4] S.G. PAWAR, Influence Of Microstructure On The Corrosion Behaviour Of Magnesium Alloys, School of Materials, 2011.
[5] Y. Zheng, Magnesium Alloys as Degradable Biomaterials, 2015. doi:10.1201/b18932.
[6] K.H. Kim, J.B. Jeon, N.J. Kim, B.J. Lee, Role of yttrium in activation of "c + a" slip in magnesium: An atomistic approach, Scr. Mater. (2015). doi:10.1016/j.scriptamat.2015.06.028.
[7] G.L. Song, Corrosion behavior and prevention strategies for magnesium (Mg) alloys, in: Corros. Prev. Magnes. Alloy. A Vol. Woodhead Publ. Ser. Met. Surf. Eng., 2013. doi:10.1533/9780857098962.1.3.

[8] J. Meng, W. Sun, Z. Tian, X. Qiu, D. Zhang, Corrosion performance of magnesium (Mg) alloys containing rare-earth (RE) elements, in: Corros. Prev. Magnes. Alloy. A Vol. Woodhead Publ. Ser. Met. Surf. Eng., 2013. doi:10.1533/9780857098962.1.38.
[9] I. Johnson, H. Liu, A Study on Factors Affecting the Degradation of Magnesium and a Magnesium-Yttrium Alloy for Biomedical Applications, PLoS One. (2013). doi:10.1371/journal.pone.0065603.

IOP Conf. Series: Materials Science and Engineering **553** (2019) 012007 doi:10.1088/1757-899X/553/1/012007

The Study of the Effect of Slag Ferronickel Substitution (FNS) in Portland Concrete on Corrosion Resistance of Reinforced Steel in Chloride Environment with Cyclic Polarization Method

M Ibnu, Rini Riastuti* A S Ashari, I T Aryani
Department of Metallurgical and Materials Engineering, Faculty of Engineering Universitas Indonesia, Kampus Baru UI Depok 16424, Indonesia.

riastuti@metal.ui.ac.id

Abstract. Ferronickel slag is a by-product of the ferronickel extraction process. Smelting process that contain the oxide compounds such as SiO_2, CaO, Al_2O_3, and Fe_2O_3. The compound content is similar to the content of Portland brand commercial cement (OPC) which is commonly used as raw material for making concrete. Therefore, this research will analyze the effect of using ferronickel slag as a cement mixture on the resistance of reinforced steel in concrete. The author will analyze the corrosion behavior of various mixtures of OPC cement and ferronickel slag with variations in the use of Ferronickel slag as much as 5%, 10%, 15%, 20% with OPC cement in the process of curing concrete for 28 days then it is immersed 3.5% NaCl solution. The concrete wiil uses cyclic polarization method to analyze pitting behaviour and resistance corrosion of reinforced steel in chloride environment. The results show that the steel in the 20% slag concrete mixture added with OPC cement has resistance compared to other mixed variations.

Keywords: *Portland concrete; Corrosion resistance; 3.5% NaCl solution; Cyclic Polarization; Slag ferronickel (FNS)*

1. Introduction

Concrete is one of the most widely used construction materials in the world. Concrete is one of the important components in buildings, the life of a building one of which is influenced by the resistance of concrete to any interference both mechanical and environmental. Indonesia, especially as an archipelago country has a large sea area than land area. The condition like this makes a considerable effect on building design. The existence of aggressive ions due to the process of building damage salinity make faster [1]. In fact, the wind can carry chlorides in the air up to distances beyond 3 km [2]. Whereas the concrete structure is generally expected to survive without repair and maintenance in a long time (up to 100 years). Therefore, the concrete materials constructed as infrastructure in coastal areas should have good corrosion resistance to the presence of aggressive ions such as chloride so that innovation is required on the raw material of cement material in steel reinforcement to address this problem. Corrosion is defined as damage of a material caused by the reaction between the environment and the material itself. Penetration by chloride ions and carbonation are some of the main causes of corrosion especially in concrete [3]. Controlling of corrosion on materials will increase the service life of the material.

Therefore, it is necessary to innovate the raw material for the concrete constituent material to overcome the existing problems.

Concrete is made of the mixture of cement and aggregate. Basically, the concrete is only made of cement and aggregate such as sands. Currently, the mixture of concrete becomes more varied, for example is the usage of Ground Granulated Blast Furnace Slag (GGBS) such as Ferronickel slag. The one of innovation is using ferronickel slag to subtitute cements. Ferronickel slag is a waste or side product from alloying industry of nickel and iron [4]. 1 ton of ferronickel production can produce 14 tons of slag that has composition of SiO_2, MgO, CaO, Al_2O_3, and Fe_2O_3 [5], so that there is a lot of waste or side product that can be used. The compound content is similar to the content of Portland brand commercial cement (OPC) commonly used as raw material for making concrete.

The method that will be used in this study is Cyclic Polarization which is used to analyze reinforcement steel in concrete which is used OPC cement with mixture some cement of Slag when exposed under a corrosive environment such as chloride solution. The method will give some parameters quantitative such as current density of corrosion, corrosion rate, pitting corrosion tendency and passivation behavior which can be analyzed from the polarization curve presented [6].

2. Experimental method

2.1. Materials

The materials of concrete consist of Cement which is a mixture from Ordinary Portland Cement (OPC) and ferronickel slag cement, coarse aggregate, fine aggregate and water. The ferronickel slag cement used is a waste product from PT Antam Pomaala Tbk which have similar composition with Ordinary Portland Cement which can be shown from the result of X-ray Difraction of OPC and ferronickel slag as presented following Table 1.

Table 1. OPC and ferronickel slag chemical compositions.

	CaO	SiO_2	Al_2O_3	Fe_2O_3	SO_3	Basicity	MgO	T.Ni	Cr_2O_3
OPC	66.00	21.10	5.50	3.90	2.75	-	5.00	-	-
Slag	2.7	52.9	2.2	7.6	-	0.63	30.8	0.12	1.1

2.2. Sample preparation

Initially, the reinforcement steel was ground with a series of abranded paper (400, 600, 800, 1000, and 1200) until the oxide scale was removed and the steel surface was smooth. The steel bar was then mounted by using resin and catalyst in order to connect with copper cable for cyclic polarization test. The OPC cement and ferronickel slag as concrete raw materials were ground using a grinding machine to achieve 74 micron or 200 mesh of particle size. The raw materials then were mixed with 0.5 water/binder (w/b) ratios to be filled in cylindrical concrete moulds with a diameter of 10 cm and a length of 20 cm, meanwhile a single steel bar was symentrically positioned in the mold shown in Fig. 1. The cylindrical concrete was then immersed in water for 28 days for curing. Then, the concrete was exposed with NaCl 3.5% solution consisting of sodium chloride in 1 L of aquadest. After that, the concrete specimens in NaCl 3.5% solution were analysed using cyclic polarization.

The ferronickel slag as raw material was ground using abrasive machine to achieve 200 mesh of partikel size. Then, the reinforcement s teel was ground with a type of abranded paper (200,400, 600,800, 100 and 1200) until some oxide scale was removed and the steel was smooth. The steel bar was mounted eith using resin and catalyst. The process of mixture uses 0.5 water/binder (w/c) ratio. After the mixture is complete, the materials will be filled to cylindrical concrete moulds with a diameter of 10 cm and a length of 20 cm. A single steel bar must be symentrically positioned in the mold shown in figure. 1. The cylindrical concrete was the immersed in water for 28 days for curing in order that can get maximum strength. After immersed, the concrete was immersed again in NaCl 3.5% solutions. The concrete in NaCl 3.5% solution were analysed using cyclic polarization in different day variations.

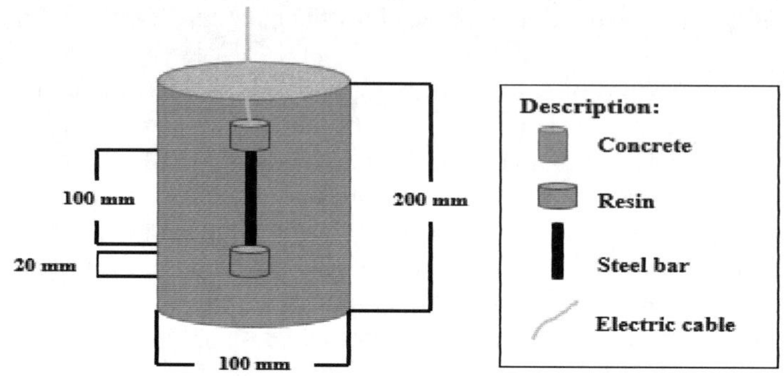

Figure 1. Dimensions of concrete for corrosion testing.

2.3. Concrete composition

The composition of Ordinary Portland Cement and Ferronickel Slag Cement used in this study are shown in Table 2.

Table 2. Formula of OPC and ferronickel slag cement as a raw material in concrete composition.

Composition of Cement	Ferronickel Slag Cement (wt. %)	Ordinary Portland Cement (wt. %)	w/c Ratio
OPC95-FNS5	5	95	0.50
OPC90-FNS10	10	90	0.50
OPC85-FNS15	15	85	0.50
OPC80-FNS20	20	80	0.50

2.4. Cyclic Polarization Test

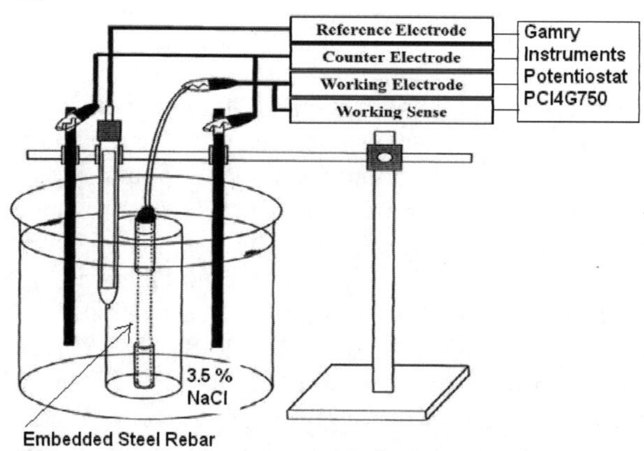

Figure 2. Devices of electrochemical measurement cell.

The Nova Autolab 2.0 with The New Electrochemistry Software from Metrohm Autolab was used for cyclic polarization tests. The electrochemical cell devices consist of the concrete as working electrode, platinum wire as a counter electrode (CE) and Ag/AgCl as reference electrode, the devices are illustrated in Figure 2. The Polarization test is carried out in 4 times test, Namely first, third, sixth, and ten days in immersed NaCl solutions. The polarization curves were determined by setting up time in *open circuit determination* (OPC) until 120 seconds and the upper and lower vortex potential until ±0.095 V.

2.5. Compressive test

The concrete mixture is poured to make compressive test specimens. The compressive strength test refers to ASTM C39/C39M-18 "Standard Test Method for Compressive Strength of Cylindrical Concrete Specimens" [9]. The loading test is carried out with the position of the specimen as shown in Figure 3.

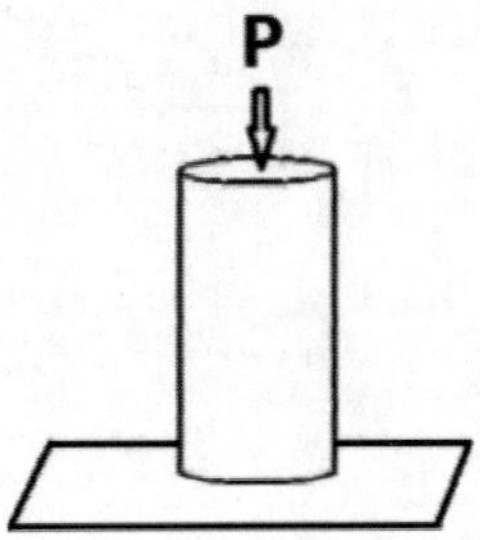

Figure 3. Compressive strength test procedure.

2.6. Carbonation test

The carbonation test was did by dripping the Phenophtalein indicator (PP) on the surfaces of the concrete steel which had been immersed in 3.5% NaCl solutions for 15 days of testing. The colour change in the PP indicator will produce a pink colour if the concrete steel non-carbonated but if the concrete show transparent colour so it is carbonated [3].

3. Results and discussion

3.1. Cyclic polarization curves

The results of reinforcement steel in beton will be shown by the orange polarization curves for OPC85-FNS5, the red curves for OPC90-FNS10, the green curves for OPC85-FNS15 and the blue curves for OPC80-FNS20. All comparison of polarization curve in testing using day variations will be shown from Figure 4.

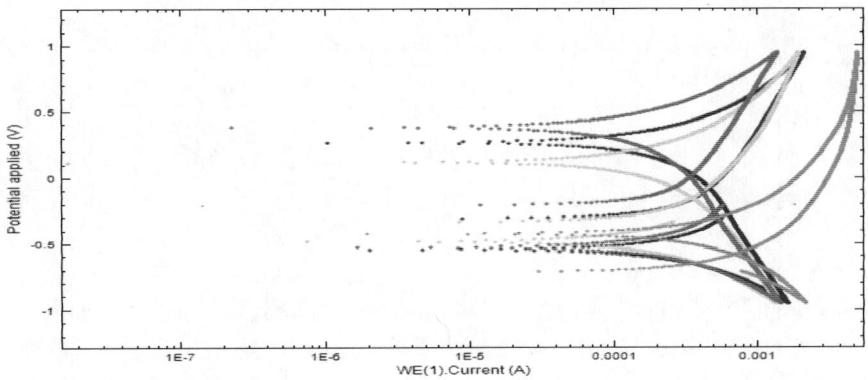

(a)

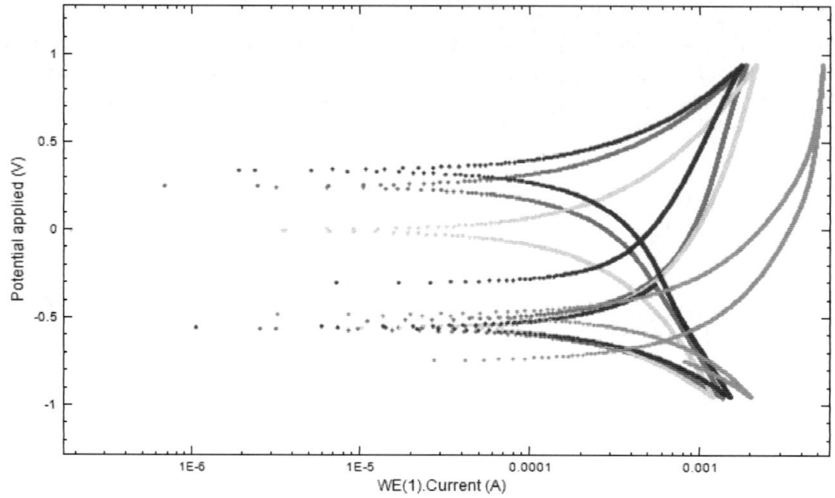

(b)

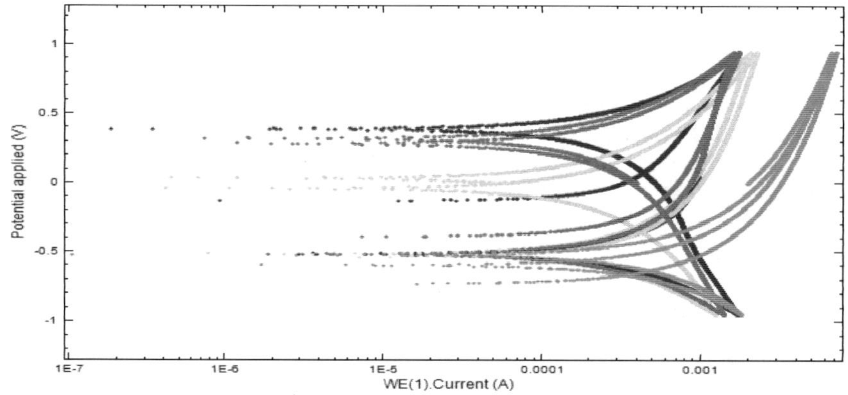

(c)

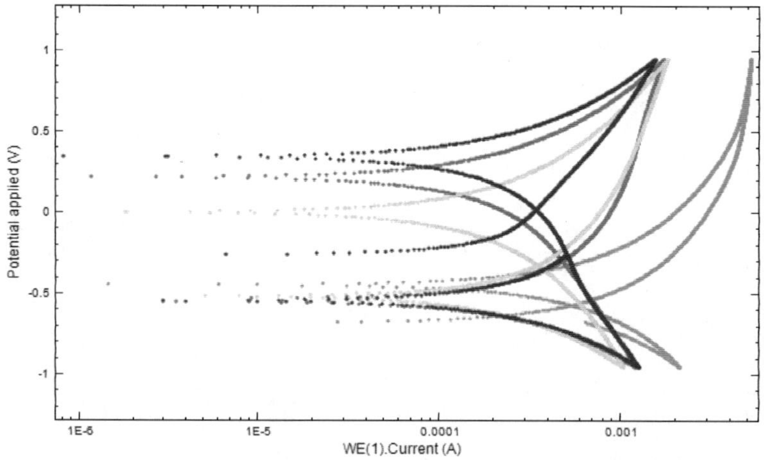

(d)

Figure 4. Comparison of cyclic polarization curve on reinforcement steel in slag and OPC mix in (a) day 1, (b) day 3, (c) day 6 and (d) day 10.

Table 3. Current density values.

Code	Immersion Time (day)	Potential Applied (V)	WE (1). Current (A)	WE (1). Current (A)
OPC95-FNS5	1	-0.775	$2.76x10^{-5}$	-0.773
	3	-0.667	$2.81x10^{-5}$	-0.666
	6	-0.727	$16.31x10^{-6}$	-0.726
	10	-0.848	$17.44x10^{-6}$	-0.846
OPC90-FNS10	1	-0.203	$1.18\ x10^{-5}$	-0.202
	3	-0.512	$10.93x10^{-6}$	-0.511
	6	-0.388	$10.33x10^{-6}$	-0.387
	10	-0.522	$9.07x10^{-6}$	-0.520
OPC85-FNS15	1	-0.329	$8.37x10^{-6}$	-0.329
	3	-0.552	$10.17x10^{-6}$	-0.55
	6	-0.535	$11.12x10^{-6}$	-0.534
	10	-0.509	$7.76x10^{-6}$	-0.507
OPC80-FNS20	1	-0.304	$8.86x10^{-6}$	-0.305
	3	-0.301	$7.2x10^{-6}$	-0.302
	6	-0.128	$0.923x10^{-6}$	-0.127
	10	-0.253	$6.57x10^{-6}$	-0.254

On the first day of testing, the cyclic polarization curve for the anodic reaction side of the OPC95-FNS5, OPC90-FNS10, OPC85-FNS15, and OPC80-FNS20 concrete steel tends to be in the right-to-left position as shown in Figure 4a. The maximum result of corrosion resistance is the OPC90-FNS10 concrete steel because it has lower current density which is 0,37 $\mu A/cm^2$ and the polarization curve is in the far-left position. Meanwhile, the minimum corrosion resistance is the OPC95-FNS5 because it has the highest current density which is about 0,87 $\mu A/cm^2$ and the polarization curve tend to right. Moreover, the tendency of a passive layer that is stable can be seen from the range potential is very large and the current meeting tends to be stable in each composition, meaning that there is no electron flow, the passive layer can keep from electron (corrosion) attacks.

On the third day of testing, the cyclic polarization curves tend to move to left compared to the others so that current density becomes but the most significant movement shown by the OPC80-FNS20. This movement make the OPC80-FNS20 has highest corrosion resistance because has the lower current density which is 0,092 $\mu A/cm^2$ and the lower corrosion resistance shown by the OPC95-FNS5. The OPC95-FNS5 steel concrete has highest current density which is 0,36 $\mu A/cm^2$ and the polarization curve still tend to right position. In this observation, there was no pitting corrosion formed in each concrete steel composition after 3 days of immersion as shown by positive loop direction which mentioned polarization curves.

On the sixth day of testing, The current density of concrete steel in each composition tends to decrease can be seen from the polarization curve which still tends to move leftward, this shows a decrease in corrosion resistance to 3.5% NaCl solution. The polarization curve OPC80-FNS20 shows the maximum results which is a value of 0.012 $\mu A/cm^2$ and the minimal results are shown by the composition of OPC95-FNS5 which is 0.208 $\mu A/cm^2$. In addition, the polarization curve above shows that the passive layer formed in the composition is still stable enough to resist the penetration of chloride which will enter the concrete. Figure.5c also shows that on the sixth day the pitting corrosion has not been formed in each composition, this is indicated by the polarization curve having a positive loop direction.

On the tenth day of testing, the cyclic polarization curves of OPC80-FNS20 steel concrete has the lowest value which means it has the best corrosion resistance. The leftmost location and the highest corrosion potential make OPC80-FNS20 composition have the highest corrosion resistance. But, the composition of OPC95-FNS5 has the lowest corrosion resistance, this already showed from the first day of immersion until the tenth day. This showed by Figure 4d if the polarization curve OPC95-FNS5 has the rightmost curve meaning that it has a large current density and OPC80-FNS20 is located on the far left so that the composition of OPC80-FNS20 is the lowest current density compared to other compositions. The current density is proportional to the electron flow. The current density is proportional to the electron flow. If there are more electron flowing can make reaction of anode became faster. So that, corrosion behaviour even easier to occur. Until the tenth day the symptoms of pitting corrosion were still not visible, it could be seen from the movement of the loop that was still positive.

The corrosion rate on third days of immersion for each concrete steel specimen increased as shown in Table 4. The highest corrosion rate was shown by the OPC95-FNS5 steel concrete which was 28,489 mm/year. The corrosion rate has increased because chloride ion between the interface of steel reinforced and concrete pore solution have increased also but the chloride threshold value still has not exceeded so that pitting corrosion still no formed [8]. The lowest corrosion rate of 0.246 mm/year is shown by concrete steel specimens in OPC80-FNS20. These results indicate that addition of 20% ferronickel slag cement will provide a higher and more stable steel reinforced corrosion resistance compared to other concrete steel composition immersed in 3.5% NaCl Solution.

Table 4. Quantitative parameter taken from cyclic polarization curve of steel reinforced for each concrete steel composition measured in 3.5% of NaCl solution for 3 days of immersed.

Code	Immersion Time (day)	βa (V/decade)	βc (V/decade)	Ecorr (mV)	Icorr (μA/cm^2)	Corrosion Rate (mm/year)
OPC95-FNS5	1	1.19	3.206	-.079	0.001617	18.875
	3	1.56	8.23	-0.74	0.00245	28.489
OPC90-FNS10	1	1.16	-2.97	-0.69	0.001915	22.248
	3	1.073	-2.97	-0.69	0.00168	19.474
OPC85-FNS15	1	0.679	-31.946	-0.54	0.000388	4.507
	3	1.142	-2.33	-0.65	0.00162	18.87
OPC80-FNS20	1	0.6283	-3.29	-0. 597	0.000721	8.37
	3	0.0915	0.0413	-0.607	$2.11x10^{-5}$	0.246

3.2. Compressive test

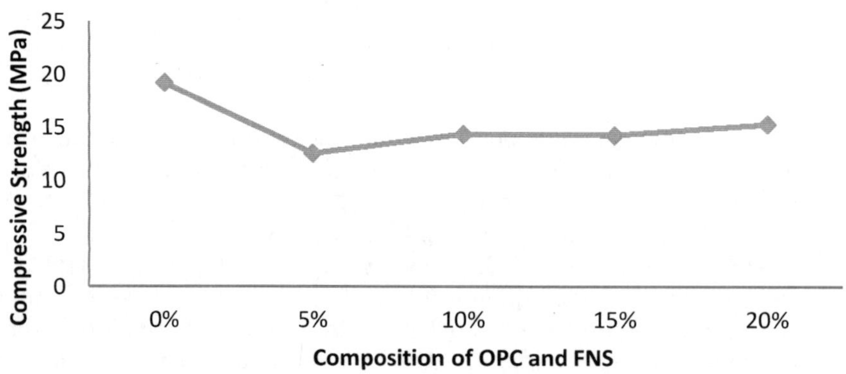

Figure 5. Compressive test result.

From Figure 5, it can be observed that the percentage of slag in each composition does not have a significant difference, it can be seen from the pattern. The maximum compressive strength for cement and ferronickel slag mixtures is in the OPC80-FNS20 composition. The difference in strength with OPC 100 is not far. This shows that the addition of slag as a substitute for cement does not significantly affect the mechanical properties of the concrete, because only a reduction in strength about 4 MPa occurs.

3.3. Carbonation test

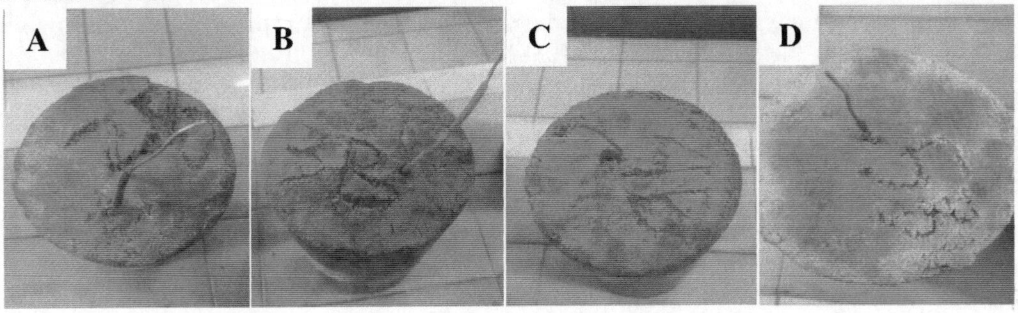

Figure 6. Carbonation test with PP indicator for concrete sample of (a) OPC95-FNS5, (b) OPC90-FNS10, (c) OPC85-FNS15, and (d) OPC80-20.

The discoloration of the concrete specimen into pink after addition of Phenophtalein indicator (PP) indicates that the carbonation reaction did not occur in the words the alkalinity of the concrete tends to be maintained so that the decreasing pH factor does not happen. The decreasing of pH will indicate breakdown of passive layer which have from concrete steel. If passive layer breakdown, the elements like CO_2 and chloride can penetrate into concrete structure so that can cause concrete corrosion [7].

4. Conclusion

The mixture of cement OPC-FNS20 concrete steel has maximum compressive strength than other composition and also provides best corrosion resistance from embedded steel bars compared to other compositions in 10 days immersion because the OPC80-FNS20 has lower current density that could hindered the movement of the electrons so that the reaction of corrosion is slower than other compositions. But in this research, no pitting corrosion was observed in carbon steel bars embedded in all of concrete steel specimen which immersed in 3.5% NaCl Solution. Otherwise, the reaction of carbonation has not occurred in all of concrete steel specimen.

Acknowledgment

The authors would like to thank University of Indonesia, especially Directorate for Research and Community Engagement (DRPM) division, for the financial support through the 2018 PITTA funding scheme No.2519/UN2.R3.1/HKP.05.00/2018 and PT Antam Pomaala Tbk as the provider of ferronickel slag as the raw material in this research.

References

[1] S. Muthulingam, and B. N. Rao, "Non-uniform time-to-corrosion initiation in steel reinforced concrete under chloride environment", *Corrosion Science*, vol. 82, pp. 304–315, May 2014, DOI: 10.1016/j.corsci.2014.01.023

[2] A. Neville, "Chloride attack of reinforced concrete: an overview", *Materials and Structures*, vol.28, pp. 63-70, Mar. 1995, DOI: 10.1007/BF02473172

[3] J. P. Broomfield, *Corrosion of Steel in Concrete, Understanding, Investigation and Repair*, 2nd ed., Jul. 2003, DOI: 10.1201/9781482265491

[4] K. Kokubu, and M. Shoya, "Guidelines for Construction Using Ferronickel Slag Fine Aggregate Concrete", *Concrete Library of JSCE No. 24,* Dec. 1994

[5] A. K. Saha, and P. K. Sarker, "Expansion due to alkali-silica reaction of ferronickel slag fine aggregate in OPC and blended cement mortars", *Construction and Building Mateials*, vol. 123, pp. 135-142, Oct. 2016, DOI: 10.1016/j.conbuildmat.2016.06.144

[6] R. Riastuti, A. Cahyadi, Y. Pratesa, and S. Siallagan, "The study of corrosion resistance ofreinforcement steel embedded in concretecomposed of commercial Portland cement and final tin slag against chloride environment", *IOP Conference Series: Materials Science and Engineering*, vol. 431, pp. 052006, Nov. 2018, DOI: 10.1088/1757-899X/431/5/052006

[7] N. Perez, *Electrochemistry and Corrosion Science*, Jan. 2004, DOI: 10.1007/b118420

[8] R. W. Revie, and H. H. Uhlig, *Corrosion and Corrosion Control: An Introduction to Corrosion Science and Engineering*, 4th ed., Mar. 2008, DOI:10.1002/9780470277270

[9] ASTM C39/C39M-18, "Standard Test Method for Compressive Strength of Cylindrical Concrete Specimens", ASTM International, West Conshohocken, PA, 2018, DOI: 10.1520/C0039_C0039M-18, www.astm.org

IOP Conf. Series: Materials Science and Engineering **553** (2019) 012008 doi:10.1088/1757-899X/553/1/012008

Effects of Poling on Electrical Properties of Flexible Piezoelectric Composites with Natural Rubber Matrix

Methee Promsawat[1,*], Kan-ubon Boonsri[1], Sukanya Samadoloh[1], Napatporn Promsawat[2] and Ekwipoo Kalkornsurapranee[1]

[1] Department of Materials Science and Technology, Faculty of Science, Prince of Songkla University, 15 Karnjanavanich Road, Hat Yai, Songkhla 90112, Thailand

[2] Synchrotron Light Research Institute, 111 University Avenue, Muang, Nakhon Ratchasima 30000, Thailand

methee.p@psu.ac.th

Abstract. This work was done to study effects of electrical applied voltage on electrical properties of flexible piezoelectric composites with natural rubber (NR) matrices. Different amounts of lead magnesium niobate titanate (PMNT) as a piezoelectric phase were added into NR matrices. NR/*x*PMNT (where x = 0, 60, 80, 100, 120 and 150 phr) composites were consequently obtained. The composites were poled under 25 kV/cm electric field for 30 mins. For all the samples, tetragonality is enhanced after poling. Dielectric and ferroelectric properties, and generated electrical voltage of unpoled and poled samples were measured. For the samples with $x \leq 80$ phr, the properties are not improved with the poling process. However, for $x > 80$ phr samples, the properties are significantly enhanced after poling. A percent increase in generated electrical voltage after poling tends to increase with increasing PMNT concentration. The overall result suggests that the NR/100PMNT composite is suitable for applications as high mechanical damage resistance piezoelectric sensors and generators.

Keywords: Piezoelectric composite, Natural rubber, PMNT, Energy harvesting

1. Introduction

Piezoelectric material is a type of smart materials that plays an important role in the technology development in the past several decades. It has been widely used as piezoelectric generators, sensors, actuators, and transducers for a wide range of applications such as hard disk drive, vehicle, medical instrument industries and etc. [1-4]. Among these applications, piezoelectric sensors and actuators have been mostly received much attention for a continuous development to reach their properties to the requirement. From searching for new piezoelectric/ferroelectric materials, lead magnesium niobate titanate [$Pb(Mg_{1/3}Nb_{2/3})_{1-x}Ti_xO_3$], had been invented. It was found that the material with x = 0.30-0.35 exhibited excellent electromechanical properties i.e. an extreme high piezoelectric coefficient (d_{33} = 600 pC/N) and a high electromechanical coupling factor (k_p = 0.7) [5, 6]. Because of its attractive piezoelectric properties, the material has been selected to use as piezoelectric sensors and actuators. However, in actual applications, they often crack and eventually fail after a large cycling number of electrical or mechanical loads is applied. This is due to a high brittleness of the materials. To solve this problem, many research investigating about the fabrication of piezoelectric materials with high mechanical failure resistance have been done. Most of them has fabricated piezoelectric materials in forms of composites, having high mechanical failure resistances [7-9]. In the preparation of flexible piezoelectric composites, piezoelectric materials were embedded in various types of polymers such as polyurethane [10, 11], epoxy [12, 13], polydimethylsiloxane [14] and polyvinylidene fluoride [15, 16]. Natural rubber (NR) has been an important economic plant of Thailand for a long time. Because its excellent mechanical properties such as a high toughness and a high elongation at break, it is selected as a subject for a wide range of applications [17]. Due to the high mechanical toughness, in this research, natural rubber was used as the matrix phase of piezoelectric composites with $Pb(Mg_{1/3}Nb_{2/3})_{0.65}Ti_{0.35}O_3$ (PMNT) particles as the dispersed phase. It is well known that for piezoelectric materials, their piezoelectric properties are enhanced after a high electric field is applied [18, 19]. Based on reviewing literature, there are no research studying electrical poling on piezoelectric composite with NR matrix.

Therefore, in this study, in the first step, NR/xPMNT (when x = 60, 80, 100, 120 and 150 phr) flexible piezoelectric composites were fabricated. The composites were then electrically poled under a high electric field. Significant electrical properties, i.e. dielectric and ferroelectric properties, and voltage generation performance, of unpoled and poled samples were measured and comparatively analyzed. The result shows that the properties of the composites with high PMNT content were enhanced after poling. The observed result is then discussed and explained in detail. We hope that this research study can develop a new flexible piezoelectric material suitable for piezoelectric sensor and generator applications.

2. Experimental method

$Pb(Mg_{1/3}Mg_{2/3})_{0.65}Ti_{0.35}O_3$ powder modified with ZnO was prepared by a columbite route technique. $MgNb_2O_6$ (columbite) powder was firstly synthesized by using MgO (99%, TOPLUS, China) and Nb_2O_5 (99.99%, TOPLUS, China) precursors. $MgNb_2O_6$ powder was then mixed with the predetermined amounts of PbO (99.5%, TOPLUS, China), TiO_2 (99%, TOPLUS, China) and ZnO (99.7%, TOPLUS, China) powders. The mixture was calcined at 900°C for 5 hr. Different amounts of PMNT powders were added into natural rubber (NR) matrices by using an internal mixer (Rheomix OS, Thermo Scientific, USA) to prepared NR/xPMNT (where x = 60, 80, 100, 120 and 150 phr, which are equivalent to 38, 44, 50, 55 and 60 wt%, respectively) mixtures. NR/PMNT composites with 1 mm thickness were formed by using a compression molding under a 2500 psi pressure at 150ºC. The NR/PMNT composites were poled under an electric field of 25 kV/cm for 30 min at 60°C. Phase, microstructure and chemical composition were characterized by an X-ray diffraction technique (XRD, Rigaku SmartLab, Japan), a scanning electron microscope (SEM, JSM-IT300LV, JEOL, Japan) and energy dispersive X-ray spectrometer (EDX, JSM-IT300LV, JEOL, Japan), respectively. Dielectric and ferroelectric properties, and generated voltages were measured by an LCR meter (GW Instek, LCR-821, Taiwan), a Sawyer-Tower circuit, and an oscilloscope (Rigol, DS1052E, USA), respectively. Effects of the electrical poling on the properties of the composites were demonstrated and discussed.

3. Results and discussion

Figure 1 shows XRD patterns of the composites with 60, 120 and 150 phr PMNT before and after poling. The main peaks (unlabeled peaks) are belonged to a perovskite phase of $Pb(Mg_{1/3}Nb_{2/3})_{0.65}Ti_{0.35}O_3$ which corresponds to ICSD file no. 159255. It is found that the reflections of pyrochlore, ZnO and unknown phases are also appeared in the XRD patterns of all the unpoled samples. It is believed that the observed pyrochlore phase is formed during PMNT preparation process, at which PbO preferably reacts with Nb_2O_5 [20], while ZnO is an additive compound of NR. The unknown phase is possibly formed from an interaction between chemical ingredients composing in the NR material. However, the unknown phase peak disappears after the samples are poled.

IOP Conf. Series: Materials Science and Engineering **553** (2019) 012008 doi:10.1088/1757-899X/553/1/012008

Enlarged views of peaks at $2\theta \approx 45°$ of corresponding patterns are shown in Fig. 1(b). For each sample, the peak is obviously split after poling. Moreover, it can be seen from the result that the ratio of the intensity of (002) peak to that of (200) peak increases after poling. This indicates the increase in the amount of the (002) plane orienting its direction parallel to the X-ray beam which is perpendicular to the sample's surface (electrodes). It is know that for a tetragonal ferroelectric an electrical polarization directs along *c* axis which is parallel to the (002) plane direction. That means that the number of the electrical polarization aligning perpendicular to sample's surface increases after poling. However, the peak intensity ratio does not change with the variation of PMNT content. The change of crystal orientation can result in an enhancement of electrical properties of the composites which will be discussed later.

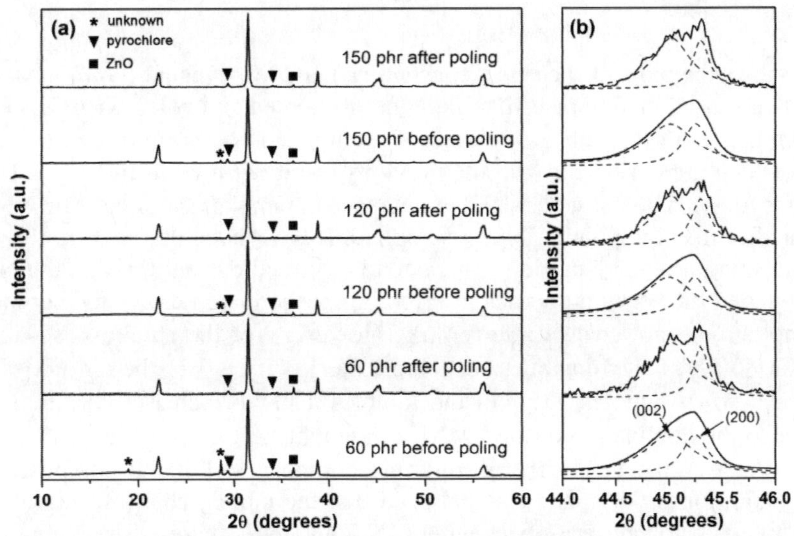

Figure 1. (a) XRD patterns before and after poling of composites with 60, 120 and 150 phr PMNT. Unmarked peaks are belonged to a perovskite phase. (b) Enlarged view of the diffraction peak at $2\theta \approx 45°$. The peaks of (b) are corresponded to those of (a).

The chemical compositions of NR/PMNT composites were measured by mean of energy dispersive X-ray spectroscopy (EDX) technique. In this experiment, the fracture surfaces of samples after tensile testing were examined. The EDX spectrum of NR/100PMNT composite is shown in Fig. 2. The observed particle contains Pb, Mg, Nb, Ti, Zn and O elements. This confirms that the presented particles are PMNT piezoelectric materials which well distributes and partially agglomerates in NR matrix. It can also be seen from the figure that the NR matrix possesses micro-voids. These voids are possibly created from pulling out of PMNT particles during the tensile testing. From the observation of microstructure of composites (the result do not show here), the amount and the size of voids increase with increasing PMNT content. This reveals that a large amount of the dispersed phase degrades the bonding performance between them and the matrix phase.

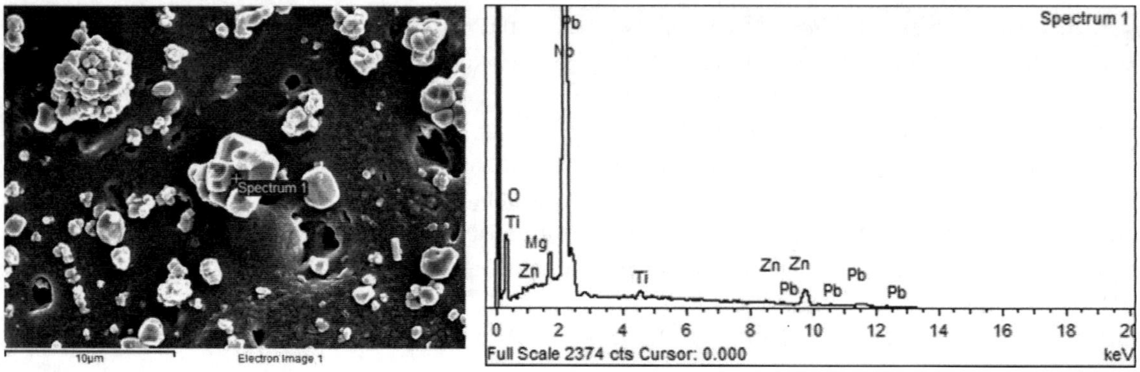

Figure 2. A chemical composition by mean of energy dispersive X-ray spectroscopy (EDS) technique of NR/100PMNT composite.

Figure 3(a) shows the plots of dielectric constant (ε_r) and loss tangent (tanδ) measured at room temperature at 200 kHz as a function of PMNT content of unpoled and poled NR/PMNT composites. It can be seen from the result that the ε_r and tanδ of an unpoled NR are found to be about 3.3 and 0.07, respectively, which are consistent with the previous result reported in Ref [21]. The properties of the NR are degraded, i.e. a decrease in ε_r and an increase in tanδ, after poling. The NR sample was seemed to be burnt after the high poling field was applied. It is believed that mobile charges included in the NR material, acting as charge carriers, conduct an electrical current through the NR bulk [22]. This causes the degraded dielectric properties after poling of the material. For the composites with x = 60-80 phr, ε_r and tanδ do not change after poling. However, for the composites with the PMNT content greater than 80 phr, the ε_r obviously develops after poling while the tanδ does not change. For both unpoled and poled samples, the ε_r of the samples tends to increase with increasing PMNT content. This is due to the domination of the PMNT phase which possesses higher ε_r as compared to the NR phase. This causes the increment of ε_r of the composites. For the composites with large amounts of PMNT additions (> 80 phr), it is believed that the mobile charges, migrating under the applied field, are trapped at interfaces between PMNT and NR phases. The trapping process is consistent with that proposed in the previous work studied by Ortiz-Serna *et al.* [23]. Moreover, the charge carriers could be trapped by space charges produced from the PMNT phase. For piezoelectric materials, space charges in forms of cation and oxygen vacancies, which are electrically negative and positive, respectively, can be formed [24]. This contributes to the reduction of electrical conductivity of the composites, which is corresponded to the less value of tanδ. Meanwhile, the ε_r of the composites is improved.

Remanent polarization (P_r) and coercive field (E_c) measured before and after poling are plotted as a function of PMNT content as shown in Fig. 3(b). The properties were measured at room temperature and at 25 Hz. For the composites with $x \leq 80$ phr, P_r does not change after poling. As x is beyond 80 phr, the P_r apparently develops after the composites are poled. For all the composites, the E_c is enhanced after poling. The enhancement of P_r of the composites is attributed to the de-aging effect. Generally, for ferroelectric materials containing charged defects, i.e. cation and anion vacancies, these defects can form defect dipoles. The stabilization of the alignment of defect dipoles in ferroelectric materials increases with time. This causes the decrease in P_r and E_c of the materials [25]. It is known that an aging state of a ferroelectric material is practically recovered, called a de-aging process, by a thermal annealing or an application of an appropriate electric field [26]. Although aging effect is dominated in ferroelectric materials in the form of ceramics, however, in this research, it is believed that the aging process also has a significant effect on ferroelectric particles.

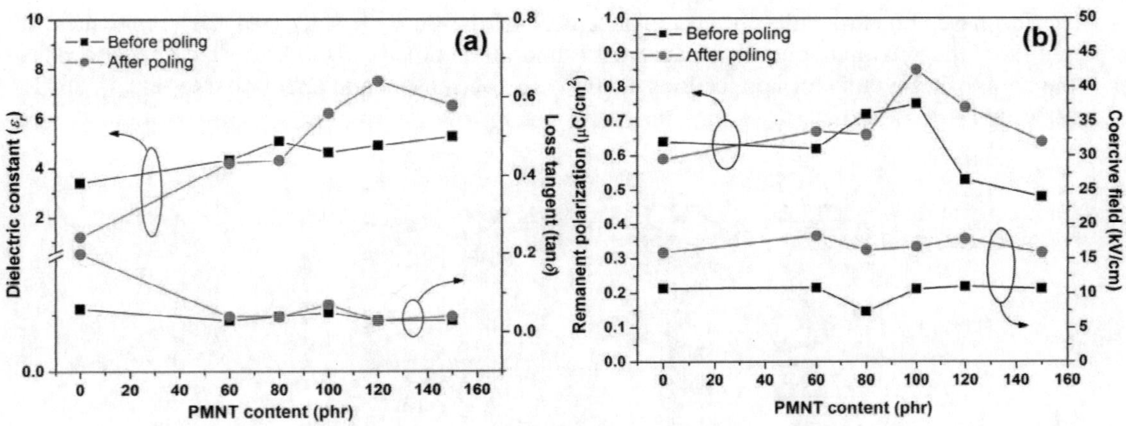

Figure 3. (a) Room temperature dielectric constant and loss tangent measured at 200 kHz and (b) remanent polarization and coercive field before and after poling of NR/PMNT composites.

Table 1. The generation of voltage of NR/PMNT composites under a 300 N impact load.

PMNT content (phr)	Open-circuit output voltage (V)		% increase in output voltage after poling
	Before poling	After poling	
0	0.24 ± 0.02	0	-
60	0.89 ± 0.07	1.08 ± 0.13	21
80	0.93 ± 0.05	1.68 ± 0.15	81
100	1.61 ± 0.05	2.42 ± 0.19	50
120	0.98 ± 0.04	2.07 ± 0.06	111
150	0.71 ± 0.03	1.97 ± 0.09	177

Average values of open-circuit output voltage generated under a 300 N impact load measured before and after poling of NR/PMNT composites are given in Table 1. The result suggests that the electric field application has a significant effect on the generation of output voltage of NR/PMNT composites. From the overall observed result, the enhanced electrical properties of NR/PMNT composites could be caused by the orientation of dipole moments inside the piezoelectric phase. From XRD result analysis, the amount of (002) plane orienting its direction to an applied field direction increases after the field is applied, as schematically shown in Fig. 4. A higher number of dipole moments remaining their directions in an applied field can promote a higher P_r of a ferroelectric. Subsequently, this is accompanied with a higher interaction strength between all the dipole moments in the material. A higher activation energy for the dipole moments to switch their directions is required, resulting in an increment of E_c. This also contributes to a higher piezoelectricity of the poled samples. Generally, for d_{33} mode, at which electric charge is generated in direction 3 (polarization direction) when the material receives a mechanical force in the direction 3. Therefore, a ferroelectric with a higher number of dipole moments aligning in direction 3 possesses a greater electric charge when a similar force is applied. Moreover, an enhancement of the voltage production of poled samples could be partially caused from a trapping of mobile charges at phase boundaries and a de-aging effect. For the trapping effect, it causes a reduction of electrical conductivity materials which can promote a separation of generated positive and negative charges. This is supported by the decrease in dielectric loss of the PMNT-added samples. For the de-aging effect, a high electric field application promotes an ordering of dipole moments of ferroelectric materials [26]. These effects subsequently lead to the enhancement of piezoelectric properties of the NR/PMNT composites. A percent increment in the output voltage of composites tends to increase with increasing PMNT content, as shown in Table 1. It is reasonably believed that an increase in piezoelectric phase content results in the enhancement of the piezoelectricity of the composites. It can be seen that the maximum output voltage was found in the poled NR/100PMNT composite. This makes the composite suitable for applications as piezoelectric sensors and energy harvesters.

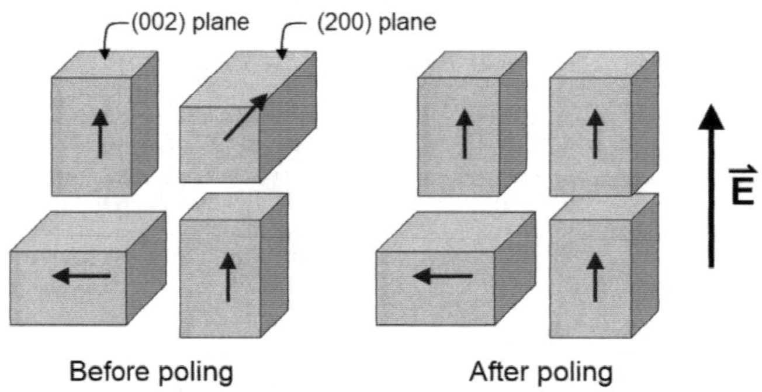

Figure 4. A schematic diagram showing the orientations of crystal structure of PMNT phase before and after poling. A blue arrow represents a direction of a dipole moment embedded in a unit cell (yellow cube). The applied field direction, which is perpendicular to the electroded-surfaces, is also shown.

4. Conclusion

Flexible piezoelectric composites with a formula of NR/xPMNT (when x = 0, 60, 80, 100, 120 and 150 phr) were successfully fabricated. Dielectric and ferroelectric properties, and voltage generation performances of composites before and after poling were measured and comparative studied. The result shows that dielectric constant and remanent polarization of composites with $x \geq 100$ phr are enhanced after poling while those of composites with $x < 100$ phr do not change after poling. For voltage generation of the composites with $x \leq 60$ phr, their output voltage values do not change after poling while those of the others raises up after the composites are poled. For all the composites, the electrical poling causes the increase in the coercive field. In addition, the percent increase in the increase in output voltage after poling tends to increase with increasing x. The improvement of these electrical properties of the composites are explained in the terms of charge trapping effect, de-aging effect and increased degree of polarization ordering. We hope that this research study leads the development of a new flexible piezoelectric composite that is suitable for applications as piezoelectric sensors and energy harvesters.

Acknowledgment

This research was supported by Natural Rubber Innovation Research Institute, Prince of Songkla University (Grant No. SCI601093S).

References

[1] O. Noclanc, P. Gaucher, and G. Calvarin, "Structural and dielectric properties of $Pb(Mg_{1/3}Nb_{2/3})O_3$-$PbTiO_3$ ferroelectric solid solutions around the morphotropic boundary," *Journal of Applied Physics*, vol.79, pp. 4291-4297, Jan. 1996 (doi: 10.1063/1.361865)

[2] B. Noheda, D.E. Cox, G. Shirane, J.A. Gonzalo, L.E. Cross, and S.-E. Park, "A monoclinic ferroelectric phase in the $Pb(Zr_{1-x}Ti_x)O_3$ solid solution," *Applied Physics Letters*, vol.74, pp. 2059-2061, Apr. 1999 (doi: 10.1063/1.123756)

[3] M. Alguero, A. Moure, L. Pardo, J. Holc, and M. Kosec, "Processing by mechanosynthesis and properties of piezoelectric $Pb(Mg_{1/3}Nb_{2/3})O_3$-$PbTiO_3$ with different

compositions," *Acta Materialia*, vol.54, pp. 501-511, Jan. 2006 (doi: 10.1016/j.actamat.2005.09.020)

[4] P. Moetakef and Z.A. Nemati, "Synthesis of pyrochlore free PMN-PZT ceramics via a seeding method," *Sensors and Actuators A: Physical*, vol.141, pp. 463-470, Feb. 2008 (doi: 10.1016/j.sna.2007.10.005)

[5] M. Promsawat, A. Watcharapasorn, Z.G. Ye, and S. Jiansirisomboon, "Enhanced dielectric and ferroelectric properties of $Pb(Mg_{1/3}Nb_{2/3})_{0.65}Ti_{0.35}O_3$ ceramics by ZnO modification," *Journal of the American Ceramic Society*, vol.98, pp. 848-854, Mar. 2015 (doi: 10.1111/jace.13372)

[6] M. Promsawat, Z.G. Ye, and A. Watcharapasorn, "Enhancement of piezoelectric properties of $Pb(Mg_{1/3}Nb_{2/3})_{0.65}Ti_{0.35}O_3$ ceramics by ZnO modification," *Materials Letters*, vol.205, pp. 126-129, Jun. 2017 (doi: 10.1016/j.matlet.2017.06.070)

[7] J. Li, J. Claude, L.E. Norena-Franco, S.I. Seok, and Q. Wang, "Electrical energy storage in ferroelectric polymer nanocomposites containing surface-functionalized $BaTiO_3$ nanoparticles," *Chemistry of Materials*, vol.20, pp. 6304-6306, Oct. 2008 (doi: 10.1021/cm8021648)

[8] P. Guggilla, A.K. Batra, and M.E. Edwards, "Electrical characterization of $LiTaO_3$:P(VDF-TrFE) composites," *Journal of Materials Science*, vol.44, pp. 5469-5474, Oct. 2009 (doi: 10.1007/s10853-009-3753-8)

[9] X. Huang, L. Xie, P. Jiang, G. Wang, and F. Liu, "Electrical, thermophysical and micromechanical properties of ethylene-vinyl acetate elastomer composites with surface modified $BaTiO_3$ nanoparticles," *Journal of Physics D: Applied Physics*, vol.42, pp. 245407/1-10, Nov. 2009 (doi: 10.1088/0022-3727/42/24/245407)

[10] W.K. Sakamoto, P. Marin-Franch, and D.F. Das-Gupta, "Characterization and application of PZT/PU and graphite doped PZT/PU composite," *Sensors and Actuators A: Physical*, vol.100, pp. 165-174, Sep. 2002 (doi: 10.1016/S0924-4247(02)00042-0)

[11] H. Khanbareh, K. de Boom, B. Schelen, R.B.N. Scharff, C.C.L. Wang, S. van der Zwaag, and P. Groen, "Large area and flexible micro-porous piezoelectric materials for soft robotic skin," *Sensors and Actuators A: Physical*, vol.263, pp. 554-562, Aug. 2017 (doi: 10.1016/j.sna.2017.07.001)

[12] S.A. Wilson, G.M. Maistros, and R.W. Whatmore, "Structure modification of 0–3 piezoelectric ceramic/polymer composites through dielectrophoresis," *Journal of Physics D: Applied Physics*, vol.38, pp. 175-182, Jan. 2005 (doi: 10.1088/0022-3727/38/2/001)

[13] Z. Zeng, L. Gai, A. Petitpas, Y. Li, H. Luo, D. Wang, and X. Zhao, "A flexible, sandwich structure piezoelectric energy harvester using PIN-PMN-PT/epoxy 2-2 composite flake for wearable application," *Sensors and Actuators A: Physical*, vol.265, pp. 62-69, Oct. 2017 (doi: 10.1016/j.sna.2017.07.059)

[14] I. Babu, and G. de With, "Highly flexible piezoelectric 0-3 PZT-PDMS composites with high filler content," *Journal of Physics D: Composites Science and Technology*, vol.91, pp. 91-97, Jan. 2014 (doi: 10.1016/j.compscitech.2013.11.027)

[15] S.H. Ji, J.H. Cho, Y.H. Jeong, J.H. Paik, J.D. Yun, and J.S. Yun, "Flexible lead-free piezoelectric nanofiber composites based on BNT-ST and PVDF for frequency sensor applications," *Sensors and Actuators A: Physical*, vol.247, pp. 316-322, Aug. 2016 (doi: 10.1016/j.sna.2016.06.011)

[16] I. Jung, Y.H. Shin, S. Kim, J.Y. Choi, and C.Y. Kang, "Flexible piezoelectric polymer-based energy harvesting system for roadway applications," *Applied Energy*, vol.197, pp. 222-229, Jul. 2017 (doi: 10.1016/j.apenergy.2017.04.020)

[17] S. Kohjiya and Y. Ikeda, "Chemistry, manufacture and applications of natural rubber," United Kingdom: Woodhead Publishing Limited, 2014.

[18] X.-H. Du, U. Belegundu, and K. Uchino, "Crystal orientation dependence of piezoelectric properties in lead zirconate titanate: Theoretical expectation for thin films," *Japanese*

Journal of Applied Physics Part 1, vol.36, pp. 5580-5587, Sep. 1997 (doi: 10.1143/JJAP.36.5580)

[19] H. Fu and R. E. Cohen, "Polarization rotation mechanism for ultrahigh electromechanical response in single-crystal piezoelectrics," *Nature*, vol.403, pp. 281-283, Jan. 2000 (doi: 10.1038/35002022)

[20] A. Mergen and W.E. Lee, "Fabrication, characterization and formation mechanism of $Pb_{1.83}Mg_{0.29}Nb_{1.71}O_{6.39}$ pyrochlore," *Journal of European Ceramic Society*, vol.17, pp. 1033-1047, 1997 (doi: 10.1016/S0955-2219(96)00187-2)

[21] S. Salaeh, N. Muensit, P. Bomlai, and C. Nakason, "Ceramic/natural rubber composites: influence types of rubber and ceramic materials on curing, mechanical, morphological, and dielectric properties," *Journal of Materials Science*, vol.46, pp. 1723-1731, Mar. 2011 (doi: 10.1007/s10853-010-4990-6)

[22] G.C. Psarras, E. Manolakaki, and G.M. Tsangaris, "Electrical relaxations in polymeric particulate composites of epoxy resin and metal particles," *Composites Part A: Applied Science and Manufacturing*, vol.33, pp. 375-384, Mar. 2002 (doi: 10.1016/S1359-835X(01)00117-8)

[23] P. Ortiz-Serna, M. Carsí, B. Redondo-Foj, and M.J. Sanchis, "Electrical conductivity of natural rubber–cellulose II nanocomposites," *Journal of Non-Crystalline Solids*, vol.405, pp. 180-187, Dec. 2014 (doi: 10.1016/j.jnoncrysol.2014.09.026)

[24] G.H. Haertling, "Ferroelectric ceramics: history and technology," *Journal of the American Ceramic Society*, vol.82, pp. 797-818, 1999 (doi: 10.1111/j.1151-2916.1999.tb01840.x)

[25] Y.A. Genenko, J. Glaum, M.J. Hoffmann, and K. Albe, "Mechanisms of aging and fatigue in ferroelectrics," *Materials Science and Engineering B*, vol.192, pp. 52-82, Feb. 2015 (doi: 10.1016/j.mseb.2014.10.003)

[26] K. Carl and K.H. Haerdtl, "Electrical after-effects in $Pb(Ti,Zr)O_3$ ceramics," *Ferroelectrics*, vol.17, pp. 473-486, Mar. 1978 (doi: 10.1080/00150197808236770)

Magnetization Dynamic Analysis of Square Model CoFe and CoFeB Ferromagnetic Materials Using Micromagnetic Simulation

Qoimatul Mustaghfiroh[1], Candra Kurniawan[1,2], Dede Djuhana[1*]
[1]Departement of Physics, Faculty of Mathematics and Natural Science, Universitas Indonesia, KampusBaru UI Depok 16424, Indonesia
[2]Research Center for Physics, Indonesian Institute of Sciences (LIPI), Puspiptek Office Area Tangerang Selatan 15314, Indonesia

dede.djuhana@sci.ui.ac.id

Abstract. In this study, dynamic magnetization of square model CoFe and CoFeB ferromagnetic materials were observed using micromagnetic simulation based on LLG equation. The geometrical side size was varied from 50 to 500 nm with the thickness of 5 nm and 10 nm. For simulation process, the used damping factor was 0.05 and the cell size of $2.5\times2.5\times2.5$ nm^3 was used with respect to exchange length of CoFe and CoFeB. The external magnetic fields were applied in in-plane and out-plane direction to generate magnetic hysteresis loop. It is found that the coercivity decreased as square size increased for both in-plane and out-plane magnetization direction. The coercivity were around 40 to 200 mT for in-plane field magnetization of CoFe. The coercivity tends to constant at 40 mT in diameter less than 100 nm and zero coercivity for diameter greater than 100 nm for out-plane field magnetization. Compared to CoFe, the coercivity in out-plane field is higher than in-plane field in CoFeB square. It is also observed that the switching time and nucleation field increased as the size increased in out-plane direction of both CoFe and CoFeB. The results showed that the different characteristics of magnetic anisotropy of both materials are important in the development of high density magnetic storage.

Keywords :CoFe, CoFeB, hysteresis, micromagnetic, magnetic anisotropy

1. Introduction

Previous publications showed that the Fe-based nanoelements were highly used because of the cheaper costs, easier to make, and easier modified [1]. Currently, many nanostructures based on CoFe and CoFeB ferromagnetic materials were developed because of some interesting properties. The CoFe based materials has unique properties that are widely used for applications requires high flux density such as data storage and electronic devices [2]. Soft magnetic CoFe thin film with high anisotropy 2.7 $\times 10^3$ J/m^3, high Curie temperature, good permeability has been paid great attention [3]. To get material with large anisotropy more than 2.7 x 10^3 J/m^3, an additive material Boron was added to the CoFe structure to be CoFeB thin film [4]. By influence of Boron element, it was found that important characteristics of perpendicular magnetic anisotropy (PMA) was emerged [5]. Therefore the dynamic magnetization process in CoFe and CoFeB become an interesting topic and can be studied by experiments and micromagnetic simulation [6]. The PMA materials raise great interest due to their potential in magnetic memory device as high-density non-volatile memory. The non-volatile memory reduces power consumption than volatile that requiring 50% power [7]. The arrangement of magnetic moment in PMA is perpendicular so that made PMA memory become small in size, high capacity, and no thermal fluctuation due to the superparamagnetic effect [8]. The superparamagnetic effect make material has zero coercivity and the information saved will be lost. Compared to longitudinal type memory, PMA can achieve more than 100 times capacity with smaller size memory [5].

When material size in the range of nanometer scale, different characteristics of magnetic, optic, electric, and mechanic will be obtained compared to bulk material [9]. These different characteristics caused by the higher surface to volume ratio of materials when in nanometer scale than bulk scale. Reactivity of material determined by the number of atom existed in the surface area. Quantum effect dominate the nanometer scale material to restrict electron movement in particle and give change to mechanical hardness, electric conductivity, and magnetization [10]. Therefore, the size dependence of nanometer size materials to its characteristics is high compared to the bulk material counterpart.

In this work, we have investigated the magnetic behavior of squared model CoFe and CoFeB thin film and explored the PMA behavior by micromagnetic simulation. Hysteresis loop provide magnetic characteristics so that intrinsic magnetic behavior can be explored from CoFe and CoFeB materials. The difference values of coercivity in in-plane and out-plane field was determined the PMA behavior. Therefore, comparable magnetic characteristics between CoFe and CoFeB will be obtained through the size optimization (thickness and diameter).

2. Simulationmethod

The PMA behavior were studied by micromagnetic simulations using Object Oriented Micromagnetic Framework (OOMMF) based on Landau Lifshitz Gilbert (LLG) equation [11]. Micromagnetic simulation used the approximation to real material by discretization of cell size less than exchange length. In this work we use 2.5×2.5×2.5 $nm^3$of cell size for both CoFe and CoFeB. For best rapprochement of simulation and experiment, we use square shape thin film. The phenomenological damping term (α) that accounts for the dissipation and lead to parallel alignment of the magnetic moment were set to be $\alpha = 0.05$. The external magnetic field was given in two directions along x-axis (in-plane) and z-axis (out-plane) with magnitude of 1000 mT for CoFe and 2000 mT for CoFeB to obtain the hysteresis curves.

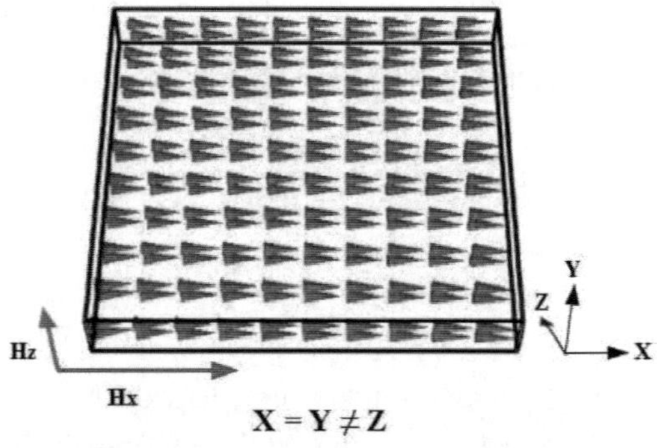

Figure 1. Geometry of thin film CoFe and CoFeB square given in-plane and out-plane field.

The geometrical side size of thin film CoFe and CoFeB were considered in this work from 50 nm to 500 nm. The material was fixed as thin film with thickness varied of 5 nm and 10 nm. Material parameter used for CoFe are magnetic saturation $M_s = 1.44 \times 10^6$ A/m, anisotropy constant $K = 2.7$ x

10^3 J/m^3, and exchange constant A = 13 x 10^{-12} J/m based on Piao, et al [3]. On the other hand, material parameter of CoFeB based on Chaves – O'Flynn, et al was set magnetic saturation M_s = 9.57 x 10^5 A/m, anisotropy constant K = 3.86 x 10^5 A/m, and exchange constant A = 13 x 10^{-12} J/m [12]. CoFeB has uniaxial anisotropy in (001) direction. The detail geometry of material was shown in Figure 1.

3. Results and discussion

3.1 Hysteresis Loop

In this section,the hysteresis loop of CoFe and CoFeB materials by micromagnetic simulation are showed in Fig. 2. The different shapes of hysteresis loop obtained in in-plane and out-plane direction. In Fig. 2(a),the CoFe with out-plane field shows a lean hysteresis curve and low coercive field. On the other hand, the square loop hysteresis and higher coercivity was showed in out-plane field direction.This suggested that the magnetization of CoFe is preferred to align in (100) plane. As revealed experimentally by Tekgul et al., magnetic easy axis direction of CoFe is parallel to the film plane [13]. By adding Boron in CoFe thin film structure to be CoFeB, the coercivity in out-plane direction is higher than the in-plane one as seen in Fig. 2(b). It is seen clearly from inset in Fig. 2 that the coercivity value has also changed by Boron influence. In CoFeB materials, the coercivity is reduced in in-plane compared to out-plane field direction. Therefore, the CoFeB is preferred to align in (001) plane. According to these results, the Boron addition has been influenced the PMA behavior of CoFe [14].

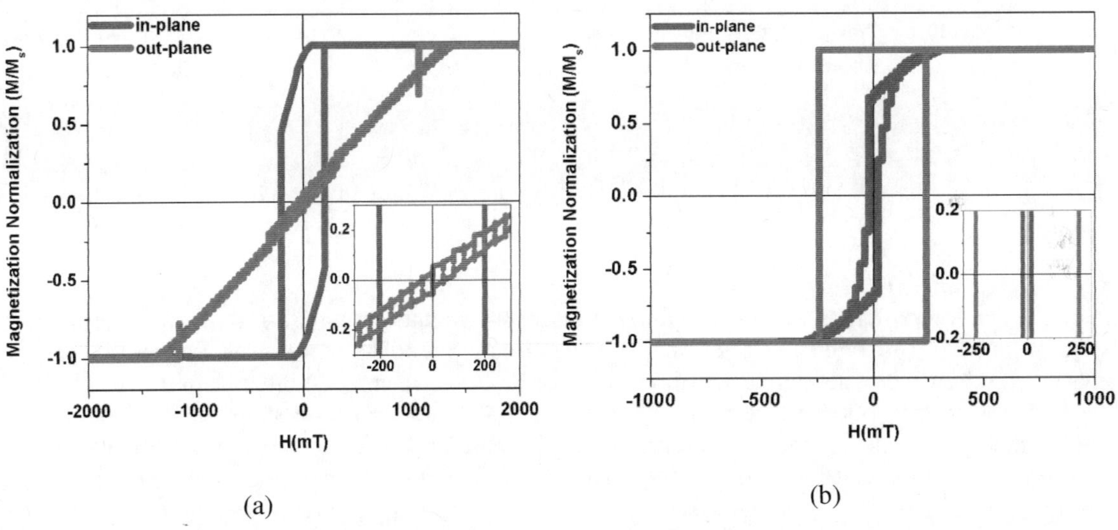

Figure 2. Hysteresis loop of (a) CoFeand (b) CoFeBsquare model given by the in-plane and out-plane magnetic field with diameter 50 nm and t = 5 nm.

3.2 Coercivity Field

Hysteresis loop provided information of magnetic properties such as the coercivity field as shown in Fig. 3. The coercivity values of CoFe and CoFeB has been achieved for in-plane and out-plane direction with diameter and thickness variation. Firstly, the CoFe thin film shows high coercivity value (H_c = 200 mT, D = 50 nm) then decreased (H_c = 40 mT, D = 500 nm) as diameter increased in in-plane direction as shown in Fig. 3(a). The reduced of coercivity field value follows the trend of coercivity after passing through the critical diameter [7]. On the other hand, the coercivity is also declining from 40 mT to 0 mT as diameter increased when out-plane field was applied. It is observed that the film thickness not significantly influenced the magnetic behavior. As expected from the hysteresis loop, it is proven that the easy axis of CoFe square was in parallel orientation to the plane. The coercivity field of CoFeB shows high coercivity (H_c = 480 mT with D = 50 nm) then decreased to 40 mT with D = 500 nm at the out-plane field direction as shown in Fig. 3(b). The H_c values fluctuate in in-plane direction on nanometer scale and tend to constant at mesoscopic scale in 40 mT. The constant coercivity values in mesoscopic scale indicate a multidomain state formation which also influenced by the anisotropy [15]. The high coercivity values in out-plane field indicated that the CoFeB has a perpendicularly easy axis and determined as PMA material. Therefore, Boron provide important role to form a PMA materials, which has high spin polarization and high stability with no heat treatment step in experiment [5].

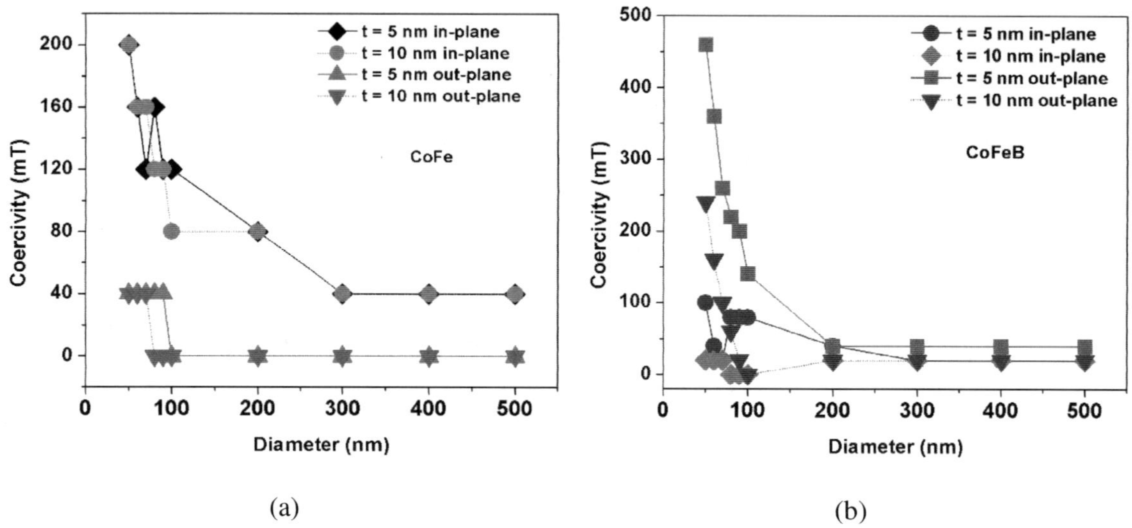

Figure 3. Coercivity field of(a) CoFeand (b) CoFeBsquare model by given in in-plane and out-plane direction.

3.3 Switching Time and Nucleation Field

The magnetization switching time is displayed in Fig.4, which was extracted from the hysteresis loop. The switching time of CoFe and CoFeB in out-plane direction tends to increase as diameter increases. These results show that with increasing film size, there are more magnetic moments in the film and the time to switch to the opposite direction from the saturation magnetization state will be longer. It is observed that the switching time of CoFe film is longer than CoFeB in the size below 200 nm. Due to the Boron addition, the CoFeB hysteresis loop has good squareness and magnetic moment will change the orientation easily. This result agree with the results from Liu et al., which showed that CoFeB has good squareness of hysteresis loop [16] .

Figure 5 shows the nucleation field of CoFe and CoFeB in hard axis direction. It is observed that the nucleation field of CoFe is higher than CoFeB up to 1080 mT. The nucleation fields to change magnetic moment direction are slightly increase as the side size in CoFe film increased, while in CoFeB film the nucleation fields are not significantly changed. In CoFeB square film with PMA behavior obtained low coercivity in high squareness hysteresis loop. It means that the nucleation field will be lower and magnetic moment orientation can be switched easily. However, the similar trend is obtained for CoFe and CoFeB film as shown in Fig. 5.The nucleation field values have a constant tendency at the mesoscopic scale (above 200 nm) as the side size increases. As explained by Quach et al., the correlation of switching time and nucleation field are proportional, which the nucleation field increased as the sweep rate increased [17].

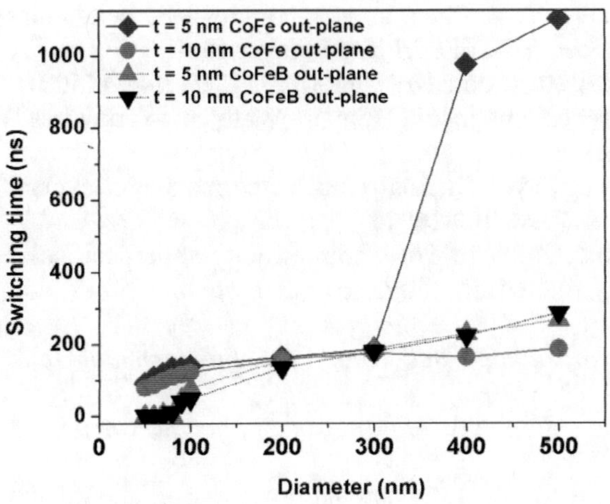

Figure 4. The switching time of CoFe and CoFeB film as side size change in out-plane direction.

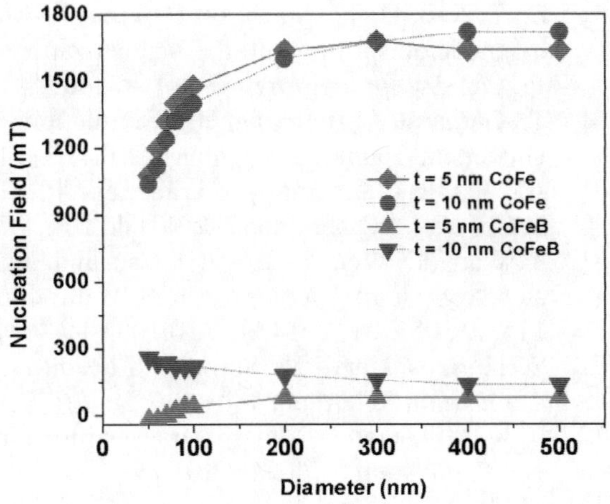

Figure 5. The nucleation field of CoFe and CoFeB film as side size change in hard axis magnetization.

4. Conclusion

In this work, thin film CoFe and CoFeB square were simulated in in-plane and out-plane magnetic field direction with varied diameter and thickness using OOMMF. The CoFe found to be in-plane magnetic anisotropy with lowest coercivity value of 40 mT. It is found that the coercivity decreased as square size increased for both in-plane and out-plane magnetization direction. The coercivity values were around 40 to 200 mT for in-plane field magnetization of CoFe. The coercivity tends to constant at 40 mT in diameter less than 100 nm and zero coercivity for diameter greater than 100 nm for out-plane field magnetization. Compared to CoFe, the coercivity in out-plane field is higher than in-plane field in CoFeB square. It is also observed that the switching time and nucleation field increased as the size increased in out-plane direction of both CoFe and CoFeB. The results showed that the different characteristics of magnetic anisotropy of both materials are important in the development of high density magnetic storage.

Acknowledgment

This work is supported by Indexed International Publication Grant (Publikasi Internasional Terindeks, PIT 9) year 2019 No. NKB-0023/UN2.R3.1/HKP.05.00/2019 through DRPM Universitas Indonesia.

References

[1] S. P. Gubin, Y. A. Koksharov, G. B. Khomutov, and G. Y. Yurkov, "Magnetic nanoparticles: preparation, structure and properties," *Russian Chemical Reviews*, vol. 74, no. 6, 2005. doi: 10.1070/RC2005v074n06ABEH000897

[2] R. S. Sundar and S. C. Deevi, "Soft magnetic FeCo alloys: alloy development, processing, and properties," *International Materials Reviews*, vol. 50, pp. 157–192, 2005. doi: 10.1179/174328005X14339

[3] H.-G. Piao, H.-C. Choi, J.-H. Shim, D.-H. Kim, and C.-Y. You, "Ratchet effect of the domain wall by asymmetric magnetostatic potentials," *Applied Physics Letters*, vol. 99, no. 19, p. 192512, 2011. doi: 10.1063/1.3658805

[4] M. Munakata, S.-I. Aoqui, and M. Yagi, "B-Concentration Dependence on Anisotropy Field of CoFeB Thin Film for Gigahertz Frequency Use," *IEEE Transactions on Magnetics*, vol. 41, no. 10, 2005. doi: 10.1109/TMAG.2005.854666

[5] D.-T. Ngo, D.-T. Quach, Q.-H. Tran, K. Møhave, T.-L. Phan, and D.-H. Kim, "Perpendicular magnetic anisotropy and the magnetization process in CoFeB/Pd multilayer films," *Journal of Physics D: Applied Physics*, vol. 47, no. 445001, 2014. doi: 10.1088/0022-3727/47/44/445001

[6] T. Gerhardt, "Micromagnetic Simulations of Ferromagnetic Domain Walls in Nanowires," Universität Hamburg, Hamburg, 2014.

[7] A. Meo, P. Chureemart, S. Wang, and R. Chepulskyy, "Thermally nucleated magnetic reversal in CoFeB/MgO nanodots," 2017. doi: 10.1038/s41598-017-16911-3

[8] Y. Wang, D. Wei, K.-Z. Gao, J. Cao, and Fulin Wei, "The role of inhomogeneity of perpendicular anisotropy in magnetic properties of ultra thin CoFeB film," *American Institute of Physics*, vol. 115, no. 053901, 2014. doi: 10.1063/1.4863139

[9] S. Horikoshi and N. Serpone, *Microwaves in Nanoparticle Synthesis: Fundamentals and Applications*. German: Wiley-VCH, 2013.

[10] P. K. Mukherjee, "Nanomaterials with Immense Potential," *Journal of Applicable Chemistry*, vol. 5, no. 4, pp. 714–718, 2016.

[11] M. J. Donahue and D. G. Porter, *OOMMF User's Guide*, 1.0. National Institute of Standard and Technology, Gaithersbug, MD., 1999.

[12] G. D. Chaves-O'Flynn, G. Wolf, D. Pinna, and Andrew D. Kent, "Micromagnetic study of spin transfer switching with a spin polarization tilted out of the free layer plane," *Journal of Applied Physics*, vol. 117, 2015. doi: 10.1063/1.4907241

[13] A. Tekgul, M. Alper, H. Kockar, and M. Haciismailoglu, "The effect of ferromagnetic and non-ferromagnetic layer thicknesses on the electrodeposited CoFe/Cu multilayers," *Journal of Materials Science Materials in Electronics*, vol. 26, pp. 2411–2417, 2015. doi: 10.1007/s10854-015-2699-7

[14] J. Sinha *et al.*, "Influence of boron diffusion on the perpendicular magnetic anisotropy in Ta|CoFeB| MgO ultrathin films," *American Institute of Physics*, vol. 117, no. 043913, 2015. doi: 10.1063/1.4906096

[15] A. P. Guimaraes, *Principles of Nanomagnetism*. New York: Springer Dordrecht Heidelberg London New York, 2009.

[16] Y. Liu, L. Hao, and J. Cao, "Effect of annealing conditions on the perpendicular magnetic anisotropy of Ta/CoFeB/MgO multilayers," *AIP Publishing*, vol. 6, no. 045008, 2016. doi: 10.1063/1.4947132

[17] D.-T. Quach *et al.*, "Minor hysteresis patterns with a rounded/sharpened reversing behavior in ferromagnetic multilayer," *Scientific Report*, vol. 8, no. 4461, 2018. doi: 10.1038/s41598-018-22810-y

Investigation of Dynamic Magnetization in FePt and FePd Disk Ferromagnets Using Micromagnetic Simulation

Ummaira Fadhilah[1], Candra Kurniawan[1,2], and Dede Djuhana[1*]

[1] Department of Physics, FMIPA Universitas Indonesia, UI Depok 16424, Depok, Indonesia

[2] Research Center for Physics, Indonesian Institute of Sciences (LIPI), Puspiptek Office Area 15314, Tangerang Selatan, Indonesia

dede.djuhana@sci.ui.ac.id

Abstract. In this study, we have observed the magnetic hysteresis loop of the highly anisotropic ferromagnetic materials FePt and FePd with disk model by micromagnetic simulation method based on the Landau-Lifshitz-Gilbert (LLG) equation. We used disk shaped model with varied size from 50 to 500 nm, thicknesses of 5 and 10 nm, and damping constant $\alpha = 0.05$. The cell size of $2.5 \times 2.5 \times 2.5$ nm^3 was used and the in-plane and out-plane fields were applied to the materials. The results showed that the hysteresis loop has a large coercivity when the external in-plane field and close to zero when the external out-of-plane field was applied. This characteristic was similar as typical of the material's hysteresis loops given the field toward the hard-axis. However, coercivity still observed in materials with size below $\leq$ 100 nm with ranging values between 20 and 80 mT. From the results, a certain value of the coercivity field appeared in out-plane applied field indicated a perpendicular magnetic anisotropy (PMA) behaviour in FePt and FePd ferromagnets. Moreover, the nucleation field was shifted as the material's size varied. The results showed that the size affected the magnetic properties of the FePt and FePd thin layers.

Keywords: FePt, FePd, hysteresis loop, micromagnetic simulation, PMA

1. Introduction

Magnetic devices have been widely explored for data storage media (magnetic memory devices) and magnetic sensor applications. Magnetic recording with high storage density and smaller size is the reason why the magnetic devices has always been developed to realize a storage media with fast data processing capabilities and low energy consumption (smaller, faster, and efficient) [1]. Data storage devices capacity is increased to store more information (bits) by increasing its area density inside a memory disc. Initially, the process of storing data on magnetic media is carried out by longitudinal recording method. The bits on a disc change in parallel or anti-parallel to the direction of the head that moves relative to the disc. To increase its density, the size of the data bit must be made in small size. However, the thermal effect of the ferromagnetic substance known as the superparamagnetic effect appeared because of influence of thermal fluctuations. To overcome the superparamagnetic effect, perpendicular magnetic recording on the materials with perpendicular magnetic anisotropy (PMA) was used to increasing the bit density in recording media [2]. Compared with in-plane magnetic anisotropy (IMA) material, materials with PMA properties are widely used for recorders and data storage. By using perpendicular magnetic recording technology with aligned the magnetic recording bits perpendicular to the disk plate, it is expected that the superparamagnetic effect will be decrease and higher storage densities would be achieved.

Another way to avoid thermal instability is using materials which has higher magnetic anisotropy constant [3,4]. One of the most promising candidates for perpendicular magnetic recording media due to the large uniaxial magnetocrystalline anisotropy are FePt and FePd alloys that can reach as high as 10^7 J/m^3 [5]. Pt- and Pd-based alloys are selected materials that are used for magnetic recording media because of their perpendicular magnetic anisotropy (PMA) properties [6]. FePt nanoparticles have almost the same percentage of Fe and Pt atoms and an important class of magnetic nano materials. The anisotropy constant of FePt is 7×10^6 J/m^3, which is one of the greatest values among all known hard magnetic materials [7]. This high uniaxial anisotropy makes FePt overcome thermal fluctuations even in very small practices [8]. The anisotropy constants of FePd is 2.6×10^6 J/m^3 which is smaller than FePt but larger than Co-Cr-based alloys that commonly used in recording media [9].

In order to have better understanding of the magnetic properties, we have investigated the characteristics of hysteresis loops from FePt and FePd disk-shaped model with different sizes and thicknesses using micromagnetic simulation. Perpendicular magnetic anisotropy properties was observed by applied magnetic field on the in-plane direction and out-of-plane direction.

2. Simulation Procedure

This study was carried out using a public micromagnetic simulation software OOMMF (Object Oriented Micromagnetic Framework) [10]. The OOMMF simulation software was built based on the Landau-Lifshitz-Gilbert (LLG) equation which is a fundamental part of time-dependent computational micromagnetics. In this simulation, the FePt and FePd with the disk shaped were used with variations in diameter d from 50 nm to 100 nm for nanometer order and 200 nm to 500 nm for mesoscopic order with two variations of thickness t of 5 and 10 nm. The cell size of discretization was fixed to be $2.5 \times 2.5 \times 2.5$ nm^3, and the damping constant α of 0.05. The material parameters for the micromagnetic simulation are set for FePt [11] $M = 1.14 \times 10^6 Am^{-1}$, the anisotropy constant $K = 6.6 \times 10^6 J/m^3$, the exchange constant $A = 10 \times 10^{-12} Jm^{-1}$ and FePd [12] $M = 1.06 \times 10^6 Am^{-1}$, the anisotropy constant $K = 2.6 \times 10^5 J/m^3$, and the exchange constant $A = 6.9 \times 10^{-12} Jm^{-1}$. Then, the FePt and FePd materials with illustrated as Figure 1 are simulated in two different states (i) the system is simulated with magnetic field in in-plane direction and (ii) out-of-plane direction with an external magnetic field -1000 mT to 1000 mT.

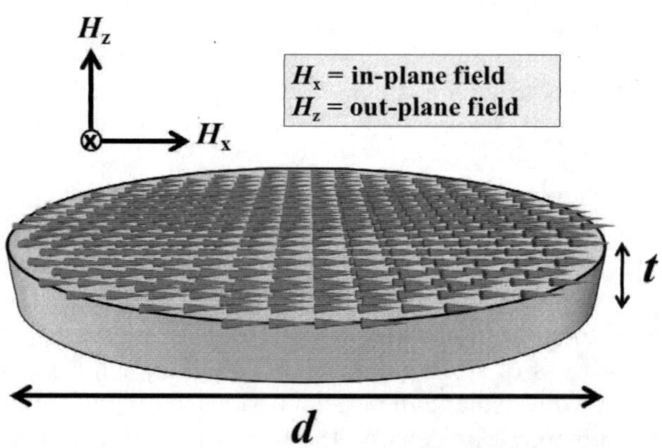

Figure 1. The dimension and geometry of FePt and FePd disk ferromagnets. The length size ***d*** was varied from 50 nm to 500 nm with external field is applied in-plane and out-of-plane direction.

IOP Conf. Series: Materials Science and Engineering **553** (2019) 012010 doi:10.1088/1757-899X/553/1/012010

3. Results and discussion

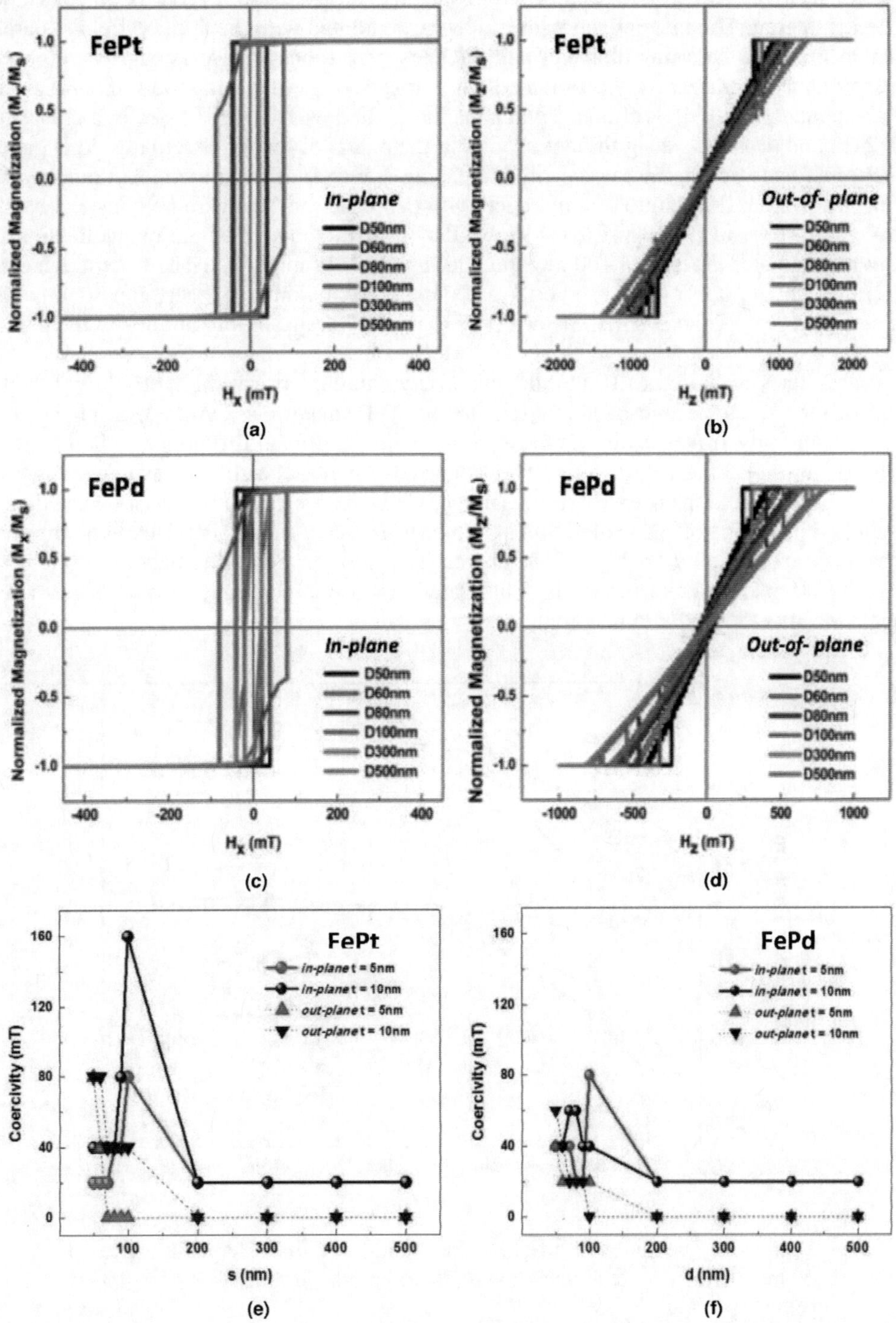

Figure 2. (**a,b,c,d**) Hysteresis loops for FePd and FePt disk-shaped with different lengths and thicknesses. The magnetic field was applied in the in-plane direction and out-of-plane (perpendicular) direction. (**e and f**) Coercivity H_c for FePd and FePt disk-shaped as a function of diameter d and thicknesses .

The magnetic behavior of a ferromagnetic material is characterized by the hysteresis loop. The observation of hysteresis curve is explained by using the magnetization process on FePt and FePd ferromagnetic material. The magnetization curves were simulated with the applied field in parallel and perpendicular direction. By using diameter and thickness variation, hysteresis curves are observed as shown in Figures 2. A straight loop is obtained when magnetic field was applied in parallel direction (a,c) while a slanted loop (b,d) is obtained when magnetic field was applied in perpendicular direction.

Figure 2 (a) and (c) shows a significant change in the coercivity when external field is given in in-plane direction. The coercivity increases with increasing diameter of the materials in range *d* 50 nm to 100 nm but start from *d* 200 nm to 500 nm coercivity decreases and tends to be constant at 20 mT as shown as Figure 2 (e) and (f). Figure 2 (e) shows that a drastic change of the magnetization loops is observed in the diameter *d* between 100 nm and 200 nm for FePt materials. The fluctuation coercivity value with range about 20 mT was observed in FePd material in diameter *d* between 70 nm and 80 nm as shown as Figure 2 (f). It shows that the effect of diameter variation only applies in the nano region which is < 100 nm.

When the thickness is raised to 10 nm, the coercivity value is observed slightly larger than ***t*** 5 nm with a significant large H_c value of about 160 mT for FePt materials. On the other hand, for FePd materials, the coercivity values is also large but there is no significant differences when the thickness is raised to 10 nm. So, the results show that coercivity increase with increasing thickness of the materials. The hysteresis loops with different diameter showed nearly zero coercivity when the external field was applied in the out-of-plane direction as shown as Figure 2 (b) and (d). However, the coercivity still observed in nano region *d* < 100 nm. The coercivity values have ranging between 40 to 80 mT for FePt and 40 to 60 mT for FePd materials. This values indicates perpendicular magnetic anisotropy (PMA) behavior in FePt and FePd ferromagnets.

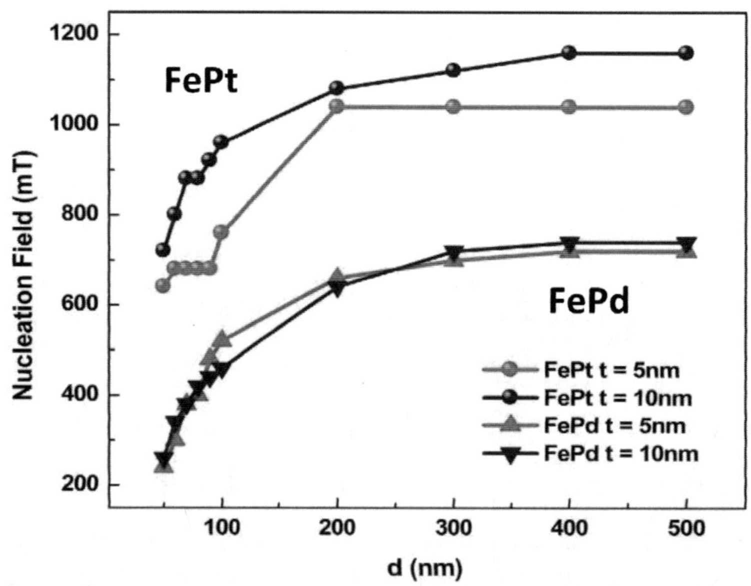

Figure 3. The nucleation field of FePt and FePd disk-shaped material as diameter function with variation ***d*** 50-500 nm and ***t*** 5 nm (red) and 10 nm (blue).

The magnitude of the field when a magnetization value decreases is known as the nucleation field which is the initial magnetic field that makes the domain structure no longer parallel [13]. The results show that nucleation field H_n are proportional to the diameter and thickness as shown as Figure 3, the nucleation field increase with increasing diameter and thickness. But, from the figure for FePt materials, the nucleation field does not show a significant values and tends to be constant when it comes to mesoscopic region. It is indicating that the diameter variation only affects the shifting of the nucleation field below *d* 200 nm. According to previous research from Shima (2002) [14], the nucleation field also increase with increasing diameter and thickness and tends to be constant at large size.

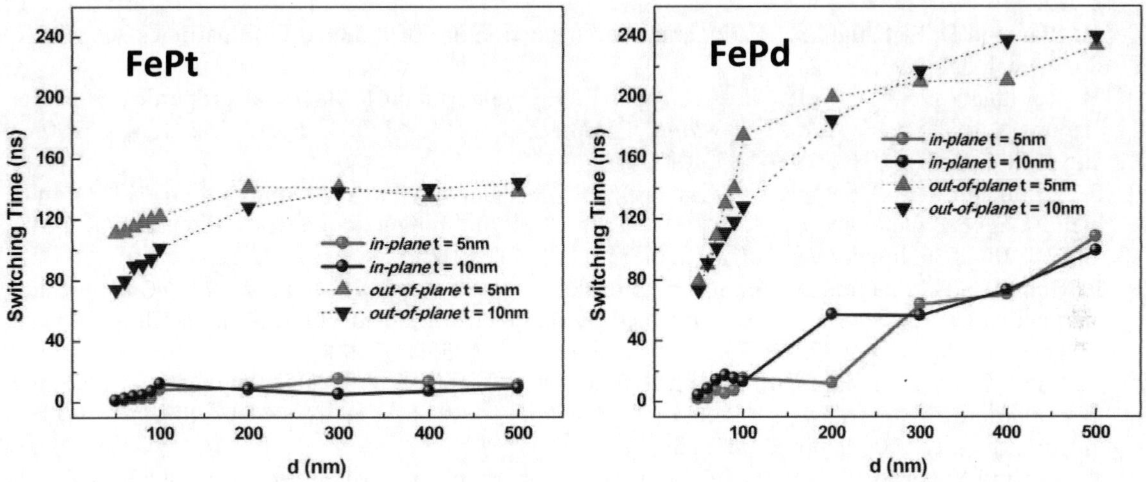

Figure 4. The switching time of FePt disk-shaped material with variation ***d*** 50-500 nm and ***t*** 5 nm (red) and 10 nm (blue) with magnetic field was applied in the in-plane direction and out-of-plane direction.

The next observations show the change in the reversal time that needed for each diameter. The amount of time which is needed to achieve a positive saturation state from a negative saturation is called switching time. Figure 4 (a) and (b) show the switching time for FePt and FePd materials as function of diameter. When the external field was applied in out of plane direction the switching time increase with increasing diameter but decrease with increasing thickness. But something that interesting comes in mesoscopic region, at *d* 300 nm for both FePt and FePd materials, the switching time increase with increasing thickness. Meanwhile, when the external field was applied in in-plane direction the switching time also increase with increasing diameter and thickness, but up to diameter *d* 100 nm, in mesoscopic region the switching time tends to fluctuate.

4. Conclusion

Hysteresis loops of FePt and FePd with disk-shaped model was observed at different diameter and thicknesses by micromagnetic simulation. It is found that two different shapes of hysteresis loops are influenced by the direction of the applied field. The results showed that the size of materials affected the magnetic properties of FePt and FePd thin layers. The highest coercivity of 160 mT was obtained for FePt , when the external field was applied in the in-plane direction. The coercivity decreased as the size of materials increased but tends to be constant in mesoscopic scale. The nucleation field was shifted to a larger field with increasing diameter and thickness. FePt materials requires a larger nucleation field compared to FePd materials. Meanwhile, along with increasing diameter, the switching time also increase because of materials with large diameter takes longer time to reverse the orientation of the momen magnet in opposite direction.

Acknowledgement

This work is supported by Indexed International Publication Grant (Publikasi Internasional Terindeks, PIT 9) year 2019 No. NKB-0023/UN2.R3.1/HKP.05.00/2019 through DRPM Universitas Indonesia.

References

[1] B. Tudu and A. Tiwari, "Recent Developments in Perpendicular Magnetic Anisotropy Thin Films for Data Storage Applications," *Vacuum*, vol. 146, pp. 329–341, 2017, https://doi.org/10.1016/j.vacuum.2017.01.031.

[2] Constantin Lefter and Mihai Dimian, "Micromagnetic analysis of magnetization behavior in Permalloy nanoparticles for data storage applications," *11th Intrnational Conf. Dev. Appl. Syst.*, 2012.

[3] B. Bian and D. E. Laughlin, "Fabrication and nanostructure of oriented FePt particles," *J. Appl. Phys.*, vol. 87, p. 9, 2000.

[4] B. Rellinghaus, S. Stappert, M. Acet, and E. F. Wassermann, "Magnetic properties of FePt nanoparticles," *J. Magn. Magn. Mater.*, vol. 266, pp. 142–154, 2003, https://doi.org/10.1016/S0304-8853(03)00465-7.

[5] R. A. Lukaszew, A. Cebollada, C. Clavero, and J. M. Garcia-Martin, "Highly ordered FEPT and FePd magnetic nano-structures: Correlated structural and magnetic studies," *Phys. B*, vol. 384, pp. 15–18, 2006, https://doi.org/10.1016/j.physb.2006.05.025.

[6] F. Bonell, S. Murakami, Y. Shiota, T. Nozaki, T. Shinjo, and Y. Suzuki, "Large change in perpendicular magnetic anisotropy induced by an electric field in FePd ultrathin films," *Appl. Phys. Lett.*, vol. 98, p. 232510, 2011, https://doi.org/10.1063/1.3599492

[7] S. Sun, "Recent Advances in Chemical Synthesis, Self-Assembly, and Applications of FePt Nanoparticles," *Adv. Mater.*, vol. 18, pp. 393–403, 2006, https://doi.org/10.1002/chin.200616239.

[8] K. Sato and Y. Hirotsu, "Structure and magnetic property changes of epitaxially grown L10-FePd isolated nanoparticles on annealing," *J. Appl. Phys.*, vol. 93, p. Number 10, 2003, https://doi.org/10.1063/1.1568531.

[9] J. R. Skuza, C. Clavero, K. Yang, B. Wincheski, and R. A. Lukaszew, "Microstructural, magnetic anisotropy, and magnetic domain structure correlations in epitaxial FePd thin films with perpendicular magnetic anisotropy," *IEEE Transaction on Magnetics*, vol. 46, issue 6, 2010, doi: 10.1109/TMAG.2009.2039923.

[10] M. J. Donahue and D. G. Porter, *OOMMF User's Guide*, 1.0. Gaithersburg, MD: National Institute of Standards and Technology, 1999.

[11] X. Hu, P. Wu, and J. Yuan, "Exchange-coupled Fe3O4/L10-FePt bilayer films by controlled oxidation of Fe/Pt multilayer," *Thin Solid Films*, vol. 517, pp. 2602–2605, 2009, https://doi.org/10.1016/j.tsf.2008.10.040.

[12] H. Niedoba and M. Labrune, "Magnetic bubbles and stripe domains in nanostructured FePd elements," *J. Magn. Magn. Mater.*, vol. 321, pp. 2178–2186, 2009, https://doi.org/10.1016/j.jmmm.2009.01.007

[13] G. S. D. Beach and A. E. Berkowitz, "Co-Fe Metal/Native-Oxide Multilayers: A New Direction in Soft Magnetic Thin Film Design I. Quasy-Static Properties and Dynamic Response," *IEEE Trans. Magn.*, vol. 41, p. No. 6, Jun. 2005, doi: 10.1109/TMAG.2005.847631.

[14] T. Shima, K. Takanashi, Y. K. Takahashi, and K. Hono, "Preparation and magnetic properties of highly coercive FePt fims," *Am. Inst. Phys.*, vol. 81, p. 1050, Aug. 2002, https://doi.org/10.1063/1.1498504.

IOP Conf. Series: Materials Science and Engineering **553** (2019) 012011 doi:10.1088/1757-899X/553/1/012011

Brownian Dynamics Simulation of Colloidal Gels as Matrix for Controlled Release Application

Hira Meidia[1*], Danial Irfachsyad[2], Dennis Gunawan[3]
[1] Department of Electronic Engineering, Faculty of Engineering and Informatics, Universitas Multimedia Nusantara, Kampus UMN, Jl Raya Boulevard, Gading Serpong, Tangerang 15810, Indonesia *
[2] Department of Physics Engineering, Faculty of Engineering and Informatics, Universitas Multimedia Nusantara, Kampus UMN, Jl Raya Boulevard, Gading Serpong, Tangerang 15810, Indonesia
3 Department of Informatics, Faculty of Engineering and Informatics, Universitas Multimedia Nusantara, Kampus UMN, Jl Raya Boulevard, Gading Serpong, Tangerang 15810, Indonesia

hira.meidia@umn.ac.id

Abstract. The form of crystallization colloidal gels is important as matrix for controlled release application. In this work, we use the Brownian Dynamics simulation to study the formation of gels by varying the inverse Debye length. We choose a fixed volume fraction, $\phi = 0.1$ and a fixed quenched temperature at room temperature, while the inverse Debye length, κ, is varied. To ensures that the simulations cover the fluid-phase region down to the unstable phase region above the critical coagulation concentration, the inverse Debye length is varied between $\kappa = 120\sigma^{-1}$ to $250\sigma^{-1}$. It shows that at the inverse Debye length $\kappa = 250\sigma^{-1}$ the gel forms by colloidal particles that can support the active ingredient by forming long range network.

Keywords: Computer Simulation, Brownian Dynamic, Colloidal, Gels, Inverse Debye Length

1. Introduction

One of the application of controlled release system is on the drug delivery. The development of this controlled drug delivery systems has greatly helped in improving patient compliance and drug effectiveness by reducing in the frequency of dosage administration and also give less side effects. The major goal of controlled release systems is to supply therapeutic levels of a drug agent at a physical site in the body for a prolonged period [1]. Thus the design and development of controlled release systems has become a key issue in current research. The controlled release system consists of the matrix and the active agent, the rate of drug release can be influenced by the drug concentration in the matrix. The potential of dependal-M loaded hydrogel matrix has been used in the rehydration therapy and the hydrogel with lesser cross linker amount has a higher drug loading capacity [2]. It has been reported that polymer nanoparticles are the responsive materials for drug delivery applications [3,4].

In this work, we study the aggregation of colloidal particles in an ionic solution by using Brownian Dynamic simulation. According to Dickinson, the Brownian Dynamics is a good approach in assessing the role of colloidal interaction during development of a particle gel network [5]. The colloidal gel is a special type of aggregation process when a space-filling network or a percolating structure is formed that can be stable for a considerable time. In the gelation process, the colloidal particles form aggregates which grow rapidly in size until they fill the available space. Further restructuring is then only possible by connecting these aggregates and it results in a structure spanning the whole system, known as a gel. The formation of gels structures promotes stabilization of the components as it prevents a phase separation over the life of the products application. In principle, the controlled release of drug can be obtained by varying system design parameters. Therefore, we simplify the model by describing the colloidal particles as spherical. We are going to report the effect of inverse Debye length in the colloidal particle gels that possible to use as a matrix of controlled release application. We use the DLVO potential to describe the interaction between the colloidal particles in ionic solution. It is shown that the percolating network of long-lived inter particle bonds between identical particles is shown at inverse Debye length $\kappa = 250\sigma^{-1}$ and $\phi = 0.1$.

2. Simulation Procedure

In this work, we use Brownian Dynamic simulation. The system is simulated in a three dimensional box of volume V. The model consists of N spherical particles and the volume fraction, ϕ, of the system is,

$$\phi = \frac{\pi N \sigma^3}{6V} \tag{1}$$

where σ is the diameter of the colloidal particles. The periodic boundary conditions are applied in all directions to simulate a macroscopic system. In our simulation, we set the diameter of the particles, σ, the viscosity, η, and the energy, k_BT, are set to unity.

$$k_BT = 1$$
$$\eta = 1$$
$$\sigma = 1$$

Time is in units of the Brownian relaxation time, τ_R. This is defined as the time for a particle to diffuse a distance on the order of one diameter in the absence of other forces and is given as,

$$\tau_R = \frac{\eta\sigma^3}{k_BT} \tag{2}$$

For a particle of one micron in diameter suspended in water ($\eta = 0.001$ Ns/m^2) at room temperature, the relaxation time is in the order of 0.24 s. The trajectories of the interacting particles were calculated using the position Langevin equation in finite time steps

$$r_i(t + \Delta t) = r_i(t) + \frac{\Delta t}{m\xi}\mathbf{F}_i^C(t) + \Delta \mathbf{r}_i^B \tag{3}$$

$\Delta \mathbf{r}_i^B$ is the displacement resulting from the Brownian force and is also taken to have a mean average of zero. It has the mean square value of $6k_BT\Delta t/3\pi\eta\sigma$, where Δt is the time step used in the simulation. We set the time step for the simulation at $4.71\text{x}10^{-6}\tau_R$. This corresponds to the mean square displacement of $10^{-6}\sigma^2$.

The conservative force is calculated from the DLVO (Derjaguin and Landau, Verwey and Overbeek) interaction. The force exerted by particle j on particle i at a distance of r_{ij} is given by,

$$\mathbf{F}_i(r_{ij}) = -\frac{\partial U_{DLVO}(\mathbf{r}_{ij})}{\partial \mathbf{r}_i} \tag{4}$$

$$U_{DLVO}(r_{ij}) = U_{el}(r_{ij}) + U_{vdw}(r_{ij}) \tag{5}$$

$$U_{vdw}(r_{ij}) = -\frac{A_H}{12}\left[\frac{\sigma^2}{r_{ij}^2-\sigma^2}+\frac{\sigma^2}{r_{ij}^2}+2\ln\left(\frac{r_{ij}^2-\sigma^2}{\sigma^2}\right)\right] \tag{6}$$

$$U_{el}(r_{ij}) = \pi\varepsilon_r\varepsilon_0\psi_0^2\sigma\ln\left(1+e^{-\kappa[r_{ij}-\sigma]}\right) \tag{7}$$

In this study, we use the parameters suitable for colloidal particles ($\sigma = 1$ µm, $A_H = 0.1$ eV, $\psi_o = 17.7$ mV) immersed in water ($\eta = 8.854\text{x}10^{-4}$ Pa.s) at room temperature in the presence of a monovalent salt. The equation for the electrostatic repulsion is valid for large value of the inverse Debye length, κ, we choose κ for 250. The surface potential is set to values that give sufficiently high maximum barrier as we are mainly interested in the study weak aggregation. We use the step-strain method by applying a sudden shear displacement and measure the stress relaxation as the gel network is trying to release the stress. In a very small deformation, a colloidal suspension has a stress response which increases linearly with the strain amplitude. Under this condition, the changes of the stress, σ_{xy}, as a function of time can be written as [6]

$$\sigma_{xy}(t') = \int_{-\infty}^{t'} G(t'-t)\dot{\gamma}(t)dt \tag{8}$$

$G(t)$ is called the relaxation modulus of the suspension and $\dot{\gamma}$ is the shear rate. We measure the mechanical strength by imposing a sudden strain of γ at the first step between $t'=-\Delta t$ and $t'=0$. The shear rate is then equal to $\gamma/\Delta t$ and the shear stress becomes,

$$\sigma_{xy}(t') = \int_{t_0-\Delta t}^{t_0} G(t'-t)\left(\frac{\gamma}{\Delta t}\right)dt \tag{9}$$

We modify the equation by calculating the average value of the relaxation modulus between $t_0 - \Delta t$ and t_0,

$$\int_{t_0-\Delta t}^{t_0} G(t)dt = \Delta t\, G(t_0 - \Delta t + c\Delta t)$$

$$= \Delta t\, G(t_0 - (1-c)\Delta t) \tag{10}$$

where $0 \leq c \leq 1$. It can then be written as,

$$\sigma_{xy}(t') = \frac{\gamma}{\Delta t}\Delta t\, G(t' - (t_0 - (1-c)\Delta t)) \tag{11}$$

$$\approx \gamma G(t')$$

for $t_0 = 0$ and $t' \gg \Delta t$.

In the simulation, the relaxation modulus is the unknown quantity and we compute the stress tensor by through the virial equation,

$$\sigma_{\alpha\beta} = \rho k_B T - \frac{1}{V}\sum_{i=1}^{N-1}\sum_{j=i+1}^{N}\frac{r_{\alpha ij}\, r_{\beta ij}}{r_{ij}}\frac{dU_{ij}}{dr_{ij}} \tag{12}$$

where V is the volume of the simulation box and U_{ij} is the interaction potential between i and j and $r_{\alpha ij}$ is the α-component of the inter particle vector. In the limit of t approaching zero, the relaxation modulus give the high frequency limit of the storage modulus, G_∞. The storage modulus at infinite frequency can also be calculated using the equation proposed by Zwanzig and Mountain[7],

$$G_\infty = \rho k_B T + \frac{2\pi\rho^2}{15}\int_0^\infty dr\, g(r) \frac{d}{dr}\left(r^4 \frac{dU_{ij}}{dr}\right) \tag{13}$$

In the long time limit, the modulus will decay and reach some equilibrium value. In a fluid system, the stress is expected to decay immediately to zero, i.e. relax completely, after a short period of time. A complete relaxation may not occur within the length of the simulation in a system which exhibits a solid-like response.

3. Results and discussion

In this work, the interaction between particles and its aggregation were calculated based on Derjaguin-Landau-Verwey-Overbeak (DLVO) theory [8,9]. The DLVO theory showed that the primary minimum of colloidal particles interaction potential was raised from the mutual van der Waals attraction where it is not accessible at low electrolyte concentrations because of a large energy barrier [10]. When the electrolyte concentration is raised above the critical coagulation concentration, the barrier height drops down to zero, leading to colloidal flocculation and precipitation. In this model, we simulate with a fixed volume fraction $\phi = 0.1$ and the inverse Debye length is varied between $\kappa = 120\sigma^{-1}$ to $250\sigma^{-1}$. This simulation model shows that the behavior of the system is very sensitive to the ionic concentration. The concentration of the ions in the solution determines the strength of the repulsive electrostatic interaction, which can be calculated using the DLVO equation. A weaker repulsion result in the van der Waals attraction becomes more dominant and the effective interaction becoming more attractive as shown in Figure 1.

We plot the configurational energy per particle as a function of time in Figure 2. It shows that for systems with κ less than $220\sigma^{-1}$, the equilibrium states are reached almost immediately on the simulation timescale with no aggregates detected. The systems are essentially still in the fluid phase after the quenches. This is confirmed in Figure 3, which show the configurations in the late stages, taken at time $471.24\tau_R$. In Fig. 3(a), the configuration for $\kappa = 200\sigma^{-1}$ is above the unstable region. The configuration demonstrates that the suspension stays dispersed throughout the simulation.

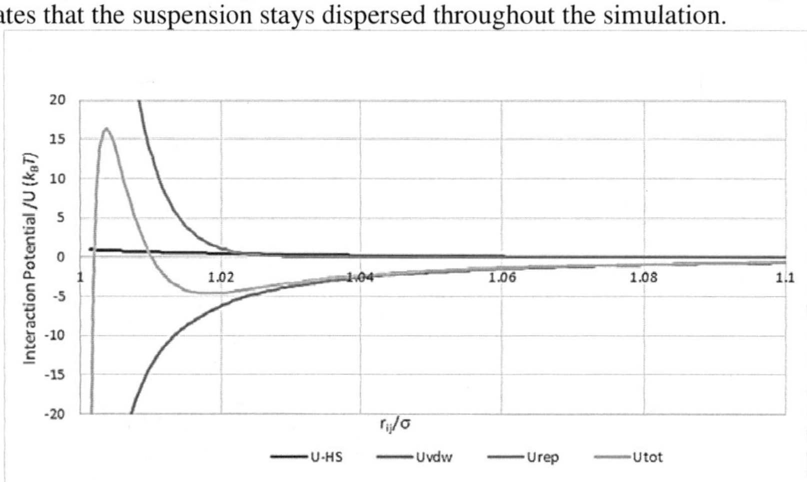

Figure 1. The profile of the DLVO, the attractive perturbation potentials as a function inter particles distance at $\kappa = 250\sigma^{-1}$.

IOP Conf. Series: Materials Science and Engineering **553** (2019) 012011 doi:10.1088/1757-899X/553/1/012011

At $\kappa = 220\sigma^{-1}$, the plot of the interaction potential as a function of time displays an inflection point at intermediate times. Beyond this point, the interaction potential decreases rapidly with time until it levels out at larger times. This suggests a dramatic departure towards the formation of a more compact structure in the system for $\kappa > 220\sigma^{-1}$. The interaction potential then shows a slow decrease until the end of the simulation and because equilibrium has not been reached.

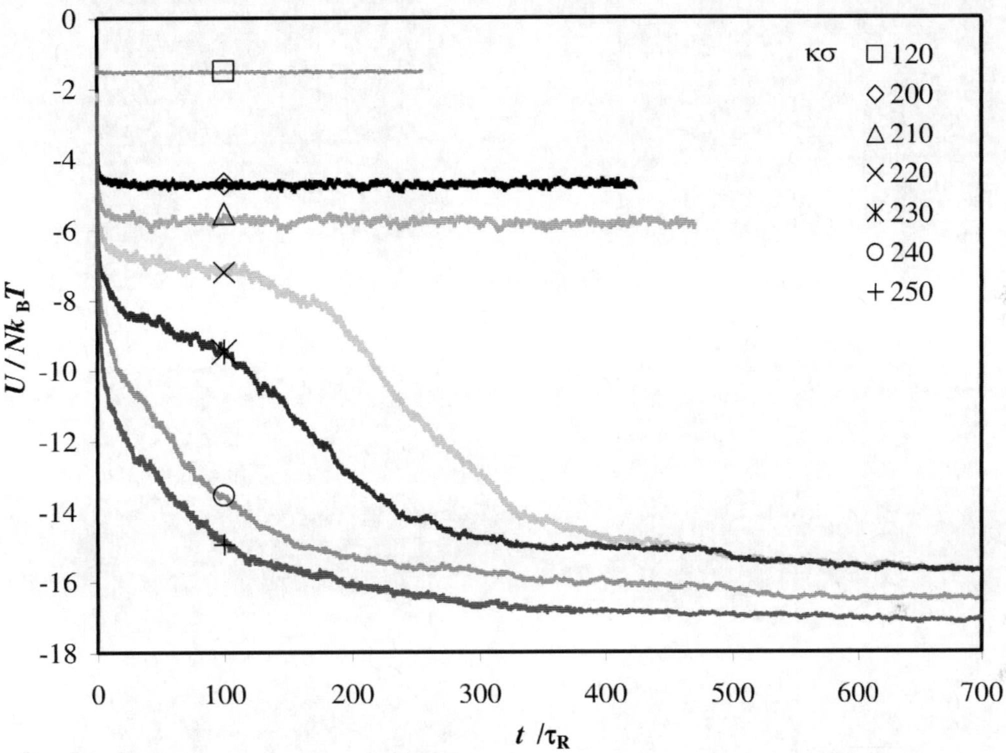

Figure 2. The average configurational energy as a function of time at different κ.

We also show that the transition relates to the onset of crystallization in the gel structure. In Figure 3, we observe the changes in the resulting gel structures as κ is varied. As we increase κ to $220\sigma^{-1}$, we start to see the formation of a gel in which the suspension forms a percolating network that fills the entire simulation box. At $\kappa > 220\sigma^{-1}$, the particles with low number densities have the short-range, hard-sphere-like interaction and the system jammed to form disordered. When κ is smaller the inter-particle interactions are featured with long-range repulsion, the particles can spontaneously assemble into disorder structure. The gel formed at lower κ is seen to have larger voids surrounding it, which suggests a denser gel is formed in this system (Figure 3b).

IOP Conf. Series: Materials Science and Engineering **553** (2019) 012011 doi:10.1088/1757-899X/553/1/012011

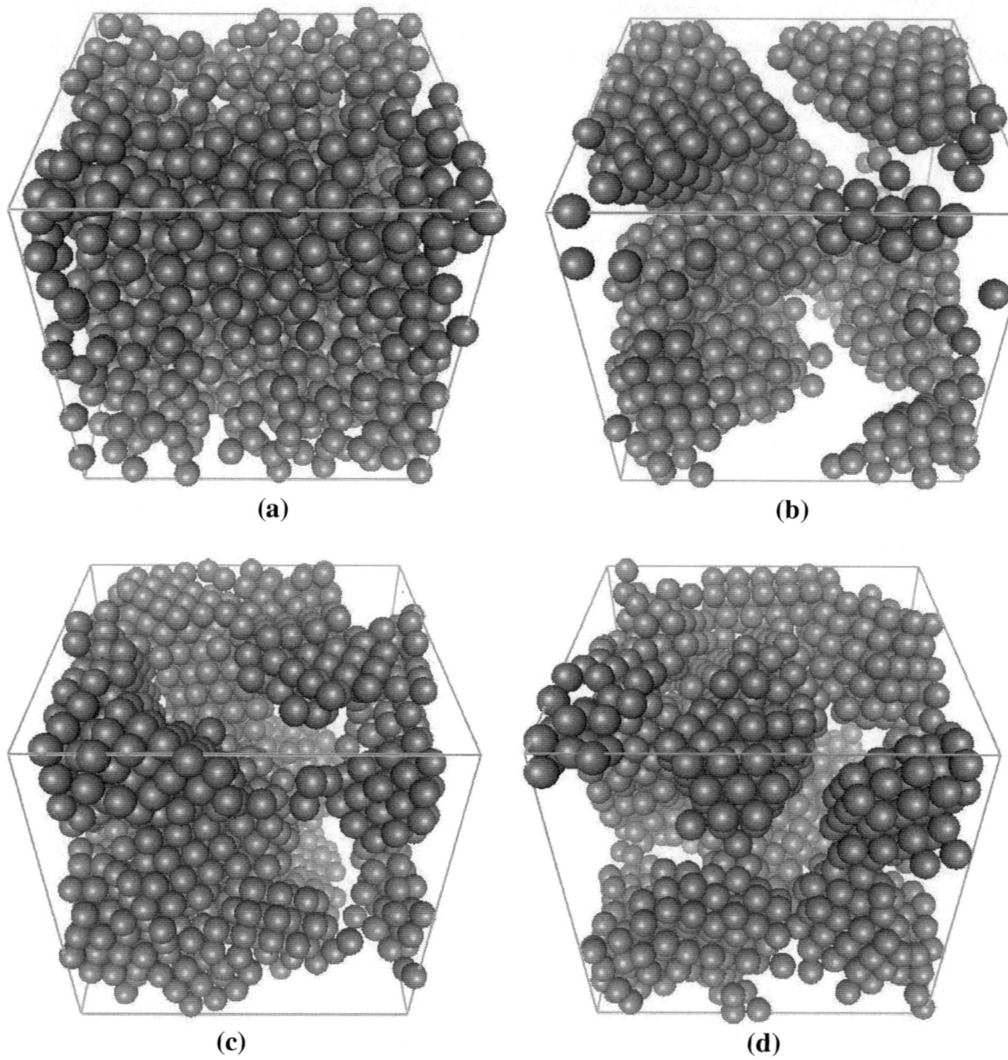

Figure 3. The pictures of the resulting structures taken at times $471.24\tau_R$ and $\phi = 0.1$ for $\kappa =$ (**a**) $200\sigma^{-1}$, (**b**) $220\sigma^{-1}$, (**c**) $230\sigma^{-1}$ and (**d**) $250\sigma^{-1}$.

4. Conclusion

We have found that at the inverse Debye length $\kappa = 250\sigma^{-1}$ the particles form percolating gel compare with others inverse Debye length. It shows that the structural evolution of colloidal gels from the long-range to the short-range orders. The colloidal particles form interconnecting network that is suitable to support active ingredients in controlled release applications.

Acknowledgments

The authors would like to acknowledge funding provided by the Ministry of Research, Technology and Higher Education of the Republic of Indonesia (research grant - PUPT Kemen Ristek Dikti 2018) and Universitas Multimedia Nusantara.

IOP Conf. Series: Materials Science and Engineering **553** (2019) 012011 doi:10.1088/1757-899X/553/1/012011

References

[1] J.R. Weiser, W. Mark Saltzman, "Controlled Release for Local Delivery of Drugs: Barriers and Models", *Journal of Controlled Release*, vol.190, pp.664-673, 2014 [doi: 10.1016/j.jconrel.2014.04.048].

[2] R. Dwivedi, A.K. Singh, A. Dhillon, "pH-Responsive Drug Release from Dependal-M Loaded Polyacrylamide Hydrogels", *Journal of Science: Advanced Material and Devices*, vol. 2, pp. 45-50, 2017 [https://doi.org/10.1016/j.jsamd.2017.02.003]

[3] C.K. Balavigneswaran, N. Misra, "Polymeric Gels for the Controlled Drug Delivery Applications," *Polymeric Gels*, pp 331-356, 2018 [doi:10.1016/b978-0-08-102179-8.00013-2].

[4] R. K. Saini, L.P. Bagri, A.K. Bajpai, A Mishra, "Responsive Polymer Nanoparticles for Drug Delivery Applications" in *Stimuli Responsive Polymeric Nanocarriers for Drug Delivery Applications*, vol.1, pp.289-320, 2018 [doi:10.1016/b978-0-08-101997-9.00015-1].

[5] E. Dickinson, "Structure and Rheology of Colloidal Particle Gels: Insight from Computer Simulation", *Advances in Colloid and Interface Science*, vol.199-200, pp. 114-127, 2013 [doi: 10.1016/j.cis.2013.07.002]

[6] R.J. Hunter, *Foundations of Colloid Science*, Vol **1**, Oxford University Press, Oxford, 1987

[7] R. Zwanzig and R.D. Mountain, "High-Frequency Elastic Moduli of Simple Fluids", *J. Chem. Phys*, vol.43, pp. 4464, 1965 [doi: 10.1063/1.1696718]

[8] G. Trefalt, F.J. Montes Ruiz-Cabello, M. Borkovec, "Interaction forces, hetero aggregation and deposition involving charged colloidal particles", *J. Phys. Chem. B*, vol.118, pp.6346-6355, 2014 [doi: 10.1021/jp503564p]

[9] F.J.M. Ruiz-Cabello, P. Maroni, M. Borkovec, "Direct Measurements of Forces between Different Charged Colloidal Particles and Their Prediction by the Theory of Derjaguin, Landau, Verwey, and Overbeek (DLVO)", *J. Chem. Phys.*, vol.138, pp.234705, 2013 [doi: 10.1063/1.4810901].

[10] Brian Vincent, "Early (pre-DLVO) studies of particle aggregation", *Advances in Colloid and Interface Science*, vol.170, Issues 1–2, pp.56-67, 2012 [doi; 10.1016/j.cis.2011.12.003].

Investigation of Notch Dept Effect on Domain Wall Depinning in Ferromagnetic Nanowires by Micromagnetic Simulation

Candra Kurniawan[1,2], Bambang Soegijono[1], Dede Djuhana[1*]
[1] Department of Physics, Faculty of Mathematics and Natural Sciences Universitas Indonesia, Kampus Baru UI Depok, 16424, Indonesia
[2] Research Center for Physics, Indonesian Institute of Sciences (LIPI), Bld. 440-442 Puspiptek Office Area, South Tangerang 15314, Indonesia

dede.djuhana@sci.ui.ac.id

Abstract. Utilization of topological nanostructure such as magnetic domain wall (DW) as a future generation of non-volatile memory as racetrack memory has been attracted more researchers due to potential to achieve higher speed of data read/writing and capacity. However, completed understanding of DW dynamics was still need to be improved by advanced analysis from the theoretical/simulation or experimental methods. In this study, the effect of geometrical notch dept on domain wall depinning in Permalloy (Py) nanowires by micromagnetic simulation method have been investigated. The varied double notch dept from 10 to 90 nm in 200 nm of wire width were used and the transverse type DW was triggered by nanosecond current pulse to observe the critical depinning current (J_d). It is observed that the depinning current was increased as the notch dept increases. However, the magnitude of J_d have a maximum values for all notch length variation. The increasing of notch length has shifted the maximum J_d to the larger notch dept values. This behavior indicated that the optimum notch design was needed to obtain higher speed and lower depinning energy in the development of domain wall based devices.

Keywords : Micromagnetic, domain wall, depinning, notch dept, nanowire.

1. Introduction

Due to the potential application as future non-volatile memories and logics, utilization of topological nanostructure such as magnetic domain wall (DW) has been an attractive research interest for the last decades [1-4]. The purpose of this research topic is to achieve higher speed of data read/writing and capacity. Another objective one was to continue the Moore's law in storage media [5]. One of the interesting proposals about magnetic non-volatile memory that utilized DWs is the concept of racetrack memory by Parkin in 2008 [6, 7]. The racetrack memory concept uses the train of DWs that move simultaneously on the long magnetic media such as 2D or 3D nanowires by electrical current or magnetic field. Previous researches show that the accurate and faster DW creep motion could be achieved by spin polarized current better than by magnetic field [8-10]. However, completed understanding of DW dynamics was still need to be improved by advanced analysis from the theoretical/simulation or experimental methods.

To optimally control the DW in the nanowire, there are some focuses on the utilization of geometrical notch as an artificial pinning potential. Most of the studies were intended to explain the dynamics of DW structure when passes or move out from the notch [11-14]. However, there is still a lack of discussion about the geometrical structure of notch in the nanowires. Piao et al. has studied the effect of asymmetrical notch in nanowire to obtain selective spin behavior using alternate magnetic field [15]. Gao et al. also studied the effect of asymmetrical notch effect to the DW random behavior during depinning out from the notch by experimental measurement and micromagnetic simulation [16]. Our previous studies explained the effect of notch length to the DW depinning behavior by electrical current and magnetic field pulse [17, 18].

This study has investigated the effect of geometrical notch dept on DW depinning driven by nano-second pulse current in Permalloy (Py) nanowires by micromagnetic simulation method. The critical current magnitude called depinning current (J_d) was observed for every notch dept parameter and the influenced depinning time (t_d) is discussed.

2. Micromagnetic simulation

A long and thin rectangular shape nanowire was modeled in this simulation. The length and thickness of the wire were 2000 nm and 5 nm, respectively. Double symmetrical triangular notch was placed at the center of the wire as shown in Figure 1. The varied double notch dept (d) from 10 to 90 nm and notch size (s) of 50, 100, and 200 nm in 200 nm of wire width were used in the simulation. Typical Permalloy material parameters are used as an input, such as magnetization saturation $M_s = 8 \times 10^5$ A/m, exchange constant $A = 13 \times 10^{-12}$ J/m, and zero magnetocrystalline anisotropy [19].

The initial transverse type domain wall (DW) was placed at the notch center with head-to-head domain structure. The DW was released in the ground state condition for 1 ns before the electrical current pulse injected for 1 ns on the x+ direction. The raising and lowering rate of current pulse was about 0.01 ns. The varied magnitude of the spin polarized current was performed to investigate the depinning current (J_d) that defined as the latest part of DW move out from the notch area. The electrical current pulse and the spin current direction are also illustrated in the Figure 1.

The simulation procedure was performed by micromagnetic simulation software package OOMMF based on the Landau-Lifshitz-Gilbert (LLG) Equation [20-22] with the spin transfer torque extension as provided by IBM [23]. The dynamic of current induced domain wall was described by the modified LLG equation as follows,

$$\frac{d\mathbf{m}}{dt} = -\gamma\mathbf{m}\times\mathbf{H}_{eff} + \alpha\mathbf{m}\times\frac{d\mathbf{m}}{dt} - (\mathbf{u}\cdot\nabla)\mathbf{m} + \beta\mathbf{m}\times\left[(\mathbf{u}\cdot\nabla)\mathbf{m}\right] \quad (1)$$

where α is the Gilbert damping constant, γ is the gyromagnetic ratio, and β is non-adiabatic constant. The spin drift velocity **u** is defined as $\mathbf{u} = PJ_e\mu_B / eM_s$, where J_e is the electrical current density, P is the spin polarization, and μ_B is Bohr magneton. The ratio of β/α was set to be 4 with α damping constant of 0.01. The dimension of the finite difference cuboid unit cell was $5 \times 5 \times 5$ nm^3, with respect to the Permalloy exchange length value of 5.6 nm [5].

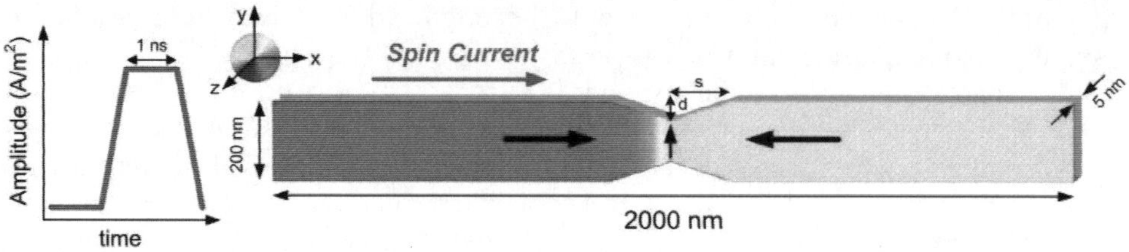

Figure 1. Geometry of the rectangular Permalloy nanowire with symmetrical triangular notch. The initial transverse type DW and head-to-head spin structure was defined. The current pulse was injected for 1 ns in the *x*+ direction and the amplitude represent the current magnitude. Color disks represent the spin direction in *xy*-axis.

3. Results and discussion

The effect of notch dept to DW depinning current (J_d) is showed in Figure 2. It is observed that the depinning current increased as the notch dept increases. However, the magnitude of J_d have maximum values for all notch length variation. The increasing of notch length (s) has shifted the maximum J_d point to the larger notch dept values. For notch length of 50 nm, the maxium value of J_d is obtained at notch dept (d) of 40 nm with the current value of 1.43×10^{12} A/m^2. For notch length of 100 and 200, local maxium J_d value are obtained at notch dept of 50 nm and 70 nm, respectively. This behavior shows that generally the notch dept value is proportional to pinning strength at nanowire. The pinning potential is needed for controlling the DW motion through the wire so that the read and write process for the non-volatile memory application can be done precisely. Therefore, the accurate process of notch shaping process is crucial in the design of nanowire based DW devices [24].

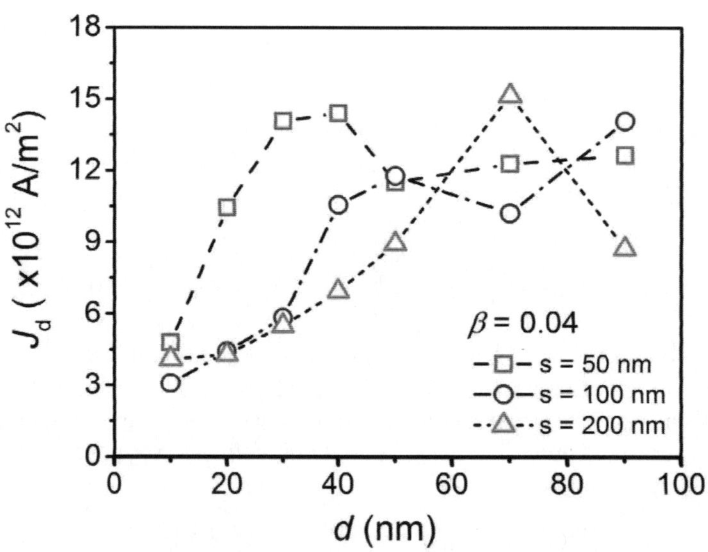

Figure 2. The effect of notch dept (d) to depinning current density (J_d) with respect to notch size (s) of 50 nm, 100 nm, and 200 nm represent with red-rectangle, blue-circle, and magenta-triangle, respectively. All simulations were performed in non-adiabatic regime with β/α ratio of 4.

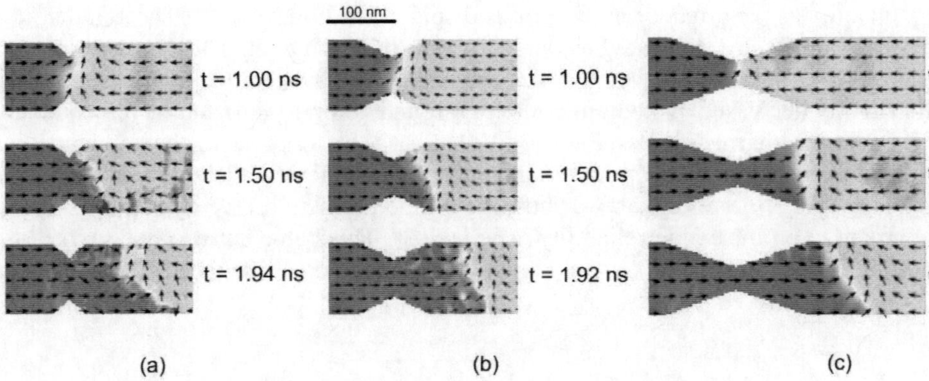

Figure 3. The dynamic of domain structure during depinning process at local maximum value of J_d at different notch size (s), the antivortex DW structure is formed during depining at (a) s = 50 nm, d = 40 nm, but kept transverse at (b) s = 100 nm, d = 50 nm, and (c) s = 200 nm, d = 70 nm.

The Figure 3 displayed the dynamic of domain structure during depinning process at local maximum value of J_d at different notch size (s). It is showed that the DW depinning motion was accompanied by changing DW structure to anti-vortex for smaller notch size but kept the transverse structure for larger notch size, which are agreed with our previous results [18]. At first nanosecond after the DW release without external current, the DW structures are shown to be asymmetric compared to notch center. This affected by the asymmetrical energy landscape at certain notch geometry [3]. Then the nanosecond current pulse was applied to the DW and depicted that the DW was shifted to the x+ direction smoothly until depinned out from the notch before the applied current current was removed. These phenomena are correlated to the DW depinning speed which represent by depinning time (t_d) measured from the first time electrical current was applied until DW edge move out from the notch completely.

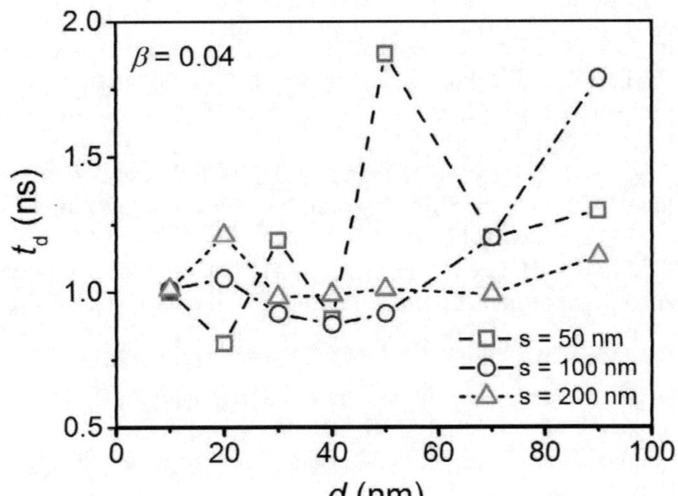

Figure 4. The effect of notch dept (d) to DW depinning time (t_d) with respect to notch size (s) of 50 nm, 100 nm, and 200 nm represent with red-rectangle, blue-circle, and magenta-triangle, respectively.

IOP Conf. Series: Materials Science and Engineering **553** (2019) 012012 doi:10.1088/1757-899X/553/1/012012

The depinning time (t_d) as the effect of notch dept is displayed in Figure 4. It is showed that in the s value of 50 nm, the t_d values are fluctuated as the increasing of notch dept. The depinning time are swinging from the value of 0.8 ns to 1.8 ns for d = 20 nm and 50 nm, respectively. However at larger s, the t_d values are slightly decreased at medium size of d (around 30 or 40 nm) and increased at larger notch dept. The t_d values are sharply increased for s = 100 nm at notch dept larger than 40 nm. But for s = 200, the changing of t_d values at d larger than 30 nm are relatively slightly increases and still around the range of 1 ns. The simulation results above showed that the predictive DW dynamics at specific notch size are important to produce high reliability DW devices. This behavior also indicated that the optimum notch design was needed to obtain higher speed and lower depinning energy in the development of domain wall based devices.

4. Conclusion

In this study, the effect of geometrical notch dept on domain wall depinning in Permalloy (Py) nanowires by micromagnetic simulation method have been investigated. The varied double notch dept were used and the transverse type DW was triggered by nanosecond current pulse to observe the critical depinning current (J_d). It is observed that the depinning current was increased as the notch dept increases. However, the magnitude of J_d have local maximum values for all notch length variation. The increasing of notch length has shifted the maximum J_d to the larger notch dept values. This behavior shows that generally the notch dept value is proportional to pinning strength at nanowire. It is also showed that the DW depinning motion was accompanied by changing DW structure to anti-vortex for smaller notch size but kept the transverse structure for larger notch size. However, the fluctuated values of depinning time as the notch dept increased made the DW dynamic become complicated. This behavior indicated that the optimum notch design was needed to obtain higher speed and lower depinning energy in the development of domain wall based devices.

Acknowledgment

This work is funded by Indexed International Publication Grant (Publikasi Internasional Terindeks, PIT 9) year 2019 No. NKB-0023/UN2.R3.1/HKP.05.00/2019 through DRPM Universitas Indonesia. CK want to acknowledge to Saintek Scholarship year 2018 from Ministry of Research, Technology and Higher Education, the Republic of Indonesia for scholarship support.

References

[1] A. Thiaville, Y. Nakatani, F. Piéchon, J. Miltat, and T. Ono, “Transient domain wall displacement under spin-polarized current pulses,” Eur. Phys. J. B, vol. 60, no. 1, pp. 15–27, Nov. 2007. DOI: 10.1140/epjb/e2007-00320-3.

[2] M. Hayashi, L. Thomas, R. Moriya, C. Rettner, and S. S. P. Parkin, “Current-Controlled Magnetic Domain-Wall Nanowire Shift Register,” Science, vol. 320, no. 5873, pp. 209–211, Apr. 2008. DOI: 10.1126/science.1154587.

[3] D. Djuhana, H.-G. Piao, S.-H. Lee, D.-H. Kim, S.-M. Ahn, and S.-B. Choe, “Asymmetric ground state spin configuration of transverse domain wall on symmetrically notched ferromagnetic nanowires,” Appl. Phys. Lett., vol. 97, no. 2, p. 022511, 2010. DOI: 10.1063/1.3459965.

[4] D. Petit et al., “High efficiency domain wall gate in ferromagnetic nanowires,” Appl. Phys. Lett., vol. 93, no. 16, p. 163108, 2008. DOI: 10.1063/1.3005586.

[5] J. M. D. Coey, Magnetism and magnetic materials. Cambridge: Cambridge University Press, 2010. DOI: 10.1017/CBO9780511845000.

[6] S. S. P. Parkin, M. Hayashi, and L. Thomas, “Magnetic Domain-Wall Racetrack Memory,” Science, vol. 320, no. 5873, pp. 190–194, Apr. 2008. DOI: 10.1126/science.1145799.

[7] M. Hayashi, L. Thomas, C. Rettner, R. Moriya, and S. S. P. Parkin, “Direct observation of the coherent precession of magnetic domain walls propagating along permalloy nanowires,” Nat. Phys., vol. 3, no. 1, pp. 21–25, Jan. 2007. DOI: 10.1038/nphys464.

[8] G. S. D. Beach, M. Tsoi, and J. L. Erskine, "Current-induced domain wall motion," J. Magn. Magn. Mater., vol. 320, no. 7, pp. 1272–1281, Apr. 2008. DOI: 10.1016/j.jmmm.2007.12.021.
[9] S. Fukami, T. Suzuki, N. Ohshima, K. Nagahara, and N. Ishiwata, "Micromagnetic analysis of current driven domain wall motion in nanostrips with perpendicular magnetic anisotropy," J. Appl. Phys., vol. 103, no. 7, p. 07E718, 2008. DOI: 10.1063/1.2830964.
[10] D. Djuhana, H.-G. Piao, S.-C. Yu, S. K. Oh, and D.-H. Kim, "Magnetic domain wall collision around the Walker breakdown in ferromagnetic nanowires," J. Appl. Phys., vol. 106, no. 10, p. 103926, 2009. DOI: 10.1063/1.3264642.
[11] L. K. Bogart, D. Atkinson, K. O'Shea, D. McGrouther, and S. McVitie, "Dependence of domain wall pinning potential landscapes on domain wall chirality and pinning site geometry in planar nanowires," Phys. Rev. B, vol. 79, no. 5, Feb. 2009. DOI: 10.1103/PhysRevB.79.054414.
[12] S. Lepadatu, A. Vanhaverbeke, D. Atkinson, R. Allenspach, and C. H. Marrows, "Dependence of Domain-Wall Depinning Threshold Current on Pinning Profile," Phys. Rev. Lett., vol. 102, no. 12, Mar. 2009. DOI: 0.1103/PhysRevLett.102.127203.
[13] V. D. Nguyen et al., "Elementary depinning processes of magnetic domain walls under fields and currents," Sci. Rep., vol. 4, p. 6509, Oct. 2014. DOI: 10.1038/srep06509.
[14] H. Y. Yuan and X. R. Wang, "Domain wall pinning in notched nanowires," Phys. Rev. B, vol. 89, no. 5, Feb. 2014. DOI: 10.1103/PhysRevB.89.054423.
[15] H.-G. Piao, J.-H. Shim, M. Yang, L. Pan, and D.-H. Kim, "Selective Behavior of Spin Wave Propagation in Asymmetrically Modulated Ferromagnetic Nanowires," IEEE Trans. Magn., vol. 51, no. 11, pp. 1–6, Nov. 2015. DOI: 10.1109/INTMAG.2015.7156928.
[16] Y. Gao et al., "Depinning of domain walls in permalloy nanowires with asymmetric notches," Sci. Rep., vol. 6, no. 1, Dec. 2016. DOI: 10.1038/srep32617.
[17] D. Djuhana, B. Soegijono, H.-G. Piao, S. K. Oh, S.-C. Yu, and D.-H. Kim, "Oscillatory transformative domain wall inner structure of the depinning domain wall around a notched ferromagnetic wire," J. Korean Phys. Soc., vol. 63, no. 3, pp. 654–658, Aug. 2013. DOI: 10.3938/jkps.63.654.
[18] D. Djuhana, C. Kurniawan, and D.-H. Kim, "Micromagnetic study of domain wall depinning driven by nanosecond current pulse in notched Permalloy nanowires," Curr. Appl. Phys., vol. 18, no. 2, pp. 236–240, Feb. 2018. DOI: 10.1016/j.cap.2017.11.005.
[19] S.-H. Huang and C.-H. Lai, "Domain-wall depinning by controlling its configuration at notch," Appl. Phys. Lett., vol. 95, no. 3, p. 032505, 2009. DOI: 10.1063/1.3187530.
[20] M. J. Donahue and D. G. Porter, "OOMMF User's Guide, Version 1.0," National Institute of Standards and Technology, Gaithersburg, MD, Interagency Report NISTIR 6376, Sep. 1999.
[21] J. E. Miltat and M. J. Donahue, "Numerical Micromagnetics: Finite Difference Methods," in Handbook of Magnetism and Advanced Magnetic Materials. Edited by Helmut Kronmuller and Stuart Parkin. Volume 2: Micromagnetism., vol. 2, 3 vols., John Wiley & Sons, Ltd., 2007, pp. 1–23. DOI: 10.1002/9780470022184.hmm202.
[22] T. L. Gilbert, "Classics in Magnetics A Phenomenological Theory of Damping in Ferromagnetic Materials," IEEE Trans. Magn., vol. 40, no. 6, pp. 3443–3449, Nov. 2004. DOI: 10.1109/TMAG.2004.836740.
[23] S. Zhang and Z. Li, "Roles of Nonequilibrium Conduction Electrons on the Magnetization Dynamics of Ferromagnets," Phys. Rev. Lett., vol. 93, no. 12, Sep. 2004. DOI: 0.1103/PhysRevLett.93.127204.
[24] R. L. Stamps et al., "The 2014 Magnetism Roadmap," J. Phys. Appl. Phys., vol. 47, no. 33, p. 333001, Aug. 2014. DOI: 0.1088/0022-3727/47/33/333001.

IOP Conf. Series: Materials Science and Engineering **553** (2019) 012013 doi:10.1088/1757-899X/553/1/012013

Compression Method in JPEG Standard

Sofian Rizal[1*]
[1] Computer Sciences, Facility and Program Department, PUSTEKSAT LAPAN, Jl. Cagak Satelit, Km. 04 Ranca Bungur, Bogor, 16310, Indonesia

sofian.rizal@lapan.go.id

Abstract. Data compression is to minimize the communication and storage. The bandwidth and storage required for a given communication system. Data compression is obviated by wideband technologies such as optical fiber communication networks, packet radio and optical storage disk. Images can generate data in such high volumes that any transmission media can be quickly overwhelmed. Personal computers and workstations has led to an increasing demand to store images in compressed from on magnetic or optical disks. The massive efforts taking place today to adopt national and international standards for signal compression. Example is the JPEG (Joint Photographic Experts Group) standard for image compression. Compression in the image from RGB in JPEG standard converted to YCbCr. Each Y, Cb, Cr divided in block 8x8 pixel. The image 8x8 pixel then transformed use DCT. Image from DCT quantized yield DC and AC component. DC component processed use DPCM (Differential Pulse Code Modulation) use Huffman table yield AC bit stream. Merge DC component and AC component yield JPEG bit stream. To get image from JPEG bit stream first use Huffman decoding. From Huffman decoding dequantized. Then transform use IDCT. Image from IDCT merge block 8x8 pixel. From block 8x8 pixel transformed YCbCr to RGB. Then get the image. Process compression is consist of lossy compression and lossless compression. Lossy compression is compression with change of data but not change meaning of information in there. Lossless compression is compression not to change data.

Keywords : JPEG, Lossy, Lossless, DCT, Quantization

1. Introduction

The aim of this work is to minimize byte the bandwidth and storage. The word signal refers to a continuous time and continuous amplitude waveform will view a signal in its more general sense as a function of time. Where time may be continuous or discrete and where the amplitude or values of the function may be continuous or discrete and may be scalar or vector-valued. Data viewed as a discrete time signal [1]. A signal may have a continuous or discrete amplitude. If the signal has both continuous time and continuous amplitude. It is called a continuous signal or an analog signal. If the signal has both discrete time and discrete amplitude say it is a discrete signal or digital signal or is data in the narrow sense. The ASCII code (American Standard Code for Information Interchange) for characters is a good example of discrete or digital data is a sequence.

Transforming the ASCII sequence into one of these alternative representations is an example of coding data. Signals must be communicated over a digital communication channel such as a modem or packet network or stored in a digital channel such as a standard computer memory. It may be however that either the original signal is analog or that it is digital but not in an appropriate form for

IOP Conf. Series: Materials Science and Engineering **553** (2019) 012013 doi:10.1088/1757-899X/553/1/012013

the transmission or storage in the given channel example for octal data and a binary channel. The original signal must be coded or transformed into a form suitable for the channel. This general operation will be referred to as signal coding. The original signal must be transformed or coded into a form suitable for the channel. This general operation will be referred to as signal coding. Signal compression or data compression codes are used to reduce the amount of memory needed to store a message or to reduce the amount of bandwidth needed to transmit data [2]. Different data require different techniques to remove the redundancy and to identify the redundancy in the data. Data compression viewed as a means for efficient representation of a digital source of data such as image. Data compression is the process of converting an input data stream into another data stream that has a smaller size.

1.1 Lossy Compression
Lossy compression techniques involve some loss of information and data have been compressed using lossy techniques generally cannot be recovered or reconstructed exactly. When the compressed stream is decompressed then the result is not identical to the original data stream. Figure 1 shows an example where a long decimal number becomes a shorter approximation after the compression-decompression process [3, 4].

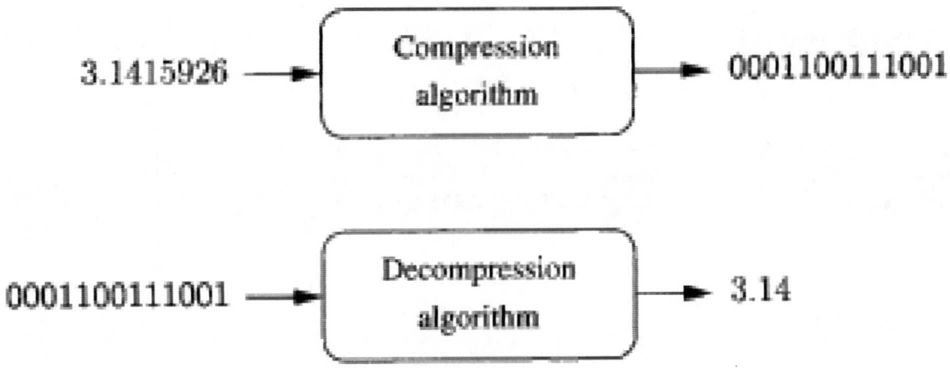

Figure 1. Lossy compression algorithms (courtesy of Ida Mengyi Pu).

1.2 Lossless Compression
If data have been losslessly compressed then the original data recovered exactly from the compressed data. For example in Figure 2. Lossless compression techniques are used when the original data of source are so important that cannot afford to lose any details.

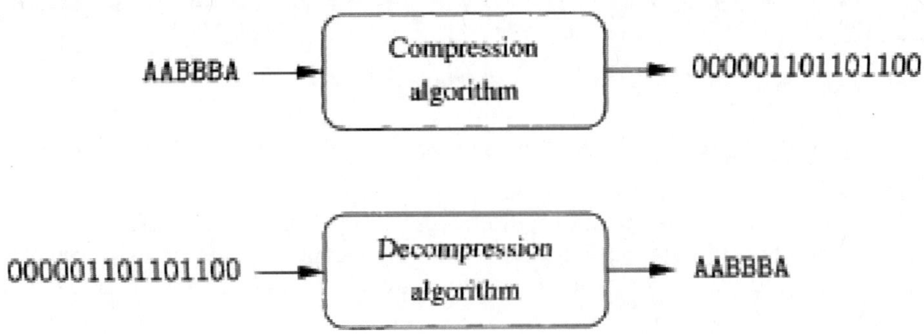

Figure 2. Lossless compression algorithms (courtesy of Ida Mengyi Pu).

2. Experimental method
The experimental method compression in JPEG (Joint Photographic Experts Group) in figure 3.

2.1 Transformation RGB to YCbCr

The eye is very sensitive to small changes in luminance which is why it is useful to have color spaces that use Y as one of their three parameters. The luminance Y varies between 0 and 255 then will be renormalized between 16 the black threshold to 235. The chrominance presence or absence of blue (Cb) and red (Cr) varies between -127.5 and 127.5 but will be renormalized between 16 and 240 with 128 as zero value [5]. There are several YCbCr sampling formats such as 4:4:4, 4:2:2, 4:1:1 and 4:2:0 [6]. Transforming RGB to YCbCr is done by:

$$Y = \frac{77}{256}R + \frac{150}{256}G + \frac{29}{256}B \tag{1}$$

$$Cb = -\frac{44}{256}R - \frac{87}{256}G + \frac{131}{256}B + 128 \tag{2}$$

$$Cr = \frac{131}{256}R - \frac{110}{256}G + \frac{21}{256}B + 128 \tag{3}$$

While the opposite transformation is

$$R = Y + 1.371(\mathrm{Cr} - 128) \tag{4}$$

$$G = Y - 0.698(\mathrm{Cr} - 128) - 0.336(\mathrm{Cb} - 128) \tag{5}$$

$$B = Y + 1.732(\mathrm{Cb} - 128) \tag{6}$$

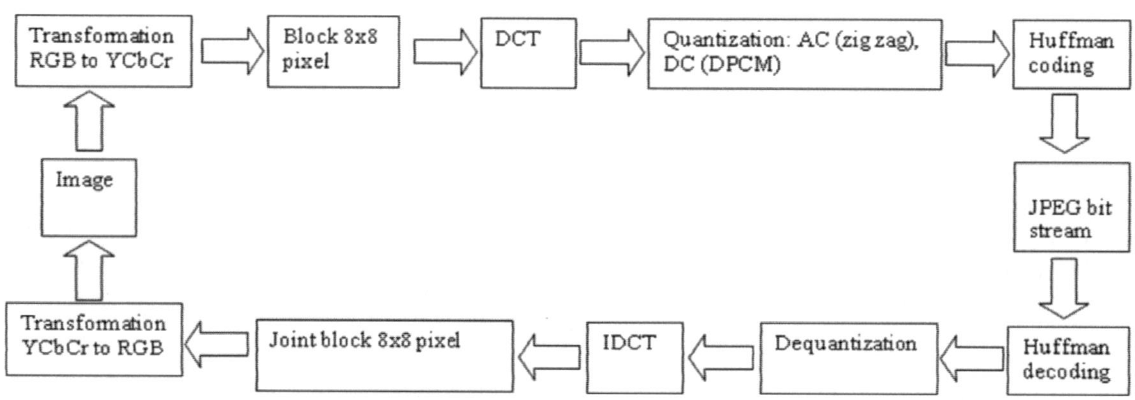

Figure 3. The experimental method compression in JPEG standard (courtesy of Author).

2.2 Block Pixel

The pixels of each color component are organized in groups of 8 x 8 pixels. If the number of image rows or columns is not a multiple of 8 then the bottom row and the rightmost column are duplicated as many times as necessary. Using ranges 8 x 8 pixels would achieve a good quality for the decoded image [7].

2.3 DCT and IDCT

The commonly used discrete cosine transform (DCT) is two dimensional and can be described by an 8x8 transform matrix as equation 7. The inverse DCT (IDCT) as equation (8) [8].

IOP Conf. Series: Materials Science and Engineering **553** (2019) 012013 doi:10.1088/1757-899X/553/1/012013

$$G_{ij} = \frac{1}{4} C_i C_j \sum_{x=0}^{7} \sum_{y=0}^{7} \left(P_{xy}\, cos \frac{(2x+1)i\pi}{16} \cos \frac{(2y+1)j\pi}{16} \right)$$

Where

$$C_f = \frac{1}{\sqrt{2}} \text{ if } f = 0, C_f = 1 \text{ if } f > 1$$

and $0 \leq i,j \leq 7$

(7)

$$G_{ij} = \frac{1}{4} C_i C_j C_{ij} \sum_{x=0}^{7} \sum_{y=0}^{7} \left(P_{xy}\, cos \frac{(2x+1)i\pi}{16} \cos \frac{(2y+1)j\pi}{16} \right)$$

Where

$$C_f = \frac{1}{\sqrt{2}} \text{ if } f = 0, C_f = 1 \text{ if } f > 1$$

and $0 \leq i,j \leq 7$

(8)

2.4 *Quantization*

The formula for quantization is

$$Quantized\ Value\ (i,j) = \frac{DCT(i,j)}{Quantum\ (i,j)} \tag{9}$$

During decoding then the dequantization formula operates in reverse.

DCT(i,j) = Quantized value (i,j) * Quantum(i,j) (10)

Errors generated in the high frequency components during dequantization normally don't have a serious effect on picture quality [9].

2.5 *Huffman Coding and Decoding*

Encoding generalized to the lossy compression called differential encoding. For example DPCM (Differential Pulse Code Modulation). The Huffman coding deals with the 63 AC (Alternating Current) coefficients. The absolute values of which are in general sensibly smaller than the absolute value of the dominant DC (Direct Current) coefficient. The encoding concerns the sequence of the non-zero coefficient in the zigzag reading of the quantized scheme. Huffman coding conveniently begin with a hierarchy of 10 categories for the non-zero coefficients [10].

Non-zero coefficient occurring in the sequential reading of a quantized scheme can be characterized by three parameters that is the number of zeros which separate it from its non-zero predecessor, category and number within the category.

JPEG from the average statistics of a large set of images with 8 bit precision. Remark on two particular symbols that is end of block (EOB) indicates the end of the non-zero coefficients in the sequence of the 63 AC coefficients to be encoded. The code word will be 1010 and zero run list (ZRL) indicates the outcome of the integer 0 preceded by a block of 15 zeros [11-13].

Huffman decoding compressed file by sliding down the code tree for each symbol is simple. The compressed file has to be read bit by bit and the decoder has to advance a node in the code tree for each bit. The method of this section uses preset partial decoding table. These table depend on the Huffman code used but not on the data to be decoded. The compressed file is read in chunks of k bits each and the current chunk is used as a pointer to a table. The code word permits the reconstruction of the matrix of the quantized values. Dequantization means multiplication with the divisor scheme that has been used for quantization. The loss of information due to quantization becomes apparent. The decompression will be finished after transforming the matrix of dequantized values back and making the necessary rounds.

3. Results and discussion

Noise can change data image. Small or big noise result from difference between the pixel of the original image by P_i and the pixels of the reconstructed image by Q_i where $1 \le i \le n$ shows in equation 11 call Mean square error (MSE).

$$\text{MSE} = \frac{1}{n}\sum_{j=1}^{n}(P_i - Q_i)^{\,2} \tag{11}$$

Figure 4 shows original image (2018) 83 Kb from camera XENICS XTM 640-CL LWIR like grayscale.

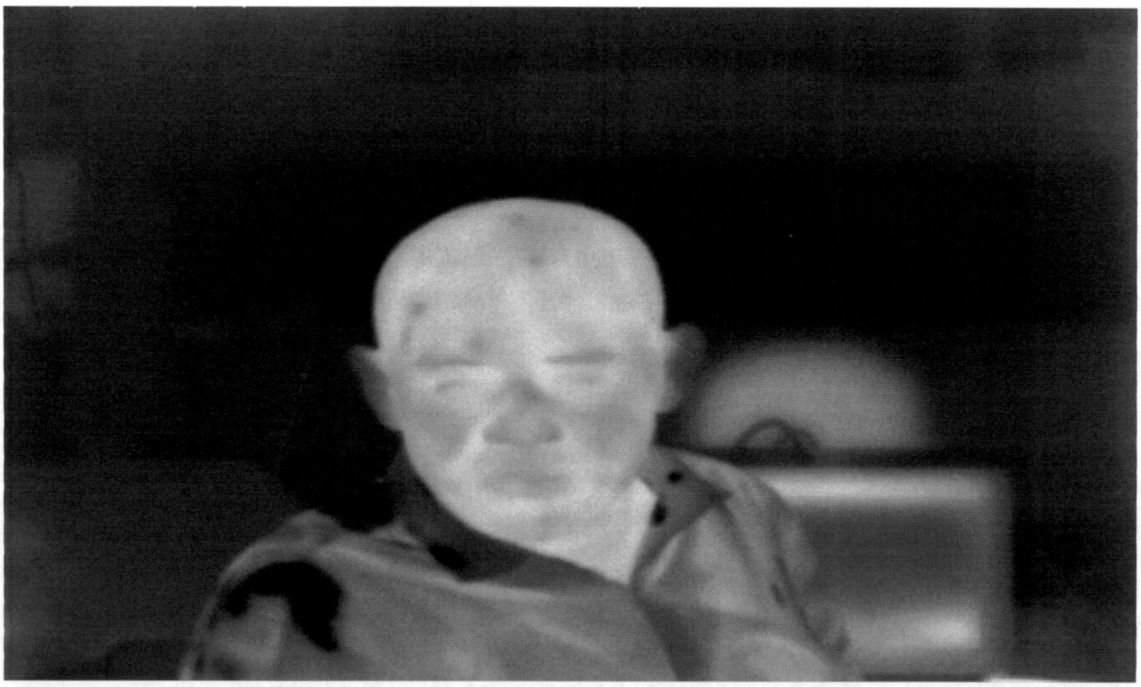

Figure 4. Original image 83 Kb from camera XENICS XTM 640-CL LWIR (courtesy of author).

Figure 5 shows representasion image 83 KB from DCT to IDCT in LabView programming. In Figure 5 that source image from RGB change to YCbCr. MSE is 2.40344 that small and quality image from restored image is good. Figure 6 shows representasion language programming from DCT to IDCT in LabView programming 2013. Source image from camera XENICS in left down size 640 x 480. Source image camera XENICS convert to source image block pixel 8x8 in left up. IDCT block 8x8 pixel in the restored image. The pixel value in right down. Figure 7 shows original image (2018) 151 Kb from camera HP ASUS Zenfone 2 like color RGB. Figure 8 shows representasion image 151 KB from DCT to IDCT in LabView programming.

Source image from camera Handphone ASUS Zenfone 2 in left down size 799 x 1422. Source image camera Handphone ASUS Zenfone 2 convert to source image block pixel 8x8 in left up. IDCT block 8x8 pixel in the restored image. The pixel value in right down. In Figure 8 that source image from RGB change to YCbCr. MSE is 1.79462 that more less than Figure 5 and quality image from restored image is more better than figure 5. Figure 8 is bigger byte than Figure 5 because Figure 5 is grayscale size 640x480 and Figure 8 is color size 799x1422. The byte in color bigger than grayscale if the same size. Figure 8 is bigger than Figure 5 because size 799x1422 in Figure 8 is bigger than size 640x480 in Figure 5.

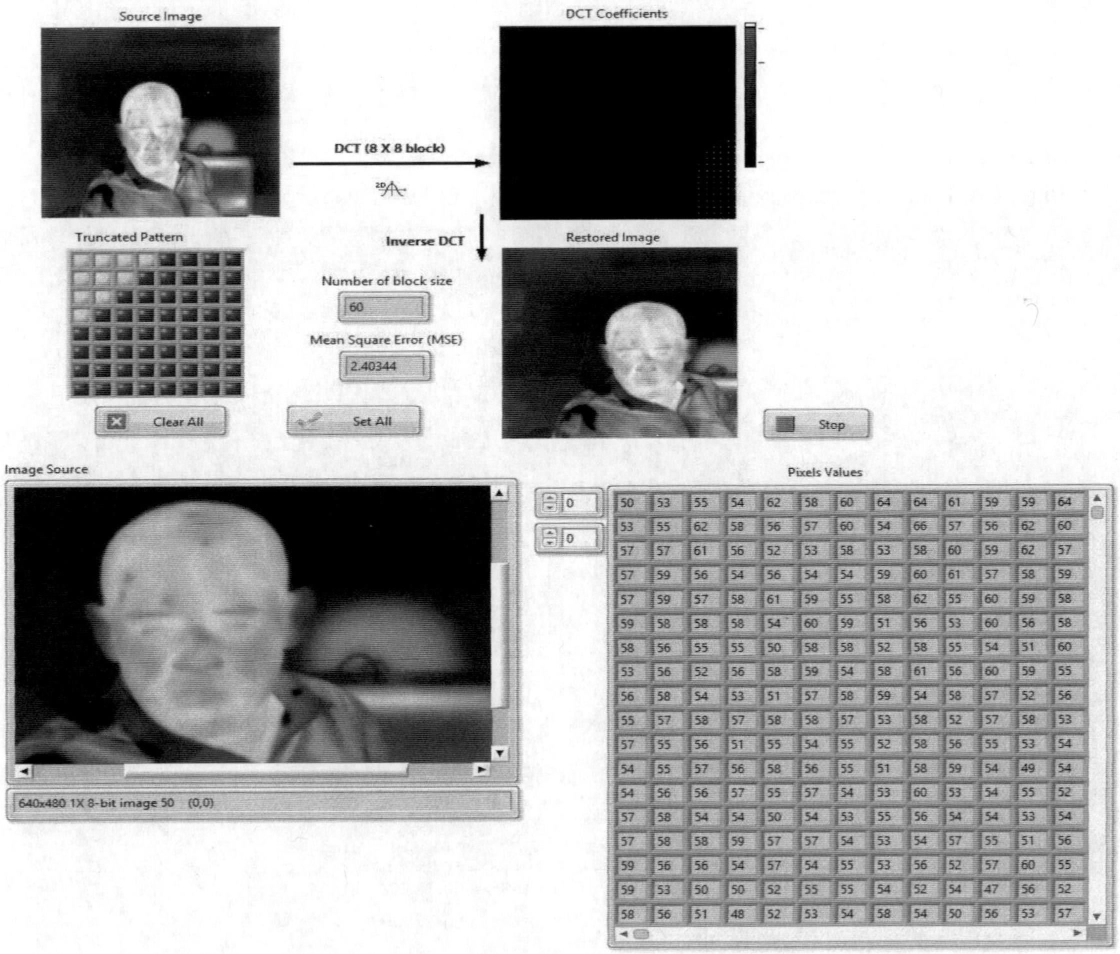

Figure 5. Structure representation image 83 Kb form DCT to IDCT.

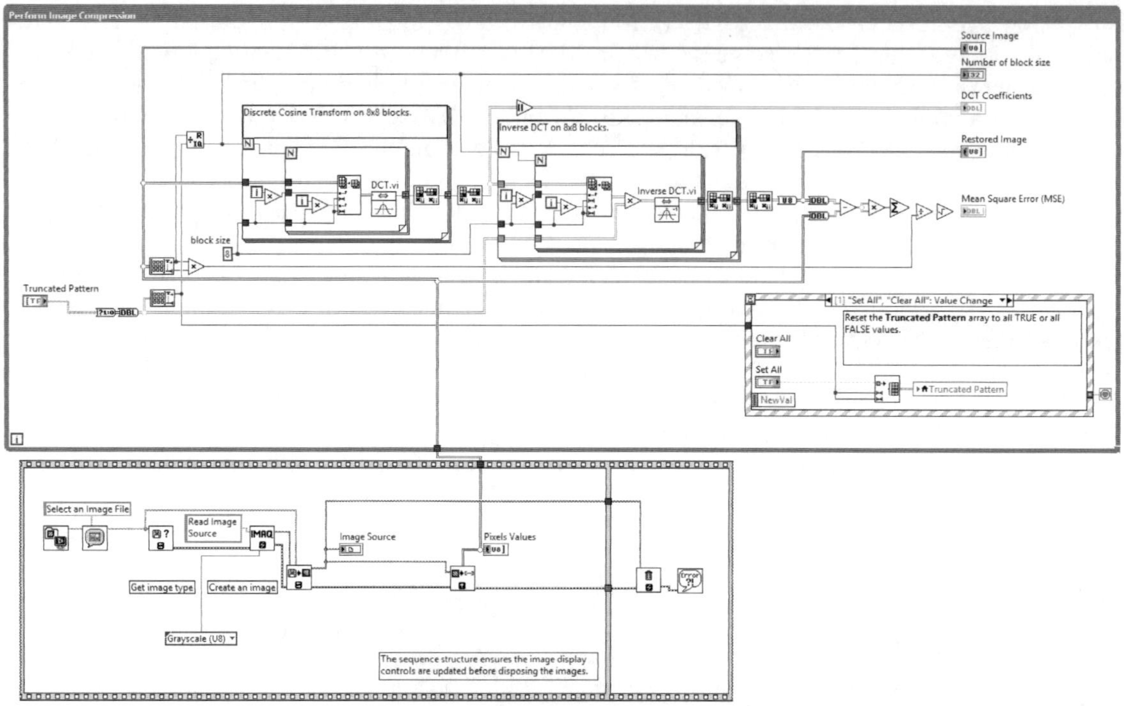

Figure 6. Language programming from DCT to IDCT in LabView programming 2013.

Figure 7. Original image 151 Kb form camera Handphone ASUS Zenfone 2.

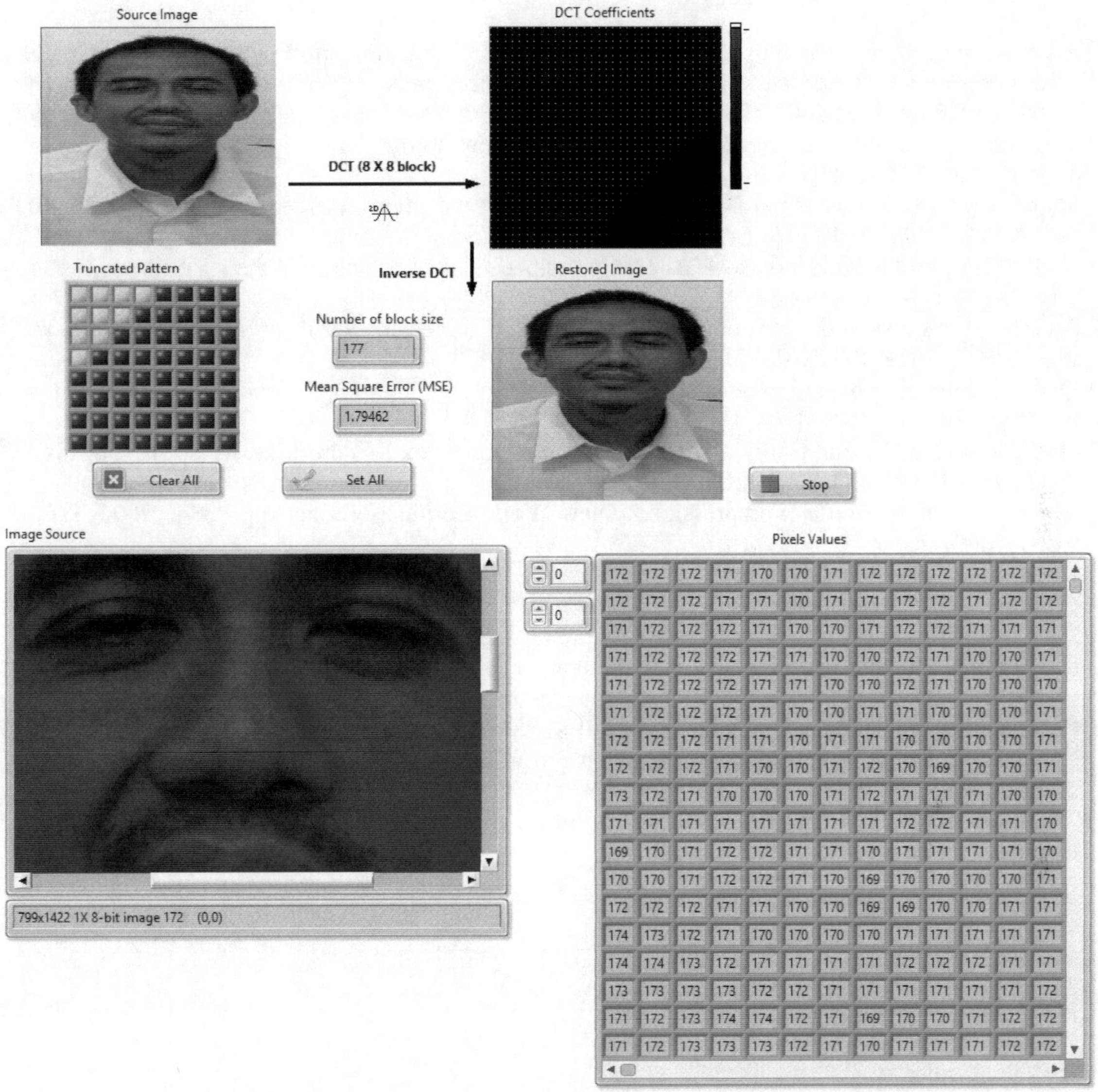

Figure 8. Structure representation image 151 Kb form DCT to IDCT.

4. Conclusion

When performing YCbCr to RGB conversion then the resulting RGB pixel values have a nominal range of 16 – 235 with possible occasional values in 0 – 15 and 236 – 255. Bigger MSE smaller quality image. The compression speed and decompression speed in JPEG standard is fast. Disadvantage for the compression method in JPEG standard is not suitable for all type of compression methods like JPEG 2000.

Acknowledgment

Thanks to Denden Ramdan, S.T and Bambang Sigit Pamadi for the support in finishing this paper.

References

[1] Allen Gersho, Robert M. Gray, "Vector Quantization and Signal Compression," United States of America Kluwer Academic Publishers Springer, pp.1, 1992 ISBN-10:0792391810 ISBN-13:978-0792391814

[2] C. Britton Rorabaugh, "Error Coding Cookbook," New York McGraw-Hill Companies, pp. 1, 1996 ISBN 0-07-911720-1 (h)

[3] Ida Mengyi Pu, "Fundamental Data Compression," Great Britain Elseiver, pp. 6, 2006 ISBN-10: 0-7506-6310-3 ISBN-13: 978-0-7506-6310-6

[4] Khalid Sayood, "Introduction to Data Compression," United States of America Elseiver, pp. 5, 2012 ISBN 978-0-12-415796-5

[5] Peter Seibt, "Algorithmic Information Theory," Berlin Heidelberg Springer-Verlag, pp. 301, 2006 ISSN 1860-4862 ISBN-10 3-540-33218-9 ISBN-13 978-3-540-33218-3

[6] David Solomon, Giovanni Motta, "Handbook of Data Compression," London Springer-Verlag, pp. 525, 2010 DOI 10.1007/10.1007/978-1-84882-903-9

[7] Mark Nelson, Jean-Loup Gailly, "The Data Compression Book," Cambridge, MA 02139, pp. 315, 1977 ISBN 1558514341

[8] David Solomon, "Data Compression," New York Springer-Verlag, pp. 334, 2004 ISBN 0-387-40697-2

[9] Yonina C. Eldar, Gitta Kutyniok, "Compressed Sensing: theory and application," Cambridge University Press, pp. 21, 2012 ISBN 978-1-107-00558-7

[10] M. Sentilkumar, V. Mathivanan, "International Journal of Advanced research in Computer Science and Software Engineering, " Vol. 6, number 5, 2016 ISSN: 2277 128X

[11] Shu Lin and Daniel J. Costello, JR., "Error Control Coding," New Jersey Prentice-Hall, Inc., Englewood Cliffs, New Jersey 07632, pp. 2, 1983 ISBN 0-13-283796-X

[12] Thomas M. Cover, Joy A. Thomas, "Elements of information Theory," Willey-Interscience, pp. 200, 2006 ISBN-13 978-0-471-24195-9 ISBN-10 0-471-24195-4

[13] Anmol Jyot Maan, "International Journal of Information and Computation Technology," Vol. 3, number 3, 2013 ISSN 0974-2239

Sensitivity of Localized Surface Plasmon Resonance (LSPR) Au Nanorod with Methylene Blue Medium Using Boundary Element Method Simulation

Miyah Sofani[1], Muhammad Sujak Mansur Kuddah[1], Miftahussur Hamidi Putra[2], Dede Djuhana [1*]
[1] Department of Physics, FMIPA Universitas Indonesia, Kampus UI Depok, Depok 16424, Indonesia
[2] Institute of Theoretical Chemistry, Ulm University, Albert-Einstein-Allee 11, D-89069 Ulm, Germany

dede.djuhana@sci.ui.ac.id

Abstract. We have calculated the sensitivity of gold nanorod (AuNR) in water solution and methylene-blue based on boundary element method with varying diameter of the nanorod, D are 20 nm, 40 nm, 60nm, and 80 nm. The aspect ratio was varied from 1.5 to 3.5 according to diameter. The dielectric function of AuNR based on Christy and Johnson experiment. To understand the sensitivity sense, we have also varied the medium refractive index is n = 1.3334 (100% water pure) and n = 1.347 (100% methylene-blue pure). The sensitivity of LSPR Au nanorod was determined by the gradient of the peak of wavelength to the refractive index medium variation for all aspect ratio. The LSPR of AuNR consisted of a longitudinal and transversal mode. The longitudinal mode appeared higher wavelength than the transversal mode in LSPR spectra and the peak of wavelength increased as the aspect ratio increased while in transversal mode the peak of wavelength relatively was constant. Furthermore, the sensitivity in longitudinal mode increased as the aspect ratio increased whereas the sensitivity in transversal decreased as the aspect ratio increased. Increasing the sensitivity in longitudinal mode related to red-shift as the nanorod volume increased and the refractive medium index change.

Keywords: Sensitivity, Localized Surface Plasmon, Methylene Blue, The Refractive Medium Index

1. Introduction

The optical characteristic of plasmonic nanoparticles can be adjusted by changing their shape, size, composition of the medium refractive index and structure [1]. Research in the field of nanotechnology shows the results of new products with better performance than large size materials. The LSPR of extinction spectra yielded from accumulated absorption and scattering spectra [2]. The scattering and absorption spectra depend on the shape, size and the refractive index. The ratio of refractive index increases by increasing the particle size such as the variant of diameter [3].

The most applications of noble metal nanoparticles, namely Au, Ag, and Cu referred to the presence of local surface plasmon resonance (LSPR), for instance, biomedical field [4], textile industry [5], energy [6], electronics [7], and agriculture [8]. Currently, many of sensor application use LSPR phenomenon due to their frequency can be tuned by changing of surrounding medium. The important factor to determine the quality of sensor by calculating the sensitivity factor, that is the magnitude of LSPR shift divided the refractive index change. Nanorod sensitivity is a value that represents the ability of nanorods to stimulate the environmental refractive index around nanorod [3]. The sensitivity value is very important especially in nanoparticle applications as sensors. Therefore, knowledge of the sensitivity of Au nanorod is very important in its use as a sensor [9].

In this study aimed calculated the sensitivity of Local Surface Plasmon Resonance (LSPR) phenomenon occurring in Au material from the rod model immersed in two different medium, water and methylene blue and analyzing the LSPR shift pattern from pneumatic on the aspect ratio curve to Au material from the rod model.

2. Simulation Procedure

We have calculated the extinction spectra of Au nanorod using public MNPBEM (Metallic Nano-Particle Boundary Element Method) simulation based on the boundary element method [10]. In this simulation, we varied the aspect ratio from 1.5 to 3.5 with increment 0.5 and the diameter D was 20nm, 40nm, 60nm, and 80nm. The length of Au nanorod followed the aspect ratio, as shown in Figure 1(a). The dielectric function of AuNR was obtained from Johnson and Christy experiment. To calculate the sensitivity, we used two mediums, first the refractive index water n_w = 1.334 and methylene blue n_{mb} = 1.347. Then, we applied the electric field to Au-nanorod in parallel and perpendicular direction, as presented in Figure 1(b).

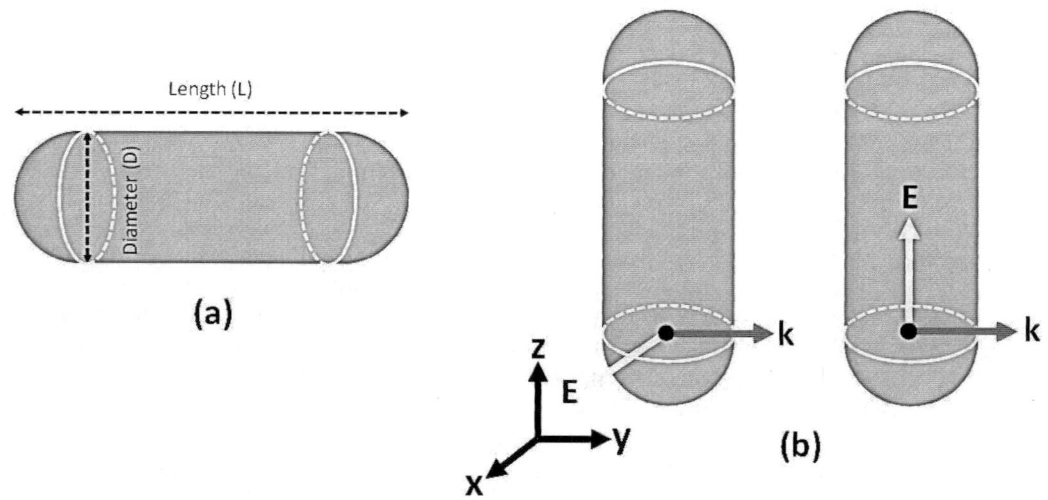

Figure 1. (a) Geometry and dimension Au nanorod. The aspect ratio is determined by length and diameter of rod and (b) Polarization direction of electric field.

From this simulation, we generated LSPR spectra such as extinction curve as the function of wavelength. Next, we can determine the sensitivity of Au nanorod with different medium based on the gradient value.

3. Results and discussion

The extinction curve of Au nanorod both water and methylene blue medium as the wavelength were presented in Figure 2. It was observed two modes such as transverse and longitudinal mode, as commonly found in nanorod shape. In the longitudinal mode, the peak of extinction curve shifted to higher wavelength or red-shift as the aspect ratio increased while the peak slightly shifted to lower wavelength or blue-shift [11].

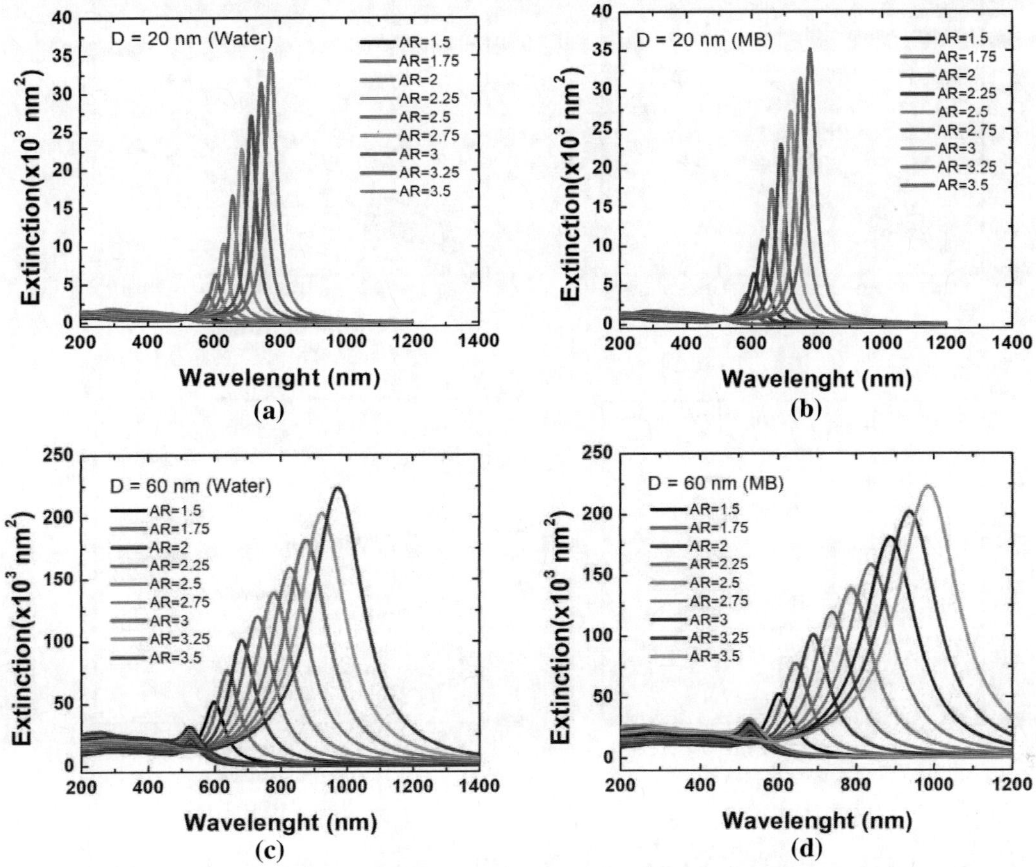

Figure 2. The extinction of wavelength (a,c) with pure water medium; (b,d) with methalyne blue medium by variations of aspect ratio.

The peak of extinction curve happened in visible-near infrared range. To understand, we have also plotted the peak of extinction curve versus the aspect ratio for two medium, as shown in Figure 3. From this figure, we observed shifting the LSPR spectra shift to lower energy in longitudinal mode and higher energy in transverse mode. Interestingly, we have observed the peak of extinction curve moved from visible to near infrared at the aspect ratio around 3 for the case small diameter. On the contrary for large diameter, the peak shifted from visible to near-infrared around the aspect ratio 1.75 [12].

Further, we have also calculated the sensitivity of Au nanorod with two different medium, water, and methylene blue. Figure 4 showed the sensitivity calculation of Au nanorod both longitudinal and transverse mode for the aspect ratio variation. In the case longitudinal mode, the sensitivity value increased as the aspect ratio increased while the sensitivity decreased as the aspect ratio increased in transverse mode. As the figure, we observed the sensitivity of Au nanorod exhibited fluctuation in small diameters such as D = 20nm and D = 40nm. But for large diameters, such as D = 60nm and 80nm was consistently increasing as the aspect ratio increased. Different in transverse mode, the sensitivity decreased following the aspect ratio increased. From this result, it showed the sensitivity of Au nanorod was originated from longitudinal mode where the peak of the extinction curve significantly changed as the aspect ratio increased. Shifting the peak in large diameter was more than small diameter compares to in transverse mode.

IOP Conf. Series: Materials Science and Engineering **553** (2019) 012014 doi:10.1088/1757-899X/553/1/012014

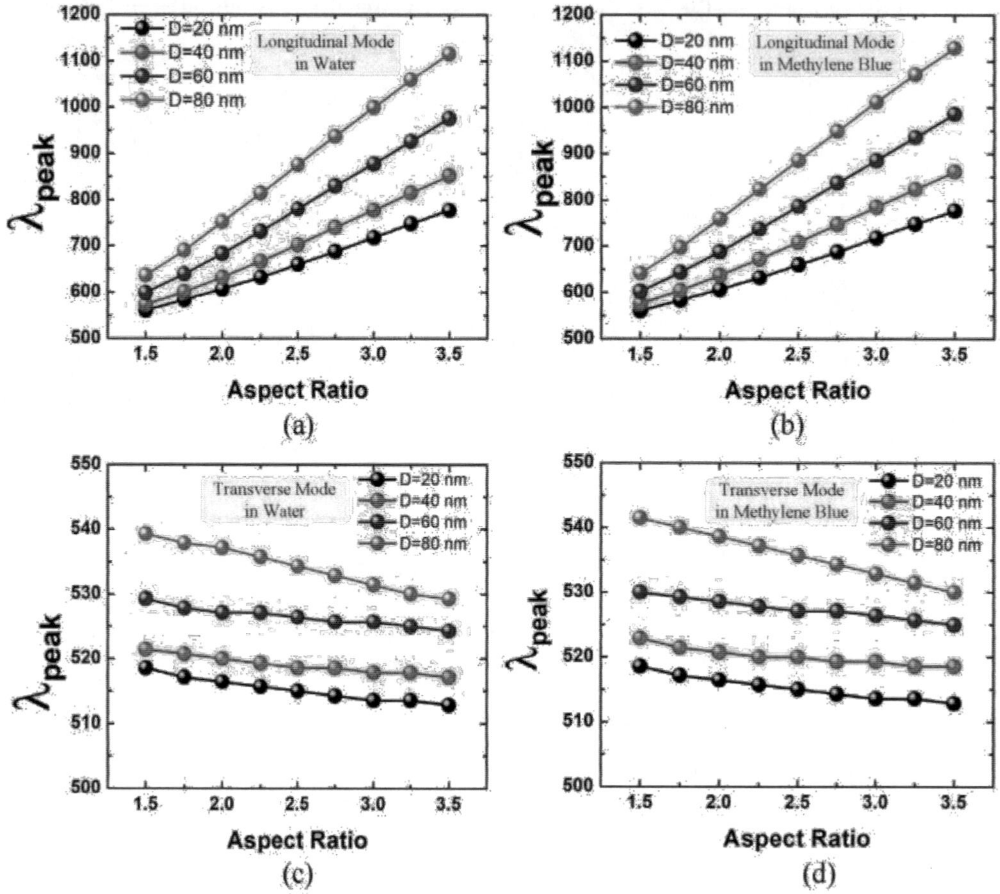

Figure 3. The LSPR peak of aspect ratio (a) longitudinal mode; (c) transverse mode with pure water medium and (b) longitudinal mode; (d) transverse mode with methylene blue by variations of diameters.

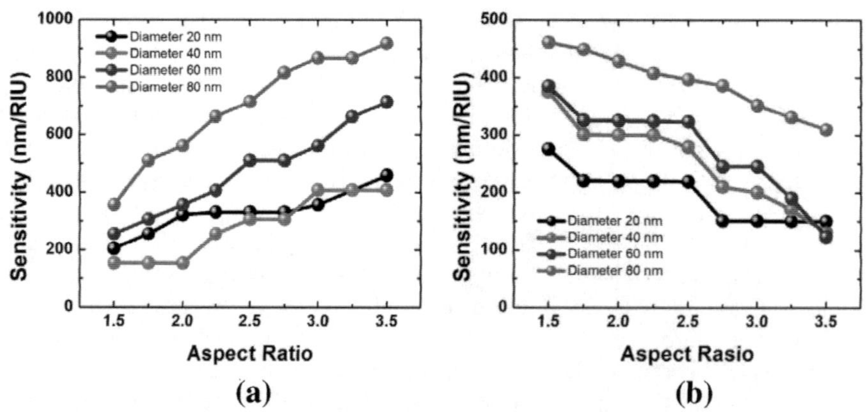

Figure 4. Sensitivity of Au nanorod for (a) longitudinal mode and (b) transversal mode.

The result calculated the sensitivity of Au nanarod gave two different polarization. We have also determined the limit range aspect ratio of gold nanorod that occurs in visible range. Then it can be seen the refractive index, diameter, and aspect ratio will produce high sensitivity values for each of the different polarization directions [13].

4. Conclusion

The sensitivity of gold nanorod shaped have been simulated by MNPBEM packaged. The LSPR phenomenon with the nanorod form, namely the transverse mode and the longitudinal mode. It shows, for longitudinal mode the sensitivity with diameter 20nm was 103,40523 RIU/nm, for 40nm was 156,46238 RIU/nm, for 60nm was 228,02452 RIU/nm, and for 80nm was 272,10904. On transverse mode the sensitivity with diameter 20nm was -61,09517 RIU/nm, for 40nm was -75,66319 RIU/nm, for 60nm was -110,62851 RIU/nm, and for 80nm was -112,783012 RIU/nm. The aspect ratio of the length to the variation of diameter affect the shift in resonant wavelength. The variation of environmental refractive index around nanorods also change the peak wavelengths of extinction of Au nanorods. The increasing in environmental refractive index leads to the peak wavelength of the domain with the longitudinal mode of the Au nanorod undergoing a shift to red wavelength (redshift). While the peak wavelength of extinction with the transversal mode of Au nanorod shifts to the blueshift. Sensitivity of the Au nanorod when the variation of the aspect ratio and the environmental refractive index increases, the sensitivity value also changes to greater, and is affected by the increasing diameter for the longitudinal mode.

Acknowledgment

This work was funded by Hibah Publikasi Internasional Terindeks (PIT) 9 Tahun 2019 through DRPM Universitas Indonesia under Grant Number : NKB-0023/UN2.R3.1/HKP.05.00/2019.

References

[1] Mahmoud A. Mahmoud Mostafa El-Sayed, “Different Plasmon Sensing Behavior of Silver and Gold Nanorods pp. 1541–1545, Apr. 2013. DOI: 10.1021/jz4005015

[2] C. Kelly, K. L E. .. Zhao, L. L; Schatz, G. C., “The Optical Properties of Metal Nanoparticles: The Influence of Size, Shape, and Dielectric Environment,” *J Phys Chem*, vol. 107, pp. 668–677, 2003. DOI: 10.1021/jp026731y

[3] K. X. Chen H Yang Z, Ni W, and Wang J., “Shape and Size Dependent Refractive Index Sensitivity of Gold Nanoparticles,” *Langmuir*, vol. vol 24, pp. 5233–5237, May 2008. DOI: 10.1021/la800305j

[4] L. A. Dykman and N. G. Khlebtsov, “Optical properties and biomedical applications of plasmonic nanoparticles,” vol. 111, pp. 1–35, Jan. 2010. DOI: 10.1021/jp066539m

[5] F. Zhang, X. Wu, Y. Chen, and H. Lin, “Application of silver nanoparticles to cotton fabric as an antibacterial textile finish,” vol. 10, pp. 496–501, Aug. 2009. DOI: 10.1007/s12221-009-0496-8

[6] M. Nortarianni, K. Veernon, A. Chou, M. Aljada, J. Liu, and N. Motta, “Plasmonic effect of gold nanoparticles in organic solar cells,” vol. 106, pp. 23–27, Aug. 2014. DOI: 10.1016/j.solener.2013.09.026

[7] M. Shur, “AlGaN/GaN plasmonic terahertz electronic devices,” vol. 486, Mar. 2014. doi:10.1063/1.3291101

[8] S. Sabir, M. Arshad, and S. K. Chaudhari, “Zinc Oxide Nanoparticles dor Revolutionizing Agriculture: Synthesis and Application,” vol. 2014, 2014. DOI:10.1155/2014/925494

[9] K. Y. Yang M et al Bruck H. A, and Rasooly, “Gold nanoparticle based enhanced chemiluminescene immunosensor for detection of staphylococcal Entetrotoxin B (SEB) in food,” *Int J Food Micribiol*, vol. 133, pp. 265–271, Aug. 2009. doi: 10.1016/j.ymeth.2013.08.031

[10] Andreas Trugler, *Optical Properties of Metalic Nanoparticles Basic Principle and simulation*, vol. 232. 2015. DOI 10.1007/978-3-319-25074-8

[11] Min Kyung Oh, Seong Kyu Kim, Hyo Jin Baik, and Sungho Park, "Multiple surface plasmon resonances in silver and copper nanorods," *J Master Chem*, vol. 21, p. September 2011. DOI: 10.1039/C1JM13613E

[12] Jie Cao, Tong Suna, and Kenneth T. V. Grattana, "Gold nanorod-based localized surface plasmon resonance biosensors: A review," *Sens Actuators*, pp. 332–351. doi: 10.1016/j.snb.2014.01.056

[13] S. L. M. A. El-Sayed, "Simulation of the Optical Absorption Spectra of Gold Nanorods as a Function of Their Aspect Ratio and the Effect of the Medium Dielectric Constant," Apr. 1999. DOI: 10.1021/jp990183

Investigation the exponential decay of LSPR Ag nanorod-coupled with side by side configuration using boundary element method

Nesy Fitria[1], Muhammad Sujak Mansur Kuddah[1], Dede Djuhana[1*]
[1] Department of Physics, FMIPA Universitas Indonesia, Kampus UI Depok, Depok 16424, Indonesia

dede.djuhana@sci.ui.ac.id

Abstract. In this study, we have systematically investigated LSPR spectra of Ag nanorod-coupled with side of side configuration by a public optical modelling, MNPBEM. The electromagnetic field was applied in perpendicular and parallel direction. The simulation showed the extinction peaks shifted to red-shift in longitudinal mode while shifted to blue-shift in transverse mode as the gap distance increased. The electric field distribution showed increasing in area between nanorod in transverse mode as the gap distance reduced. Contrary, the electric field distribution relatively was constant in longitudinal mode. The electric field increased at end of Ag nanorod-coupled. Furthermore, we have also calculated the different value of Ag coupling and single nanorod to the aspect ratio of the gap distance and the length of nanorod. Interestingly, it was found following the exponential behaviour in longitudinal mode with the decay length related to the electric field distribution as the gap distance increased.

Keywords : LSPR, Plasmon coupling, side by side configuration

1. Introduction

For recent years, the development of nanoparticles research has brought many advantages to various application such as sensor, bioimaging, cancers [1]. Noble particles such as Au, Ag, Cu, and alloys based on noble particle shows a resonance collective conduction electron that known as localized surface plasmon resonance (LSPR) [1,2]. The spectra of LSPR depends on shape, size, composition, medium, and structure [2,3]. One of part LSPR issue relates to the coupling configuration and the gap distance since useful to enhance the sensitivity and intensity of LSPR [2]. Several works have been reported about the coupling configuration in noble particle [4-7]. However, there are few studies about the effect of the gap distance in Ag nanorod-coupled and the decay length of coupling configuration.

In this study, we have investigated LSPR of Ag nanorod-coupled with side by side configuration based on optical simulation, Metallic Nano-Particle Boundary Element Method (MNPBEM). The length of nanorod and the gap distance of coupling are varied. The incident wave is applied on Ag nanorod-coupled in perpendicular and parallel polarization. We find the different LSPR spectra Ag nanorod-coupled and single nanorod to the aspect ratio the gap distance and the length of nanorod showed an exponential behavior and followed plasmon rule phenomenon.

2. Simulation method

In this study, we carried out the public of optical modeling MNPBEM to investigate the exponential decay of LSPR Ag nanorod-coupled with side by side configuration based on boundary element method [4,8]. The side by side configuration of Ag nanorod-coupled and the incident wave application were illustrated in Figure 1. The diameter D of Ag nanorod was fixed in 20 nm and the length L was varied from 30 nm, 60 nm, and 100 nm or the ratio of the diameter and length were 1.5, 3, and 5, respectively. The gap distance of Ag nanorod-coupled, s was from 5nm to 210nm. The dielectric function of Ag nanorod was used by Palik's experiment [5] and the refractive index of medium was water n_m=1.334. Then, we applied the incident wave in perpendicular and parallel direction of Ag nanorod configuration to produce the spectra, such as such as absorption, scattering, and extinction cross section as the function of wavelength.

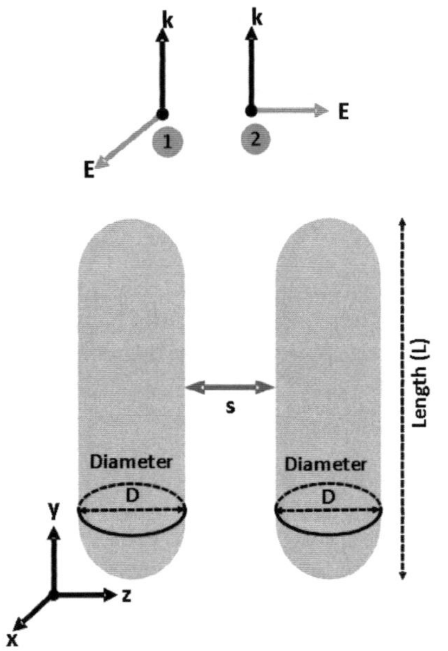

Figure 1. The configuration of Ag nanorod-coupled with side by side. The diameter D is 20 nm and the length L are 20 nm, 60 nm, and 100 nm. The gap distance of Ag nanorod s is varied from 5 nm to 200 nm. The incident wave is applied in perpendicular and parallel direction.

3. Results and discussion

The LSPR spectra of Ag nanorod-coupled with side by side configuration for varying the gap distance from 5 nm to 210 nm for the aspect ratio 3 was shown in Figure 2. For this purpose, we only used the extinction cross section as the function of wavelength. There were two mode such as longitudinal and transverse mode, as commonly was found in nanorod shape [1]. The extinction peaks of both longitudinal dan transverse mode shifted as the gap distance changed. The transverse mode shifted to blue-shift while the longitudinal mode shifted to red-shift as the gap distance increased. More, we found LSPR spectra of Ag nanorod-coupled intensity showed higher than LSPR intensity of single nanorod. For the case s = 210nm, the peak of extinction around λ = 660nm closed to the peak of single nanorod around λ = 661nm in longitudinal mode while in transverse mode, the peaks showed same value both Ag nanorod-coupled and single nanorod around λ = 360nm. Higher LSPR intensity of Ag nanorod-coupled was affected by the surface charge polarization and the dipole moment. Increasing the surface charge of Ag nanorod-coupled caused the LSPR intensity also increased [9,10].

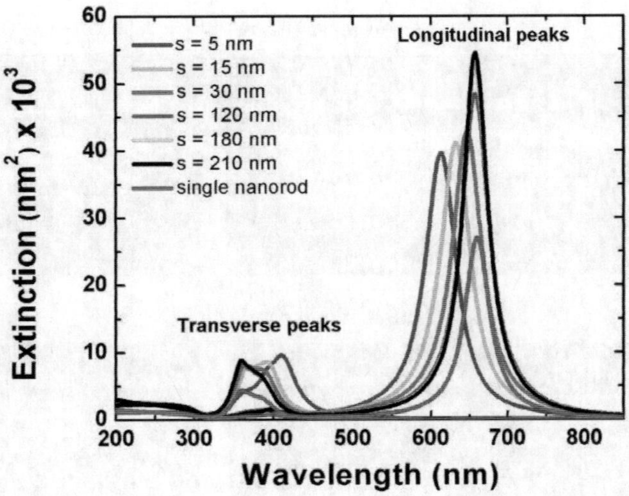

Figure 2. The extinction of Ag nanorod-coupled side by side configuration for s = 5, 15, 30, 120, 180, and 210 nm in the case the aspect ratio is 3 and the single nanorod. The peak of transverse mode is blue-shift and the peak of longitudinal mode is red-shift.

Next, we have also presented the LSPR of Ag nanorod-coupled in the image of electric field distribution for the aspect ratio 3 both transverse and longitudinal mode, as shown in Figure 3. As the figure, we found the coupling interaction was stronger as the gap distance reduced [3]. The value of electric field tended to increase between Ag nanorod-coupled from 6V/m to 12V/m as the gap distance reduced from s = 30nm to s = 5nm in transverse mode. Opposite in longitudinal mode, the value of electric field was not change. The value of electric field showed increasing at end of Ag nanorod-coupled around 30V/m. Based on Figure 3, the coupling interaction of Ag nanorod-coupled was more attractive in transverse mode and tended to blue-shift as the gap distance increased in transverse mode. In longitudinal model, the surface charge at end of Ag nanorod-coupled increased that contributed to the LSPR intensity became higher than single nanorod and tended to red-shift as the gap distance increased.

Further study, we have also investigated the relation of the extinction peak of Ag nanorod-coupled and single nanorod as the aspect ratio between the gap distance and the length of nanorod. It was found the LSPR peak of coupling configuration to a single nanorod followed an exponential decay behavior in longitudinal mode, as reported by Sayed and Huang [6],

$$\Delta\lambda / \lambda_0 = A \exp(-s / L\tau) \tag{1}$$

where $\Delta\lambda = \lambda - \lambda_0$ was the different the LSPR peak between the coupling and single nanorod. A was constant value, s was the gap distance of coupling, L was the length of nanorod, and τ was the decay length. The curve of $\Delta\lambda / \lambda_0$ vs. s / L and the exponential fitting for the aspect ratio AR = 1.5, 3, and 5 was presented in Figure 4. For the exponential fitting, we applied the approximation $y = A\exp(-x / k)$ with $x = s / L$ and $k = \tau$. The result of exponential fitting was shown in Table 1.

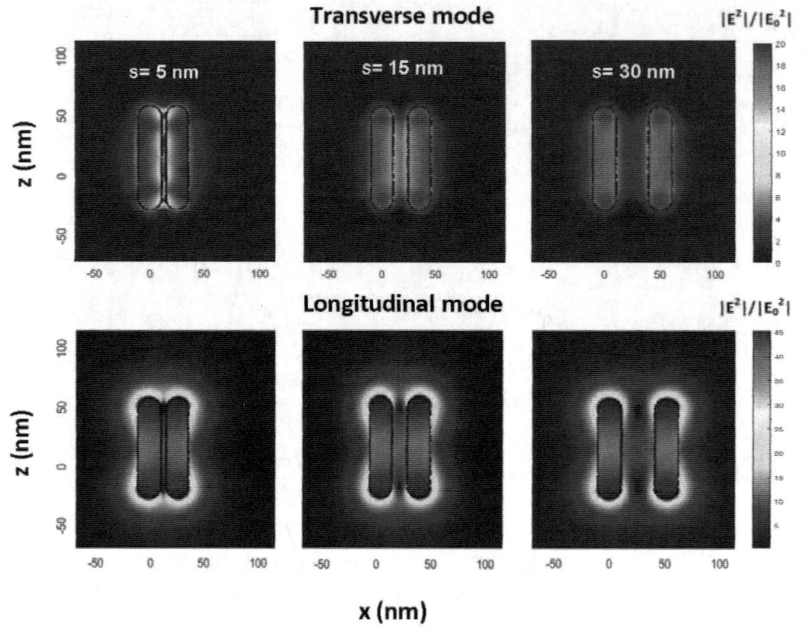

Figure 3. The electric field distribution of Ag nanorod-coupled in transverse mode (above) and longitudinal mode (below) for the aspect ratio 3 and s = 5nm, 15nm, and 30nm. The bar color presents the rasio of electric field

Table 1. The result of exponential fitting of $\Delta\lambda / \lambda_0$ vs. s / L for the aspect ratio AR=1.5, 3, and 5.

Aspect ratio (AR)	Constant (A)	Decay length (τ)
AR = 1.5	A = 0.14	$\tau = 0.27$
AR = 3	A = 0.23	$\tau = 0.20$
AR = 5	A = 0.19	$\tau = 0.10$

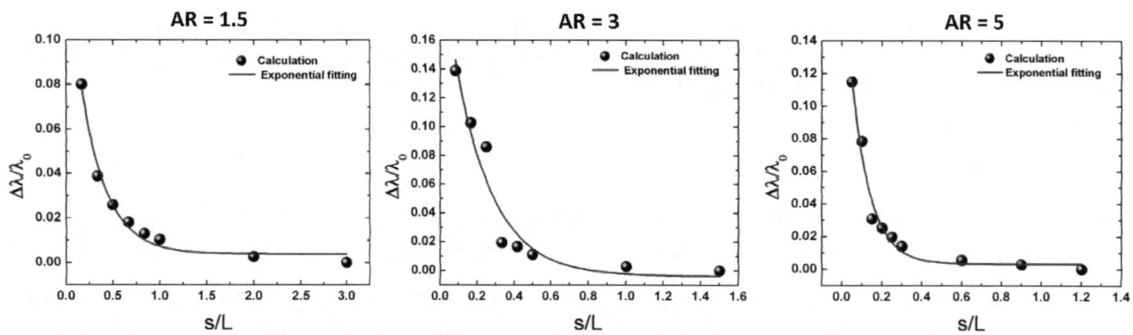

Figure 4. The curve of $\Delta\lambda / \lambda_0$ vs. s / L and the exponential fitting for the aspect ratio AR = 1.5, 3, and 5 of Ag nanorod-coupled in longitudinal mode.

Based on Table 1 and Figure 4, we showed our calculation following the exponential behavior and fitted to the exponential function. Interestingly, the decay length exhibited decreasing as the aspect ratio increased and the constant A varied around 0.14 to 0.23. The decay length τ related to the electric field intensity reduced as the gap spacing increased [7]. The decay length was around 0.10 to 0.27 and the constant A was around 0.14-23 agree with others results [6].

4. Conclusion

In conclusion, we have performed LSPR of Ag nanorod-coupled with side by side configuration. LSPR spectra of Ag nanorod-coupled showed two mode, longitudinal and transverse mode. The peak of LSPR shifted to red-shift in longitudinal mode while blue-shift in transverse mode as the gap distance increased. In longitudinal mode, the different of LSPR peak between the coupling configuration and single nanorod to the ratio between the gap distance and length of nanorod showed an exponential behaviour. The dipole-dipole interaction contributed to enhance the surface charge and increased the LSPR intensity of Ag nanorod-coupled than a single nanorod.

Acknowledgment

The authors would like to thank DRPM Universitas Indonesia for funding this work through Hibah PIT 9 Tahun 2019 under Grant Number: NKB-0023/UN2.R3.1/HKP.05.00/2019.

References

[1] Mahmoud A. Mahmoud and Mostafa A. El-Sayed, "Different Plasmon Sensing Behavior of Silver and Gold Nanorods," *J Phys Chem Lett*, vol. 4, pp. 1541–1545, 2013, http://dx.doi.org/10.1021/jz4005015

[2] Kyeong Seok Lee and Mostafa A. El-Sayed, "Gold and Silver Nanoparticles in Sensing and Imaging: Sensitivity of Plasmon Response to Size, Shape, and Metal Composition," *J Phys Chem B*, vol. 110, pp. 19220–19225, 2006, http://dx.doi.org/10.1021/jp062536y

[3] B. Lamprecht *et al.*, "Metal Nanoparticle Gratings: Influence of Dipolar Particle Interaction on the Plasmon Resonance," *Phys Rev Lett*, vol. 84, no. 20, pp. 4721–4724, 2000, http://dx.doi.org/10.1103/PhysRevLett.84.4721

[4] Ulrich Hohenester, "Simulating electron energy loss spectroscopy with the MNPBEM toolbox," *Comput. Phys. Commun.*, vol. 185, no. 3, pp. 1177–1187, Mar. 2014, http://dx.doi.org/10.1016/j.cpc.2013.12.010

[5] Edward D. Palik, *Handbook of Optical Constants of Solids*. Academic Press, 0-12-544420-6

[6] Prashant K. Jain, Wenyu Huang, and Mostafa A. El-Sayed, "On the Universal Scaling Behavior of the Distance Decay of Plasmon Coupling in Metal Nanoparticle Pairs: A Plasmon Ruler Equation," *Nano Lett*, vol. 7, no. 7, pp. 2080–2088, Jun. 2007, http://dx.doi.org/10.1021/nl071008a

[7] Srdjan S. Aćimović, Mark P. Kreuzer, María U. González, and Romain Quidant, "Plasmon Near-Field Coupling in Metal Dimers as a Step toward Single-Molecule Sensing," *ACS Nano*, vol. 3, no. 5, pp. 1231–1237, Apr. 2009, http://dx.doi.org/10.1021/nn900102j

[8] F. J. Garcı´a de Abajo, "Retarded field calculation of electron energy loss in inhomogeneous dielectrics," *Phys. Rev. B Vol. 65 115418* http://dx.doi.org/10.1103/PhysRevB.65.115418

[9] J. Aizpurua, Garnett W. Bryant, Lee J. Richter, F. J. García de Abajo, Brian K. Kelley, and T. Mallouk, "Optical properties of coupled metallic nanorods for field-enhanced spectroscopy," *Phys Rev B*, vol. 71, no. 23, p. 235420, Jun. 2005, http://dx.doi.org/10.1103/PhysRevB.71.235420

[10] Prashant K. Jain, Susie Eustis, and Mostafa A. El-Sayed, "Plasmon Coupling in Nanorod Assemblies: Optical Absorption, Discrete Dipole Approximation Simulation, and Exciton-Coupling Model," *J Phys Chem B*, vol. 110, no. 37, pp. 18243–18253, 2006 http://dx.doi.org/10.1021/jp063879z

The Thickness Effect on LSPR Spectra of Au-Nanorod Coated TiO_2 and SiO_2 Studied by MNPBEM

Muhammad Sujak Mansur Kuddah[1], Miftahussurur Hamidi Putra[2], Dede Djuhana[1*]
[1] Department of Physics, FMIPA Universitas Indonesia, Kampus UI Depok, Depok 16424, Indonesia
[2] Institute of Theoretical Chemistry, Ulm University, Albert-Einstein-Allee 11, D-89069 Ulm, Germany

dede.djuhana@sci.ui.ac.id

Abstract. We have studied the effect on localized surface plasmon resonance (LSPR) of Au-nanorod (AuNR) coated TiO_2 (Au@TiO_2) and SiO_2 (Au@SiO_2). The length and diameter of AuNR are 70 nm and 20 nm, respectively. For this purpose, then AuNR was coated by TiO_2 and SiO_2 with thickness varying from 2nm to 20nm with increment 2nm. The LSPR exhibited two modes, transverse mode (TM) and longitudinal mode (LM) as commonly was found in a rod-shaped particle. Interestingly, the peak of extinction red-shift as the thickness increased both TM and LM. Increasing volume and refractive index around AuNR contributed to red-shift, therefore the rate of decreasing energy TiO_2 faster than SiO_2. Furthermore, we have also observed the peak of LSPR extinction curve of Au@TiO_2 have a similar trend to Au@SiO_2 as the thickness increased, but its beginning to decrease at ~8nm. In the LM, for a given thickness, we found the peak of extinction curve Au@TiO_2 show more than one peak present around ~2-3nm and ~8nm. However, only one peak happened in Au@SiO_2. In a sense, Au coated TiO_2 can be fully utilized to enhance photocatalytic activity under visible light of AuNR and useful in photocatalytic activity, on the other hand, SiO_2 plays an insignificant role for a certain limit thickness under visible light – near-infrared.

Keywords: MNPBEM, LSPR, refractive index, longitudinal mode, transverse mode

1. Introduction

In the recent decade, the inquiry of noble metals such as Au, Ag, and Cu has a gorgeous thought controlled the electron oscillate driven by electric field so-called localized surface plasmon resonance (LSPR). This phenomenon strongly depends on how to controlled the size, shape, morphology, and material used [1] and occurred with prerequisite is the metal size smaller than photon wavelength [2]. The noble metallic was attracting to study because of their unique properties and great potential to applied in technology and shows fascinating surface plasmon in visible range, then mostly used as sensor [3,4], biological imaging [3], catalysis/photocatalysis [5,6], and drug delivery [5]. Its has been well-known coating surrounding the Au-nanorod (AuNR) will influence the LSPR effect. Isabel et al. did gold nanosphere and nanorod shape coated SiO_2 and both showed the optical spectra shift to a longer wavelength from visible to near-infrared [7]. Also Wiebke et al. showed gold nanorod coated SiO_2 more stable than uncoated [5]. The silica has several advantages as a material veneer to enhance the thermal, mechanical, and colloidal stability [5]. The AuNR coated SiO_2 are good material core-shell structure for sensing application, photoacoustic imaging, and increased the efficiency of the drug delivery process. On the other side, gold coated TiO_2 is essential to deeper investigation cause titania is a common photocatalyst [8-10], abundant in the earth [11], non-toxicity [8,11], friendly surrounding, chemical inertness [8], preserve aggregation of metal nanoparticles and low cost. The titania was found in 1969 as a fascinating material for photochemical and photocatalysis application due to TiO_2 has a greater band gap (>3.0eV) and was found in three phase such as anatase, rutile, and brookite [12,13]. Rutile is more stable in bulk material and anatase are good water adsorption and has a longer electron-hole pairs lifetime for a surface reaction. For photocatalysis application under visible light is still poor to utilized due to the exciton generated by visible light lower than its band gap, therefore the electron-hole pair recombined rapidly and and no affect photocatalysis [8,10]. The researcher has done many ways to fully utilized titania for photocatalysis. The common ways are doping or coating with a metal or nonmetal [9]. This modification aims to modify the hot electron injection from metal to the conduction band of TiO_2 and also the local electric field of incident wavelength can be fully utilized the photocatalysis efficiency enhancement [8], this electron which is initiated by decay surface plasmon resonance [10,11,14].

The main objective of this study, firstly, is determined qualitatively how the LSPR spectra of extinction spectrum AuNR affected by the TiO_2 and SiO_2 coating, respectively. The TiO_2 was used within anatase phase which is as good water adsorption. Secondly, calculated the tunability of maximum thickness AuNR coated TiO_2 and SiO_2 under visible light – near-infrared (NIR).

2. Simulation procedure

The systematic investigation of thickness effect on LSPR spectra of Au-nanorod coated TiO_2 and SiO_2 was performed by MNPBEM (Metallic Nano-Particle Boundary Element Method) public software [15] based on boundary element method [16]. The geometry of Au-nanorod (AuNR) with coating TiO_2 and SiO_2 were presented in Figure 1. The length (*L*) and diameter (*D*) of AuNR are 70nm and 20nm, respectively or the aspect ratio was 3.5. The thickness (*t*) of TiO_2 and SiO_2 were varied from 2nm to 20nm with increment 2nm. The dielectric function of AuNR was obtained from Johnson and Christy experiment [17]. The refractive index of TiO_2 based on Bodurov's work [18] and SiO_2 was adopted Gao's experiment [19]. For calibrating, we used the Sellmeier equation to recalculate the refractive index of TiO_2 in the range 0.251μm-1.5μm.

$$n^2(\lambda) = 1 + \frac{B\lambda^2}{\lambda^2 - C} \tag{1}$$

where n is a refractive index, λ is a wavelength in μm, and coefficient B=4.6796 and C = 0.040086. The nanoparticle immersed in water with refractive index n = 1.33. Then, we applied the plane wave in parallel and perpendicular to long axis of AuNR with respect to the wavelength 400nm to 1100nm. From this simulation, we produced spectra such as extinction spectrum as the wavelength function and near-field map.

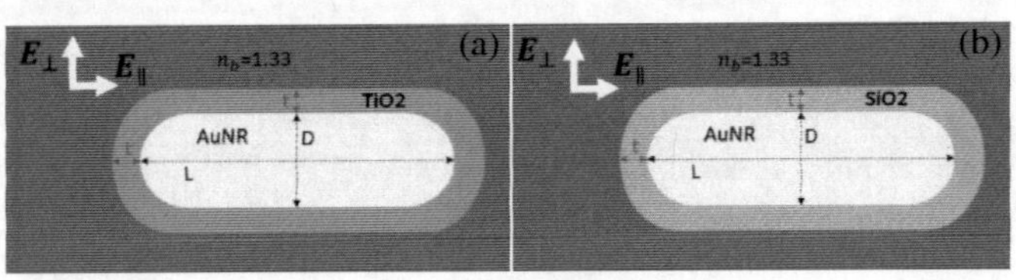

Figure 1. The schematic diagram of **(a)** AuNR@TiO_2 and **(b)** AuNR@SiO_2.

IOP Conf. Series: Materials Science and Engineering **553** (2019) 012016 doi:10.1088/1757-899X/553/1/012016

3. Results and discussion

The LSPR spectra of AuNR pure and coated TiO_2 and SiO_2 for the thickness variation in the extinction curve, as shown in Figure 2. It was found two resonance modes such as transverse and longitudinal modes, as usually happened in rod shape [1,3,20]. For the case longitudinal modes, the LSPR spectra shifted to longer wavelength or red-shift as the thickness increased both AuNR coated TiO_2 and SiO_2. The LSPR spectra shifting observed in the range visible-NIR, as shown as in Figure 2(a)-right and Figure 2(b)-right. For transverse mode case, the spectral slowly shifted to red-shift as the thickness increased and all LSPR shifting still occurred in visible range, as depicted in Figure 2(a)-left and Figure 2(b)-left. Further, we have also observed the extinction curve of AuNR coated TiO_2 in longitudinal mode around 500nm-700 nm. It was observed that some multiple peaks were formed with low amplitude. For the thickness around 2nm–6nm as labeled (i), it was originated from the interband transition from 5d electron to conduction band excited by the wavelength 550nm-600nm (2.25-2.07 eV) [21], thus this peak also correlated with the peak in transverse mode greater 600nm. This result similar to Winsemius's work that Au has a limit absorption around ~640nm (~1.94eV) called interband transition [21]. Then, above the thickness 6 nm, others peak began to form in LSPR spectra. This peak was a higher order in LSPR spectra of AuNR such as quadrupole and hexapole mode as indicated (ii).

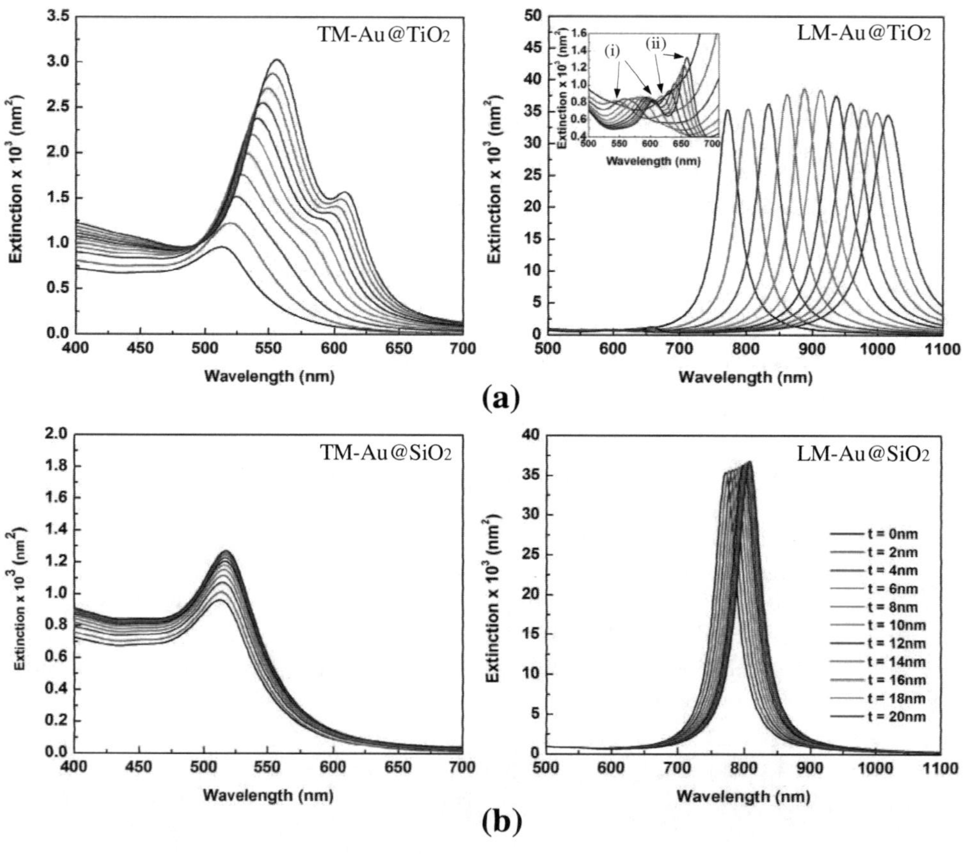

Figure 2. The extinction spectrum vs wavelength as thickness variation of **(a)** AuNR coated TiO_2 in transverse mode (TM) and longitudinal mode (LM) **(b)** AuNR coated SiO_2 in transverse mode (TM) and longitudinal mode (LM). In inset figure which (i) and (ii) are interband transition and higher order like quadrupole/hexapole mode, respectively.

Next, we have also plotted the peak value of the extinction curve of AuNR related to the thickness variation extract from the longitudinal mode, as shown as in Figure 3. As Figure 3, the peak value of AuNR coated TiO_2 was higher than coated SiO_2. It was understood the peak value of AuNR coated TiO_2 higher than SiO_2 since the refractive index of TiO_2 larger than SiO_2. Large shifting the LSPR spectra of AuNR coated TiO_2 widely used in photocatalytic activity under visible range [8,11] whereas small shifting the LSPR spectra of AuNR coated SiO_2 has potentially improved the stability of Au nanoparticle [7,22]. Then, we have also analyzed the LSPR shifting corresponded to the peak value, as previous plotted in Figure 3. Interestingly, the growth of LSPR shifting AuNR coated TiO_2 and SiO_2 followed the exponential growth function [23,24], such as

$$\Delta\lambda(t) = m(n_c - n_s)[1 - \exp(-2t/\delta)] \tag{2}$$

where m is the change of wavelength per refractive index unit (RIU), n_c is refractive index of material coating, in this case $n_c = 2.51$ for TiO_2 and $n_c = 1.47$ for SiO_2, n_s is refractive index medium ($n_s = 1.33$), δ is the growth length, and t is the thickness of the coating. Then, we have carried out a fitting process to determine the value of m and δ based on this function. The value of AuNR coated TiO_2 are m = 452 nm/RIU and $\delta = 64.8$ nm and for SiO_2 are m = 360 nm/RIU and $\delta = 31.4$ nm. From this result showed the growth length of Au coated TiO_2 was higher than coated SiO_2. The peak and fitting value of LSPR shifting AuNR coated TiO_2 and SiO_2 was presented in Figure 3.

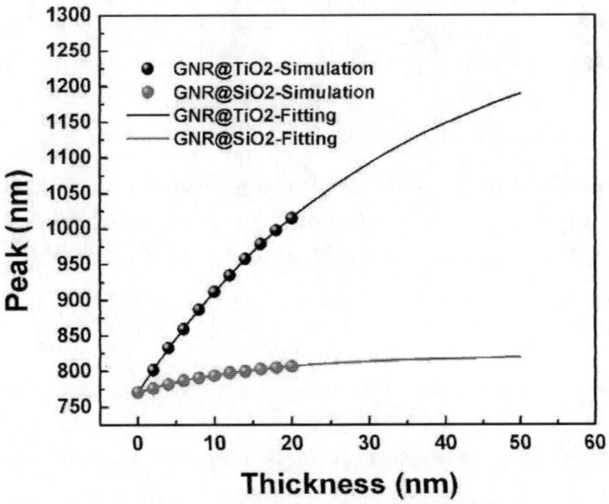

Figure 3. LSPR peak versus the thickness from the data simulation and fitting growth function.

Furthermore, we have also produced the near-electric field image of LSPR AuNR coated TiO_2 and SiO_2 both longitudinal and transverse mode for the thickness from 2nm to 20nm. The near-electric field image was shown in Figure 4(a) for Au@TiO_2 and Figure 4(b) for Au@SiO_2 both longitudinal and transverse mode. For longitudinal mode, the near-electric field of Au@TiO_2 and Au@SiO_2 appeared dawn among interface Au/TiO_2 and Au/SiO_2 at around 8nm. It means that the magnitude of the electric field increased as the thickness increased [9]. At the interface between Au and TiO_2 (e.g.,

IOP Conf. Series: Materials Science and Engineering **553** (2019) 012016 doi:10.1088/1757-899X/553/1/012016

longitudinal mode and also Au coated titania shows near-electric field higher than coated silica. The transition happened at thick 8nm as well.

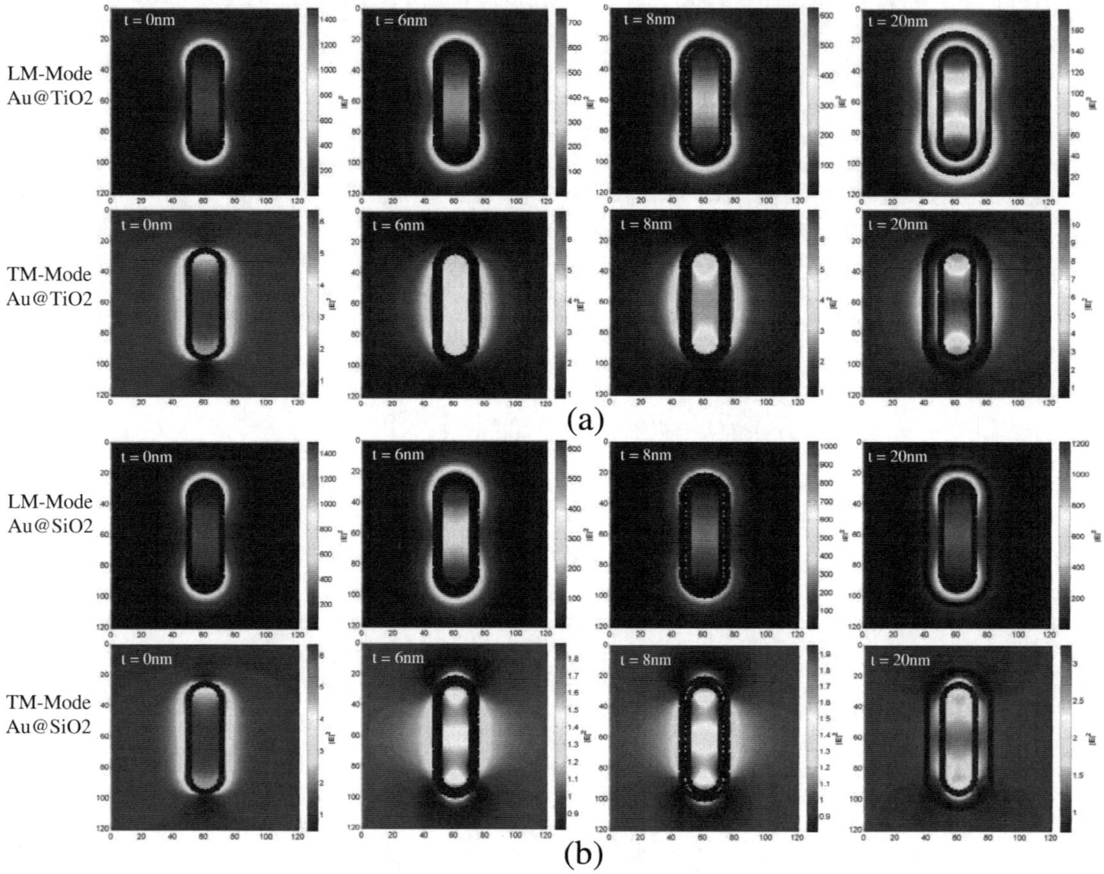

Figure 4. The near-electric field map LSPR peak in longitudinal mode and transverse mode **(a)** Au@TiO_2 and **(b)** Au@SiO_2. For thickness t= 0nm is Au pure and transition between t=6nm and t=8nm observed in the interface Au with TiO_2/SiO_2 both longitudinal and transverse mode.

IOP Conf. Series: Materials Science and Engineering **553** (2019) 012016 doi:10.1088/1757-899X/553/1/012016

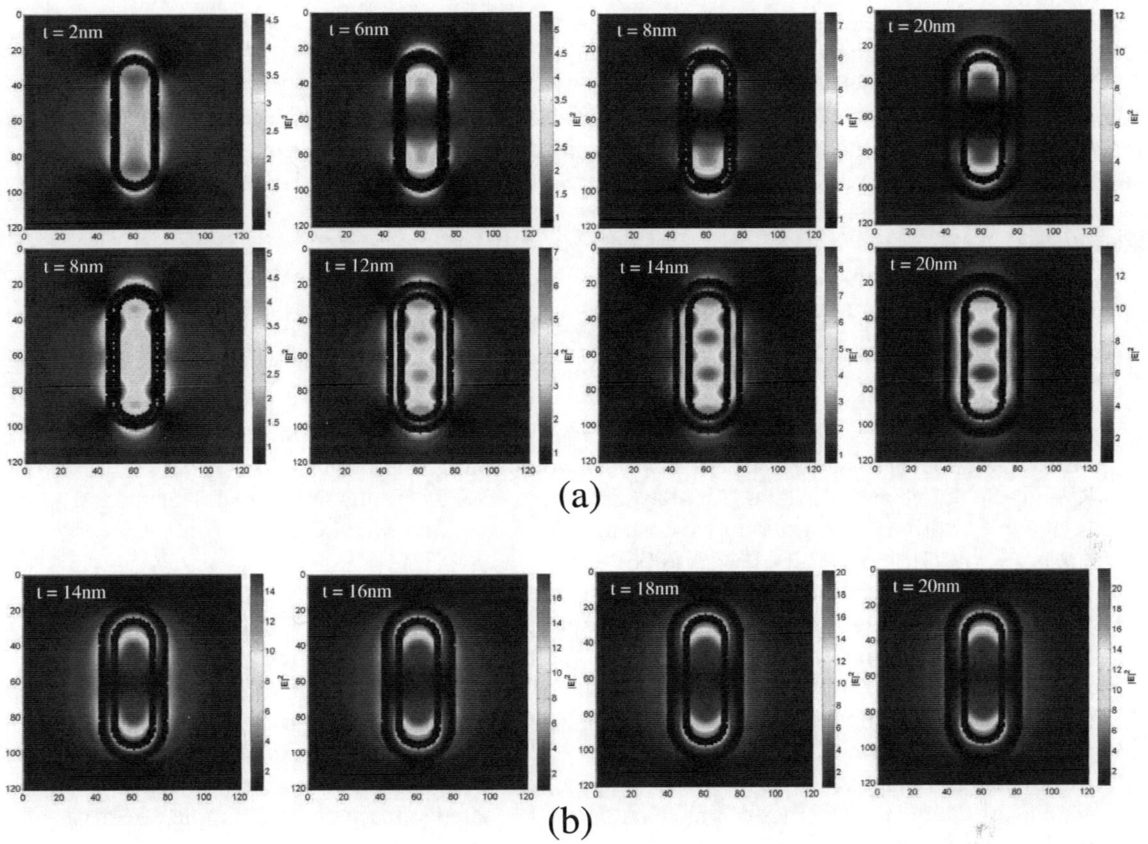

Figure 5. The distribution near-electric field of multiple peaks for longitudinal mode **(a)** interband transition (upper row) and higher order (lower row) **(b)** interband transition for transverse mode.

Thus, we have also produced the near-electric field distribution of the multiple peaks from the inset Figure 2(a) for longitudinal mode and Figure 2(b) for transverse mode as shown in Figure 5. The multiple peaks display a similar near-electric field pattern for longitudinal mode thick 2nm-20nm (Figure 5(b)) so that it's a supporting confirmation as the interband transition of Au nanoparticle [9]. In Figure 5(a)-lower row is a higher order transition from quadrupole to hexapole happened for thickness above 8nm, where the energy quadrupole higher than hexapole. This transition exists for a big particle like increasing volume [26]. The energy transition (quadrupole-hexapole) appear at the interface between gold and titania coincide with decreasing the intensity of LSPR.

4. Conclusion

In conclusion, we have performed AuNR coated TiO_2 and SiO_2, respectively by MNPBEM simulation. The result suggests that LSPR spectra of AuNR coated SiO_2 more stable than uncoated and the shifting of LSPR spectra can be tuned by increasing the thickness of TiO_2 under visible light-near infrared. The shifting AuNR coated SiO_2 was much smaller than AuNR coated TiO_2, about 807 ± 12 nm for AuNR@SiO_2 and 1015 ± 81.3 nm for AuNR@TiO_2 with thick 20nm. According to the fitting method, the maximum thickness of AuNR@SiO_2 around 42 ± 14.1 nm and unseen for AuNR@TiO_2 under visible light-NIR. Based on our result, AuNR coated SiO_2 is the best candidate to control the stability of LSPR gold nanorod under visible light – NIR. On the other hand, for photocatalytic enhancement, the TiO_2 within anatase phase is good material coating aim to improved the photocatalytic activity under visible light-NIR. For future projects, it can be used to consider the effective thickness of Au@TiO_2 or Au@SiO_2 in conducting an experiment.

Acknowledgment
This work was funded by Indexed International Publication Grants (PIT) 9 Year 2019 through DRPM Universitas Indonesia through grant contract number: NKB-0023/UN2.R3.1/HKP.05.00/2019.

References

[1] S. Link, M. B. Mohamed, and M. A. El-Sayed, "Simulation of the Optical Absorption Spectra of Gold Nanorods as a Function of Their Aspect Ratio and the Effect of the Medium Dielectric Constant," *J Phys Chem B*, vol. 103, pp. 3073–3077, 1999.

[2] STEFAN A. MAIER, *PLASMONICS: FUNDAMENTALS AND APPLICATIONS*. New York: Springer Science+Business Media LLC, 2007.

[3] Mahmoud A. Mahmoud and Mostafa A. El-Sayed, "Different Plasmon Sensing Behavior of Silver and Gold Nanorods," *J Phys Chem Lett*, vol. 4, pp. 1541–1545, 2013.

[4] Kyeong Seok Lee and Mostafa A. El-Sayed, "Gold and Silver Nanoparticles in Sensing and Imaging: Sensitivity of Plasmon Response to Size, Shape, and Metal Composition," *J Phys Chem B*, vol. 110, pp. 19220–19225, 2006.

[5] Wiebke Albrecht, Tian-Song Deng, Bart Goris, Marijn A. van Huis, Sara Bals, and Alfons van Blaaderen, "Single Particle Deformation and Analysis of Silica-Coated Gold Nanorods before and after Femtosecond Laser Pulse Excitation," *Nano Lett*, vol. 16, pp. 1818–1825, 2016.

[6] S. Link, Z. L. Wang, and M. A. El-Sayed, "Alloy Formation of Gold-Silver Nanoparticles and the Dependence of the Plasmon Absorption on Their Composition," *J Phys Chem B*, vol. 103, pp. 3529–3533, 1999.

[7] Isabel Pastoriza-Santos, Daniel Gomez, Jorge Pe´rez-Juste, Luis M. Liz-Marza´n, and Paul Mulvaney, "Optical properties of metal nanoparticle coated silica spheres: a simple effective medium approach," *Phys Chem Chem Phys*, vol. 6, pp. 5056–5060, 2004.

[8] Na Zhou *et al.*, "TiO2 coated Au/Ag nanorods with enhanced photocatalytic activity under visible light irradiation," *Nanoscale*, vol. 5, p. 4236, 2013.

[9] Guo-Ying Yao, Qing-Lu Liu, and Zong-Yan Zhao, "Studied Localized Surface Plasmon Resonance Effects of Au Nanoparticles on TiO2 by FDTD Simulation," *catalysts*, vol. 8, p. 236, 2018.

[10] Xiang-Chao Ma, Ying Dai, Lin Yu, and Bai-Biao Huang, "Energy transfer in plasmonic photocatalytic composites," *Light Sci. Appl.*, vol. 5, 2016.

[11] Wenguang Fan and Michael K. H. Leung, "Recent Development of Plasmonic Resonance-Based Photocatalysis and Photovoltaics for Solar Utilization," *molecules*, vol. 21, 2016.

[12] Reyes-Coronado, G Rodr´ıguez-Gattorno, M E Espinosa-Pesqueira, C Cab, R de Coss, and G Oskam, "Phase-pure TiO2 nanoparticles: anatase, brookite and rutile," *Nanotechnology*, vol. 19, 2008.

[13] Tim Luttrell, Sandamali Halpegamage, Junguang Tao, Alan Kramer, Eli Sutter, and Matthias Batzill, "Why is anatase a better photocatalyst than rutile? - Model studies on epitaxial TiO2 films," *Sci. Rep.*, vol. 4, 2014.

[14] Muhammad Shahid Arshad, Špela Trafela, Kristina Žužek Rožman, Janez Kovač, Petar Djinović, and Albin Pintar, "Determination of Schottky barrier height and enhanced photoelectron generation in novel plasmonic immobilized multisegmented (Au/TiO2) nanorod arrays (NRAs) suitable for solar energy conversion applications," *J. Mater. Chem. C*, 2017.

[15] Andreas Trugler, *Optical Properties of Metalic Nanoparticles Basic Principle and simulation*, vol. 232. Switzerland: Springer International Publishing, 2016.

[16] F. J. Garcı´a de Abajo, "Retarded field calculation of electron energy loss in inhomogeneous dielectrics," *Phys. Rev. B*, vol. 65, p. 115418, 2002.

IOP Conf. Series: Materials Science and Engineering **553** (2019) 012016 doi:10.1088/1757-899X/553/1/012016

[17] P.B. Johnson and R. W. Christy, "Optical Constants of the Noble Metals," *Phys. Rev. B*, vol. 6, no. 12, pp. 4370–4379, Dec. 1972.

[18] I. Bodurov, I. Vlaeva, A. Viraneva, T. Yovcheva, and S. Sainov, "Modified design of a laser refractometer," *Nanosci. Nanotechnol.*, vol. 16, pp. 31–33, 2016.

[19] L. Gao, F. Lemarchand, and M. Lequime, "Refractive Index Determination of SIO2 Layer in the UV/VIS/NIR Range: Spectrophotometric Reverse Engineering on Single and Bi-Layer Designs," *J. Eur. Opt. Soc. - Rapid Publ.*, vol. 8, 2013.

[20] Hye-Mi Bok, Kevin L. Shuford, Sungwan Kim, Seong Kyu Kim, and Sungho Park, "Multiple Surface Plasmon Modes for Gold/Silver Alloy Nanorods," *Langmuir*, vol. 25, no. 9, pp. 5266–5270, Mar. 2009.

[21] P. Winsemius, M. Guerrisi, and R. Rosei, "Splitting of the interband absorption edge in Au: Temperature dependence," *Phys. Rev. B*, vol. 12, 1975.

[22] Chuanliu Wu and Qing-Hua Xu, "Stable and Functionable Mesoporous Silica-Coated Gold Nanorods as Sensitive Localized Surface Plasmon Resonance (LSPR) Nanosensors," *Langmuir*, vol. 25, no. 16, pp. 9441–9446, 2009.

[23] Linda S. Jung, Charles T. Campbell, Timothy M. Chinowsky, Mimi N. Mar, and Sinclair S. Yee, "Quantitative Interpretation of the Response of Surface Plasmon Resonance Sensors to Adsorbed Films," *Langmuir*, vol. 14, pp. 5636–5648, 1998.

[24] Lihua Qian, Bo Shen, Gaowu W. Qin, and Biswajit Das, "Widely tuning optical properties of nanoporous gold-titania core-shells," *J. Chem. Phys.*, vol. 134, 2011.

[25] José Luis Montañ-Priede, Ovidio Peña-Rodríguez, and Umapada Pal, "Near-Electric-Field Tuned Plasmonic Au@SiO2 and Ag@SiO2 Nanoparticles for Efficient Utilization in Luminescence Enhancement and Surface-Enhanced Spectroscopy," *J Phys Chem C*, vol. 121, pp. 23062–23071, 2017.

[26] Amit Bansal and S.S.Verma, "Optical response of noble metal alloy nanostructures," *Phys. Lett. A*, vol. 379, pp. 163–169, 2015.

Study of Domain Wall Propagation in FePt and FePd Nanowires Driven by Sub-nanosecond Pulse of Magnetic Field by Micromagnetic Simulation

Sulaiman Hawibowo[1], Candra Kurniawan[1,2], and Dede Djuhana[1*]

[1] Department of Physics, Faculty of Mathematics and Natural Science Universitas Indonesia, Depok 1624, Indonesia
[2] Research Centre for Physics, Indonesian Institute of Sciences (LIPI), Puspiptek Office Area, Tangerang Selatan 15314, Indonesia

dede.djuhana@sci.ui.ac.id

Abstract. In this study, we investigated the domain wall propagation behavior in FePt and FePd nanowires by micromagnetic simulation based on LLG equation. A rectangular shape nanowire model was generated with a fixed length of 2000 nm and varied thicknesses of 2.5 nm and 5 nm. The wire widths were varied of 50 nm, 100 nm, and 150 nm. The 500 ps length of magnetic-pulse was applied in-plane to a transverse type domain wall. We observed the Walker Breakdown field (H_{WB}), which means the maximum domain wall velocity before it was abruptly decreased. The transverse DW structure was maintained below H_{WB} while vortex/antivortex wall structure was formed in the higher magnetic field. Interestingly, the H_{WB} value of both FePt and FePd were decreased as the width and thickness of the nanowires increased. We have also analyzed the changing of the demagnetization and exchange energy for the domain wall structure transition from transverse wall to vortex/ anti-vortex wall.

Keywords: micromagnetic, nanowire, Walker Breakdown, domain wall, transverse wall, vortex wall.

1. Introduction

The new effective data storage devices such as Magnetic Random Access Memory (MRAM) have become one of the interesting development non-volatile memory technology that was trying to establish for electronics equipment with the utilization of electron spin [1]. The spin direction manipulation process provides a new paradigm for data storage processes in the development of spintronics-based technology [2,3]. Another important invention idea was the Racetrack Memory which was developed by Stuart Parkin from IBM [4,5]. In perpendicular magnetic recording, magnetic bits were built perpendicularly to the magnetic thin layer film. Depending on the external field, DW can move with or without changing its inner structure along on nanowire [6,7]. Materials that have the properties of perpendicular magnetic anisotropy (PMA) are expected to reduce the superparamagnetic effect and become a important improvement for the development of technology by utilizing the dynamics of domain-wall magnets, considering the needs for storage media that requires the development of devices that are smaller, faster, and more efficient [8].

Several studies of thin film layers on FePt and FePd alloys show very large uniaxial magnetocrystalline anisotropy and can be applied to an anisotropic axis perpendicularly to the film plane. These properties made it suitable for applications in particular high-density magnetic recording media [9,10]. To correlate the film microstructure with its magnetic properties, it should be considered with typical structural defects [11]. The crystallographic structure of FePd undergoes tetragonal distortion induces strong PMA (Perpendicular Magnetic Anisotropy) in a thin layer parallel to the film plane [12,13].

In this study, the materials used are alloy materials of FePt and FePd nanowires with rectangular geometries that are varied in thickness and width. Then, it will be observed the characteristics of the domain-wall propagations when triggered with a magnetic field that is varied when triggered with a magnetic field pulse for 0.5 ns. Domain wall will move with a certain speed until it reaches the field of Walker Breakdown (H_{WB}). This method can be carried out with a micromagnetic simulation so it can be further analyzed the influence of magnetic energy on the speed and structure of the domain wall.

2. Simulation Method

The simulation process of domain-wall propagation was performed by using public software Object Oriented Micromagnetic Framework (OOMMF) based on the LLG equation (Landau-Lifshitz Gilbert) [14,15]. This domain-wall propagation simulation used a thin rectangular geometry of nanowire. A magnetic pulse of 500 ps length was applied on the nanowire in the x-direction. We have varied the magnitude of a magnetic pulse to drive DW along the nanowire. The wire sizes used a fixed length (L) = 2000 nm, thickness (t) 2.5 nm and 5 nm, and width varied to 50 nm, 100 nm, and 150 nm. The cell size used was 5 x 5 x t nm^3 where the t value indicates the thickness of the wire and the damping constant value was $\alpha = 0.05$. The material parameters of FePt using the magnetization saturation Ms, exchange constant A, and anisotropy constant K of 1.14×10^6 A/m, 1.0×10^{-11} J/m, and 6.6×10^6 J/m^3, respectively [16]. While, The parameter of FePd the magnetization saturation Ms, exchange constant A, and anisotropy constant K of 1.03×10^6 A/m, 6.9×10^{-12}, and 2.6×10^5 J/m^3, respectively [17]. The geometry of the nanowire form used in this simulation is shown in Figure 1 as the initial transverse wall (TW) was located at the center of the nanowire.

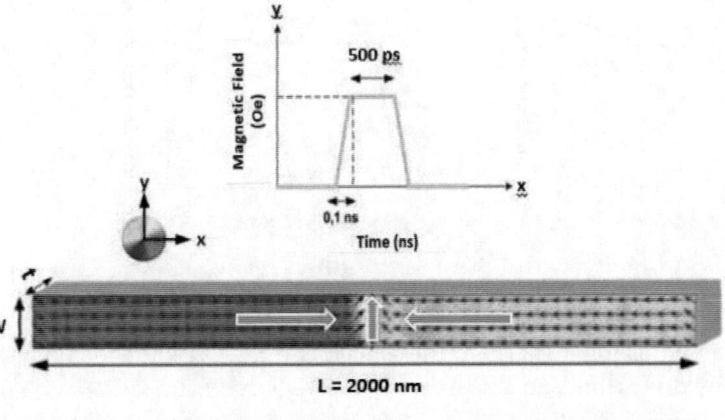

Figure 1. The geometry and dimension of FePt and FePd nanowire with a transverse wall structure at the center of a nanowire. A 500 ps time-length magnetic pulse was applied along the nanowire. The color bar is the magnetization direction.

3. Results and discussion

The domain-wall propagation is a change of the domain-wall position along the nanowire. When a magnetic field pulse was given, the position of the domain-wall would change, which means that the amplitude variation by an active pulse was expressed of the magnitude of the external magnetic field. Each of materials was given an external magnetic field ranging from B = 20 Oe to B = 300 Oe. The

IOP Conf. Series: Materials Science and Engineering **553** (2019) 012017 doi:10.1088/1757-899X/553/1/012017

domain-wall velocity profile will increase linearly with increasing low external magnetic field as shown in Fig. 2. This velocity will continue to increase until it reached the maximum value or critical field known as the Walker Breakdown [18,19]. Meanwhile, after a larger H_{WB}, the domain-wall velocity will be decreased. It is observed that the domain wall velocity decrease as the thickness and width of the geometry of the nanowire increased. This results exhibited a similar trend to other reported results. On the other hand for FePt material, when the thickness is 5 nm, the Walker Breakdown phenomenon is observed by giving a larger external magnetic field or more than 250 Oe. But interestingly, the change of the domain-wall structure occurred when the external magnetic field was giving around 9 Oe. It might be caused by the different anisotropy properties of those materials. Table 1 showed H_{WB} value for the width and thickness variation for FePt and FePd nanowires. The table showed that the DW velocity was sensitive to the thickness rather than the width of the nanowire.

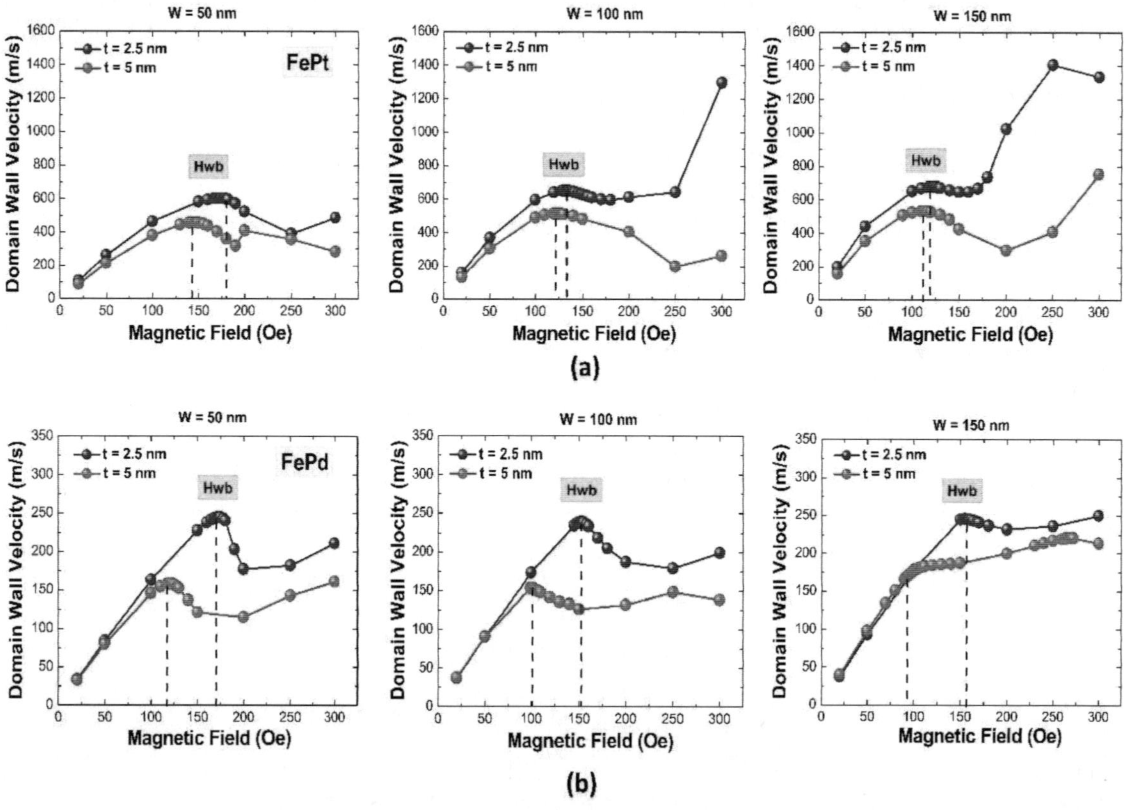

Figure 2. DW velocity as the magnitude of the magnetic pulse with the width and thickness variation for (a) FePt and (b) FePd nanowires. The critical magnetic field is known as the Walker breakdown field (H_{WB}) denoted by dotted-line.

Table 1. The Walker breakdown field of FePt and FePd nanowires concerning the width and thickness variation.

Material	Thickness	Width	H_{WB}	Material	Thickness	Width	H_{WB}
FePt	2.5 nm	50 nm	174 Oe	FePd	2.5 nm	50 nm	174 Oe
		100 nm	133 Oe			100 nm	154 Oe
		150 nm	123 Oe			150 nm	156 Oe
	5 nm	50 nm	148 Oe		5 nm	50 nm	122 Oe
		100 nm	123 Oe			100 nm	101 Oe
		150 nm	112 Oe			150 nm	93 Oe

Then, we also observed the change of the inner structure based on the simulation results as shown in Fig. 3. It shows the magnetization images of the domain-wall propagation in the given thickness of $t = 5$ nm and width of $W = 100$ nm. The results showed that when the external magnetic field is below the Walker Breakdown field (B = 100 Oe), the domain-wall structure still maintains the transverse wall structure, whereas the external magnetic field is greater than the Walker Breakdown field (B = 250 Oe), the domain-wall structure changed from transverse to the vortex / antivortex-wall structures as shown by Figure 3. For FePt material, when the thickness of nanowire is 5 nm the domain-wall structure was changed from the transverse wall into the antivortex-wall because the DW core nucleation appeared and by the difference in time of several seconds will be followed by the appearance of nucleated vortex-wall when t = 1.3 ns. This nucleated DW core was starting from the top and bottom edge of the nanowire. When the time took a longer than 1.6 ns it will be followed by the disappearance of the nucleated anti-vortex wall, which means the magnetic field pulse returns to zero.

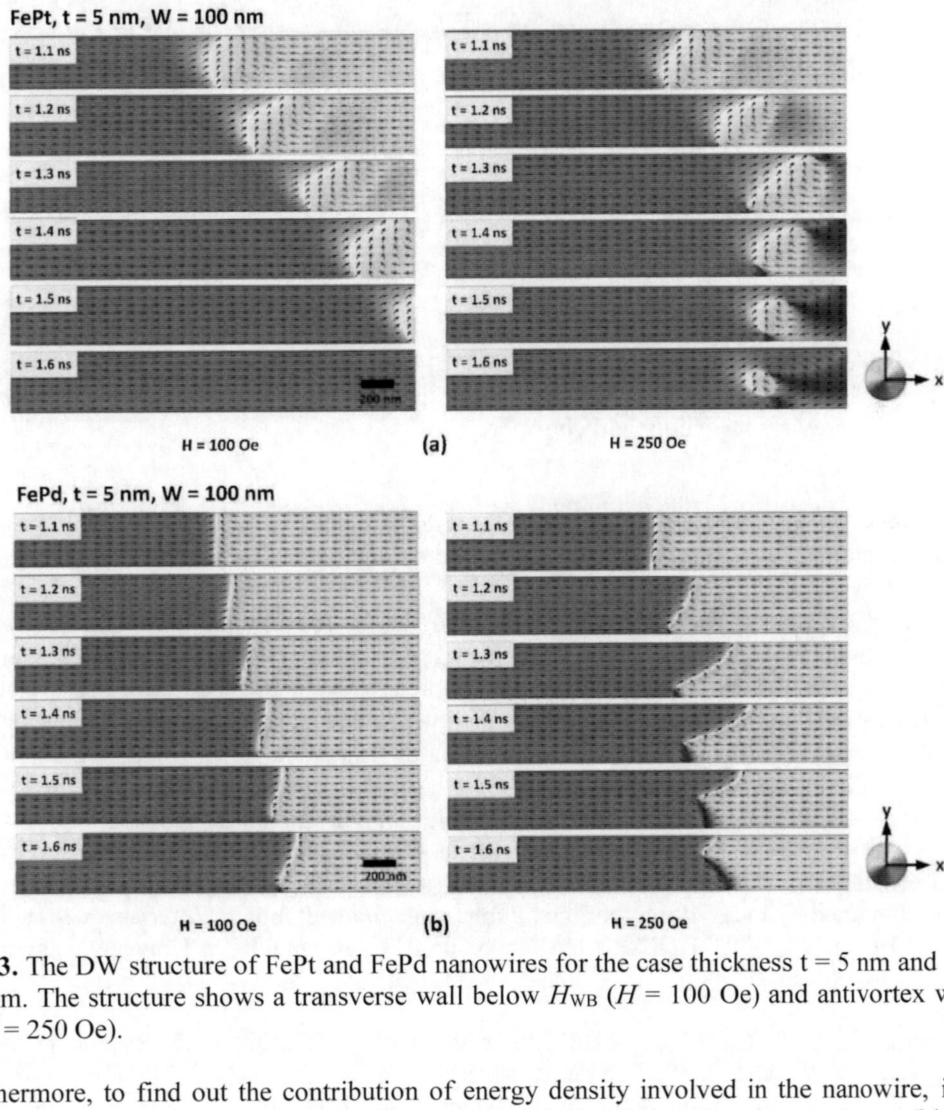

Figure 3. The DW structure of FePt and FePd nanowires for the case thickness t = 5 nm and width W = 100 nm. The structure shows a transverse wall below H_{WB} (H = 100 Oe) and antivortex wall after H_{WB} (H = 250 Oe).

Furthermore, to find out the contribution of energy density involved in the nanowire, it can be represented by the changing of demagnetization energy (red color) and exchange energy (blue color) as shown in Figure 4. The demagnetization energy density for both materials was dominant than the exchange energy. The larger width and thickness in the nanowire geometries produced higher energy density. When the external magnetic field was given, the demagnetization energy will increase followed by increasing nanowire volume. However, for external magnetic fields around the Waker Breakdown field, the demagnetization energy density begins to decrease, which means that the interaction between magnetic moments is not uniform anymore or anti-parallel so that the domain-wall structure will change from transverse-wall to the vortex/antivortex wall. On the other hand, the energy exchange will gradually increase, which means there is a change in the domain-wall structure to the vortex/antivortex-wall where the direction of spin interaction of the neighboring magnetization is become anti-parallel or get into maximum energy. Compared to the DW velocity of FePt and FePd in Fig 2, the energy density also did not abruptly decreased as it was found in Permalloy [20] but slowly decreased as the magnitude of magnetic pulse increased.

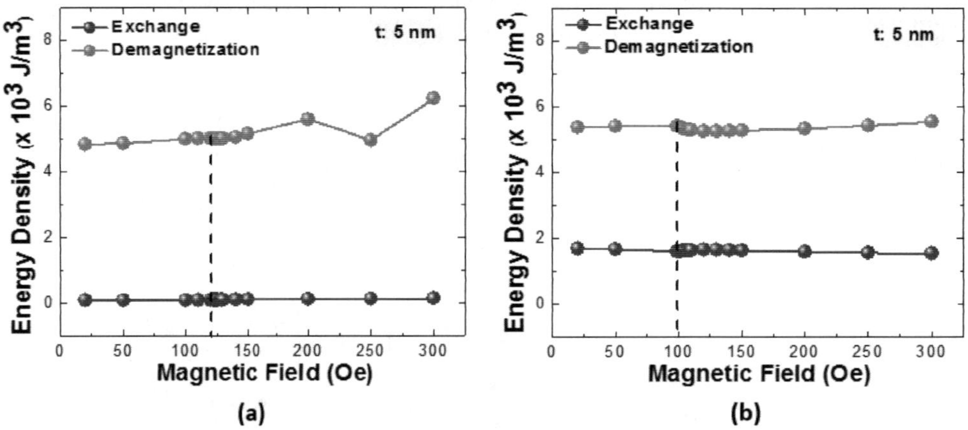

Figure 4. Magnetic energy properties of (a) FePt and (b) FePd nanowires for the case thickness $t = 5$ nm and width $W = 100$ nm. The energy curve shows an increasing exchange energy and a decreasing demagnetization energy for magnetic field above H_{WB}.

4. Conclusion

In this study, we have observed the characteristics of the domain wall propagation in the FePt and FePd nanowires by micromagnetic simulation. When the external magnetic field increased, the velocity of the domain-wall will remain increased until the critical field (H_{WB}). This critical field also relies on nanowire geometries. The thicker and wider the nanowire is, the H_{WB} field will be decreased. Below the critical field, the domain transverse-wall structure was observed. When the external magnetic field has exceeded the WB field, the domain-wall structure was changed from transverse to vortex/antivortex-wall. During propagation, the demagnetization energy increased followed by the increasing of the external magnetic field before reaching the H_{WB} field so it would be maintained as the transverse-wall structure. Meanwhile, after the H_{WB} field, the vortex/antivortex structure has appeared from the edges of nanowire as followed as increasing the energy exchange.

Acknowledgment

This work is supported by Indexed International Publication Grant (Publikasi Internasional Terindeks, PIT 9) year 2019 No. NKB-0023/UN2.R3.1/HKP.05.00/2019 through DRPM Universitas Indonesia.

References

[1] R.C. Sousa, I.L. Prejbeanu, "Non-volatile magnetic random access memories (MRAM)," *C.R. Physique*, vol.6, 2005. DOI: 10.1016/j.crhy.2005.10.007

[2] Baibich, M.N., Broto, J.M., Fert, A., Nguyen Van Dau, F., Petroff, F., 1988. Giant magnetoresistance of Fe/Cr magnetic superlattice. *Phys. Rev.* Lett. 61, 2471. DOI:https://doi.org/10.1103/PhysRevLett.61.2472

[3] Binasch, G., Grünber, P., Saurenbach, F., Zinn, W., 1989. Enhanced magnetoresistance in layered magnetic structures with antiferromagnetic interlayer exchange. *Phys Rev B* **39**, 4828. DOI:https://doi.org/10.1103/PhysRevB.39.4828

[4] S. S. P. Parkin, M. Hayashi, and L.Thomas, "Magnetic Domain-Wall Racetrack Memory," *Science*, vol. **320**, no. 5873, pp. 190-194, Apr.2008. DOI: 10.1126/science.1145799

[5] S. S. P. Parkin, "Data in the Fast Lanes of Racetrack Memory," *Sci. Am.*, vol. **300**, no. 6, pp. 76-81, Jun. 2009. DOI:10.1007/978-981-10-2720-8

[6] R. D. Michael and M.J. Donahue, Head to head domain wall structures in thin magnetic strips, *IEEE Trans. Magn.* 33 (1997), 4167. DOI: 10.1109/20.619698

[7] Y. Nakatani, A. Thiaville, and J. Miltat, J, "Determination of Domain Wall Depinning and Driving Currents in Doped Permalloy Structures, "Magn. Mag. Matter. 290 (2005), 750. DOI: 10.1109/TMAG.2010.2042578

[8] B. Tudu, A. Tiwari, 2017. Recent developments in Perpendicular Magnetic Anisotropy Thin Films for Data Storage Applications. *Vacuum* **146**, 329–341. DOI:https://doi.org/10.1016/j.vacuum.2017.01.031

[9] Ovanov O.A., Solina L.V., Demshina, V.A., *Phys. Met. Mettallogr* **35** (1973), 81. DOI: https://doi.org/10.1063/1.1669327

[10] Shima, T., Takanashi, K., Hono, K., 2002. Preparation and magnetic properties of highly coercive FePt films. *Appl. Phys. Lett.* **81**. DOI: https://doi.org/10.1063/1.1498504

[11] Bi, M., Wang, X., Lu, H., Zhang, L., Deng, L., Xie, J., 2017. Thickness dependence of magnetization reversal mechanism in perpendicularly magnetized L10 FePt films. **428**, 412–416. DOI: 10.1016/j.jmmm.2016.12.126

[12] J. R. Skuza, C. Clavero, K. Yang, B. Wincheski, R. A. Lukaszew, 2010. Microstructural, Magnetic Anisotropy, and Magnetic Domain Structure Correlations in Epitaxial FePd Thin Films With Perpendicular Magnetic Anisotropy. *IEEE Trans. Magn.* **46**, 1886–1889. DOI: https://doi.org/10.1109/TMAG.2009.2039923

[13] Skomski, R., Kashyap, A., Zhou, J., 2005.*Scr. Mater.* **53**, 389–394. DOI: https://doi.org/10.1016/j.scriptamat.2005.04.036

[14] M. J. Donahue and D.G Porter, 'OOMMF User's Guide, Version 1.0." National Institute of Standards and Technology, Gaithersburg, MD, Interagency Report NISTIR 6376, Sep. 1999. DOI: https://doi.org/10.6028/NIST.IR.6376

[15] T.L. Gilbert, "Classics in Magnetics a Phenomenological Theory of Damping in Ferromagnetic Materials," *IEEE Trans. Magn.*, vol. **40**, no. 6, pp. 3443-3449, Nov. 2004. DOI: 10.1109/TMAG.2011.2145366

[16] Hu, X., Wu, P., & Yuan, J. (2009). Exchange-coupled Fe3O4/L10-FePt bilayer films by controlled oxidation of Fe/Pt multilayer. *Thin Solid Films, 517*(8), 2602-2605. DOI: 10.1016/j.tsf.2008.10.040

[17] Niedoba, H., & Labrune, M. (2009). Magnetic bubbles and stripe domains in nanostructured FePd elements. *Journal of Magnetism and Magnetic Materials,* 321(14), 2178-2186. DOI: https://doi.org/10.1016/j.jmmm.2009.01.007

[18] D.Djuhana, H.-G. Piao, J. –H. Shim, S.-H. Lee, S.-M. Ahn, and D.-H. Kim, Interaction of antiparallel transverse domain walls in ferromagnetic nanowire, J. Nanosci. Nanotechnol. vol **11**, (2011), pp. 6237-6240(4). DOI: https://doi.org/10.1166/jnn.2011.4485.

[19] N.L. Schryer, L.R. Walker, 1974. The motion of 180° domain walls in uniform dc magnetic field. *J Appl Phys*, **45**, 5406. DOI: 10.1063/1.1663252

[20] Y. Nakatani, A. Thiaville, and J. Miltat, Head to head domain walls in nano-strips: a refined phase diagram, J. Magn. Magn Mater**.** **290**, 750. (2005). DOI: 10.1016/j.jmmm.2004.11.355

Effect of FeV Residual Powders as an Innovation of Thermo Reactive Deposition Process (TRD) with Material Balance Method to Carbide Surface Characteristics on SUJ2 Tool Steel

Reza Fawazul[1], Myrna Ariati Mochtar[2*]
[1,2] Department of Metallurgical and Materials Engineering, Faculty of Engineering, Universitas Indonesia, Kampus UI Depok, Depok 16424, Indonesia

myrna@metal.ui.ac.id

Abstract. Thermal Reactive Deposition (TRD) using pack cementation method with vanadium as carbide former has been examined to form the carbide layer on SUJ2 tool steel. In this study, effect of addition recycled FeV powder to new FeV powder to the product quality. Calculation analysis resulted 4 variations of FeV powder ratio to be studied i.e. 100% new FeV powder; 100% residual FeV powder; new powder FeV : residual FeV powder 50%:50%; New FeV powder : residual FeV powder 25%:75%. Electron microscope (SEM) and optical microscope (OM) analysis showed that the carbide layer formed in each sample had the same relative layer thickness of about 18μ. X-ray diffraction (XRD) characterized showing that there were a vanadium carbide compound in the formed layer. Hardness test and wear test resulted a relatively similar hardness value of about 1800 HV and $3x10^{-5}$ mm^3/m. It can be concluded that the use of residual FeV powder to the new FeV powder can be applied because the test results with various ratios that have been done have relatively similar results and meet the hardness application specification of 1700 HV.

Keywords: Thermal Reactive Deposition, vanadium carbide, SUJ2 steel, residual FeV powder, FeV powder ratio

1. Introduction

Thermo-reactive Deposition (TRD) is the process of depositing carbide-forming elements on carbon-containing surface. This layer have high hardness, good wear resistance, low friction coefficient, oxidation resistance and good corrosion resistance [1-3]. When compared to CVD and PVD, this TRD process has the advantage of not requiring high space to run the process, so that equipment is simpler and reduces production costs [4]. One of the TRD methods is *pack cementation* which is done by integrating carbide, substrate, Al_2O_3, and NH_4Cl forming powder into furnace with temperature between 950°C until 1100°C. Catalysts fill with carbide-forming powder to produce a carbide layer which will diffuse into the substrate and help impact to get even results. After the process is carried out in a certain period, the substrate is cooled. This process will produce layers with a very high bond between the diffusion layer and the substrate [5].

The vanadizing process aims to deposit vanadium to the material substrate. During the heating process, vanadium atoms diffuse into the surface of the material/substrate. Counter Diffusion of carbon in the material/substrate which will be fundamentally used for the material/substrate and the shape of vanadium carbide as a result of the interaction of carbon to the vanadium layer. From the vanadium carbide for the solid surface layer of vanadium carbide, the vanadium carbide layer formed will have a good performance and does not accentuate the shaft in its layer [6]. Vanadium carbide can achieve hardness of up to 3000 HV and has oxidation resistance at high temperatures (up to 750°C). The reaction is in Equations 1 to 4 [3,7]:

$$NH_4Cl_{(s)} \rightarrow NH_{3(g)} + HCl_{(g)} \quad (1)$$
$$2NH_{3(g)} \rightarrow N_{2(g)} + 3H_{2(g)} \quad (2)$$
$$2HCl_{(g)} + V_{(s)} \rightarrow VCl_{2(g)} + H_{2(g)} \quad (3)$$
$$VCl_{2(g)} + C_{(s)} \rightarrow VC_{(s)} + Cl_{2(g)} \quad (4)$$

But during its application, generally there is always residual powder after the surface hardening process ends. This residual powder is no longer used in the process. This research will examine the effect of FeV residual powder (recycled FeV powder) from TRD results on SUJ2 tool steel into FeV new powder in TRD process, on the coating thickness, hardness, microstructure, and elements distribution in carbide layer formed. So that the utilization of residual FeV powder is expected to save production costs from the motor vehicle industry.

2. Experimental Method

This research uses SUJ2 steel samples which have been cleaned and mashed, ferrovanadium powder (FeV), Alumina powder (Al_2O_3), and NH_4Cl powder. Before the research was carried out, it was necessary to do a normal vanadizing process which aims to obtain FeV residual powder which will be analyzed for the composition of the elements between the new powder FeV (powder before the process) and the remaining powder FeV (powder obtained after the process is carried out). This analysis was carried out by XRF analysis which aims to determine the calculation approach to be carried out to add a number of new powder FeV to the residual FeV powder.

After testing is complete and the product is obtained, some characterization of the material will be carried out to observe the carbide layer formed. Characterization carried out is microstructure observation with Optical Microscope (OM), layer morphology observation and element distribution by Scanning Electron Microscope - Energy Dispersive Spectroscopy (SEM-EDS), observation of compound composition with X-Ray Diffraction (XRD), layer hardness test and wear resistance layer.

3. Results and Discussion

Table 1. XRF Analysis Result of New FeV Powder, in wt%

Element	Mg	Al	V	Fe	Si	P	Ni	Mo
%	18,75	29,43	44,69	5,71	0,97	0,23	0,035	0,012

Table 2. XRF Analysis Result of Residual FeV Powder, in wt%

Element	Mg	Al	V	Fe	Si	P	Ni	Mo
%	5,78	32,27	49,49	11,23	1,01	0,12	0,08	0,02

Calculation analysis is performed to determine the percentage difference of the most important elements in powders such as vanadium before and after the TRD process. So we can find out how much new powder FeV needs to add to the residual FeV powder. Table 1 and Table 2 are the results of XRF analysis which shows the levels of elements in the new powder FeV and the recycled powder of FeV in the same series of processes.

IOP Conf. Series: Materials Science and Engineering **553** (2019) 012018 doi:10.1088/1757-899X/553/1/012018

For the calculation to be carried out, we want to know how many grams of vanadium are contained in each powder. It is known for the new powder FeV analyzed at XRF that is weighing 10.5376 grams, while for the remaining powder FeV analyzed in XRF weighing 11.2518 grams. Calculations to find out how many grams of vanadium are contained in each powder is done by calculating the usual chemical equilibrium in equation below:

$$\%\ element = \frac{element\ grams}{total\ grams} x\ 100\% \tag{5}$$

So it can be obtained for:

new FeV powder →

$$\%\ element = \frac{element\ grams}{total\ grams} x\ 100\% \tag{6}$$

$$44.69\% = \frac{element\ grams}{10{,}5376\ gr} x\ 100\% \tag{7}$$

Vanadium are obtained = 4.709 grams

residual FeV powder →

$$\%\ element = \frac{element\ grams}{total\ grams} x\ 100\% \tag{8}$$

$$49.49\% = \frac{element\ grams}{11{,}2518} x\ 100\% \tag{9}$$

Vanadium are obtained = 5.5685 grams

From these two results, there are 4.709 grams of vanadium in 10.5376 grams of FeV new powder and there are 5.5685 grams of vanadium in 11.2518 grams of residual FeV, it can be concluded that vanadium levels are not reduced because the results obtained from these calculations show the vanadium obtained in the new powders did not show a decrease in residual FeV powder. This can be seen from the distance between the total weight of the two powders, namely 10.5376 grams and 11.2518 grams, and also the distance between the weight of vanadium obtained from the calculation of 4.709 grams and 5.5685 grams having the same range of 0.7 grams.

From the conclusions obtained based on this calculation, a research will be conducted on how the difference in the use of residual FeV powder on SUJ2 steel. Wheteher its obtained the different results from those using only FeV new powder or have the same results. Also, because it is assumed that the vanadium level does not change, it will also be done mixing new powder and FeV residual powder and will also see whether it will have different results or not.

Since it is assumed that the vanadium content in FeV residual powder is not reduced, it is estimated that the experiments to be carried out will have results that are not much different. The experiment will be carried out with the following variables:
New FeV powder, Residual FeV powder 100%, New FeV powder 50% : residual FeV powder 50%, New FeV powder 25% : residual FeV powder 75%.

3.1 Analysis of the Use Effect Recyled FeV Powder on the Carbide Layer Morphology

The carbide layer morphology resulting from vanadizing process for various variations of FeV powder ratios is shown in Figure 1. The morphology of this carbide layer is obtained from the results of SEM testing. In the results of these tests, it showed that the carbide layer is formed in each variation of the use of FeV powder and with similar relatif layer morphology.

In the SEM test results, also obtained the thickness of the layers formed in the observed area. From the results of these tests, obtained results with carbide layer thickness size that is not significantly different. This is consistent with the hypothesis that the use of residual powder will be obtained by products with results that will not be much different from products that use 100% new FeV powder without recycled FeV powder mixture.

IOP Conf. Series: Materials Science and Engineering **553** (2019) 012018 doi:10.1088/1757-899X/553/1/012018

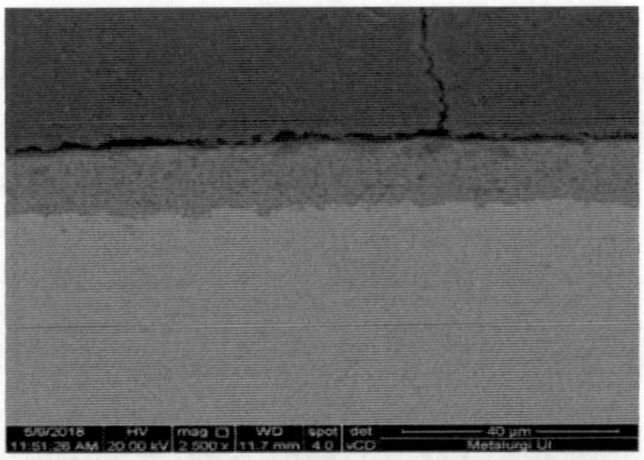
(a)

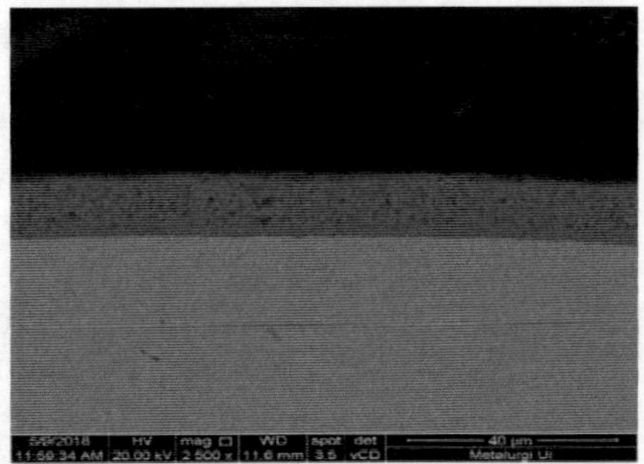
(b)

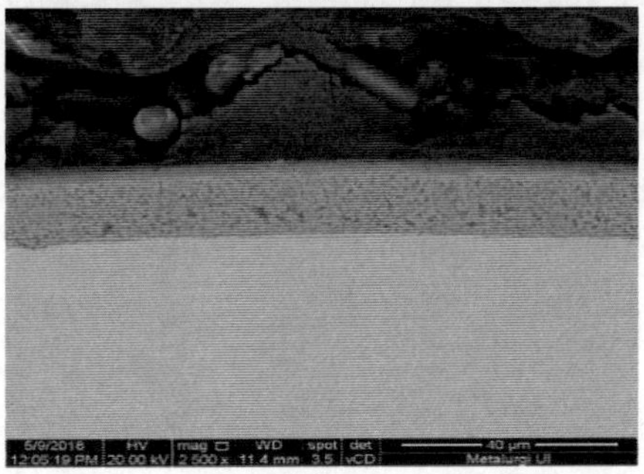
(c)

(d)

Figure 1. Morphology of carbide layers (a) new FeV, (b) 100% Residual FeV, (c) New FeV 50% : Residual FeV 50%, (d) New FeV 25% : Residual FeV 75%, resulting from the TRD process for various variations of FeV powder ratios on SUJ2 tool steel

3.2 Analysis of the Use Effect Recyled FeV Powder on the Carbide Layer Homogeneity

This analysis is conducted to find out whether there are differences in the thickness of the layers formed from each variable. This analysis is shown in Figure 2. which is a combination of 4 images with different positions taken from the same variable. For 1 picture there will be 3 measurements so that the total for 1 variable will get 12 measurements which will be averaged and compared to other variables whether there is a significant difference in the thickness of the carbide layer formed or not.

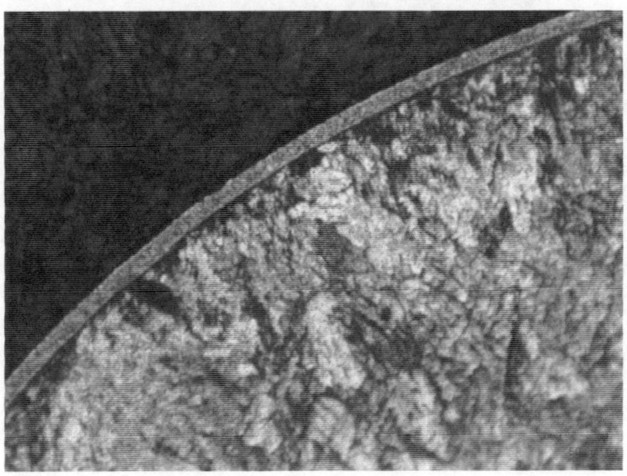

Figure 2. The homogeneity of carbide layers resulting from the vanadizing process for various variations of FeV powder ratios in tool steel SUJ2 TRD Process Results.

From the results of these observations obtained the average of each variable is as follows:

Table 3. Average Thickness of each variable.

Powder	Average Thickness (μm)
New FeV powder	18.0125
Residual FeV 100% powder	17.9283
New FeV powder 50% : Residual FeV powder 50%	18.2792
New FeV powder 25% : Residual FeV powder 75%	18.2092

From the results of these calculations as shown on the table, it showed the carbide layer formed has an average thickness that is not much different. So the use of FeV residual powder that is done will not have much different results from products that only use FeV powder.

3.3 Element Distribution Analysis

In the SEM (Scanning Electron Microscope) test that has been carried out, there are also EDS (Energy Dispersive Spectroscopy) for determine the distribution of chemical elements in the observed area. This analysis is carried out by drawing a straight line starting from the carbide layer which is formed to the base of the material being tested. This analysis aims to determine the distribution of elements in the carbide layer to the material base of each variable.

In Figure 3. shows the results of linescan SEM-EDS analysis that have been done. The results of the analysis showed that there was no difference in the distribution of elements between products using residual FeV powder with products that only used FeV new powder. From the graphs obtained, the vanadium element found in the layers is very high. This shows that the carbide layer is formed in each variable.

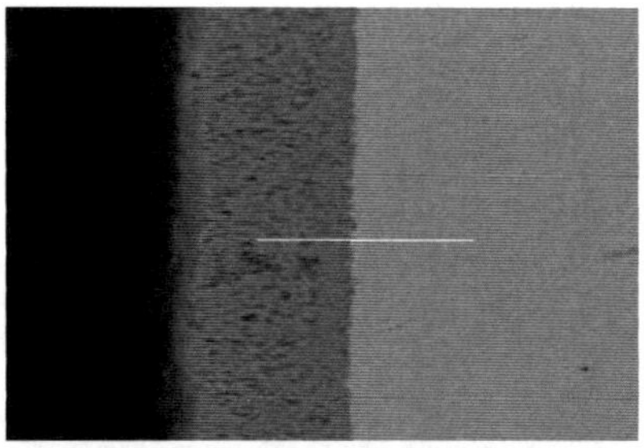

(a)

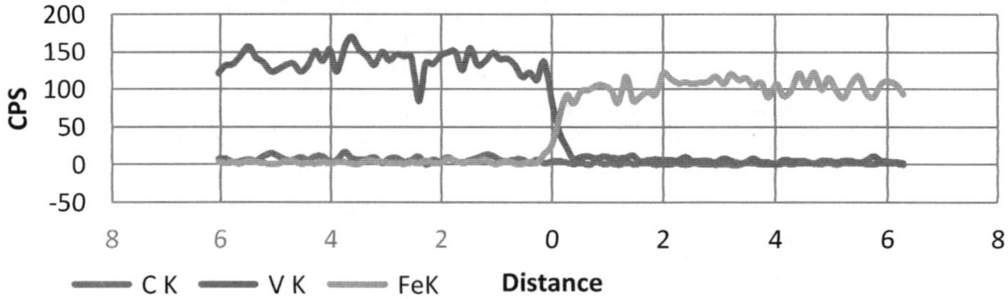

(b)

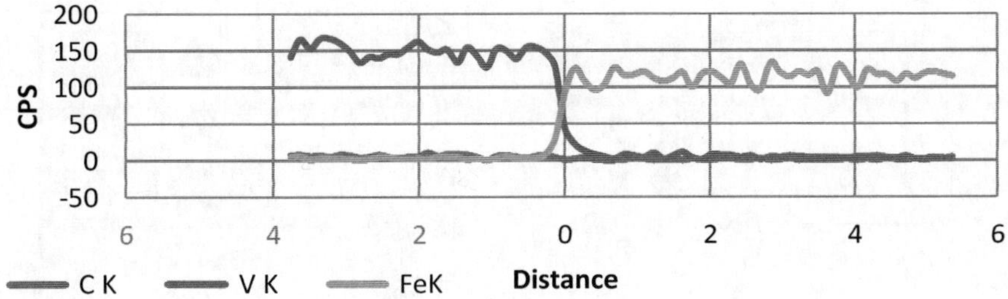

(c)

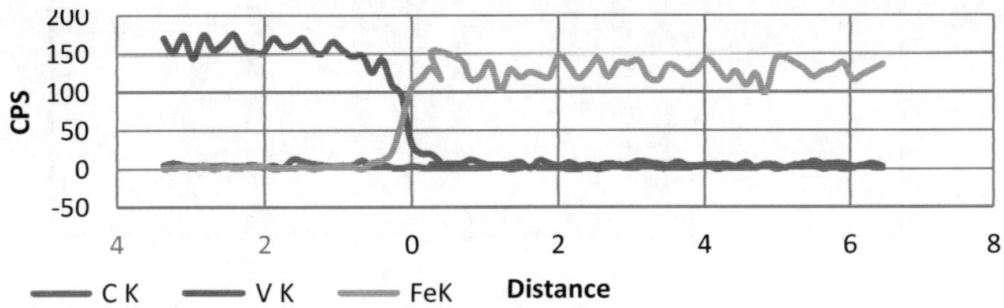

(d)

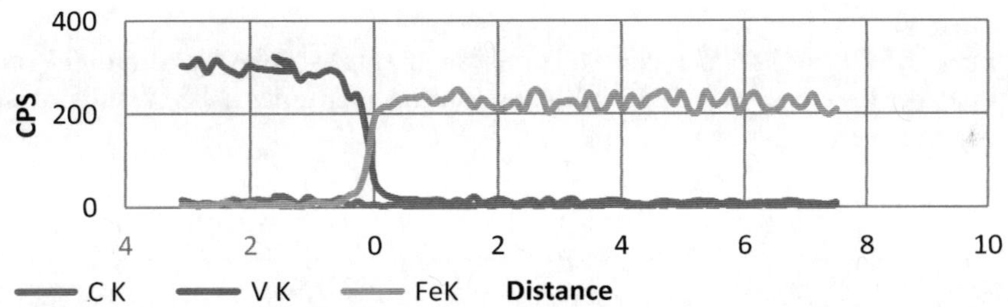

(e)

Figure 3. Linescan carbide layer results example from the vanadizing process for various variations of FeV powder (a). The graphics show the EDS result of : new FeV (b), residual FeV 100% (c), new FeV 50% : residual 50% (d), new FeV 25% : residual FeV 75% (e) on Tool Steel SUJ2 TRD Process Results.

3.4 Compound Composition Analysis

Observation of the composition of compounds contained in the sample analysed by XRD (X-Ray Diffraction). Observations obtained from this XRD will then be processed by the XRD application, one of which is X'Pert HighScore Plus. The results of these observations are shown in Figure 4 as follows.

In Figure 4. there are at least 3 peaks which represent the dominant compounds found in Vanadizing's SUJ2 Steel. When matched with the vanadium XRD database/literature for vanadium compounds, these 3 peaks are vanadium carbide compounds. The detected vanadium carbide compounds are V6C5 and V8C7.

The detected Vanadium Carbide compounds were also found in the study of Aliakbar Ghadi [4] and Bayu Mahardika [8] who also carried out surface hardening with the TRD method and used FeV powder. This shows that the use of FeV residual powder applied in this study can produce carbide layers containing vanadium carbide compounds.

IOP Conf. Series: Materials Science and Engineering **553** (2019) 012018 doi:10.1088/1757-899X/553/1/012018

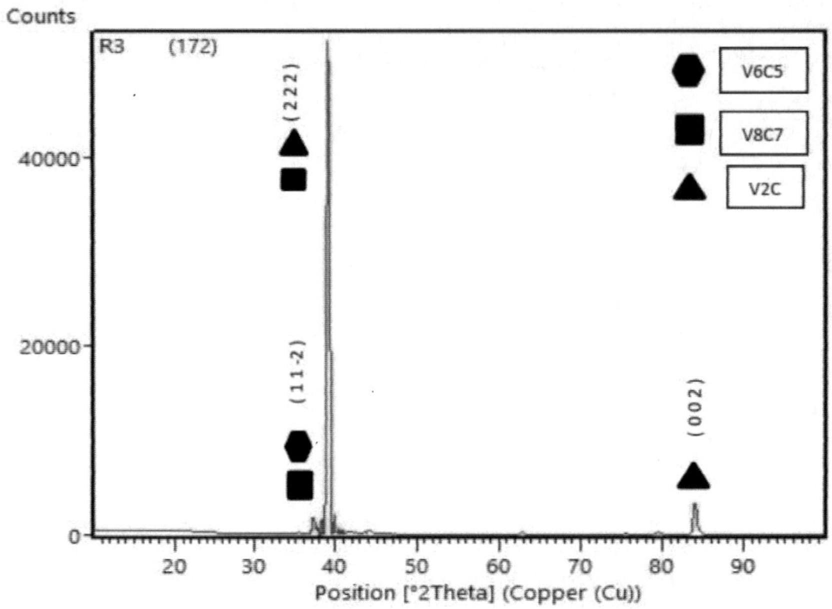

Figure 4. Observation Results of XRD on SUJ2 Steel Results of Vanadizing Treatment.

3.5 Hardness Analysis

Testing the hardness of the SUJ2 Steel sample from vanadizing is done by Microvickers where the hardness results can be directly available from the measuring instrument. The results of the hardness testing can be seen in Figure 5. belows.

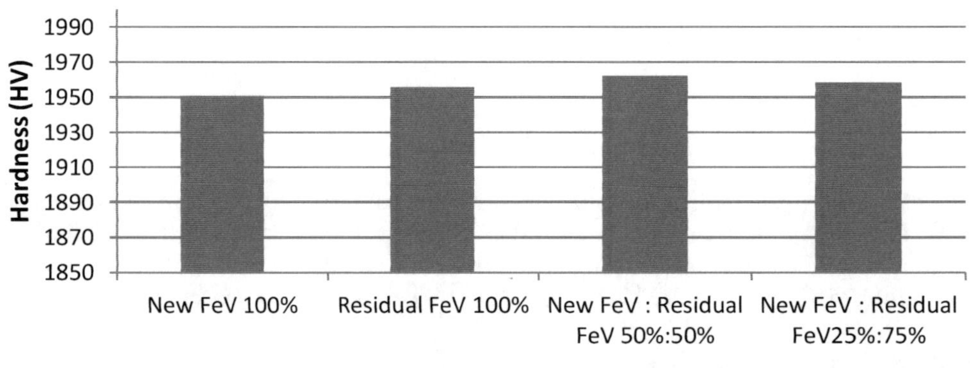

Figure 5. Results of Testing Hardness in SUJ2 Steel Vanadizing Results (HV)with Various FeV Powder.

The hardness testing was done on the SUJ2 Steel pin by giving each 1 point on 10 samples for 1 variable. The results of this hardness test obtained the average hardness of each sample variable that was not significantly different.

3.6 Wear Analysis

This test uses the Ogoshi method where the test object obtains the frictional load of the rotating ring. From the width of the abrasive material gap, it can be obtained for the wear rate of each sample are:

Table 4. Abrasive Width and Wear Rate in SUJ2 Steel Vanadizing Results with Various FeV Powder.

FeV Samples	Abrasive Width (mm)	Wear Rate (mm^3/m)
New FeV Powder	1.3	1.0866×10^{-5}
Residual FeV Powder	1.2	1.4363×10^{-5}
New FeV Powder 50% : Residual FeV Powder 75%	1.05	$5{,}7252 \times 10^{-5}$
New FeV Powder 25% : Residual FeV Powder 75%	1.1	6.5826×10^{-5}

From the table we could see the wear rate was not significantly different. So it can be concluded that the use of recycled FeV powder has a sample quality that is not so far from the sample that only uses new FeV powder.

4. Conclusion

Based on the results of research on the use of residual FeV powder as an innovation of the Thermo Reactive Deposition (TRD) process with the Matierals Balance method on the characteristics of carbide layers in SUJ2 tool steel it can be concluded that the effect of using residual FeV powder with various ratios on new powder FeV results in the quality of carbide coating has a thickness that is relatively the same as the carbide layer produced by products that only use FeV new powder. The calculation approach used to determine the addition of FeV new powder to residual FeV powder is carried out using the usual chemical equilibrium calculation method. The calculation is done after obtaining data on the percentage of vanadium in the new powder FeV and recycled FeV powder. The quality of the products produced by the use of recycled FeV powder with various ratios that have been carried out has relatively similar results with products that only use FeV new powder both from the results of carbide layer morphology, layer homogeneity, hardness, and wear resistance.

Acknowlegment

This research is fully funded by PITTA 2018 given by Ministry of Higher Education of Republic of Indonesia and Directorate Research and Public Services of University of Indonesia for financial support and project administration respectively to conduct this research under PITTA Project for 2018 fiscal. And also in terms of research equipment supported by PT FSCM Manufacturing Indonesia.

References

[1] X.S. Fan, Z.G. Yang, C. Zhang, Y.D. Zhang, "Thermo-reactive deposition processed vanadium carbide coating: growth kinetics model and diffusion mechanism", *Surface & Coating Technology,* volume 208, 2012, pp 80-86. doi: 10.1016/j.surfcoat.2012.08.010

[2] T. Arai, "Thermoreactive Deposition/Diffusion Process for Surface Hardening Steels," *ASM Handbook Vol 04, Heat Treating,* ASM International, 1991, pp. 1000-1003.

[3] N. Tecnolog and D.Fecha, "Vanadium carbide coatings produced on gray cast iron using the thermo-reactive deposition / diffusion technique," *Ingeniera Mecanica Tecnologia Y Desarrollo, volume 5,* 2015, pp. 333-338.

[4] A. Ghadi, M. Soltanieh, H. Saghafian, and Z. G. Yang, "Investigation of Chromium and Vanadium Carbide Composite Coatings on CK45 Steel by Thermal Reactive Diffusion" *Surface and Coating Technology,volume 289,* 2016, pp 1-10.

[5] F. Fazlalipour, M.Aghair-Khafri, "Vanadium carbide coatings on die steel deposited by the thermo-reactive diffusion technique", *Journal of Physics and Chemistry of Solids, volume 69,* 2008, pp 2465-2470. doi: 10.1016/j.jpcs.2008.04.040

[6] A. Borisova, Y. Borisov, E. Shavlovsky, I. Mits, L. Castermans and R. Jongbloed, "Vanadium Carbide Coatings : Deposition Process and Properties," *Reutte, volume 2,* 2001, pp 452-468.

[7] C. Oliveira, R. Riofano and L. Casteletti, “Formation of carbide layers on AISI H13 and D2 steels by treatment in molten borax containing dissolved both Fe–Nb and Fe–Ti powders,” *Material Letters , volume 59*, 2005, pp. 1719-1722. doi: 10.1016/j.matlet.2005.01.052

[8] M. A. Mochtar, W. N. Putra, B. Mahardika, “Optimizing the dual elemental thermal reactive deposition time in carbide layer formation on SUJ2 tool steel,” *AIP Conference Proceeedings,* 2018, pp 1-9. doi: 10.1063/1.5038294

The Effect of Thermo Reactive Diffusion (TRD) Processing Time with Ferrochromium Powder on Carbide Layer Characteristics on SUJ 2 Tool Steel Substrate

Annida Jihan Maulida[1], Myrna Ariati[2*]
[1,2]Departement of Metallurgy and Materials Engineering, University of Indonesia, Depok, 16424, Indonesia

myrna@metal.ui.ac.id

Abstract. In this study, the formation of chromium carbide layer on SUJ 2 tool steel substrate as a surface treatment by thermo reactive diffusion process with pack cementation method was observed. The powder used are Al_2O_3, NH_4Cl, and FeCr as carbide forming elements with the composition of 37%, 3%, 60% for each. TRD process performed in rotating vacuum furnace for 4, 6, and 8 hours on 980°C. The results show time variation effects on the carbide layer. Thickenss carbide layer on SUJ2 after TRD process observed by Optical Microscope. The morphology of carbide layer observed with SEM shows a smooth and dense layer with almost no porosity formed. $Cr_{23}C_6$ and Cr_7C_3 found on the layer by XRD and hardness value obtained using Vickers microhardness testing for each samples are 1738.62, 1738.72, and 1739.64 HV. The wear rates are inversely proportional with the hardness value where the values are 0.000698, 0.000658, and 0.000627 mm^3/m.

Keywords: Thermo Reactive Diffusion, Chromium Carbide, Pack Cementation

1.Introduction

Thermo Reactive Diffusion is a method to form a very hard carbide, nitride, or carbonitride layer which has an excellent wear resistance on a carbon steel substrate [1]. From this process, hard ceramic coating's expected to formed, which can be used to increase the service life of engine components, cutting tools, and forming tools, usually consist of carbide or carbonitride [2] with a layer that has a better and more uniform adhesion propeties [3]. TRD coating that has been done, mostly using Vanadium as Carbide Forming Elements to produce VC (Vanadium Carbide) coating with a great hardness surface. But, its resistance from oxidation in high temperature still below Chromium [1]. Carbon diffusion process takes place in two stages [4], carbons from the inside of metal move to the surface by heating process so that the surface has relatively higher consentration of carbon and CFE inside the medium reacts with carbon in the surface to form a carbide layer. Kinethic growth of carbide layer itself can be explained by equation below

$$l^2 = K\,t = K_0 \exp\left(-\frac{Q_{TRD}}{RT}\right) t \ [5] \qquad (1)$$

Where l is thickness (cm), t is time (s), K is growth rate constant (J/mol), T is Temoerature (K), and R is gas constant (8.31415 J/mol K). In this experiment, the effect of different processing time with TRD method at 980°C in 4, 6, and 8 hours towards carbide layer characteristics will be studied. Processing time variable is used to determine the most optimum one in this process.

IOP Conf. Series: Materials Science and Engineering **553** (2019) 012019 doi:10.1088/1757-899X/553/1/012019

2. Experimental and Method

Cylinder shaped SUJ 2 steel with diameter 1 cm is cleaned with barrel polishing process. SUJ 2 steel chemical composition can be seen on Table 1. Cylinder shaped SUJ 2 that has been cleaned weighed with pin and mixed with Al_2O_3, NH_4Cl, and FeCr powder by comparison 1:1 for sample : powder, with powder composition on Table 2. Sample and powder are mixed inside the pot and installed into the furnace, vacuum process using Argon gas with 1.5 bar pressure. Vacuum process carried out until -92 KPa reached. Pre-heating process takes place at 400 dan 800°C for 5 minutes each, followed by TRD process for 4, 6, and 8 hours and cooled by furnace cooling process. Sample will be characterized using Optical Microscope (OM) to find out the homogenity of carbide layer that can be seen from the thickness value (measured by infinite analyzer), SEM-EDS to know the morphology and elements distribution of carbide layer, XRD to estimate compounds formed on the surface, wear test to determine wear resistance after TRD treatment, and hardness measurement using Micro Vickers.

Table 1. Chemical Composition of SUJ 2 Steel.

Fe	C	Si	Mn	P	S	Ni	Cr	Mo	Ti	Cu	Al	V
Bal	1.03	0.24	0.28	0.013	0.006	0.06	1.68	0.004	0.007	0.05	0.1	0.006

Table 2. Powder Coating Composition.

Al_2O_3	37% 300 gram x 37% = 111 gram
NH_4Cl	3% 300 gram x 3% = 9 gram
FeCr	60% 300 gram x 60% = 180 gram

3. Result and Discussion

3.1 The Effect of Processing Time Variable On Layer Thickness

From TRD process using Chromium as carbide forming elements, the chromium carbide layer obtained has a compact and smooth surface with low porosity which can be seen on Figure 1. The layer obtained has uniform thickness on sample (a), (b) and (c). The boundary layer and surface clearly seen without any additional layer between them which has a similarity with F.E Castillejo's experiment [7] so that we can say that there is no interdiffusion layer formed. It also can be proved from the linescan data that shows no additional elements found on carbide layer and substrate as seen on the graph in Figure 2 and 3. The longer the time process, the thicker the layer formed [8] thickness value from 4, 6, and 8 hours time process increases in sequence. For 4 hours processing time, thickness value obtained is 13μm, for 6 hours processing time, the thickness value is 13.7μm, and for 8 hours processing time, the thickness values is 17.2μm.

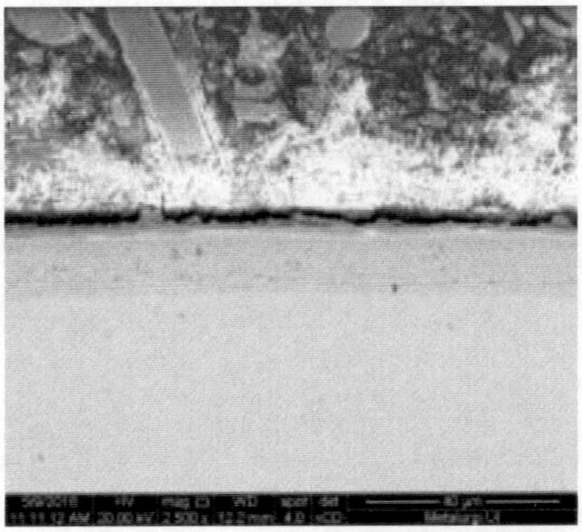

(a)

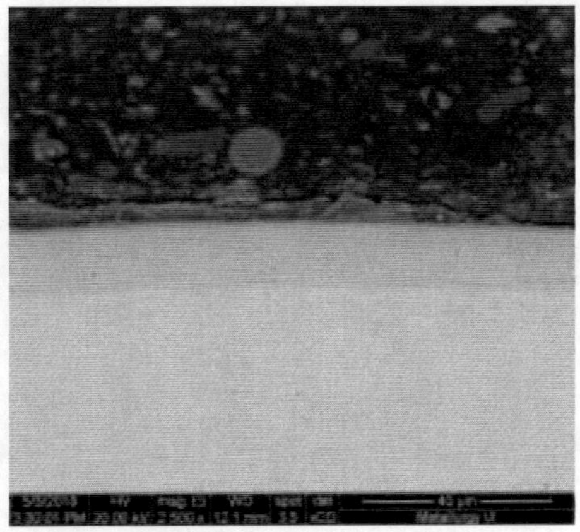

(b)

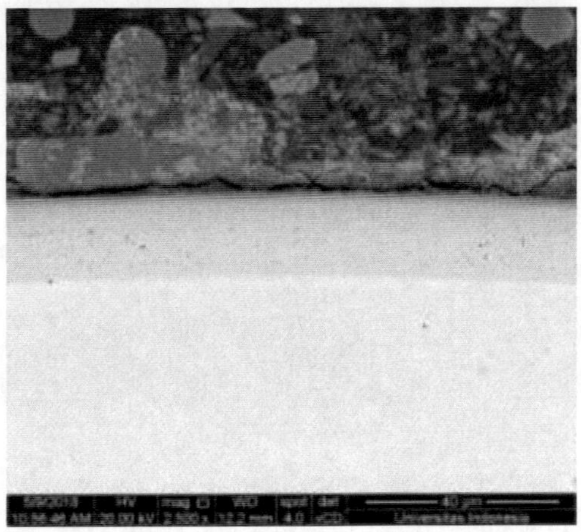

(c)

Figure 1. (a) Carbide Layer t = 4 hrs, (b) Carbide Layer t = 6 hrs, (c) Carbide Layer t = 8 hrs.

In Figure 2 and 3 can be seen the result of EDS linescan characterization which shows elements distribution throughout the carbide layer and surface of the substrate

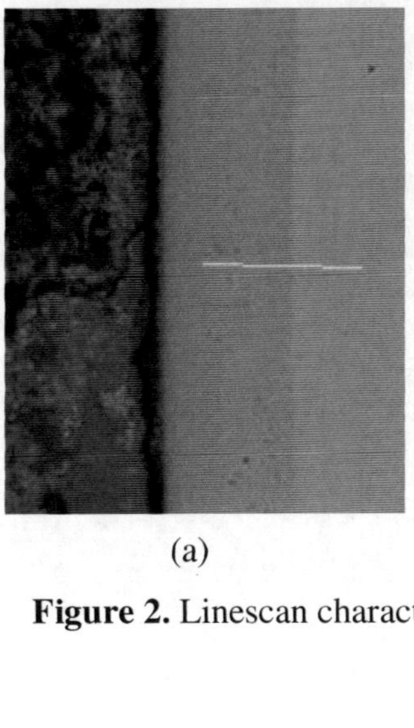

(a)

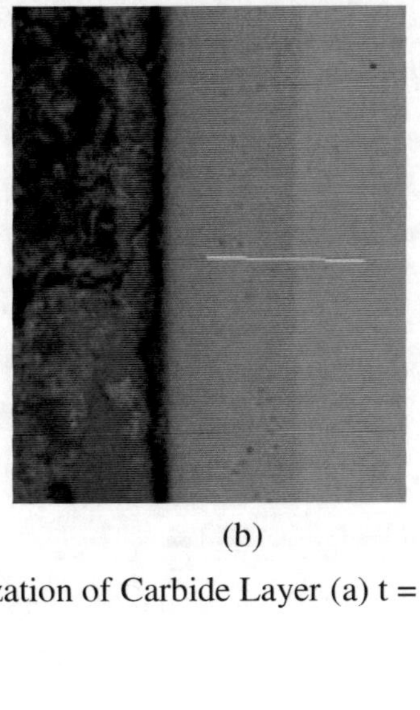

(b)

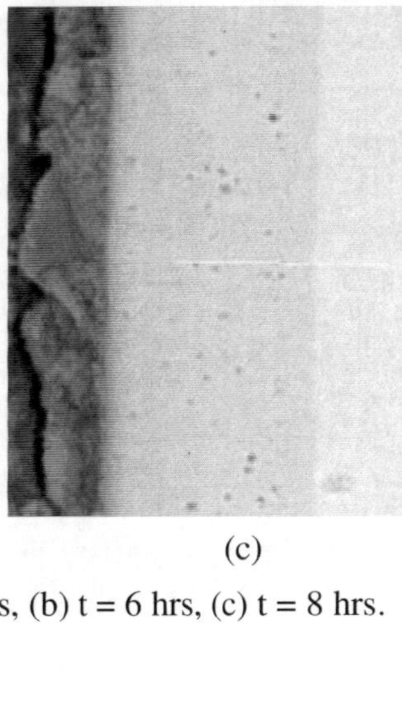

(c)

Figure 2. Linescan characterization of Carbide Layer (a) t = 4 hrs, (b) t = 6 hrs, (c) t = 8 hrs.

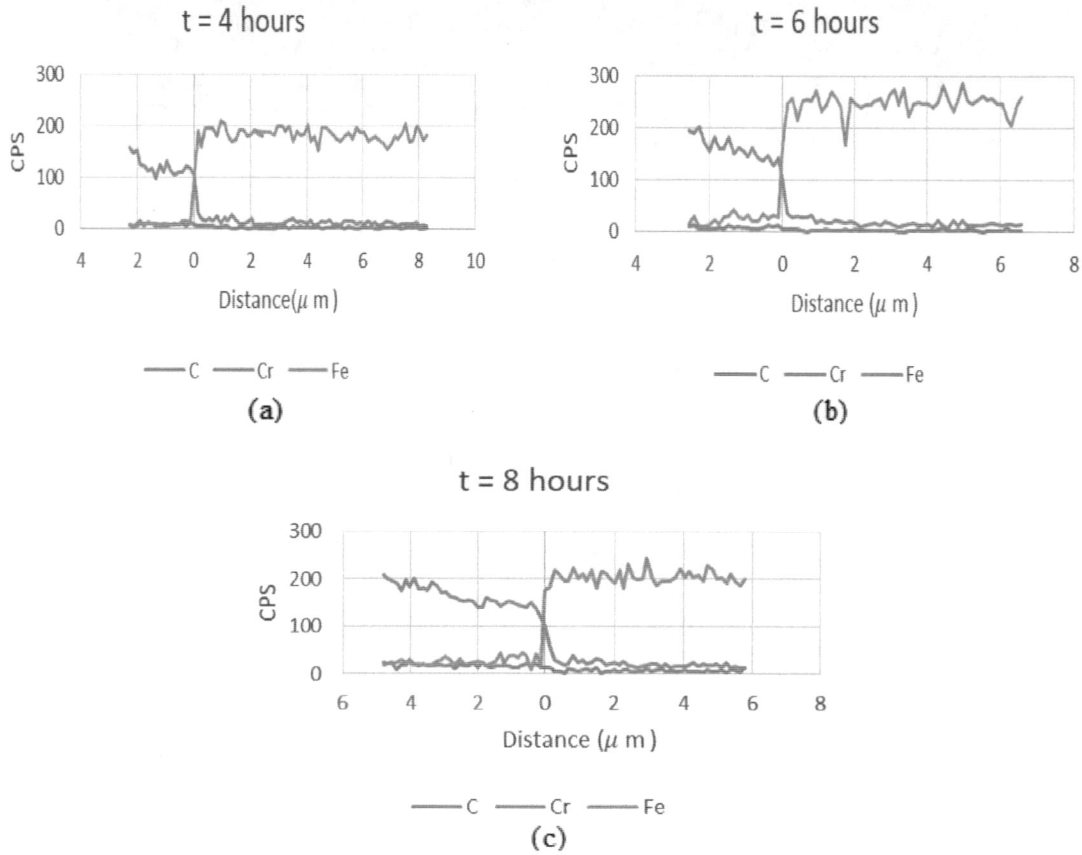

Figure 3. CPS VS Distance Graph (a) t = 4 hrs, (b) t = 6 hrs, (c) t = 8 hrs.

From the linescan graph above, can be seen the tendency that occurs to all samples are the same. Cr content is decreasing from layer to substrate and so the carbon content, even not really significant because the carbon on the surface already bond with carbide forming elements on the layer. It's proven that the diffusion direction moves from the middle of the substrate to the surface, when the heating process happens, the carbon will gather in the surface area of the substrate. The carbon gathered then reacts with carbide forming elements powder and forming Chromium Carbide layer.

3.2 Analysis of The Compound Formed On The Layer

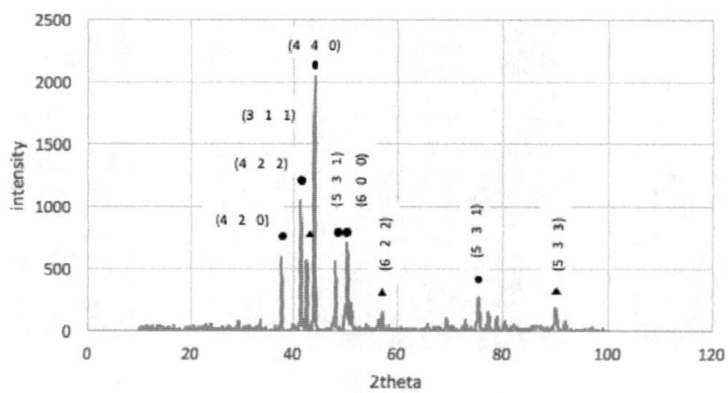

Figure 4. XRD Characteristics Graph.

From XRD characterization that already converted with X'pert Highscore Plus software, the peaks as seen on Figure 4 was obtained. From the graph above, the compounds identified as chromium carbide found in the form of $Cr_{23}C_6$ and Cr_7C_3 that also found in F. E. Castillejo's experiment [7], which performs Thermo Reactive Diffusion process using AISI D2 tool steel as the substrate, so it can be proved that the carbide layer is formed in this experiment.

3.3 The Effect of Processing Time On Thickness Value and Homogenity of Layer Formed

Based on the experiment by Fernandes [8], it's stated that the longer the processing time, the thicker the carbide layer formed. It is the result from the diffusion process that runs longer. From the measurement process, layer thickness obtained from 4, 6, and 8 hours processing times are 13, 13.8, 17.6 μm. Thickness value between 4 and 6 hours processing time are not that different, meanwhile for 8 hours processing time has the thickest carbide layer with a significant difference as can be seen in Figure 5. It supports the theory explained before that the longer the process, the carbide layer formed will also higher. The significant difference between thickness value of 8 hours processing time with the others might be because it is the most optimum processing time in this process.

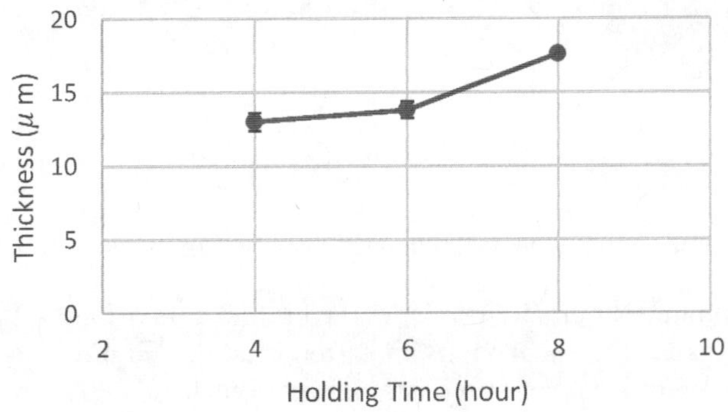

Figure 5. Layer Thickness vs Holding Time.

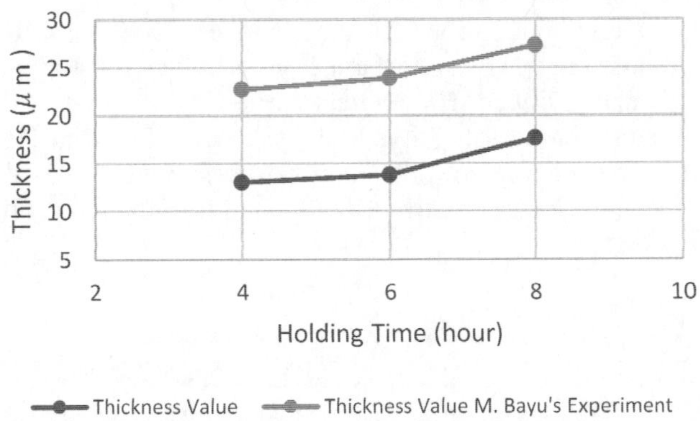

Figure 6. Comparison Layer Thickness.

Thickness increases alongside with increasing processing time in accordance with equation $d^2 = Kt$ [3] where d is layer thickness, K is diffusional coefficient, and t is processing time. The equation shows the longer the processing time, the thicker the layer. Thickness value variation can be made as a layer homogeneity reference. Based on the experiment by M. Bayu [9], coating process using powder mixture of FeCr and FeV with TRD, thickness increases alongside with increasing processing time. The difference of thickness value obtained can be influenced by powder composition used in the TRD process. Where M. Bayu using powder contains Vanadium which affinity towards carbon is higher than

IOP Conf. Series: Materials Science and Engineering **553** (2019) 012019 doi:10.1088/1757-899X/553/1/012019

Chromium so that the layer formed also thicker. From the collected data, carbide layer obtained is quite homogeneous and can be proved by thickness measurement result on 3 points for 4 different areas of sample's surface that shows insignificant difference for each parameters. Layer homogeneity can also be supported by morphology condition that shows compact surface with low porosity.

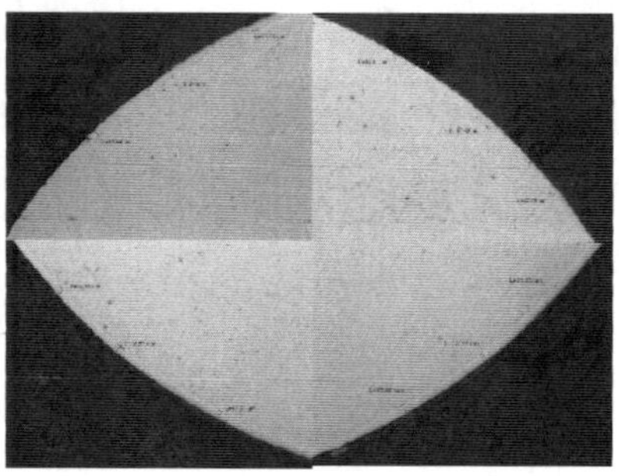

(a)

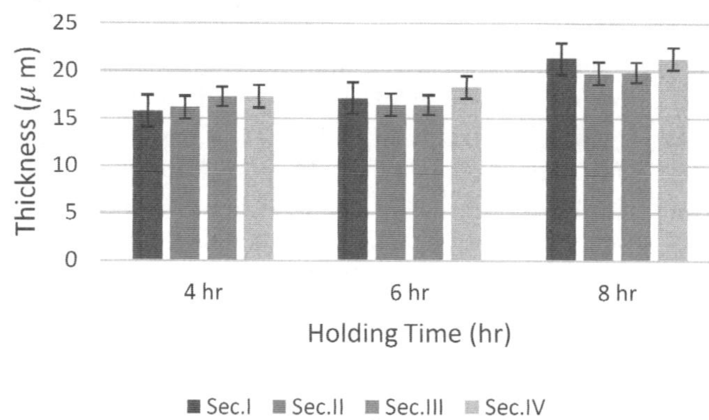

(b)

Figure 7. (a) Homogenity Measurement (b) Homogenity Measurement Result.

Figure 7 shows result from measurement process on layer thickness using OM and Infinite Analyzer Software to determine the value for each points in different areas. Figure 7 shows thickness value distribution taken from 4 different areas for every sample where each area measured three times on three different points, generates an average thickness value. For each parameter, layer thickness obtained have relatively similar values.

3.4 The Effect of Processing Time On Hardness Value

Coating process with TRD method applied for each sample takes place at 980°C with processing time parameters are 4, 6, and 8 hours. Figure 8 shows that hardness values for 4, 6, and 8 hours processing time are 1738.62, 1738.72, and 1739.64 HV. It can be seen that hardness value didn't change significantly for different amount of processing time, it shows that holding time is not affect significantly towards hardness value, the same as found on F.S. Chen's experiment [1]. It is also show that different processing time will not affect any change in carbide compound forming so that the hardness value will also not be affected.

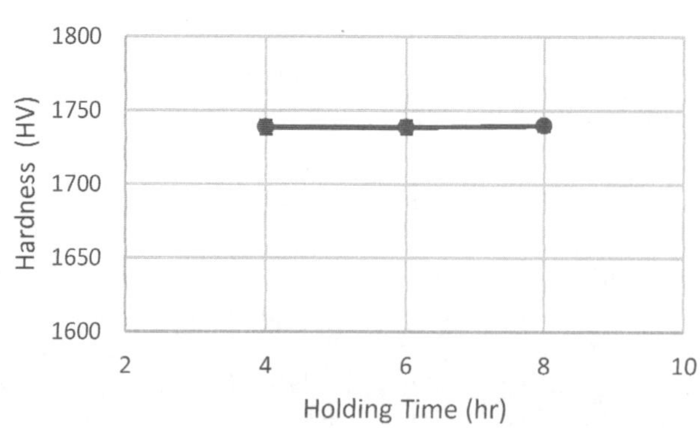

Figure 8. Hardness Value vs Holding Time.

Based on XRD characterization, the compounds found in the form of $Cr_{23}C_6$ and Cr_7C_3, from G. Cios's experiment, it is said that chromium carbide can be classified based on its hardness value. $Cr_{23}C_6$ has hardness value around 976-1750 HV and for Cr_7C_3 the hardness value is 1336-2200 HV [10] so there is a resemblance between the theory and what we found from this experiment because the data obtained from this experiment is still in the same range with the theory.

3.5 The Effect of Processing Time Towards Wear Resistance

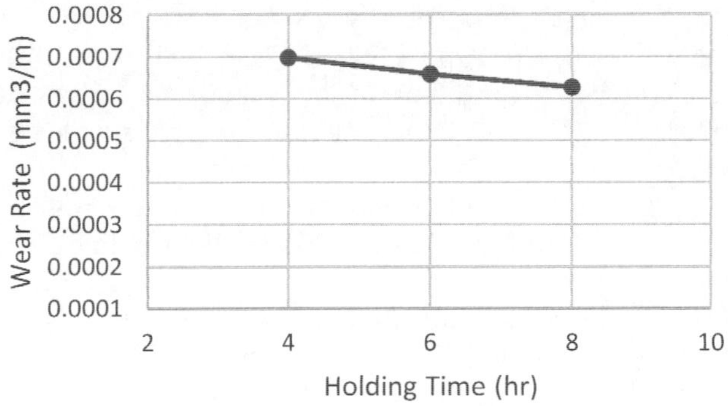

Figure 9. Wear Rate vs Holding Time.

Wear test was done using Ogoshi testing method with 12 kg load, 400 m sliding distance, and 2.38 m/s velocity. From the data obtained, the wear rate tendency is reduced alongside with the increasing of processing time. Figure 9 shows that for 4, 6, and 8 hours processing time, the wear rate values are 0.000698, 0.000658, dan 0.000627 mm^3/m. Wear rate value has no significant difference, it can be explained from the insignificant hardness values for each different sample that also affects the wear rate values.

4.Conclusion

Hardness values obtained for 4, 6, and 8 hours processing time are 1738.62, 1738.72, 1739.64 HV that occupies minimum hardness specification for silent chain pin which is 1700 HV and wear rate values are 0.000698, 0.000658, 0.000627 mm^3/m. Increasing processing time for Thermo Reactive Diffusion resulting thickness value of 13 μm, 13.8 μm, dan 17 μm. Carbide layer thickness increases due to diffusion process that going on continuously along with the increasing of processing time that can be explained by $d^2 = Kt$ equation. Carbide layer from Thermo Reactive Diffusion process using FeCr powder as carbide forming elements produce Chromium Carbide coating with smooth and uniform surface and carbide compounds on the layer found in the form of $Cr_{23}C_6$ and Cr_7C_3. For every processing time variable, layer homogeneity measured are uniform. The longer the processing time, the better the homogeneity. The optimum processing time for TRD in this process is 8 hours.

Acknowledgement

Authors would like to thank to Ministry of Higher Education of Republic of Indonesia and Directorate Research and Public Services of University of Indonesia for financial support and project administration respectively to conduct this research under PITTA Project for 2018 fiscal. The Author also would like to thank PT Federal Superior Chain for their support on this research. The Assistance of PT Federal Superior Chain for Thermo Reactive Deposition process and Equipment is gratefully acknowledged.

References

[1] F.-S. Chen, P.-Y. Lee, M.-C. Yeh, "Thermal reactive deposition coating in a fluidized bed furnace", *Mater. Chem. Phys.*, volume 53, 1998, pp 19–27, doi: 10.1016/S0254-0584(97)02048-8

[2] T.H Liu, C.T Su, T.P Cho, "Effect of chromium content on the dry machining performance of magnetron sputtered CrxC coating", *Mater Sci Eng*, volume 364, 2004, pp A188–A97, doi: 10.1016/j.msea.2003.08.010

[3] Z. J. Shan, Z. G. Pang, F. Q. Luo, F. D. Wei, "Surface & coatings technology kinetics of V (N , C) and Nb (N , C) coatings produced by V – Nb – RE deposition technique", *Surf. Coat. Technol*, volume 206 (19–20), 2012, pp 4322–4327, doi: 10.1016/j.surfcoat.2012.04.057

[4] M. A. Khafri, F. Fazlalipour, "Vanadium carbide coatings on die steel deposited by thermo-reactive diffusion technique", *Physics and Chemistry of Solids*, volume 69, 2008, pp 2470, doi: 10.1016/j.jpcs.2008.04.040

[5] Z.B. Wang, J. Lu, K. Lu, "Chromizing behaviors of a low carbon steel processed by means of surface mechanical attrition treatment", *Acta Materialia*, volume 53, 2005, pp 2081 – 2089, doi:10.1016/j.actamat.2005.01.020

[6] W. D. Callister, D. G. Rethwisch, "Material Science and Engineering: An Introduction", 9th ed., John Wiley and Sons Inc, 2009, pp 327

[7] F.E Castillejo, D. M. Marulanda, J. J. Olaya, J. E. Alfonso, "Wear and corrosion resistance of niobium-chromium carbide coatings on AISI D2 produced through TRD", *Surf. Coat. Technol*, volume 254, 2014, pp 104-111, doi : 10.1016/j.surfcoat.2014.05.069

[8] F. A. P. Fernandes, T. L. Christiansen, K. V. Dahl dan Somers, "Growth of Vanadium Carbide by Halide-Activated Pack Diffusion", 2013

[9] A. M. Myrna, N. P. Wahyuaji, M. Bayu "Optimizing the dual elemental thermal reactive deposition time in carbide layer formation on SUJ2 tool steel"**,** AIP Conference Proceedings 1964, 020012 (2018) doi: 10.1063/1.5038294

[10] G. Cios, P. Bala, M. Stepien, K. Gorecki, "Microstructure Of Casti Ni-Cr-Al-C Alloy", *Archives of Metallurgy and Materials,* volume 60, 2015, 145-148, doi: 10.1515/amm-2015-0022

The Effect of Neodymium on the Solidification Process of Al-5Zn-0,5Cu Alloy Sacrificial Anode

Yudha Pratesa[1], Adimas Habib Iqbal[1], Deni Ferdian[1*]
[1] Department of Metallurgical and Materials Engineering, Faculty of Engineering Universitas Indonesia, Kampus Baru UI Depok 16424, Indonesia

deni@metal.ui.ac.id

Abstract. The effect of the addition of neodymium rare earth on Al-5Zn-0.5Cu alloy was investigated with Optical Microscope (OM), Differential Scanning Calorimetry (DSC), and Cyclic Polarization. The content variable of neodymium tested was 0.1wt%, 0.3wt%, dan 0.5wt%. Observation with OM was conducted to see the changes of the SDAS and the precipitate formation. DSC was used to identify the phase transformation and solidification process of the intermetallic phase. Cyclic Polarization was used to know the corrosion characteristics of Al-5Zn-0.5Cu-xNd. The presence of neodymium formed precipitates on the grain boundary which made shorter SDAS. Addition of neodymium as an alloying element of Al-5Zn-0.5Cu sacrificial anode may decrease pitting corrosion resistance and reduce the size of the loop. This behavior will be more preferable for Al Anode because the indication of more uniform attack.

Keywords: Solidification, Rare Earth, Neodymium, Sacrificial Anode, Aluminium Alloy

1. Introduction

Pautasso et al. (1998) explain the tendency for hydrogen embrittlement (HE) and stress corrosion cracking (SCC) on HSS steel structures as a result of overprotection [1]. Overprotection occurs because the potential value of Al-Zn-In reaches -1.15 V vs. SSC [2]. The properties of aluminum sacrificial anodes include high current efficiency, low density, and long life usage [3]. Pure aluminum sacrificial anodes cannot be used because the sacrificial anode will immediately form a passive γ-Al_2O_3 layer on its surface [4]. Sacrificial anodes that form a passive layer will make the corrosion attack which should attack the sacrificial anode to attack the supposedly protected structure. This indicates that the formation of a passive layer on the sacrificial anode will make the sacrificial anode not work properly. It is also known that the addition of Zn will damage the stability of Al_2O_3 passive layer forming which will facilitate the process of corrosion [5]. The addition of the Zn element will destroy the passive layer on the anode surface with its second phase particles. These second phase particles have a certain volume fraction known as β-phase [6]. Studies show that rare earth metals as micro-alloying exhibit a beneficial impact on the mechanical properties of aluminum alloys. Studies are showing that Nd is primarily distributed in the form of intermediate compound AlCuNd which gives a resistive force on grain boundaries and increased the performance of Al 2A70 alloys at high temperatures [7]. The α-Al phases tend to corrode without the protection of the Al_2O_3 layer. The corrosion product layer grows with the dissolution of the η-Zn and α-Al phases. The corrosion resistance decreases as the immersion

IOP Conf. Series: Materials Science and Engineering **553** (2019) 012020 doi:10.1088/1757-899X/553/1/012020

time increases due to the destruction of the Al_2O_3 protective layer, selective corrosion of alloys, and the formation of defects in the corrosion product layer [8]. According to Minliang Su *et al.*, the as-cast HPDC AlNd44 alloys exhibit relatively good corrosion resistance strength, which is due to the presence of Al-Nd intermetallic phases that converge at the grain / dendritic boundary and act as a corrosion barrier. After annealing at temperatures above 573K, the corrosion rate increases with changes in the microstructure, i.e., the disintegration of reticular structures near the grain / dendritic boundaries [9].

2. Experimental method

2.1 Sample Preparation

The raw materials for this test were high purity of Al, Zn, Cu, and Nd. Each of element was melted with Resistance Furnace at temperature 750^{0}C for 30 minutes after preheated at 400^{0}C to make Al-Zn-Cu alloy. The melting of metals is aided by a graphite stirrer to allow the melt to homogenize appropriately. When the molten aluminum is completely dissolved, the dross was cleaned, and the metals were cast into the graphite mold. Neodymium was inserted by using Al-10 Nd master alloy. This master alloy was made at temperature 800^{0}C for 30 minutes. a The process of adding Nd to the melt must be done very quickly due to the high reactivity of neodymium with oxygen. Casting process of Al-5Zn-0.5Cu-xNd starts with melting the Al-5Zn-0.5Cu alloy at 720^{0}C and held for 10 minutes. Afterward, Al-10Nd is added to the melt at 730^{0}C. The melting metals is homogenized for 5 minutes and held for 10 minutes. Then the Al-5Zn-0.5Cu-xNd alloy is poured into the mold.

2.2 Characterization of Al-5Zn-0.5Cu-xNd Samples.

In this stage, each sample is characterized using Optical Microscope (OM), Differential Scanning Calorimetry (DSC) and Cyclic Polarization with metroohm potentiostate (CP). OM is used to see the grain boundary shape, the precipitate formed, and the length of the SDAS. DSC is used to see the phase transformations that have formed during the solidification process. This DSC experiment was measured with 10°C/minutes scanning rate. Polarization test is used to determine the corrosion properties of alloys. Cyclic polarization was conducted in 3.5% Sodium Chloride solution at ambient temperature. A standard three-electrode system was used for the potentiodynamic test. Graphite as a counter electrode and Ag/AgCl electrode as a reference. The OCP procedure was done for 15 minutes before starting the experiment.

3. Results and discussion

3.1 Differential Scanning Calorimetry (DSC).

DSC test result for Al-5Zn-0.5Cu-0.5Nd showed in Figure 1. The result showed the cooling process from temperature 670^{0}C to 450^{0}C. The cooling speed of this solidification process is 10°C / min. Fig 1 showed the starting point of the formation of the α-Al phase at 622◦C. Then, a great release of energy happened until temperature 620^{0}C (2), so the formation of the α-Al phase occurs in this temperature. If we compare with the Al 5Zn 0.5Cu result, the α-Al phase solidification is slightly shifting from 625 to 622 ^{0}C.

3.2 Optical Microscope

Microstructural observations were performed using Zeiss Primotech optical microscope on Al-5Zn-0,5Cu material with variations of 0.1 Nd, 0.3 Nd, and 0.5 Nd addition. The sample is an as-cast sample. . Microstructures of Al-5Zn-0,5Cu alloys, Al-5Zn-0,5Cu-0.1Nd, Al-5Zn-0,5Cu-0.3Nd, and Al-5Zn-0,5Cu-0,5Nd is shown in Figures 2. Based on the microstructural observations in Figure 2, clear grain boundaries with lighter colors than their matrices can be observed. The precipitate may also be visible in white on the grain boundary. 0.5 w.t% Nd addition had more precipitates at grain boundaries than neodymium addition of 0.1 and 0.3 w.t%.This phenomena suggests that the formation of precipitates

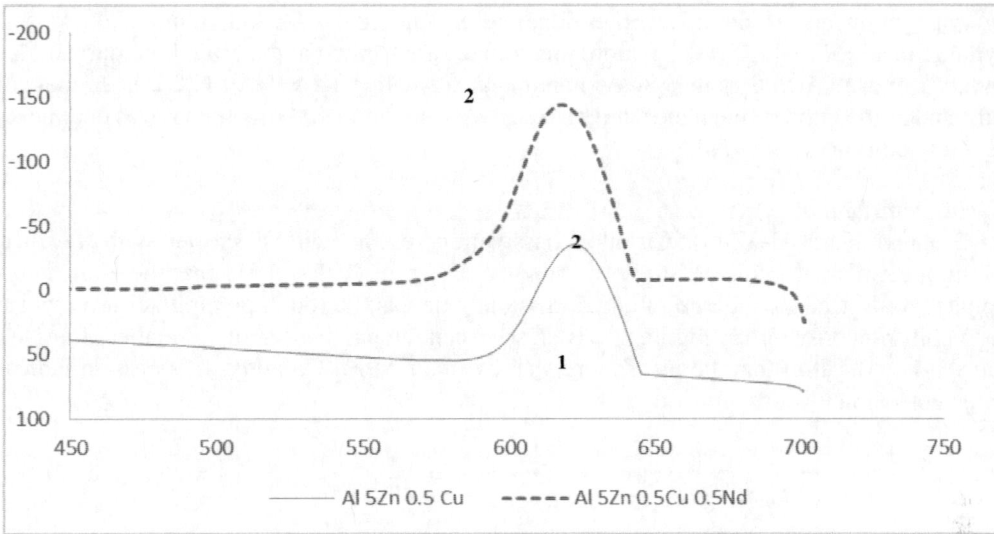

Figure 1. A slightly shifting of α-Al phase peak.

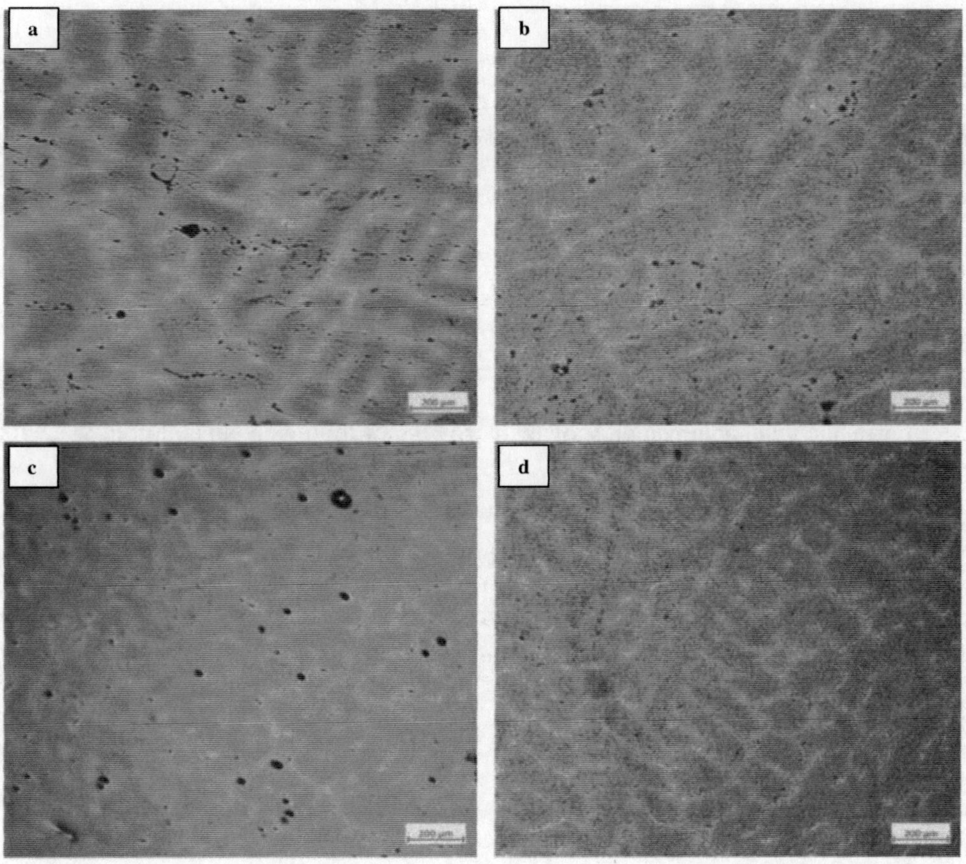

Figure 2. As-cast microstructure of a) Al-5Zn-0,5Cu, b) Al-5Zn-0,5Cu-0,1Nd, c) Al-5Zn-0,5Cu-0,3Nd, and d) Al-5Zn-0,5Cu-0,5Nd (Magnification 50x).

at the grain boundary is directly proportional to the number of neodymium additions, as more neodymium is added to the alloy, the more precipitates are formed at the grain boundaries. The size of Secondary Dendrite Arm Spacing (SDAS) area is also smaller 0.5 wt% of Nd. It is an indication that the Nd acted as the grain refinement. The 0% Nd SDAS is around 200 μm, then it was decline to around 65μm after addition of 0.5% Nd.

3.3 Cyclic Polarization
Figure 3 showed the Al-5Zn-0,5Cu alloy passivation region will be shorter with the increase of neodymium element alloying. A study by Junguan He et al. [10] showed that the elongated form of precipitate would tend to undergo pitting corrosion, whereas the round precipitates are more likely to undergo uniform corrosion. In Figure 3, it is clearly seen the passive region is smaller after Neodinium addition. It is indication that the pit is more shallow than Al Zn Cu alloy. It is also an indication Nd make the corrosion is more uniform

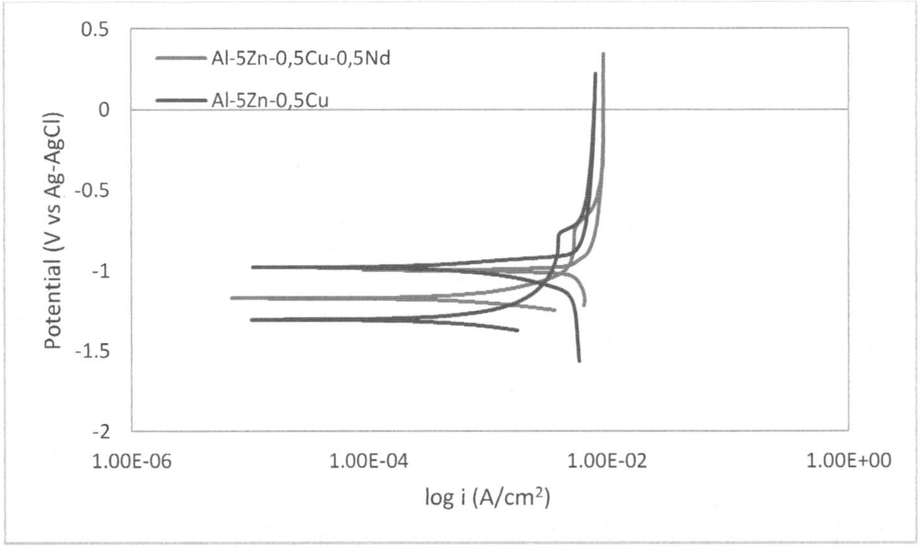

Figure 3. Cyclic polarization showed narrowing of loop with Nd addition.

4. Conclusion
The formation of precipitates at the grain boundaries is directly proportional to the number of neodymium additions, the more neodymium added to the alloy, the more precipitates formed at the grain boundaries. The function of adding neodymium elements is as a grain refiner. The effect of the addition of neodymium elements as much as 0.1%, 0.3%, and 0.5% decreases the length of SDAS. The addition of neodymium decreased pitting corrosion resistance and make the loop smaller. It is a good indication for the shallower pit which is more beneficial for sacrificial anode

Acknowledgment
We would like to give our thanks and appreciation to Hibah PITTA 2018 Nomor: 2413/UN2.R3.1/ HKP.05.00/2018 for supporting the funding in this research

References
[1] J. Pautasso, B. Victor, H. L. E. Guyader, and V. Debout, "Low Voltage Cathodic Protection For High Strength Steel Part I - Definition of A New Aluminum Galvanic Anode Material," *NACE Int.*, no. 725.

[2] DNV, "Cathodic Protection Design," *Rev. Perform. high strength steels used offshore*, no. June, 2017.

[3] S. M. A. Shibli, B. Jabeera, and R. Manu, "Development of high performance aluminium alloy sacrificial anodes reinforced with metal oxides," *Mater. Lett.*, vol. 61, no. 14–15, pp. 3000–3004, 2007.

[4] T.-M. Tsai, "Protection of Steel Using Aluminum Sacrificial Anodes in Artificial Seawater." p. 21, 1996.

[5] S. M. A. Shibli and V. S. Gireesh, "Activation of aluminium alloy sacrificial anodes by selenium," *Corros. Sci.*, vol. 47, no. 8, pp. 2091–2097, 2005.

[6] A. Muazu and S. . Yaro, "Effects of Zinc Addition on the Performance of Aluminium as Sacrificial Anode in Seawater," *J. Miner. Mater. Charact. Eng.*, vol. 10, no. 2, pp. 185–198, 2011.

[7] X. Ren, L. Peng, G. Huang, R. Zong, H. Xie, and J. Li, "Effect of Nd on microstructure and properties of 2A70 alloy," *J. Alloys Compd.*, vol. 731, no. 2, pp. 1014–1021, 2018.

[8] Z. Cao, G. Kong, C. Che, and Y. Wang, "Influence of Nd addition on the corrosion behavior of Zn-5%Al alloy in 3.5wt.% NaCl solution," *Appl. Surf. Sci.*, vol. 426, pp. 67–76, 2017.

[9] M. Su *et al.*, "Al-Nd intermetallic phase stability and its effects on mechanical properties and corrosion resistance of HPDC Mg-4Al-4Nd-0.2Mn alloy," *J. Alloys Compd.*, vol. 691, pp. 634–643, 2017.

[10] J. He, J. Wen, and X. Li, "Effects of precipitates on the electrochemical performance of Al sacrificial anode," *Corros. Sci.*, vol. 53, no. 5, pp. 1948–1953, 2011.

Characteristics of Products of Thermo Reactive Deposition Surface Treatment on JIS-SUJ2 Steel using Fe-Cr Coating Powder with Process Temperature Variation

Muhammad Panji Wiguna[1], Myrna Ariati[1*]

[1]Department of Metallurgical and Materials Engineering, Faculty of Engineering Universitas Indonesia, Kampus Baru UI Depok 16424, Indonesia

myrna@metal.ui.ac.id

Abstract. In this study the effect of temperature variation of the thermo-reactive deposition (TRD) process on carbide layer formation on SUJ2 steel substrates by pack cementation using Ferrochromium powder as carbide former was studied. This process was carried out on SUJ2 steel substrates at temperatures of 900, 980 and 1060 °C for 6 hours. The effect of temperature on layer thickness, homogenity and hardness was studied. Results show that the higher the temperature the thicker the layer formed on the substrate surfaces. Furthermore the XRD results show that the layer formed by this process is chromium carbide. The average microhardness of the layer for 3 process temperature variations is around 1750 HV while the wear rate is around 7 x 10^{-4} mm^3 / m.

Keywords: Thermo reactive deposition, SUJ2 steel, chromium carbide

1. Introduction

Thin hard carbide coating deposition is one common surface treatment applied to steels to improve life time of tools, dies, and other mechanical parts. As of now, three methods of producing hard carbide are widely used: chemical vapor deposition (CVD), physical vapor deposition (PVD) and thermo-reactive diffusion (TRD) [1]. CVD and PVD, however, have a number of drawbacks, namely high plant investment costs and the requirement of a vacuum or very controlled atmosphere. To obtain the same surface hardness as the CVD and PVD process the TRD process could be an easier, environment-friendly and lower-cost alternative compared to the other processes [2].

TRD process can be done on parts by three methods: molten salt bath [3,4], fluidized bed [5,6] and pack cementation [7,8]. In this process, the carbon and nitrogen in steel substrate diffuse into a deposited layer with carbide forming and nitride forming element such as vanadium, niobium, tantalum, chromium, molybdenum or tungsten. Diffused carbon or nitrogen reacts with carbide and nitride forming elements in the deposited layer to form carbide or nitride layer which is solid and with metallic bonding [9].

TRD coating process can be carried out using chromium. Chromium has good oxidation and corrosion resistance; furthermore CrC has superior hardness, strong adhesion properties and wear resistance which can replace electroplated metal as protection for the mold [5,10]. Several coatings using chromium produce good characteristics for heavy-wear applications such as high hardness and toughness, high Young modulus, and high melting temperatures suitable for applications at relatively high temperatures [5,10]. The aim of this study is to produce hard carbide coating on substrate steel by pack cementation. Ferrochromium powder is used to form chromium carbide layers in SUJ2 steel.

2. Experimental method

In this study the coating process is carried out on JIS-SUJ2 steel substrates, with a sample diameter of 1.00 mm and a height of 6.07 mm. The chemical composition of SUJ2 steel is shown in Table 1, all samples were prepared.

Table 1. Chemical composition of SUJ2 Steel (wt %).

Element	**Fe**	**C**	**Si**	**Mn**	**Ni**	**Cr**
Composition	Bal	1.03	0.24	0.28	0.06	1.68

The mass ratio of steel and powder used was 1:1 where the mass of SUJ2 steel is 300 grams. In addition to carbide-forming powders, Al_2O_3 powder as an anti-sintering agent and NH_4Cl as a catalyst was respectively added 37% and 3% of the amount of powder used. The TRD process was carried out by the cementation pack method in a powder mixture consisting of 180g ferrochromium, 111g Al_2O_3, and 9g NH_4Cl on Table 2. The process was carried out in a furnace for 6 hours at temperatures of 900, 980 and 1060 °C.

Table 2. Powder Coating Composition.

Material	**Percentage (%)**	**Amount (300 g x percentage)**
Al_2O_3	37%	111 grams
NH_4Cl	3%	9 grams
FeCr	60%	180 grams

The cross-section of the carbide coating produced on the sample surface was observed, and layer thickness measured, by optical microscopy, using Nital (2% HNO3 and 98% alcohol) as etchant. Vickers microhardness testing was performed to measure coating layer microhardness with load of 100g and holding time of 10 s.

Wear resistance testing of the coated SUJ2 steel samples was carried out using Ogoshi high speed universal wear testing machine with sliding distance of 400m, load of 120 N and 2.38 m/s sliding speed. The revolving disk used for testing had a radius and thickness of 14.87 and 3.53 mm, respectively. Identification of phases formed on the coating was carried out by XRD analysis with with 2θ varying from 10° to 100°, while identification of chemical composition was carried out by EDS linescan in conjunction with SEM.

3. Results and discussion

3.1. The influence of treating temperature on the formation of carbide layer

The effect of treatment temperature on layer thickness is shown in Figure 1. This shows that the average thickness of the carbide layer increases with an increase in treatment temperature from 900 to 1060°C. The correlation can be expressed in the Arrhenius equation:

$$\frac{d^2}{t} = K = K_0 exp\left(\frac{-Q}{RT}\right) \quad (1)$$

where d represents layer thickness (m), t is time (s), K_0 is the pre-exponential constant, K is the diffusion coefficient (m / s^2), Q is the activation energy in joules and R is the ideal gas constant (8.314 J / mol K).

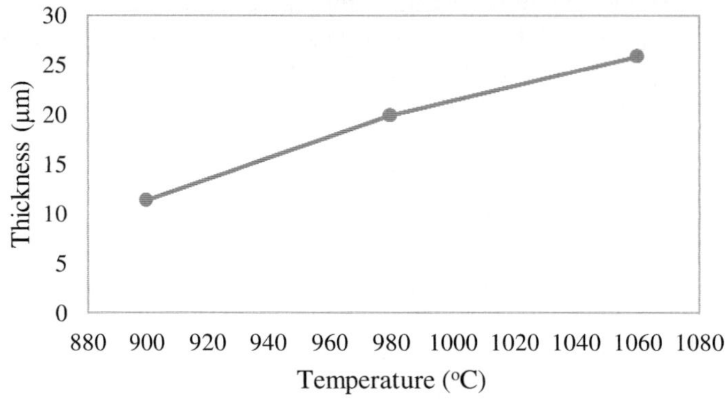

Figure 1. Influence of Temperature on Thickness.

As can be seen from the equation, changes in temperature theoretically will affect the rate of diffusion of carbon and carbide elements [3,5]. The increase in temperature will accelerate the reaction of the carbide layer formation. As a result, the thickness of the formed coating layer increased with the increase of treatment temperature.

3.2. Influence of Temperature on Hardness

The effect of temperature on the average hardness of the formed layer is shown in Figure 2. From the figure it can be seen that temperature does not have a significant effect on the hardness of the resulting layer. In this study, the hardness for samples treated at temperatures 900, 980 and 1060 °C are 1746.8, 1751.6 and 1753.4 HV, respectively.

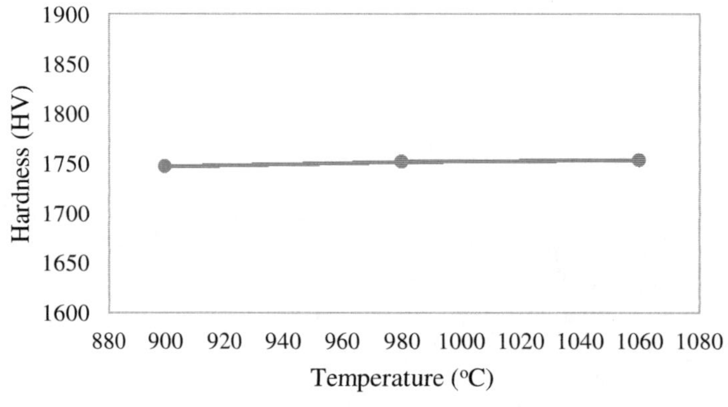

Figure 2. Influence of Temperature on layer Hardness.

3.3. Characterization of Carbide Layer

In Figure 3, the results of optical microscopy on SUJ2 steel cross section coated with Fe-Cr powder at 980°C for 6 hours are shown. There is a clear difference between the substrate and the surface carbide coating without the formation of an interdiffusion layer. The same results were obtained by X.S Fan *et al.* where in the study no interdiffusion layers were formed between the substrate and the formed coating. However, the aforementioned study used a vanadium-based carbide former, V_2O_5, to form the carbide layer and the lack of the interdiffusion layer formation was caused by the low solubility of vanadium and carbon in austenite at the TRD process temperature [4].

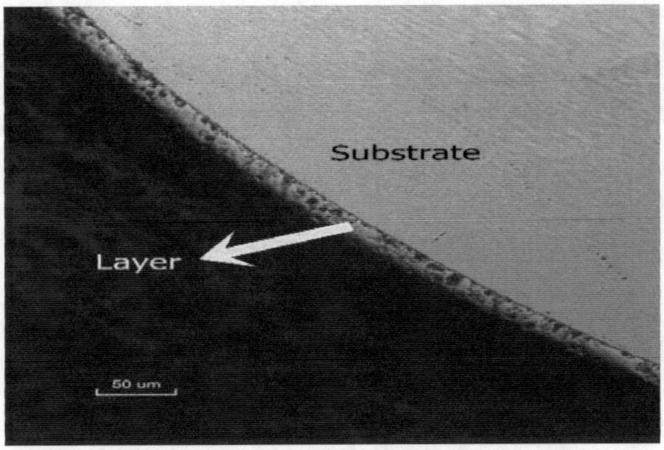

Figure 3. Cross section optical micrograph.

Figure 4 compares thickness of carbide coatings under different process temperatures. The linescan results further show the diffusion capability of chromium based carbide former with respect to the process temperature. The higher the process temperature, the faster the speed of diffusion to form carbides from this element as shown in Figure 5. This can be seen from the CPS of the element which increases as process temperature increases.

Furthermore, based on calculation of elemental increase using EDS-linescan, in Figure 5 it is shown that Fe content from the layer to the substrate tends to increase. Conversely, the chromium content tends to decrease. Meanwhile chromium content has increased significantly in the substrate/coating interface area.

Figure 6 shows the result of the carbide layer compound chemical composition analysis with a temperature of 1060°C. It can be seen that the compounds found in the layer consist of chromium carbides. The chromium carbides detected are $Cr_{23}C_6$, Cr_7C_3, and Cr_3C_2.

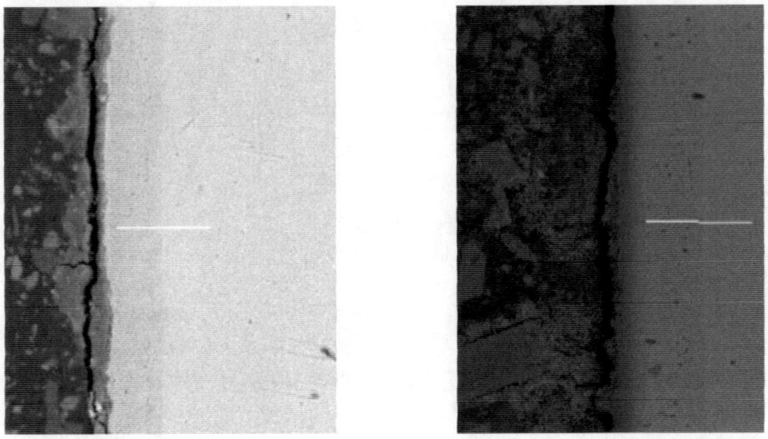

(a) (b)

Figure 4. Cross-section of sample at: (a) 900°C and (b) 980°C.

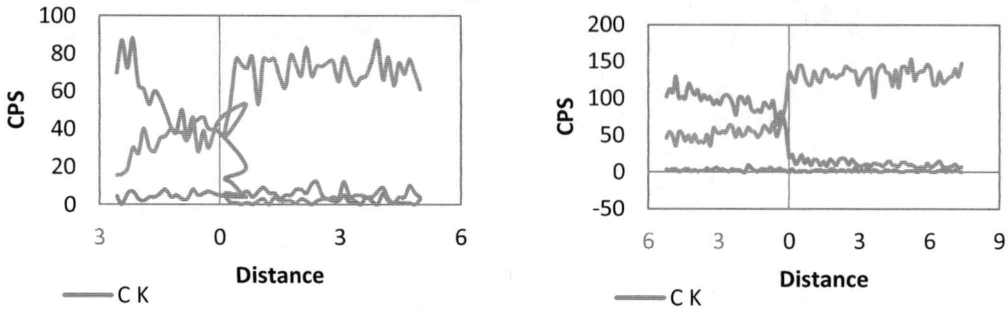

(a) (b)

Figure 5. Linescan analysis of samples: (a) 900°C and (b) 980°C.

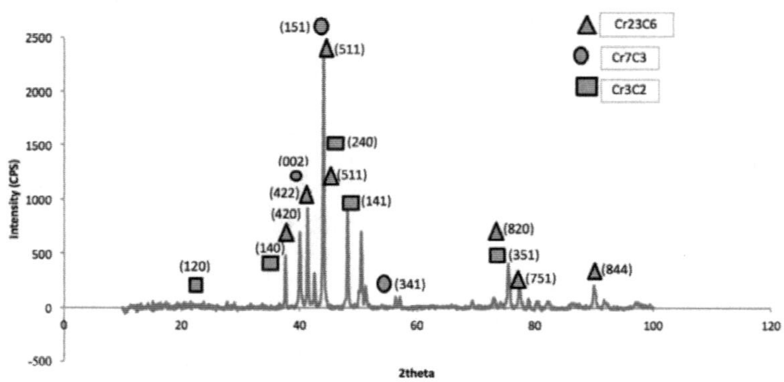

Figure 6. XRD Pattern for 1060°C process.

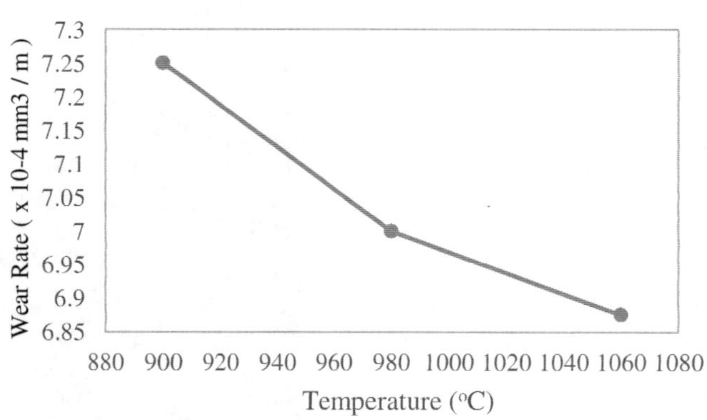

Figure 7 Influence of Temperature on wear rate.

3.4. The effect of temperature on wear properties

Figure 7 shows the linear relationship between the wear rate and the TRD process temperature. The higher the process temperature, the better the wear resistance of the material. At process temperature of 900°C, the wear rate as measured is 7.25 x 10^{-4} mm^3 / m. The wear rate decreases at process temperatures of 980 and 1060°C with a wear rate of 7 x 10^{-4} mm^3 / m and 6.875 10^{-4} mm^3 / m. This corresponds to the hardness values measured on each TRD process temperature.

The highest wear rate is seen in the layer formed at 900°C TRD process temperature with an average hardness of 1587.6 HV. Meanwhile the lowest wear rate is seen at 1060°C, which has the highest hardness value among the other samples. Thus, it can be concluded that the wear rate is inversely proportional to the average hardness value.

4. Conclusion

The coating of chromium carbide to SUJ2 steel substrate by means of pack cementation process performed in this study may be summarized as follows: wear rate is inversely proportional with the average hardness value from the experiment; process temperature has no significant effect on the average hardness where hardness values obtained tend to be similar at 1746.8 HV at 900°C, 1751.6 HV at 980°C and 1753.4 HV at 1060°C. This also impacts wear rate which shows a similar trend in all three process variations at approximately 7 x 10^{-4} mm^3/m. The higher the process temperature, the thicker the carbide layer formed, where increasing temperature speeds up diffusion where the average thickness obtained at temperatures of 900, 980 and 1060 °C are 11.35 19.89 and 25.86 μm, respectively. With FeCr as carbide former, compounds found in the layer consisted of chromium carbides ($Cr_{23}C_6$, Cr_7C_3 and Cr_3C_2).

Acknowledgment

The authors would like express their gratitude to the Ministry of Higher Education of Republic of Indonesia as well as the Directorate of Research and Public Services of Universitas Indonesia for financial support and project administration, respectively, to conduct this research under the PITTA Project Grant 2018. The authors are also grateful to PT Federal Superior Chain for their support to this study, in particular, PT Federal Superior Chain's assistance in the Thermo Reactive Deposition process and equipment is sincerely acknowledged.

References

[1] F. Chen, P. Lee and Y. M.Chyi, "Thermal reactive deposition coating of chromium carbide on die steel in a fluidized bed furnace," Materials Chemistry and Physics, vol. 53, pp. 19-27, 1998.

[2] M. Biesuz and V. M. Sglavo, "Chromium and Vanadium Carbide and Nitride Coatings Obtained by TRD Techniques on UNI 42CrMoS4 (AISI 4140) Steel," Surface & Coatings Tecnology, p. 8, 2015.

[3] G. Khalaj, A. Nazari, S. Khoie, M.J. Khalaj and H. Pouraliakbar, "Chromium carbonitride coating produced on DIN 1.2210 steel by thermo-reactive deposition technique: Thermodynamics, kinetics and modeling," *Surface & Coatings Technology,* vol. 225, pp. 1-10, 2013.

[4] X. Fan, Z. Yang, C. Zhang, Y. Zhang and H. Che, "Evaluation of vanadium carbide coatings on AISI H13 obtained by thermo-reactive deposition/diffusion technique," *Surface & Coatings Technology,* vol. 2015, pp. 641-646, 2010.

[5] F. Chen, P. Lee and M. Yeh, "Thermal reactive deposition coating of chromium carbide on die steel in a fluidized bed furnace," *Material Chemistrys and Physics,* pp. 19-27, 1998.

[6] P. King, A. Brownrigg, J. Long and R. Reynoldson, "Fluidized bed CrN coating formation on prenitrocarburized plain carbon steel," *Journal of Materials Engineering and Performance,* vol. 13, pp. 431-438, 2004.

[7] S. Sen, "A study on kinetics of CrxC-coated high-chromium steel by thermo-reactive diffusion technique," *Vacuum,* vol. 79, pp. 63-70, 2005.

[8] U. Sen, "Friction and wear properties of thermo-reactive diffusion coatings against titanium nitride coated steels," *Materials and Design,* vol. 26, pp. 167-174, 2005.

[9] T. Arai and S. Harper, ASM Handbook Volume 4 Heat Treating, Ohio: ASM International, p.607, 1992.

[10] S. Sen, "Influence of Chromium Carbide Coating on Tribological Performance of Steel", pp. 63-70, Turkey, 2006

IOP Conf. Series: Materials Science and Engineering **553** (2019) 012022 doi:10.1088/1757-899X/553/1/012022

Study of Temperature and Holding Time of Second Stage in Two-stages Thermal Upgrading of Lateritic Nickel Ore

B Suharno[1*], A Sa'adah[1], A Shofi[1] and F Nurjaman[2]
[1]Department of Metallurgical and Materials Engineering, Faculty of Engineering Universitas Indonesia, Kampus Baru UI Depok 16424, Indonesia
[2]Research Unit for Mineral Technology, Indonesian Institute of Science, Lampung 35361, Indonesia

suharno@metal.ui.ac.id

Abstract. A two-stage thermal upgrading process with the addition of 10 wt.% sodium sulfate and sodium chloride followed by magnetic separation was conducted for treatment low-grade nickeliferous laterite ore to produce a ferronickel product. The aim of this experiment was to investigate the effect of reduction temperature and holding time in two-stage thermal upgrading process, which was carried out from 950°C to 1150°C for 30 to 90 minutes, into the grade and recovery of nickel also the morphology of ferronickel particle. A wet magnetic separation with 500 gauss magnet was conducted to produce ferronickel in concentrate. Increasing the reduction temperature of nickel laterite increased the nickel grade and recovery. The nickel grade and recovery also increased along with the longer holding time in the reduction process. Nevertheless, it decreased the iron recovery.

Keywords : Low-grade Nickel, Two-stage Thermal Upgrading, Ferronickel, Particle Growth, Sodium Sulfate, Sodium Chloride

1. Introduction

More than 65% of the nickel consumption is used to make stainless steel [1,2]. The demand ferronickel has been increased in the last few years due to the rapid development of stainless steel industries [3]. The continuous depletion of sulfide ore deposits forced the interest of the industry to focus on the utilization and start developing nickeliferous laterite deposits as one of the alternative raw material in nickel production [4].

Nickeliferous laterite ore was not desirable as a major source of nickel since the nickel concentration is lower than saprolitic nickel ore which is typically difficult to process by physical beneficiation methods [1]. It requires more energy intensive and complex processing to concentrate nickel from laterite ore [5]. Innovation development related to the process for increasing the nickel grade of lateritic ore was obtained by conducting a pre-treatment process of selective reduction roasting at the temperature below 1200°C using two-stage thermal upgrading mechanism and the addition of additives followed by magnetic separation [6]. A various operating condition such as temperature and holding time was performed to increase the growth of metal particle size which is favorable when magnetic separation is carried out.

The temperature influences metal enrichment on the nickel laterite ore. The high temperature increased the percentage of nickel grade in the ferronickel compound due to the wustite reduction rate simultaneously with the growth of ferronickel particles, increased to the formation dissolved nickel in taenite (γ-Fe-Ni) [7]. Longer holding time in the reduction process has a significant impact on nickel enrichment, especially at high temperature (700°C) [8,9]. In the reduction process, the percentage of nickel grade increases with the length of holding time [10].

In this study, the function of temperature and holding time in the reduction of low-grade nickel ore using 10 wt. % sodium sulfate (Na_2SO_4) and sodium chloride (NaCl) on promoting the growth of metallic ferronickel to increase the recovery and grade of nickel in the ferronickel concentrate was investigated clearly.

2. Experimental method

Nickel lateritic ore in this research is from Sulawesi, Indonesia. A 100 grams of raw materials were crushed into less than 147 μm and drying at 105°C for prior to the characterization analysis was performed. From X-Ray Fluorescence (XRF) analysis, the chemical composition of lateritic ore was 1.4 Ni - 50.5 Fe - 2.68 Cr - 16.5 Si - 4.86 Al - 1.81 Mg. The palm kernel shell was used as the reductant which contains 77% of fix carbon and 21% ash.

The drying process was conducted at 105°C into the as-received ore by using an oven for 24 h to remove the free water. The dried lateritic ore and reductant were crushed become less than 149 μm. The 50 grams of fine lateritic ore, 22 grams of reductant, and 10 wt. % of additives (Na_2SO_4 and NaCl) were thoroughly mixed for preparing pellets.

The pellets were dried at 100°C for 2 h, then it was charged into a crucible which made from graphite. The crucible was put in a muffle furnace at a certain temperature and holding time. The process consists of a pre-heating process (first stage) at an initial temperature of 500°C with holding time for 60 and 90 minutes. It was continued with the reduction process (second stage) at a final temperature of 950, 1050, and 1150°C with holding time for 30, 60, and 90 minutes.

After the reduction process, the reduced pellets were quenched in distilled water. The cooled-reduced pellets were ground to less than 74 μm using a vibrating mill before it was analyzed by using X-Ray Diffraction (XRD) to observe the transformation of mineral or phase compound after the reduction process. The magnetic separation was carried out to the finely reduced pellets by dissolved it in distilled water with 1:10 ratio and separated the concentrate and tailing using 500 gauss magnet. The iron and nickel content in concentrate and tailing was analyzed by using Inductively Coupled Plasma-Optical Emission Spectrometry (ICP-OES). A Scanning Electron Microscopy With Energy Dispersive X-Ray Spectroscopy (SEM-EDS) was performed to identify the chemical compound and to investigate its distribution after the magnetic separation process.

3. Results and discussion

3.1 Effect of Temperature on Selective Reduction Process

Figure 1 shows the result of the reduction and magnetic separation process of the lateritic pellets ore with the addition of 10 wt.% additives (Na_2SO_4 and NaCl). The reduction process of lateritic ore was conducted at 950 to 1150°C for 30 minutes. The preheating temperature was fixed at 500°C for 60 minutes. From figure 1, the Ni and Fe grades increased from 2.03% and 46.36% to 3.62% and 72.45% respectively, and the recovery of Ni and Fe also increased. The higher the temperature, the more troilite (FeS) was formed which increased the nickel grade. The reaction between Na_2SiO_3 and olivine was also more considerable which increased the liberation of nickel and iron. In addition, nickel and iron oxide chlorination process was more considerable and led to higher grades and recoveries, both for iron and nickel [11].

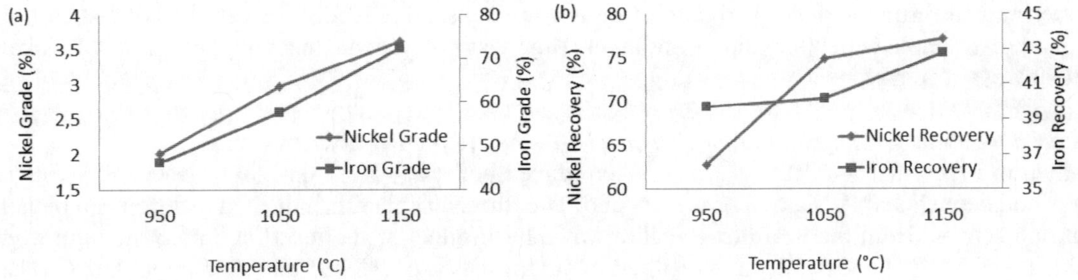

Figure 1. Effect of temperaturewith the addition of 10 wt.% Na_2SO_4 and NaCl on (a) nickel and iron grades, (b) nickel and iron recoveries.

The reduction of nickel laterite at the temperature of 1150°C resulted in highest nickel grade and recovery rather than at temperatures of 950 and 1050°C, which was 3.62% and 77.15%, respectively. The grade and recovery of Ni and Fe increased as the temperature further increased from 950 to 1150°C. It was agreed with Li et al. (2012) that increasing temperature from 800 to 1200°C in the selective reduction process with the presence of additives increased the nickel grade, iron grade, nickel and iron recovery [12]. In addition, the increase in nickel grade and recovery due to the increasing metallization of iron and nickel, more FeO and NiO are reduced to Fe and Ni metals [10]. Increasing temperatures also increase the aggregate activation of ferronickel, molecular movement of reactive gases. Increasing CO/CO_2 ratio in systems affecting Fe and Ni particle growth and reaction rate that have a direct impact on increasing nickel and iron [13,14].

3.2 Effect of Holding Time on Selective Reduction Process

The results of the reduction and magnetic separation reduced laterite pellet at reducing time from 30 to 90 minutes is shown in figure 2. The preheating temperature was fixed at 500°C for 90 minutes and the second stage temperature at 1150°C.

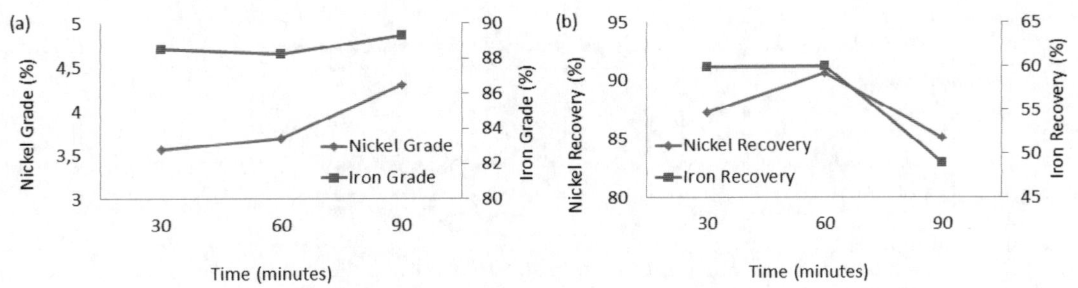

Figure 2. Effect of holding time with the addition of 10 wt.% Na_2SO_4 and NaCl on (a) nickel and iron grades, (b) nickel and iron recoveries.

The results showed that prolonged reducing periods favorable for the Ni and Fe grades. From Figure 2 (a), the Ni and Fe grades of the magnetic separation increase from 3.57% and 88.51% to 4.31% and 89.34%, respectively. The longer holding time in the reduction process resulted in the increasing concentration of reducing gas which increased the reduction of iron and nickel oxide, thus it affected the increasing of nickel and iron grades in concentrate [8]. An increasing in nickel grade from 3.69% to 4.31% at a holding time of 60 to 90 minutes could be attributed to the decreasing recovery of Fe in concentrate from 60.05% to 48.89%. The Ni and Fe recoveries initially increased to a maximum value and then decreased after 60 minutes. It could explain that with increasing the length of holding time, more FeS were formed and went into tailing when the magnetic separation process carried out. From the experimental results with the variation of holding time at the reduction process (second stage), it was found that sample with 90 minutes of reducing time was optimal due to the highest grade of nickel and the lowest of iron recovery.

3.3 The Phase Transformation of Lateritic Nickel Ore during the Reduction Process

As shown in Figure 3, the XRD patterns of the reduced pellet at 950°C that SiO_2, NaCl, FeS, FeNi, fayalite, and nepheline ($Na_2Mg_2Si_2O_7$) were found. The formed nepheline indicated that iron and nickel liberation occurred from the lizardite. Fayalite was a by-product of the reaction due to the high iron content. Trevorite ($NiFe_2O_4$) reduction resulted in the formation of FeNi at temperature of 950°C. The trevorite phase reduced to NiO and Fe_3O_4 at temperature of 900°C [15]. Magnetite was found at all temperatures. Based on the thermodynamic analysis, the magnetite reduction reaction with the solid carbon has a more negative free energy compared with the reduction of CO gas. However, at the temperature from 950 to 1150°C, the available solids carbon has been reduced due to boudouard reaction, thus the magnetite reduction was inhibited. At a temperature of 1050°C, the intensity of NaCl and SiO_2 was decreased because both of them were reacted to form HCl and Na_2SiO_3. The FeS as the result of a reaction between FeO with Na_2S and NiS was also observed at this temperature, as expressed in equation (1) and (2).

$$Na_2S_{(s)} + FeO_{(s)} + SiO_{2(s)} \rightarrow Na_2SiO_{3(s)} + FeS_{(s)} \qquad \Delta G^{\theta} = -24{,}821 \text{ Kcal} \qquad (1)$$
$$NiS_{(s)} + FeO_{(s)} + CO_{(g)} \rightarrow FeS_{(s)} + Ni_{(s)} + CO_{2(g)} \qquad \Delta G^{\theta} = -6{,}628 \text{ Kcal} \qquad (2)$$

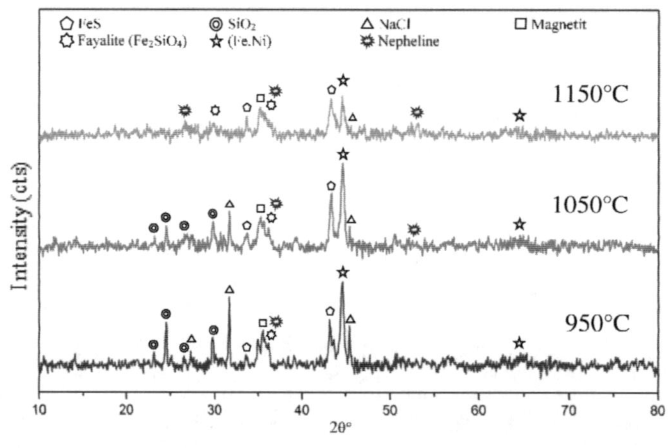

Figure 3. The XRD patterns of reduced lateritic pellets at 950, 1050 and 1150°C temperatures for 30 minutes with 10 wt.% of additives (Na_2SO_4 and NaCl).

Figure 4 shows the microstructures of a sample with temperature of 1050°C by using SEM-EDS. Figure 4 shows the size of the ferronickel grains observed at the temperature of 1050°C, which was 28.5 μm. This phenomenon was due to a grain growth mechanism from the formation of the Fe-FeS liquid phase eutectic system at temperature of 998°C. Nevertheless, the morphological analysis of grain size on ferronickel for the temperature of 950°C was not conducted due to the very small size and it was unfavorable when the magnetic separation was carried out.

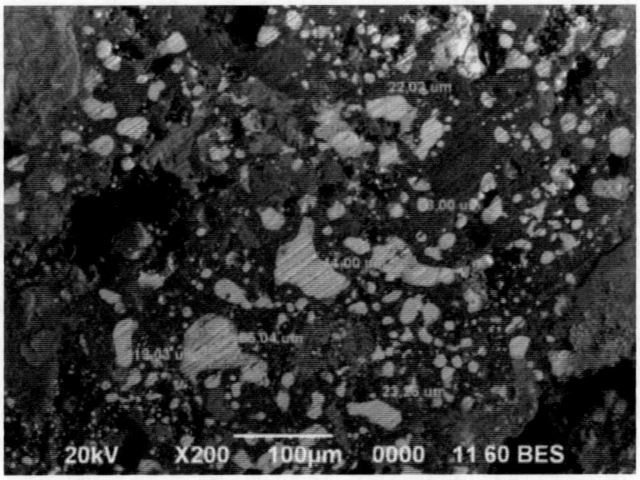

Figure 4. SEM-EDS picture of a sample with the temperature of 1050°C.

4. Conclusion

Increasing the reduction temperature improved the grade and recovery of nickel due to the maximum nickel liberation from the lizardite phase. The longer holding time increased the grade and recovery of nickel but decreased the iron recovery at the concentrate due to the increasing amount of FeS, which decreased the recovery of iron. The growth of FeNi particles increased with the higher reduction temperature.

Acknowledgment

This experiment was funded by DRPM UI as PITTA grants research (No. of contract: 2395/UN2.R3.1/HKP.05.00/2018) thus the authors would like to say thank you.

References

[1] G. M. Mudd, "Global trends and environmental issues in nickel mining: sulfides versus laterites,"*Ore Geology Reviews*, vol. 38, pp. 9–26, 2010.

[2] J. Kim, G. Dodbiba, H. Tanno, K. Okaya, S. Matsuo, T. Fujita, "Calcination of low-grade laterite for concentration of Ni by magnetic separation,"*Minerals Engineering,*vol. 23, pp. 282-288, 2010.

[3] R. R. Moskalyk, A. M. Alfantazi, "Nickel laterite processing and electrowinning practice,"*Minerals Engineering*,vol. 15, pp. 593–605, 2002.

[4] C. A. Pickles, "Microwave heating behaviour of nickeliferous limonitic laterite ores,"*Minerals Engineering*, vol. 17, pp. 775–784, 2004.

[5] J. Kyle, "Nickel Laterite Processing Technologies - Where to Next," *Murdoch University Repository*, 2010.

[6] R. Elliott, F. Rodrigues, C. A. Pickles, J. Peacey, "A Two-stage Thermal Upgrading Process for Nickeliferous Limonitic Laterite Ores,"*Canadian Metallurgical Quarterly*, 2015.

[7] D. Q. Zhu, Y. Cui, K. Vining, S. Hapugoda, J. Douglas, J Pan, G. L. Zheng, "Upgrading low nickel content laterite ores using selective reduction followed by magnetic separation,"*International Journal of Mineral Processing*, vol. 106-109, pp. 1-7, 2012.

[8] F. O'Connor, W. H. Cheung, M. Valix, "Reduction roasting of limonite ores: effect of dehydroxylation,"*International Journal of Mineral Processing*, vol. 80, pp. 88-99, 2006.

[9] J. D. Schulz and B. P. Faulkener, US Patent 4195986 (1980).

[10] B. Li, H. Wang, Y. Wei, "The reduction of nickel from low-grade nickel laterite ore using a solid-state deoxidisation methode," *Minerals Engineering*, vol. 24, pp. 1556-1562, 2011.

[11] S. Zhou, Y. Wei, B. Li, H. Wang, B. Ma, C. Wang, "Chloridization and reduction roasting of high-magnesium low-nickel oxide ore followed by magnetic separation to enrich ferronickel concentrate,"*Metallurgical and Materials Transactions B*, vol. 47B, pp. 1-9, 2015.
[12] G. Li, T. Shi, M. Rao, T. Jiang, Y. Zhang, "Beneficiation of nickeliferous laterite by reduction roasting in the presence of sodium sulfate,"*Minerals Engineering,*vol. 32, pp. 19-26, 2012.
[13] Y. J. Li, Y. S. Sun, Y. X. Han, P. Gao, "Coal-based reduction mechanism of low-grade laterite ore,"*Transactions of Nonferrous Metals Society of China*, vol. 23, pp. 3428-3433, 2013.
[14] H. Donghua, Z. Jianliang, M. Rui, C. Mingming, "Thermal behaviors and growth of reduced ferronickel particles in carbon-laterite composites,"*Rare Metals*, vol. 30, pp. 681-687, 2011.
[15] W. Mayangsari, A. B. Prasetyo, "Phase transformation of limonite nickel ores with Na_2SO_4 addition in selective reduction process,"*International Conferenceon Advanced Materials Science and Technology*, 202, 2017.

Microstructure, Phases Transformation and Hardness of Sintered Fe-Mo-Si Alloys Prepared by Powder Metallurgy Technique

Ciswandi[1,2], Rini Riastuti[1], Agus Sukarto Wismogroho[2], Bambang Hermanto[2], Toto Sudiro[2]*

[1]Department of Metallurgical and Materials Engineering, Faculty of Engineering, Universitas Indonesia, Kampus Baru UI Depok 16424, Indonesia

[2]Research Center for Physics, Indonesian Institute of Sciences, Kawasan Puspiptek Serpong, Tangerang Selatan 15310, Indonesia

cis.wandi26@gmail.com

Abstract. The transition-metal silicides are attractive materials for high-temperature applications due to its mechanical properties and resistance to oxidation-corrosion at high temperature. In this paper, the development of silicides alloy based on ferromolybdenum (Fe-Mo) and ferrosilicon (Fe-Si) lumps as starting materials were discussed. The Fe-Mo and Fe-Si lumps were manually crushed into small granules. Afterwards, the small granule of Fe-Mo and Fe-Si were separately milled into fine powder and were then mixed by using Wet-High Energy Milling technique for 4h and 2h, respectively. The powders were compacted and then sintered at 1200°C for 2h in a vacuum. The microstructure of sintered alloys was observed by using SEM-EDX. Meanwhile, the formation of silicides phases was identified by using XRD. The Vickers hardness testing was performed to measure the hardness of sintered alloys. According to the results, the $MoSi_2$ phase was detected in the (Fe-Si)-30(Fe-Mo) alloy. When the Fe-Si concentration was decreased to 40 % at, the FeMoSi phase was formed. The detail results of this study are presented and discussed in this paper.

Keywords : Fe-Mo-Si alloys; powder metallurgy; sintering; microstructure; hardness

1. Introduction

Transition-metal silicides constitute intermetallic compounds of inorganic solids have attractive physical and technical properties, such as high melting temperature (much greater than 2000°C), good oxidation and corrosion resistance, relatively low density, high electrical conductivity, high strength and excellent creep behavior at elevated temperature [1, 2]. Therefore, these materials are widely used in the different and modern engineering applications as gas turbines, combustion chamber components, missile nozzles, diesel engine glows plugs and industrial gas burners and furnaces [3].

One of the attractive transition-metal silicide materials is $MoSi_2$. It was considered as a candidate material to substitute nickel-based super-alloys for aircraft applications [4]. This is because $MoSi_2$ have moderate density and high melting-point temperature, namely 6.24 g/cm^3 and 2050°C, respectively [5]. Si element in the alloys plays a role in promoting the formation of glassy protective silica (SiO_2) scale at high temperatures. This compound will protect the alloys from the oxidizing and aggressive environments. Meanwhile, Mo element content in the alloys can prevent embrittlement caused by Oxygen and Nitrogen [1]. Moreover, $MoSi_2$ has a good ductility and toughness, resistance to corrosion and fatigue crack growth at high-temperature that is better than the nickel-based super-alloy [4]. In the previous study, $MoSi_2$ was successfully applied as a coating material on the molybdenum-based substrate using a thermal spray method [6]. On the other hand, the manufacturing $MoSi_2$ has also been successfully created by the self-propagating high-temperature service (SHS). The other processing techniques, often with two or more combination methods, including arc-melting and casting, mechanical alloying, hot pressing, reaction sintering, spark plasma sintering, combustion synthesis, and solid-state displacement reactions were used [7].

The unique of other intermetallic silicide compounds is iron disilicides ($FeSi_2$). In contrast to the $MoSi_2$, $FeSi_2$ is generally used for high-temperature electrical applications, such as thermoelectric, semiconductor and magnetic materials [8, 9]. Furthermore, the $FeSi_2$ compound have also high resistance to oxidation, low vapor pressure, lack of toxicity, lower resistivity and a constituent element that is widely available in the Earth [9]. In the previous study, the nanostructure of $FeSi_2$ material was successfully synthesized using Czochralski growth and chemical vapor transport method, powder metallurgy and a rapid eco-process [9-11].

Up to now, most researches on $FeSi_2$, $MoSi_2$ and other types of alloys still utilize high purity powder as the starting material to synthesis the aforesaid alloys. Though, the purity powder is quite expensive. Therefore, from the view point of material cost, the exploration of other sources of material is needed to fabricate silicides alloy. Ferro-lumps materials are considerably low cost material compared to high purity powder that has been commonly used in steel fabrication. Previously, our groups have utilized the lumps materials, such as ferrochrome (Fe-Cr), ferroboron (Fe-B) and ferrosilicon (Fe-Si) lumps [12-14] for coating applications. But there is no report in silicides alloys preparation using lumps materials as the starting materials.

A powder metallurgy technique consisting of powder mixing, compaction and sintering was widely used for preparing the compact alloys. Firstly, two or more kinds of powders were alloyed using a mechanical milling, either under wet or dry milling condition. During MA, the collision of balls mill and powder can reduce the powder particle size from micro to nano. It can also be used to synthesis new or amorphous phases. With the small particle size, it is expected that the powder compaction produces more dense, smooth and uniform structure. In order to avoid the sample oxidation, sintering process can be conducted in a vacuum or inert atmosphere [15].

In this study, Fe-Mo-Si alloys were developed using a powder metallurgy method using ferromolybdenum (Fe-Mo) and ferrosilicon (Fe-Si) lumps as the starting materials.

2. Experimental method

In the present study, the Fe-Mo-Si alloys were synthesized from Fe-Si and Fe-Mo lumps with the chemical composition as presented in Table 1 and 2, respectively. The raw materials were manually crushed into small granules. Afterwards, the small granule of Fe-Mo and Fe-Si were separately milled and then mixed by using Wet-High Energy Milling technique for 4h and 2h, respectively at oscillation frequency of about 700/min in order to obtain a fine powder. Based on the broadening of the XRD peaks, it was known that the crystallite size of Fe-Si and Fe-Mo powder are less than 360nm.

Table 1. Chemical composition of Ferrosilicon (Fe-Si) lumps (at%).

Fe	Si	Al
13.7	85.12	1.18

Table 2. Chemical composition of Ferromolybdenum (Fe-Mo) lumps (at%).

Fe	Mo	Si	Cu	P	S	C
49.14	45.45	3.88	0.57	0.12	0.23	0.61

The Fe-Mo-Si alloys preparation procedures were carried out according to the following steps. Firstly, the various compositions of Fe-Si and Fe-Mo powders as presented in Table 3 were mixed in the 70ml steel vial for 2h using Wet-high Energy milling technique with the ball mill (a diameter of around 5mm) and powder ratio of 5:1. Secondly, 1g of mixed powder was put in a diameter of 1cm of steel dies and compacted with a pressure of about 7metric-tons for 5 minutes. The powder compaction produces of a pellet sample with a diameter of about 1cm and a thickness of around 0.3 cm. Then, the alloys were sintered in an alumina tube furnace at 1200°C for 2 h and vacuum atmosphere. The sample heating rate was 20°C/min from room temperature to 1200°C. The sintered alloys were polished using SiC papers from #500 until #1500 grades.

Table 3. The various composition of Fe-Mo-Si powder (at%).

Samples	Fe-Si	Fe-Mo
1	100	-
2	70	30
3	40	60
4	-	100

For the density measurement, an Archimedes's principles were applied. The mass measurement was carried out using an electronic balance and calculated using the following formula

$$\rho = \frac{A}{A-B} \times \rho_o \quad (1)$$

Where, ρ is a density of sample (g/cm^3), A is mass of sample in an air (g), B is mass of sample in a liquid (g) and $\rho_o$is a density of liquid at a certain temperature (g/cm^3). In this study, aquades was used as immersing medium and measured at 30°C.

The microstructure of sintered alloys was observed by using Scanning Electron Microscope equipped with Energy Dispersive X-ray spectrometer (SEM Hitachi SU3500-EDX Horiba). The formation of silicides phases of the sintered alloys was identified by using X-ray Diffractometer (XRD SmartLab Rigaku) with Cu Kα radiation at 40kV; 30mA. The hardness of sintered alloys was measured by micro-Vickers hardness tester (Leco Microhardness Tester LM 100AT). The sample indentation was carried out using 500kgf in load for 15s.

3. Results and discussion

The obtained results in the development of Fe-Mo-Si alloys using Fe-Mo and Fe-Si lumps as starting materials can be explained as follows:

3.1 Phase composition of the Fe-Mo-Si alloys

Fig. 1 shows the X-ray diffraction patterns of Fe-Mo-Si alloys sintered at 1200°C. The results reveal that the formation of intermetallic silicides phases is depended on the alloy composition. The 100(Fe-Si) alloy is composed of Si and $FeSi_2$ phases (Fig. 1a). The present of Si phase is in good agreement with Fe-Si binary phase diagram which shows that the Si phase appears, if the Si content is higher than 65 at% [16]. With the decrease of Fe-Si content in the alloy, instead of FeMo phase, the new phases as $MoSi_2$ and FeSi are found in the (Fe-Si)-30(Fe-Mo) alloy as shown in Fig. 1b. The absence of Si phase in this alloy suggests that Fe and Mo elements were reacted with Si to form $MoSi_2$ and FeSi phases, according to the following reaction:

$$Fe + Si \rightarrow FeSi \qquad \Delta G_{1200^oC} = -76{,}77\ kJ/mol \quad (2)$$

$$Mo + 2Si \rightarrow MoSi_2 \qquad \Delta G_{1200^oC} = -177{,}03\ kJ/mol \quad (3)$$

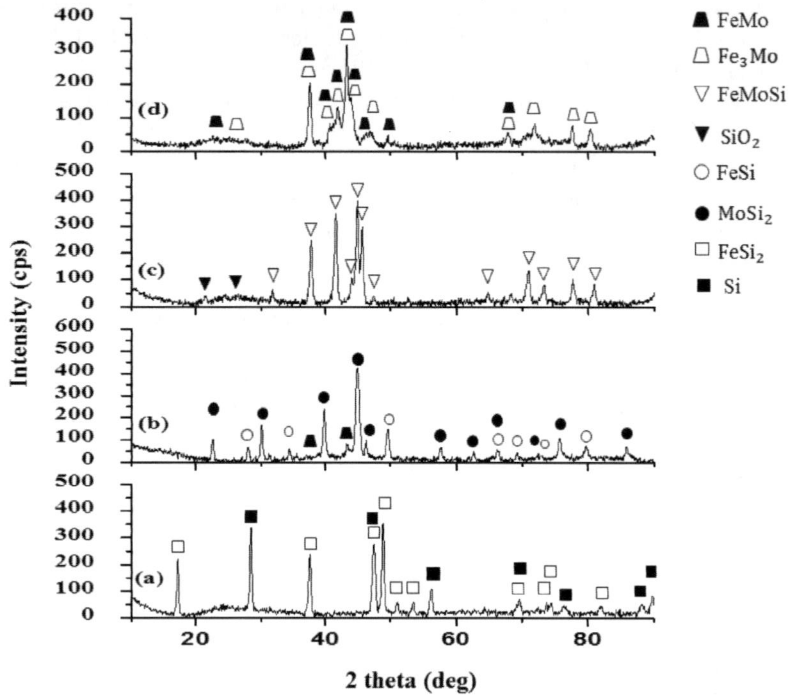

Figure 1. X-ray diffraction patterns of (a) 100(Fe-Si), (b) (Fe-Si)-30(Fe-Mo), (c) (Fe-Si)-60(Fe-Mo), and (d) 100(Fe-Mo)

It can be seen that the Gibbs free energy formation of both reactions is negative. This infers that the above reaction is liable to occur. In addition, it is interesting to note that the diffraction peaks of $FeSi_2$ phase is not observed and the $MoSi_2$ phase was detected as the main phase in the (Fe-Si)-30(Fe-Mo) alloy. Accordingly, it is reasonable to assume that the $MoSi_2$ and FeSi phases formation may also proceed through the reaction below:

$$FeSi_2 + Mo + Si \rightarrow MoSi_2 + FeSi \qquad \Delta G_{1200^oC} = -126{,}32\ kJ/mol \quad (4)$$

The Gibbs energy value of reaction 2-4 is calculated based on the entrophy and entalphy values obtained from literature data [17]. In contrast, the formed phase in the (Fe-Si)-60(Fe-Mo) alloy is totally different compared to that of 100(Fe-Si) and (Fe-Si)-30(Fe-Mo) alloys. The FeMoSi phase was detected as the main constituent of the alloy (see Fig. 1c). This evidence suggests that the lump elements was completely reacted, forming FeMoSi phase at sintering temperature of 1200°C.

$$(Fe - Si) + (Fe - Mo) \rightarrow FeMoSi \qquad at\ 1200^o C \quad (5)$$

Meanwhile, the 100(Fe-Mo) alloyis composed mainly of FeMo and Fe_2Mo_3 phases (Fig. 1d). As shown in the Experiment, the Fe-Mo lump composition contains a small concentration of Si. However, the silicides formation could not be observed. This is due to the low content of Si (around < 2 at%) in Fe-Mo lumps, as shown in Table. 1. According to Fe-Si and Si-Mo binary phases, Si element doesn't go through solubility to Fe or Mo element to form intermetallic phase when the Si content is below than 2 at% [16, 18]. Other important results that can be seen from XRD patterns in Fig. 1 are the increase of background noise at diffraction angle of around 20° to 30° in the 100(Fe-Si), (Fe-Si)-60(Fe-Mo), and 100(Fe-Mo) alloys. That is a typical reflection of amorphous phases [19, 20].

Due to the high content of Si and Mo, probably the amorphous Si and/or Mo oxides are formed in the sintered alloys.

3.2 Surface microstructure of Fe-Mo-Si alloys
In order to confirm the XRD results, the surface morphology of sintered alloys was studied using Scanning Electron Microscope (SEM) equipped with Energy Dispersive X-ray Spectrometer (EDX). The BSE comp mode of SEM images of Fe-Mo-Si alloys sintered at 1200°C for 2 h are presented in Fig. 2.

It can be seen that the alloy composition affects the typical microstructure of sintered alloys. Some areas with different contras and brightness associated with phase structure different can be observed in the Fe-Mo-Si alloys, corroborating the results of XRD Analysis. For the 100(Fe-Si) alloy, the microstructure consists of two distinguished areas: gray and dark areas. According to the results of EDX semi-quantitative analysis, the gray area of point 1 and dark gray area of point 2 are composed of (26 at % Fe, 66 at % Si) and (1.48 at % Fe, 95.35 at % Si), respectively. The Fe and Si atomic ratio of gray area is near to that of $FeSi_2$. The dark area is Si-rich which is believed to be Si phase as detected by XRD analysis (Fig. 1a).

On the contrary, the rough microstructure is seen in the (Fe-Si)-30(Fe-Mo) alloys. There are three types of microstructure in this alloy, namely light (point 3), gray (point 4) and black areas (point 5), which are composed of (29.09 at % Fe, 26.06 at % Mo, 49.85 at % Si), (45.10 at % Fe, 1.09 at % Mo, 53.80 at % Si), and (13.97 at% Fe, 7.04 at % Mo, 79.00 at % Si), respectively. Based on the aforesaid atomic ratio, the light area is suspected to be $MoSi_2$ and gray area is believed to be FeSi. While the dark area is rich in Si, composing mainly of Si phase. It can be seem that the fraction of Si phase area is remarkably decreased compared to that of in 100(Fe-Si) alloy.

As the Fe-Mo content in the alloy increases to 60%, the sintered alloys have a smoothest, dense and more uniform microstructure compared to the other alloy compositions. From the BSE comp image as shown in Fig. 2c, two different areas can be observed: gray area matrix and black precipitates. The result of EDX point analysis of point 7 in Figure. 2c indicates that the matrix is composed of 36.37 at % Fe, 29.75 at % Mo, 24.95 at % Si, and some amount of O, suspected to be FeMoSi phase, as indicated by XRD analysis. Meanwhile, the black precipitates as seen in point 6 of Figure. 2c, contains 11.46 at % Fe, 5.42 at % Mo, 19.42 at % Si, and 63 at % O which should be amorphous SiO_2 as detected by XRD measurement (see Fig. 1c).

In the 100(Fe-Mo) alloys, the microstructure presents the matrix marked by gray region (point 8). Based on the result of EDX point analysis, the gray region is consisted of 49.26 % Fe, 48.57 % Mo, and 2.17 % Si (in at%) which is probably composed of FeMo and Fe_3Mo phases in the XRD analysis. The fine pores seem to be found at the grain boundaries. This could be related to not fully sinter of the samples. Based on the results as presented in Fig. 2, the different in microstructure of the Fe-Si alloys with the increase of Fe-Mo content might be strongly affected by solubility of Mo in the Fe-Si phase.

3.3 Density and Vickers hardness of Fe-Mo-Si alloys
The density of Fe-Mo-Si alloys sintered at 1200°C for 2 h is presented in Figure. 3. Table 4 shows the literature data of density and hardness of constructed phases of sintered alloys. According to Fig. 3, the density of the alloys containing 0, 30, 60 and 100(Fe-Mo) is 3.43; 5.75; 6.82; and 8.18, respectively. Evidently, the density of the Fe-Mo-Si alloys increases almost linearly with the decrease of Si content. These results are in a good agreement with the results of a previous study [21]. In comparison to the literature data as shown in Table 4, the results of density measurement strongly suggest that the phase fraction affects the final end density of the alloy.

IOP Conf. Series: Materials Science and Engineering **553** (2019) 012023 doi:10.1088/1757-899X/553/1/012023

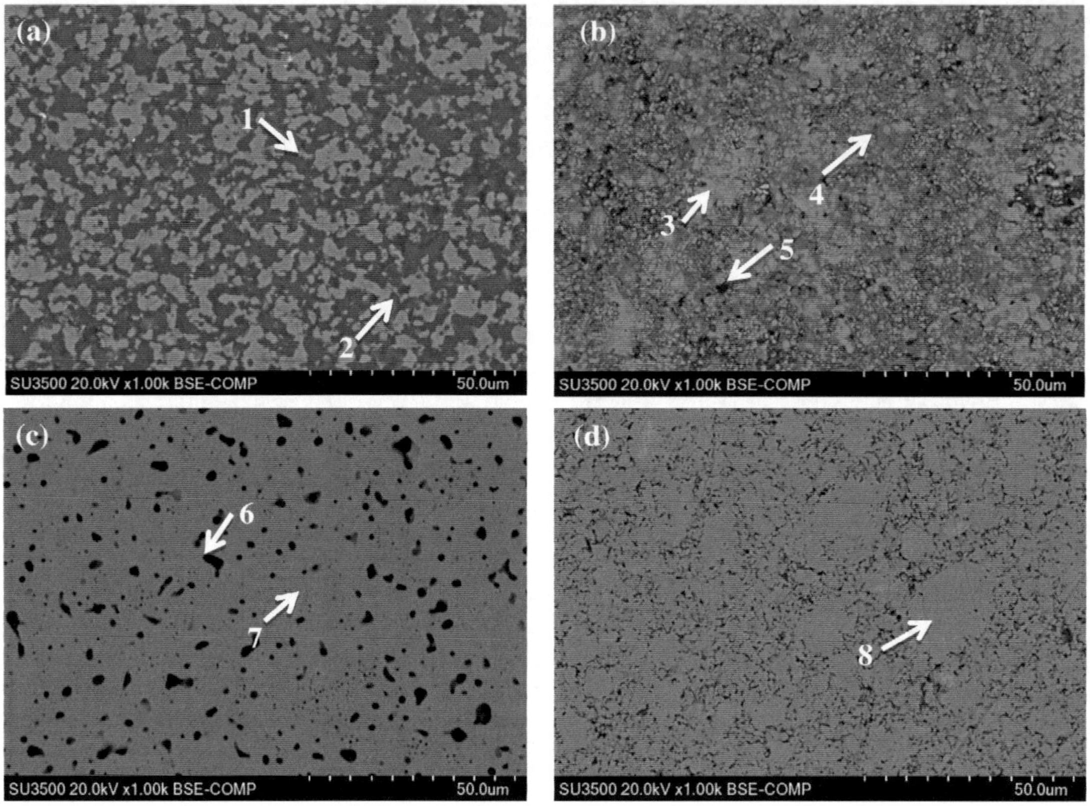

Figure 2. BSE comp images of (a) 100(Fe-Si), (b) (Fe-Si)-30(Fe-Mo), (c) (Fe-Si)-60(Fe-Mo), and (d) 100(Fe-Mo) sintered alloy.

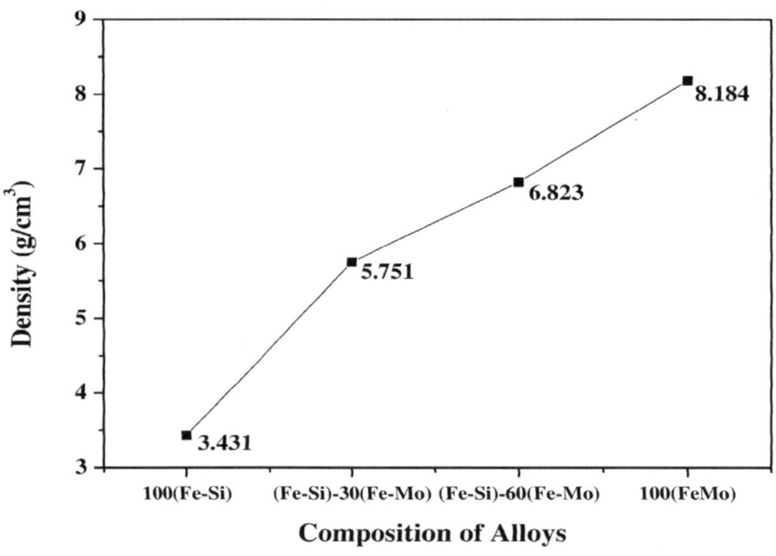

Figure 3. Density of sintered Fe-Mo-Si alloys at 1200°C for 2 h.

Table 4. Density and hardness of Fe-Mo-Siphases.

No.	Phases	Density (g/cm^3)	References	Hardness (Hv)	References
1	Si	2.328	[22]	560	[26]
2	$FeSi_2$	4.930	[23]	572.4	[26]
3	FeSi	5.100	[24]	936	[26]
4	$MoSi_2$	6.240	[5]	876.92	[27]
5	FeMoSi	Not available	-	Not available	-
6	FeMo	9.000	[25]	931.98	[28]
7	Fe_3Mo	Not available	-	Not available	-

Fig. 4 show the Vickers hardness of Fe-Mo-Si alloys sintered at 1200°C. The hardness value as presented in Fig. 4 is an average value of 5 times measurement which carried out randomly in the samples.

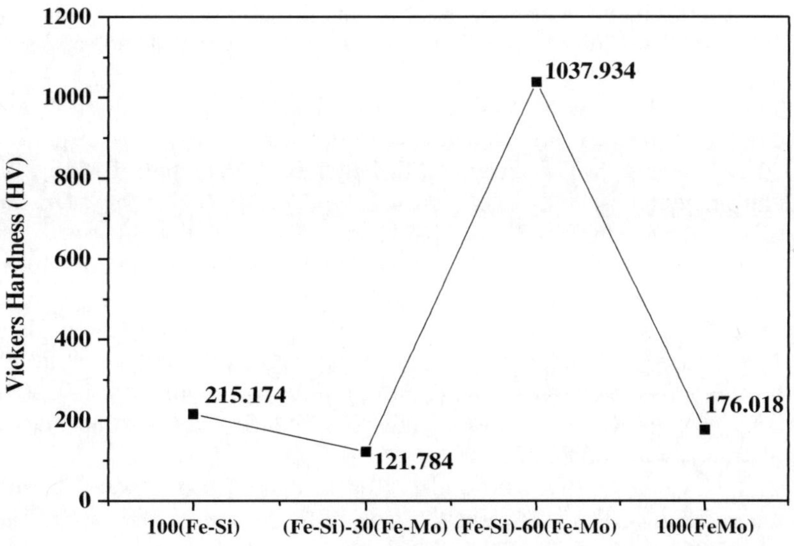

Figure 4. Vickers hardness of the sintered Fe-Mo-Si alloys at 1200°C for 2 h.

The obtained results show that the hardness of the samples is varied with the decrease of Si content in the alloy. For 100(Fe-Si) and 100(Fe-Mo) alloys, the measured hardness is 251.174 HV and 176.018 HV, respectively. While the hardness of (Fe-Si)-30(Fe-Mo) and (Fe-Si)-60(Fe-Mo) alloys is about 121.787 HV and 1037.934 HV, respectively. It can be seen that the hardness of (Fe-Si)-60(Fe-Mo) alloy is remarkably high compared to the other composition. In General, the material hardness was affected by microstructure and the phase composition of the alloys. Accordingly, a dense and relatively homogeny microstructure of (Fe-Si)-60(Fe-Mo) alloy increases the hardness value of the alloy. Furthermore, the presence of ceramic SiO_2 also increases the hardness of this composition. The present results are in good agreement with the results of a previous study [12] which found that more homogeneous and denser structure have the higher hardness. On the contrary, the rough microstructure of (FeSi)-30(FeMo) alloy leads to the lowest hardness value among all alloy composition. It is important to note that the hardness of 100(Fe-Si), 30(Fe-Mo) and 100(Fe-Mo) is smaller compared to the hardness of Si, FeSi, $FeSi_2$, $MoSi_2$ and FeMo as shown in Table 4. This indicates that the aforesaid alloys seem to be not fully sintered.

4. Conclusions
The Fe-Mo-Si alloys are successfully synthesized by powder metallurgy technique using Fe-Si and Fe-Mo lumps as starting materials at sintering temperature of 1200°C. The phase formation and surface microstructure of sintered Fe-Mo-Si alloys are depended on the alloy composition. The (Fe-Si)-60(Fe-Mo) alloy exhibits more homogenous and dense microstructure compared to the other composition, leading to the highest sample hardness.

Acknowledgements
This work was financially supported by Program Unggulan 2018 of Indonesian Institute of Sciences and Insinas RistekDikti (058/P/RPL-LIPI/INSINAS-1/III/2018) for providing characterization facilities

References
[1] F. Chu, D. J. Thoma, K. J. McClellan, and P. Peralta, "Mo_5Si_3 single crystal : physical properties and mechanical behaviour," *Journal of Materials Science and Engineering*, A26. 1999
[2]P. S. Tantri, A. K. Bhattacharya, and S. K. Ramasesha, "Synthesis and properties of $MoSi_2$ based engineering ceramics," *Proc. Indian Acad. Sci. (Chem. Sci)*, vol.113, pp. 633-649, Oct. – Dec. 2001
[3] Z. Yao, J. Stiglich, and T. S. Sudarshan, "Molybdenum silicide based materials and their properties," *Journal of Materials Engineering and Performance,* vol. 8 (3), 291-304, Jun. 1999
[4] M. C. Kushan, Y. Uzunonat, S. C. Uzgur and F. Diltemiz, " Potential of $MoSi_2$-Si_3N_4 composite for aircraft gas turbine engines," in *Recent Advances in Aircraft Technology, Dr. Ramesh Agarwal (Ed.),* ISBN: 978-51-0150-5, InTech,2 012, 98-116.
[5] G. Ouyang, P. K. Ray, M. J. Kramer, and M. Akinc, "Pressureless sintering of Mo-Si-B alloys with Fe additive," *Journal of Materials Engineering and Performance,* vol. 26 (5), 2417, May. 2017
[6] S. Weis, T. Uhlig, G. Wagner, T. Lampke, W. Bauer and A. Moldenhauer, "High-temperature corrosion and radiation characteristics of thermal sprayed molybdenum disilicide-based coatings," *IOP Conf. Series: Materials Sciences and Engineering,* vol. 118, 1-7. 2016
[7] C. –L. Yeh, and J. –A. Peng,"Combustion synthesis of $MoSi_2$-Al_2O_3 composites from thermite-based reagents," *Metals,*vol. 6, 235, 2016
[8] H. S. Kim, W. Liu, G. Chen, C. –W. Che, and Z. Ren, "Relationship between thermoelectric figure of merit and energy conversion efficiency," *Proc. The National Acad. Sci. United States of America*, vol. 112,8205-8210, Jul. 2015
[9] C. Kloc, E. Arushanov, M. Wendl, H. Hohl, U. Malang and E. Bucher," Preparation and properties of FeSi, α-FeSi, and β-FeSi single crystals, " *Journal of Alloys and Compounds,*vol. 219, 93-96, 1995
[10] H. Inoue, T. Kobayashi, M. Kato, and S. Yoneda," A low-cost production method of $FeSi_2$ powder generation thermoelectric modules, " *Journal of Electronics Materials,* Dec. 2015
[11] M. Shibuya, M. Kawata, Y. Shinohara, and M. Ohyanagi," Eco-fabrication process and thermoelectric properties of β-$FeSi_2$, " *Trans. Mat. Res. Soc. Japan,* vol. 40(3), 219-222, 2015
[12] Ciswandi, D. Aryanto, Irmaniar, A. Tjahjono, and T. Sudiro, " The effect of annealing on structure and hardness of (Fe-Cr)-50 at.% Al coatings synthesized by mechanical alloying," *AIP Conf. Proc. Series*1964, 020005-1-020005-7, 2018
[13] R. Y. Sundawa, D. Aryanto, A. S. Wismogroho, and T. Sudiro," Microstructure and phase composition of Fe-B-Al coatings on low carbon steel prepared by using mechanical alloying technique," *IOP Conf. Series : Journal of Physics: Conf. Series* 817, 2017

IOP Conf. Series: Materials Science and Engineering **553** (2019) 012023 doi:10.1088/1757-899X/553/1/012023

[14] D. Aryanto and T. Sudiro," Preparation of ferrosilicon-aluminium coating using mechanical alloying technique: study of thermal annealing on their structural characteristics, " *Journal of Surface and Coatings Technology* ,vol. 337, 35-43, Dec. 2018

[15] M. J. B Suleiman, A. M. A. B. Ahmad, R. B. Ibrahim, M. B. Mohamad, N. B. A. Kasim, M. R. B. D. A. Kadir, S. B. Muhamad, Y. Itoh, K. Hanada, and T. Shimizu," Effect of sintering conditions on mechanical properties and microstructure of Titanium alloy produced by metal injection moulding (MIM), " *Advanced Materials Research* ,vol. 686, pp. 164-169, Mar. 2013

[16] I. Ohnuma, S. Abe, S. Shimenouchi, T. Omori, R. Kainuma, and K. Ishida, "Experimental and thermodynamic studies of the Fe-Si binary system," *ISIJ International*, vol.52, pp.540-548, Dec. 2011

[17] Ihsan Barin. *Thermochemical Data of Pure Substances*, Third ed. Reading, VCH Verlagsgesellshaft mbH, Weinheim, 1995

[18] B. Paul, S. P. Chakraborty, J. Kishor, I. G. Sharma and A. K. Suri"Studies on synthesis and characterization of Mo based in situ composite by silicothermy co-reduction process, " *Metalurgical and Materials Transaction B,* Apr. 2011

[19] E. A. Okoronkwo, P. E. Imoisili, S. A. Olubayode, and S. O. O. Olusunle,"Development of silica nanoparticles from corn cob ash, " *Advances in Nanoparticles,* vol. 5, 135-139, May. 2016

[20] Y. Li, H. Yu, X. Huang, Z. Wu, and M. Chen,"A simple sythesis method to prepare a molybdenum oxide hole-transporting layer to efficient polymer solar cells, " *Royale Society of Chemistry* ,vol. 7, 7890-7900, Jan. 2017

[21] V. Kumar, H. Mehdi, and A. Kumar, "Effect of silicon content on the mechanical properties of aluminium alloy, " *International Research Journal of Engineering and Technology (IRJET)*, vol.02, 1320-1330, Jul. 2015

[22] I. J. Majhi, S. K. Sahoo, S. C. Patnaik, B. Sarangi, and A. Behera, "Effect of pouring temperature in Al-16Si-1% Al_2O_3 hypereutectic alloys," *IJSRD – International Journal for Scientific research and Development*, vol. 6, 698-700, 2018

[23] K. Takakura, and H. Ohyama, "Hole mobility of *p*-type β-$FeSi_2$ thin films grown from Si/Fe multilayers," *Journal of Applied Physics*, vol. 97,093716-1 – 093716-5, 2005

[24] Y. Shin, D. A. Tuan, Y. Hwang, T. V. Cuong, and S. Cho,"Formation and ferromagnetic properties of FeSi thin films," *Journal of Applied Physics*, vol. 113, 17C306-1 – 17C306-3, Apr. 2013

[25] J. Arvidsson, Iron and molybdenum containing pellets, European Patent Application 11190836.4 (2011)

[26] V. Milekhine, M. I. Onsoien, J. K. Solberg, and T. Skaland, "Mechanical properties of FeSi, $FeSi_2$, and Mg_2Si," *Intermetallics*, vol. 10, 743-750, Apr. 2002

[27] S. K. Ramasesha, P. S. Tantri, K. B. Anup, "$MoSi_2$ and $MoSi_2$ – based materials as structural ceramics," *Metals Materials and Processes*, vol. 52, pp.181-190, Jan. 2000

[28] Y. S. Borisov, A. L. Borisova, E. A. Astakhov, A. N. Burlachenko, Z. G. Ipatova, and V. F. Gorban, "Detonation coatings of composite powder of ferromolybdenum-silicon carbide produced using method of mechanical-and-chemical synthesis," *The Paton Welding Journal*,25-32, Mar. 2014

Reduced Remanence to Coercivity Increase Ratio of $Pr_{11,76x}Dy_xFe_{82}B_6$ (x = 0; 1; 2; 3) Permanent Magnets

Nanang Sudrajat[1,2], Wisnu A.[3)], Azwar Manaf[1*)]
[1]Departemen of Physics, Faculty of Mathematics and Natural Sciences, Universitas Indonesia Depok, 16424, Indonesia
[2] Research Center for Electronics and Telecommunications, Indonesian Institute of Sciences, Bandung 40135 Indonesia
[3]Centre for Science and Technology of Advanced Materials, Kawasan Puspiptek Serpong, Tangerang Selatan15314, Indonesia

azwar@ui.ac.id

Abstract. Dysprosium substituted Pr-Fe-B permanent magnet samples with $Pr_{11,76\text{-}x}Dy_xFe_{82}B_6$ (x = 0, 1, 2 and 3) composition were made up by compacting the crystalline fine powders prepared through mechanical milling and successive high power ultrasonic irradiation. The (Pr,Dy)-Fe-B alloys with designated compositions were obtained from arc melted the charges that consisted of respectively Dy-Fe, Fe_3B alloys and pure Fe and Pr metals. Results of the xrd analysis are permanent magnetic samples confirmed that the mass of the compound fraction formed is 90.07%. A change in lattice parameter of $Pr_2Fe_{14}B$ phase occurred when Pr was partially substituted by Dy. Furthermore, the effect of dy partial substitution on magnetic properties can increase the coercivity value, but with a decrease in saturation magnetic. Dy = 3 substitution produces coercivity and remanence values of 160 kA / m and 0.275 T respectively in the permanent magnet category.

Keywords : permanent magnets, NdFeB, coercivity, sintered magnets, magnetic properties

1. Introduction

Praseodymium Iron Boron is a magnetic alloy that promises for permanent magnet application and supporting electronic devices, this is because this material has the same crystal structure with NdFeB, also the same coercivity and remanence on room temperature [1]. NdFeB is a superior permanent magnet to replace SmCo in electric vehicle applications [2] and until recently no one compound that replaced it. NdFeB has a low currie temperature while the application of an electric motor requires high room temperature coercivity. Recently, a newly developed process called the grain boundary diffusion process (GBDP), provides a suitable solution for the problems mentioned above by placing Dy between NdFeB grains [3] . Although the saturation magnetization (Ms) and the maximum energy product ((BH)max) of Nd-Fe-B magnets decreases with increasing Dy content, Dy is usually used for the improvement of the coercivity in Nd-Fe-B magnets. [4]. In particular in the application of permanent magnet, high coercivity field (Hc) is a very basic characteristic because Hc is one of the main parameters for maximum energy product (Bhmax) besides magnetic remanence. [5] A lot of research that attempts to increase the coercivity field include adding Dy to NdFeB compound [6]. In this experiment dysprosium was substituted to PrFeB to increase the coercivity field. Samples were analyzed using X-Ray powder diffraction (XRD), surface morphology using optical microscopy, and magnetic loop hysteresis curves using Permagraph Magnet-Phisyk which were carried out at room temperature. So the purpose of this experiment is to determine the effect of phase changes and microstructure on changes in the intrinsic properties of alloyed magnetic materials $Pr_{11.76-x}Dy_xFe_{82}B_6$.

2. Experimental Method

Dysprosium is substituted for Paraesidimyum Iron Boron ($Pr_{11.76-x}Dy_xFe_{82}B_6$) with several variations of composition x = 0, 1, 2, and 3. The samples $Pr_{11.76-x}Dy_xFe_{82}B_6$ are made by Pr,Dy, Fe and B commercial ingots and melted using arc melting in the Argon environment, then the ingots are crushed and milled using planetary Ballmill for 6 hours with a revolution speed of 500 RPM and an evolution speed of 1000 RPM to get the permanent magnetic powder. After being powdered, it was compressed at a pressure of 5 tons and synthesized in a vacuum furnace 1100 ° C for 2 hours, stoichiometry calculations for the substitution variation of the day are shown in Table 1.

Table 1. Calculations of sample stoichiometry $Pr_{11.76-x}Dy_xFe_{82}B_6$.

composition	Samples name	Mass of raw materials (gram)			
x		Pr	Dy	Fe	B
0	$Pr_{11.76}Fe_{82}B_6$	3.9446	0	10.9009	0.1544
1	$Pr_{10.76}Dy_1Fe_{82}B_6$	3.5968	0.3855	10.8637	0.1538
2	$Pr_{9.76}Dy_2Fe_{82}B_6$	3.2515	0.7684	10.8267	0.1533
3	$Pr_{8.76}Dy_3Fe_{82}B_6$	2.9084	1.1487	10.7900	0.1528

Phase analysis of the samples was measured using an XRD device which was used by the Brucker brand with an X-ray Cu anode anode (λ = 1.5406 Å) with a step size of 0.01. Quantitative analysis was carried out using a general structure analysis system (GSAS) software. Surface morphology using the JEOL JED 350 SEM brand, and magnetic properties testing using a Permagraph Magnet-Physik 255 with a range of -0.6 to 0.6 tesla.

3. Results and Discussions

The X-ray diffraction pattern (XRD) of the $Pr_{11.76-x}Dy_xFe_{82}B_6$ alloy sample for x = 0 - 3 is shown in Figure 1. The results of the qualitative analysis using the Match program found that all samples consisted of the neodymium iron boron phase. The peaks of the XRD pattern of all samples are very dominant corresponding to the ref data of JCPDS 96-100-8719 forming the tetragonal structure $Pr_2Fe_{14}B$. But all samples have three phases, $Pr_2Fe_{14}B$, Fe, and Pr.

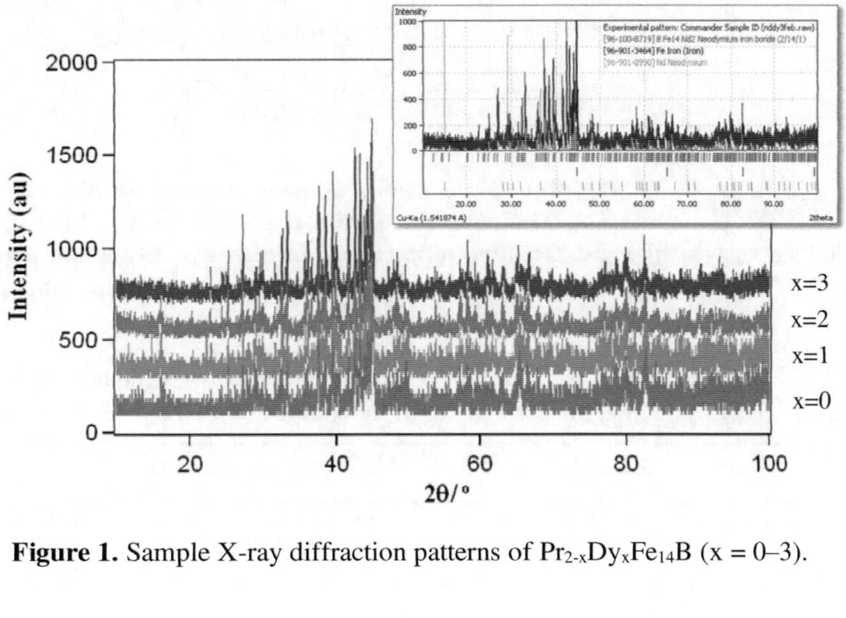

Figure 1. Sample X-ray diffraction patterns of $Pr_{2-x}Dy_xFe_{14}B$ (x = 0–3).

The result of phase identification shows that the arc melting process has succeeded in forming the dominant phase $Pr_{11.76-x}Dy_xFe_{82}B_6$ in the composition x = 0 - 3. It can be seen in Figure 1 that the greater the composition of Dy, the number of minor phase contents decreases, so further analysis is needed. determine the phase change, the amount of mass fraction formed, and the influence of Dy substitution into the Pr atom.

In Figure 2 shows the results of the improvement of X-ray diffraction patterns from the sample $Pr_{11.76-x}Dy_xFe_{82}B_6$ with variations in composition (x = 0 - 3). Figure 2 (a-d) is the result of the refinement of the XRD pattern for x = 0-3 which has formed the Bragg diffraction peak with the dominant phase following the structure of $Pr_{11.76-x}Dy_xFe_{82}B_6$. Qualitative and quantitative analysis refers to Open Database Crystallography with card numbers (JCPDS: 00-043-0002), (JCPDS: 00-033-0664), and (JCPDS: 00-037-1493) for phase $Pr_{11.76}Fe_{82}B_6$ respectively , Fe, and Pr.

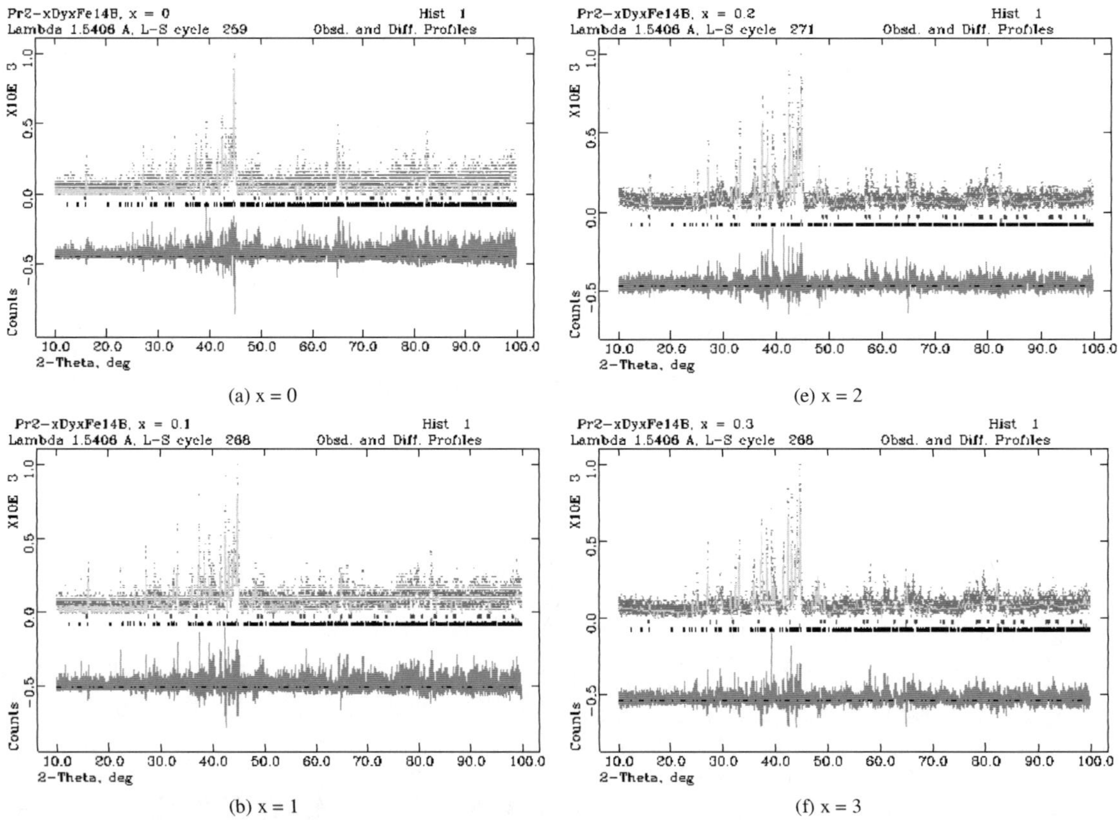

Figure 2. Results of sample X-ray diffraction pattern refinement $Pr_{11.76-x}Dy_xFe_{82}B_6$ (x = 0 - 3).

A complete summary of the results of the refinement of the X-ray diffraction pattern from the Pr11.76-xDyxFe82B6 sample with variations in composition (x = 0 - 3) for all samples is shown in Table 2.

Table 2. Structure parameter values, fit criteria (Rwp), goodness of fit (χ2) and the phase mass fraction formed in the sample $Pr_{11.76-x}Dy_xFe_{82}B_6$ with variations in composition (x = 0 - 3).

Sample	Phase	Lattice parameter (Å)			V	ρ	Fraction	R_{wp}	χ^2
(*x*)		*a*	*b*	*C*	($Å^3$)	(g/cm³)	wt%	(%)	
0	$Pr_{11.76-x}Dy_xFe_{82}B_6$	8.822(2)	8.822(2)	12.261(3)	954.4(6)	7.524	81.11	19.26	2.28
	Fe	2.8716(4)	2.8716(4)	2.8716(4)	23.68(1)	7.832	18.61		
	Pr	3.730(5)	3.730(5)	11.211(3)	135.0(5)	10.399	0.28		

	$Pr_{11.76-x}Dy_xFe_{82}B_6$	8.821(1)	8.821(1)	12.243(2)	952.7(5)	7.537	87.42		
1	Fe	2.8721(4)	2.8721(4)	2.8721(4)	23.69(1)	7.829	12.25	17.36	2.29
	Pr	3.743(4)	3.743(4)	11.103(2)	134.7(3)	10.295	0.32		
	$Pr_{11.76-x}Dy_xFe_{82}B_6$	8.815(1)	8.815(1)	12.222(1)	949.7(3)	7.561	90.72		
2	Fe	2.8721(3)	2.8721(3)	2.8721(3)	23.69(1)	7.829	9.11	9.20	1.52
	Pr	3.715(4)	3.715(4)	11.146(2)	133.2(4)	10.535	0.18		
	$Pr_{11.76-x}Dy_xFe_{82}B_6$	8.810(1)	8.810(1)	12.205(2)	947.4(3)	7.580	90.80		
3	Fe	2.8746(1)	2.8746(1)	2.8746(1)	23.75(8)	7.808	9.12	9.76	1.59
	Pr	3.706(1)	3.706(1)	11.406(7)	135.6(6)	10.695	0.72		

Figure 2 and Table 2 show that the results of refinement of X-ray diffraction patterns have very good fitting quality according to the criteria of fit (Rwp) and goodness of fit ($\chi 2$). Rwp is the ratio of the weight of the difference between the pattern of observation and calculation of XRD (ideal value of Rwp <10%). Whereas $\chi 2$ (chi-squared) is the ratio of the XRD pattern observations that are comparable to expectations.

In Figure 3 is a sample observation using a 200x magnification optical microscope and shows that the distribution of the grain boundary of the composition x = 0 - 0.3 has very good and uniform particle homogeneity across the sample surface but there is another element in the compound boundary with the estimated Fe and Pr regardless of the results of the XRD analysis. The diameter of the grain decreases relative to the addition of Dy metal.

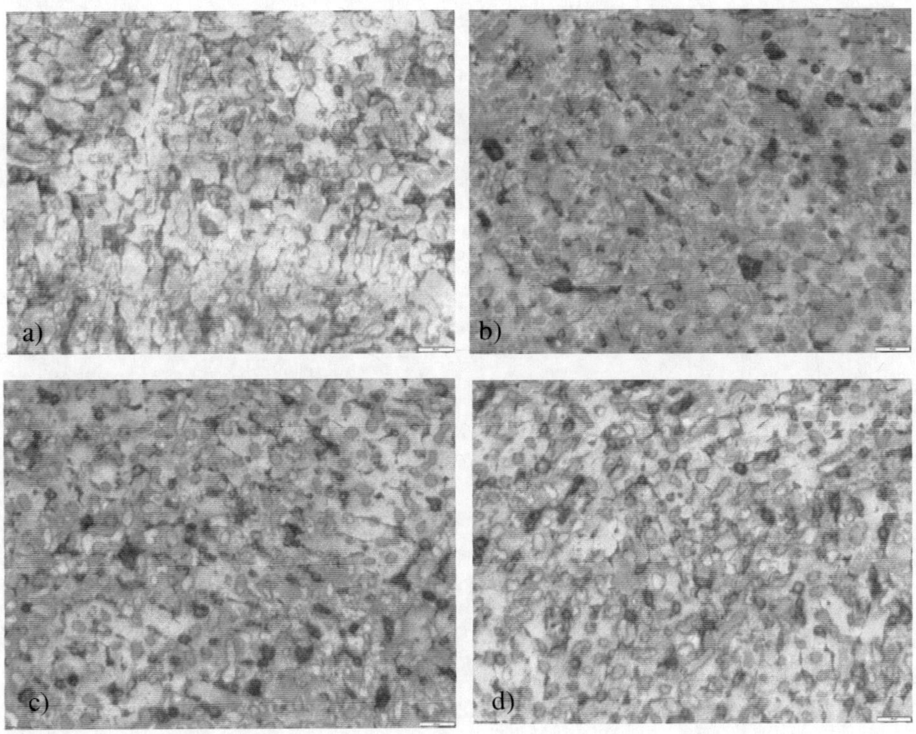

Figure 3. The morphology of the $Pr_{11.76-x}Dy_xFe_{82}B_6$ particles observed using an optical microscope a) x = 0; b) x = 1; c) x = 2; and d) x = 3.

In addition, the presence of Dy can influence the magnetic properties of this system. In Figure 4 shows the hysteresis curve measured by magnetic properties after sintering 1100^0 celcius for 2 hours using permagraph Magnet-Phisyk. All samples have a high magnetic saturation value, and the value of the remanent magnet is not less than half the value of magnetic saturation. Samples with the composition x = 4 have remanent and coercivity values in the category of permanent magnets [7], to

IOP Conf. Series: Materials Science and Engineering **553** (2019) 012024 doi:10.1088/1757-899X/553/1/012024

increase the characteristic value in the category of permanent magnet requires proper sintering and annealing process for grain growth and grain boundary [8,9].

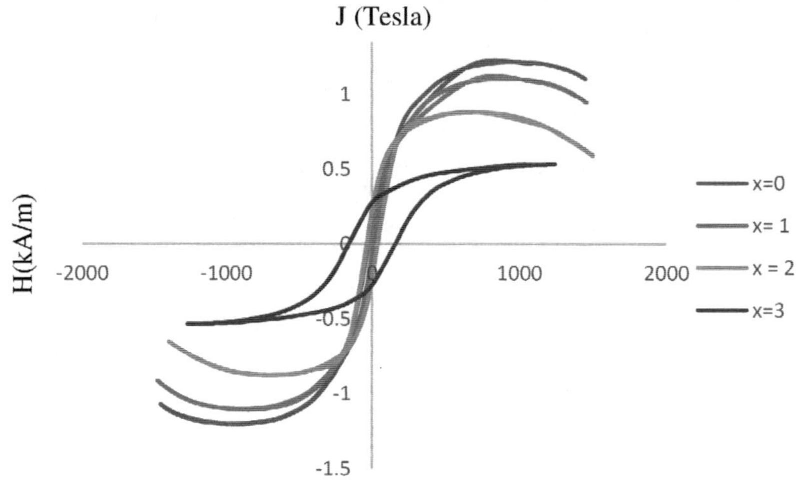

Figure 4. Hysteresis curve of the $Pr_{11.76-x}Dy_xFe_{14}B$.

The addition of Dy substitution to compount composition causes an increase of coercivity but decreases of magnetic saturation [10,11] as shown in Figure 5.

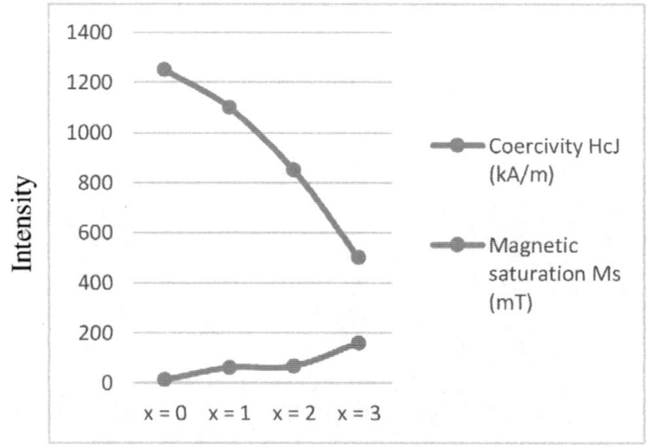

Figure 5. Ratio between saturation and coercivity.

Figure 5 shows the ratio between saturation magnetic and coercivity with the addition of Dy, the significant coercivity value increases while the saturation magnetic significantly decreases, and Dy x = 4 has reached the saturation and coercivity values of the permanent magnet category characterized by a high remanent value, also Mr = 0.275 Tesla. It can be said that the coercivity value is inversely proportional to the saturation magnet or the remanent magnet.

IOP Conf. Series: Materials Science and Engineering **553** (2019) 012024 doi:10.1088/1757-899X/553/1/012024

4. Conclussion

Successfully made by dy-sintered permanent magnet PrDyFeB by mixing the weight percent of elements of Pr, FeDy and FeB alloys to form Pr11.76-xDyxFe82B6 compound with variations x = 0-3. The results of the XRD analysis showed that there were still three phases of PrDyFeB, Fe and Pr with a weight percent fraction of about 90% PrDyFeB. All experimental results have shown high saturation magnetic values and substitution of x = 3 has been able to show the properties of permanent magnets with a remanent value of 0.275 T and a coercivity value of 160 kA / m, still requiring an annealing process to increase these values.

Acknowledgement

This experiment was conducted at the Faculty of Mathematics and Natural Sciences and is supported by the grand program PITTA Universitas Indonesia 2018.

References

[1] Y. He, X. Bao, and Q. Zhou, "Magnetic field stability of PrFeB magnets developed by GBD for cryogenic permanent magnet undulators," *J. Rare Earths*, vol. 36, no. 4, pp. 385–389, 2018.
[2] M. Sagawa, S. Fujimura, N. Togawa, H. Yamamoto, and Y. Matsuura, "New material for permanent magnets on a base of Nd and Fe (invited)," *J. Appl. Phys.*, vol. 55, no. 6, pp. 2083–2087, 1984.
[3] L. ZHOU, J. LI, X. CHENG, T. LIU, X. YU, and B. LI, "Dy gradient and coercivity in grain boundary diffusion processed Nd-Fe-B magnet," *J. Rare Earths*, vol. 35, no. 6, pp. 559–566, 2017.
[4] C. L. Harland and H. A. Davies, "Magnetic properties of melt-spun Nd-rich NdFeB alloys with Dy and Ga substitutions," *J. Alloys Compd.*, vol. 281, no. 1, pp. 37–40, 1998.
[5] S.-Y. Park, J.-H. Kim, K.-M. Lim, S.-Y. Ahn, K. R. Choi, and C.-S. Kim, "Microstructure and magnetic properties of (Nd,Dy)FeB and (Nd,Pr)FeB strips," *J. Korean Phys. Soc.*, vol. 66, no. 11, pp. 1755–1760, 2015.
[6] W. Xiaoli, Z. Lina, D. Kaihong, C. Shengli, S. Yongcong, and L. Musen, "Influence of Dysprosium Distribution on Properties of Sintered and Aged Dy-doped NdFeB Permanent Magnets," *Rare Met. Mater. Eng.*, vol. 45, no. 2, pp. 309–314, 2016.
[7] G. H. Yan, R. J. Chen, Y. Ding, S. Guo, D. Lee, and A. R. Yan, "The preparation of sintered NdFeB magnet with high-coercivity and high temperature-stability," *J. Phys. Conf. Ser.*, vol. 266, no. 1, 2011.
[8] I. A. Shah, T. Abbas, Z. Ullah, N. Hassan, and A. Rauf, "Sintering Effects on Structural and Magnetic Behaviours of NdFeB Magnets," *Armen. J. Phys.*, vol. 8, no. 4, pp. 185–190, 2015.
[9] G. P. Wang, W. Q. Liu, Y. L. Huang, S. C. Ma, and Z. C. Zhong, "Effects of sintering temperature on the mechanical properties of sintered NdFeB permanent magnets prepared by spark plasma sintering," *J. Magn. Magn. Mater.*, vol. 349, pp. 1–4, 2014.
[10] Y. TAKADA, Y. KANEKO, N. MIYAMOTO, and M. HIRAOKA, "Correlation between Dy Distribution and Coercivity of Dy - diffused NdFeB Sintered Magnets," *J. Japan Soc. Powder Powder Metall.*, vol. 60, no. 3, pp. 117–120, 2013.
[11] S. qing Hu, K. Peng, and H. Chen, "Influence of annealing temperature on the Dy diffusion process in NdFeB magnets," *J. Magn. Magn. Mater.*, vol. 426, pp. 340–346, 2017.

Synthesis and Deposition of High Crystallinity Cu_2ZnSnS_4 (CZTS) Using Dip Coating Method

Adlan Mizan[1], Fathur Rahman[2], Badrul Munir[3]*, Achmad Subhan[4]
[1 2 3*]Departement of Metallurgical and Materials, Faculty of Engineering, Universitas Indonesia, Depok-West Java 16424
[4]Metallurgy Departement of Lipi, Banten-West Java 15314

bmunir@ui.ac.id

Abstract. Kesterite Cu_2ZnSnS_4 (CZTS) thin film was successfully obtained using simple dip coating Synthesis of CZTS consist of mixing copper chloride, zinc chloride, tin chloride, and thiourea into a mixture of ethanol and ethanolamine to make CZTS suspension. Soda lime glass was used as a different substrate for dip coating method. Dip coating method uses 3 cycles of dip and drying at 170°C to remove ethanol and ethanolamine and then the sample was annealed at 550°C in argon atmosphere. X-ray diffraction was used to check the formation of kesterite before and after annealing. After annealing the diffraction pattern shown high crystallinity of kesterite with little secondary phase. Using Scherrer equation we can predict the crystallite size of CZTS before and after annealing. The scanning electron microscope was used to show the morphology of the CZTS before and after annealing. It is shown that although the grain was distributed homogenously a void can be seen throughout the surface. Optical properties of samples were checked before and after annealing using UV-vis Spectrometer. The result shows that dip coating method could create a good properties CZTS.

Keywords: Annealing Effect, CZTS, Dip-coating, Ethanolamine, High Crystallinity.

1. Introduction

With the recent increasing use of sustainable energy as main source of the energy, the study around photovoltaic material has significantly increased. One of the commonly researched photovoltaic material was Cu_2ZnSnS_4 (CZTS) [1]. CZTS was selected because it has a relatively good band gap of 1,4-1,5 eV, cheap production, non-toxic and the material source is abundant [2].

In the solar cell application device, CZTS was used as the absorber layer because CZTS have the absorber coefficient of 10 4 cm-1 in visible spectrum range [2]. And as n-type semiconductor like silicone but because the absorber coefficient is a lot higher than silicone the film did not require high thickness. So it reduces the cost of raw materials [3].

Many processes have been conducted to create CZTS including solid-state method [4]. But the chemical method is often selected because it creates CZTS with high purity, low cost and simple [5]. Many papers synthesized CZTS using successive ion layer adsorption reaction (SILAR) method but the complex method and require a lot of time are slowing the production of CZTS [6]. Another method that has been used was spin coating method of Cu-Zn-Sn precursor then annealed with sulfur atmosphere [7]. Spin coating method creating relatively good homogeneity of CZTS because of centrifugal energy can create evenly distributed precursor of CZT. But comes the problem of the film not sticking in the spin coater if the contact angle was too high, the film would just slip off the glass and did not create good evenly distributed CZTS [8]. In addition, although many researchers have successfully created a perfect variable for sulphurization, it is still relatively hard to control the environment of sulphurization if we did not use trial and error method.

Dip coating method offers the benefits of SILAR method with already CZTS precursor formed. Almost every precursor will stick to the substrate using dip coating method [9]. This paper will discuss the performance of CZTS using dip coating method and compared the use of annealing CZTS at 540 °C with the performance characterized by X-ray diffraction (XRD), ultraviolet-visible (UV-vis), Energy dispersive spectroscopy (EDS) and field emission scanning electron microscope (FE-SEM).

2. Method and material

The dip-coating method to create CZTS also shown in the Fig. 1. The synthesis of precursors was done by dissolving each one chloride elemental constituents in solvent and stabilizer [10]. Whereas ethanol was used as solvents and ethanolamine as a stabilizer. The volume fraction of ethanol compared to ethanolamine was 70% and 30% of the total volume respectively. The molar value of the constituent elements are and the volume of the solvent is adjusted with theoretical calculations of 0.075 M, 0,050 M, 0,050M and 0,23 M for Cu^{2+}, Sn^{2+}, Zn^{2+}, and S^{2-} concentration respectively. The optical and morphological properties of CZTS which are found in conditions of Cu poor (Cu / (Sn + Zn) <1) and Zn rich (Zn / Sn> 1) [11,12].

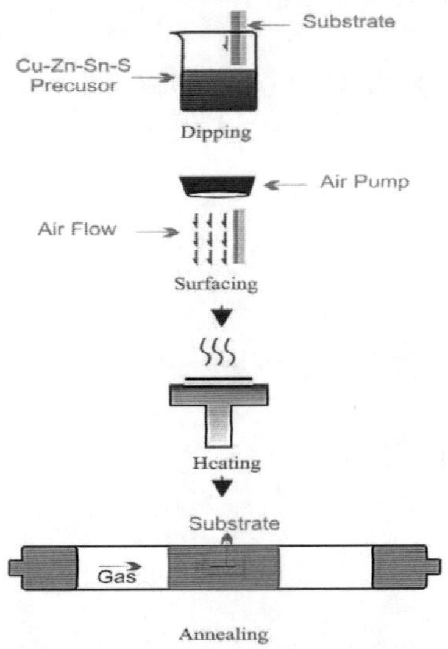

Figure 1. Dip coating process.

After the precursor was made the deposition technique that been used is dip-coating with 3 cycles dipping and drying at 150 C for 10 minutes each cycle continued with final drying at 200°C for 10 minutes for removing all ethanolamine. For each dipping, the surface was homogenized by introducing airflow vertically. Lastly, the sample was annealed at 540°C in Argon atmosphere for 15 minutes to make crystalline structure of CZTS.

The samples then characterized before and after annealing by PANanalytical X-ray diffraction with an inert atmosphere to check the diffraction pattern of samples. AMETEK scanning electron microscope and

IOP Conf. Series: Materials Science and Engineering **553** (2019) 012025 doi:10.1088/1757-899X/553/1/012025

EDS are used to check the morphology and the chemical composition of the samples. Shimadzu UV-vis under atmospheric pressure to check the optical properties of the samples.

3. Result and Discussion

3.1. Visual result

It is shown in Fig. 2 That comparison of the visual result of CZTS before and after being annealed. Fig 2a. Shown that before annealing the color of presumably CZTS is mostly black, this phenomenon happens because there is still no formation of crystalline CZTS and only form a paint-like layer.

In other hand Figure 2b. Shown the CZTS after being annealed, and shown to have a slightly different tone of black almost like grey [3]. The different gradations of color possibly change the outcome properties of CZTS.

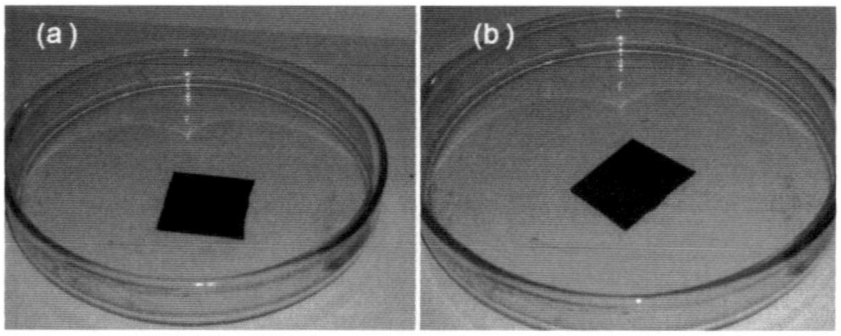

Figure 2. CZTS (a) before annealing and (b) after annealing.

3.2. Diffraction Result

Diffraction result of the CZTS was carried out using X-ray diffraction shown in Fig. 3 and Fig. 4. With comparison with 00-026-0575 reference in UCDD PDF2. Fig. 3 shown the diffraction patterns of CZTS before annealing was being used, it is shown that although the crystallinity was still very low the peak with the highest intensity already formed. This means the CZTS was there but not yet formed to be crystal kesterite structure.

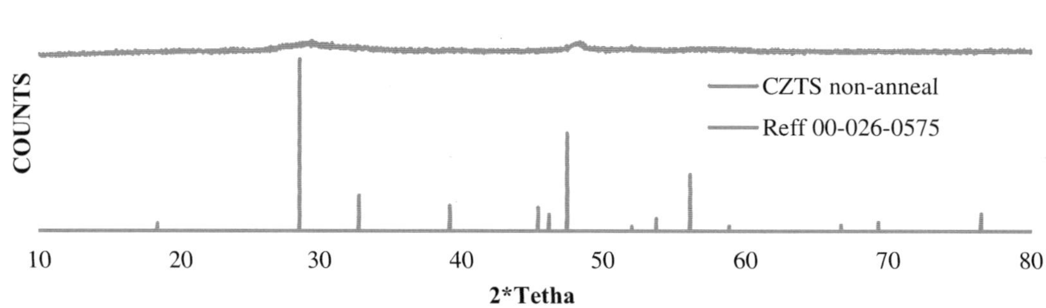

Figure 3. The diffraction pattern of non-annealed CZTS compared to reference 00-026-0575 (Kesterite).

On the other hand, the diffraction pattern of CZTS after annealing is shown in Fig. 4. It is clearly shown that crystallinity was higher compare to CZTS that did not go through the annealing process. The CZTS sample shown to have matched with the diffraction pattern of kesterite reference. This proves that CZTS samples were almost made entirely by kesterite CZTS crystal structure.

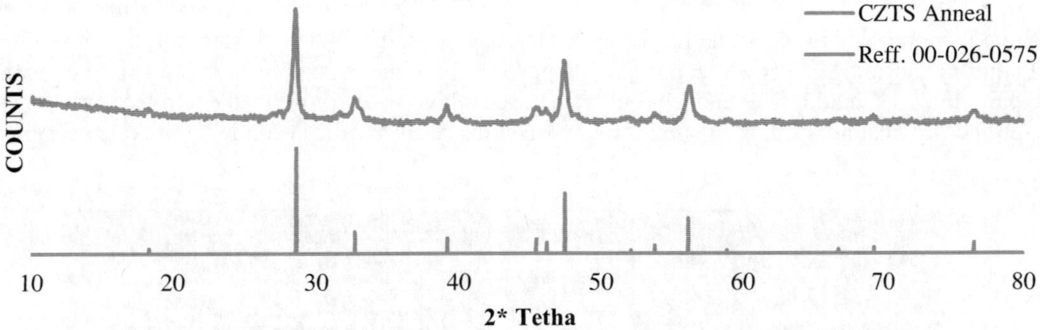

Figure 4. The diffraction pattern of annealed CZTS compared to reference 00-026-0575 (Kesterite).

These diffraction patterns hinted that the synthesis was successfully carried out and resulting in high crystallinity that almost identical to the reference [13].

3.3. *Crystallite Size*

Using X'pert highscore we can calculate the FWHM (full width at half maximum). This need to be done in order to calculate crystallite size using Scherrer equation with FWHM. Scherer equation is shown in equation 1. With K is the shape factor and the value are 0,89, λ is 1,54 Å, B is FWHM and D is the crystallite size [14].

$$D = \frac{K\lambda}{B\frac{1}{2}\cos\Theta} \quad (1)$$

All peaks of CZTS with and without annealing were labeled and calculated. The crystallite size is the average of crystallite size from each peak. Table 1 shows that the crystallite size of the annealed sample was a lot higher with almost 3-time multiplier compared to CZTS without annealing process. This concludes the successful process of CZTS synthesis.

Table 1. Crystallite size of CZTS sample

Process	D(nm)
CZTS Non-annealing	8,01
CZTS annealing	24,81

3.4. *Morphology Result*

The morphology of CZTS was then checked using scanning electron microscope to check not only the grain but also the distribution of the surface of CZTS. Distribution result will show if this process is applicable to making CZTS thin film [15]. Fig. 5 show the CZTS before and after it was annealed. Fig. 5a shows a blurry image over the substrate, this means that there is still a lot of impurities considering this process may still be left some Cl^- and some ethanolamine since it is not fully evaporated [16]. Using ImageJ we are able to obtain the average size of the particle was around 400 nm which was acceptable considering the CZTS has not yet gone through the annealed process. The homogeneity of the non-annealed sample shows that although the image was blurry we can safely assume that it is not fully distributed because there are still a lot of color tones in it means the voids can still be made.

Fig. 5b show CZTS after being annealed, the blurry part was completely removed. This means the

IOP Conf. Series: Materials Science and Engineering **553** (2019) 012025 doi:10.1088/1757-899X/553/1/012025

impurities that habit in CZTS without annealing was completely removed. From imageJ we can calculate the average ratio size of particle CZTS, it was shown that CZTS annealed have a particle size of around 700 nm which is higher from CZTS that has not yet been annealed. This phenomenon can be caused by grain growth during annealing effect which normally happens around the annealing sample [17]. The distribution of the particle was spread homogenously but we can see a hole in every bit of the surface. This means that although the dip coating method created the homogenous result it still needs to improve to eliminate the void.

Figure 5. Morphology result of CZTS (a) without annealing and (b) with annealing.

3.5. Chemical Result

The chemical result of CZTS with annealing process and without annealing process was carried out using electron dispersive spectroscopy (EDS). EDS was tested to see the confirmation on the chemical result in impurities from CZTS without annealing process and also to see the loss of impurities in CZTS after being annealed. Fig. 6 shown EDS result of CZTS before going through an annealing process and it is confirmed that CZTS before annealing have high impurities such as Chlorine, Oxygen, and Nitrogen because the hydrocarbon material from ethanolamine is not fully evaporated yet and the chlorine from the salt used to make CZTS was not yet to be removed.

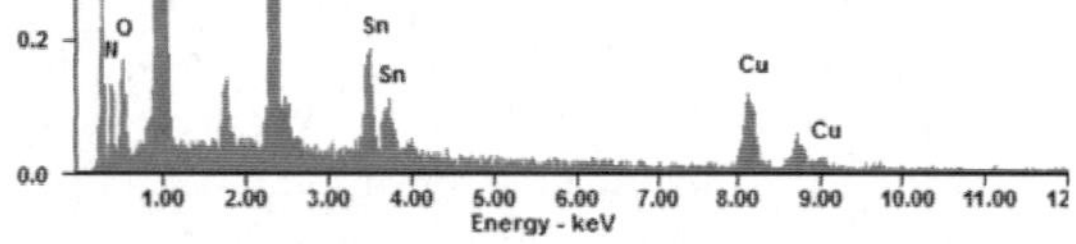

Figure 6. The chemical result of CZTS with annealing.

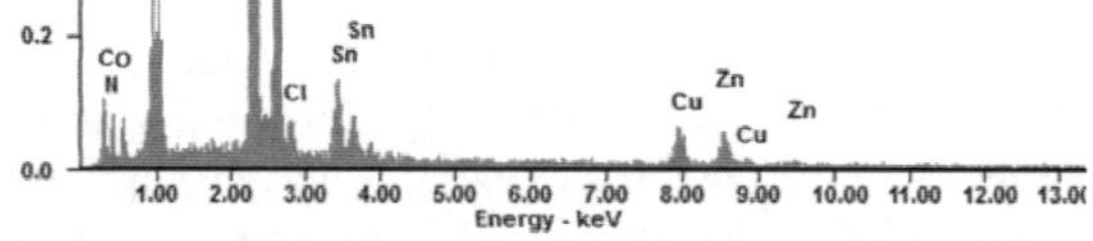

Figure 7. The chemical result of CZTS without Annealing.

CZTS without annealing was not the perfect candidate for CZTS and not yet be ready to be a semiconductor. In other hands, the chemical result of CZTS after annealing are shown in Fig. 7 are not showing the chlorine impurities at all which mean that the impurities were completely removed. This proves that the crystalline structure of CZTS was almost pure made of CZTS and that is a good indication of a good absorber CZTS.

3.6. Optical Result

The optical result of CZTS was carried out using UV-vis spectroscopy. CZTS with annealing process and without annealing process was placed in the holder and the optical properties (absorbance, transmittance, and reflectance) was calculated in the specific wavelength [18].

Fig. 8 show the comparison absorbance optical properties between CZTS with annealing and without annealing process. CZTS without annealing process shown to have almost linear absorbance from all the wavelength. This means the crystalline structure that responded to specific optical properties has yet to be formed. CZTS without annealing process was acting more like black paint compared to an absorber layer. In contrary CZTS with annealing process create specific optical properties when in contact with a specific wavelength. This phenomenon proves that the crystalline structure/semiconductor layer was already formed because it acts in a specific order.

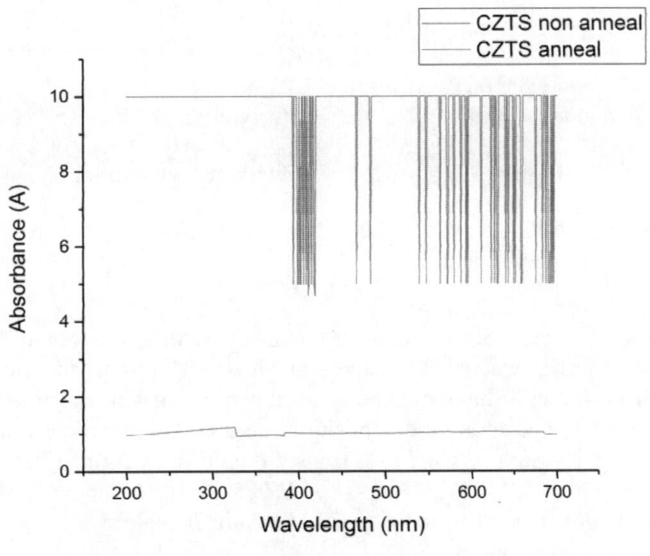

Figure 8. Absorbance optical result of CZTS.

If we calculate the average of absorbance in CZTS after annealing process we can calculate the average transmittance using beer's law in equation 2. With %T is a transmittance in percentage and A is absorbance value [19].

$$\%T = 10^{(2-A)} \tag{2}$$

Since the average absorbance in all wavelength is ~1 A it can easily be calculated that transmittance of the absorber layer was 10%. The value of this transmittance was good since we want the transmittance of the absorber layer was as low as possible. This concludes this CZTS can deliver for being absorber layer.

The other properties we can calculate for semiconductor was the band gap of the absorber layer. The band gap was used to indicate how many energy it takes to turn the semiconductor material from isolator to conductor. Fig. 9 shown the band gap of CZTS using tauc's plot. The value of band gap of this CZTS was around 1,36 eV which is still acceptable range for CZTS absorber layer in theory [20]. This proves that the CZTS semiconductor absorber layer was formed perfectly with this process.

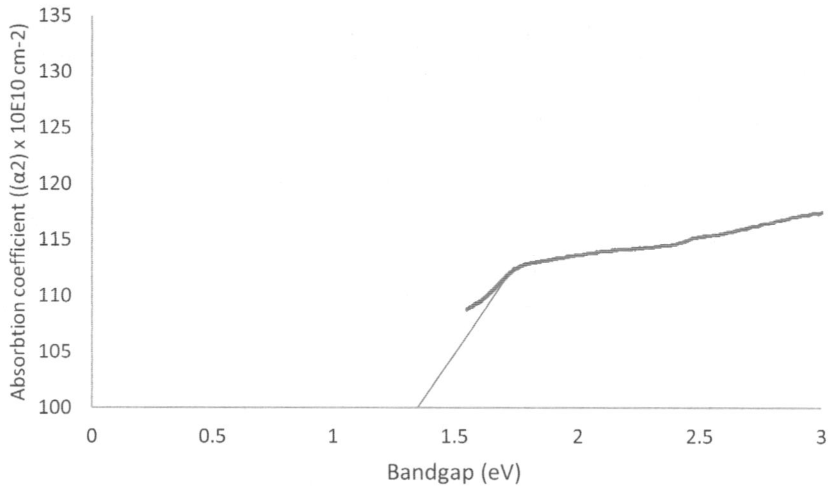

Figure 9. Tauc plot of CZTS after annealing.

Overall the characterization leading towards to the formation of CZTS was confirmed by optical, diffraction pattern, morphological and chemical.

4. Conclusion

The characterization of CZTS synthesized by using metal-ETA complex is resulted in good properties of CZTS with crystallinity kesterite structure after the annealing with almost identical peaks with reference. The dip coating method is shown to have created a homogenous structure across the surface with a little void which can be improved by perfecting the deposition method. Optical measurement resulted in proving that CZTS are present in the structure with low transmittance optical properties that indicating a good absorber layer. The band gap that has been found for CZTS using this method was 1,36 eV which later prove that this method creates a good properties CZTS. And annealing process will highly improve all properties that require to make a good CZTS.

References

[1] W. Environmental, U. Czts, and T. Film, "Solar Cell Without Environmental Pollution," *Energy Convers.*, pp. 2874–2879, 2003.

[2] H. Katagiri *et al.*, "Development of CZTS-based thin film solar cells," *Thin Solid Films*, vol. 517, no. 7, pp. 2455–2460, 2009.

[3] W. Wang *et al.*, "Device characteristics of CZTSSe thin-film solar cells with 12.6% efficiency," *Adv. Energy Mater.*, vol. 4, no. 7, 2014.

[4] V. A. Akhavan *et al.*, "Colloidal CIGS and CZTS nanocrystals: A precursor route to printed photovoltaics, "*Journal of Solid State Chemistry*, 2012, vol. 189, pp. 2–12.

[5] N. M. Shinde, P. R. Deshmukh, S. V. Patil, and C. D. Lokhande, "Aqueous chemical growth of Cu2ZnSnS4(CZTS) thin films: Air annealing and photoelectrochemical properties," *Mater. Res. Bull.*, vol. 48, no. 5, pp. 1760–1766, 2013.

[6] J. Henry, K. Mohanraj, and G. Sivakumar, "Electrical and optical properties of CZTS thin films prepared by SILAR method," *J. Asian Ceram. Soc.*, vol. 4, no. 1, pp. 81–84, 2016.

[7] S. K. Swami, A. Kumar, and V. Dutta, "Deposition of kesterite Cu2ZnSnS4 (CZTS) thin films by spin coating technique for solar cell application," *Energy Procedia*, vol. 33, pp. 198–202, 2013.

[8] A. G. Kannan, T. E. Manjulavalli, and J. Chandrasekaran, "Influence of Solvent on the Properties of CZTS Nanoparticles," *Procedia Engineering*, vol. 141, pp. 15–22, 2016.

[9] G. Rajesh, N. Muthukumarasamy, E. P. Subramaniam, S. Agilan, and D. Velauthapillai, "Synthesis of Cu2ZnSnS4thin films by dip-coating method without sulphurization," *J. Sol-Gel Sci. Technol.*, vol. 66, no. 2, pp. 288–292, 2013.

[10] P. Prabeesh, P. Saritha, I. P. Selvam, and S. N. Potty, "Fabrication of CZTS thin films by dip coating technique for solar cell applications," *Mater. Res. Bull.*, vol. 86, pp. 295–301, 2017.

[11] B. Munir, B. Eko Prastyo, D. Marta Nurjaya, E. Muslih, and S. Karim, "High crystalline Cu2ZnSnS4 semiconductor prepared from low toxicity ethanol-based precursors," *AIP Conference Proceedings*, vol. 1788. p. 30022, 2017.

[12] B. Munir, B. E. Prastyo, and E. Y. Muslih, "Synthesis and characteristic of Cu 2 ZnSnS 4 thin film prepared by appropriate non-stoichiometry precursor," *IOP Conf. Ser. Earth Environ. Sci.*, vol. 105, no. 1, p. 12128, 2018.

[13] C. Malerba *et al.*, "CZTS stoichiometry effects on the band gap energy," *J. Alloys Compd.*, vol. 582, pp. 528–534, 2014.

[14] A. Monshi, M. R. Foroughi, and M. R. Monshi, "Modified Scherrer Equation to Estimate More Accurately Nano-Crystallite Size Using XRD," *World J. Nano Sci. Eng.*, vol. 02, no. 03, pp. 154–160, 2012.

[15] M. Zhou, Y. Gong, J. Xu, G. Fang, Q. Xu, and J. Dong, "Colloidal CZTS nanoparticles and films: Preparation and characterization," *J. Alloys Compd.*, vol. 574, pp. 272–277, 2013.

[16] B. S. Pawar *et al.*, "Effect of complexing agent on the properties of electrochemically deposited Cu2ZnSnS4 (CZTS) thin films," *Appl. Surf. Sci.*, vol. 257, no. 5, pp. 1786–1791, 2010.

[17] S. W. Shin *et al.*, "Studies on Cu2ZnSnS4 (CZTS) absorber layer using different stacking orders in precursor thin films," *Sol. Energy Mater. Sol. Cells*, vol. 95, no. 12, pp. 3202–3206, 2011.

[18] Q. Guo *et al.*, "Fabrication of 7.2% efficient CZTSSe solar cells using CZTS nanocrystals," *J. Am. Chem. Soc.*, vol. 132, no. 49, pp. 17384–17386, 2010.

[19] E. J. Einholm, "Introduction to Spectroscopy," *J. Magn. Reson. Ser. A*, vol. 121, no. 1, pp. 93–94, 1996.

[20] M. Jiang, R. Dhakal, Y. Li, P. Thapaliya, and X. Yan, "Cu2ZnSnS4 (CZTS) polycrystalline thin films prepared by sol-gel method," *37th IEEE Photovolt. Spec. Conf.*, vol. 4, pp. 001283–001286, 2011.

[21] P. K. Sarswat, S. Kar, and M. L. Free, "Bifacial photodetector using CZTS absorber material," *2012 International Conference on Emerging Electronics*, 2012.

[22] M. P. Suryawanshi *et al.*, "CZTS based thin film solar cells: a status review," *Mater. Technol. Adv. Perform. Mater.*, vol. 28, no. 1, pp. 98–109, 2013.

[23] T. J. Pundsack, B. D. Chernomordik, A. E. Béland, E. S. Aydil, and D. A. Blank, "Excited-state dynamics in CZTS nanocrystals," *J. Phys. Chem. Lett.*, vol. 4, no. 16, pp. 2711–2714, 2013.

[24] V. Kheraj, K. K. Patel, S. J. Patel, and D. V. Shah, "Synthesis and characterization of Copper Zinc Tin Sulphide (CZTS) compound for absorber material in solar-cells," *Journal of Crystal Growth*, vol. 362, no. 1, pp. 174–177, 201

Separation of Cerium from Neodymium Hydroxide Concentrate Through Decomposition and Leaching Using HNO_3

MV Purwani[1*], Triyono[1], M Setyadji[1], RA Amiliana[1],
[1]Center for Accelerator Science and Technology
National Nuclear Energy Agency, Jalan Babarsari PO Box 6101 ykbb
Yogyakarta 55281

purwani@batan.go.id

Abstract. The thermal decomposition or calcination of neodymium hydroxide concentrate product by pilot plant of REOH into CeO_2, La_2O_3 and Nd hydroxide concentrate in Center for Accelerator Science and Technology has been performed. From the result of start-up and validation was obtained neodymium hydroxide concentrate which still containing of impurities especially Ce. The purpose of this research to separate Ce from $Nd(OH)_3$ concentrate by thermal decomposition to form Nd oxide (Nd_2O_3). Calcination temperature was 1000°C, and variation of time. The result of calcination were dissolved using aqueous HNO_3 with molarity variation and the level of leaching was five times. From the results, it can be stated that calcination could convert $Nd(OH)_3$ concentrate to Nd_2O_3 concentrate. Cerium hydroxide was easy dissolve in aqueous HNO_3 while CeO_2 couldn't. The longer of calcination time, the formed of Nd_2O_3 is more perfect. The results of 3 hours calcination for the %weight of La increase until 12,50%, Ce increase to 10,42% and Nd increase to 52,08%. The composition of La in solid of leaching products use 1 M HNO_3 with five leaching times is 3,57%, while Ce is 50.26% and Nd is 0.36%. The kinetic calcination and leaching followed reaction control.

Keywords: neodymium hydroxide concentrate, separation, calcination, leaching, aqueous HNO_3

1. Introduction

Neodymium hydroxide concentrate was a REOH (rare earth hydroxide) process products at pilot plant of REOH to form CeO_2, La_2O_3 and Nd hydroxide concentrate in Center for Accelerator Science and Technology that containing REEs namely Ce, La, Pr, Y, Sm and Gd [1]. There is a growing demand of the high-tech rare earth metals such as Nd, Pr and Dy in daily life because their numerous usages in various electronic and mechanical devices like computers, audio-visual components, magnetic separators, cranking motor of automobiles, military and aerospace system [2-6]. Cerium oxide (CeO_2) has many applications, such as solid electrolytes in solid oxide fuel cells, catalysts, optical additives. For example, catalysts based on CeO_2 are widely used as efficient oxidation systems in heterogeneous catalysis such as in oxidation of CO in oxygen-rich and in hydrogen-rich atmospheres [7].

For the separation of Ce of La and Nd. neodymium hydroxide concentrate should be calcined and leached, so that the solution is formed. Leaching was a liquid-solid extraction which is the process of extraction of a soluble constituent (solute) on the solid compound by using the solvent. The leaching of

Nd using mineral acid were investigated for substantial recovery of Nd [8-12]. The leaching reaction that occurs could be between base and an acid. The base will react with the acidic solution to form salt and water [13]. Leaching of the rare earth metals with various organic and inorganic acids have been investigated by some researchers [9-11]. K Stonea et al., who studied comparison of H_3PO_4, HCl, $HClO_4$ and HNO_3 usage leading to the leaching, had proven the leaching process can use H_3PO_4, HCl, and HNO_3. Using of HNO_3, resulting in the greatest leaching efficiency [14].

The obtained cerium oxide, depending on the synthesis procedure, had the following values of calcination temperature [15]. Leonid et al. was produced of high porosity nanoparticles of cerium oxide that was calcined at 550°C [7]. Cerium is largely used in various applications such as phosphor lamps, polishing media, automotive three way emission catalyst, nickel metal hydride batteries and ceramics and is always found as a mixture with most of the REEs (Rare Earth Elements) and numerous transition metals [3,16]. Leaching of metals present in such devices was previously reported using sulfuric acid [17-19]. The separation of transition metals from lanthanides was carried out by hydrometallurgical techniques, REEs precipitating under the form of sulphate salts [20,21]. The separation of cerium (III) from lanthanum(III), neodymium(III) and praseodymium (III) can be done by oxidation and liquid-liquid extraction using ionic liquids [22].

Process optimization and kinetics for leaching of cerium, lanthanum and neodymium elements from iron ore waste's apatite by nitric acid. The maximum REE leaching recoveries of 66.1%, 56.8% and 51.7% for Ce, La and Nd, respectively were achieved at the optimum leaching condition with 18% nitric acid concentration, 0.06 solid to liquid ratio and 38 min leaching time [23]. Extraction of Nd and Fe are quantitative using dilute CH_3COOH (0.4 M) solution and the result of the extraction was 89.76% of Nd [24]. The purpose of this work is to combine thermal decomposition or calcination and leaching to recover Ce, La and Nd component from neodymium hydroxide concentrate. Thermal decomposition or calcination to form Nd oxide (Nd_2O_3) containing Ce and La oxide and leaching using aqueous HNO_3. Neodymium oxide and La_2O_3 dissolved in aqueous HNO_3 while Ce oxide (CeO_2) was insoluble in aqueous HNO_3.

2. Experimental method

2.1. Materials and Reagents

Neodymium hydroxide concentrate was a REOH process products at pilot plant of REOH to form CeO_2, La_2O_3 and Nd hydroxide concentrate in National Nuclear Energy Agency using nitric acid solution.

2.2. Thermal decomposition or calcination conditions

The calcination experiment was carried out in furnace. The prepared sample was calcined at 1000 °C for 1, 2, and 3 hours.

2.3. Leaching procedure conditions

The results of calcination were leached or dissolved in HNO_3. The leaching experiment was carried out in a three necked round bottom flask which was placed on a heating mantle (set at desired temperature). The solution was kept under agitation using an externally connected agitator. The nitric acid concentration was studied at the concentration range of 0.2 M – 1 M. The ratio of neodymium hydroxide concentrate weight with volume of nitric acid (w/v) was 2.5. Time of leaching was 5 minutes with stirring speed 150 rpm. The stage of leaching was five times. The sample is analyzed using XRF.

3. Results and discussion

3.1. Calcination of Neodymium hydroxide concentrate

3.1.1. Effect of time of calcination on weight fraction and conversion of Solid Weight. The calcination was conducted at various temperatures. The weight conversion calculation uses following formula [1].

IOP Conf. Series: Materials Science and Engineering **553** (2019) 012026 doi:10.1088/1757-899X/553/1/012026

$$\alpha = \frac{m_0 - m}{m_0 - m_f} \quad (1)$$

Where m_0 is the initial mass, m is the mass at time t, and m_f is the final mass when it has turned into Nd oxide. The weight of each fraction was characterized using XRF (X-ray fluorescence spectrometry), so we can calculate the conversion of the compound. Effect of increasing time is the decreasing of weight fraction, but the weight conversion of α is increase [25,26]. The reaction that occurred in this process can be seen in this following equation and the relationship between calcination time and weight fraction and conversion is shown at Figure 1.

$$RE(OH)_{3(s)} \rightarrow RE_2O_{3(s)} + H_2O_{(g)} \quad (2)$$

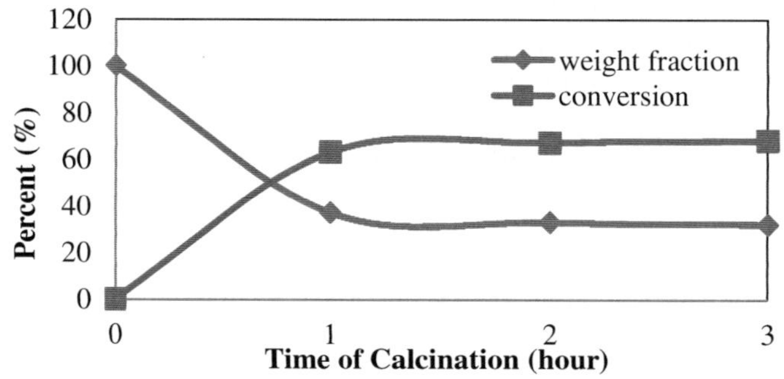

Figure 1. The relationship between calcination time and weight fraction and conversion.

3.1.2. Effect of time of calcination on %weight of RE element and %weight of RE compounds. The calcination was conducted at various temperatures too. The %weight of RE element after calcination is shown in Figure 2 and the %weight of RE compounds is shown in Figure 3. The %weight of RE element and RE compounds is increase with increasing calcination time. But for water and hydroxide are decreasing. In calcination for 1 to 2 hours the RE oxide that is formed, but it is not perfect yet because it still contains H_2O and hydroxide. In 3-hour calcination, the %weight of La is increase from 3.50% to 12.50%, Ce content increase from 3.33% to 10.42% and %weight of Nd increase from 16.67% to 52.08%. All water content in Nd hydroxide became vapor. Before calcination, the %weight of La compound ($La(OH)_3$) was 5.47%, and %weight of Ce compound ($Ce(OH)_3$) was 4.55%. After calcination, the %weight of La compound (La_2O_3) was 14.66%.

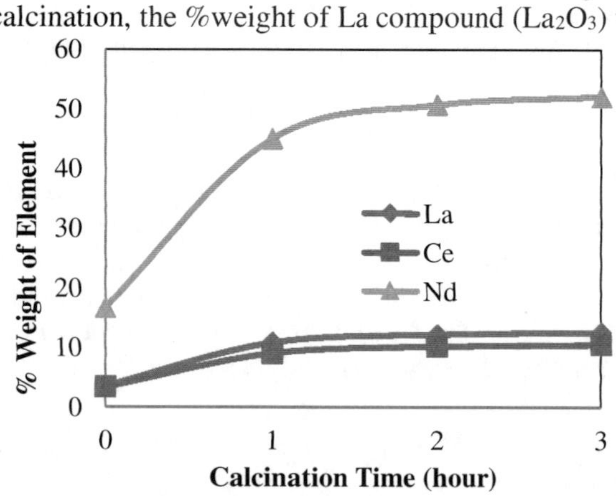

Figure 2. The relationship between calcination time and %weight of RE element.

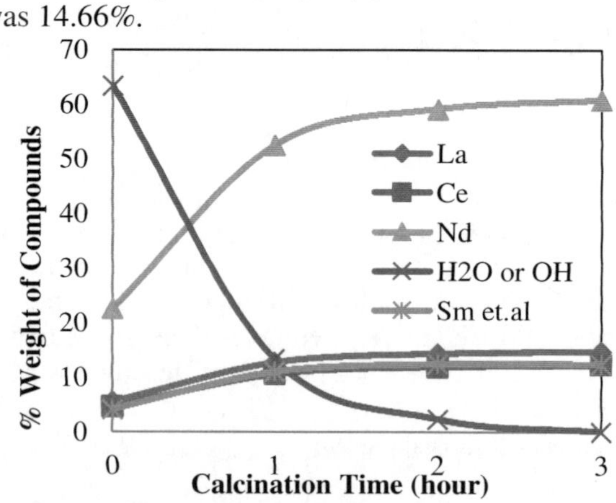

Figure 3. The relationship between calcination time and %weight of RE compound.

3.2. Leaching for separation of Ce, La and Nd

3.2.1. Effect of HNO_3 molarity on leaching efficiency of Ce, La and Nd. The leaching of La and Nd was investigated at varying HNO_3 concentration in the range of 0.2 M –1 M while keeping other parameters constant. The leaching efficiency of La and Nd was increased with the increase in the HNO_3 concentration as shown in Figure 4. The leaching efficiency of La is 11.16% and the leaching efficiency of Nd is 30.72%, while Ce could not leach

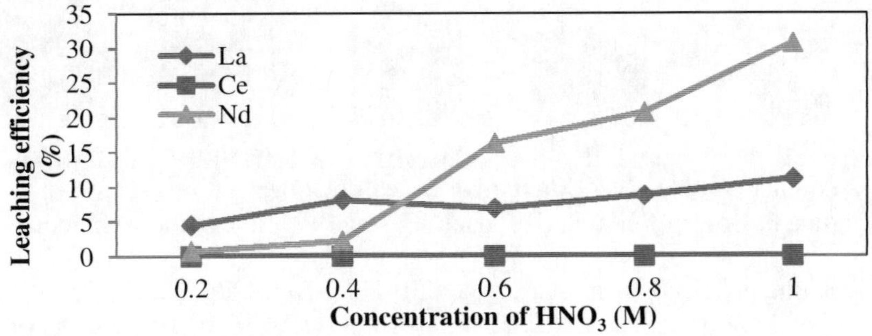

Figure 4. Effect of nitric acid concentration on leaching efficiency of La, Ce, and Nd.

3.2.2. Effect of leaching stage on leaching efficiency of La at various concentration of HNO_3. The stage of leaching was investigated at the range of I – V (1st – 5th) at varying HNO_3 concentration in the range of 0.2 M until 1 M. The leaching efficiency of La is increased with increase in the HNO_3 concentration as shown in Figure 5 and the leaching efficiency of La at the range of I – V (1st – 5th) in 0.2 M is 4.46% - 5.9%, while the leaching efficiency of La in 0.4 M is 8.13% - 7.98%, in 0.6 M is 10.89% - 10.53%, in 0.8 M is 8.87% - 12.78%, and the last in 1 M is 13.16% - 15.21%.

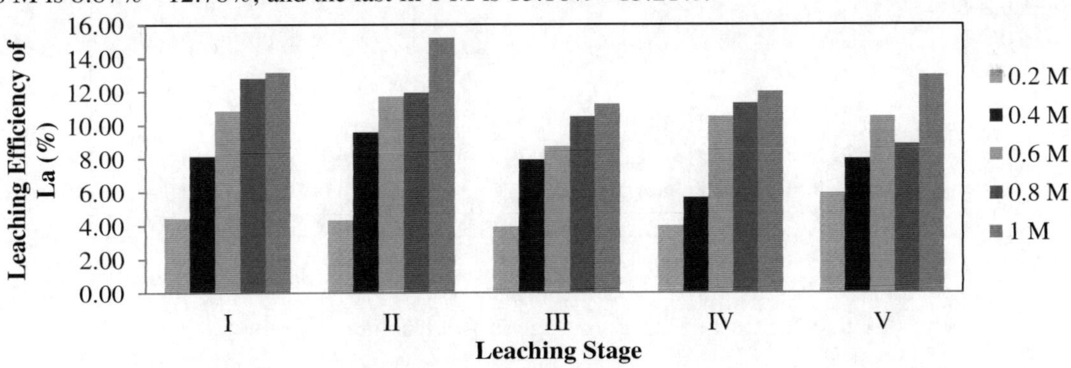

Figure 5. Effect of leaching stage on La leaching efficiency.

3.2.3. Effect of cumulative leaching stage on leaching efficiency of La at various concentration of HNO_3. The cumulative leaching efficiency of La was the amount of La leaching efficiency at each stage (I - V). The relationship between cumulative stage and cumulative leaching efficiency was presented in Figure 6.

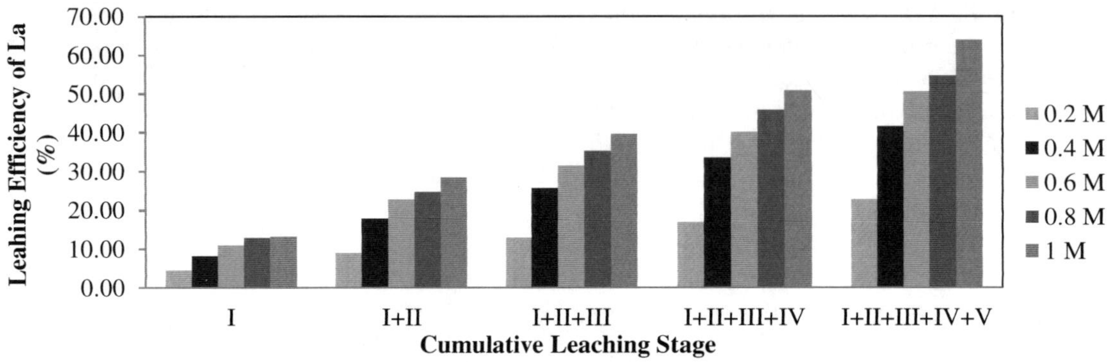

Figure 6. Effect of cumulative leaching stage on La leaching efficiency.

The cumulative leaching efficiency is increase when the HNO_3 molarity is increase too. The cumulative leaching efficiency of La at 0.2 M is 22.62%, at 0.4 M is 41.54%, at 0.6 M is 50.66%, at 0.8 M is 54.62%, and at 1 M is 63.84%.

3.2.4. Effect of leaching stage on leaching efficiency of Nd at various concentration of HNO_3. The leaching efficiency of Nd was increased with the increase of HNO_3 concentration as shown in Figure 7. The leaching efficiency of Nd at the range of I – V (1st – 5th) in 0.2 M is 0.75% - 5.9%, in 0.4 M was 2.23% - 7.98%, in 0.6 M is 15.86% - 18.02%, and in 0.8 M is 15.87% - 24.27%. Meanwhile the leaching efficiency of Nd at the range of I – III (1st – 3rd) in 1 M is 30.72% - 33.36%.

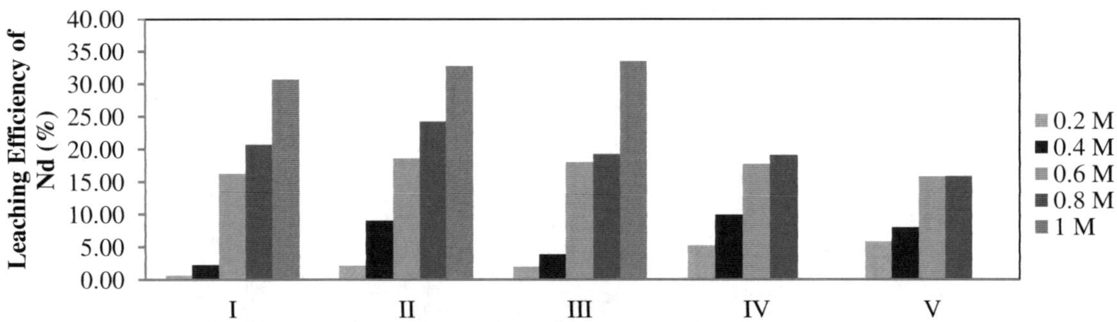

Figure 7. Effect of leaching stage on Nd leaching efficiency.

3.2.5. Effect of cumulative leaching stage on leaching efficiency of Nd various concentration of HNO_3. The cumulative leaching efficiency of Nd was the amount of Nd leaching efficiency at each stage (I - V). The relationship between cumulative stage and cumulative leaching efficiency was presented in Figure 8. The cumulative leaching efficiency is increase when the increasing of HNO_3 molarity. The cumulative leaching efficiency of Nd at 0.2 M is17.40%, at 0.4M is 34.83%, at 0.6 M is 88.21%, and at 0.8 M is 97.65%. The cumulative leaching efficiency of Nd at 1 M was 99.44% at cumulative stage I – III (1st – 3rd), so the leaching of Nd only required 3 stage

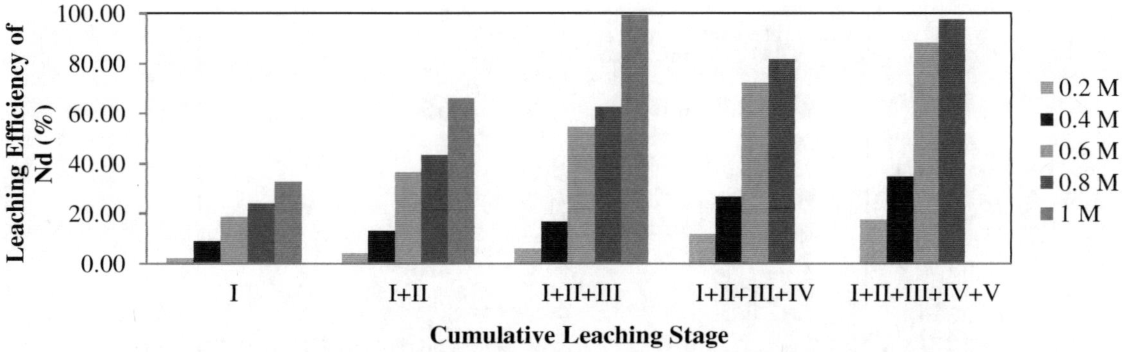

Figure 8. Effect of cumulative leaching stage on Nd leaching efficiency.

3.3. Kinetic model

3.3.1. Thermal decomposition or calcination kinetic. Kinetic decomposition was studied by applying unreacted shrinking core model. The influence of different temperature regions upon the thermal behavior of chemical compounds can provide kinetic parameters indicating change in the reaction pathway. The extent of conversion, *X*, is defined by the following equation [27].

$$X = \frac{m_0 - m}{m_0 - m_f} \quad (3)$$

where *m* is the mass of the sample at a given time *t*, m_0 and m_f refer to masses at the beginning and the end of process. The result of the conversion at calcination of Nd Hydroxide concentrate can be seen in Figure 9.

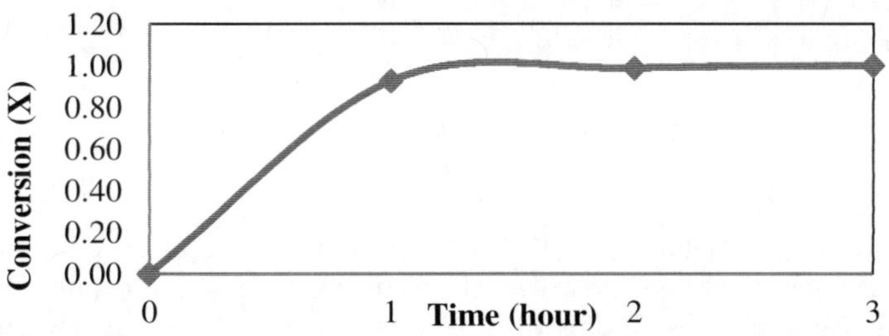

Figure 9. Relation between time and fractional conversion at calcination of $Nd(OH)_3$ concentrate.

The mechanisms occurring and mathematical models of thermal decomposition followed reaction control or ash diffusion control [25,26,28–30]. If the calcination process was controlled by the chemical reaction, the following expression can be used to describe the leaching kinetics of the process.

$$1 - (1 - X)^{\frac{1}{3}} = k_1 t \quad (4)$$

Similarly, if the was controlled by diffusion, the following expression of the shrinking core model can be used.

$$1 - 3(1 - X)^{\frac{2}{3}} + 2(1 - X) = -k_2 t \quad (5)$$

The using of these equation can be seen on Figure 10.

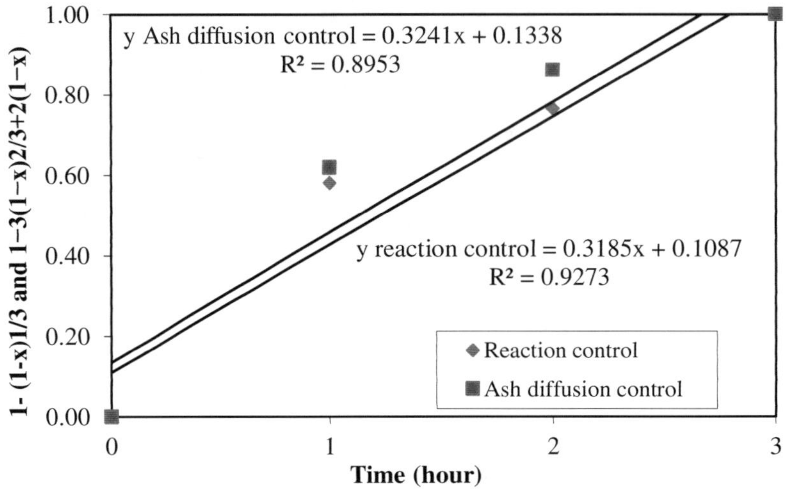

Figure 10. The relationship between time and shrinking core equation at calcination process.

The equations line linearity (R^2) of the relationship between time (t) with f (X) for the diffusion control and reaction control were shown in Table 1.

Table 1. The equation of the relationship between t and f (X) at calcination process.

	Ash diffusion control	Reaction control
Equation	$Y = 0.3241x + 0.1338$	$Y = 0.3185x + 0.1087$
Linearity (R^2)	$R^2 = 0.8953$	$R^2 = 0.9273$

Table 1 shows that the linearity (R^2) of Reaction control was greater than ash diffusion control. It can be concluded that calcination was controlled by reaction.

3.3.2. Kinetic of leaching. The major models that have been developed for kinetics of non-catalytic liquid–solid reactions are the shrinking (including shrinking particle and shrinking core), homogeneous, grain, uniform pore and random pore models [31]. The homogeneous, grain, uniform pore and random pore models applied to the porous solid–liquid system that were not suitable to our leaching process. The shrinking core is suitable to the reaction between an initially non-porous particle and a reagent, during which the unreacted core gradually shrinks. In this model, a solid product layer is formed around the unreacted grains. Same with shrinking core, the shrinking particle model also suitable to the reaction between an initially non-porous particle and a reagent, but no solid is produce [32].

In shrinking core model, the leaching process is controlled either by the diffusion of reactant through the solution boundary, or through a solid product layer, or by the surface chemical reaction rate [33]. If the leaching process is controlled by the chemical reaction, equation (4) can be used to describe the leaching kinetics of the process. Similarly, if the diffusion of the reagent through a product layer is the slowest step, the equation (5) of the shrinking core model can be used.

Figure 11 presents the fractional conversion during the leaching. As shown in Figure 11, Nd hydroxide concentrate was leached out to $Nd(NO_3)_3$ solution after the reaction of 25 min at room temperature, with 0.8 mol/L HNO_3 and the solid is Ce oxide concentrate. Meanwhile the relation between time (t) and the fractional conversion equation ($x = f(x)$) for the leaching process was shown in Figure 12.

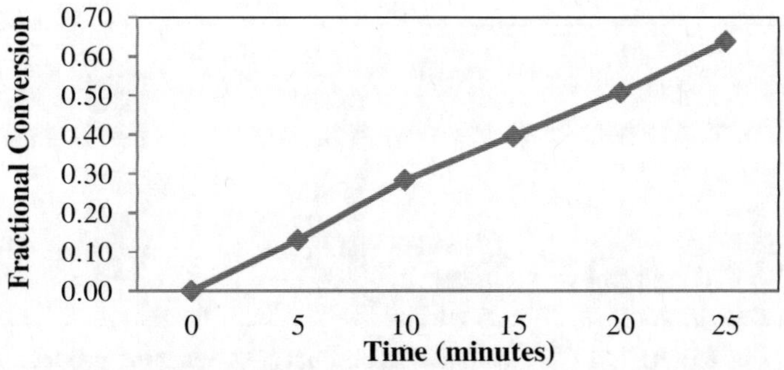

Figure 11. Relation between time and fractional conversion at leaching of Nd hydroxide concentrate.

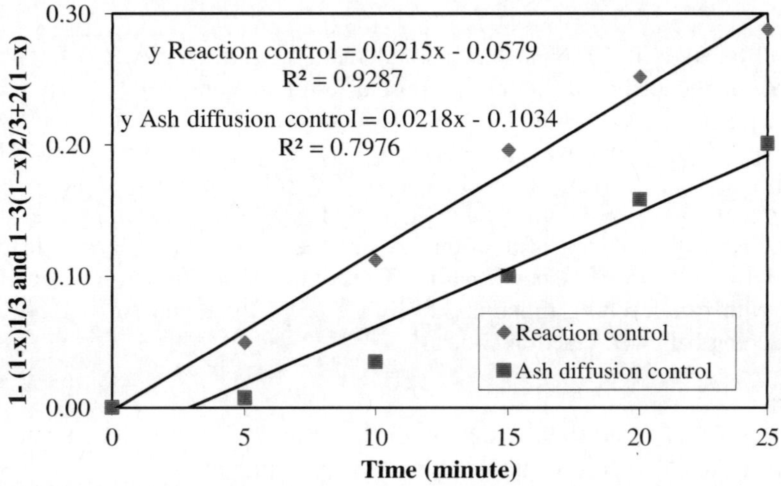

Figure 12. The relationship between time with shrinking core equation at leaching process

The equations line linearity (R^2) of the relationship between time (t) with f (X) for the diffusion control and reaction control at leaching process was shown in Table 2.

Table 2. The equation of the relationship between t and f (X) at leaching process

	Ash diffusion control	Reaction control
Equation	$y = 0.0218x - 0.1034$	$Y = 0.0215x - 0.0579$
Linearity (R^2)	$R^2 = 0.7976$	$R^2 = 0.9287$

From Table 2, the linearity (R^2) of reaction control was greater than ash diffusion control, it can be concluded that leaching was controlled by reaction

3.4. The composition of solid product

The composition of raw material (REOH), product of calcination and product of leaching was shown at Figure 13. The %weight of Ce in REOH is 3.33%, after calcination increased to 9.00% and after leaching increased to 50.26%. The %weight of La in REOH is 3.50%, after calcination increased to 6.29% and after leaching decreased to 3.57 %. The %weight of Nd in REOH is 16.67%, after calcination increased to 45.01% and after leaching decreased to 0.36 %. It can be concluded that Ce difficult to dissolve in HNO_3, but Nd and La easy dissolve in HNO_3.

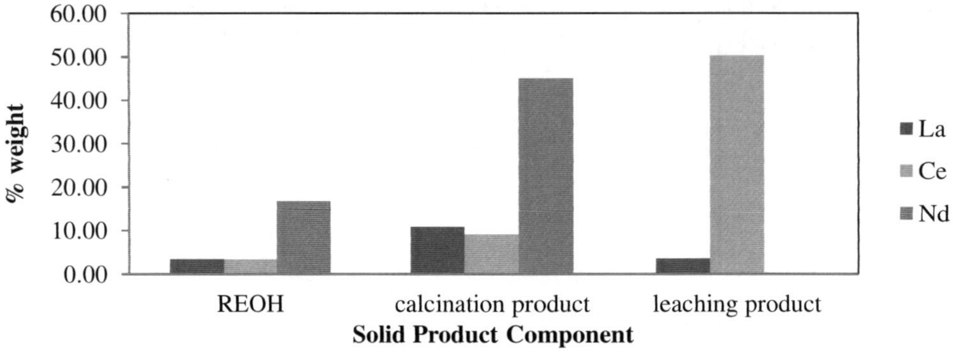

Figure 13. The composition of solid product.

4. Conclusion

From the results of the study it can be stated that calcination could convert $Nd(OH)_3$ concentrate to Nd_2O_3 concentrate as evidenced by the solubility properties in aqueous HNO_3. Cerium hydroxide was easy dissolve in aqueous HNO_3 while CeO_2 could not soluble in aqueous HNO_3. The more of calcination time, the Nd_2O_3 formed was more perfect. The concentration La in $Nd(OH)_3$ concentrate before calcination is 3.50%, Ce is 3.33% and Nd is 16.67%. The results of 3 hours calcination, the %weight of La is 12.50 %, Ce is 10.42 % and Nd is 52.08 %. After leaching the result of 3 hour calcination, HNO_3 molarity was very influent on leaching La and Nd, but Ce can't leach. The greater molararity of HNO_3 increase the efficiency leaching. The composition of La in solid of leaching product using 1 M HNO_3 with five leaching times is 3.57%, Ce is 50.26 % and Nd is 0.36 %. The kinetic calcination and leaching followed reaction control.

Acknowledgment

Sincere thanks to DIPA PSTA and INSINas 2018 for financial support. The authors are also thankful to the technicians that were involved with this work, mainly Suprihati.

References

[1] L. Berry, J. Galvin, V. Agarwal, and M. S. Safarzadeh, *Miner. Eng.*, vol.109, pp. 32–41, 2017, DOI: 10.1016/j.mineng.2017.02.007.
[2] S. Massari and M. Ruberti, *Resour. Policy*, vol.38, no. 1, pp. 36–43, 2013, DOI: 10.1016/j.resourpol.2012.07.001.
[3] K. Binnemans *et al.*, *J. Clean. Prod.*, vol.51, pp. 1–22, 2013, DOI: 10.1016/j.jclepro.2012.12.037.
[4] E. Antolini and J. Perez, *Int. J. Hydrogen Energy*, vol.36, no. 24, pp. 15752–15765, 2011, DOI: 10.1016/j.ijhydene.2011.08.104.
[5] M. Kul, Y. Topkaya, and I. Karakaya, *Hydrometallurgy*, vol.93, no. 3–4, pp. 129–135, 2008, DOI: 10.1016/j.hydromet.2007.11.008.
[6] L. V. Resende and C. A. Morais, *Miner. Eng.*, vol.23, no. 3, pp. 277–280, 2010, DOI: 10.1016/j.mineng.2009.12.012.
[7] L. V Gorobinskii and G. Y. Yurkov, vol.100, pp. 134–138, 2007, DOI: 10.1016/j.micromeso.2006.09.047.
[8] J. Pana and R. William, pp. 171–178, 2013, DOI: 10.1007/s10163-012-0105-6.
[9] H. Yoon *et al.*, vol.30, no. 9, pp. 1–6, 2013, DOI: 10.1007/s11814-013-0259-5.
[10] C. Rao, Y. Pontikes, K. Binnemans, and T. Van Gerven, *Miner. Eng.*, vol.76, pp. 20–27, 2015, DOI: 10.1016/j.mineng.2015.01.005.
[11] Y. Wu, X. Yin, Q. Zhang, W. Wang, and X. Mu, *Resour. Conserv. Recycl.*, vol.88, no. 100, pp. 21–31, 2014, DOI: 10.1016/j.resconrec.2014.04.007.
[12] X. Yang, J. Zhang, and X. Fang, *J. Hazard. Mater.*, vol.279, pp. 384–388, 2014, DOI:

10.1016/j.jhazmat.2014.07.027.
[13] C. Sasikumar, D. S. Rao, S. Srikanth, N. K. Mukhopadhyay, and S. P. Mehrotra, *Hydrometallurgy*, vol.88, no. 1–4, pp. 154–169, 2007, DOI: 10.1016/j.hydromet.2007.03.013.
[14] K. Stone, A. M. T. S. Bandara, G. Senanayake, and S. Jayasekera, *Hydrometallurgy*, vol.163, pp. 137–147, 2016, DOI: 10.1016/j.hydromet.2016.03.014.
[15] A. Martinez-arias, J. A. Anderson, J. C. Conesa, and J. Soria, no. Iii, pp. 1325–1330, 2000.
[16] I. B. De Lima and W. L. Filho, *Highlights on Rare Earths: Research Advances, Possibilities, Challenges, and Trends Based on Authors' Findings and Views*. Elsevier Inc., 2016, DOI: 10.1016/B978-0-12-802328-0.00026-7.
[17] L. Pietrelli, B. Bellomo, D. Fontana, and M. Montereali, *Waste Manag.*, vol.25, no. 2 SPEC. ISS., pp. 221–226, 2005, DOI: 10.1016/j.wasman.2004.12.013.
[18] R. Panda *et al.*, *J. Ind. Eng. Chem.*, vol.20, no. 4, pp. 2035–2042, 2014, DOI: 10.1016/j.jiec.2013.09.028.
[19] V. Innocenzi and F. Vegliò, *J. Power Sources*, vol.211, pp. 184–191, 2012, DOI: 10.1016/j.jpowsour.2012.03.064.
[20] L. Pietrelli, B. Bellomo, D. Fontana, and M. R. Montereali, *Hydrometallurgy*, vol.66, no. 1–3, pp. 135–139, 2002, DOI: 10.1016/S0304-386X(02)00107-X.
[21] L. Eduardo, O. Carmo, and M. B. Mansur, *J. Power Sources*, vol.195, no. 11, pp. 3735–3741, 2010, DOI: 10.1016/j.jpowsour.2009.12.071.
[22] M. Gras, N. Papaiconomou, E. Chainet, F. Tedjar, and I. Billard, *Sep. Purif. Technol.*, vol.178, pp. 169–177, 2017, DOI: 10.1016/j.seppur.2017.01.035.
[23] A. Ferdowsi and H. Yoozbashizadeh, *Trans. Nonferrous Met. Soc. China (English Ed.*, vol.27, no. 2, pp. 420–428, 2017, DOI: 10.1016/S1003-6326(17)60048-7.
[24] S. S. Behera and P. K. Parhi, *Sep. Purif. Technol.*, vol.160, pp. 59–66, 2016, DOI: 10.1016/j.seppur.2016.01.014.
[25] G. Zhang, T. Guo, H. Zheng, and Z. LI, *J. Fuel Chem. Technol.*, vol.44, no. 6, pp. 674–679, 2016, DOI: 10.1016/S1872-5813(16)30031-7.
[26] G. Zhan, J. X. Yu, Z. G. Xu, F. Zhou, and R. A. Chi, *Trans. Nonferrous Met. Soc. China (English Ed.*, vol.22, no. 4, pp. 925–934, 2012, DOI: 10.1016/S1003-6326(11)61266-1.
[27] C. J. Kim *et al.*, *Hydrometallurgy*, vol.146, pp. 133–137, 2014, DOI: 10.1016/j.hydromet.2014.04.003.
[28] A. Amiri, G. D. Ingram, A. V. Bekker, I. Livk, and N. E. Maynard, *Adv. Powder Technol.*, vol.24, no. 4, pp. 728–736, 2013, DOI: 10.1016/j.apt.2013.01.016.
[29] T. Wanjun, L. Yuwen, Y. Xi, and W. Cunxin, *Ind. Eng. Chem. Res.*, vol.43, no. 9, pp. 2054–2059, 2004, DOI: 10.1021/ie030418g.
[30] Y. Altaş and H. Tel, *J. Nucl. Mater.*, vol.298, no. 3, pp. 316–320, 2001, DOI: 10.1016/S0022-3115(01)00655-9.
[31] K. Li, J. Chen, D. Zou, T. Liu, and D. Li, *J. Rare Earths*, 2018, DOI: 10.1016/j.jre.2018.05.015.
[32] K. Liu, Q. Chen, Z. Yin, H. Hu, and Z. Ding, *Hydrometallurgy*, vol.125–126, pp. 125–136, 2012, DOI: 10.1016/j.hydromet.2012.06.001.
[33] M. Li, C. Wei, S. Qiu, X. Zhou, C. Li, and Z. Deng, *Hydrometallurgy*, vol.104, no. 2, pp. 193–200, 2010, DOI: 10.1016/j.hydromet.2010.06.001.

The Role of Heat Treatment on Characteristics of Mg-Al-Sr Reinforced Nano-SiC Composite

Vioni Dwi Sartika[1],Achmad Fauzi Trinanda[1],Rama Aditya Syarif[1],Ivan Orlando Limouswan[1],Anne Zulfia Syahrial[1*]
[1]Department of Metallurgy and Materials Engineering, Faculty of Engineering Universitas Indonesia, KampusBaru UI Depok 16424, Indonesia

anne@metal.ui.ac.id

Abstract. The effect of heat treatment on microstructure and mechanical properties of SiC/Mg-Al-Sr alloy matrix composite was investigated. The heat treatment involved annealing at 425, 375, and 325°C for 4 hours, water quenching then ageing at room temperature for 48 hours. The heat treatment process caused dissolution ofs Mg_2Si precipitated and decomposition of nano-SiC particles. The composites with 0.15 Vf-% nano-SiC which was treated at 425°C has reached the optimum hardness due to precipitate of the intermetallic phase. Besides, the hardness of the treated composites still has good ductility. The fracture surface of the composite consisted of numerous dimples as a result of micro voids nucleated at the interface between the matrix and SiC particles and second-phase particles as well. The fracture of heat treated composite was similar to microvoids which were nucleated mainly on nano-SiC particles as obtained during fractography analysis

Keywords : Magnesium composite, Mg-Al-Sr, Nano-SiC, Heat treatment

1. Introduction

The transportation industries are facing some challenges, such as the increasing demand for economical use of energy resources and stricter control over emissions to lower the climate change effect. This condition triggered the industries to seek lightweight materials as an alternative to conventional materials. The use of lightweight materials in vehicles become a solution to cut the energy consumption because it takes less energy to accelerate a lighter object than a heavier one, lightweight materials offer great potential for increasing vehicle efficiency. One of promising material to overcome this problem is magnesium[1]. Magnesium is material which has high specific strength with low density and good mechanical properties[2]. Commonly, magnesium being strengthened by adding alloys of other elements[1]. It is hard to do a cold-working process due to the HCP crystal structure of this metal[3]. Aluminum is an element that is commonly used as magnesium alloy. Another element like strontium can be added to Mg-Al alloys to enhance the strength at elevated temperature [4]. The mechanical properties of this alloy can be improved with the addition of various types of reinforcement such as SiC particles[5]. This particles have a good interface with the magnesium alloy matrix and can be distributed uniformly[6]. Another way to improve the mechanical properties of this alloy is through a heat treatment process to modify the intermetallic phases in the matrix. Intermetallic phases such as $Mg_{17}Al_{12}$, Al_4Sr, Mg_2Sr, MgSr and $Mg_{17}Sr_2$ in magnesium matrix could enhance the strength[7]. These intermetallic phases can retard the movement of dislocations, so it needs higher energy for deformation to occur.

In the previous work, magnesium composite was produced using stir casting route. The chemical composition of matrix is Mg-1.25Al-0.75Sr wt% and addition of nano-SiC was varied by 0.05; 0.10; 0.15; 0.20; 0.25 Vf%[8]. In this work, the composites were heat treating in 3 variable temperatures, 325, 375, 425°C for 4 hours. The aims of this experiment are to analyse the effect of age-hardening heat treatment on microstructure, mechanical properties and fracture mechanism of composites.

2. Experimental method

2.1 Synthesis of Mg-Al-Sr reinforced nano-SiC composite

The stir casting process is used to make the specimens. Magnesium ingot (99.9 wt%) was used as the matrix of the composite. The alloying element was added by using Al-15Sr alloy, then combined with various compositions of the nano-SiC particles (0.05; 0.10; 0.15; 0.20; and 0.25 Vf-%). In the making, magnesium was heated to its molten state in 860°C, then Al-15Sr in the form of a rod was added and stirred for 2 minutes. Before poured in the molten magnesium, nano-SiC was preheated in a muffle furnace for an hour in 900°C to remove moist and increase surface tension. The composite is poured into the mould which had been preheated in 900°C for 1 minute and coated with zirconia to reduce thermal shock. The Mg, Al and Sr content of all casts were analysed using Optical emission spectroscopy (OES) (the chemical compositions are shown in Table 1).

Table 1. Chemical composition of Mg-Al-Sr/Nano-SiC composite[8].

Element	Vf% Nano-SiC					
	0%	0.05%	0.10%	0.15%	0.20%	0.25%
Mg	96.3	96	96	96.5	95.8	95.3
Al	3.54	3.78	3.83	3.28	3.89	4.48
Be	<0.0001	0.0001	<0.0001	<0.0001	0.0002	0.0001
Cu	<0.0010	0.003	<0.0010	<0.0010	0.0168	0.0015
Mn	0.0645	0.0391	0.0477	0.0485	0.0524	0.0259
Zn	<0.0010	0.0343	0.0264	0.0339	0.0789	0.026
Ag	<0.0010	<0.0010	<0.0010	<0.0010	0.0047	<0.0010
Ca	0.0024	0.0031	0.0024	0.003	0.0031	0.0096
Cd	<0.0010	0.0011	<0.0010	<0.0010	0.0019	0.001
Sn	0.0327	0.0543	0.0291	0.0279	0.0834	0.0545
Sr	>0.0008	>0.0008	>0.0008	>0.0008	>0.0008	>0.0008

2.2 Heat Treatment

Each composite that is obtained through stir casting process then being heat treated at three different temperatures (325°C, 375°C and 425°C). The holding time for each temperature variable is 4 hours and the temperature rises at 10°C/minute with preheating for about 15 minutes.

2.3 Mechanical Properties

The composites, which had been heat treated, were prepared and mechanical properties tested (hardness& wear resistance). Someparts are cut into 2×3(cm) and ground to obtain a fine surface then tested by Brinell Hardness at room temperature. Weartest is conducted using the pin-on-disk machine by sliding a specimen under dry condition.

IOP Conf. Series: Materials Science and Engineering **553** (2019) 012027 doi:10.1088/1757-899X/553/1/012027

2.4 Metallography

All casts microstructurally characterised to observe the grain size of matrix alloy using an optical microscope (OM)and scanning electron microscope (SEM). They were cut, polished using alumina and ethanol, then etched using Nital reagent. EDX and XRD analysis are conducted to confirm the phases formed in the cast.

2.5 Fractography

The impact test sample was further analysed the fracture surface by a scanning electron microscope (SEM). The composite with 0,15vf% nano SiC is selected as a comparative date to see the fracture behaviour of the composite after heat treatment.

3. Results and discussion

3.1 Hardness and Wear Resistance

In the previous work, it was stated that the increasing of nano-SiC could enhance the hardness of composites. However, after certain addition, there is no significant effect since the tendency of particles to aglomerate[8]. In this work, heat treatment process improved the hardness value of composites, as a temperature increased as seen in figure.1. The highest hardness was found in the composite which had been heat treated at 425°C with the addition of 0.15 Vf% nano-SiC, that is 75 BHN as present with the blue line in figure.1. This hardness is 47.67% greater than magnesium alloy without reinforcement and heat treatment.

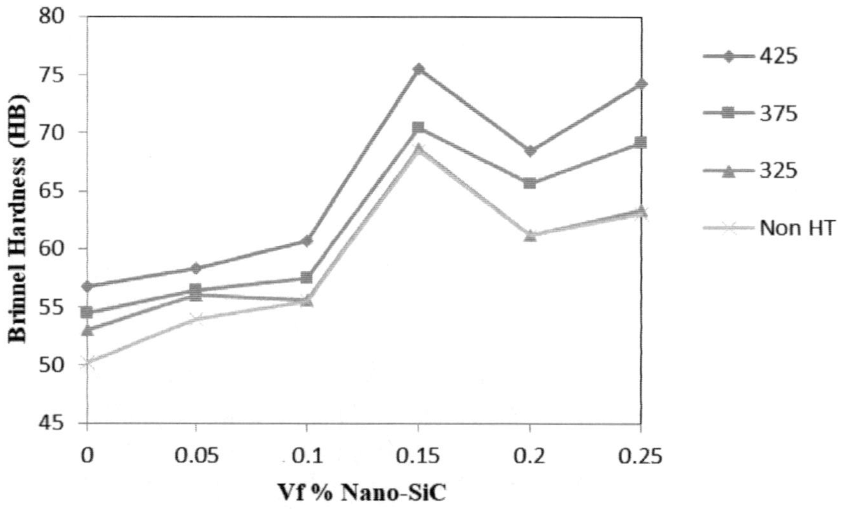

Figure 1. The hardness value of composites.

Usually, the hardness of material relating to its wear resistance. Abrasive testing gives the wear rate which shows the lowest value. It is matched with the hardness given in figure.1. In some conditions, it is still observed the fluctuation wear rate in figure.2. Another contributing factor which increases the wear rate is the debris formed and stick on the rotating pin, so it could erode the sliding surface of the composite.

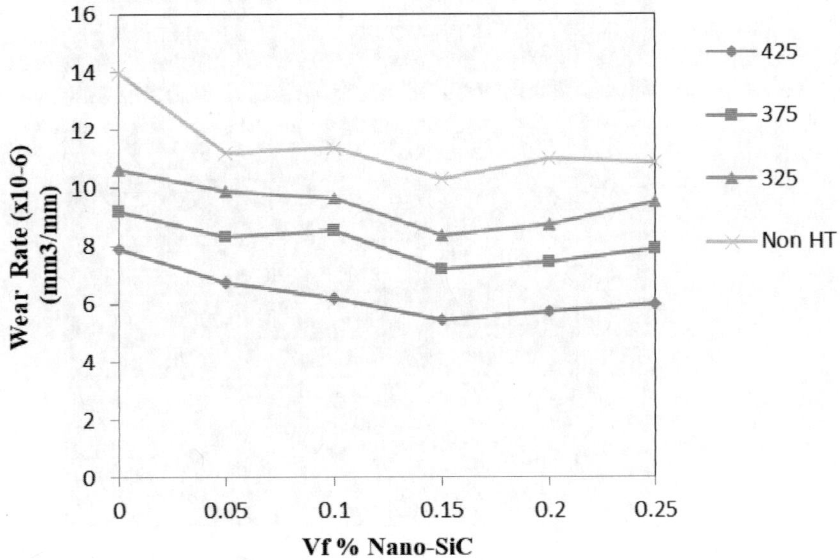

Figure 2. Wear rate of composites.

3.2 Microstructure

The observation of microstructure using optical microscope is conducted on Mg-Al-Sr/0.15 Vf% nano-SiC before[8] after heat treatment process, since it gives the highest hardness value and wear resistance. The microstructure of the two composites consists of α-Mg, $Mg_{17}Al_{12}$, MgAlSr, which shown in Figure.3.

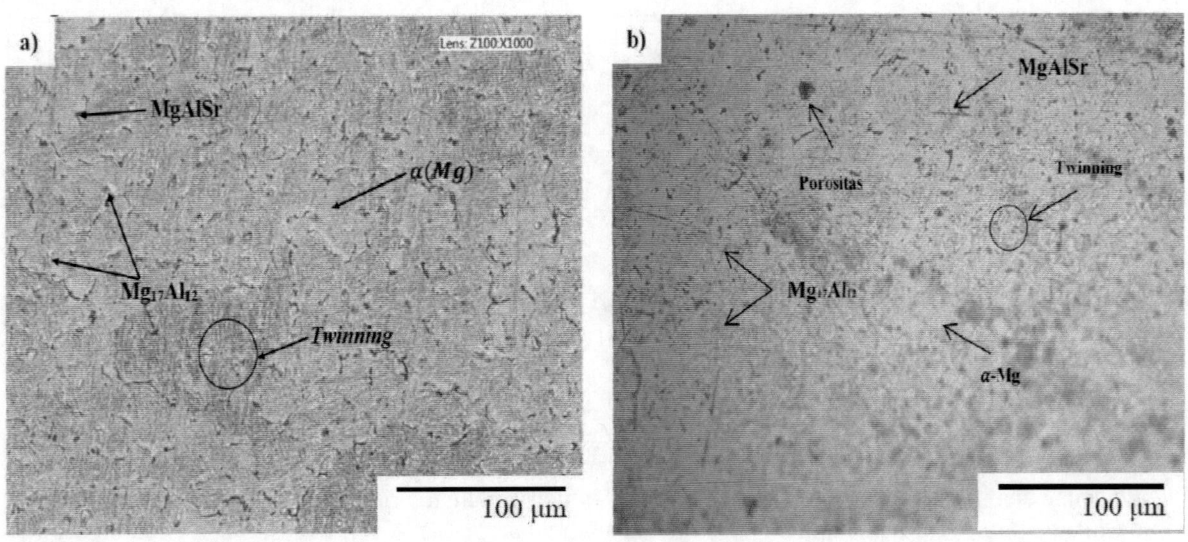

Figure 3. The microstructure of Mg-Al-Sr/0.15 Vf% nano-SiC, (a) before[8] and; (b)after heat treatment process.

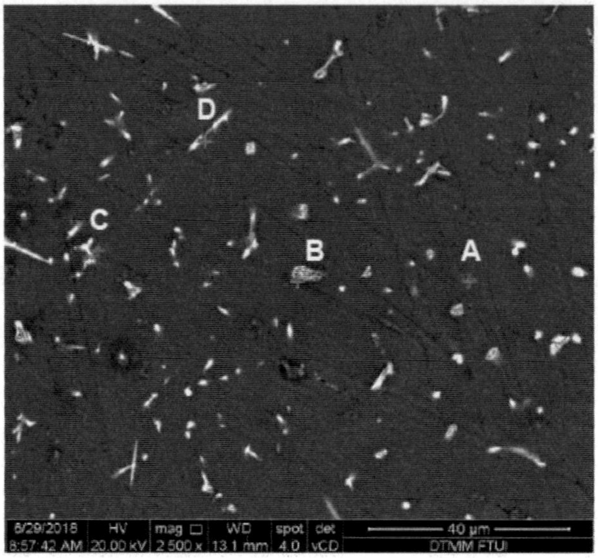

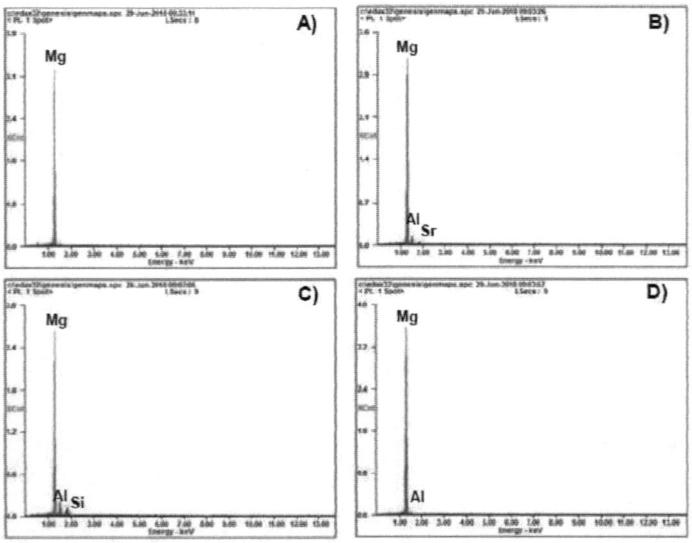

Titik	Wt (%)			
	Mg	Al	Sr	Si
A	100	0	0	0
B	82.26	16.2	1.54	0
C	70.88	21.73	0	7.39
D	95.09	4.91	0	0

Figure 4. Microstructural analysis using SEM-EDX.

Microstructure characterisation using SEM found that the average size of intermetallic phases before heat treatment is about 7.777μm[8]. On the other hand, there is significant change observed after heat treatment process which reduces the average size of intermetallic phases. The measurement shows the size is about 2.990 μm. EDX analysis which conducted on composite after heat treatment reveals the phases formed at point A, B, C, D in figure 4 are α-Mg,Al_4Sr, Mg_2Si, MgAl. According to Xia et al[9], Mg_2Si could found in magnesium composite as the result of a reaction between Mg and SiC particles. It indicates that the improvement of hardness after heat treatment is caused by the finer precipitates and existence of Mg_2Si phase.

3.3 XRD analysis

Another way to see the phases that are formed after heat treatment is through XRD (*X-ray Diffraction)* analysis as shown in Figure 5.

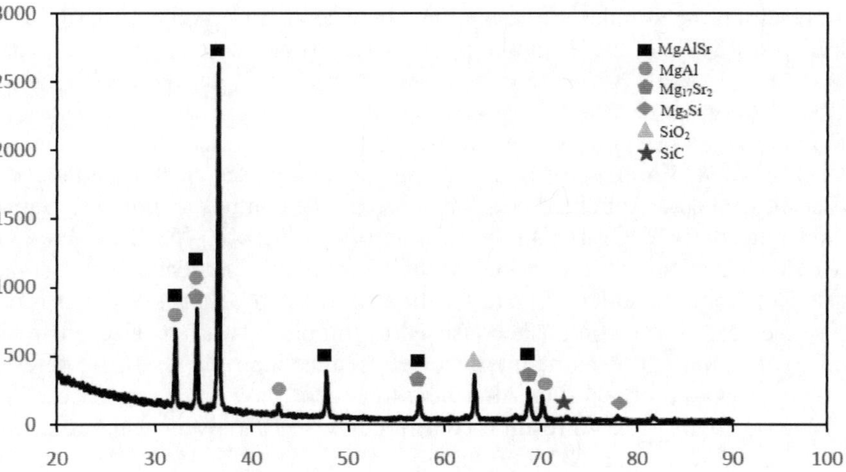

Figure 5. XRD analysis result

XRD analysis in figure. 5 found the MgAl, $Mg_{17}Sr_2$ compound. This compound has resulted after heat treatment process. This precipitate gave strengthening effect to the matrix, so that the hardness of the composite enhanced after the heat treatment process[10]. Then, the XRD analysis found Mg_2Si compound at 2 theta position of 78.1. Mg_2Si compound is formed from the interfacial reaction between nano-SiC and Mg matrix[11]. Another compound such as SiC and SiO_2 are found at 2 theta position of 71 and 63.2 respectively. SiO_2 compound is formed on the surface of SiC within the pre-heating process before casting. Li et al[9] reported that pre-oxidation of nano-SiC at 973-1273K in 2 hours could produce the SiO_2 layer on the surface. This layer could enhance the adhesion force between the matrix and reinforce.

3.4 Fractography

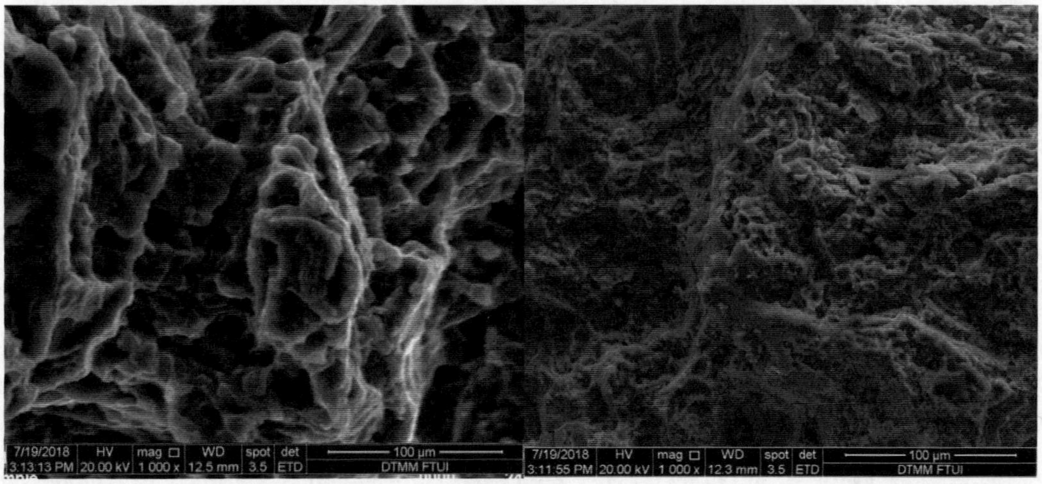

Figure 6. SEM image of fracture surface obtained for magnesium composite, (a) before heat treatment and; (b) after heat treatment process.

Figure 6 shows the fracture surface of the non-heat treated and heat treated composite. Numerous dimples of various size are seen on the fracture surface of both composites. The nucleation of microvoids were resulting the dimples at the interface between matrix and SiC-particles and second-phase particles and/or by SiC-particles irregular multiphase particles cracking and its coalescence[12].

4. Conclusion

The effect of Heat Treatment on Mg-Al-Sr/nano-SiC to its microstructure and mechanical propertieswere investigated. From work discussed in this paper, the heat treatment enhances the hardness of composites. Theoptimum hardness is reached by the composite which contained 0.15Vf% reinforcement and treatedat 425°C. The highest value of hardness is 75 BHN. Furthermore, heat treatment decreases wear rate of composites with the optimum is reached by composite with composition is 0.15Vf% reinforcement which was treated at 425°C, that is 5.46×10^{-6} mm^3/mm.The fracture surface of the heat treated compositeconsisted of dimples of various sizes. The dimples were a result of the microvoids nucleation at the interface between the matrix and SiC particles and second-phase particles.Microstructural observation after heat treatment shows the existence of some phases such asα-Mg, Al_4Sr, Mg_2Si, MgAl. This result is confirmed by XRD analysis which found the existence of MgAl, MgAlSr,Mg_2Si, $Mg_{17}Sr_2$, SiC, SiO_2 phases.

Acknowledgment

Authors would like to express their gratitude for financial support from the Ministry of Research and Higher Education, Republic of Indonesia, through PKM Grant.

References

[1] M. Gupta and N. Sharon, *Magnesium, Magnesium Alloys, and Magnesium Composites*. Hoboken, N.J.: Wiley, 2013, pp. 1-2. doi:10.1002/9780470905098

[2] C. Xu, J. Zhang, S. Liu, Y. Jing, Y. Jiao, L. Xu, L. Zhang, F. Jiang, M. Zhang and R. Wu, "Microstructure, Mechanical and Damping Properties of Mg–Er–Gd–Zn Alloy Reinforced with Stacking Faults", *Materials &Design*, vol. 79, pp. 53-59, 2015. doi:10.1016/j.matdes.2015.04.037

[3] M. Gupta and N. Sharon, *Magnesium, Magnesium Alloys, and Magnesium Composites*. Hoboken, N.J.: Wiley, 2013, pp. 118. doi:10.1002/9780470905098

[4] F. Jinping, X. Bingshe, W. Shebin, L. Lu and F. Zhiyong, "Effect of Sr/Al Ratio on Microstructure and Properties of Mg-Al-Sr Alloy", *Rare Metal Materials and Engineering*, vol. 41, no. 10, pp. 1721-1724, 2012. doi:10.1016/S1875-5372(13)60013-2

[5] K. Deng, K. Wu, Y. Wu, K. Nie and M. Zheng, "Effect of Submicron Size SiC Particulates on Microstructure and Mechanical Properties of AZ91 Magnesium Matrix Composites", *Journal of Alloys and Compounds*, vol. 504, no. 2, pp. 542-547, 2010. doi:10.1016/j.jallcom.2010.05.159

[6] P. Poddar, V. Srivastava, P. De and K. Sahoo, "Processing and Mechanical Properties of SiC Reinforced Cast Magnesium Matrix Composites by Stir Casting Process", *Materials Science and Engineering: A*, vol. 460-461, pp. 357-364, 2007. doi:10.1016/j.msea.2007.01.052

[7] A. Janz, J. Gröbner, D. Mirković, M. Medraj, J. Zhu, Y.Chang and R. Schmid-Fetzer, "Experimental Study andThermodynamic Calculation of Al-Mg–Sr PhaseEquilibria", *Intermetallics*, vol. 15, no. 4, pp. 506-519,2007. doi:10.1016/j.intermet.2006.09.001

[8] Z. Anne, D. Vioni, "The Role of Nano-SiC on Characteristics of Mg-Al-Sr/Nano-SiC Composites Produced by Stir Casting Route", *IOP Publishing Journal*, 2018.

[9] Xia, C., Xianliang, Z., Delai, O., & Tongcai, "Effects of Heat Treatment on Microstructure and Corrosion Behavior of SiC p /Al-6wt%Mg Composites", *Rare Metal Materials and Engineering*, vol *45*no.12, pp. 3085–3088, 2016. doi:10.1016/S1875-5372(17)30058-9

[10] Kvačkaj, T., Bidulská, J., Fujda, M., Kočiško, R., Pokorný, I., & Milkovic, O. (2009). Nanostructure Formation and Properties in Some Al Alloys after SPD and Heat Treatment. *Materials Science Forum*, *633-634*, 273–302. doi:10.4028/www.scientific.net/MSF.633-634.273

[11] C. Li, Z. Wang, M. Zha, C. Wang, H. Yu, H. Wang and Q. Jiang, "Effect of Pre-Oxidation Treatment of Nano-SiC Particulates on Microstructure and Mechanical Properties of SiC/Mg-8Al-

1Sn Composites Fabricated by Powder Metallurgy Combined with Hot Extrusion", *Materials*, vol. 9, no.12, p. 964, 2016. doi:10.3390/ma9120964

[12] Glogovský, M., Fujda, M., Vojtko, M., Zubko, P., & Škrobian," Effect of heat treatment on microstructure and mechanical properties of extruded SiC/6061 composite". *Acta Metallurgica Slovaca*, vol 21, no.1, 2015. doi:10.12776/ams.v21i1.547

Effect of Sodium Carbonate on the Reduction Process of Nickel Slag from Sulawesi

Arum Cahyanurani Setyabrata[1], Ahmad Maksum[1,2], Agus Budi Prasetyo[1], Bambang Priyono[1], Johny Wahyuadi Soedarsono[1*]
[1] Center of Minerals Processing and Corrosion Research, Department of Metallurgical and Materials Engineering, Universitas Indonesia, Depok 16424, Indonesia
[2] Department of Mechanical Engineering, Politeknik Negeri Jakarta, Depok 16425, Indonesia

jwsono@metal.ui.ac.id

Abstract. Nickel slag is one of the results of the nickel ore smelting process which is still a waste and has not been used optimally. Considering the amount of nickel slag waste in Indonesia, the effort to utilize it into a more valuable product becomes very important. Therefore, in order to extract the precious metal content in nickel slag, this study aims to analyze the effect of adding sodium carbonate in the reduction process of nickel slag. In this study, the reduction process is carried out using coal as a reducing agent, where the variation of the reduction temperature is 800, 900, and 1000°C; while the variation of the ratio between nickel slag and additives is 1:1, 1:2, and 2:1. The XRD analysis results show that the temperature rise up to 1000 °C and the addition of sodium carbonate will increase the amount of sodium magnesiosilicate up to 29.4% and also hematite up to 25.1%. The same thing happened in samples with a ratio of 1:1 and 1:2 where the amount of sodium magnesiosilicate increased from 29.4% to 30.0% and hematite from 25.1% to 28.8%.

Keywords: Additive, Nickel slag, Reduction, Sodium carbonate, Waste

1. Introduction

The production of ferro-nickel in Indonesia mostly comes from processing laterite ore, where some researchers have succeeded in increasing the concentration of nickel through a reduction process[1–3]. In addition to nickel, nickel ore processing also produces a side product called nickel slag which still contains other valuable metals such as lanthanum and yttrium[4]. Nickel sulphide ore smelting firstly recovered by increasing sulphur product called as matte [5-6]. Sulphide iron in the initial ore is oxidized in two stages that are calcination and oxygen flash smelting which then followed by the conversion of sulphide iron to iron oxide and SO_2. Iron oxide then fluxed with silica and other combination of oxide mineral that ends with the produce of byproducts called as slag[7].

This causes the high silica content on slag nickel where it usually contain 20% silica in the form of amorphous silica or metal silicates and fayalite[4]. Unfortunately, the precious metals in fayalite are strongly bonded in the silica iron matrix and difficult to separate. Therefore, in this study sodium carbonate is added to break the silica iron matrix bond and separating silica so that the precious metals content in the nickel slag will increase.

2. Experimental method

2.1 Materials
The nickel slag sample used in this study originated from Sulawesi and was characterized in previous studies [8], where the Si oxide content was 25.02% (based on XRF) and in the form of forsterite, olivine, quartz, and fayalite as seen in the XRD pattern, Figure 1.

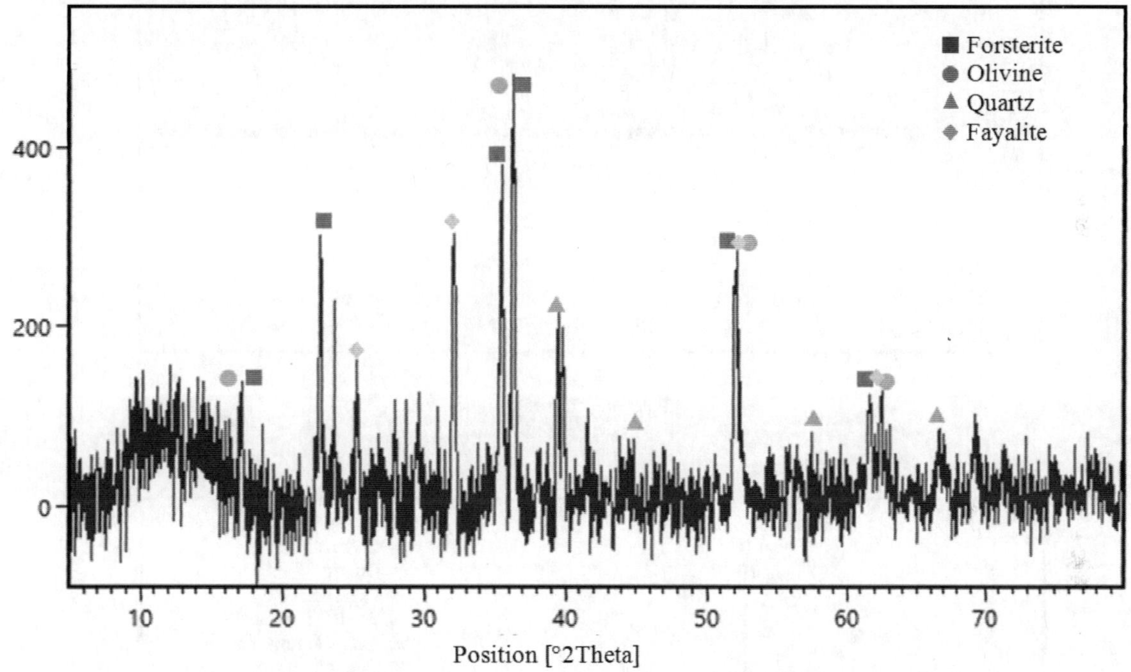

Figure 1 XRD results of initial sample (nickel slag) [8].

2.2 Experiment
To determine the effect of temperature, three samples each containing a mixture of 20gr nickel slag, coal 2gr, and sodium carbonate 20gr were reduced for one hour at different temperatures namely 800, 900, and 1000°C, respectively. Meanwhile, three other samples carried out variations in the weight ratio of nickel slag with sodium carbonate, namely 1: 1, 1: 2, and 2: 1, and heated at a temperature of 1000°C for 1 hour. After the reduction process, the samples were taken to BATAN for XRD testing and the results were analyzed using the highscore plus software

3. Results and discussion

The result of XRD pattern analysis can be seen on Figure 2, and its semi-quantitative analysis is separated to Figure 3 until Figure 5. Figure 2a and Figure 3 show that the percentage of quartz tends to increase with increasing temperature, while a percentage of forsterite tends to decrease. This indicates the possibility of the decomposition of forsterite into quartz due to the presence of the reducing agent. In this reduction without sodium carbonate, the optimum decomposition of forsterite into quartz occurs at the highest experimental temperature of 1000°C.

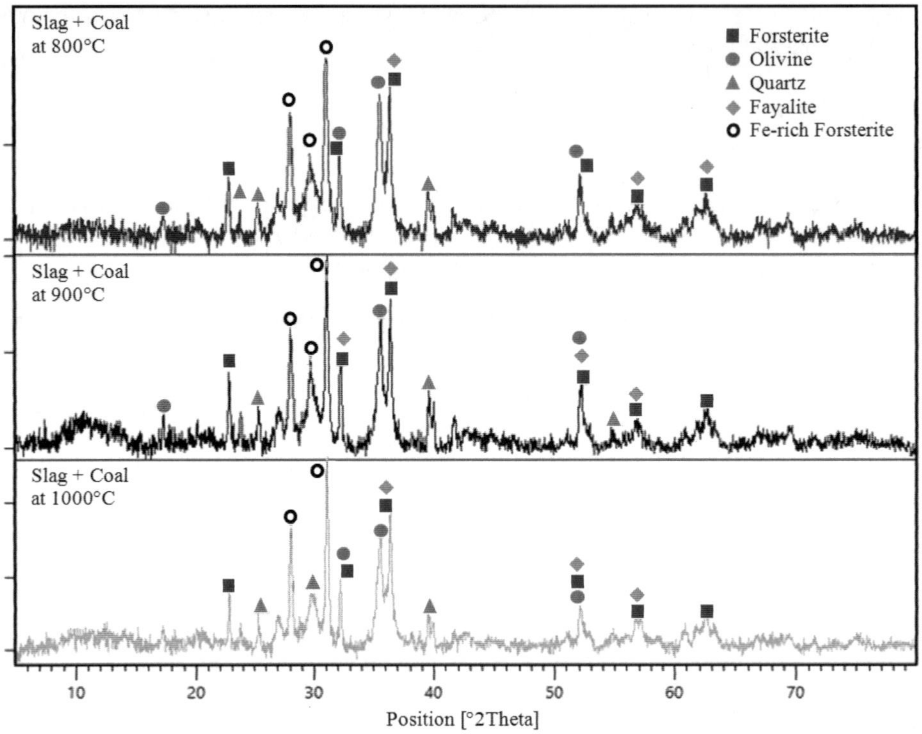
Slag + Coal
at 800°C
Slag + Coal
at 900°C
Slag + Coal
at 1000°C
Forsterite
Olivine
Quartz
Fayalite
Fe-rich Forsterite
10
20
30
40
50
60
70
Position [°2Theta]

(a)

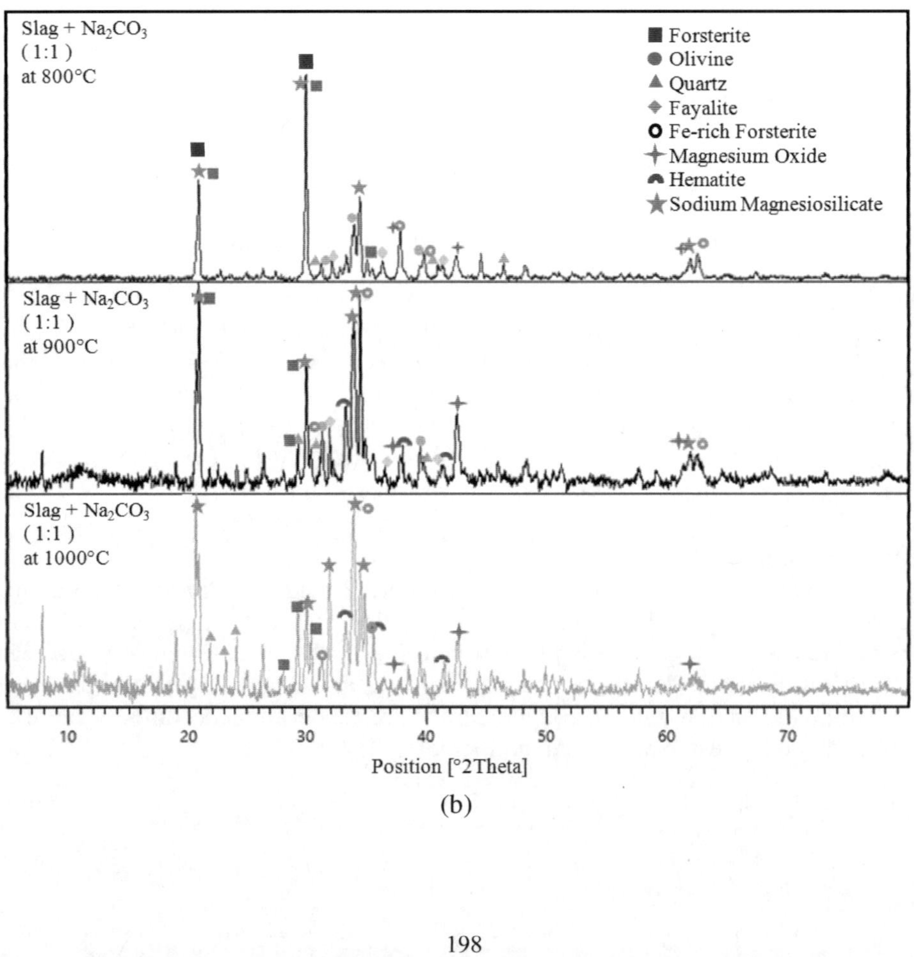
Slag + Na_2CO_3
(1:1)
at 800°C
Slag + Na_2CO_3
(1:1)
at 900°C
Slag + Na_2CO_3
(1:1)
at 1000°C
Forsterite
Olivine
Quartz
Fayalite
Fe-rich Forsterite
Magnesium Oxide
Hematite
Sodium Magnesiosilicate
10
20
30
40
50
60
70
Position [°2Theta]

(b)

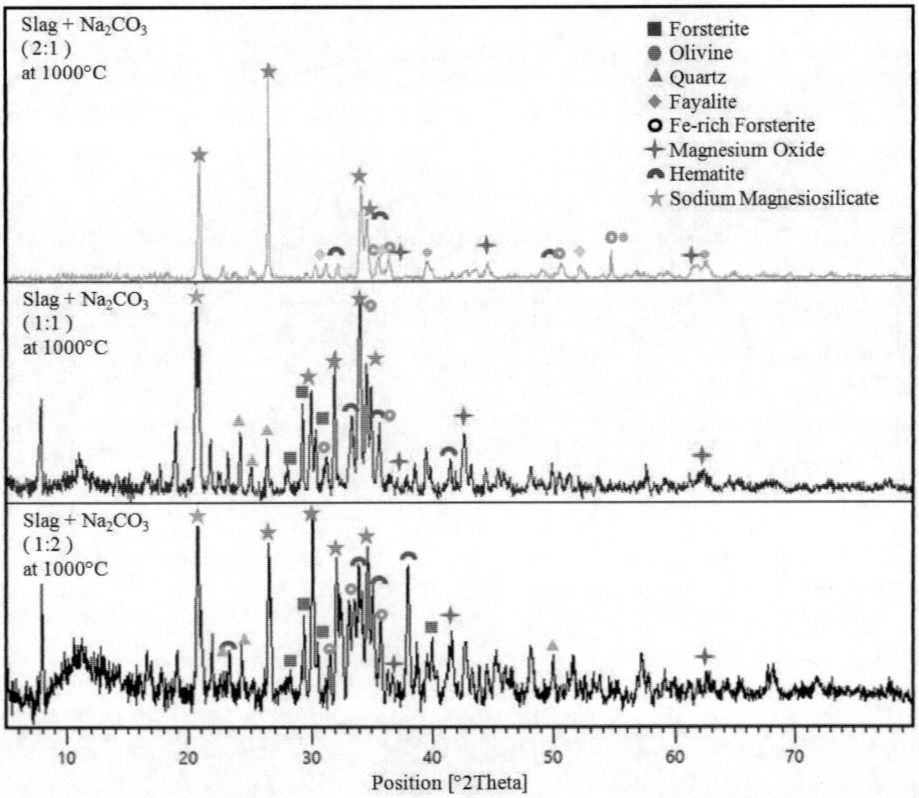

(c)

Figure 2. XRD patterns of nickel slag after reduction (a) without sodium carbonate at various temperatures; (b) with sodium carbonate (ratio 1:1) at various temperatures; (c) in various sodium carbonate ratios at 1000°C.

In Figure 2a there is also no significant change in compound. This is presumably due to the strong bond of iron silicate compounds from nickel slag so that they cannot be reduced directly by using coal only. This assumption is strengthened from the results of experiments using sodium carbonate as an additive as seen in the XRD results in Figure 2b and Figure 4. Figure 2b shows the presence of other dominant compounds namely Magnesiosilicate (Na_2MgSiO_4) and hematite, although the iron silicate compounds are still present. This can also be seen in Figure 2c that as the addition of sodium carbonate increases, more hematites are formed. ccording to the semi-quantitative results in Figure 5, Fe-rich forsterite is reduced from 18.5% to 14.9% by reducing the amount of carbonate sodium, ie from 1:1 ratio of nickel / sodium carbonate to 2:1.

Based on semi-quant result of XRD as shown in Table 7, it can be seen where at 800°C, the most dominant phase is Forsterite (Mg2SiO4) with the content of 25.9%. In the increase of temperature, it is seen that the Hematite began to form. Temperature of 900°C and 1000°C shows the presence of Hematite and Sodium Magnesiosilicate with the value of 10.6%, 25.1% and 36.2% and 29.4% respectively.

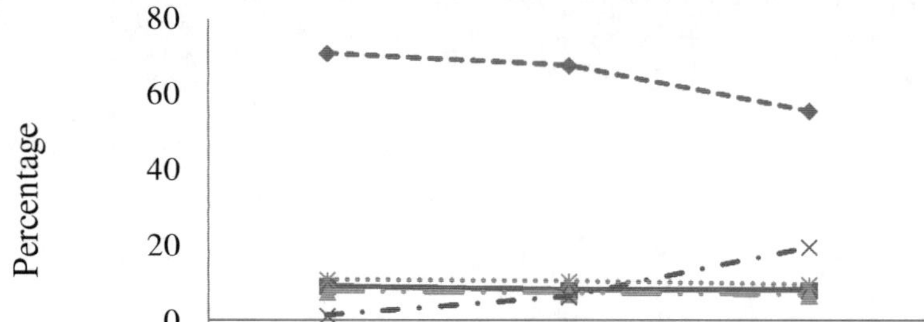

	800°C	900°C	1000°C
Fe – rich Forsterite	71	67.8	55.7
Olivine	9.2	8.1	8.2
Forsterite	7.6	7.2	6.8
Quartz	1.4	6.4	19.7
Fayalite	11	10.5	9.7

Figure 3. XRD semi-quantitative analysis of nickel slag after reduction without sodium carbonate at various temperatures.

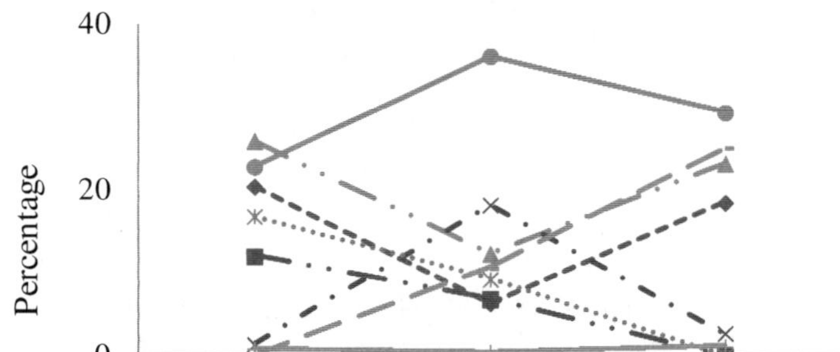

	800°C + Na2CO3	900°C + Na2CO3	1000°C + Na2CO3
Fe – rich Forsterite	20.4	6.3	18.5
Olivine	12	6.8	0
Forsterite	25.9	12.3	23.2
Quartz	1.3	18.2	2.7
Fayalite	16.8	9.2	0
Sodium Magnesiosilicate	22.8	36.2	29.4
Magnesium Oksida	0.8	0.4	1.1
Hematit	0	10.6	25.1

Figure 4. XRD semi-quantitative analysis of nickel slag after reduction with sodium carbonate (ratio 1:1) at various temperatures.

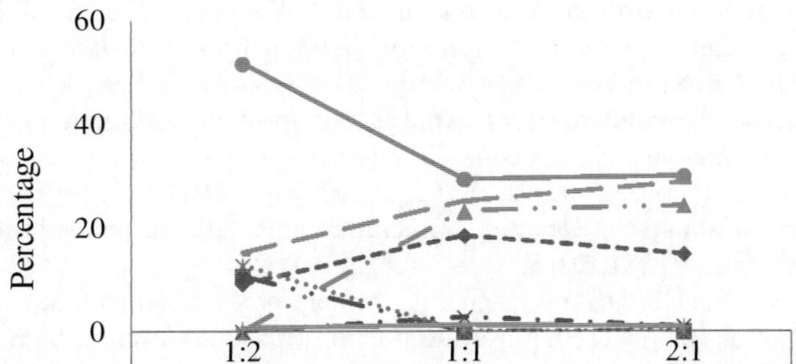

	1:2	1:1	2:1
Fe – rich Forsterite	9.2	18.5	14.9
Olivine	10.7	0	0
Forsterite	0	23.2	24.3
Quartz	0	2.7	1
Fayalite	12.7	0	0
Sodium Magnesiosilicate	51.7	29.4	30
Magnesium Oksida	0.6	1.1	1
Hematit	15.1	25.1	28.8

Figure 5. XRD semi-quantitative analysis of nickel slag after reduction in various sodium carbonate ratios at 1000°C.

Figure 5 shows the semi-quant XRD result of slag nickel that had been added by Na_2CO_3 in different weigh ratio of nickel slag and Na_2CO_3. In 2:1 ratio shows that there is still fayalite and hematite with the value of 12.7% and 15.1% respectively. On the other side, with the increase of Na_2CO_3 content in the ratio to 1:2 it can be seen there is no fayalite found. Figure 5 also informs that the higher the weight ratio between nickel slag and Na_2CO_3, the higher the hematite content that is 25.1% and 28.8% respectively.

4. Conclusion

Nickel slag used in this study is an amorphous compound in which silicate compounds dominate and become a barrier to the separation of precious metals.
Sodium carbonate added in the tin slag reduction process can accelerate the decomposition of Fe-rich forsterite ($FeMgSiO_2$) into Sodium Magnesiosilicate (Na_2MgSiO_4), Magnesium Oxide (MgO), and Hematite (Fe_2O_3)..

Acknowledgment

The authors also wish to thank the Ministry of Research Technology and Higher Education and Directorate of Research and Community Engagement, Universitas Indonesia, under PITTA Research Grants Contract No. 2460/UN2.R3.1/HKP05.00/2018 for the financial support of this research.

References

[1] A. Maksum, M. K. E. Husein, S. Permana, A. Rustandi, and J. W. Soedarsono, "A preliminary study on the reduction of limonite ore by using rice husk as a reducing agent," *IOP Conf. Ser. Mater. Sci. Eng.*, vol. 316, no. 12050, 2018.

[2] R. Rahmat and A. Maksum, "Effects of Na2SO4 addition on the selective reduction of limonite ore from Southeast Sulawesi Effects of Na2SO4 addition on the selective reduction of limonite ore from Southeast Sulawesi," *IOP Conf. Ser. Earth Environ. Sci.*, vol. 105, 2017.

[3] R. Hidayanti, S. Permana, A. Maksum, and J. Wahyuadi, "Study on the effects of temperature in the carbothermic reduction laterite ore using palm kernel shell as reducing agent," *IOP Conf. Ser. Earth Environ. Sci.*, vol. 105, 2017.

[4] I. Perederiy, "Dissolution of Valuable Metals from Nickel Smelter Slags by Means of High Pressure Oxidative Acid Leaching," University of Toronto, 2011.

[5] F. Huang, Y. Liao, J. Zhou, Y. Wang, and H. Li, "Selective recovery of valuable metals from nickel converter slag at elevated temperature with sulfuric acid solution," *Sep. Purif. Technol.*, vol. 156, pp. 572–581, 2015.

[6] Z. Guo, D. Zhu, J. Pan, and F. Zhang, "Mineralogical Characteristics and Preliminary Beneficiation of Nickel Slag from Reduction Roasting-Ammonia Leaching," *Minerals*, vol. 7, no. 6, p. 98, 2017.

[7] K. Yan, Y. Guo, L. Fang, L. Cui, F. Cheng, and T. Li, "Decomposition and phase transformation mechanism of kaolinite calcined with sodium carbonate," *Appl. Clay Sci.*, vol. 147, no. July, pp. 90–96, 2017.

[8] R. H. Pangaribuan, J. Patrick, A. B. Prasetyo, A. Maksum, and B. Munir, "The effect of NaOH (natrium hydroxide) to slag nickel pyrometallurgy in different temperature and additive ratio," *E3S Web Conf.*, vol. 67, no. 3052, 2018.

The Influence of Spray Distance of Thermal Sprayed Aluminum (TSA) on its Corrosion Resistance and Bond Strength

Rini Riastuti[1], Bravola A H Bancin[1], Zamri Rahmat[1], and Sonia T Siallagan1[1]
[1] Metallurgy and Materials Engineering Department, Faculty of Engineering, Universitas Indonesia, Depok, West Java, Indonesia

bravolabancin@gmail.com

Abstract. This study investigates the ideal spray distance to result in a high quality aluminum coating on 316L stainless steel substrates. Thermal spray process is used using Aluminum as the coating material, well-known as TSA (Thermal Sprayed Aluminum. To get high coating quality must do the coating using proper parameters, one of them is the spraying distance. In this investigation is want to observe the influence of the spray distance on corrosion resistancea and its bond strength. the spray distance variants 10, 20, and 30cm. To evaluate the corrosion resistance and the bond strength of thermal sprayed aluminum, the researchers apply a salt spray and mechanical pull-off test. The result shows that the spray distance of 20cm produce the strongest corrosion resistance and the highest bond strength, 12.5 MPa.

Keywords : Corrosion resistance; Electric arc spraying; 316L stainless steel substrates; Aluminum; Bond strength

1. Introduction

Stainless steel substrates are, abundantly, used in such industries as oil and gas, chemical, and automotive because they have excellent corrosion resistance, good ductility, high mechanical strength at room temperature, and are relatively low cost [1]. Despite high corrosion resistance, in environments containing aggressive ions such as Cl-ions, this steel may still risk local corrosion [2]. Enhanching the corrosion resistance and life time of these steel, one can coat them with paints or metals such as Al, Ni, Cr, and Mo [3].

To coat stainless steel substrates, the researchers applied a thermal spray method. The method was used to improve the properties of the stainless steel against abrasive environment, corrosive environment and application in high temperature. The procedure of thermal spray includes, first, spraying molten metal particles on the surface of the object and, secondly, quenching which causes solidification. This coating technique will provide protection from the negative effects of chemicals, atmosphere, brine, and high-temperature environments [3,4].

Among several thermal spray techniques is electric arc spraying. This technique is applied to this investigation and involves two wires as the melted coating materials and then sprayed on the surface of the object. Unlike other thermal spray techniques that commonly utilize heat from combustion gases to melt coating materials, electric arc spraying exploits the thermal energy of the arc generated by a contact between two electrically-charged metal wires [3,4]. Of the many measures classified as electric arc spraying, Thermal Spray Aluminum (TSA) is the one applied to this study. The advantages of TSA are relatively low cost, simple, having no curing time, having an operating temperature up to 538°C, providing sacrificial anodes, and protecting objects for around 30 years in corrosive environments. This technique is widely used to such high corrosive environments as platforms, tanks, and risers in the oil and gas industry [5,6].

The research gap is obtained from Performance and Microstructure Analysis of 99.5% Aluminium Coating by Thermal Arc Spray Technique [5]. There, Muhamad Hafiz Abd Malek, Nor Hayati Saad, Sunjadi Kiyai Abas, N.R. Nik Roselina, and Noriyati Mohd Shah investigate the aluminum bond strength, its ductility, hardness, as well as porosity and find that overall performance of the aluminum coating is generally good. Nevertheless, they do not base those four features on the spray distance, as this research does, in exploring the aluminum coating best performance. In addition, lecturers and researchers interested in this investigation as well as experts at PT Thermic Coating Industries also contribute to determining the topic of this research.

This study explores the ideal spray distance to yield the best aluminum coating on stainless steel substrates. To find out the quality of the coating, this study involves a salt spray and mechanical pull-off test. Before undergoing those two tests, the surface of the sample is cleaned with thinner, heated to 80-90°C, and Al_2O_3-blasted.

2. Experimental method

2.1 Materials

The research materials are comprised of stainless steel substrates, aluminum wires, thinner solution, and aluminum oxide abrasives. The type of stainless steel substrates in this research is 316L (SS 316L) with dimensions of 150x70x5 mm. The table below gives information about the chemical composition of 316L stainless steel substrates.

Table 1. The chemical composition of SS 316L.

Elements	C	Si	Mn	P	S	Ni	Cr	Mo	Ni
(%)wt	0.025	0.6	0.9	0.035	0.001	10.1	16.7	2.04	0.03

Aluminum wires were involved in the process of electric arc spraying. These wires contain 99.5% Al and are 1.6 mm in diameter. Thinner and C # 24 aluminum oxide abrasives were, respectively, used in the process of cleaning and blasting.

2.2 Method

In this sub-chapter, the researchers explain the stages of investigation including the surface preparation, coating, the corrosion resistance test, and the bond strength test.

First, the surface preparation began with cleaning surface contaminants such as dust, oil, and rust with thinner. The surface then underwent a preheating process at a temperature by 80°C-90°C, controlled with an infrared digital thermometer. Subsequently, the surface was blasted in a blasting cabinet with a spray distance of 10-15 cm and a 4-5 bar air pressure to obtain an Sa 3 clean level in accordance with ISO 8501-1. After the blasting process was conducted, the sample should enter the coating process. Samples must not be left in an open room for more than 4 hours, otherwise the sample must be re-blasted.

Next, the surface of stainless steel substrates was coated with aluminum. This process was carried out through electric arc spraying. To deal with electric arc spraying, the researchers used an OSU hesller, its consuming two 1.6 mm-diametered wires of metallized aluminum. The process required a voltage of 27-29 V and a current of 150-200 A, and was taking place under atmospheric condition inside a closed container to reduce noise and dust pollution. The thickness of the coating layer on each sample is 90-100 μm while the spray dístance parameters are 10, 20, and 30 cm. Before do the coating, all preparation must be in proper results, the cleanliness and the surface roughness. In the process, electric arc spraying exploits the thermal energy of the arc generated by a contact between two electrically-charged metal wires. The melted Aluminum wires are sprayed by compressed air and help the Aluminum bonded to the substrate (SS 316L). The coating results are done and without finishing process represent to the actual applicaton in Oil & Gas platform.

The corrosion resistance of the layers was then examined through a salt spray test according to ASTM B117-16 standard. In this process, the surface of the object was first scratched with a knife diagonally to form an X. Afterwards, the samples were placed in a salt spray cabinet at 30° against the vertical line. This test spent 96 hours to finish. Next, an ASTM D 1654-standardized visual observation was conducted to evaluate whether any layer was peeled after the scratched area was scraped for the second time.

To test their coating layer's bond strength, the samples underwent a mechanical pull-off test. Before this evaluation took place, the surfaces of both samples and dolly were cleaned. the dolly surface then coarsened with a sandpaper. The surface of sample was attached to the dolly using Araldite glue. The dolly position was kept and designed so it unable to move until the glue was completely dry. The curing time was 24 hours. In operation, the pull-off test device utilized a perpendicular force to pull the dolly until it detached from the samples' surface or reached its maximum limit, 22 MPa by using a tool for rotating the screw inside the device. Below is the picture of the device and the process of mechanical pull-off test.

Figure 1. Mechanical pull-off test.

3. Results and discussion

3.1 Visual Observation

Figure 2 shows the results of visual observation of coatings at spray distances of 10, 20, and 30 cm. On this image, particulate aluminum obviously spread over the entire surface of each sample. The white gloss layers also do not exhibit any particular traits. In addition, the appearance of the coating layers does not indicate any sticking failures. As well as that, no layer peels occur in all samples. These points imply that the spray distance has no effect on the layers' view.

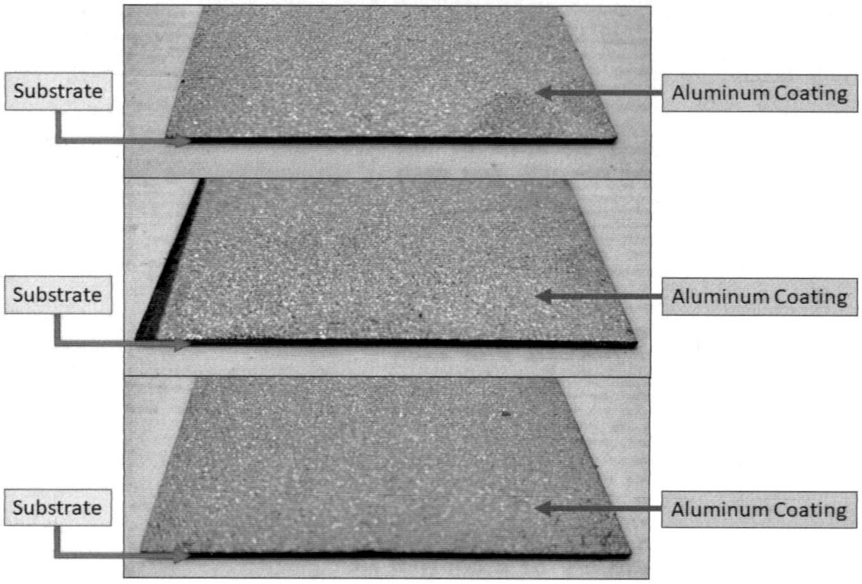

Figure 2. The visual observation of aluminum coatings at spray distances of: (a) 10 cm, (b) 20 cm, and (c) 30 cm.

3.2 Corrosion Resistance Test

The next stage was the corrosion resistance test. The aluminum layers were tested through a salt spray test. Before undergoing the salt spray test, as shown on Figure 3, the aluminum coatings looked shiny and clean while after passing this test, as shown on Figure 4, its color changed wholly at all spray distances. The color entire change indicates a response to destructive environments and protection capability.

Figure 3. The pre-salt spray test samples.

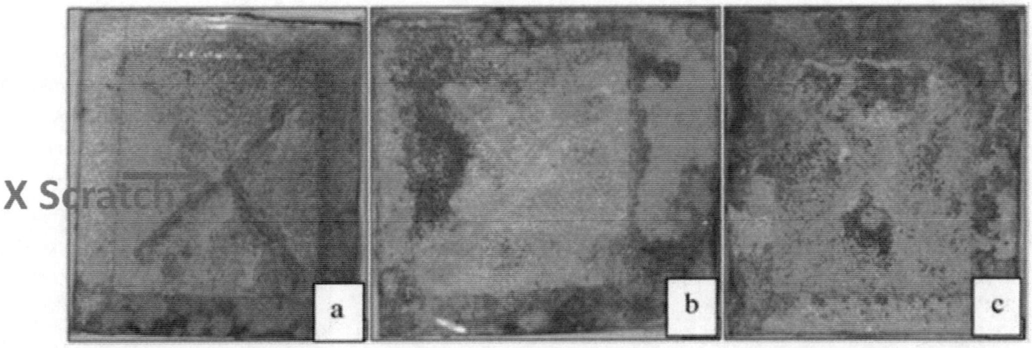

Figure 4. The post-salt spray test samples at spray distances of; (a) 10 cm, (b) 20 cm, and (c) 30 cm.

This observation indicates that as long as the aluminum coatings are attached entirely to the samples' surface and have no sticking failures, they will keep protecting the surface from aggressive ion attacks in the corrosive environment. The aluminum layers act as a barrier between the coated objects and destructive surroundings. As well as that, the coatings also served as a sacrificial anode and cathodic protector [8].

Among all the spray distances, 20cm yields the best corrosion resistance. On the contrary, the corrosion resistance is relatively low at further spray distance, 30cm, as the porosity becomes higher. This high porosity is triggered by the pace of the bigger particulates which faster leads to the surface of the object that, then, drives air trapped when quenching. So happens at closer spray distance. This distance brings about microcracks which, in turn, produces porosity within the layer [11,12]. All these imply that aluminum coatings can provide long-term protection which is much better than painting [9]. The same results are shown in research conducted by Oladijo (2016) and Wang (2016).

3.3 Bond Strength Test

The bond strength of the aluminum coatings on the stainless steel substrates' surface determines the resistance power of the object in question. This aluminum-coated object tends to be durable when the coating retains on the surface as it keeps the object against all destructive factors. The bond strength value of each sample is shown on the graph below.

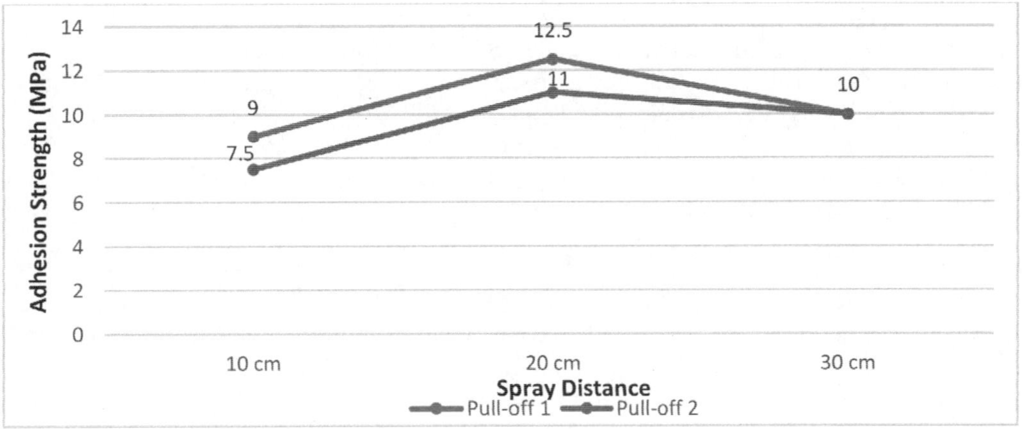

Figure 5. The trend of pull-of test result.

Figure 5 shows that the bond strength of the coating layers at a spray distance of 20cm, 12.5 MPa, is greater than that at 10 and 30cm. That of 20cm has proved to have the best interlocking strength because at this distance, particulate aluminum penetrate well on the stainless steel surfaces coarsened through the blasting process. Meanwhile, the bond strength of those at 10cm is not as good as that at 20cm due to its closeness. This nearness leads to overheating and internal stress in the aluminum coating layers [10]. Like the layers' bond strength at 10cm spray distance, that at 30cm also has a lower adhesive force than that at 20cm. The reason is that at this spray distance, the larger particulates accelerate to the surface of the object so that they do not penetrate properly. Because of this poor penetration, the interlocking between the coating layers and the surfaces of the substrates becomes weak. This investigation results are not different from those obtained by Gedzevicius dan M Hafidz Abd Malek, that the best spray distance for thermal spray methods, particularly, electric arc is 15-25cm [13].

4. Conclusion

The aluminum spray distance of 10 cm are not optimum enough because the heat of the electric arc spraying can lead to overheating and make the internal stress to the coating results, it can lead to lower bond strength (9 MPa) and lower corrosion resistance due to in this distance can bring microcracks inside the layer and then produce porosity or coating failure within the layer. If the distance are further can results the interlocking mechanism and the corrosion resistance become weak due to the larger particulates accerate to the surface due to its kinetic force and causing poor penetration and leading the high porosity to the coating results. In fact, All the spray distance does not indicate any coating failure and all the surface are covered with Aluminum which the coating also served as a sacrificial anode and cathodic protector, so this implied that the spray distance 10 cm, 20 cm, and 30 cm are may be used. Based on this investigation above, to get high quality of the coating, the 20cm spray distance yields the strongest corrosion resistance and the highest bond strength, 12.5 MPa, among others.

Acknowledgment

This research is fully funded by PITTA UI year 2018 under the contract number of 2520/UN2.R3.1/HKP.05.00/2018 and, in terms of research equipment, supported by Thermic Coating Industries and Gunanusa Utama Fabricators, Inc.

References

[1] Saedi K., M. Neikter, J. Olsen, Zhijian James Shen, and F. Akhtar. 316L stainless steel designed to withstand intermediate temperature. Materials and Design 135, p.1-8 (2007).

[2] Cai, Baoping, Yonghong Liu, Xiaojie Tian, Fei Wang, Hang Li, and Renjie Ji. An experimental study of crevice corrosion behaviour of 316L stainless steel in artificial seawater. Corrosion Science, 52:3235–3242, (2010).

[3] Davis, J.R., Handbook of Thermal Spray Technology. ASM International (2004). p. 54-76

[4] Oerlikon Group: Oerlikon Metco, An introduction to thermal spray issue 6 [online] from https://www.oerlikon.com (2016). [accessed on May 29th 2018]

[5] Malek, Muhamad Hafiz Abd , Nor Hayati Saad, Sunhaji Kiyai Abas, N.R. Nik Roselinaaand Noriyati Mohd Shah. Performance and Microstructure Analysis of 99.5% Aluminum Coating by Thermal Arc Spray Technique. The Malaysian International Tribology Conference, Procedia Engineering 68:558 – 565, (2013).

[6] Oerlikon Group : Oerlikon Metco, BRO-0007.7 - Electric Arc Wire Spray Solution [online] from https://www.oerlikon.com (2016). [Accessed on May 29th 2018].

[7] Prepared Under the Direction of the ASM International Handbook Committee. ASM Handbook. Volume 5, Surface Engineering. Metals Park, Ohio :ASM International, (1994).p.1446-1471.

[8] Tailor, Satish. Thermal Spray Aluminum (TSA) Coating. India : Metallizing Equipment Co. Pvt. Ltd. Jodhpur, p.1-10 (2016).

[9] Jones, Denny. Principles and Prevention of Corrosion, Second Edition. New York: Prentice-hall, Inc, pp.1-35,198-229, (1996).

[10] Muhamad Hafiz Abd Malek et al. Thermal Arc Spray Overview. IOP Conf. Ser.: Mater. Sci. Eng.46 012028. (2013).

[11] O.P. Oladijo, B.A. Obadele, A.M. Venter, L.A. Cornish, African Corrosion Journal, 2:37-45(2016)
[12] Y. Wang, K.Y. Li, F. Scenini, J. Jiao, S.J. Qu, Q. Luo, J. Shen, The effect of residual stress on the electrochemical corrosion behavior of Fe-based amorphous coatings in chloride-containing solutions, Surface & Coatings Technology (2016), doi:10.1016/j.surfcoat.2016.05.03
[13] Gedzevicius, I, and A.V. Valiulis. Analysis of Wire Arc Spraying Process variables on coatings properties. 12th International Scientific Conference: Achievements in Mechanical & Materials Engineering, 2009.

IOP Conf. Series: Materials Science and Engineering **553** (2019) 012030 doi:10.1088/1757-899X/553/1/012030

Role of Zn Substitution on Improvement of Electromagnetic and Microwave Absorption Characteristics of $BaTi_{(1-x)} Zn_xO_3$ (x= 0; 1/3; ½; 2/3; 1)

Yuda Bakti Zainal[1,2], Tony Kristiantoro[3], Dedi[3], Azwar Manaf[1*]

[1] Postgraduate Program of Materials Science, Faculty of Mathematics and Natural Sciences, Universitas Indonesia, Jl. Salemba Raya No.4, Jakarta 10430, Indonesia
[2] University of Jenderal Achmad Yani, Jl. Terusan Jenderal Sudirman Po.Box 148 Cimahi, 40531, Indonesia
[3] Indonesian Institute of Sciences, Research Centre for Electronics and Telecommunication, Bandung, 40135, Indonesia

azwar@ui.ac.id

Abstract. In the current study, we have explored the electromagnetic and microwave absorption characteristics, especially the permeability, permittivity and reflection loss of magnetoelectric $BaTi_{(1-x)}Zn_xO_3$ with x= 0; 1/3; ½; 2/3; 1. The composition was selected to include the effect of structural change when Ti is substituted with Zn in $BaTi_{(1-x)}Zn_xO_3$ material system. When x = 0, the Zn free $BaTiO_3$ posesses a tetragonal phase with lattice parameters a = b = 3.996361 Å and c = 4.029751 Å. The lattice parameters changed when x > 0 and phase transformation as well as material structure change occurred when x = 1. In addition, the effect of the geometric effect of replacing Ti ions with Zn which has a bigger ion radius will increase the absorption of electromagnetic waves.

1. Introduction

Microwave Absorbing Materials (MAM) are currently being explored by researchers for applications in high-speed wireless communication systems to reduce electromagnetic waves that affect human health [1-3]. For military applications MAM is usually used to avoid radar detection [1, 2]. Good candidates that meet the requirements of MAM are those materials that have high complex electrical permittivity (ε'-jε'') and complex magnetic permeability (μ'- jμ'') which contribute to increasing the reflection loss value (RL) [3]. Property like complex permitivity and permeability reflect the electromagnetic waves absorption characteristics of MAM.

Research on electromagnetic wave absorbing materials based on $BaTiO_3$ is very unique considering that this material has perovskite structure and poor ferromagnetic properties [4]. A good MAM is to tune the combination of these two properties [5, 6] in obtaining a macthing impedance which provides a better absorption effect. The preparation methods employed for the fabrication of $BaTiO_3$ based material, mostly are a sol gel, coprecipitation, conventional ceramic techniques, In-situ chemical oxidative polymerization and there are still few using solid state reaction [6].

Barium titanate ($BaTiO_3$) is a dielectric compound and has a small complex permeability. Despite dielectric constant pertaining in $BaTiO_3$ relatively large, its performance for MAM is found still relatively low [4]. In this study, we report results of our current studies on the improvement of absorbing capability of zinc doped barium titanate of $BaTi_{(1-x)}Zn_xO_3$ (BTZ). The x is selectively selected being 0; 1/3; ½; 2/3; 1.

2. Experimental method

Stoichiometry quantities of standard research-grade $BaCO_3$, TiO_2, ZnO precursors were mixed and wet milled in a high energy milling (Shaker Mill), with the addition of alcohol until 60 % volume for 10 hours. All mixed precursors were mechanically alloyed produced heavy deformed powder materials. The powder materials of $BaTiO_3$ composition were then dried and sintered at 1200 ^{0}C for 3 hours. Whereas dried powders of $BaTi_{2/3}Zn_{1/3}O_3$ and $BaTi_{1/2}Zn_{1/2}O_3$ compositions were sintered at 1100 °C and those of $BaTi_{1/3}Zn_{2/3}O_3$ and $BaZnO_3$ compositions were sintered at 1000 °C. The $BaTi_{(1-x)}Zn_xO_3$ compound has been prepared using solid state reaction.

The characterization of polycrystalline samples was characterized by X-Ray Diffraction (Brukker D8 Advance). XRD data of polycrystalline samples were refined by GSAS for structural analysis. Microwave absorption properties were measured by Vector Network Analyzer (Advantest R3770 300 kHz - 20GHz]. While complex permittivity $(\varepsilon' - j\varepsilon")$ and permeability $(\mu' - j\mu")$ values were calculated by Nicolson-Ross-Weir (NRW) method.

3. Results and discussion

All samples with $BaTi_{(1-x)}Zn_xO_3$ compositions were successfully characterized by X-ray diffraction (XRD). Figure 1 compares all diffraction patterns of mechanically alloyed $BaTi_{(1-x)}Zn_xO_3$ powders which show typical of crystalline sample. Obviously, the diffraction pattern of all samples indicates a similar pattern, but with an additional indication that the diffraction peaks are shifted due to partial substitution of Zn.

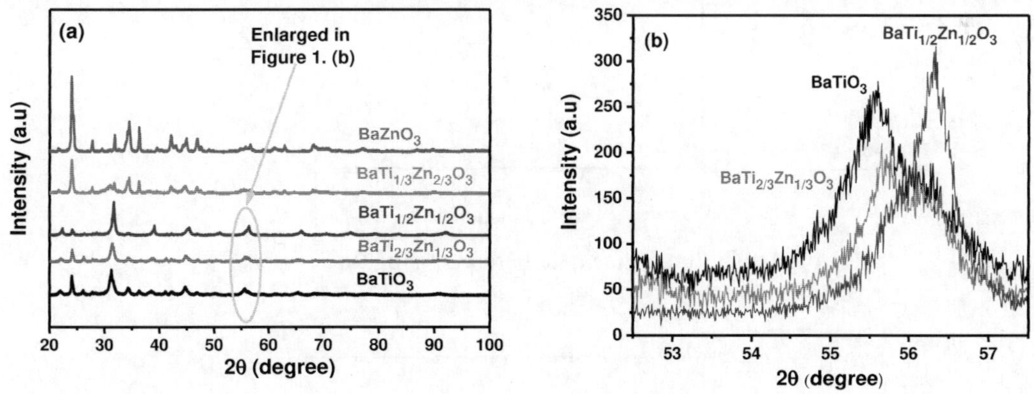

Figure 1. (a) XRD spectrum of BaTi(1-x)ZnxO$_3$ (x= 0; 1/3; ½; 2/3; 1) and (b) Shift in peak intensity of BaTi(1-x)ZnxO$_3$ x= 0; 1/3; ½; 2/3; 1).

The formation of a single phase is confirmed from peaks identification of the pattern which match well with ICSD number 98-007-8876 that all patterns correspond to $BaTiO_3$ phase. Figure 1. (b) compares a diffraction peak of (1 2 1) which clearly shown peaks shifting when Zn partially substituted Ti ion in $BaTiO_3$ phase. With increasing the Zn ratio, the peaks shifted to the right direction because the lattice parameter changes. The change must be due to the different in ionic radius between Zn^{2+} (74 pm) and Ti^{4+} (53 pm) [7].

IOP Conf. Series: Materials Science and Engineering **553** (2019) 012030 doi:10.1088/1757-899X/553/1/012030

Table 1. Lattice data parameters of $BaTi_{(1-x)}Zn_xO_3$.

No.	x	$BaTi_{(1-x)}Zn_xO_3$ Sample	Lattice Parameters			
			a (Å)	b (Å)	c (Å)	α, β, γ (degree)
1	0	$BaTiO_3$	3.996	3.996	4.030	$\alpha = \beta = \gamma = 90°$
2	1/3	$BaTi_{2/3}Zn_{1/3}O_3$	4.009	4.009	4.032	$\alpha = \beta = \gamma = 90°$
3	1/2	$BaTi_{1/2}Zn_{1/2}O_3$	4.047	4.047	4.042	$\alpha = \beta = \gamma = 90°$
4	2/3	$BaTi_{1/3}Zn_{2/3}O_3$	5.843	11.801	11.940	$\alpha = \beta = \gamma = 90°$
5	1	$BaZnO_3$	5.833	11.376	12.585	$\alpha = \beta = \gamma = 90°$

All data of the diffraction patterns are then refined by GSAS in which the refined pattern is determined by Goodness of Fit (GOF) and weighted profile R-factor (Rwp) as the acceptance indicators. Analysis of refined data have resulted in lattice parameters and crystal system of $BaTi_{(1-x)}Zn_xO_3$ as summarized in Table1. In addition, the dimension of cell allows for calculation of the cell volume of $BaTi_{(1-x)}Zn_xO_3$ which then plotted in Figure 2. Data in Table 1 indicate that there is a limit of the Zn to substitute Ti while still maintaining the shape of the tetragonal structure. Since the ionic radii of Zn^{2+} is much higher than that of Ti^{4+}, the a and c parameters changed with x. However, at x = 2/3 and 1 the crystal structure changes from the tetragonal to orthorhombic.

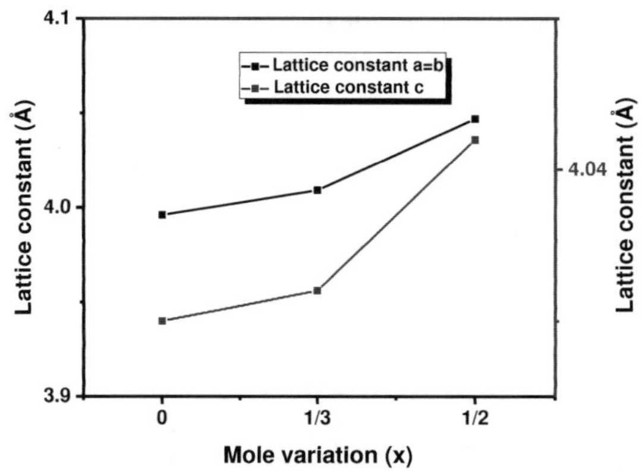

Figure 2. Plot lattice data parameters of $BaTi_{(1-x)}Zn_xO_3$.

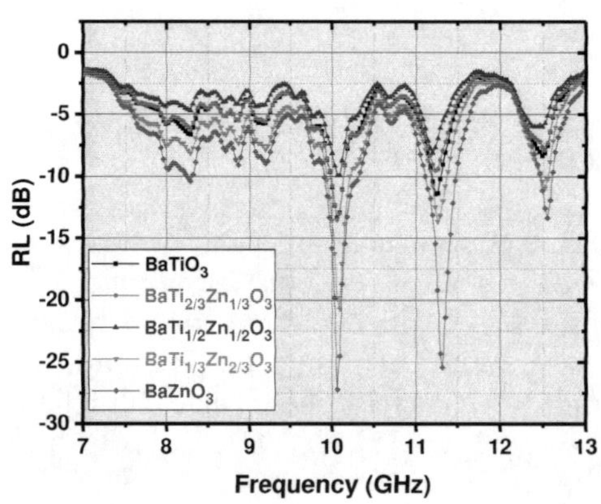

Figure 3. Reflection Loss of $BaTi_{(1-x)}Zn_xO_3$.

Values of RL, which were calculated from the relation $RL = 20log|(Z_{in} - Z_0)/(Z_{in} + Z_0)|$ [6, 8], in the frequency range 7 – 13 GHz for $BaTi_{(1-x)}Zn_xO_3$ (x = 0; 1/3; ½; 2/3; 1) samples are compared in Figure 3. Z_{in} and Z_0 are respectively impedance of the sample and electromagnetic wave in air. The absorption spectrums clearly indicate that the incoming electromagnetic waves entering the samples were absorbed partially in which the RL values are below 0 dB in the frequency range but at a certain frequency, the RL value even reduced further. The RL value at peak frequencies 8.29 GHz, 8.86 GHz, 9.19 GHz, 10.06 GHz, 11.32 GHz and 12.55 GHz increases with a decrease in x from x = ½ to x = 0. However, a significant increase in RL was obtained in samples with x = 2/3 and 1 where the crystal system has changed from cubic to orthorhombic. The change in the crystal system must also be followed by a change in dielectric constant since there is a discrepancy in ionic size between Zn^{2+} and Ti^{4+}.

The RL value of $BaTi_{1/3}Zn_{2/3}O_3$ and $BaZnO_3$ are 20.69 dB and 27.20 dB respectively at a frequency 10.06 GHz being the largest value within the frequency range. Hence, the highest microwave absorption up to 95.64% was obtained in $BaZnO_3$ at a frequency 10.06 GHz. The presence of Zn ion as a doping agent increases permittivity value [9]. This might be the reason for such RL enhancement. In Table 3, the electromagnetic characteristics of $BaTi_{(1-x)}Zn_xO_3$ samples are summarized which including the value of complex permittivity and permeability calculated based on NRW. The RL value is determined by both the magnetic and dielectric properties in addition to resistivity which overall determine the total impedance of material. $BaTi_{(1-x)}Zn_xO_3$ sample with x = 2/3 and x = 1 must have the great dielectric value, because in the electric field the frequency of the dielectric constant can increased the electrical energy loss [4]. It is also worthy to note that $BaTi_{(1-x)}Zn_xO_3$ possesses the negative value of imaginary permeability (μ'') due to the substitution effects that enhance the level of the microwave absorption [10].

Table 2. Microwave Absorption characteristics of $BaTi_{(1-x)}Zn_xO_3$.

No.	BZT	Frequency 10.06 (GHz)			Frequency 11.32 (GHz)		
	x	Permittivity (ε_r)	Permeability (μ_r)	Absorbtion (%)	Permittivity (ε_r)	Permeability (μ_r)	Absorbtion (%)
1	x=0	1.87+j2.10	4.31-j1.35	78.47	0.30+j1.87	2.53+j0.20	60.51
2	x=1/3	1.80+j2.07	4.26-j1.42	78.85	0.43+j1.67	2.18-j0.29	56.45
3	x=1/2	1.02+j1.74	3.64-j2.47	68.17	0.56+j1.36	1.78-j1.41	44.54
4	x=2/3	2.21+j0.10	2.92-j1.64	89.30	-0.83+j1.68	1.48+j2.07	71.09
5	x=1	2.05-j0.01	2.82-j1.06	95.64	1.36-j0.35	1.00-j1.35	91.49

4. Conclusion

The role of Zn substitution in $BaTi_{(1-x)}Zn_xO_3$ with composition x = 0; 1/3; ½; 2/3; 1 is to change both phase and dielectric properties. The $BaTi_{(1-x)}Zn_xO_3$ with x = 0, 1/3 and ½ are a single phase material with a tetragonal crystal system which change to an orthorhombic crystal system at x = 2/3 and 1. The change of crystal system brought the change in dielectric constant and microwave absorption characteristics of the BZT. The microwave absorption characteristics are found improved in BZT from which the absorption of more than 89 and 95 % is respectively obtained in samples with x = 2/3 and 1.

Acknowledgment

The authors gratefully acknowledge the support of Postgraduate Program of Materials Science Universitas Indonesia for the research facilities. We are also thankful for the financial support provided partially by the Directorate of Research and Community Service Universitas Indonesia. Our thanks also go to the Research Centre for Electronics and Telecommunication, Bandung, Indonesian Institute of Science Indonesia Ministry of Research, Technology and Higher Education for the XRD and VNA facilities.

References

[1] Afghahi S S S, Jafarian M and Atassi Y 2016 *J. Magtm. Magtcs. Mater.* **419** 62 DOI: 10.1016/j.jmmm.2016.06.003
[2] Pande A, Gairola P, Sambyal P, Gairola S P, Kumar V, Singh K and Dhawan S K 2017 *Mater. Chem. Phys.* **189** 22 DOI: 10.1016/j.matchemphys.2016.12.045
[3] Luo H, Gong R, Wang X, Nie Y, Chen Y and Harris V G 2015 *J. Alloys Compd.* **646** 345 DOI: 10.1016/j.jallcom.2015.05.208
[4] Chen X, Wang G, Duan Y and Liu S 2008 *J. Alloys Compd.* **453** 433 DOI: 10.1016/j.jallcom.2006.11.092
[5] Wang X, Li Q, Su Z, Gong W, Gong R, Chen Y, and Harris V G 2015 *J. Alloys Compd.* **643** 111 DOI: 10.1016/j.jallcom.2015.04.122
[6] Xiang J, Hou Z, Zhang X, Gong L, Wu Z, and Mi J 2018 *J. Alloys Compd.* **737** 412 DOI: 10.1016/j.jallcom.2017.12.047
[7] Chen D, Luo F, Zhou W, and Zhu D 2018 *J. Alloys Compd.* **757** 348 DOI: 10.1016/j.jallcom.2018.05.116
[8] Meng F, Wang H, Huang F, Guo Y, Wang Z, Hui D and Zhou Z 2018 *Composites* **137** 260 DOI: 10.1016/j.compositesb.2017.11.023
[9] Liu Z, Xing Z, Wang H, Xue Z, Chen S, Jin G and Cui X 2017 *J. Alloys Compd.* **727** 696 DOI: 10.1016/j.jallcom.2017.08.175
[10] Chandra Babu Naidu K, and Madhuri W 2017 *Mater. Chem. Phys.* **187** 164 DOI: 10.1016/j.matchemphys.2016.11.062

Synthesis and Characterization of Biochar from Crab Shell by Pyrolysis

Muhamad Nasir[1*], Triannisa Rahmawati[1], Fitri Dara[1]
[1] Development Unit for Clean Technology, Indonesia Institute of Sciences (LIPI), Cisitu-Sangkuriang, Bandung 40135, Indonesia

mnasir71@yahoo.com

Abstract. Carapace, which is part of the crab shell exoskeletons were analysed through structural and chemical methods. This material was compared between grinded and ungrinded process. This two processes resulted product called biochar. SEM-EDS, FT-Ir, XRD, TG-DSC, AAS, and PSA were conducted to characterized biochar. The SEM images showed different in biochar particles size. The composition of biochar differ for each pyrolyzed temperature, where Ca was the dominant element and C was the poor element. FT-Ir spectra provided different spectra for biochar prepared at 500-600 °C and biochar prepared at 700-900 °C. XRD data showed that calcite and lime-based biochar depend on pyrolyzed temperature. Thermal analysis generated that biochar decomposed into three main stages. Otherwise, biochar through grinded step give resemble result as like biochar without grinded step, but has smaller particle size, which means grinded process did not give significant effect both structural and chemical composition, but give any effect on particle size. Furthermore, these study indicate that biochar potentially can be used as adsorbent.

Keywords: Biochar, crab shell, grinded, lime, pyrolysis

1. Introduction

Every year, around 6 to 8 million tonnes of waste crab, lobster, and shrimp shell are produced globally, where in developing countries, waste shells are often just dumped in landfill or the sea [1]. The abundant source of waste shell should be used, because the shell consist of some material that beneficial for many fields. But unfortunately, the potential value of such shell for chemical industry is being deficiented. But now, many researcher have great attention on this material. Some material from crustacean like crab, shrimp and wood lice being study now.

Crustacea are one of omnifarious group of aquatic animals, that lives at diffenent places, like in tropical to artic water, extending into freshwater and in some onto land for part of their life history, and the shoreline to the deep ocean [2]. For crustacea, the main important part is exoskeleton.

The exoskeleton is the outer part of animal that has function to stabilizes animal body and as protection against predators. The crustaceans exoskeleton consist of biomineralized stucture which construct by organic matrix, along with inorganic mineral, that mostly contain calcium carbonate [3]. Calcium carbonate is widely used as filler [4], extender material in paper [5], cosmetic, medicine, paint, plastic, sealant, food, ceramics, textile, adhesive, etc. In previous paper, calcium carbonate has been produced from waste marble powder (WMP) by calcination-dissolution-precipitation (CDP) method [6]. While in animal, cacium carbonate is material that made exoskeleton of animal harder, for example in crab shell exoskeleton.

The exoskeleton of crab shell is a natural composite consist of highly mineralized chitin-protein fibers arranged in a twisted plywood or Bouligand pattern [7], where it was (snow crab) consist of 34.2% dw protein and amino acids, 17.1% dw fat, and 28.5% dw (calcium, phosporous, and magnesium as major materials), and yother materials 21.2% bw [8]. This exoskeleton divided into three parts: carapace, claw, finger. Here, carapace was used as raw material of the study.

Carapace is the hard upper shell of tortoise, crustacean, or arachnid. In crustaceans, the function of carapace is as protective cover. This part of crab shell can be used as raw materials to produce biochar after pyrolysis applied because that is the most abundant part in crab shell. Combined with its application in soils, the production of biochar have been suggested as promising for mitigating climate change [9, 10], enhancing crop production [11] the waste disposal [12], and as simultaneously amending soil [13, 14]

Previously, researcher have been investigate SEM, X-ray micro diffraction, optical microscopy to decipher the functional role of calcium carbonate in lobster [9-11], identify the composition of the exoskeleton of two crustacea (lobster and crab) [12], investigate calcium rich biochar from the pyrolysis of crab shell for phosphorus removal [13]. But in present study, the as overarching objective was to compare the effect of insertion grinded and ungrinded process of biochar derived from carapace of crab shell. SEM-EDS, FT-Ir, TG-DSC, XRD, AAS, and PSA were conducted to analysed the sample characteristics.

2. Experimental method

2.1. Materials and biochar preparation

Crab shell waste was collected from Grogol Village, Gunung Jati District, Cirebon Regency. The shell was separated based on their parts (Carapace, claw, finger, etc). On this study carapace was used as raw materias. The carapace was washed with clean water, after that dried in the oven at 90 °C for 24 h.

The carapace was placed in mortal and pounded roughly. Then, ground with grinder, finally sieved through 200 meshes (0.74 μm) before use. Carapace powder that passes 200 meshes sieving machine was collected. A crucible with around 6 gram of carapace powder was put in furnace for pyrolysis. Each of carapace powder was pyrolysis at 500 to 900 °C using Nabertherm GmbH Bahnhoftstr 20.28865. at at heating rate 10 °C/min and keep for 1.5 h before cooling to room temperature. Further, the sample produced from this step was called biochar. The sample denoted for pyrolysis temperatur 500 to 900 as follow: B500 (biochar pyrolysis at 500 °C), B600, B700, B800, B900 respectively with interval temperature 100 °C. Each of biochar produced from pirolysis was weighned to calculate powder loss when pyrolysis occured.

Another sample was prepared with the same procedure, but between sieving and pyrolysing, carapace was grinded using high energy milling (HEM) for 15 h, then pyrolyzed at 500 °C (BG500). For grinded proses, the apparatus controlled as follow: 1 h grinded (on), 10 minutes rest grinded (off).

2.2. Characterization of biochar

The morphology of carapace powder and biochar were captured using scanning electron microscopy (SEM JEOL JSM IT-300) at 20.0 kV accelerating voltage for scanning the surface, whereas the electron dispersive spectroscopy (EDS) used to monitored the % composition of sample. Thermogravimetric-differential thermal analysis (TG-DTA) is used to quantitatively determine the mineral constituent, and also organic compounds that might be consist in the samples. The sample was measured by using TGA 209 Lybra (Netzsch), where the sample was heated from 40 to 900 °C for 10 °C/min of heating rate under N_2 gas. Diffraction peak of biochar and carapace 200 mesh were recorded by x-ray diffraction (XRD) Bruker D8 ADVANCE at ambient temperature. Analysis was performed on peaks from the 2θ range between 10° to 70° with the rotating anode CuKα tube (λ = 1.5418 Å).

Particle size was calculated using particle size analyzer (PSA) Beckman Coulter$_{TM}$ LS I3 320 Laser Diffraction, where the powder were dispersed in DMF. Fourier transform infrared (FT-Ir) Thermo Scientific Nicolet iS5 was taken to observe functional group in the sample. Atomic absorpsion spectroscopy (AAS) Agilent Technologies 200 Series AA was conducted to measure calsium and magnesium content. In this processes, wet destruction was applied to sample using HNO_3 and H_2O_2 overnight, continued with digestion using Advanced Microwave Digestion System Ethos at 1800 watt, 200 °C for 15 minutes, where 15 minutes was time to reach that temperature.

3. Results and discussion

3.1. Pyrolysis temperature variation and mineral analysis of biochar

Yield of crab shell pyrolyzed at various temperature were taken. It was shown that the yield was decreased from 66.21 % for BG500 and 66.36% for B500 to 43.07% for B900. As temperature pyrolysis increase, % yield of biochar getting lower. It was happened because $CaCO_3$ decomposed to be CaO and CO_2, this carbon dioxide was release to the atmospher caused low of % yield. For BG500 dan B500 also showed quite similar data. In which for grinded process the % yield was 66.21%, and for ungrinded % yield 66.36%. The EDS was conducted to measure material composition. While AAS taken to compare composition data with EDS and provided data quantitatively. The detailed yield for each treatment are summarized in Table 1.

Table 1. The yield percentage and particle size.

Treatment	% Yield	Particle Size (μm) (PSA data) Mean	Mode	Interval
200 mesh	-	1.01	1.75	0.09-292.04
Grinded 15 h, pyrolysis 500 °C	66.21	23.22	18.00	0.34-1041.05
500 °C	66.36	9.19	5.36	0.09-133.75
600 °C	64.86	155.60	153.80	0.04-1377.20
700 °C	45.66	17.64	19.76	0.04-634.19
800 °C	45.74	12.48	13.61	0.04-634.192
900 °C	43.07	90.73	168.90	0.04-863.88

Table 2. Summary of EDS and AAS result for elemental analysis.

Treatment	Composition (wt.%)				Other	Composition (wt.%)	
	Ca	C	O	Mg	(Na, Cl, Al) (wt.%)	Ca	Mg
Method	EDS					AAS	
200 mesh	27.97	24.19	43.83	1.46	2.55	13.29	3.26
Grinded 15 h, pyrolysis 500 °C	24.90	17.83	52.34	2.03	2.90	24.78	4.68
500 °C	27.78	15.52	48.84	1.92	5.94	24.32	4.63
600 °C	22.16	16.58	49.88	3.23	8.15	28.96	5.11
700 °C	31.49	9.13	54.84	2.66	1.88	40.61	6.35
800 °C	41.04	6.79	48.06	1.88	2.23	38.05	6.20
900 °C	39.20	6.46	46.83	4.12	3.39	45.14	6.49

Energy dispersive spectroscopy detected that the mineral content varies for each treatment. Calcium, carbon, oxygen, phosporus, magnesium present as major components of crab shell and biochar, whereas sodium, aluminium, chloride exist as minor components. This result is in tune that

IOP Conf. Series: Materials Science and Engineering **553** (2019) 012031 doi:10.1088/1757-899X/553/1/012031

calcium is the highest percentage of element, followed by magnesium [15]. From this result, as a function of pyrolysis temperature increased, the total calcium tend to increased (27.97 to 41.04 wt.%), except for B600 and B900, where calcium in B600 lower then B500, and B900 lower then B800; oxygen around 50 wt.%; and carbon 24.19 for B500, BG500 to 6.46 wt.% for B900.

Atomic absorption spectroscopy was conducted to measure elemental analysis of magnesium and calcium. The resulted data showed that there was not orderliness for magnesium, while for calcium, as temperature pyrolysis increased, the total calcium in the sample increase, except for calcium prepared at 800 °C. This measurement indicated that there was relationship between EDS and AAS calculation, namely there was not orderness both in EDS and AAS for magnesium as pyrolysis temperature increased, and there was an increasing tendency of calcium as pyrolysis temperature increased. Table 2 summarizes EDS and AAS results.

3.2. Biochar morphology and particle size

The SEM image afforded that there was not significant surface morphology from crab shell, BG500, B500 to B900. From SEM scanning, the all sample seems like powder, layer stacking and still in micrometer size. Crab shell 200 mesh typically has bigger size than grinded crab shell for 15 h. Furthermore, after pyrolysis treatment applied, the size of particle seems like smaller (Figure 1). But, by validating with PSA measurement, actually, after pyrolysis applied, the interval size of particle became wider. The mean, mode, size interval also varied, and there was not any trend. Table 1 presented PSA measurment result.

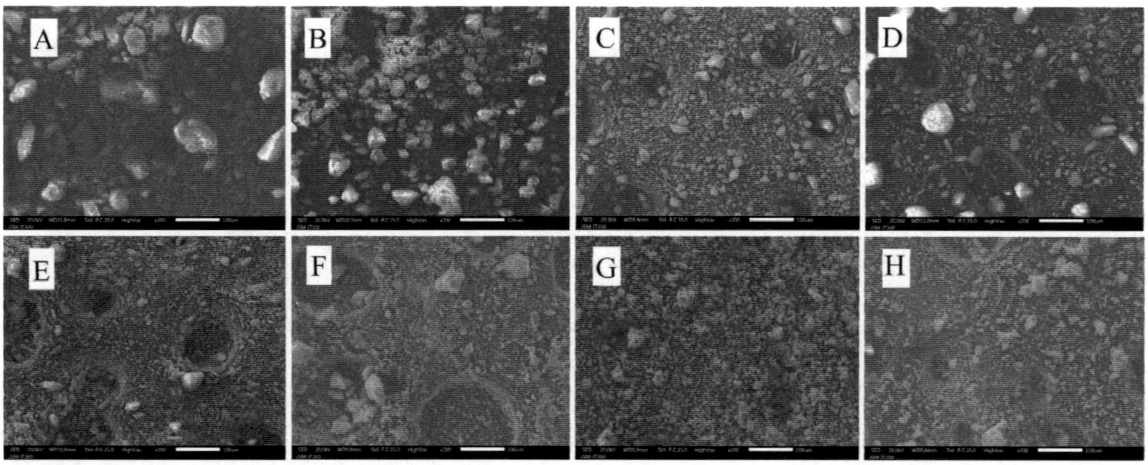

Figure 1. SEM analysis of crab shell and biochar, (a) crab shell, (b) crab shell grinded 15 h, (c) BG500, (d) B500, (e) B600, (f) B700, (g) B800, (h) N900.

3.3. TG-DSC analysis

Crab shell and biochar treated at 500 to 900 °C showed a quite different weight loss at some temperature (Figure 2, 3, 4). For crab shell and biochar, the first decomposition occurred at ambient temperature, which was signified to the volatilization such solvent and include endothermic process. The second stage at around 380-475 °C is exothermic process, it resulted two different appearance which attributed to organic matrix. For crab shell, BG500, B500, B600 this decomposition process do not appear, but appear at B700 to B900. On this stage, % mass loss 14.44%, 14.44%, and 14.77% for B700, B800, and B900 respectively. Whereas the mass loss associated to calcite thermal decomposition take placed at around 475-900 °C. At under pyrolysis treatment 600 °C, it involved high % mass loss calcite, from 36.39% to 38.38%, while for higher temperature, its content of calcite decreased slight with temperature increased. These thermal analysis showed that, the spectra of BG500 is typical to B500. It mean, grinded did not change thermal properties. The thermal decomposition of calcite is endothermic.

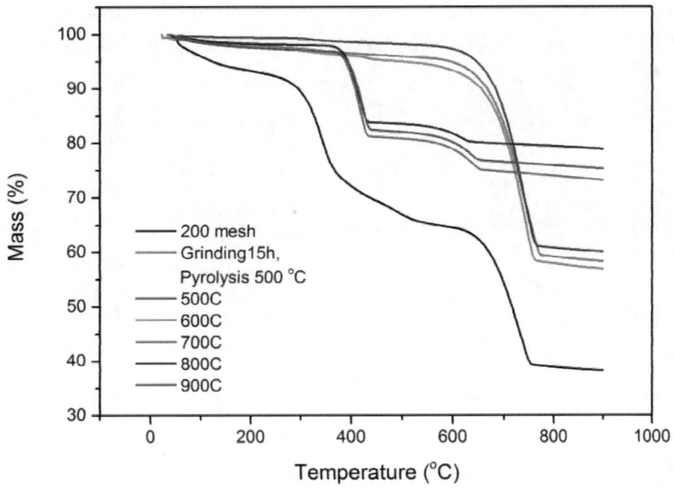

Figure 2. TGA of crab shell and biochar at different pyrolysis temperature.

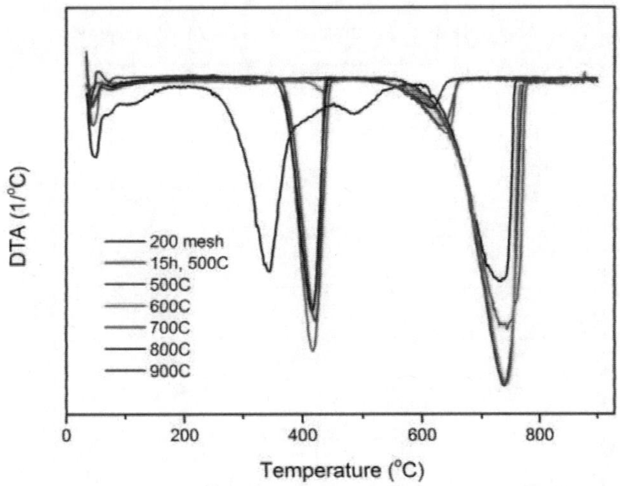

Figure 3. DTG of crab shell and biochar at different pyrolysis temperature.

3.4. XRD analysis

Crystalization of crab shell and biochar pyrolysis at temperature under 600 °C mainly constructed with calcite in which evidance with strong peak at 29.6° and slowly decreased at pyrolysis temperature above 700 to 900 °C. 23.2° also assign calcite peak, but in weak peak intensity [16]. Meanwhile, as peak 29.6° decreased, peak at 34.1° increased significanly at B700, B800, B900. The peak suggested that calcite has been undergo thermal decomposition to another crystal form. This crystal form known as lime. The thermal decomposition of calcite shown in Figure 5. The other peak that showed lime were at 50.8°, and 18.0°.

The crystallinity of all sample also determined. The crystalinity from crab shell to B600 gradually increase, and decline for B700, further increase slowly from B700 to B900. The crystallinity of crab shell, BG500, B500, B600, B700, B800, and B900: 65.9%, 79.5%, 79.1%, 80.6%, 74.3%, 77.5%, 78,6% respectively. The reduced % crystallinity from B600 to B700 because the transformation crystallinity form from calcite to lime. This XRD result conclude that biochar prepared at bellow 600 °C was calcite-based biochar, and at above 700 °C resulted lime-based biochar [17].

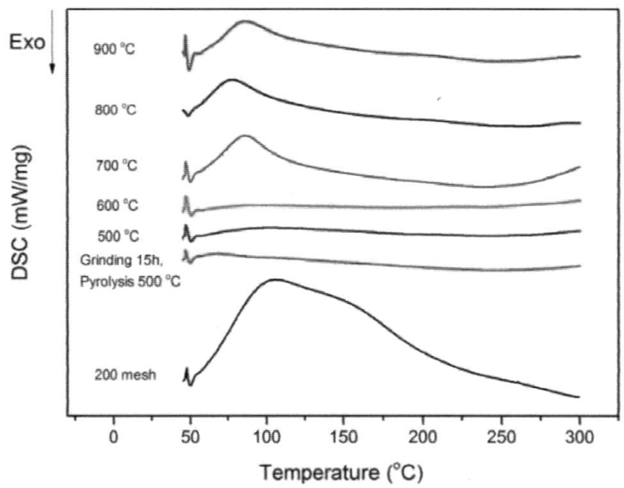

Figure 4. DSC of crab shell and biochar at different pyrolysis temperature.

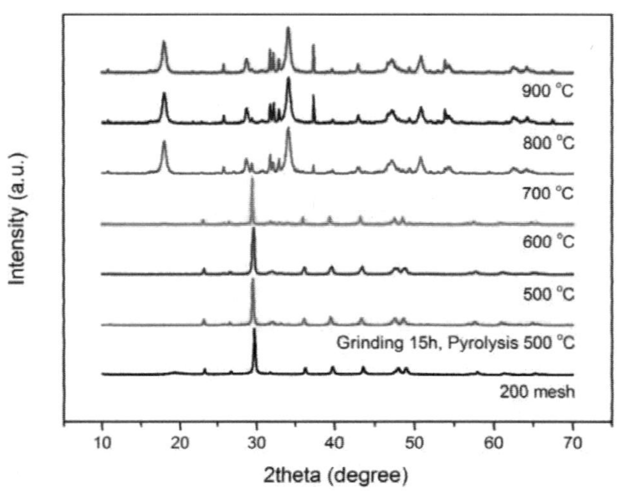

Figure 5. XRD data of crab shell and biochar at different pyrolysis temperature.

3.5. FT-Ir Analysis

The FT-Ir analysis of crab shell and biochar with different pyrolysis temperature showed spesific band of calcite and lime (Figure 6). The calcite band occured around 1450 cm^{-1}, 874 cm^{-1}, 712 cm^{-1} for crab shell, BG500, B500, B600. This peak is in a good aggrement with calcite typical absorption peaks around 1433, 874, and 713 cm^{-1} [16]. The other literature resulted that calcite ($CaCO_3$) in crab shell occured at 1796, 1456, and 874 cm^{-1} [18].

IOP Conf. Series: Materials Science and Engineering **553** (2019) 012031 doi:10.1088/1757-899X/553/1/012031

On the other hand, lime band represented at 2642 cm^{-1}, it was suggested that $Ca(OH)_2$ formed. This band appeared to crab shell which pyrolyzed at above 700 °C. Together with XRD data, the FT-Ir data showed that calcite is the main calcium crystal for biochar prepared under 600 °C, while at pyrolysis temperature above 700 °C, lime is formed and the most abundant calcium crystal form. However, the other band appeard for CH, CH_2, CH_3 at 2925-2850 cm^{-1}.

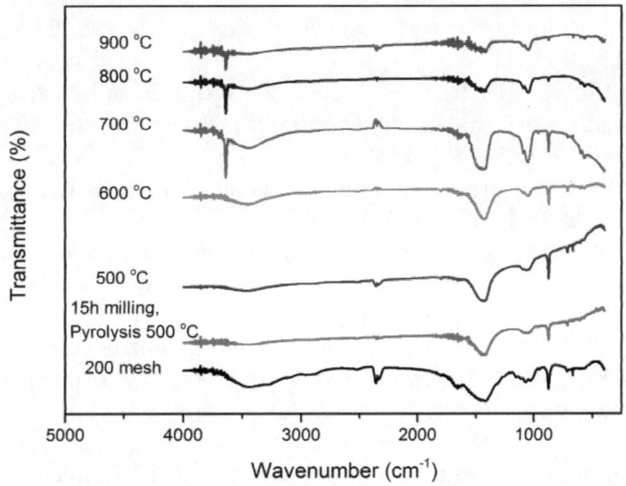

Figure 6. FT-Ir of crab shell and biochar at different pyrolysis temperature.

4. Conclusions

This study presented that there was not any effect on the insertion grinded proses for 15 h before pyrolysis. Furthermore, from FT-Ir and thermal analysis the grinded did not give different significant chemical or physical properties of biochar. .XRD analysis showed that at bellow 600 °C, it produced calcite-based biochar, while at above 700 °C produced lime-based biochar.

This data also conclude that as themperature pyrolysis increased, % yield of biochar gradually decrease. From elemental analysis showed that calcium is the major component, and carbon is the minor componen, where EDS and AAS calculation give linear trend for magnesium and calcium element

Acknowledgements

This research acknowledge to Ministry of Research, Technology, and Higher Education of the Republic of Indonesia and Indonesia Institute of Science.

References

[1] Yan N and Chen X 2015 Don't waste seafood waste *Nature* **524** 155-157 DOI 10.1038/524155a

[2] Penn J W, Caputi N, Lestang S D, Johnston D, Kangas M, and Bopp J 2018 Crustacean fisheries *Earth Systems and Environmental Sciences* DOI: 10.1016/B978-0-12-409548-9.09577-4

[3] Meldrum F C 2003 Calcium carbonate in biomineralisation and biomimetic chemistry *International Materials Reviews* **48** 187-224 DOI: 10.1179/095066003225005836

[4] Karakas F and Celik M S 2012 Effect of quantity and size distribution of calcite filler on the quality of water borne paints *Progress in Organic Coatings* **74** 555-563 DOI: 0.1016/j.porgcoat.2012.02.002

[5] Lopez-Periago A M, Pacciani R, Garcia-Gonzalez C, Vega L F and Domingo C 2010 A breakthrough technique for the preparation of high-yield precipitated calcium carbonate *J. Supercrit. Fluids* **52** 298–305 DOI: 10.1016/j.supflu.2009.11.014
[6] Erdogan N and Eken H A 2017 Precipitated calcium carbonate production, synthesis and properties, *Physiochem. Probl. Miner. Process* **53** 57-86 DOI: 10.5277/ppmp170105
[7] Chen P-Y, Albert Y-M, Joana M and Marc A M 2008 Structure and mechanical properties of crab exoskeletons *Acta Biomaterialia* **4** 587-596 DOI: 10.1016/j.actbio.2007.12.010
[8] Maria A L Y, Maria V M, Susana A P and Julia L H 2011 Chemical composition of snow crab shells (*chionoecetes* opilio) *CyTA-Journal of Food* **9** 265-270 DOI: 10.1080/19476337.2011.596285
[9] Cusack D F, Axsen J, Shwom R, Hartzell-Nichols L, White S and Mackey KRM 2014 An interdisciplinary assessment of climate engineering strategies *Frontiers in Ecology and the Environment* **12** 280–287 DOI: 10.1890/130030
[10] Smith P 2016 Soil carbon sequestration and biochar as negative emission technologies *Global Change Biology* pp. 1-10 DOI: 10.1111/gcb.13178
[11] Dickinson D, Balduccio L, Buysse J, Ronsse F, van Huylenbroeck G and Prins W 2015 Cost-benefit analysis of using biochar to improve cereals agriculture *Global Change Biology Bioenergy* **7** 850–864 DOI: 10.1111/gcbb.12180
[12] Jeffery S, Bezemer T M, Cornelissen G, Kuyper T S W, Lehmann J, Mommer L, Sohi S P, Voorde T F J V D, Wardle D A and Groenigen J W V 2015 The way forward in biochar research: targeting trade-offs between the potential wins *Global Change Biology Bioenergy* **7** 1-13 DOI: 10.1111/gcbb.12132
[13] Gul S, Whalen J K, Thomas B W, Sachdeva V, Deng H Y 2015 Physico-chemical properties and microbial responses in biochar-amended soils: mechanisms and future directions *Agriculture, Ecosystems and Environment* **206** 46–59 DOI: 10.1016/j.agee.2015.03.015
[14] Puga A P, Abreu C A, Melo L C A and Beesley L 2015 Biochar application to a contaminated soil reduces the availability and plant uptake of zinc, lead and cadmium *Journal of Environmental Management* **159** 86–93 DOI: 10.1016/j.jenvman.2015.05.036
[15] Bobelmann F, Romano P, Fabritius H, Raabee D and Epple M 2007 The composition of the exoskeleton of two crustacea: the american lobster *Homarus americanus* and the edible crab *Cancer pagarus Thermochemica Acta* **463** 65-68 DOI: 10.1016/j.tca.2007.07.01
[16] Jimoh O A, Otitoju T A, Hussin H, Ariffin K S and Baharun N 2017 Understanding the presipitated calcium carbonate (PCC) production mechanism and its characteristics in the liquid-gas system using milk of lime (MOL) suspension *S. Afr.J.Chem.* **70** 1-7 DOI: 10.17159/0379-4350/2017/v70a1
[17] Dai L, Tan F, Li H, Zhu N, He M, Zhu Q, Hu G, Wang L, and Zhao J 2017 Calcium-rich biochar from the pyrolysis of crab shell for phosphorus removal *Journal of Environmental Management* **198** 70-74 DOI: 10.1016/j.jenvman.2017.04.057
[18] Gbenor O P, Adeosun S O, Lawal G I and Jun S 2016 Role of $CaCO_3$ in the physicochemical properties of crustacean-sourced structural polysaccharides *Materials Chemistry and Physics* **184** 203-209 DOI: 10.1016/j.matchemphys.2016.09.043

Effect of Sn-addition on the properties of the biomedical Ti-17Nb-6Ta alloy.

Alaa Keshtta [1*], Mohamed A.-H. Gepreel[1]
[1] Department of Materials Science and Engineering, Egypt–Japan University of Science and Technology, New Borg El-Arab City, Alexandria 21934, Egypt.

alaa.keshtta@ejust.edu.eg

Abstract. The effect of Sn-addition (0, 1.5, and 3 Sn, at.%) to the biomedical Ti-17Nb-6Ta alloy has been investigated in this study. The three alloys were proved using XRD analysis to be β-type alloys. Microstructural analysis using optical microscope showed martensite lathes in all alloys and proved that Sn addition stabilized β phase and suppresses the martensite formation during quenching. Micro-hardness results showed a slight increase with 1.5% Sn adding but superior addition of Sn with 3% at. has negligible effect on the hardness compared to Sn-free alloy. The compressive yield stress for the three alloys located between 400 to 500 MPa, and the stain values increased with the increasing of Sn percentages in the TNT alloy.

Keywords : β Ti-alloys, Ti-17Nb-6Ta alloy, Sn-addition, phase stability, biomedical alloy.

1. Introduction

Recently, researchers gave a great attention for new biocompatible materials as a solution of allergic problems in some commercial biomedical alloys. These alternatives opened the door for new alloy generations which are nickel-free ones, these alloys target vital elements to reach the highest biocompatibility [1,2]. Recent statistics showed that nearly two third of the patents in US over 13 years period in the last two decades went to biomedical sector[3].

The elements added to titanium to form an alloy show different effects on the stability of the titanium phases in projected alloy. The alloying elements of titanium alloys are classified into three different categories; α-stabilizers, β-stabilizers, and neutral elements. In the first type, α-stabilizers elements, the added element is working on increasing the temperature range at which α-phase could present. On the other hand, β-stabilizers elements are those elements that contribute to lower the temperatures of β-phase field. Neutral elements do not show a significant effect on stabilizing any of the two phases [4]. Sn was proved to be generally weak α-stabilizers but recent studies showed that in β-type alloys Sn acts like a weak β-stabilizing element [5].

The TNT alloy was developed recently by Gepreel et al, [6] and showed many promising mechanical properties (such as controllable strength, high ductility and excellent cold workability) [7,8]. High corrosion resistance besides excellent biocompatibility were confirmed by cytotoxicity tests [9].The purpose of this study is to investigate the effect of Sn adding to Ti-17Nb-6Ta (TNT) alloy on the microstructure and mechanical properties. Sn was added to the TNT with two different percentages and compared with the original TNT alloy.

2. Experimental method

2.1 Alloy design considerations

Ti-17Nb-6Ta (TNT) alloy has $\overline{Bo}$, $\overline{Md}$ and e/a values very close to that of GUM metals[10]. Here, $\overline{Bo}$ is defined as the average bond order between alloying element and Ti atom, and $\overline{Md}$ is the average d-orbital energy level and (e/a) is the valence electrons per atom ratio of the alloy [5]. Sn is known to be high biocompatible element for developing new Ti-alloys. Also, Sn is known as α-stabilizing element but it acts as a weak β-stabilizing element in β type alloys. The addition of Sn to TNT alloy will not affect the e/a value but decrease both the $\overline{Bo}$ and $\overline{Md}$ values and may result in higher β-phase stability, as shown in Table 1. This will give a room to control the phases stability and so the mechanical properties of this TNT alloy system. In this study, Sn is added to TNT alloy with amounts of 1.5 and 3 at % (named hereafter as TNT-1.5 and TNT-3, respectively).

Table 1. $\overline{Bo}$, $\overline{Md}$ and e/a values of TNT, TNT-1.5 and TNT-3 alloys.

Alloy	$\overline{Bo}$	$\overline{Md}$ (eV)	e/a
TNT	2.86377	2.44813	4.23
TNT-1.5	2.856165	2.442925	4.23
TNT-3	2.84856	2.43772	4.23

2.2 Material preparation

The three alloys were produced using arc melting furnace with electromagnetic stirrer option in argon atmosphere. Then cold rolling deformation was carried out on the cast ingot using rolling mill to ~30% of ingot thickness. After that the rolled ingot was cut into smaller specimens using precision cutting saw under lubricant cooling. Solution treatment was carried on the samples at 900° C for 10 min followed by ice-water quenching. Oxide layers were removed using sandpapers. Specimens were mounted for microstructure analysis then grinded with abrasive papers up to 2400 grit and followed by polishing. Specimens were followed by chemical etching as immersed in etching solution of 10% HF (40% conc.) and 20% HNO3 (70% conc.) [11]. Samples were prepared for XRD analysis and hardness testing by grinding them up to 1000 grit. Also, samples for compression tests were prepared using immerged wire cutting to cut cylindrical specimens with 6mm diameter and 7.5mm height. Then, these samples were re-solution treated at 900° C for 10 min followed by ice-water quenching to insure of removing any heat effect occurred while cutting the samples. After that, oxide layers were removed using manual grinding set. Also, a special set was used in order to get a fine parallel surfaces in compression samples.

2.3 Material testing

The optical micrographs of the alloys were observed using optical microscope. Existed phases of the alloys were investigated through Cu Kα X-ray diffraction with conditions 30 kV and 30 mA at a scan rate of 1 degree/min using (XRD-6100, Shimadzu, Japan) diffractometer. The micro-hardness was measured by (Shimadzu HMV-FA) tester under the load of 5 N and holding for 15 seconds. Every specimen was measured 5 times, and then the mean values were calculated. Also, the error deviations were obtained. Compression tests were operated using (Shimadzu AGS-X, 100 kN) universal testing machine in air at room temperature with stain rate $5*10^{-4}$ sec^{-1}. The compression test was stopped at 50% engineering strain.

3. Results and discussion

The microstructure of the solution treated TNT, TNT-1.5 and TNT-3 alloys are showed in Figure 1. The martensite lathes are observed in the three alloys with different amounts. For TNT martensite lathes are clearly appeared inside the grains and at the grain boundaries as presented in Figure 1(a), almost everywhere. Also, in TNT-1.5 the martensite lathes are observed but with less amounts than in TNT, as shown in Figure 1(b). This suggests and confirms the previous idea that Sn acts as β stabilizing element in this β-type TNT alloy. In the same way for TNT-3 alloy, the martensite lathes also appeared with less amount but scattered in random areas because of the confirmed higher β-phase stability in this alloy as shown in Figure 1(c),(d). These results have a good agreement with XRD ones shown in Figure 2 as discussed below.

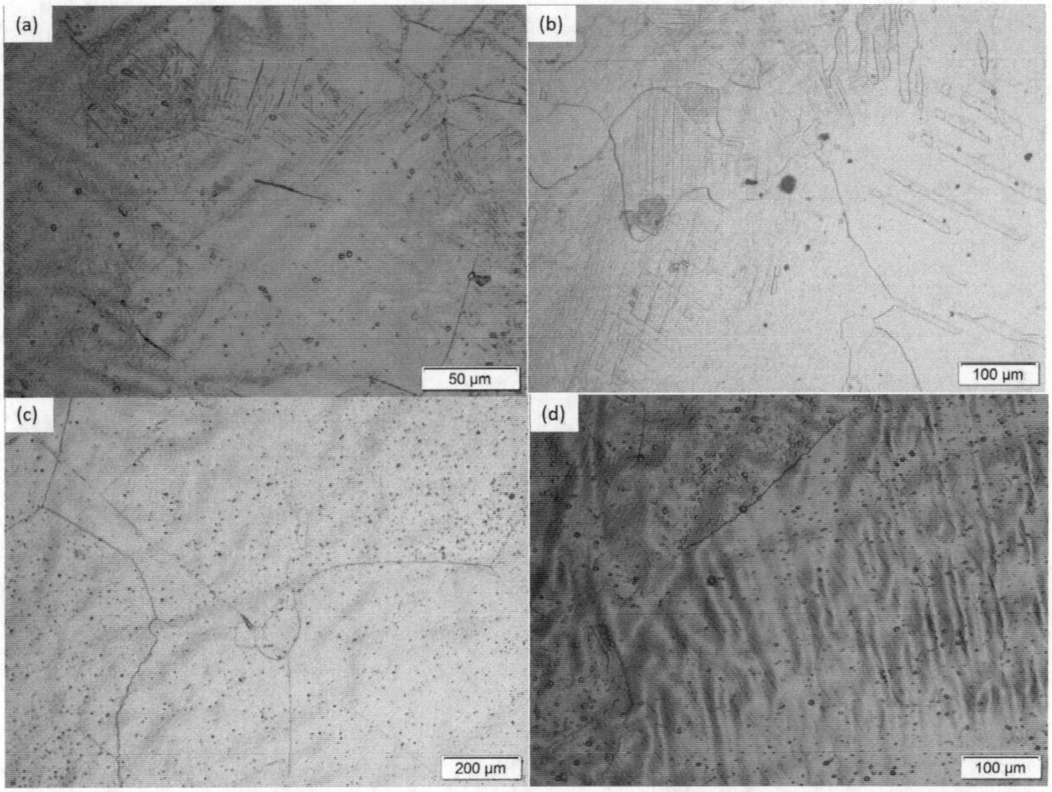

Figure 1. Optical micrographs obtained for (a)TNT, (b) TNT-1.5 and(c),(d) TNT-3 alloys in the solution treated condition.

The XRD patterns showed the changes in phases existed in the solution treated TNT, TNT-1.5 and TNT-3 alloys. Main Bragg reflections are related mainly to the α" martensite phase in addition to small amounts of β-phase in the solution treated TNT alloy. The α" martensite phase was formed during the quenching from austenitizing temperature. However, the β-phase is the predominant phase in the TNT-3 alloy. Besides the β-phase, the presence of α" martensite phase is also observed in this TNT-3 alloy. Martensite peaks of TNT-1.5 alloy are moderate between TNT-3 and TNT alloys with the predominance of β-phase. This confirms that the Sn-addition have a proven influence on stabilizing β-phase in the TNT β-type alloy through both microstructural and XRD investigations.

The micro-hardness change with the addition of Sn to the TNT alloy in the solution treated condition is measured and reported in Figure 3. The hardness value increased by the addition of 1.5%at. Sn from around 190 HV in TNT alloy to 220 HV in TNT-1.5 alloy. But with further addition of 3%at. Sn, in TNT-3 alloy, had almost no effect on the hardness and same values recorded as TNT alloy. Although the β-phase stability in TNT-3 alloy is higher than TNT, both showed the same hardness. The reason behind this is unclear at the moment and further detailed study is needed.

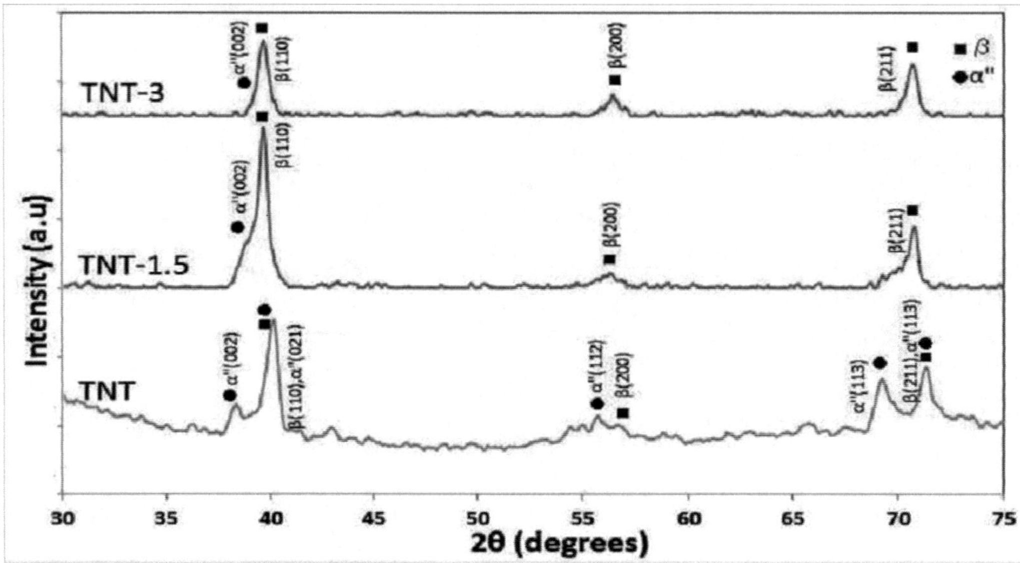

Figure 2.XRD patterns of TNT, TNT-1.5 and TNT-3 alloys in the solution treated condition.

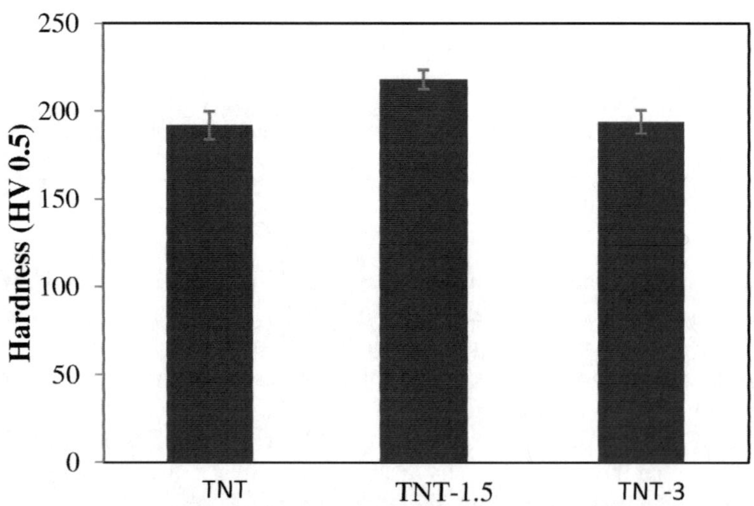

Figure 3. Micro-hardness changes of TNT, TNT-1.5 and TNT-3 alloys.

Compression tests were carried out on the solution treated samples of the alloys and the true stress – true strain curves were plotted in Figure 4. From this figure, it is clear that the elasticity/elastic strain of both alloys are high since the apparent yielding of the alloys is at compression true stain of more than 10%. Also, a double-yielding-like phenomena is clearly observed in the TNT-3 alloy. In addition, the alloys plastic yielding is in the range between 450 and 550 MPa. The lowest yielding strength is observed in TNT alloy and the highest is in TNT-1.5 alloy in good agreement with the hardness trend. The ultimate compression strength is showing the same trend; TNT-1.5 > TNT > TNT-3 with values 750, 700 and 600 MPa, respectively. The alloys show ultimate compression strength at a compression true strain of ~40%, while the test was stopped at ~50% true strain without specimens failure.

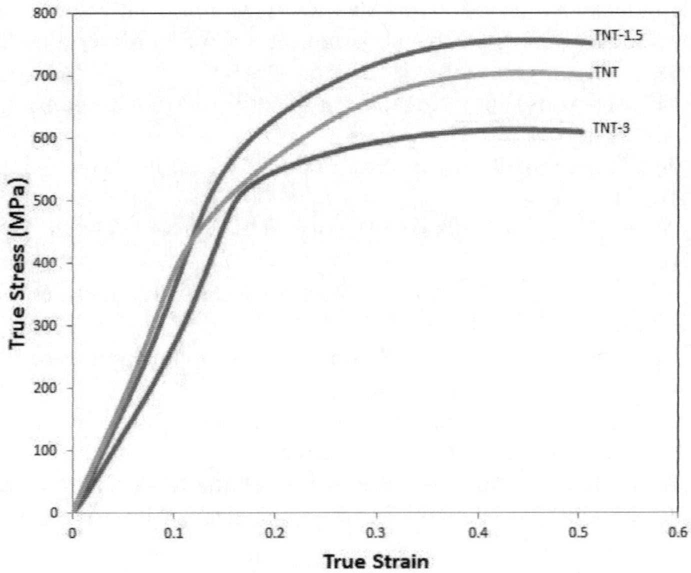

Figure 4. True stress - true strain compression curves for TNT, TNT-1.5 and TNT-3 alloys in the solution treated condition.

A similar study by the research group was done on TNT alloy with addition of Zr element instead of Sn element [12].That study done on Ti-17Nb-6Ta-3Zr alloy. Zr-addition did not show any significant effect on stabilizing β-phase compared to the original TNT alloy, on the contrary, Sn-addition in this study proved to act like β-stabilizing element. But, for both Zr and Sn addition the microstructure and XRD results agreed that β-phase is the predominant phase, beside lower amount of α" martensite phase. Hardness measurements showed that Zr-addition to TNT alloy resulted in relatively higher hardness values compared with values of Sn-addition.

4. Conclusion

Sn addition to the Ti-17Nb-6Ta (TNT) alloy stabilizes β-phase and suppresses the α" martensite phase formation during quenching from austentizing temperature zone. Hardness, Yield stress and ultimate compression stress increased by the addition of 1.5%at. Sn (TNT-1.5) alloy, however, further Sn addition (i.e., 3% Sn in TNT-3 alloy) proved to have no effect on the yield stress and ultimate compression stress as well .

Acknowledgment

The authors gratefully acknowledge the Missions Sector-Higher Education Ministry, Egypt, for the financial support of this work.

References

[1] Fu, J., et al., Novel Ti-base superelastic alloys with large recovery strain and excellent biocompatibility. Acta biomaterialia, 2015. 17: p. 56-67.
[2] Konopatsky, A., et al., Ternary Ti-Zr-Nb and quaternary Ti-Zr-Nb-Ta shape memory alloys for biomedical applications: Structural features and cyclic mechanical properties. Materials Science and Engineering: A, 2017. 702: p. 301-311.

[3] Jani, J.M., et al., A review of shape memory alloy research, applications and opportunities. Materials & Design, 2014. 56: p. 1078-1113.
[4] Peters, M. and C. Leyens, Titanium and titanium alloys: Fundamentals and applications. 2003: Wiley.
[5] Abdel-Hady, M., K. Hinoshita, and M. Morinaga, General approach to phase stability and elastic properties of β-type Ti-alloys using electronic parameters. Scripta Materialia, 2006. 55(5): p. 477-480.
[6] Gepreel, M.A.H., New Ti-Alloy with Negative and Zero Thermal Expansion Coefficients. Key Engineering Materials, 2011
[7] Hussein, A.H., et al., Biocompatibility of new Ti–Nb–Ta base alloys. Materials Science and Engineering: C, 2016. 61: p. 574-578.
[8] Keshtta, A.M. and M.A.-H. Gepreel. Shape Memory Effect in New Ti-Nb-Ta Alloy. in Materials Science Forum. 2017.
[9] Saleh, Y.E., M.A. Gepreel, and N.K. Allam, Functional Nanoarchitectures For Enhanced Drug Eluting Stents. Scientific Reports, 2017. 7.
[10] Saito, T., et al., Multifunctional Alloys Obtained via a Dislocation-Free Plastic Deformation Mechanism. Science, 2003. 300(5618): p. 464-467.
[11] Keshtta, A. and M.A.-H. Gepreel, Superelasticity Evaluation of the Biocompatible Ti-17Nb-6Ta Alloy. Journal of Healthcare Engineering, 2019. 2019: p. 8.
[12] Nyamuchiwa, K., et al. Microstructure and Mechanical Properties Change with Cold Deformation of the Biomedical Ti-17Nb-6Ta-3Zr Alloy. in Key Engineering Materials. 2018: Trans Tech Publ.

IOP Conf. Series: Materials Science and Engineering **553** (2019) 012033 doi:10.1088/1757-899X/553/1/012033

Synthesis and Characterization of Bi_2Te_3 Thermoelectric Material Using the Cold Pressing and Sintering Method

Tony Kristiantoro[1,2], Dedi[2], Vivi Fauzia[1*]
[1] Departemen Fisika, Fakultas MIPA, Universitas Indonesia, Kampus Baru UI Depok 16424, Indonesia
[2] Research Centre for Electronics and Telecommunications – LIPI, Bandung 40135 Indonesia

vivi@sci.ui.ac.id

Abstract. Research on thermoelectric material that could convert heat into electricity is expanding rapidly in the effort to improve efficiency. Bismuth telluride (Bi_2Te_3 or BT) is one of the most promising thermoelectric materials in converting waste heat into electrical energy; however, oxidation that occurs in the sintering process often decreases efficiency. In this study, we propose the use of activated carbon in the sintering of Bi_2Te_3, which is prepared using cold pressing to reduce the oxidation effect. Our results show Bi_2Te_3 possessing a polycrystalline single-phase rhombohedral form with lattice parameters of a = 4.3841 Å and c = 30.4816 Å. Covering the BT samples with activated carbon during the sintering process improved the power factor of Bi_2Te_3 from $0.775 \times 10^{-3} W.m^{-1}K^{-2}$ to $1.0125 \times 10^{-3} W.m^{-1}K^{-2}$. The improved thermoelectric properties exhibited in the carbon-covered BT samples may be attributed to the absence of an oxidation process occurring during sintering, as indicated by the energy dispersive spectrometry (EDS).

Keywords: Thermoelectric, Bi_2Te_3, Power factor, Seebeck coefficient

1. Introduction

Thermoelectric material becomes increasingly important in its use as an alternative source of harvesting energy and for energy conversion. The efficiency of thermoelectric materials is determined by the dimensionless figure of merit, $ZT = S^2\sigma T/\kappa$, where S, σ, κ, and T represent the Seebeck coefficient, electrical conductivity, thermal conductivity, and temperature, respectively [1]. Bi_2Te_3 (BT) has optimal thermoelectric properties and is the best material for thermoelectric cooling and thermoelectric power generation at near room temperature [2]. The crystal structure of BT is hexagonal or rhombohedral with the space group of R-3m [3].

The thermoelectric properties of BT-based alloys are influenced by the synthesis method. Besides the use of directed conventional growth from the melt, BT alloys are typically prepared with powder processing that is followed by spark plasma sintering (SPS), hot pressing, or hot extrusion. These methods are expensive and face scale-up challenges [4–10]. In this experiment, we propose the use of cost-effective cold pressing and sintering. However, oxidation occurring in the sintering process often causes a decrease in efficiency. Therefore, we propose the use of activated carbon in the sintering process to reduce the oxidation effect.

2. Experimental method

Commercial BT powder was determined to weigh about 0.5 g before samples were loaded into a steel mold and compacted into rod shapes (5 mm × 12 mm), as shown in Figure 1. Then, one sample was covered with activated carbon in an alumina boat, sintered at 470 °C for 3h in the furnace and then naturally cooled to room temperature.

Measurements of the Seebeck coefficient (S) and electrical resistivity (ρ) were determined using a commercial measurement system (Linseis LSR-4, Germany). The microstructures of samples were observed by X-ray diffraction (XRD D8 Advance, Bruker, Germany) at room temperature using CuKα radiation (λ = 1.5406 Å). The morphology and composition of the samples were observed using field-emission scanning electron microscopy (FESEM) and electron dispersive spectrometry (EDS) (JEOL JSM-IT300, Japan).

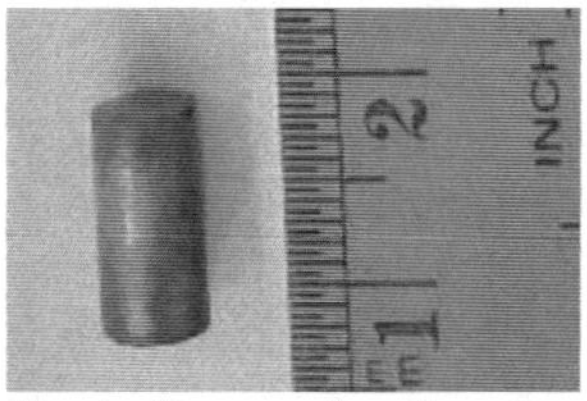

Figure 1. The Bi_2Te_3 bulk sample after the sintering process.

3. Results and discussion

XRD patterns of carbon-covered BT (C-BT), non-covered BT (NC-BT), and commercial BT samples are shown in Figure 2. Using a database Bi_2Te_3 pattern (ICSD#98-004-8203), we determined that all peaks exhibited in samples matched with the peaks of standard patterns, and no peaks were of the secondary phase. The diffraction peaks were indexed to reflect the rhombohedral phase of Bi_2Te_3 with the lattice constants of a = 4.3841 Å and c = 30.4816 Å.

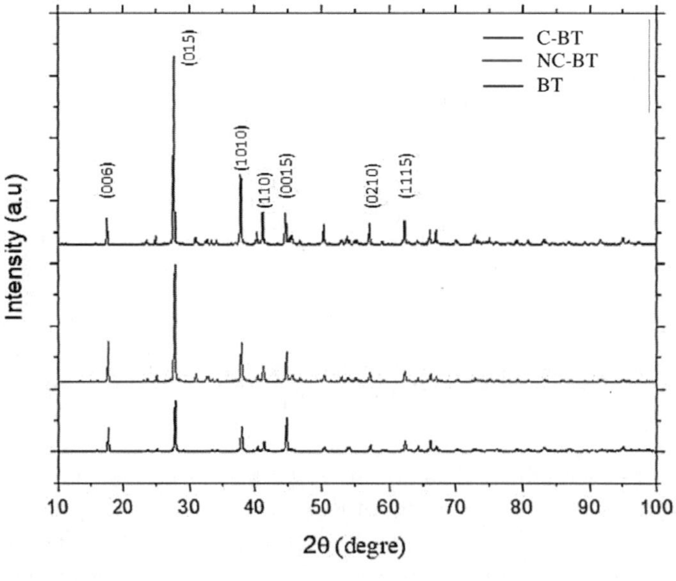

Figure 2. XRD patterns of non-covered BT (NC-BT), carbon-covered BT (C-BT), and commercial BT powder.

The morphologies of NC-BT, C-BT, and commercial BT powders are presented in Figure 3 (a,c,e). The material compositions of all samples were then confirmed by the EDS, as shown in

IOP Conf. Series: Materials Science and Engineering **553** (2019) 012033 doi:10.1088/1757-899X/553/1/012033

Figure 3 (b,d,f). From Figure 3b, it is clear that the NC-BT sample contains oxygen, indicating the occurrence of oxidation during the sintering process.

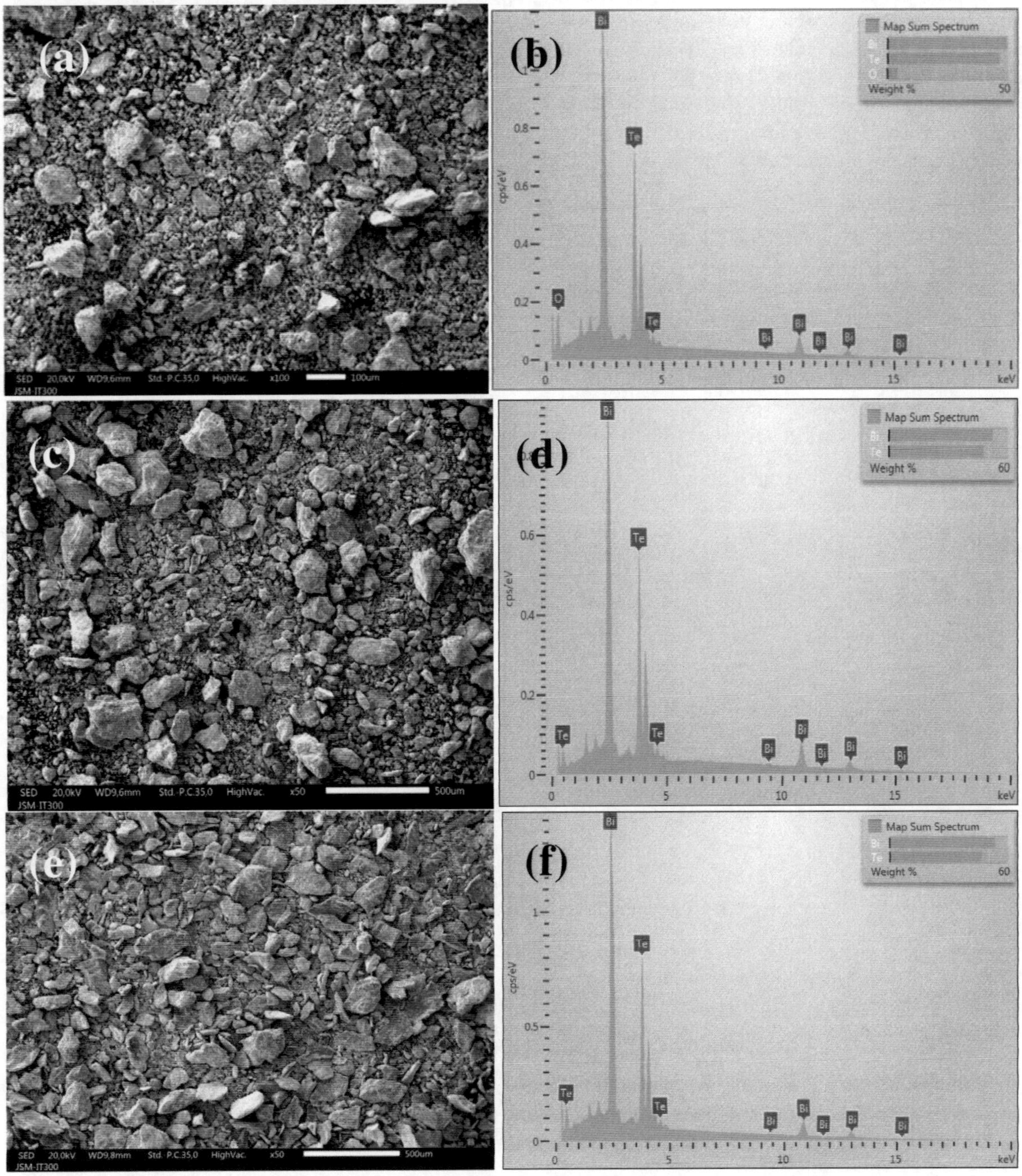

Figure 3. FESEM images and energy dispersive spectrometry (EDS) of non-covered BT (NC-BT) (a, b), carbon-covered BT (C-BT) (c, d), and commercial BT powder (e, f).

Figure 4 shows the temperature-dependent thermoelectric properties of samples that contain the Seebeck coefficient, S (a); the power factor, $S^2\sigma$ (b), and electrical resistivity, ρ (c). The negative value of the Seebeck coefficient shows that both samples are *n*-type semiconductors, and the Seebeck coefficient of C-BT is higher than that of the NC-BT samples. Figure 4 also shows that the Seebeck coefficient of each sample increases, from -117.7 to -147 μV/K, as the temperature increases from room temperature to 200 °C. Meanwhile, the ρ values of both samples also increase from 13.81 to 26.72 μΩ.m as the temperature increases from 30 °C to 200 °C. This is a typical characteristic of semiconductor materials. The electrical resistivity, ρ, of C-BT is lower than that of the NC-BT samples. The higher S and lower ρ values of the C-BT sample results in the power factor, $S^2\sigma$, of the C-BT sample being higher than that of the NC-BT sample, as shown in Figure 4(c). The C-BT sample exhibiting better thermoelectric properties may be attributed to the absence of an oxidation process during sintering, as indicated by the EDS.

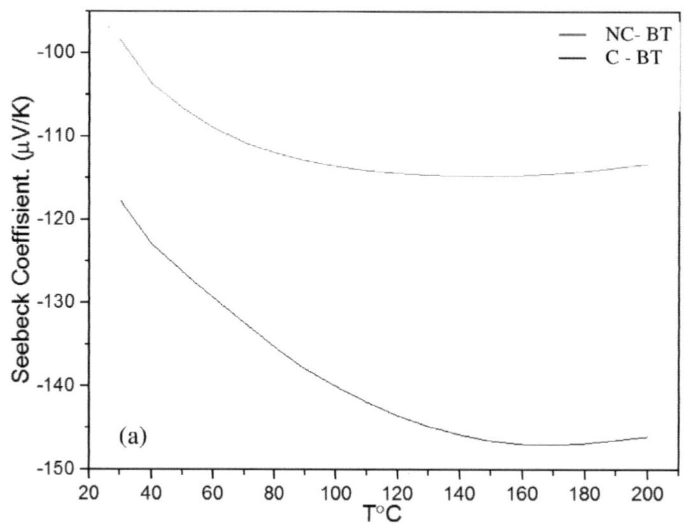

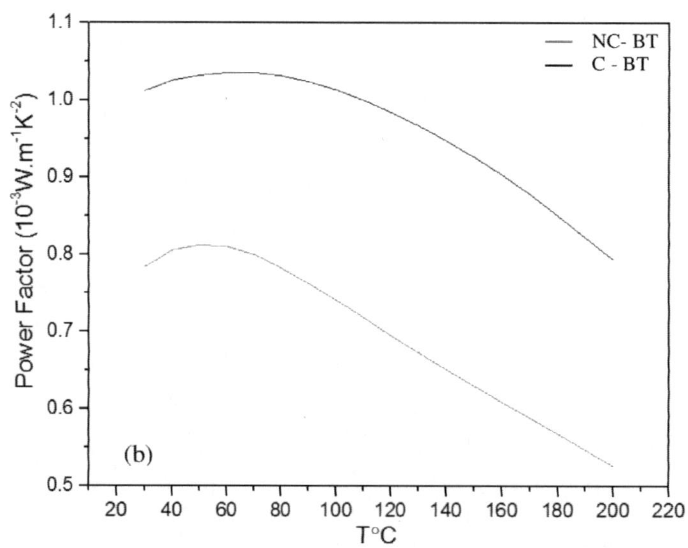

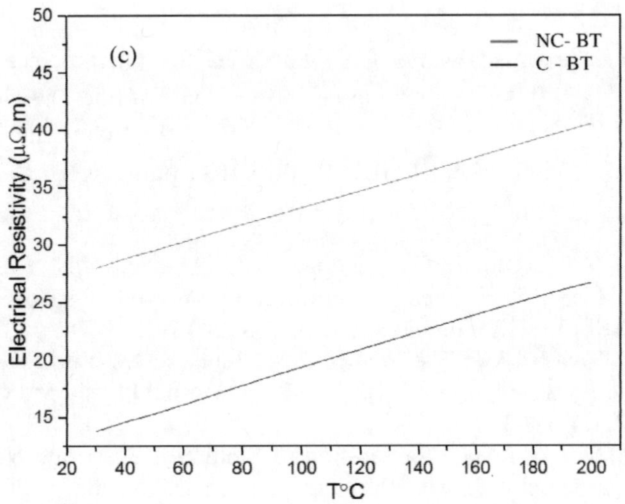

Figure 4. Temperature-dependent thermoelectric roperties, including (a) Seebeck coefficient S; (b) power factor $S^2\sigma$, and (c) electrical resistivity ρ of NC-BT and C-BT samples.

To compare our results with previously published results, a summary of the thermoelectric properties of Bi_2Te_3 at room temperature is shown in Table 1. Based on this table, it is evident that the power factor of the C-BT sample is higher than that of Bi_2Te_3 samples prepared using microwave sintering, but it is still lower than that of Bi_2Te_3 prepared by the spark plasma sintering method. This may be due to the microstructure properties or because the conductivity of our samples is not yet optimal. These results show that covering Bi_2Te_3 samples with activated carbon during the sintering process is an alternative, low-cost method of improving the thermoelectric properties of Bi_2Te_3.

Tabel 1. Thermoelectric properties of Bi_2Te_3 prepared by different sintering methods.

Preparation Method	Seebeck Coefficient (μV/K)	Electrical Resistivity (μΩ.m)	Power Factor (10^{-3}W.m^{-1}K^{-2})	References
Spark Plasma Sintering	192.15	9.85	2.125	[11]
Microwave Sintering	40.3	25.26	0.410	[12]
Carbon Covered	117.7	13.81	1.0125	Our work

4. Conclusion

We propose the simple and low-cost alternative method for sintering bulk BT samples of covering the BT samples with activated carbon during the sintering process, which our study shows can improve the power factor of Bi_2Te_3 from 0.775×10^{-3}W.m^{-1}K^{-2} to 1.0125×10^{-3}W.m^{-1}K^{-2}. The improved thermoelectric properties of carbon-covered BT samples may be due to the absence of an oxidation process during sintering, as indicated by the EDS.

Acknowledgments
Technical support was provided by the Core Facilities for Thermoelectric Laboratory, Research Centre for Electronics and Telecommunication, Indonesian Institute of Sciences. This research was financially supported by Thematic Research Funding from the Indonesian Institute of Sciences and Grant PITTA 2019 No. 2329/UN2.R3.1/HKP .05.00/2019 from Universitas Indonesia.

References
[1] Dedi, P.C. Lee, C.H. Chien, G.P. Dong, W.C. Huang, C.L. Chen, C.M. Tseng , S.R. Harutyunyan, C.H.Lee, Y.Y. Chen, "Thermoelectric power of single crystalline lead telluride nanowire," Applied Physic Letter, no.1-5, vol. 103, 2013
[2] DM. Rowe,"CRC Handbook of Thermoelectrics," 1st ed., New York: CRC Press, 1995 p. 239.
[3] S. Nakajima, "The crystal structure of $Bi_2Te_{3-x}Se_x$, "Journal Physic and Chemical Solids, vol. 24, pp. 479-485, March 1963
[4] K. Kishimoto and T. Koyanagi, "Preparation of Sintered Degenerate N-Type PbTe with a Small Grain Size and its Thermoelectric Properties," Journal Applied Physic, vol.92, pp. 2544–2549, 2002
[5] Y. Gelbstein, Z. Dashevsky, and M.P. Dariel, "High performance n-type PbTe-based materials for thermoelectric applications ," Physica B, no. 1-4,vol. 363, pp.196–205, 2005
[6] J. Yang, R. Chen, X. Fan, W. Zhu, S. Bao, and X. Duan," Microstructure control and thermoelectric properties improvement to n-type bismuth telluride based materials by hot extrusion," Journal Alloys and Compounds, vol. 429, pp. 156–162, 2007
[7]P. Zhu, Y. Imai, Y. Isoda, Y. Shinohara, X. Jia, and G. Zou, "Composition-dependent thermoelectric properties of PbTe doped with Bi2Te3," Journal Alloys and Compdounds, vol. 420, pp. 233–236, 2006
[8] Y. Dong, M. McGuire, A. Mailk, and F.J. DiSalvo, "Transport properties of undoped and Br-doped PbTe sintered at high-temperature and pressure $\geq$4.0 GPa,"Journal of Solid State Chemistry, vol. 182, pp. 2602–2607, 2009
[9] J. Gellman *et al.*, "Thin film thermoelectric devices with high room-temperature figures of merit," *Nature*, vol. 55, no. 1, pp. 1–8, 2001
[10] D. Vasilevskiy, R.A. Masut, and S. Turenne, "Thermoelectric and Mechanical Properties of Novel Hot-Extruded PbTe n-Type Material," Journal of Electronic Materials, Vol. 41, No. 6, 2012
[11] D. Li, R.R. Sun, X.Y. Qin, "Improving thermoelectric properties of p-type Bi2Te3-based alloys by spark plasma sintering," Progress in Natural Science: Materials International 21, pp. 336–340, 2011
[12]J. A. Zavala, D. Vasilevskiy, S. Turenne, and R. A. Masut, "Microwave Sintering of Bi2Te3- and PbTe-Based Alloys: Structure and Thermoelectric Properties," Journal of Electronic Materials, Vol. 42, No. 7,pp. 1992-1998, 2013.

Mechanical Properties and Microstructure at Stainless Steel HAZ from Dissimilar Metal Welding After Heat Treatment Processes

Raden Dadan Ramdan[1*], M Agus Kariem[1], Oki Neswan[2], Fadhil Wiriawan[1], Rochim Suratman[1], Bambang Widyanto[1] , Riza Wirawan[3]

[1] Faculty of Mechanical and Aerospace Engineering, Institut Teknologi Bandung, Jl. Ganesa no 10 Bandung 40132, Indonesia
[2]Faculty of Mathematic and Natural Science, Institut Teknologi Bandung, Jl. Ganesa No 10 Bandung 40132, Indonesia
[3]Jakarta State University, Department of Mechanical Engineering, Jl. Rawamangun Muka,Jakarta 13220, Indonesia

dadan@material.itb.ac.id

Abstract. Metallurgy of HAZ at both metals after dissimilar metal welding is important to be evaluated, because differences in metallurgical characteristics due to heating effect on bothHAZ might take place after the process. For the case of austenitic stainless steel, metallurgical aspects that should be considered after welding are the formation of chromium carbide (sensitization) and delta ferrite phase. In the present research, focus is given on the HAZ microstructure and hardness both after welding and post-heat treatment process. Microstructure observation was carried out by optical microscope and Scanning Electron Microscope (SEM), whereas hardness testing was carried out by micro-Vickers method. From this study, it can be concluded that the increase in heat treatment temperature can reduce the volume fraction of delta ferrite and chrome carbide precipitation, whereas hardness at the HAZ region decreases after the heat treatment.

Keywords : dissimilar metal welding, post heat treatment, sensitization

1. Introduction

Dissimilar metal welding is a welding process involving two different metals to be joined. It has been applied in various applications including important components in the turbine application. During welding process, the weld area experiences fusion whereas the area of the base metals close to the weld area are still in the solid state condition. However this area affected by heat during the process and is called as heat affected zone (HAZ) [1-3]. Generally heat involved in the HAZ causing microstructure evolution and phase transformation as well in many cases. This metallurgical effect of welding on the HAZ might cause different condition on different metals, i.e. one metal has good influences while the other has worse condition. Therefore finding the optimum condition on both metals to be joined should be find out before performing dissimilar metal welding.

Generally various important welding metallurgies that should be evaluated after welding process are at HAZ. These including, phase transformation [4], microstructure evolution [5], the formation of secondary phase [6] and segregation [7]. For the case of dissimilar metal welding these phenomena might present in the different conditions on different metal, and these could be the opposite condition. In carbon steel, one of condition that should be avoided is the transformation to martensite phase at the HAZ [8,9], whereas for austenitic stainless steel, sensitization that causing grain boundary corrosion [10,11] is an example of phenomenon that should be avoided.

In the previous work, we have evaluated the effect of heat input during welding processes by gas metal arc welding (GMAW) method [12]. In the present work, focus was given on the metallurgical phenomena at the HAZ of stainless steel, which including the formation of sensitization. Heat treatment after welding is commonly performed to eliminate the formation of sensitization [13], and therefore in the present work it was performed evaluation of heat treatment after dissimilar metal welding. Evaluation on the specific effect of post heat treatment on dissimilar metal welding by GMAW method is rarely found in the literature. And therefore this studied is expected to contribute in enriching the knowledge on this specific field of engineering.

2. Experimental method

Dissimilar metal welding in the present works was performed by gas metal arc welding (GMAW) method. Type of carbon steel and austenitic stainless steel used as the base metal is A53 and SS304 respectively, whereas SS308 was used as the filler material. Welding control was performed by varying current, voltage and speed of welding. After welding, heat treatment was performed in the normal atmosphere at temperature of 1010°C and 1120°C and held at this temperature for 13 minutes. Several characterizations were performed after the process, including micro structure observation, hardness measurement and elemental composition characterization. General to detail microstructure observation was performed by optical microscope method, and focus is given on the metallurgical features at HAZ of stainless steel. This observation was supported by hardness test using micro-Vickers method at the same area. In order to confirm the type of metallurgical feature observed by optical microscope and hardness test, scanning electron microscope and energy dispersive spectroscopy (EDS) were also performed, and focus was given on suspicious feature related with sensitization mechanism at the HAZ of austenitic stainless steel.

3. Results and discussion

Figure 1 shows example of microstructure after welding process, showing the weld area and HAZ of austenitic stainless steel. From this figure it can be seen austenite phase at the base metal, and dendritic structure at weld area. At the HAZ area it can be observed the relatively bold line of grain boundary which is delta ferrite in the form of ferrite stringer that is normally formed following rolling orientation and due to heat effect during welding process.

On the other hand Figure 2 shows example of microstructure after post heat treatment processes at temperature 1010°C (Figure 2(a)) and 1120°C (Figure 2(b)). From both figures it can be seen that dendrite structure that was formed after welding process at the weld area had been transformed to the equiaxed grain of austenite after both heat treatment processes. In addition delta ferrite is almost diminished after both heat treatment process, as bold line around the HAZ or fusion line cannot be observed. However minor features in the form of dot feature as shown by the red arrow is considered as the remaining delta ferrite of the sample heat treated at 1010°C. For the case of sample heat treated at 1120°C, this feature is completely diminished. The appearance of delta ferrite is associated with the presence of sensitization as the chromium carbide exist at the boundary between delta ferrite and austenite phase. At high temperature such as the present heating temperature, chromium carbides are dissolved into austenite matrix, and therefore effect of sensitization can be diminished. Another point to be considered is, heat treatment also influences the grain size at the HAZ area, as is shown in the Figure 3. This figure show the data for the sample with different heat input condition, and it can be observed both heat treatment in the present work increase the grain size and higher as the heating temperature increases.

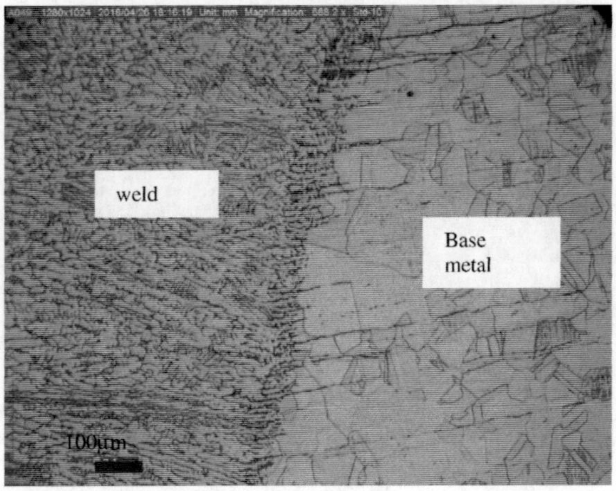

Figure 1. Example of microstructure after welding processes.

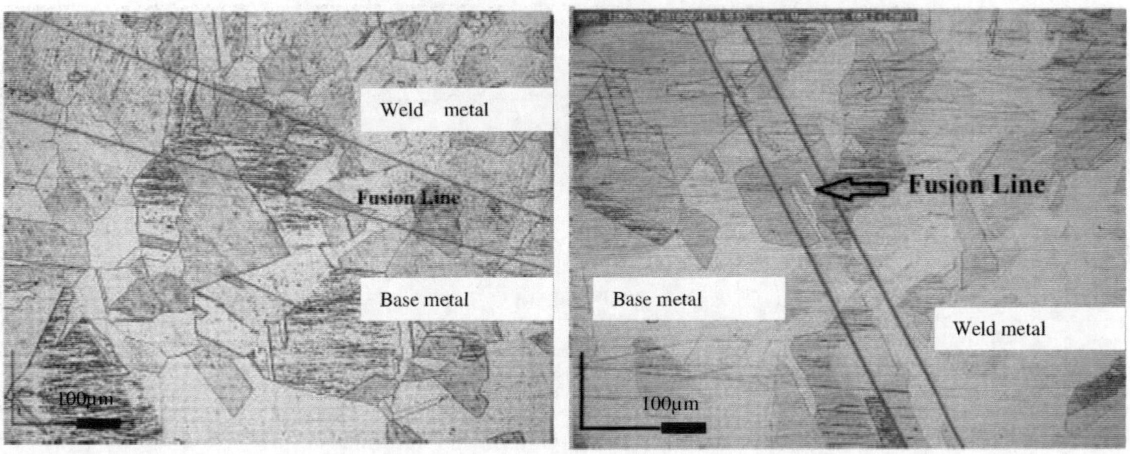

Figure 2 Example of microstructure after post heat treatment process (a) heating temperature 1010°C, (b) heating temperature 1120°C.

Figure 4 shows the effect of heat treatment on hardness at base metal of stainless steel (BM), HAZ and weld metal. It can be seen that for as weld specimen, HAZ show highest hardness of all, however after heat treatment, weld metal become more dominant especially for the sample heat treated at 1010°C. The highest hardness of HAZ is considered due to the formation of chromium carbide at the boundary between austenite phase and delta ferrite. After heat treatment, this carbide is dissolved as previously mentioned and as also shown in the Figure 3, grain growth also takes place during heat treatment. Therefore the hardness after heat treatment is decreased at all area.

IOP Conf. Series: Materials Science and Engineering **553** (2019) 012034 doi:10.1088/1757-899X/553/1/012034

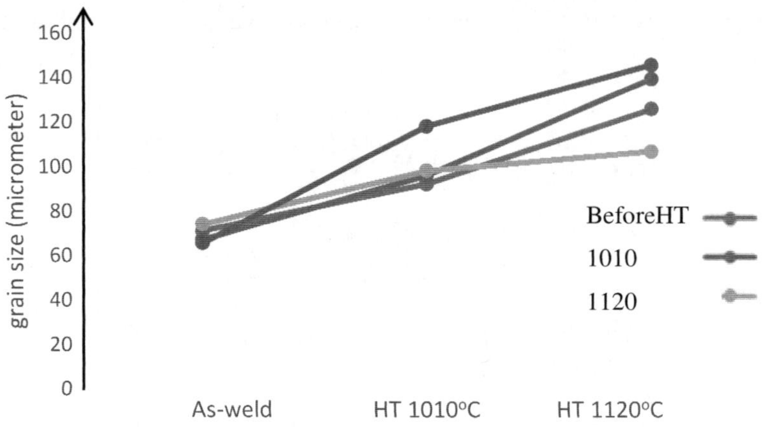

Figure 3. Effect of heat treatment on the grain size.

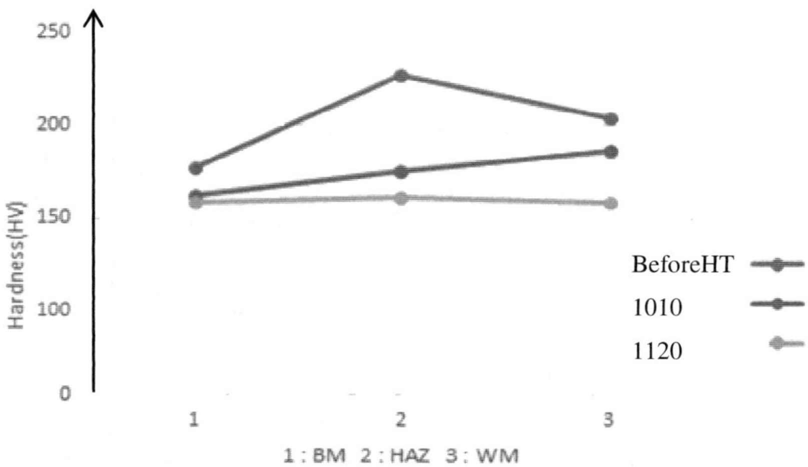

Figure 4. Effect of heat treatment on hardness.

In addition to the above data, Figure 5 shows scanning electron micrograph of as weld sample and after heat treatment. As weld sample show continuous line of delta ferrite which is in accordance with data in the Figure 1, whereas heat treated sample show delta ferrite in the form of discontinuous dot feature which also indicate partial dissolvement of carbide after heat treatment. By comparing Figure 5(b) and (c) it can also be observed that sample heat treated at temperature of 1120°C show less delta ferrite as compared with sample heat treated at 1010°C. This condition also suggests longer holding time is required in order to fully diminish delta ferrite as well as carbide in the HAZ of stainless steel. On the other hand, as confirmation on the former data about the occurrence of sensitization, energy dispersive spectroscopy (EDS) characterization was performed on the as weld sample. Characterization was performed by point mode characterization on the small boundary feature at the grain boundary between delta ferrite and austenite (Figure 6(a), and also on the surrounding area (Figure 6(b) covering the feature in the Figure 6(a). It was observed higher carbon concentration for the small feature in the Figure 6(a) (the mass % about 12%) as compared with surounding area (mass% about 6%). This fact indicates the formation of carbide at the grain boundary of austenite and delta ferrite as a result of sensitization at the HAZ of austenitic stainless steel during welding process.

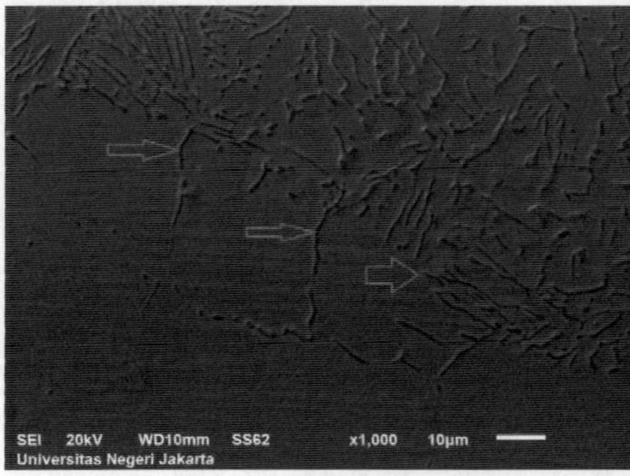

(a)

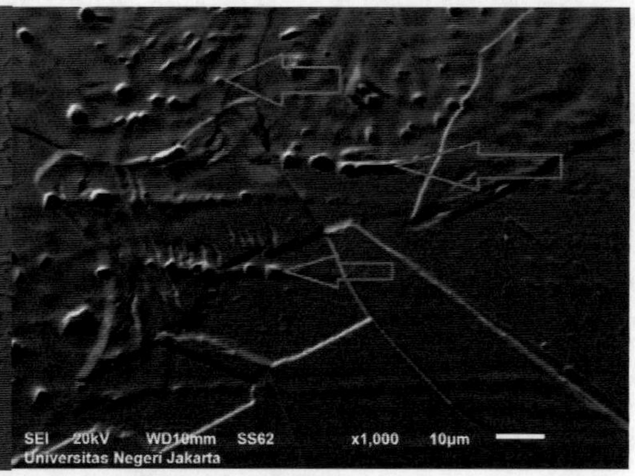

(b)

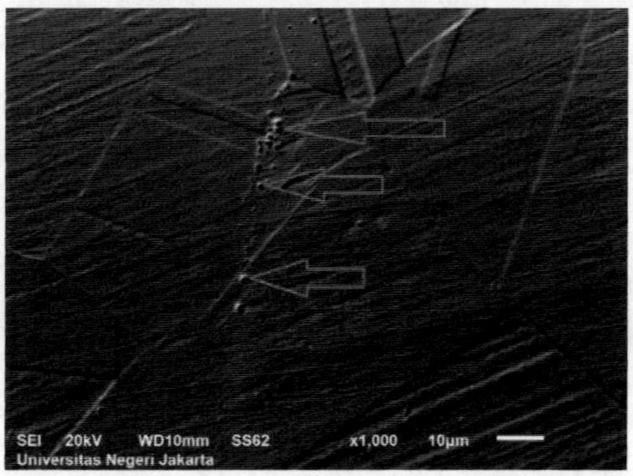

(c)

Figure 5. SEM observation of: a) *as-weld* , b) heat treated at1010°C, c) heat treated at 1120° C.

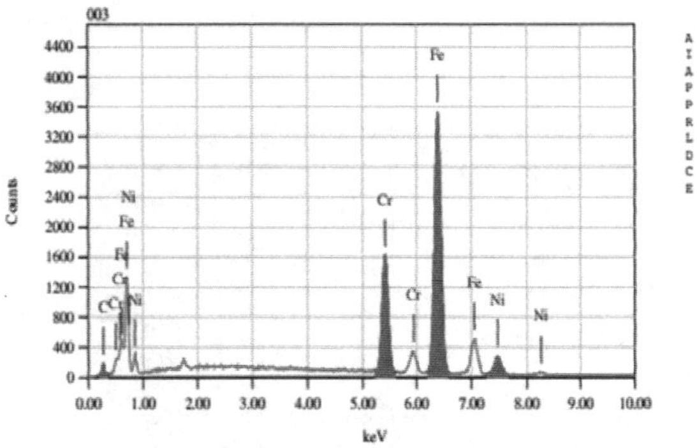

ZAF Method Standardless Quantitative Analysis
Fitting Coefficient : 0.2020

Element	(keV)	Mass%	Error%	Atom%	Compound	Mass%	Cation	K
C K	0.277	12.27	0.33	39.13				3.1888
Cr K	5.411	17.75	0.15	13.08				21.8782
Fe K	6.398	63.73	0.21	43.72				68.6470
Ni K	7.471	6.25	0.32	4.08				6.2861
Total		100.00		100.00				

(a)

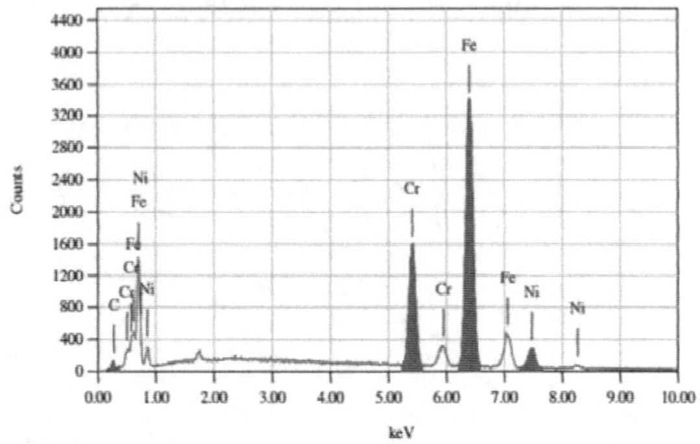

ZAF Method Standardless Quantitative Analysis
Fitting Coefficient : 0.2049

Element	(keV)	Mass%	Error%	Atom%	Compound	Mass%	Cation	K
C K	0.277	6.78	0.38	25.06				1.6538
Cr K	5.411	18.75	0.16	16.01				22.1132
Fe K	6.398	67.25	0.23	53.47				69.3017
Ni K	7.471	7.22	0.37	5.46				6.9314
Total		100.00		100.00				

(b)

Figure 6. Energy dispersive spectroscopy (EDS) results of: a) one point feature at the grain boundary austenite shown by the number, b) area shown by the photo.

4. Conclusion

Sensitization occurs after welding at the HAZ of stainless steel with the presence of chromium carbide at the grain boundary between delta ferrite and austenite phase. On the other hand heat treatments applied in the present research are able to partially diminish sensitization. In order to completely diminish the sensitization, heat treatment should be performed at longer holding time. Before heat treatment HAZ has the highest hardness, and highest decreasing of hardness observed after heat treatment at the HAZ.

Acknowledgment

The present author would like to express our gratitude to the ministry of research and higher degree education for the funding of the present research.

References

[1] Zhaohan Zhang, Zhibin Wang, Wenxian Wang, Zhifeng Yan, Min Ding, "Microstructure evolution in heat affected zone of T4003 ferritic stainless steel", *Materials & Design*, Vol. 68, pp 114-120, March 2015 (DOI: 10.1016/j.matdes.2014.12.018)

[2] M. Mohammadijoo, J. Valloton, L. Collins, H. Henein, D. G. Ivey, "Characterization of martensite-austenite constituents and micro-hardness in intercritical reheated and coarse-grained heat affected zones of API X70 HSLA steel", *Materials Characterization*, Vol. 142, pp. 321-331, August 2018 (/10.1016/j.matchar.2018.05.057)

[3] Bevis Hutchinson, Jacek Komenda, Gregory S. Rohrer, Hossein Beladi, "Heat affected zone microstructures and their influence on toughness in two microalloyed HSLA steels", *Acta Materialia,* Vol. 97, pp. 380-391, 15 September 2015 (10.1016/j.actamat.2015.05.055)

[4] Xuanwei Lei, Shi Dong, Jihua Huang, Jiang Yang, Xingke Zhao, "Phase evolution and mechanical properties of coarse-grained heat affected zone of a Cu-free high strength low alloy hull structure steel", *Materials Science and Engineering: A*, Vol.718, pp. 437-448, March 2018 (10.1016/j.msea.2018.02.008)

[5] Xinjie Di, Min Tong, Chengning Li, Chen Zhao, Dongpo Wang, "Microstructural evolution and its influence on toughness in simulated inter-critical heat affected zone of large thickness bainitic steel", *Materials Science and Engineering: A*, Vol. 743, pp. 67-76, January 2019 (10.1016/j.msea.2018.11.070)

[6] Seonghoon Jeong, Bongyoon Kim, Joonoh Moon, Seong-Jun Park, Changhee Lee, "Influence of κ-carbide precipitation on the microstructure and mechanical properties in the weld heat-affected zone in various FeMnAlC alloys", *Materials Science and Engineering: A*, Vol. 726, pp. 223-230, May 2018 (10.1016/j.msea.2018.04.088)

[7] Liqing Zhai, Yuichi Miyahara, Hiroshi Abe, Yutaka Watanab, "Segregation behavior of phosphorus in the heat-affected zone of an A533B/A182 dissimilar weld joint before and after simulated thermal aging", *Journal of Nuclear Materials*, Vol. 452(1–3), pp.133-140, September 2014 (10.1016/j.jnucmat.2014.05.005)

[8]. H. Wang, W. Woo, D. K. Kim, V. Em, S. Y. Lee, "Effect of tailored martensitic transformation in a thick weld: Residual stresses mitigation, heterogeneous microstructure, and mechanical properties", *Materials Characterization*, pp. 345-355, Vol.144, October 2018 (10.1016/j.matchar.2018.07.025)

[9] Xiang Luo, Xiaohua Chen, Tao Wang, Shiwei Pan, Zidong Wang, "Effect of morphologies of martensite–austenite constituents on impact toughness in intercritically reheated coarse-grained heat-affected zone of HSLA steel", *Materials Science and Engineering: A*, Vol. 710, pp.192-199 , January 2018 (10.1016/j.msea.2017.10.079)

[10] Subodh Kumar, A. S. Shahi, "Studies on metallurgical and impact toughness behavior of variably sensitized weld metal and heat affected zone of AISI 304L welds", *Materials & Design*, Vol. 89, pp. 399-412, January 2016 (10.1016/j.matdes.2015.09.145)

[11] R. Demarque, "Evaluation of the effect of the thermal cycle on the characteristics of welded joints through the variation of the heat input of the austhenitic AISI 316L steels by the GMAW process" *Science and Technology of Materials*, Vol 30(1), pp. 51-59, November 2018 (10.1016/j.stmat.2018.09.001)

[12] R.D Ramdan, A.L. Koswara, Surasno, R.Wirawan, F. Faturohman, B. Widyanto, R.Suratman, "Metallurgy and mechanical properties variation with heat input,during dissimilar metal welding between stainless and carbon steel", IOP Conference Series: Materials Science and Engineering, 307 2018, pp 012056 (10.1088/1757-899X/307/1/012056)

[13] M. J. Marques, A. Ramasamy, A. C. Batista, J. P. Nobre, A. Loureiro, "Effect of heat treatment on microstructure and residual stress fields of a weld multilayer austenitic steel clad", *Journal of Materials Processing Technology*, Vol. 222, pp. 52-60, August 2015 (10.1016/j.jmatprotec.2015.03.004)

Effect of Interlaminated Carbon and Basalt Fiber Reinforced Hybrid Composites on Mode I Fracture Toughness

I.D.G Ary Subagia[1*], A. Herman Yuwono[2], Yonjig Kim[3]

[1]Department of Mechanical Engineering, Udayana University (UNUD), Bukit Jimbaran Badung Bali – Indonesia 80361
[2]Department of Metallurgical and Materials Engineering, Faculty of Engineering University Indonesia Kampus Baru UI Depok, West Java-Indonesia (16424)
[3]Division of Mechanical Design Engineering, Chonbuk National University, Jeonju 561-756, South Korea.

arsubmt@me.unud.ac.id

Abstract. The fracture toughness and failure behavior of interlaminated hybrid composites were investigated. Hybrid interlaminated carbon and basalt fibers have been manufactured by fluid injection molding. The compact tension (CT) from hybrid composite panel has been used for Mode-I tension tests based on ASTM-D5045 standards. The experiment was aimed at investigating the stress intensity factor (K_{IC}) of interlaminated hybrid composites having different stacking of carbon and basalt fibers. The experimental results showed that hybrid composites with configuration CBCBC (B2 code) have maximum average values about 5.8% and 2.0% higher than the C_4 and B1 configurations, respectively. Replacing 20%, 30%, and 40% carbon by basalt fibers led to decrease in stress intensity factors (K_{IC}) by 8.87%, 7:56%, and 5.46%, respectively. Additionally, SEM measurements showed that the dominant failures that occurred include delamination, pull-out fiber to epoxy and fiber splitting. Thus, the fracture toughness of composites is significantly influenced by interlaminated hybrid composite epoxies.

Keywords: Intensity factor, Fracture toughness; Failure, Hybrid composites; Interlaminated

1. Introduction

Hybrid composites are a combination of reinforcements in single matrices that were developed due to the challenges involved in obtaining new materials. [1, 2]. They offer numerous advantages such as superior stiffness and strength, ease of fabrication, corrosion resistance, light weight and low costs [2]. Therefore they have been widely employed in automobile components, boats, sport equipment, etc. [3]. However, composite materials have the disadvantages of brittleness and low toughness; thus, they have been widely investigated in recent years. During the last two decades, natural fibers have been commonly used for the strengthening of composites. Basalt fibers are a new type of fibers in composite reinforcement, and they are a natural resource from volcanic rock. Most combinations of carbon and basalt fibers have resulted in improved mechanical, thermal, and environmental properties. Basalt fiber is analogous to glass fiber produced by the melting of volcanic rock at a temperature of 1700 °C [4-6]. Basalt fibers have mechanical properties similar to glass fibers. Moreover, they are non-toxic, eco-friendly and easy to recycle. There have been several studies focused on the mechanical properties of

IOP Conf. Series: Materials Science and Engineering **553** (2019) 012035 doi:10.1088/1757-899X/553/1/012035

basalt fibers as a reinforcement on composites [4-13].

As described above, fracture toughness is an important structural property of composite materials. Fracture toughness is the ability of composites containing a crack to resist propagation fracture. Cracks on a material cause a decrease in its strength [14]. In many cases, increasing the crack length on materials significantly decreases their static strength. This is usually the case in brittle materials, hence they have low fracture toughness. Laffan, et.al [15] reviewed the fracture toughness associated with translaminar failure modes of continuously reinforced laminated composites. They presented measurements on the toughness linked with failure modes, and showed that several steps need to be taken before a resistance curve can be fully characterized. Evaluation and standardization in terms of the linear elastic fracture mechanics as well as the elastic-plastic fracture mechanics parameters have been previously studied [16]. Chamnho et.al [17] studied the fracture toughness based on the analytical model with an aim to calculate the mode I fracture toughness of a multidirectional laminate material. Additionally, the fracture toughness of hybrid composites have also been studied [18-21].

In this work, the interlamination of carbon and basalt fibers has been performed with the purpose of improving the toughness behavior of carbon fiber reinforced polymers (CFRP). This has not been studied previously for mode I fracture toughness of hybrid composites. Additionally, our aim is to also investigate the resistance of an interlamination of basalt fibers on carbon fibers to loads capable of causing brittle or ductile crack extension.

In order to realize this, CT specimens were prepared according to the ASTM D 5045 standards. The critical stress intensity factor (K_{IC}) has been calculated by the compliance method to predict the fracture toughness of hybrid composites. The failures of hybrid composites after the tests have been analyzed using a scanning electron microscope (SEM).

2. Experimental method

2.1 Materials

In this work, two different phases of woven fibers have been employed for the strengthening of composites. Carbon plain weave fabrics (C120-3K), produced by Hyundai Fiber Co. Ltd. (Korea), have been used as the main filler, and basalt plain weave fabrics (EcoB4-F210), fabricated by Seco-Tech (Korea), have been used as the complementary filler. A thermosetting polymer, like an epoxy resin (HTC-667C), modified with an aliphatic amine hardener, produced by Jet Korea Industrial Corporation, was used as the matrix [10]. Table 1 shows the interlaminated variation hybrid composites where C and B represent carbon fibers and basalt fibers respectively. Additionally, alcohol and wax were used as supporter materials for the injection molding processes.

2.2 Fabrication of hybrid composites

In this experiment, laminated hybrid composites, containing woven plain fabric carbon and basalt with epoxy resin matrices were fabricated, and coded as CFRP and BFRP respectively. Eighty layers of both fabrics were laminated. Basalt fabrics replaced the carbon fabrics by 10% to 40%. Hybrid composite panels were fabricated according to weight fraction in the filler (i.e. woven fabrics carbon and basalt) to matrix (i.e. epoxy resin and hardener in ratio 5:1 *wt.%.*) ratio of 60 to 40.). Basalt fabrics layered according to their weight fraction values and positioned as the core of laminates were coded as CF/BF/CF, while basalt fabrics positioned on the external laminates were coded as BF/CF/BF.

Figure 1 shows the injection molding process of interlaminated hybrid composites. The process begins with fiber lamination, and is followed by mold preparation. In this experiment a mold base of Aluminum with size 500 mm x 500 mm and thickness 5 mm was employed. Plastics bags of size 600 mm x 600 mm were prepared for wrapping the fabric laminates. The epoxy and the hardener were mixed in the weighted ratio of 100:20, and the mixture was inserted into a desiccator at a pressure of 70 cmHg for 40 min to release all the bubbles and flatten the mixture. The mixture was then injected through a vacuum pump (Air-tech Ulna G-100D) at a pressure of 80 KPa for 45 minutes. During the process, a steady density of mixtures should be controlled in order to cover all the reinforcements. Also, the conditions during the injection process should be kept steady at room temperature (28°C). As the next step, the panels are cured inside an oven at a constant temperature of 65°C for at least 2 hours (see Figure. 1e). The hybrid composite panel is then released from the mold and cut using a water cutter machine into CT shapes with geometry as shown in the inset of Figure 2 [22]. A diamond cutter was used to produce a pre-crack (a = 1 mm) in all the CT specimens.

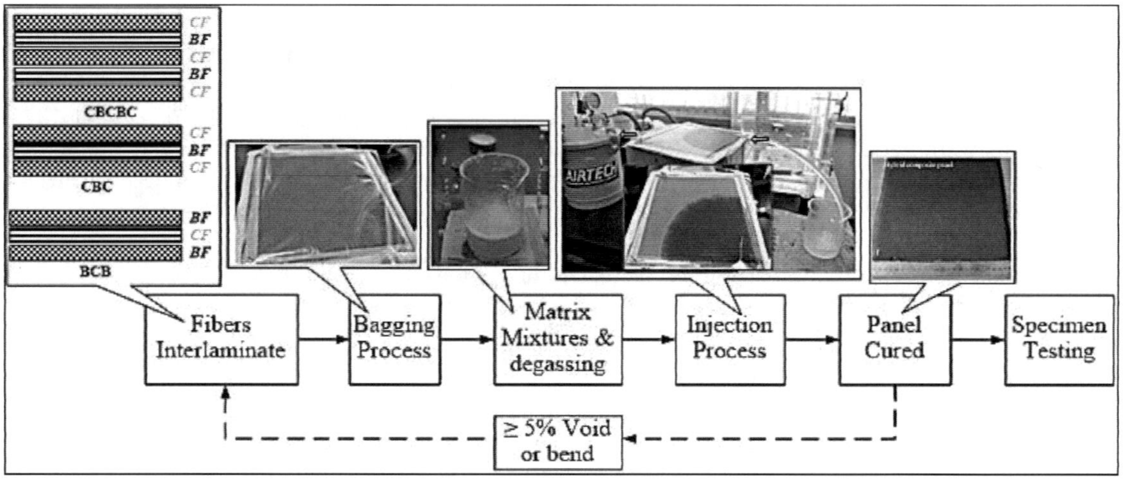

Figure 1. Schematic of Vacuum Injection molding processes.

2.3. Test setup

Figure 2 shows the tension test setup of the CT specimen. A universal testing machine (Unitec-M, R&B) with two tons of load cells and a cross head speed of 1 mm/min was employed. The crack propagation of the CT specimen was measured using a 0.25 mm clip-on gauge (epsilon, technology corp. Jackson, WY USA). The CT specimen based on the ASTM standard is shown in Figure 2b.

The fracture toughness of hybrid laminated composites for materials with thick sizes has been calculated using the formula:

$$B, a, (W-a) \geq 2.5\left(\frac{K_Q}{\sigma_{ys}}\right)^2 = K_{IC} \tag{1}$$

where K_Q is fracture toughness, σ_{ys} is the yield stress, a is the crack length, and w is the specimen width. The width of the specimen is defined as:

$$w \geq 2.5\left(\frac{K_Q}{\sigma_{ys}}\right)^2 \geq 2B \tag{2}$$

The stress intensity factor is usually expressed as:

$$K_Q = \frac{P_Q}{Bw^{0.5}} f(x) \tag{3}$$

The toughness of the CT specimen is calculated using Eq. (4)

$$f(\delta) = \frac{(2+\delta)}{(1-\delta)^{3/2}}\left[0.886 + 4.64(\delta) - 13.32(\delta)^2 + 14.72(\delta)^3 - 5.60(\delta)^4\right] \tag{4}$$

$$\delta = \left(a/w\right) \tag{5}$$

P_Q is the critical load measured by projecting a line whose slope is 5% less than the original slope of curve, $f(\delta)$ is the polynomial function, and B is the specimen thickness (mm). The crack length, a, is selected as per the rule, $0.45 < a/w < 0.55$ [23].

Table 1. Interlaminated configuration and strength fracture toughness of hybrid composites.

Laminates	Code	CT Geometry B [mm]	CT Geometry w [mm]	CT Geometry a [mm]	$x =$ (a/w)	A = $P_Q/Bw^{½}$	$f(a/w)$	K_Q (MPa.√m)
$[CFRP]_{80}$	CF	17.39	69.6	3.82	0.55	3.23	11.37	36.69
$[BFRP]_{80}$	BF	14.00	56	3.06	0.55	2.33	11.19	26.04
$C_{36}/B_8/C_{36}$	C1	17.11	68.4	3.77	0.53	3.13	11.38	35.57
$C_{32}/B_{16}/C_{32}$	C2	16.23	64.9	3.44	0.53	3.27	10.65	34.78
$C_{28}/B_{24}/C_{28}$	C3	15.85	63.4	3.38	0.54	3.13	10.74	33.69
$C_{24}/B_{32}/C_{24}$	C4	15.23	61.5	3.17	0.54	3.15	10.83	34.11
$B_{16}/C_{48}/B_{16}$	B1	15.04	60.6	3.23	0.54	3.25	10.83	35.38
$B_{16}/C_{16}/B_{16}/C_{16}/B_{16}$	B2	15.00	60.0	3.24	0.55	3.28	11.02	36.08

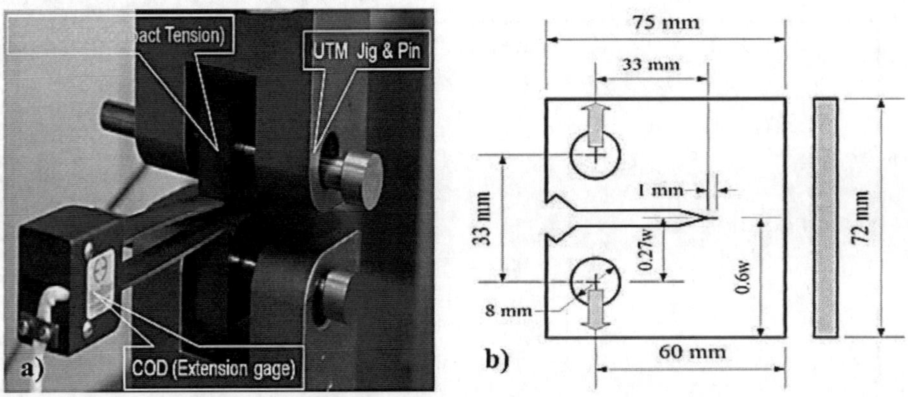

Figure 2. Geometry of Compact Tension (CT)

3. Results and discussion

3.1. Mode-I Fracture toughness behavior of Interlaminated hybrid composites

Table 1 summarizes the mode-I fracture toughness results calculated using equations 1-4. The 5% secant lines to predict the P_Q of each hybrid interlaminated composite specimen have been derived using the formula $P_{max}/P_Q \leq 1.1$, implying that $P_5 = P_Q$. Figures 3a and 3b show the tension curves of CT-CFRP and CT-BFRP respectively. The curves have been compared on the basis of the load and the displacement of the CT specimen during the tests. The crack propagation of CT-CFRP started without plastic deformation, implying that the CT-CFRP was brittle during the tension. In contrast, the CT-BFRP crack propagation started with a little plastic deformation in the initial crack, implying that it is more deformable. In this test, the loads have resulted in higher crack propagation in the CT-CFRP as compared to the CT-BFRP, but the displacement of CT-BFRP was longer than CT-CFRP. Additionally, from figures 3a and 3b it can also be observed that the CT-BFRP crack propagation is more stable than the CT-CFRP. In this case, jump crack propagation of CT-CFRP occurred during the mode-I test. This result shows that brittle materials have a higher stress intensity factor than ductile materials. The results are consistent with previous results [3, 24-26].

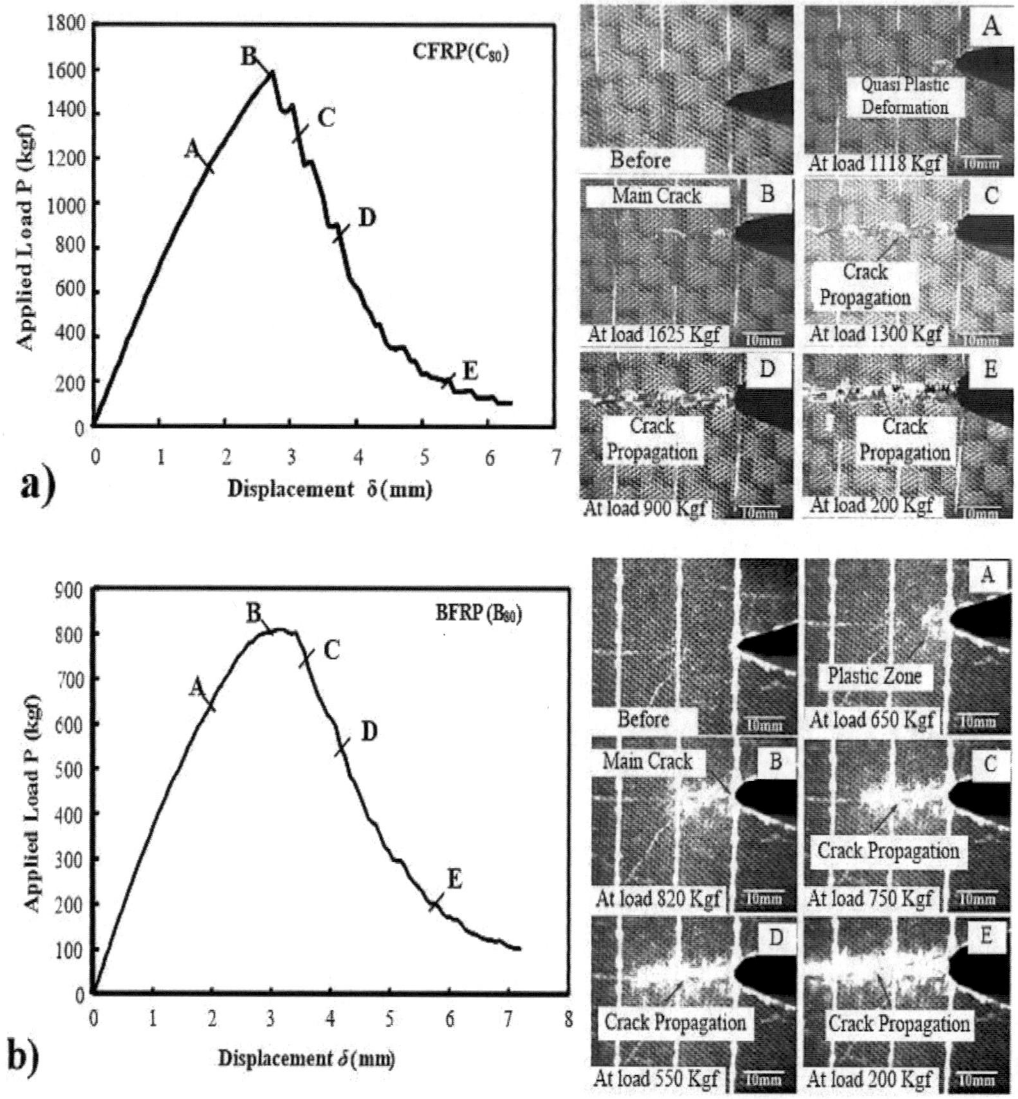

Figure 3. Load vs displacement curves of a) CT-CFRP; b) CT-BFRP.

Figures 4a and 4b show the graphic average values for the stress intensity factor of interlaminated hybrid composites tested on Mode-I. First, five interlaminated variations with basalt fiber as the core of the composite were tested, followed by three interlaminated with stacking modes CBC, BCB, and CBCBC. Figure 4a shows the average stress intensity factor for the interlaminated variation with added basalt fiber on the carbon fiber core of reinforced plastics. Clearly the stress intensity factor (K_Q) of the interlaminated hybrid composite with 20~40 wt. % basalt fiber added on the core carbon fiber gradually decreases. The results show that the interlaminated hybrid composite [C_{36}/B_8/C_{36}] (C1) has strength toughness lower than CFRP by 3.15%. Further, the basalt fabric content rises by 20%, 30%, and 40% in the CFRP, resulting in the decrease of the fracture toughness by 5.46%, 8.87%, and 7.56% respectively. This shows that the presence of interlaminated basalt fabric in the place of 10 – 40 wt. % of the carbon fibers does not significantly influence the fracture toughness of the composite during the Mode-I loading due to its brittleness. This is consistent with previous studies [19, 27-30]. However, the mechanical properties of the interlaminate during the Mode-I test show an increase in its durability, as shown before in previous tests [27, 31].

Figure 4b shows the average of the stress intensity factor for interlaminated hybrid composites with stacking sequences fibers. Clearly, the basalt fabrics laminated in the external positions sharply increase the strength of toughness. Table 1 summarizes the stress intensity factors for interlaminated hybrid composites, which are 3.589% lower than B1 for C4, and compared with CFRP, the hybrid composite (B1) is lower by about 3.56%. The hybrid composite (B2) has strength intensity factor 5.772% greater than C4, and about 1.975% higher than (B1). This suggests that hybrid interlaminated composites with alternate stacking of carbon and basalt fabrics can potentially improve the toughness properties of the composites.

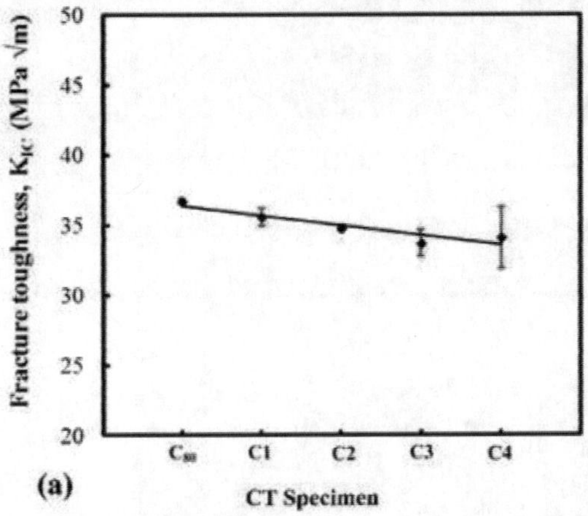

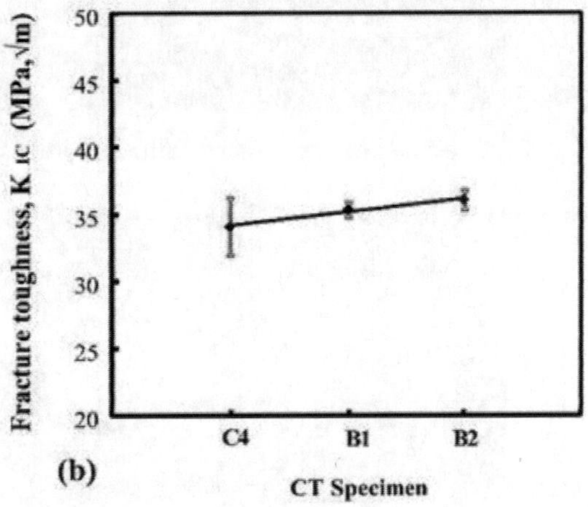

Figure 4. Average stress intensity factor values for a) Interlaminate of basalt fibers into the CFRP, b) Interlaminate fibers based on the stacking sequence (CBC, BCB and CBCBC)

According to the results of each K_{IC} with Mode-I loads, the difference between experimental and numerical methods were exploited to determine the error of the fracture toughness for various specimens. We can approximate the relationship between the experimental and numerical values of the mode I fracture toughness by the formula

$$K_{IC} = -7.0358(X) + 36.38 \qquad (MPa.\sqrt{m}) \tag{6}$$

$$X = \frac{w_B}{W_F} \tag{7}$$

where, X , w_B , and W_F are the total basalt fabric, the number of basalt fabrics, and the total number of fabrics, respectively.

The comparison of the results shows that the hybrid laminated composite with configuration $C_{27}B_{32}C_{27}$ (C4) has the highest error of 3.99%. Based on these test results we can summarize that the stress intensity factor is a function of loading, crack size, and structural geometry.

3.2. Failure behavior of Interlaminated hybrid composite

The surface morphology of an interlaminated hybrid composite epoxy resin matrix, containing interlaminated carbon-basalt fabrics reinforced on mode-I load was studied using SEM (JEOL JSM 5900). Figure 5 shows the CT specimen failures for CFRP and BFRP composites after Mode-1 tests. A flat failure surface is visible on the CFRP composite, resembling a dog-tooth (see Figure 5a). On the other hand, a rupture-like failure is visible on the BFRP composite (see Figure 5b). In fact, BFRP exhibited a failure which is more ductile than the CFRP composites, although the stress intensity factor of CFRP is higher than the BFRP composites. The propagation cracks of the basalt fabric are quite similar to glass fabric reinforced composites [20, 32].

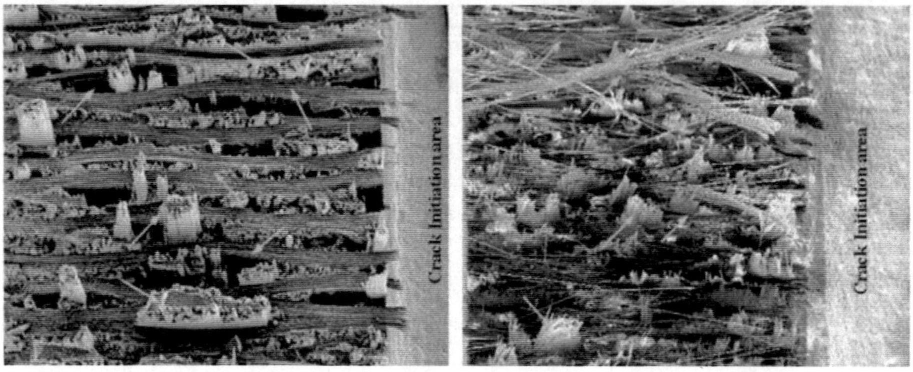

Figure 5. SEM analysis of failures a) the CFRP composites, b) the BFRP composites.

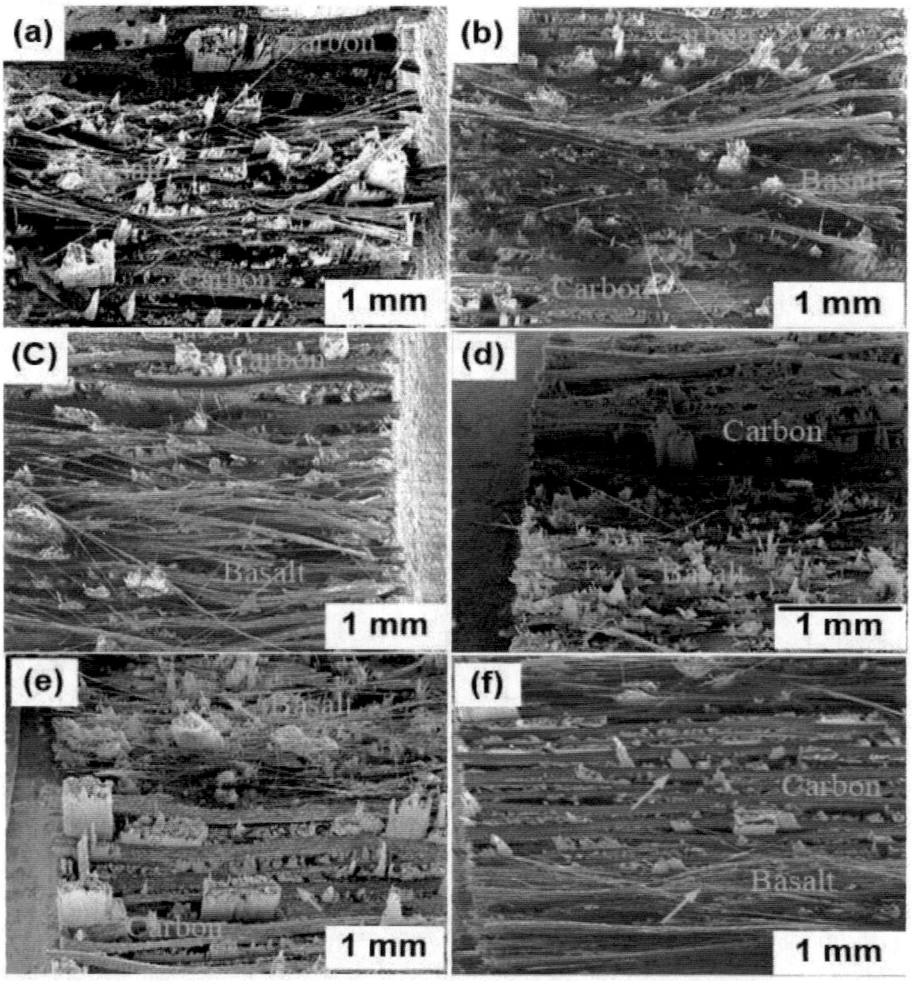

Figure 6. SEM images interlaminated hybrid composites; a) 10 wt.% Basalt Fibers, b) 20wt.% basalt fibers, c) 30 wt.% basalt fibers, d) 40 wt.% basalt fibers, e) BCB configuration, f) CBCBC configuration.

Figure 6 shows the fracture failure of an interlaminated hybrid composite epoxy matrix based on the enhancement of basalt fabric content and stacking sequence configuration. Figures 6a, b, c, and d show the failures for interlaminate basalt fibers on a CFRP composite. Each interlaminate configuration clearly has a unique pattern. Fiber pull-out, debonding, fiber unraveled, and fabrics delaminate are common features shown by the hybrid composite carbon-fabric epoxy resin matrix in post mode-I loads. Additionally, a shear hackle of the matrix is also visible on specimens C2, C3, and C4, caused by matrix cracks under maximum tension, perpendicular to the delaminate direction. This indicates that the hybridization of basalt fabrics with carbon fabric epoxy matrix significantly influences the fracture toughness of the hybrid composites, even though there is a slight decrease in strength. The failure fracture on mode-I loading with stacking sequence between the carbon and basalt fabrics (C/B/C, B/C/B and C/B/C/B/C) is shown in Figures 6e and f. Evidently many delaminates and longitudinal cracks occur in the B/C/B model, and fiber pull-out is shown by the hybrid composite in C/B/C/B/C configuration. Additionally, shear hackles influenced by shear stress during the mode-I loading test are also visible. These results are consistent with the mode I fracture toughness test in Figure 4. The hybrid composite with C/B/C/B/C stacking showed better fracture mechanics and toughness as compared to C/B/C and B/C/B staking modes.

4. Conclusions

In this study, we have successfully investigated the fracture toughness and failure characteristics of hybrid interlaminated composite epoxies containing reinforced carbon-basalt fabrics. Three types of hybrid interlaminated composite modes were studied. Vacuum resin transfer molding was used to manufacture all variations of interlaminated hybrid composites. CT was performed for Mode-I fracture toughness tests.

Hybridization between carbon and basalt fibers has been effectively exploited to modify the mechanical properties, and has contributed well towards the reduction in the cost of production of the composites. The stress intensity factor during the Mode-I test of CFRP showed a higher value than the BFRP. The K_{IC} of 20, 30, and 40 wt.% inserted CFRP has decreased linearly by 5.46%, 8.87%, and 7.56%, respectively. Hybrid interlaminated composites with added basalt fabrics show decreased strength toughness, but, at the same time, enhance the strain toughness of the hybrid composite. Furthermore, the fiber-stacked laminates have significantly influenced the stress intensity factor values of composites. This is an effective way to improve the toughness properties of the composites. Fiber pull-out, de-bonding, fiber delamination, and epoxy splitting are common failures which occur on hybrid composites after Mode-I loads. Additionally, shears hackles occur due to cracking of the matrix (epoxy) on the plane of maximum tension, perpendicular to the direction of delamination. This is indicative of local shear deformations in structures.

Acknowledgments

The authors would like to thank the Ministry of Research, Technology and Higher Education for funding this research.

References

[1] L. Mishnaevsky and G. Dai, 2014 "Hybrid carbon/glass fiber composites: Micromechanical analysis of structure–damage resistance relationships," *Computational Materials Science,* vol. 81, pp. 630-640.

[2] Yentl Swolfs, Larissa Gorbatikh, and I. Verpoest, 2014 "Fibre hybridisation in polymer composites: a review," *Composites Part A: Applied Science and Manufacturing* vol. 67, pp. 181-200.

[3] M. K. H. Y. Mohammed, Abu El-Ainin H, A. M. Hashem, 2013 "Fracture Properties of Glass Fiber Composite Laminates and Size Effect," *Scholars Journal of Engineering and Technology (SJET),* vol. 1(1), pp. 13-26.

IOP Conf. Series: Materials Science and Engineering **553** (2019) 012035 doi:10.1088/1757-899X/553/1/012035

[4] V. Fiore, G. Di Bella, and A. Valenza, 2011 "Glass–basalt/epoxy hybrid composites for marine applications," *Materials and Design,* vol. 32, pp. 2091–2099.

[5] Kunal Singha, 2012 "A short review on basalt fiber," *International Journal of Textile Science,* vol. 1 (4), pp. 19-28.

[6] V. Lopresto, C. Leone, and I. De Iorio, 2011 "Mechanical characterisation of basalt fibre reinforced plastic," *Composites Part B: Engineering,* vol. 42, pp. 717–723.

[7] Reza Eslami Farsani, Seyed Mohammad Reza Khalili, Vahid Daghighi, and Reza Fazaeli, 2011 "Creep Behavior of Basalt and Glass Fiber Reinforced Epoxy Composites,," *Journal of Mechanical Research and Application,,* vol. 3 (1), pp. 29-36.

[8] C. Colombo, L. Vergani, and M. Burman, 2012 "Static and fatigue characterisation of new basalt fibre reinforced composites," *Composite Structures,* vol. 94, pp. 1165–1174.

[9] I. D. G Ary Subagia, Tijing Leonard D, Kim Yonjig, Kim Cheol Sang, Vista Iv Felipe P, and S. H. Kyong., 2014 "Mechanical performance of multiscale basalt fiber–epoxy laminates containing tourmaline micro/nano particles," *Composites Part B: Engineering,* vol. 58, pp. 611-617.

[10] I.D.G Ary Subagia, Kim Yonjig, Tijing Leonard D, Kim Cheol Sang, and Shon Ho Kyong, 2014 "Effect of stacking sequence on the flexural properties of hybrid composites reinforced with carbon and basalt fibers," *Composites Part B: Engineering,* vol. 58, pp. 251-258.

[11] F. Sarasini, J. Tirillò, L. Ferrante, M. Valente, T. Valente, L. Lampani, P. Gaudenzi, S. Cioffi, S. Iannace, and L. Sorrentino, 2014 "Drop-weight impact behaviour of woven hybrid basalt–carbon/epoxy composites," *Composites Part B: Engineering,* vol. 59, pp. 204-220.

[12] S.-O. Lee, S.-H. Choi, S. H. Kwon, K.-Y. Rhee, and S.-J. Park, 2015 "Modification of surface functionality of multi-walled carbon nanotubes on fracture toughness of basalt fiber-reinforced composites," *Composites Part B: Engineering,* vol. 79, pp. 47-52.

[13] V. Dhand, G. Mittal, K. Y. Rhee, S.-J. Park, and D. Hui, 2015 "A short review on basalt fiber reinforced polymer composites," *Composites Part B: Engineering,* vol. 73, pp. 166-180.

[14] M. Janssen, J. Z. and, and R. J. H. Wanhill, *Fracture Mechanics* vol. 2nd Edition. Spon Press 2 Park Square, Milton Park, Abingdon, Oxfordshire, OX14 4RN: Taylor & Francis e-Library, 2005.

[15] M.J. Laffan, S.T. Pinho, P. Robinson, and A. J. McMillan, 2012 "Translaminar fracture toughness testing of composites: A review," *Polymer Testing,* vol. 31, pp. 481–489.

[16] X.-K. Zhu and J. A. Joyce, 2012 "Review of fracture toughness (G, K, J, CTOD, CTOA) testing and standardization," *Engineering Fracture Mechanics,* vol. 85, pp. 1-46.

[17] P.P. Camanho and G. Catalanotti, "Analytical Model for The Prediction of The Fracture Toughness of Multidirectional Laminates," presented at the 18TH International Conference on Composite Materials, ICC Jeju, Korea, 2011.

[18] G. Belingardi, E. G. Koricho, and A. T. Beyene, 2013 "Characterization and damage analysis of notched cross-ply and angle-ply fabric GFRP composite material," *Composite Structures,* vol. 102, pp. 237-249.

[19] M. T. Demirci, N. Tarakçıoğlu, A. Avcı, and Ö. F. Erkendirci, 2014 "Fracture toughness of filament wound BFR and GFR arc shaped specimens with Charpy impact test method," *Composites Part B: Engineering,* vol. 66, pp. 7-14.

[20] P.S. Shivakumar Gouda, S.K. Kudari, Prabhuswamy. S, and Dayananda Jawali, 2011 "Fracture Toughness of Glass-Carbon (0/90)S Fiber Reinforced Polymer Composite – An Experimental and Numerical Study," *Journal of Minerals & Materials Characterization & Engineering,* vol. 10 (8) pp. 671-682.

[21] Shay Hamer, Herman Leibovich, Anthony Green, Ron Intrater, Ron Avrahami, Eyal Zussman, Arnon Siegmann, and Dov Sherman, 2011 "Mode I Interlaminar Fracture Toughness of Nylon 66 Nanofibrilmat Interleaved Carbon/Epoxy Laminates," *Polymer composites.*

[22] I.D.G. Ary Subagia and Yonjig Kim 2014 "Tensile behavior of hybrid epoxy composite laminate containing carbon and basalt fibers," *Science and Engineering of Composite Materials.* , vol. 21 (2), pp. 211-217.

[23] ASTM-D-5045-99, 2007 "Standard test method for plane-strain fracture toughness and strain energy release rate of plastic maerials," *ASTM International, West conshohocken,* pp. PA 19428-2959, United State.

[24] M. Rokbi, H. Osmani, N. Benseddiq, and A. Imad, 2011 "On experimental investigation of failure process of woven-fabric composites," *Composites Science and Technology,* vol. 71, pp. 1375-1384, Jul 28.

[25] R. W. Li, P. Dong, B. Bai, and X. L. Wang, 2009 "Stress Distribution and Fracture Behavior of Compact Tension Beryllium Specimen," *Rare Metal Materials and Engineering,* vol. 38, pp. 95-98, Jan.

[26] M. Yahyazadehfar, A. Nazari, J. J. Kruzic, G. D. Quinn, and D. Arola, 2014 "An inset CT specimen for evaluating fracture in small samples of material," *J Mech Behav Biomed Mater,* vol. 30, pp. 358-68, Feb.

[27] I. Taketa, J. Ustarroz, L. Gorbatikh, S. V. Lomov, and I. Verpoest, 2010 "Interply hybrid composites with carbon fiber reinforced polypropylene and self-reinforced polypropylene," *Composites Part a-Applied Science and Manufacturing,* vol. 41, pp. 927-932, Aug.

[28] M. Davallo, H. Pasdar, and M. Mohseni., 2010 "Effects of Laminate Thickness and Ply-Stacking Sequence on the Mechanical Properties and Failure Mechanism of Unidirectional Glass-Polyester Composites," *International Journal of ChemTech Research CODEN(USA): IJCRG,* vol. 2, No.4, pp. 2118-2124.

[29] Mohammad T. Bashar, Uttandaraman Sundararaj, and P. Mertiny, 2013 "Mode-I interlaminar fracture behaviour of nanoparticle modified epoxy/basalt fibre-reinforced laminates," *Polymer Testing,* vol. 32, pp. 402-412.

[30] Jae Il Lim, Kyong Yop Rhee, Hyun Ju Kim, and Dong Ho Jung, 2014 "Effect of stacking sequence on the flexural and fracture properties of carbon/basalt/epoxy hybrid composites," *Carbon Letters,* vol. 15, No. 2, pp. 125-128.

[31] H. Hadavinia and H. Ghasemnejad, 2009 "Effects of Mode-I and Mode-II interlaminar fracture toughness on the energy absorption of CFRP twill/weave composite box sections," *Composite Structures,* vol. 89, pp. 303–314.

[32] M. Rokbi, H. Osmani, N. Benseddiq, and A. Imad, 2011 "On experimental investigation of failure process of woven-fabric composites," *Composites Science and Technology,* vol. 71, pp. 1375–1384.

Preliminary study on the use of rice husk as a reducing agent in iron sand reduction

Muhammad Faris Ghiyats[1], Ahmad Maksum[1,2] and Johny Wahyuadi Soedarsono[1*]
[1] Center of Minerals Processing and Corrosion Research, Department of Metallurgical and Materials Engineering, Universitas Indonesia, Depok 16424, Indonesia
[2] Department of Mechanical Engineering, Politeknik Negeri Jakarta, Depok 16425, Indonesia

jwsono@metal.ui.ac.id

Abstract. Titanium is a valuable mineral that has many advanced functions in our daily lives. This has caused the demand for this mineral to increase over the years. Researchers have long studied the use of reduction methods using coal in increasing titanium content in iron sand before further processing. Considering the sulfur content in coal which causes various excess problems in the reduction process, then in this experimental study we tried to examine the potential of rice husk as a reduction agent in the iron sand reduction process. This study includes several stages, namely roasting and reduction processes with or without the addition of Na_2SO_4, and magnetic separation. The weight ratio of iron sand and rice husk used in this study was 10: 1, while the reduction process was carried out at a temperature of 700°C for 60 minutes, followed by the magnetic separation process. Characterization of samples was carried out using XRD and ICP-MS. And the final results obtained after magnetic separation showed that the Ti content in the product tail was higher than the Ti content in the concentrate, which was 0.692% and 0.221% respectively.

Keywords: Iron sand, rice husk, reduction, titanium.

1. Introduction

In Indonesia, Titanium is one of the chemical compounds that have a large number of sources. Titanium is used in many applications in daily life. It is because the characteristic of Titanium is really forceful. Titanium is as strong as steel but much less dense. Titanium is mainly used in spacecraft, aircraft, and missiles. Due to their high tensile strength to density ratio, high corrosion resistance, and the ability to withstand the extreme temperatures, titanium is used for several advanced applications such as sensor[1], photovoltaic[2], and medical applications [3].

It has been observed that Iron sand is the part of metal mining material with the sedimentation process of sand-sized material containing iron element. It contains many oxide compounds, such as ilmenite ($FeTiO_3$), magnetite (Fe_3O_4), hematite (Fe_2O_3), and many more [4].

Pyro-metallurgy is the method that used in this study. it is very necessary to use redactor agent to be able to do the reduction process. Initially the reduction agent commonly used in this process is the use of coal, but its use causes serious pollution impacts to the environment as well as the limited resources to be renewed.

To solve the environmental issue, the mining and industrial sectors associated with the metal processing are expected to reduce coal consumption due to the decreasing the non-renewable energy consumption and greenhouse gas emissions. The alternative way that can be implemented to break this issue is using biomass energy or waste from natural resources to replace the role of coal as a reducing agent, such as rice husk.

2. Experimental method

2.1 Materials

To support this research process, there are some equipment required including furnace, thermocouple, digital scale and also followed by XRD and ICP-MS at BATAN for characterization process. Then, the materials that used in this study were iron sand from West Java as the raw material, rice husk as the redactor agent as well, bentonite for palletizing process and Na_2SO_4.

From the XRD result as described at Figure 1, the iron sand ore has hematite, magnetite, and magnesium titanate phases.

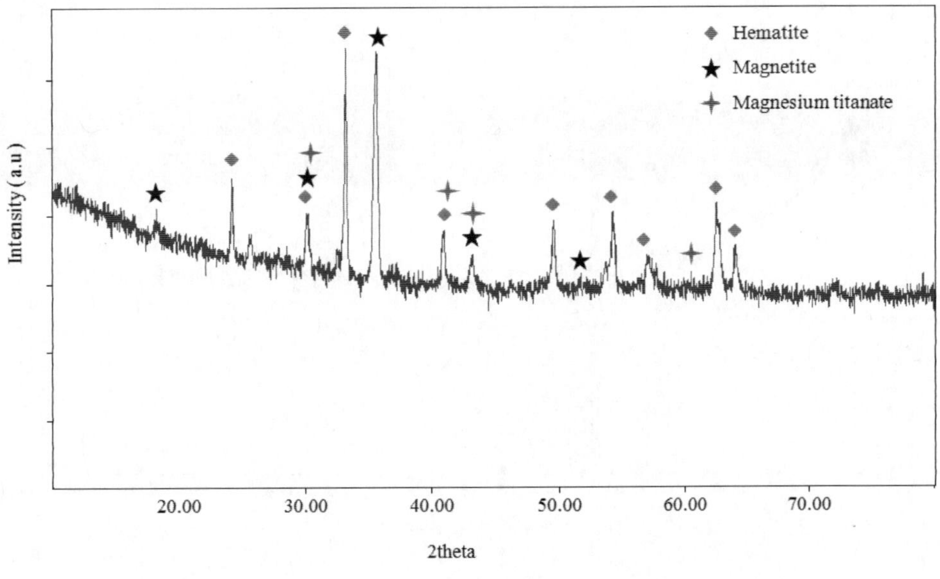

Figure 1. XRD analysis results of initial sample of iron sand.

ICP-MS is also done to investigate the quantitative data of the initial sample method, and the result as shown in Table 1 indicated that the percentages of the Fe and Ti that consisted in the initial condition of the iron sand. It clearly proved that there is Ti content in that sample even the percentage is really small.

Table 1. The contents of Ti and Fe elements in iron sand ore are measured using ICP-MS.

Element	Fe	Ti
Content (%)	51.772	2.872

2.2 Experiment

The samples were prepared with each of them were crushed up to #150. Before we add the Na_2SO_4, we did the initial characterization test to know the phase and the contention of the raw iron sand. After sample preparation finished, iron sand and bentonite and Na_2SO_4 were homogenized and mixed into 2 sample with the ratio ratio: 10:1:0 and 10:1:1 respectively. Then after that, the crucible was used for purpose of roasting process, at 700°C for 1 hour. And finally, the main process which is the reduction process was done before the XRD and ICP-MS for the characterization step were finished.

3. Results and discussion

The ICP-MS test results on samples that have undergone a roasting process, as shown in Table 2, show an increase in Fe and Ti levels. This is presumably caused due to reduced water content, volatile matter and other oxides such as Ca and Na oxides. This phenomenon is in line with the results obtained in previous studies where the roasting process is carried out on iron ore [5], [6] or limonite ore [7].

Table 2. The contents of Ti and Fe elements in iron sand after roasting process that measured by using ICP-MS.

Element	Fe	Ti
Content (%)	52.659	4.786

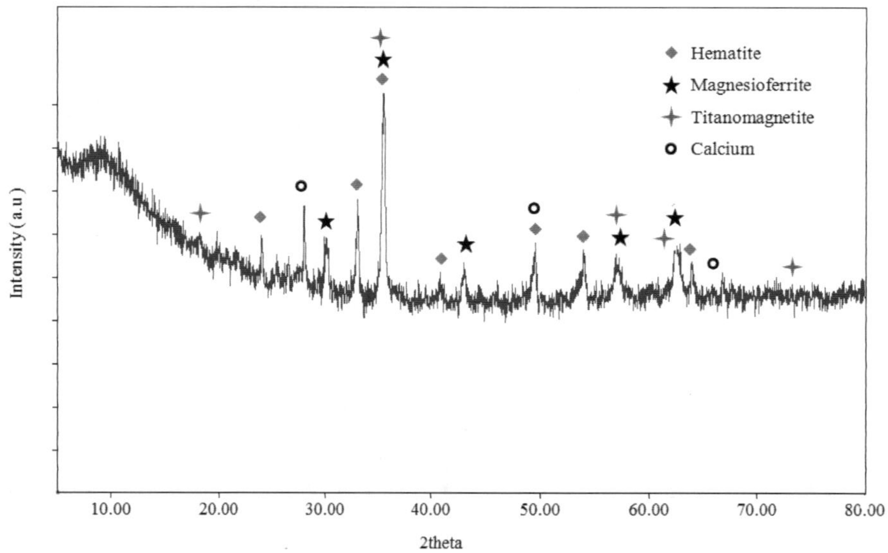

Figure 2. XRD analysis results of sample after reduction without Na_2SO_4.

The reduction process, which is the main step of this experiment proceed the characterization of the initial sample of iron sand to purposed or to order to know the changes in the compounds form that occurs after the reduction at temperature 700°C with reduction time in 1 hour. The sample is separated into two, sample 1 with the addition of Na_2SO_4 and sample 1 without the addition. And after the magnet separation, the purposed to collect the powder which stayed on the contact surface of the magnetic separator and it also called as the concentrate the results showed in the qualitative and quantitative result.

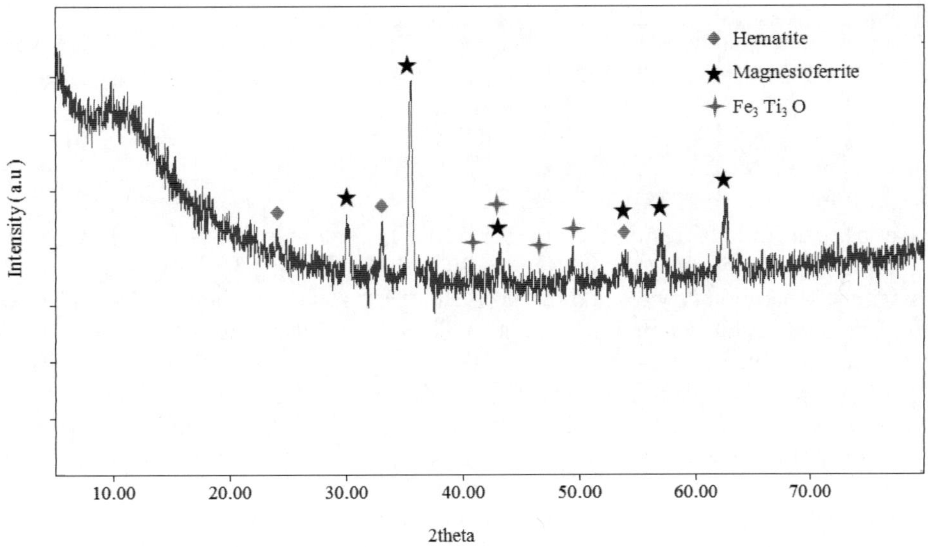

Figure 3. XRD analysis results of sample after reduction with Na_2SO_4.

From the XRD result between sample with and without addition of Na_2SO_4 as seen in Figure 2 and 3, the results show the similar pattern in both of hematite and magnetite phases and they are still presenting the highest peak of the phase. Those phases would contain the Fe and Ti within the sample. Compare to the result of the quantitative result shows the result of Ti Content and Fe with this following Table 3.

Table 3. The contents of Ti and Fe elements in iron sand after reduction process that measured by using ICP-MS.

Sample	Fe (%)	Ti (%)
Iron ore without Na_2SO_4	33.12	1.49
Iron ore with Na_2SO_4	31.02	2.9

It can be seen from the table, the sample with the addition of Na_2SO_4 has higher number of Ti compared to those who are not. This is suspected due to the ineffectiveness of the mixed rice husk to give the needed support to reduce the hematite into magnetite. The ineffectiveness is suspected due to the accumulation of high gas of rice husk after being burned.

In the quantitative test, the data shows the increasing of titanium content in the tailing after being separated by magnetic separation. The Ti content in tailing is much higher than in concentrate. Because, mostly, the element that successfully magnetized is the higher Fe content. As the result of the reduction process shows that the sample 2, or the one who proceed with the addition of the Na_2SO_4, has the higher recovery for the Ti Content, then, the researcher decided to continue the quantitative test of the sample after being separated. The results of the magnetic separation are named as the concentrate and tail. Concentrate usually has the higher Fe content and on the other hand, the tail is the one that consists higher Ti. The numbers of each result show in Table 4.

Table 4 The contents of Ti and Fe elements in iron sand after magnetic separation that measured by using ICP-MS.

Sample	Fe (%)	Ti (%)
Concentrate	47.97	0.221
Tail	34.22	0.692

It clearly can be seen from the result, the number of the Ti content in tail product is higher than the concentrate. This result can be the approval of the hypothesis that the Fe and Ti can be separated through the magnetic separation process and tail product should has the higher number of Ti compared to concentrate. The same result, where the content of Ti in the tail is greater than the content of Ti in concentrate, was also obtained by Jin-Fang L et al in his study to get Ti and Cr from raw ilmenite [8]. But some of the miscalculation error can be seen in this case, the Ti content that recorded before the magnetic separation has been done, the number of the reduction result has the higher number of Ti. At the first place, sample 2 underwent same procedure as sample 2, with the same amount of rice husk, reduction time differs them. With more time added to the process, more carbon monoxide gas would be observed for the reduction process. The sufficient amount of gas produced for the reduction reaction to occur. So, the miscalculated number or unexpected low number that quantitative test result may be caused by the error of the magnetic separations.

4. Conclusion

The reduction process done have been able to recover Ti content from iron sand while The weight of the added rice husk and the time have an optimum point that has to be reached but cannot be surpassed in order to get the best Ti concentration. The addition of Na_2SO_4 will increase the content of Ti, therefore, can be known as the result of transformation of hematite into magnetite phase during roasting reduction process and accelerated by rice husk as reducing gas agent.

Acknowledgment

Special thanks are due to Mrs. Anita (Chemical and Nuclear Material and Analysis, BATAN), Mr. Agus B Prasetyo (LIPI Metalurgi, Serpong-Indonesia) and Mr. Haidir (Politeknik Negeri Jakarta) for helpful supports. The authors also wish to thank the DRPM Universitas Indonesia under PITTA grant 2018 for the financial support of this research.

References

[1] J. Bai and B. Zhou, "Titanium dioxide nanomaterials for sensor applications," *Chem. Rev.*, vol. 114, no. 19, pp. 10131–10176, 2014.

[2] Y. Bai, I. Mora-Seró, F. De Angelis, J. Bisquert, and P. Wang, "Titanium dioxide nanomaterials for photocatalytic applications," *Chem. Rev.*, pp. 10131–10176, 2014.

[3] M. Kulkarni, A. Mazare, E. Gongadze, Perutkova, V. Kralj-Iglic, I. Milošev, P. Schmuki, A. Iglič, and M. Mozetič, "Titanium nanostructures for biomedical applications," *Nanotechnology*, vol. 26, no. 6, 2015.

[4] Y. ran Liu, J. liang Zhang, Z. jian Liu, and X. dong Xing, "Phase transformation behavior of titanium during carbothermic reduction of titanomagnetite ironsand," *Int. J. Miner. Metall. Mater.*, vol. 23, no. 7, pp. 760–768, 2016.

[5] J. W. Soedarsono, A. Rustandi, Y. Pratesa, R. D. Sulamet-ariobimo, B. H. Prabowo, and J. S. Exsaudy, "The Effects of Reduction Parameter to Composite Pelet of Iron Ore and Coal Using Single Conveyor Belt Hearth Furnace," *Adv. Mater. Res.*, vol. 842, pp. 115–119, 2016.

[6] A. Kawigraha, J. Wahyuadi, and S. Harjanto, "Thermogravimetric Analysis of The Reduction of Iron Ore with Hydroxyl Content," *Adv. Mater. Res.*, vol. 774–776, pp. 682–686, 2013.

[7] A. Maksum, M. K. E. Husein, S. Permana, A. Rustandi, and J. W. Soedarsono, "A preliminary

study on the reduction of limonite ore by using rice husk as a reducing agent," *IOP Conf. Ser. Mater. Sci. Eng.*, vol. 316, no. 12050, 2018.
[8] J. F. Lv, H. P. Zhang, X. Tong, C. L. Fan, W. T. Yang, and Y. X. Zheng, "Innovative methodology for recovering titanium and chromium from a raw ilmenite concentrate by magnetic separation after modifying magnetic properties," *J. Hazard. Mater.*, vol. 325, pp. 251–260, 2017.

Effect of the Processing Parameters on the Shrinkage Defect Using A High-Pressure Die-Casting Process

Dwi Rahmalina[1*], Hendri Sukma[1], Andry[1]
[1] Department of Mechanical Engineering, Faculty of Engineering, Universitas Pancasila, Srengseng Sawah, Jagakarsa, Jakarta 12640, Indonesia

drahmalina@univpancasila.ac.id

Abstract. Effects of the processing parameters of a high-pressure die-casting (HPDC) process on the shrinkage defect of housing block compressor (HBC) were examined in this research. Because of the reduction of defect can increase the productivity of HBC that representing a good quality product, the processing parameters determining the quality of product must be optimized. In this study, the processing parameter of high speed in the range of 3.97 to 4.97 m/s and that of fast start point in the range of 293 to 333 mm were calculated and then the HBC samples of using the Al alloy of ADC12 were produced through the HPDC process. The shrinkage defects on HBC were characterized by visual inspection, radiography testing and microstructure examination. The result showed that the best parameters for reducing shrinkage defect of HBC by the HPDC process are 4.97 m/s of high speed and 313 mm of fast start point. The optimum parameters of high speed and fast start point were verified to contribute to producing a high quality of the HBC for the applications in automotive engine.

Keywords: High pressure die casting; Fast start point; High speed; Shrinkage defect.

1. Introduction

Quality of the casting product is an important factor in producing a superior manufacture product, due its competitiveness in the market. This has motivated the manufacture world doers to compete in innovating and to strive for various ways to achieve this goal. One of the ways to increase productivity of the production line is to increase the good quality products and reduce the ratio of not good quality products (*NG In-Process*). The casting technology has been developed to optimize the parameter of the casting process. High Pressure Die Casting (HPDC) is a shape manufacturing process and has been used to produce many automotive components. The molten metal is injected into metal mold at high speed and solidified under high pressure [1-2]. The HPDC products have high dimensional precision and elevated production efficiency, also significant financial profit for automotive industries [3-6]. However, components manufactured by HPDC contain defects, such as shrinkage defects, that could decrease their mechanical properties [7,8]. Shrinkage defect is the existence of cavity or trapped air inside a product produced from cast in HPDC production line. The mechanical properties is influenced by filling process and solidification conditions, particularly depends on HPDC parameters such as the condition as pressure applied and the gate velocity [9].

IOP Conf. Series: Materials Science and Engineering **553** (2019) 012037 doi:10.1088/1757-899X/553/1/012037

Automotive component, particularly compressor is functioned as an equipment to compress the refrigerant in a car cooler system (Fig. 1). This block compressor consists of 4 main components; they are Front Housing, Front Cylinder, Rear Cylinder, and Rear Housing. This block compressor is using ADC12 material as the raw material. There is a need to decrease NG (Not Good) product ratio value of compressor dies on 10SRE11 Rear Housing model with minimum standard as much as 1% [10,11]. Therefore a deeper and thorough study on getting the method in reducing the "Not Good" product is required. One thing to achieve is by examining the injection parameter needed to reduce cavity defect on the product, especially high speed and fast start point. High speed (HS) is defined as injection speed of material into mold, while fast start point (FSP) is beginning point of implementing high speed. This research is aiming to study the effect of high speed parameter concerning shrinkage defect on RH 10SRE11 block compressor metal casting; to examine the effect of fast start point parameter toward shrinkage defect occurred on RH 10SRE11 block compressor metal casting; also to investigate the characteristic of shrinkage defect.

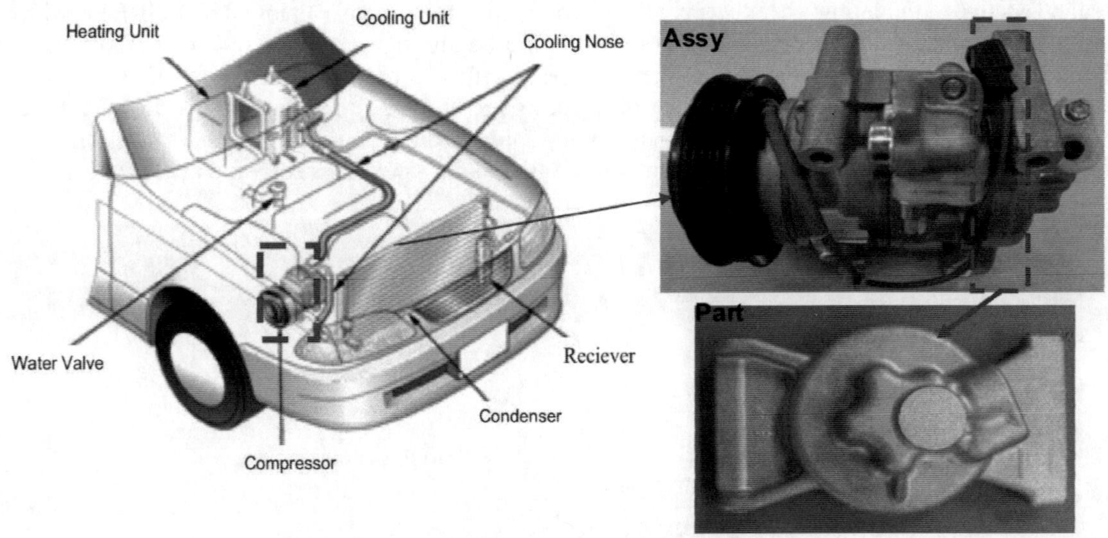

Figure 1. Housing block compressor type 10SRE11.

2. Experimental method

The block compressor was produced using aluminum alloys of ADC12. The steps of processing and research are described in two steps. First, re-calculation of casting aspects so that beginning parameter can be identified, which will be used to prove the formulation. Gating system on the cast in HPDC process consists of 3 cast cavities, as seen in Figure 2 and 3. To achieve an optimum parameter determination, a calculation of high-speed value and fast start point is conducted. Furthermore, a method to determine HPDC parameter in order to obtain optimum parameter will be discussed.

IOP Conf. Series: Materials Science and Engineering **553** (2019) 012037 doi:10.1088/1757-899X/553/1/012037

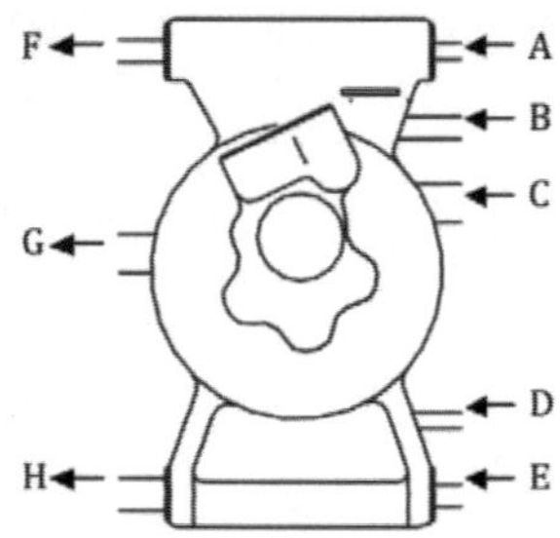

Figure 2. Input and output gating systems of Al alloy on the Cavity produced by block compressor.

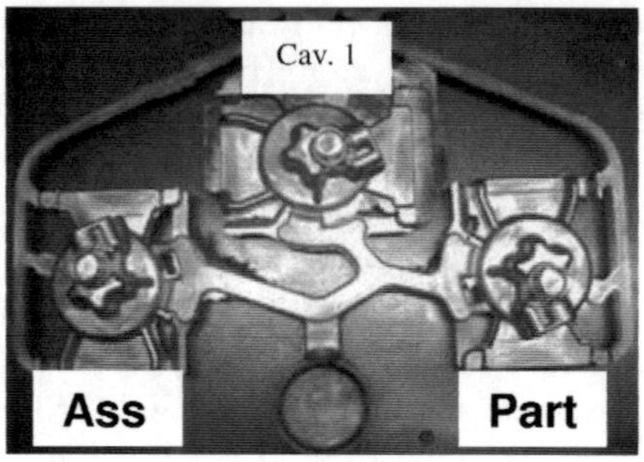

Figure 3. Number of Cavity.

Second step is conducting an experiment on production process of casting RH 10SRE11 with injection systems as shown as Fig. 4, that is by using the beginning parameter re-calculation result, particularly high speed and fast starts point. High-speed effect is presented in Figure 5, showing a schematic picture of a plunger tips movement of injecting the liquid Al material into the die cast. Fast start point (FSP) is the beginning of high-speed point application during the process of pushing the molten aluminum into the cast. Fast start point will be trialed on 3 positions; of which before gate, on the gate, and after the gate (Fig. 6).

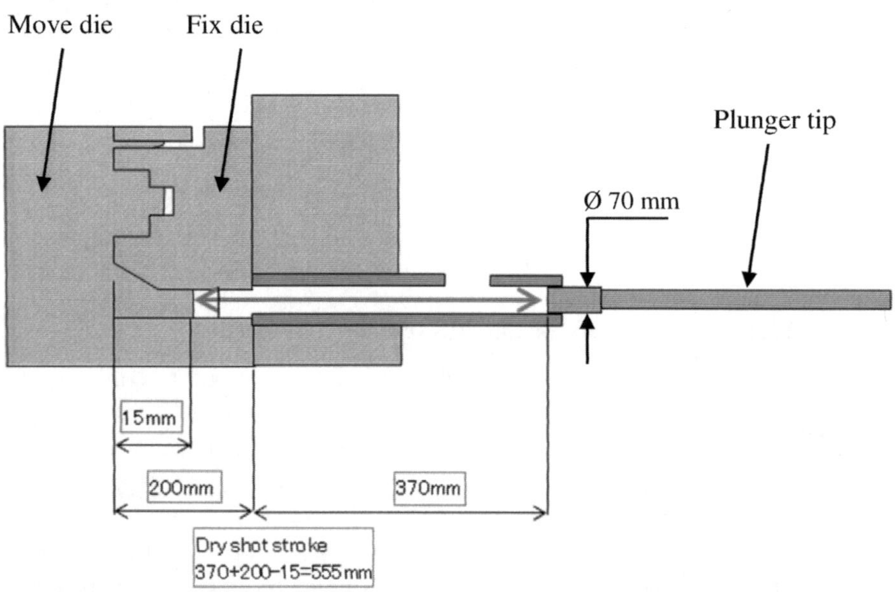

Figure 4. Injection system of molten Al alloy into metal mold using HPDC process.

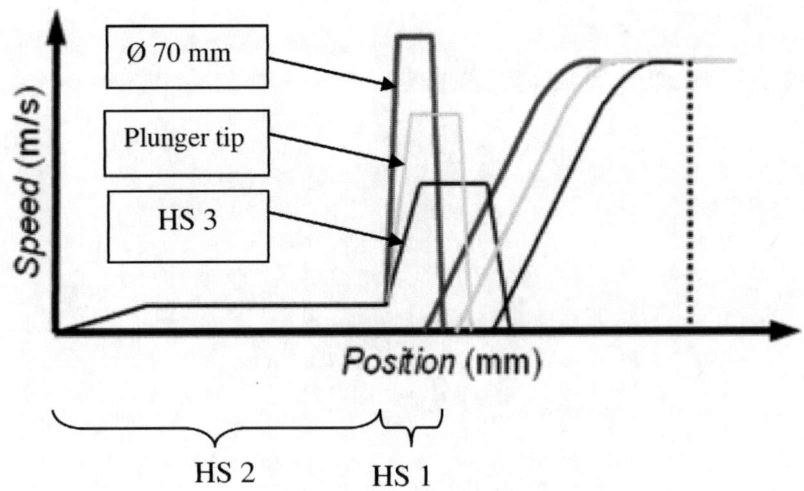

Figure 5. Illustration of high-speed variation of HPDC process on compressor product.

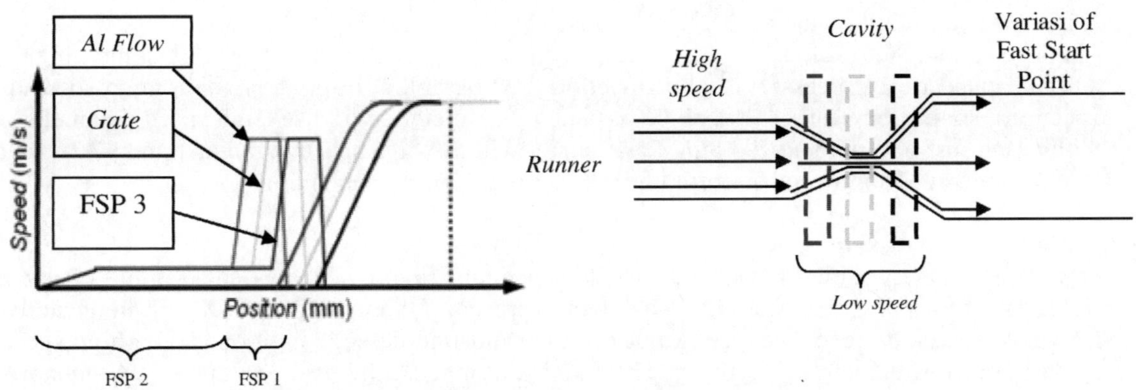

Figure 6. Fast start point on the gate of HPDC process on compressor product.

The parameters of HPDC process are cross section area of gate of 91 mm^2/ cavity; over flow (O/F) of 43.5 mm^2/cavity; diameter of plunger tip of 70 mm, cross section area of plunger tip of 3846.5 mm^2, product weight of 1600 gr (3 cavity); over flow weight of 450 gr; ρ (density melt Al) of 2.4 gr/cm^3 and charging volume of. 284.7 cm^3. Based on these parameters, charging time can be calculated as shown in equation (1).

t (*Charging time lower limit*) $$= k \cdot \left(\frac{T_i - T_f + S \times Z}{T_f - T_d} \right) \cdot X \quad (1)$$

= 0.05206 sec = 52.06 msec

t = *Charging time* (s)
k = > 0.0346 s/mm
T_i = Melting temperature on *Gate* (°C) = 632 °C
T_f = Pouring Temperature = 570 °C
T_d = Dies Temperature = 150 °C
S = *Solid rate* = 20 %

Z = Coefficient of Conversion Unit = 4.8 °C / %
X = Thickness of product = 4 mm

$$\text{Vp } (\textit{lower limit speed injection value}) = \frac{V \times W}{A_{tip} \times t} \tag{2}$$

$$= 4.27 \text{ m/s}$$

Setting lower limit = - 0.2 m/s
Speed requirement = Vp – (setting lower limit)
= 4.47 m/s

$$\text{Vg } (\textit{Gate speed}) = \frac{V_{requirement} \cdot A_{section\_tip}}{A_{section\_gate} / Cavity} \tag{3}$$

$$= 62.9 \text{ m/s}$$

$$\text{FSP } (\textit{Fast start point}) = \text{Dry shot stroke} - \text{biscuit thickness} - \frac{W_p(Part + O/F)}{\rho_{Al} \cdot A_{tip}} \tag{4}$$

$$= 313 \text{ mm}$$

Therefore from the beyond calculation, it can be concluded that high-speed value needed is 4.47 m/s and Fast start point value needed is 313 mm.

In this section, the result of the experiment of the research will be addressed by comparing the result of inspection process (visual inspection, radiographic inspection, and microstructure inspection), so that the best result shall be obtained. Parameter 1 will use High speed parameter at 3.97 m/s and fast start point at 293 mm. Parameter 2 will use high speed parameter at 3.97 m/s and fast start point at 313 mm, and so forth until all the parameter are examined.

3. Results and discussion

Below is a summary table of result of visual inspection. Figure 7 shows an example of visual checking result under a condition where a defect is present (O) and no defect (X). From the above observation, it can be seen that speed value with the most no defect condition (O) is (high speed, fast start point) parameter, respectively, (3.97 m/s, 293 mm), (3.97 m/s, 333 mm), (4.47 m/s, 293 mm), (4.97 m/s, 293 mm), (4.97 m/s, 313 mm), and (4.97 m/s, 333 mm). Whereas, parameter with least no defect (O) is (*high speed, fast start point*) parameter, respectively, (4.47 m/s, 313 mm), and (4.47 m/s, 333 mm).

Table 1. Summary of visual inspection examinations.

Fast Start Point		293 mm			313 mm			333 mm		
Cavity No.	.	1	2	3	1	2	3	1	2	3
High speed	3.97 m/s	X	X	O	X	X	X	X	X	O
	4.47m/s	X	X	O	X	X	X	X	X	X
	4.97m/s	X	X	O	X	O	X	X	O	X

Note: X = Defect is Present on compressor
O = No Defect

In the following, radiographic examination will be explained. On this radiographic examination, a comparison of void sizes resulting from each measurement is conducted. The standard used is the values below 15 mm^3 by using reference from the standard of the housing block compressor manufacturer. From the observation, it can be concluded that parameter of the smallest void value is high-speed 3.97 m/s with 333 mm fast start point. Figure 8 explains (a) is a product with minimum void value, and figure (b) is the product with maximum void value.

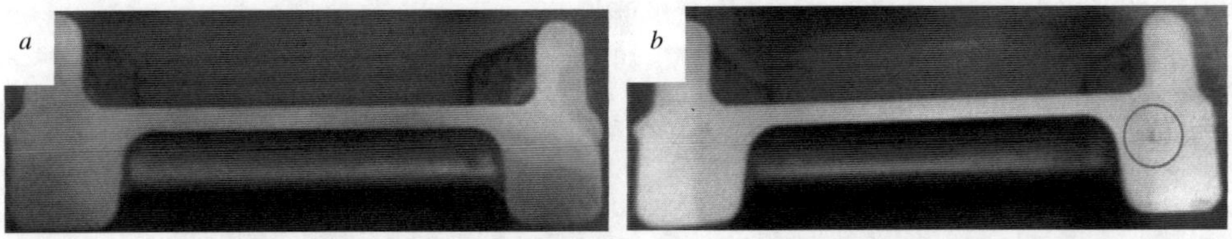

Figure 7. Visual inspection of the block compressor: (a) No defect. (b) Defect is Present.

Table 2. Summary of Radiographic analysis result (in mm^3).

Parameter FSP		293 mm			313 mm			333 mm		
Cav No.	.	1	2	3	1	2	3	1	2	3
Parameter *High speed*	3.97 m/s	56.9	29.9	16.7	50.4	29.2	15.4	39.9	20.9	6.3
	4.47m/s	35.9	36.1	21.0	49.2	42.7	9.9	35.3	28.8	24.7
	4.97m/s	21.1	12.4	12.4	24.5	13.8	10.2	24.3	23.3	13.7

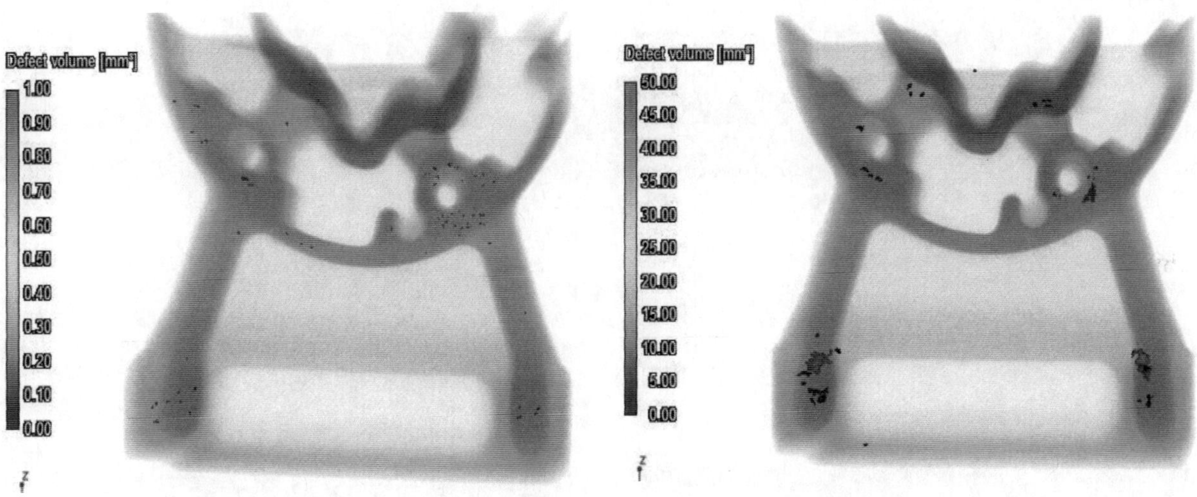

(a) (b)

Figure 8. Radiographic observation image: (a) minor defect, (b) major defect.

In the following, the result of structure micro observation will be explained. From the result, morphology shapes of a microstructure of each parameter will be compared. Figure 9 shows the sample position on which a microstructure observation will be conducted. Each of the samples is observed using 2 types of magnification, which are 100x and 500x. Figure 10 is resulted from 100x magnification. From this figure, it can be seen that the structure on figure (b) is more uniform and homogenous than the structure in figure (b). This explains that figure (b) has an evenly distributed Si Eutectic structure than figure (a). Also from figure (a), it can be seen that a micro shrinkage, which is spread almost every location in figure (b).

IOP Conf. Series: Materials Science and Engineering **553** (2019) 012037 doi:10.1088/1757-899X/553/1/012037

Figure 11 describes that the present of dendritic sizes between figure (a) and figure (b) are almost similar. Moreover, micro porosity is also can be seen, which explains that micro porosity is still formed in each of the parameters. Therefore, it can be concluded that this parameter has effect on porosity and shrinkage, which occurred on the sample.

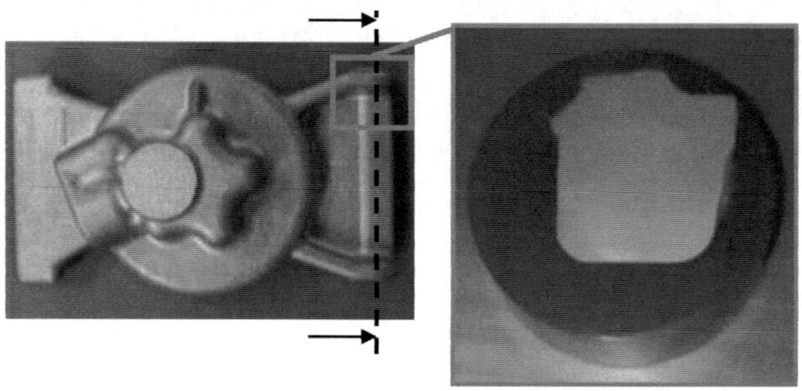

Figure 9. Sample position to examine the microstructure.

Figure 10. Microstructure of block compressor with HPDC parameters of: (a) 3.97 m/s with 293 mm, (b) 4.97 m/s with 313 mm.

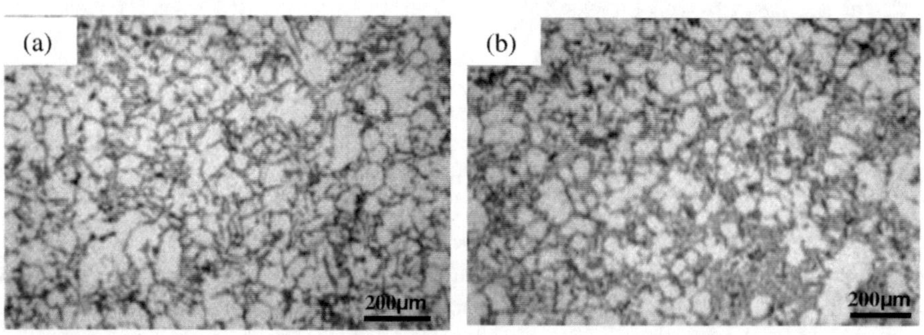

Figure 11. Microstructure of block compressor with HPDC parameters of: (a) 3.97 m/s with 293 mm, (b) 4.97 m/s with 313 mm.

4. Conclusion

From the research in previous chapter, it can be concluded that the result of visual inspection is the parameter with minor defect is 4.97 m/s high speed and 293 mm, 313 mm, and 333 mm fast start point. Radiographic analysis result is the parameter with the small void volume (under $15mm^3$) is high-speed

4.97 m/s with 293 mm and 313 mm fast start point. The result of microstructure analysis is the parameter with distributed eutectic silicon grain size is 4.97 m/s with 313 fast start point. Generally the dendritic size is not much different from each of the parameters. The best parameter for manufacturing block compressor is the parameter of 4.97 m/s high speed with 313 mm fast start point.

Acknowledgment
The authors are grateful to The State Ministry of Research, Technology and Higher Education, The Republic of Indonesia by funding this work under the PTUPT Incentive 2018.

References
[1] E. Degarmo, J.T. Black, R. A. Kohser, Materials and Process in Manufacturing, 9th Edition, Prentice Hall, USA, 2003.
[2] PN Rao, Manufacturing Technology: Foundry, Forming, and Welding, 2nd Edition, McGraw-Hill Book Company Inc, Singapore, 2001
[3] L. Wang, M. Makhlouf, D. Apelian, Aluminum die casting alloys: alloy composition, microstructure, and properties-performance relationships, Int. Mater. Rev. 40 (1995) 221-238. doi: 10.1179/imr.1995.40.6.221.
[4] S. Ji, W.C. Yang, F. Gao, D. Watson, Z. Fan, Effect of iron on the microstructure and mechanical property of Al-Mg-Si-Mn and Al-Mg-Si die cast alloys, Mater. Sci. Eng. A 564 (2013) 130-139. doi: 10.1016/j.msea.2012.11.095.
[5] X.X. Dong, L.J. He, X.S. Huang, P.J. Li, Effect of electromagnetic transport process on the improvement of hydrogen porosity defect in A380 aluminum alloy, Int. J. Hydrogen Energy 40 (2015) 9287-9297. doi: 10.1016/j.ijhydene.2015.05.160.
[6] X.X. Dong, X.S. Huang, L.H. Liu, L.J. He, P.J. Li, A liquid aluminum alloy electromagnetic transport process for high pressure die casting, J. Mater. Process. Technol. 234 (2016) 217-227. doi: 10.1016/j.matprotec.2016.03.028.
[7] Outmani, L. Fouilland-Paille, J. Isselin, M.E. Mansori, Effect of Si, Cu and processing parameters on Al-Si-Cu HPDC castings, J. Mater. Process. Technol., 249 (2017), pp. 559-569. doi: 10.1016/j.jmatprotec.2017.06.043.
[8] Y. Ling, J. Zhou, H. Nan, Y. Yin, X. Shen, A Shrinkage Cavity prediction model for gravity castings based on pressure distribution: A casting steel case, Journal of Manufacturing Processes, 26 (2017), 433-445. doi: 10.1016/j.jmapro.2017.02.017.
[9] A.R. Adamane, L. Arnberg, E.E. Fiorese, G. Timelli, F. Bonollo, Influence of injection parameters on the porosity and tensile properties of high-pressure die cast Al-Si alloys: a review. Am. Foundry Soc. 9 (2015), 43–53. doi: 10.1007/BF03355601.
[10] Toshiba, Manual for the Die Casting Machine Training Course, Toshiba Machine Co., LTD., Japan, 2007

Wet-spinning of Cellulose Acetate Reinforced with Acetylated Nano-crystalline Cellulose as Carbon Fibre Precursors

Mahyuni Harahap[1], Bongkot Hararak[1], Inam Khan[1], Surya Pandita[1], Saharman Gea[2]
[1] School of Metallurgy and Materials, The University of Birmingham, Edgbaston, Birmingham B15 2TT, United Kingdom.
[2] Department of Chemistry, Faculty of Mathematic and Natural Science, Universitas Sumatera Utara, Jl. Bioteknologi No.1 Medan 20155, Indonesia.

s.gea@usu.ac.id

Abstract. The search precursors that are bio-based, renewable and biodegradable has attracted significant attention due to increased environmental awareness. Recently, nano-crystalline cellulose (NCC) has been considered as a potential precursor to produce carbon fibre owing to its high crystallinity and excellent mechanical properties. However, its crystallinity prevents it from being dissolved in comment solvents. The main aims of this study were: (i) to acetylate NCC to improve its compatibility with cellulose acetate and (ii) to produce wet-spun cellulose acetate fibres reinforced acetylated and unacetylated NCC. In this study, rod-like particles of NCC with diameters in the range 2.0-4.8 nm and lengths of 46-114 nm were characterised using a Transmission Electron Microscope (TEM). The carbonyl (C=O) group of the acetyl group was confirmed using Fourier Transmission Infrared (FTIR), and the crystallinity of NCC was characterised using X-Ray Diffraction (XRD). The crystallinity indicates of the NCC before and after acetylation were 78% and 68% respectively. The morphology of wet-spun cellulose acetate reinforced acetylated NCC was smoother than reinforced NCC. The acetylation on NCC increased its compatibility with the polymer matrix.

Keywords : Nano-crystalline cellulose, Acetylation, Wet-spinning, Carbon Fibre.

1. Introduction

Cellulose is the biggest biopolymer extracted from plants and is also found in micro-organism such as fungi, bacteria, algae and tunicate [1]. Cellulose is a non-aromatic, organic compound with chemical formula $(C_6H_{10}O_5)_n$ connected by β-1-4 linkages and composed of D-anhydroglucopyranose units [2]. Cellulose is a semi-crystalline structure which is bonded with lignin, hemicellulose and other compounds as the matrix. A higher crystalline content can be found in nano-crystalline cellulose (NCC) and nano-fibril cellulose (NFC), which can be extracted by acid hydrolysis [3] and steam explosion [4] respectively.

Its nano-crystalline has several benefits in use for polymeric materials such as enhancing abrasion resistance, chemical resistance, hardness, glass transition temperature (*Tg*) and tensile strength. Similarly, NCC has a property of polyelectrolyte due to charge sulphate ester introduced on the surface of NCC during acid hydrolysis [5], high specific strength (7.5 GPa) [6] and high modulus (138 GPa) [7].

The utilisation of NCC as a functional material increase significantly over the past 20 years due to these properties [8]. Some examples are, (i) biomedical applications [9], printed electronics [10], fillers in nano-composites [11] and food packaging [12]. The fabrication of nano-fibres from NCC continues to attract significant attention including wet-spinning. For example, Jinfeng W *et.al.,* [13] spun NCC-silver to produce highly conductive fibres. Torres-Rendon J *et.al.,* [14] produced NCC-based fibre with high mechanical properties (33 GPa). In addition, Lundahl M *et.al.,* [15] investigated the effect of hydrogel viscosity and spinnability of NCC wet-spinning.

In this project, cellulose acetate was investigated as a carbon fibre precursor. Modifying NCC through acetylation will improve its dispersibility with the polymer matrix to produce circular and smooth wet-spun fibres. During the carbonisation process, filaments of cellulose acetate will be subjected to thermo-mechanical loading. Adding NCC to cellulose acetate is expected to increase the thermo-mechanical properties of cellulose acetate filaments.

2. Experimental method

2.1 Materials

NCC with molecular weight (M_w) of 14,700 - 27,850, a density of 1.50 g/cm^3 and a sulphate content of 246 - 261 mmol/kg was purchased from Celluforce, Canada. Cellulose acetate (M_w of 30,000 g/mol), dimethylsulfoxide (DMSO), pyridine and acetic anhydride were purchased from Sigma-Aldrich, UK.

2.2 Acetylation of NCC

The acetylation process was carried out in a gel form by dispersing a 1.5% (w/v) of NCC in deionised (DI) water in a 30 ml of pyridine using sonication for 10 minutes at 20 °C - 30 °C. The suspension was transferred to a 250 ml three-necked bottom-flask. The acetylation was commenced by adding 15 ml of acetic anhydride under reflux in a nitrogen atmosphere at 80 °C for five hour. After that, the product was precipitated using DI water and acetone (1:1) several times until pH was close to 7. The product was evaporated using a Vacuum brand CVC 3000. Finally, the solid was dried in a vacuum oven at 80 °C for five hours and coded as Ac-NCC.

2.3 Wet-spinning of cellulose acetate reinforced with NCC and Ac-NCC

A 20% (w/v) cellulose acetate in DMSO introduced to a 6% (w/v) NCC and Ac-NCC solution in DMSO. The concentration of NCC and Ac-NCC in cellulose acetate was 5%. NCC and Ac-NCC were dispersed in cellulose acetate using sonication at 24 kHz at 20 °C - 30 °C for 5 minutes. The polymer solutions were spun with condition as following: needle diameter of 0.25 mm, feed rate of 0.004 g/minute and winding speed of 0.44 m/minute. The fibre was precipitated using DI water in a coagulating bath.

2.4 Characterization

2.4.1. Transmission electron microscope. The as-received NCC was characterised using a LoJeol 1200EX transmission electron microscope (TEM). A 1% (w/v) of NCC suspension was dropped on a copper grid (200 mesh) using a pipette, and then it was stained with a drop of 2% uranyl acetate. The excess solution was absorbed with the tip of filter paper. Finally, the nano-particle of NCC was observed using an accelerating voltage of 80 kV.

2.4.2. FTIR spectroscopy. The functional group present in the NCC before and after acetylation was characterised using an FTIR Spectrometer (Nicolet 8700, Thermo Scientific). The disks were prepared by grinding potassium bromide (KBr) with NCC and Ac-NCC with a ratio of 200 mg and 2 mg

IOP Conf. Series: Materials Science and Engineering **553** (2019) 012038 doi:10.1088/1757-899X/553/1/012038

respectively. The instrument was operated in a transmission-mode in the range 400 - 4000 cm^{-1}, along with a resolution of 4 cm^{-1} and 100 scans.

2.4.3. Environmental scanning electron microscope. The morphology of NCC and Ac-NCC was studied using a Philips XL-30 FEG Environmental SEM. It was operated with an accelerating voltage of 10 kV. Before investigated, the sample was sputter-coated (SC 500 emscope) with a thin layer of gold alloy to reduce charging during analysis.

2.4.4. X-ray diffraction. X-Ray diffraction (Equinox 3000) was used to calculate the crystallinity of NCC and Ac-NCC. The instrument was performed over 2 Theta ranging from 0 to 100° with a scan rate of 0.02° sec^{-1} at 35 kV and 25 mA. The index crystallinity of samples was calculated based on the Segal method [16].

$$\%\mathrm{Crl} = \frac{I_{002} - I_{am}}{I_{002}} \times 100\% \quad (1)$$

3. Results and discussion

3.1. Transmission electron microscope

The TEM micrographs of the as-received NCC dried at 80 °C for 5 hours are presented in the Figure 1. The length and diameter distribution of NCC was obtained by calculating 115 NCC particles using image J software analysis. Selected particles were dragged and clicked 'Analyse — Measure'. Once the data were acquired and saved, the dimensions of the particles were imported for statically analysis. The NCC resembles needle-like particles with a length 46-114 nm and diameter 2.0-4.8 nm. The AFM data reported was similar to the average length and diameter distribution by the supplier which was 44-108 nm and 2.3-4.5 nm respectively. The slight different might be caused by differences of instrument used and the sample preparation.

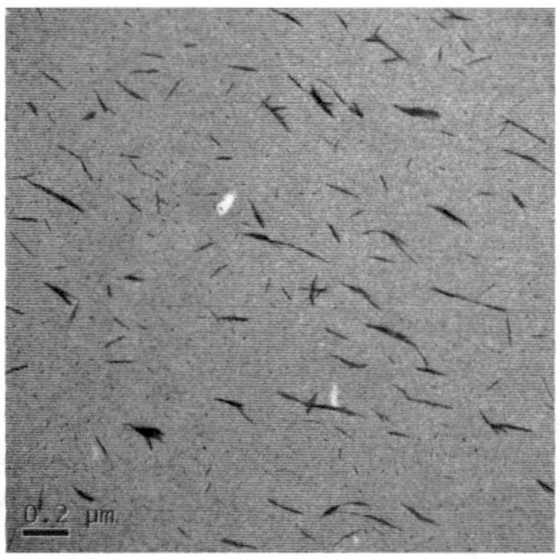

Figure 1. TEM micrograph of as-received NCC.

3.2. Acetylation of NCC

3.2.1. FTIR spectroscopy analysis. Figure 2 presents the FTIR spectra of NCC and Ac-NCC. The hydrogen bonding (O-H) in both of NCC and Ac-NCC appeared at around 3350 cm^{-1} was associated with the structure of cellulose-I [17]. The intensity of the O-H group after acetylation decreased because the hydroxyl groups in NCC were partially substituted by acetyl groups during the acetylation reaction [18]. There is a peak of C-H stretching at 2900 cm^{-1} for NCC before and after acetylation. After acetylation, a new low-intensity peak appeared around 1730 cm^{-1} which was attributed to the carbonyl (C=O) stretching vibration introduced with the acetate groups on the surface [19]. It did not appear for as-received NCC. Similarly, unreacted acetic anhydride and by-products of acetic acid in the sample was not seen. It was confirmed by the absence of a peak at 1700 cm^{-1} and between 1760 and 1840 cm^{-1} [20].

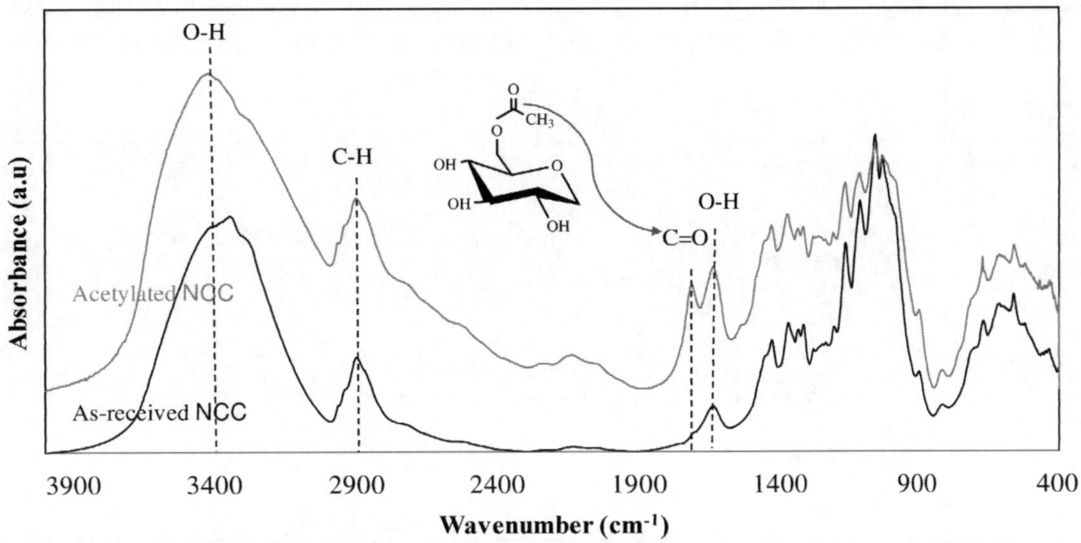

Figure 2. FTIR spectra of NCC before and after acetylation.

3.2.2. The morphology of NCC before and after acetylation. The morphology of NCC before and after acetylation dried at 80 °C for 5 hours are presented in Figure 3. The surface of as-received NCC was smooth, and it changed to be rough and flake-like after acetylation. The same behaviour was also observed from another literature [21]. In addition, Ac-NCC appeared in the form of iridescent films instead powder form. This is due to the rod-like particle of NCC has a behaviour of a chiral nematic at critical concentrations. Hence, it produces iridescent film upon drying [21].

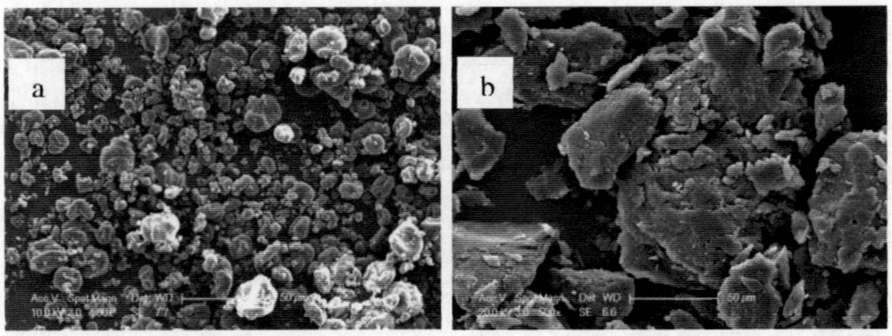

Figure 3. ESEM micrograph for (a) as-received NCC and (b) Ac-NCC.

3.2.3. X-ray diffraction of NCC. Figure 4 shows the XRD pattern of NCC before and after acetylation. From the result, it shows that NCC has 2 Theta angles at around 14.5, 16.3, 23.5 and 34.5° assigned to the typical reflection planes of 101, 10$\bar{1}$, 002 and 040 respectively which was presented as cellulose-I [22], the same result as FITR spectra. The crystallinity of NCC after acetylation decreased from 72% to 68%. The intra- and inter-molecular hydrogen bonds was disrupted by introducing an acetyl group on the surface NCC which in turn disrupts the crystal structure of NCC [23]. Table 1 summarises the decreasing of the crystallinity of NCC after acetylation. The Table shows that there are differences of the crystallinity of NCC from once source to another. It is caused the raw material used for acid hydrolysis of NCC not the same. In addition, the method for the acetylation of NCC being different.

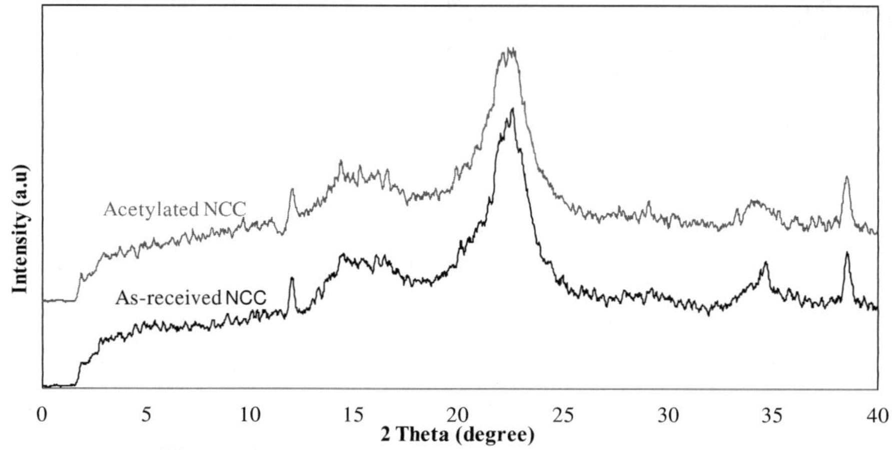

Figure 4. XRD pattern before and after acetylation.

Table 1. The crystallinity of NCC before and after acetylation.

Samples	Crystallinity (%)		
	Acetylating agent	Before acetylation	After acetylation
NCC	Acetic anhydride	72	68
NCC* (24)	Acetic anhydride	81	74
NCC* (25)	Acetic anhydride	77	68

*Data obtained from reference.

3.2.4. Thermogravimetric analysis. Figure 5 represents the TGA traces for NCC before and after acetylation which show three distinct regions. The first stage is dehydration of the sample from room temperature to 100 °C which was assumed to be prominent feature [26].

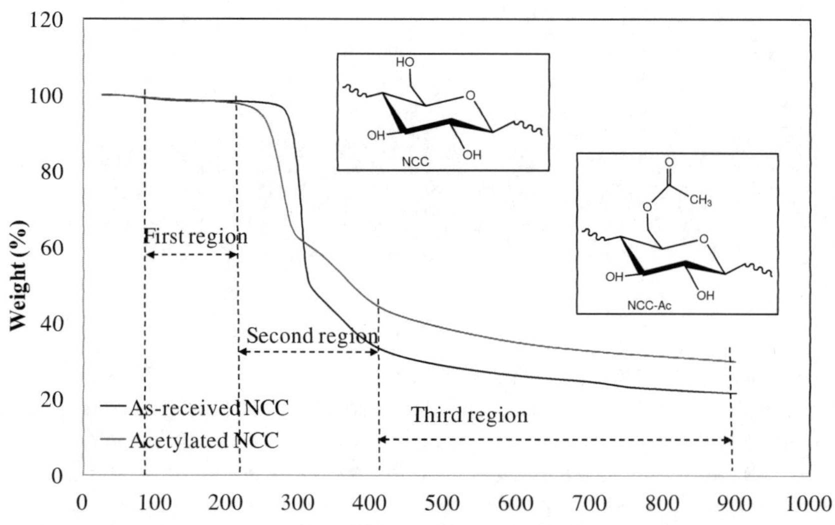

Figure 5. TGA trades for NCC and Ac-NCC with a heating rate of 10 C.minute^{-1} in argon.

The second step region in the TGA trace is the decomposition of NCC to the monomer of D-glucopyranose which is seen to be between 270 °C and 400 °C [27]. However, Ac-NCC degraded at a lower temperature between 210 °C and 400 °C. The decrease in the onset thermal degradation temperature for NCC after acetylation can be rationalised through degree of the crystallinity. It was confirmed from the XRD result. The final step region for both as-received and Ac-NCC was observed to commence around 400 °C attributing to the completing decomposition. The carbon yield of Ac-NCC is approximately 10% higher than as-received NCC.

3.3. The morphology of wet-spun fibres

The morphology and the cross-section of wet-spun cellulose acetate reinforced NCC and Ac-NCC are presented in Figure 6. Longitudinal lines are observed for both polymer fibres which might be coming from inner bone of the needle. The fibre reinforced NCC has lower compatibility with the polymer solution compared to Ac-NCC. It can be seen pores or void appeared at the cross-section of the fibres. In addition, the grooves with some particles are also seen on the surface of wet-spun cellulose acetate. On the other hand, the morphology of cellulose acetate reinforced Ac-NCC is smoother and has fewer particles. It has been reported that the presence of NCC in a polymer matrix leaded to the formation "bead fibres" and the morphology was rough due to poor compatibility [28]. The acetylation on NCC increases its compatibility with the polymer matrix. It assumes that it improves the mechanical properties of the fibre.

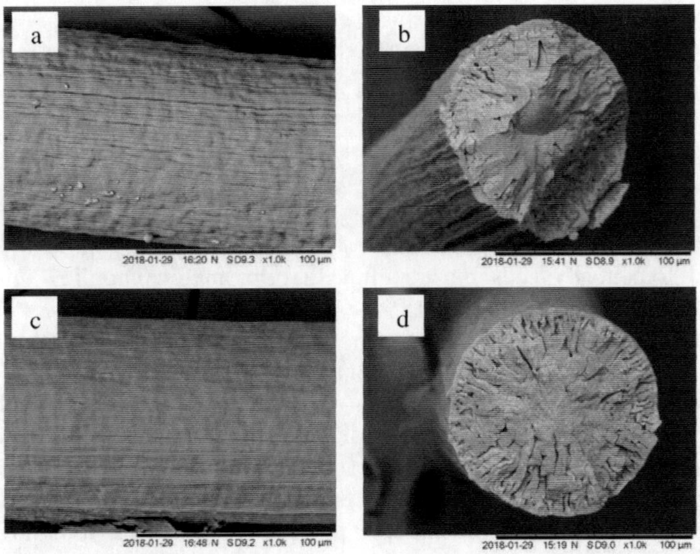

Figure 6. SEM micrographs and cross-section area of wet-spun (a-b) cellulose acetate reinforced 5% NCC and (c-d) cellulose acetate reinforced 5% Ac-NCC.

4. Conclusion

Surface acetylation of NCC in a gel-form has been achieved by introducing acetyl group on the surface of NCC using acetic anhydride as the acetylating agent and pyridine as the catalyst. The porosity of wet-spun fibres decreases and it produces circular as well as continuous fibre. The morphology of cellulose acetate reinforced Ac-NCC is smoother than those reinforced NCC. The particles are also found on the surface of the wet-spun fibre reinforced NCC. It indicates that the acetylation on NCC has improved its compatibility with cellulose acetate. The heat treatment of cellulose acetate reinforced NCC is still on-progress.

Acknowledgment

The authors would like to thank the Indonesian government through the Indonesia Endowment Fund for Education (LPDP) for supporting funded from 2016 to 2018.

References

[1] Eichhorn SJ, Young RJ. The Young's modulus of a microcrystalline cellulose. Cellulose. 2001;8(3):197–207. doi:10.1023/a:1013181804540.
[2] Sun XF, Sun RC, Fowler P, Baird MS. Isolation and characterisation of cellulose obtained by a two-stage treatment with organosolv and cyanamide activated hydrogen peroxide from wheat straw. Carbohydr Polym. 2004;55(4):379–91. doi:10.1016/j.carbpol.2003.10.004.
[3] Silvério HA, Neto WPF, Dantas NO, Pasquini D. Extraction and characterization of cellulose nanocrystals from corncob for application as reinforcing agentin nanocomposites. Ind Crop Prod. 2013;44:427–36. https://doi.org/10.1016/j.indcrop.2012.10.014.
[4] Gea S, Zulfahmi Z, Yunus D, Andriyani A, Hutapea YA. The Isolation of Nanofibre Cellulose from Oil Palm Empty Fruit Bunch Via Steam Explosion and Hydrolysis with HCl 10 %. J Phys Conf Ser. 2018;979:1–8. doi:10.1088/1742-6596/979/1/012063.
[5] Peng BL, Dhar N, Liu HL, Tam KC. Chemistry and applications of nanocrystalline cellulose and its derivatives: A nanotechnology perspective. Can J Chem Eng. 2011;89(5):1191–206. https://doi.org/10.1002/cjce.20554.
[6] Ioelovich M. Optimal Conditions for Isolation of Nanocrystalline Cellulose Particles. Nanosci Nanotechnol. 2012;2(2):9–13. DOI: 10.5923/j.nn.20120202.03.
[7] Abraham E, Deepa B, Pothan LA, Jacob M, Thomas S, Cvelbar U, et al. Extraction of nanocellulose fibrils from lignocellulosic fibres: A novel approach. Carbohydr Polym. 2011;86(4):1468–75. doi:10.1016/j.carbpol.2011.06.034.
[8] Tang J, Sisler J, Grishkewich N, Tam KC. Functionalization of cellulose nanocrystals for advanced applications. J Colloid Interface Sci. 2017;494:397–409. http://dx.doi.org/10.1016/j.jcis.2017.01.077.
[9] Jorfi M, Foster EJ. Recent advances in nanocellulose for biomedical applications. J Appl Polym Sci. 2015;132(14):1–19. DOI: 10.1002/APP.41719.
[10] Hoeng F, Denneulin A, Bras J. Use of nanocellulose in printed electronics: A review. Nanoscale. 2016;8:13131–54. DOI: 10.1039/c6nr03054h.
[11] Capadona JR, Shanmuganathan K, Trittschuh S, Seidel S, Rowan SJ, Weder C. Polymer nanocomposites with nanowhiskers isolated from microcrystalline cellulose. Biomacromolecules. 2009;10(4):712–6. DOI: 10.1021/bm8010903.
[12] Sinaga MZE, Gea S, Panindia N, Sihombing YA. The preparation of cellulose nanocomposite film from isolated cellulose of corncobs as food packaging. Orient J Chem. 2018;34(1):562–7. DOI: 10.13005/ojc/340166.
[13] Wang J, Huang S, Lu X, Xu Z, Zhao Y, Li J, et al. Wet-spinning of highly conductive nanocellulose–silver fibers. J Mater Chem C. 2017;5:9673–9. DOI: 10.1039/C7TC03217J.
[14] Torres-Rendon JG, Schacher FH, Ifuku S, Walther A. Mechanical performance of macrofibers of cellulose and chitin nanofibrils aligned by wet-stretching: A critical comparison. Biomacromolecules. 2014;15(7):2709–17. dx.doi.org/10.1021/bm500566m.
[15] Lundahl MJ, Cunha a G, Rojo E, Papageorgiou AC, Rautkari L, Arboleda JC, et al. Strength and Water Interactions of Cellulose I Filaments Wet-Spun from Cellulose Nanofibril Hydrogels. Sci Rep. 2016;6:30695. DOI: 10.1038/srep30695.
[16] Wu Q-L, Gu S-Y, Gong J-H, Pan D. SEM/STM studies on the surface structure of a novel carbon fiber from lyocell. Synth Met. 2006;156(11–13):792–5. DOI: 10.1016/j.synthmet.2006.04.007.
[17] Moharram MA, Mahmoud OM. FTIR spectrocopic study of the effect of microwave heating on the transformation of cellulose I into cellulose II during mercerization. J Appl Polym Sci. 2008;107:30–6. DOI 10.1002/app.26748.
[18] Ashori A, Babaee M, Jonoobi M, Hamzeh Y. Solvent-free acetylation of cellulose nanofibers

for improving compatibility and dispersion. Carbohydr Polym. 2014;102(1):369–75. http://dx.doi.org/10.1016/j.carbpol.2013.11.067.
[19] Lin N, Huang J, Chang PR, Feng J, Yu J. Surface acetylation of cellulose nanocrystal and its reinforcing function in poly (lactic acid). Carbohydr Polym. 2011;83:1834–42. doi:10.1016/j.carbpol.2010.10.047.
[20] Shaikh HM, Pandare K V., Nair G, Varma AJ. Utilization of sugarcane bagasse cellulose for producing cellulose acetates: Novel use of residual hemicellulose as plasticizer. Carbohydr Polym. 2009;76(1):23–9. doi:10.1016/j.carbpol.2008.09.014.
[21] De Souza Lima MM, Borsali R. Rodlike cellulose microcrystals: Structure, properties, and applications. Macromol Rapid Commun. 2004;25(7):771–87. DOI: 10.1002/marc.200300268.
[22] Liu D, Zhong T, Chang PR, Li K, Wu Q. Starch composites reinforced by bamboo cellulosic crystals. Bioresour Technol. 2010;101(7):2529–36. doi:10.1016/j.biortech.2009.11.058.
[23] Tang L, Huang B, Lu Q, Wang S, Ou W, Lin W, et al. Ultrasonication-assisted manufacture of cellulose nanocrystals esterified with acetic acid. Bioresour Technol. 2013;127:100–5. http://dx.doi.org/10.1016/j.biortech.2012.09.133.
[24] Jonoobi M, Mathew AP, Abdi MM. A Comparison of Modified and Unmodified Cellulose Nanofiber Reinforced Polylactic Acid (PLA) Prepared by Twin Screw Extrusion. J Polym Environ. 2012;20:991–7. DOI: 10.1007/s10924-012-0503-9.
[25] Jonoobi M, Harun J, Mathew AP, Hussein MZB, Oksman K. Preparation of cellulose nanofibers with hydrophobic surface characteristics. Cellulose. 2010;17(2):299–307. DOI: 10.1007/s10570-009-9387-9.
[26] Tomé LC, Brandão L, Mendes AM, Silvestre AJD, Neto CP, Gandini A, et al. Preparation and characterization of bacterial cellulose membranes with tailored surface and barrier properties. Cellulose. 2010;17(6):1203–11. DOI 10.1007/s10570-010-9457-z.
[27] Yang Z, Xu S, Ma X, Wang S. Characterization and acetylation behavior of bamboo pulp. Wood Sci Technol. 2008;42(8):621–32. DOI 10.1007/s00226-008-0194-5.
[28] Rojas OJ, Montero GA, Habibi Y. Electrospun nanocomposites from polystyrene loaded with cellulose nanowhiskers. J Appl Polym Sci. 2009;113:927–35. DOI: 10.1002/app.30011.

Production of Nanobiocontrol Agent in Molasses and Tofu Liquid Waste Media by *Streptomyces sp*. TT10

Desak Gede Sri Andayani[1], I Dewa Putu Hermida[2], Novik Nur Hidayat[3], Toni Kristiantoro[2], Desak Gede Tirta Andini[4]

[1]Research Unit for Clean Technology, Indonesian Institute of Sciences, Jl. Cisitu-Sangkuriang, Campus LIPI, Bandung 40135 Indonesia
[2]Research Center for Electronics and Telecommunication, Indonesian Institute of Sciences, Jl. Cisitu Sangkuriang, Campus LIPI, Bandung 40135 Indonesia
[3]Research Center for Biology, Indonesian Institute of Sciences, Jl. Raya Jakarta-Bogor Km 46 Cibinong 16911 Indonesia
[4] Department of Chemistry, Airlangga University, Jl. Dr. Ir. H. Soekarno Mulyorejo Campus C, Surabaya 60115, Indonesia

desakgedesa@gmail.com

Abstract. The objective of this study has processed the waste of agriculture into high economic value materials using fermentation technology by *Streptomyces sp*. TT 10 in the form of new materials in Nano size. In this study, anti-phytopathogenic agents would be produced using fermentation technology in the form of antifungal that fought fungi that often infected strawberries. The research series consisted of the fermentation process for seven days, 30°C, pH 7 on molasses and tofu liquid waste medium. The broth of fermentation was extracted using ethyl acetate. Concentrated ethyl acetate extracts were ground using a shaker mill PPF-UG to obtain the size up to 100 nm. Furthermore, Nano sizes were characterized using (particle size analyzer) PSA and (scanning electron microscope) SEM. The anti-phytopathogenic agent was analyzed by agar diffusion and microdilution. The results showed that activity of ethyl acetate extract of *Streptomyces sp.* TT10 was increased up to 80% after milling process, against pathogenic strawberry that is *Penicillium sp., Aspergillus sp., Rhodotorula sp., Kluyveromyces sp., Candida krusei, Candida pseudotropicalis* and *Fusarium oxysporum* with inhibitory diameter (mm), minimum inhibition concentrations (MIC) (ppm) and the minimum fungicidal concentration (MFC) (ppm) value (%) are respectively (82.85:50;50), (81.29;50;50), (82.85;50;50), (83.87;50;50), (74.07;50;50), (79.95;50;50), and (83.33; 50; 50).

Keywords: molasses, tofu liquid waste, fermentation technology, nanotechnology, pathogenic strawberry

1. Introduction

The need for fruits as a source of vitamins and minerals has become an urgent matter for current conditions. Tropical fruits provide many benefits for surviving of creatures in the world. Increased production of tropical fruits is very dependent on how to handle from where it grows to distribution in the field. Maintaining tropical fruits from pathogen is an effort to maintain the quality of tropical fruits [1].

Based on this, it is necessary to look for agents in the form of secondary metabolites of *Streptomyces sp.* TT10 to overcome diseases in tropical fruits (case study: strawberry fruits). *Streptomyces sp.* was known as a producer of secondary metabolites such as antimicrobial, antiviral and anti-fungi [2-7]. *Streptomyces sp.* TT10 is isolated from Indonesian volcanic soils.

The agricultural industry produces primary and secondary products as renewable resources that support other important industrial sectors outside agriculture. Agricultural secondary products that are not used in the industry still contain high nutritional value to produce secondary metabolites in the form of antifungal, antibacterial, antiviral and antioxidant. These metabolites can be applied as biological control agents in general having an inhibitory mechanism against pathogens through its metabolites [8].

The objective of this study was processed the waste of food and agricultural become high economic value materials using fermentation technology and find the formula of anti-pathogenic for tropical fruits (case studies of strawberry fruit) in the form of a new materials in Nano size from *Streptomyces sp.* strain TT10.

Nanoparticles are microscopic particles that have the same size as atoms to molecules. Nanoparticles are ultrafine particles with a length greater than 0.001 micrometers (1 nanometer) and smaller than 0.1 micrometers (100 nanometers) (ASTM). In general, nanoparticles are also known as collections of atoms that bind to each other with a diameter of less than 100 nm [9]. Nanoparticles have very small sizes or even invisible, but have a very large influence on living things and the environment. These particles help the development of technology, improve health and well-being, and have helped maintain environmental stability [10,11].

Nanoparticles have a large surface area due to the small geometry size. This large surface area causes nanoparticles to become highly reactive and easily bind to each other to form larger particles to become a macro size. To get nanoparticles, it can be through bottom up and top down methods. The bottom up method is how to form nanoparticles by making the smallest atoms or molecules. This method is generally used by nature to form nanoparticles. The top down method is how to form nanoparticles with mechanical equipment, in general with the milling process. In the process, the material will be reduced to Nano size [12,13].

One method to get Nano scale material is to use high energy milling method, which is a technique that uses collision energy between crushing balls and rotating chamber walls in a certain way. The ball in the chamber will collide with each other which causes the particle to break to reach Nano size [14].

There are several variables that must be considered to get the desired particle size, such as: type of grinder, duration of milling, temperature of grinder, type of material to be milled, number and type of grinding media material, ratio of mass of grinding and powder media, and frequency [15].

In this study Nanobiocontrol agents will be produced based on fermentation technology and nanotechnology through self-assembly or biosynthesis and top down (milling process) method in the form of new material anti-fungi that can fight pathogenic strawberries fruits.

2. Experimental method

2.1. Microorganisms

2.1.1. Secondary metabolite producer. Streptomyces sp. TT10 is used as a producer of anti-pathogenic metabolite regenerated on potatoes dextrose agar (PDA) media Difco.

2.1.2. Pathogenic Strawberry. The microorganisms used for testing were *Penicillium sp., Aspergillus sp., Rhodotorula sp., Kluyveromyces sp., Candida krusei, Candida pseudotropicalis* and *Fusarium oxysporum* are isolated from infected strawberries. All isolates were grown, maintained, and tested in potatoes dextrose agar (PDA) Difco as well as in potatoes dextrose broth (PDB) Difco media.

IOP Conf. Series: Materials Science and Engineering **553** (2019) 012039 doi:10.1088/1757-899X/553/1/012039

2.2. Sporulation Process, Fermentation Process and *Purification of Fermentation Broth.* Sporulation process of *Streptomyces sp.*TT10 was carried out on starch (10 g/L), glucose (10 g/L), trypton (5 g/L), yeast extract (2.5 g/L) and calcium carbonate (1 g/L) for 72 hours at 30°C, pH 7. Five to ten percent of sporulation media at optimum growth times (hours) are inoculated into fermentation media containing molasses and tofu liquid waste. The biosynthesis process is carried out for seven days, 30°C, pH 7, and 120 rpm. The concentration of glucose and protein was analyzed during the fermentation process. The broth of fermentation and biomass were separated using centrifuge TOMY MX-301 at 9000 rpm, 4°C for 20 minutes. The broth of fermentation is extracted using ethyl acetate organic with a ratio of 1: 1 at 28 ° C. Ethyl acetate extract was evaporation using evaporator at 50 rpm, 1atm, and 50 ° C [16-18].

2.3. Milling process
Ethyl acetate extract is converted into Nano particles by milling from 1 to 3 hours in PPF-UG shaker mill with a duration of one hour on and 10 minutes off at 70°C. The mass ratio of grinding and ethyl acetate extract is 10: 1 BPR (ball to powder weight ratio) with a frequency of 500 rpm. The results of milling were analyzed using a particle size analyzer (PSA) HORIBA LB-550 (IK 03 TP 016) and scanning electron microscope (SEM) JEOL JSM IT-300-JAPAN [19-22].

2.4. Testing of Antiphatogenic Strawberry Activities
Micro and Nano sized of ethyl acetate extracts of *Streptomyces sp.* TT10 were tested for anti-pathogenic strawberry using diffusion and dilution methods [23].

The diffusion method was carried out in a Petri dish that containing PDA media which had been homogeneously mixed with pathogenic strawberry suspension, then incubated for 72 hours at 30°C. Observations were made by measuring the diameter of the clear barriers in millimeters.

Minimum inhibition concentration (MIC) is determined by a liquid micro dilution method using PDB media in 96 well plates Nunc. Ten different concentrations of fermented extracts were tested for pathogenic strawberries. Stock solutions used for ten different concentrations (ppm): 1000; 500; 250; 125; 62.50; 31.25; 15, 63; 7.8; 3,9; 1.95 then incubated for 72 hours at 30°C. MIC is the lowest concentration that inhibits pathogenic strawberry.

After determining the MIC, the cultures of all wells that were not overgrown with strawberry pathogenic were observed in a petri dish that had been filled with PDA media and then incubated for 72 hours, 30°C. The endpoint of minimum fungicidal concentration (MFC) was defined as the lowest concentration of an antifungal agent which killed > 99.9% of the initial pathogenic strawberry population where no growth was observed. Efficiency calculated by:

$$[(A-B)/ B] \times 100 \% \quad (1)$$

Where are GA = zones of diameter (mm) after milling process, MIC (ppm) after milling process and MFC (ppm) after milling process. B = zones of diameter (mm) before milling process, MIC (ppm) before milling process and MFC (ppm) after milling process.

3. Results and discussion

3.1. Sporulation, Fermentation and Extraction of Streptomyces sp. strain TT10
The growth and formation of secondary metabolites in the form of anti-pathogenic strawberry are formed at stationary phase at 38-52 hours and reach optimum at the 50nd hour referring to Figure 1.

From pH value was showed *Streptomyces sp.* TT10 produced anti-pathogenic strawberry metabolites at acidic atmosphere. The data of concentration of glucose and protein showed that *Streptomyces sp.* TT 10 required carbon and protein sources for the growth and formation of anti-pathogenic strawberry metabolites that was indicated by a decrease of glucose concentrations from 6.39 mg / ml to 4.12 mg / ml and a decrease of protein concentrations from 0.66 mg / ml to 0.42 mg / ml (Figure 1).

The fermentation liquid which has been separated from the biomass was tested for antifungal activity against pathogenic strawberry. The table shows the antifungal metabolites was formed from the first day to the seventh day. The highest activity was reached at the third day, and then the activity was decreased [24].

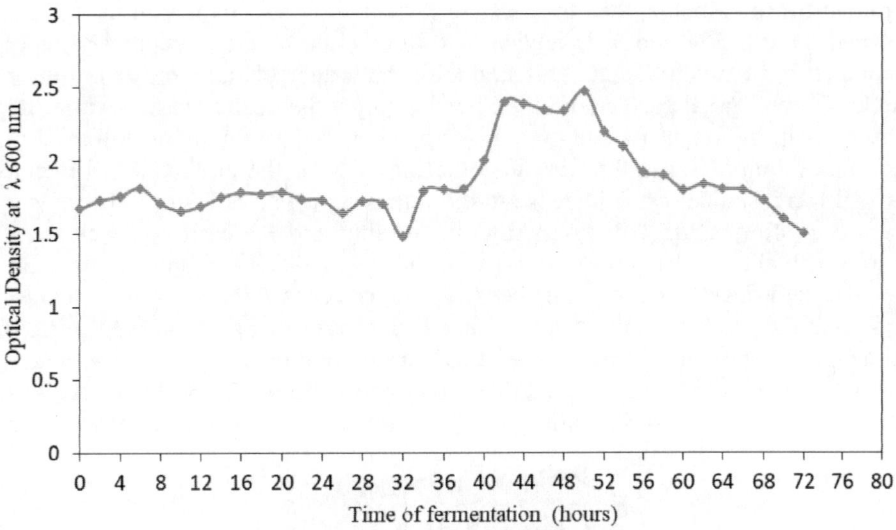

Figure 1. Growth curve of *Streptomyces sp*. TT10

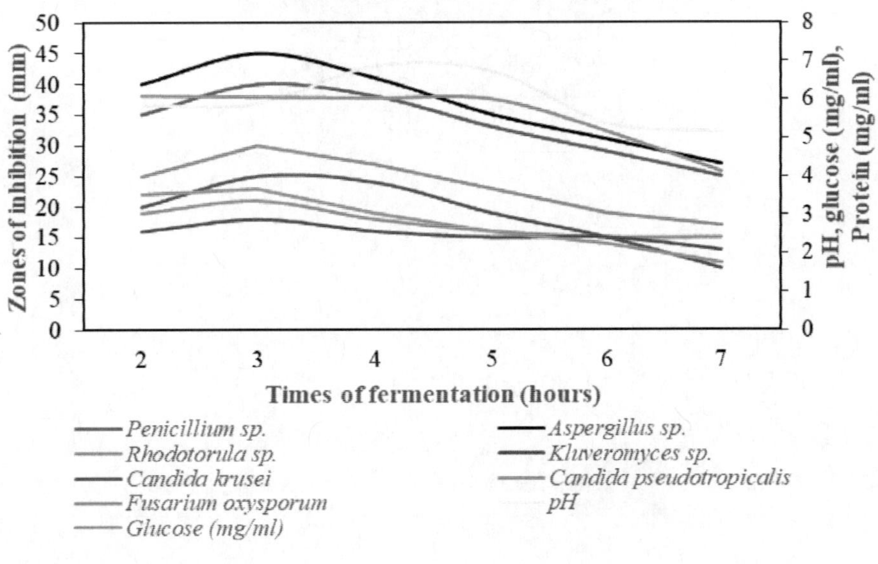

Figure 2. Data of pH, glucose, protein (mg/ml) and zones of inhibition for seven days fermentation of *Streptomyces sp*. TT10

3.2. Testing of Ethyl Acetate Extract of Streptomyces sp. TT10

In vitro test on ethyl acetate extract showed the activities of pathogenic strawberry with zones of inhibition (mm): *Penicillium sp*. (35 mm), *Aspergillus sp*. (38.75 mm), *Rhodotorula sp*. (17.5 mm), *Kluyveromyces sp*. (15.5 mm), *Candida krusei* (20.25 mm), *Candida pseudotropicalis* (19.75 mm) and

Fusarium oxysporum (30 mm) with MIC from 31.25 to 3.91 ppm and MFC from 62.5 to 7.81 ppm (Figure 2) [25].

3.3. Mechanical treatment with shaker mill against ethyl acetate extract of Streptomyces sp. TT10

After being milled for three hours, the ethyl acetate extract showed a reduction in size from micro to Nano as shown in Figure 3, 4 and 5. Likewise with the weight of the extract, the longer of milling process, the smaller the extract weight. This is due to the longer of duration of milling process, the reduced particle size will increase. The duration needed to get the desired results varies depending on the type of grinder, milling intensity, ball mass ratio: powder, and milling temperature. The greater the ratio, the less time required for milling because the intensity of the collision will increase. Milling temperatures will increase due to the increase in the intensity of ball collisions. There is a possibility that this increase in temperature will result in cold welding and increasing particle size. The same effect was also shown due to variations in the size of the steel ball. The bigger the size, the greater the impact energy that will occur. The rotation speed of the grinder is directly proportional to the energy that will be produced. The higher the rotation speed, the more energy will be generated. Ball mill grinder rotation speed is limited by critical speed. When the rotation speed is above the critical speed, the steel balls will move around the surface in the grind and will not give a shock load to the powder. Therefore, it is necessary to know the optimal speed to be able to produce the biggest impact energy [26].

Figure 3. Shaker Mill PPF-UG

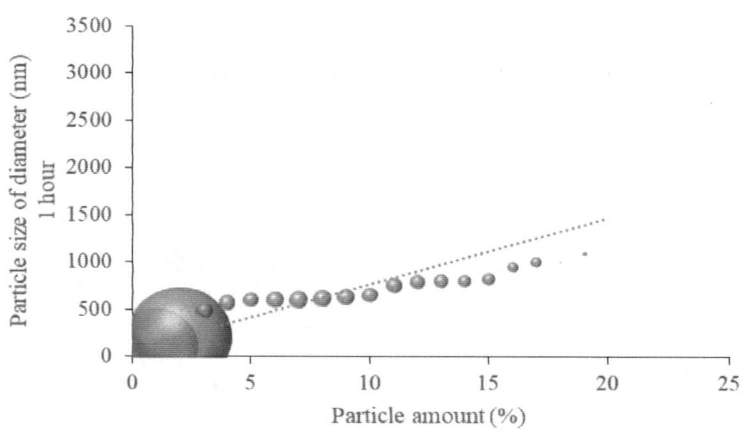

(a)

IOP Conf. Series: Materials Science and Engineering **553** (2019) 012039 doi:10.1088/1757-899X/553/1/012039

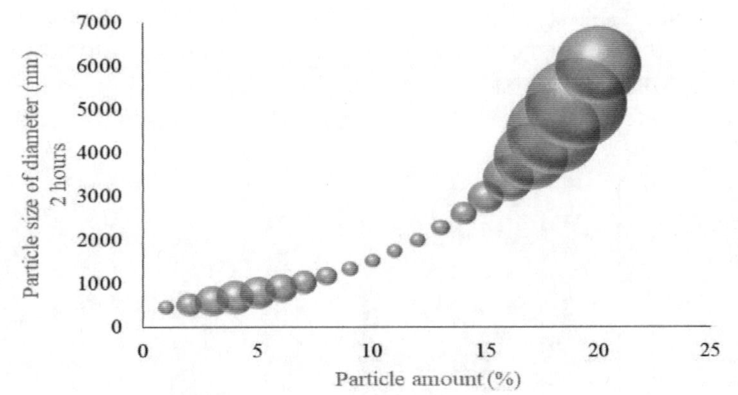

(b)

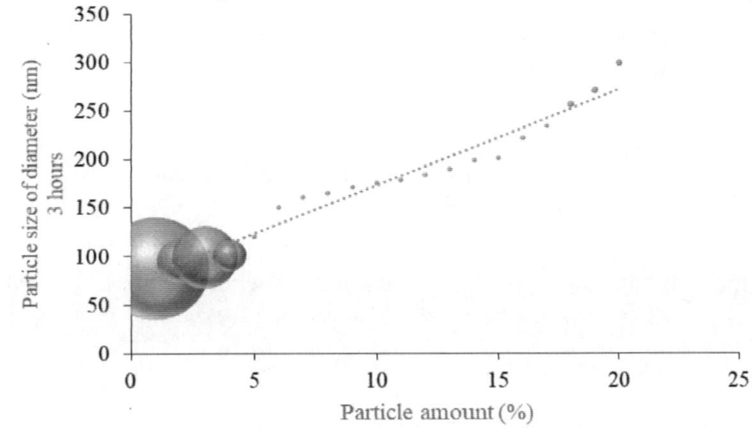

(c)

Figure 4. Size and Distribution Particles Analysis of Nano biomaterial of Ethyl Acetate Extract of *Streptomyces sp.* TT10 for (a): 1 hour milling; (b): 2 hours milling; (c): 3 hours milling

Figure 5. Nanometer analysis using SEM after 3 hours of grinding

IOP Conf. Series: Materials Science and Engineering **553** (2019) 012039 doi:10.1088/1757-899X/553/1/012039

In Figure 6, 7, 8 and Table 1, shows a clear difference in the in-vitro activity of Nano-acetate extract in Nano size was 80% higher inhibiting pathogenic strawberry compared with ethyl acetate extract before and after milling process.

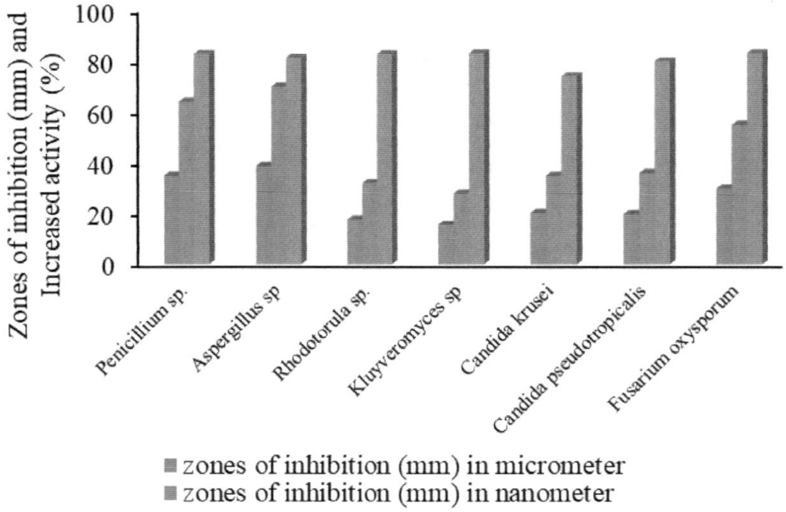

Figure 6. Increased of anti-pathogenic strawberry activities of ethyl acetate extract *Streptomyces sp.* TT10 after self-assembly processed

This is caused by smaller size changes which will increase the antifungal effect. High viscosity and wide range of tissue and high efficiency and stability of active substances in the absorption of active ingredients into the tissue enhancements so that it easily penetrates the cell lipid membrane in plant tissues.

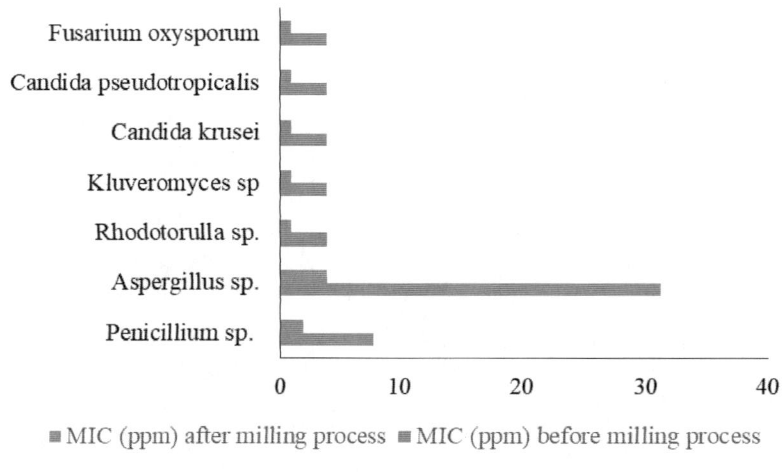

(a)

IOP Conf. Series: Materials Science and Engineering **553** (2019) 012039 doi:10.1088/1757-899X/553/1/012039

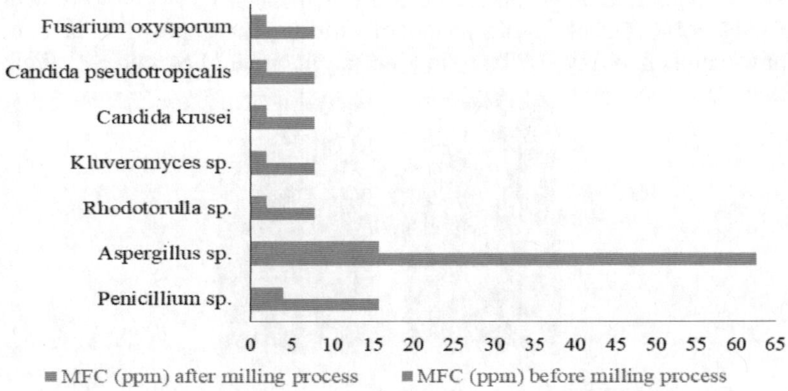

(b)

Figure 7. (a): MIC and (b): MFC of ethyl acetate extract *Streptomyces sp*. TT10 before and after milling process

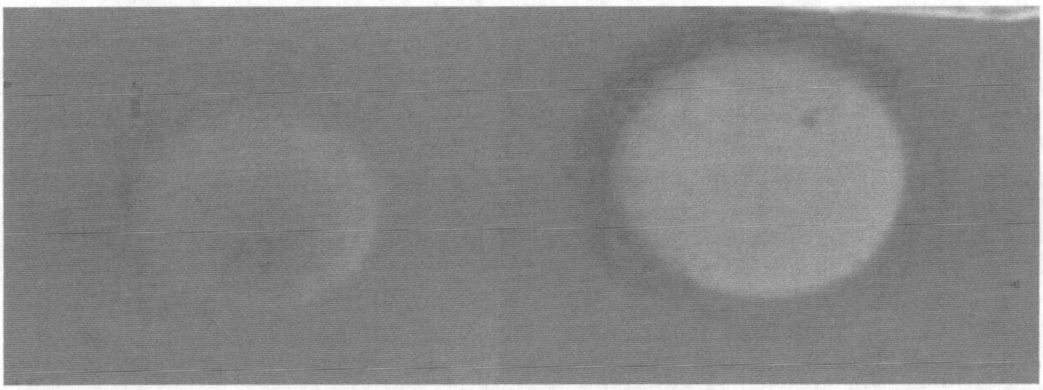

(a) **(b)**

Figure 8. Anti pathogenic activity of *Streptomyces sp* TT10 against strawberry by diffusion agar methods, (a) nanobiocontrol > 100 nm and (b) nanobiocontrol < 100 nm

Table 1. Anti pathogenic activity of *Streptomyces sp*. TT10 against strawberries by micro dilution method

K+	K-	1.95 ppm	3.91 ppm	7.81 ppm	15.63 ppm	31.25 ppm	62.5 ppm	125 ppm	250 ppm	500 ppm	1000 ppm
PDB+*	PDB	-	-	-	-	-	-	-	-	-	-

*= pathogenic strawberries = *Penicillium sp., Aspergillus sp., Rhodotorula sp., Kluyveromyces sp., Candida krusei, Candida pseudotropicalis* and *Fusarium oxysporum*

Milling process will result the newly produced material, it was assessed that the size of material particle decreases efficiently with increased grinding time. Milling process is able to produce until 100% particle without affected the chemical phases. With the optimal average size of 25 nm and a narrow particle size distribution. The spontaneous self-assembling behavior observed in nature that leads to complex, multifunctional, hierarchical structures within biological systems. Recent research undertaken to synthesize hierarchically assembled functional material have underscored the need as well as the benefit in synergistically combining top down fabrication methods with bottom up self-assembly [27-30]. Nanobiocontrol in the shape of emulsion and powder have been produced by Research Unit for Clean technology (LPTB) , Indonesian Institute of Sciences (LIPI).

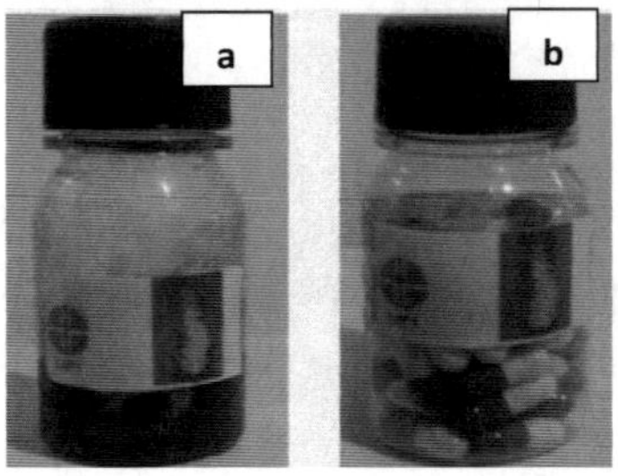

Figure 9. Laboratory scale of nanobiocontrol in emulsion (a) and powder (b) produced by LIPI

4. Conclusion

Food and agriculture industry waste, namely tofu and molasses can be used as a medium for the growth of *Streptomyces sp*. TT10 to produce bio-control agents. Ethyl acetate extract of *Streptomyces sp*. TT10 in the shape of new nanoparticles has been shown enhancement to the antifungal activities against pathogenic strawberries fruits. This top-down approach by milling process provides a new outline for the future design of nanostructured of bio-pesticide.

Acknowledgement
This research supported by Indonesian Institute of Sciences and Ministry of Research, Technology and Higher Education Republic of Indonesia.

References

[1] Q. Chaudhry, L. Castle, "Food applications of nanotechnologies: An overview of opportunities and challenges for developing countries", *Trends in Food Science & Technology*, pp. 595-603, 2011. doi:10.1016/j.tifs.2011.01.001

[2] H.K. Daima, V. Bansal, "Properties of Nanomaterials on Their Antibacterial Applications", *Nanotechnology in Diagnosis*, pp. 151-166, 2015.

[3] J. M. Raaijmakers, M. Vlami, and J. T. De Souza, Antibiotic production by bacterial biocontrol agents, *Antonie van Leeuwenhoek*, vol. 81, pp. 537–547, 2002. doi:10.1023/A:1020501420831

[4] J.B. Qureshi, R.J. Mauger, J.L. Cano, Galazzo, and M.D. Lee, "New metabolites from microbial fermentation of a *Streptomyces sp*." *J Antibiot*, vol. 54, pp. 1100–1103, 2001.

[5] A.M. Pastrana, M.J.B. Ureba, A. Aquado, and N. Capote, "Potential Inoculum Sources and Incidence of Strawberry Soil –borne pathogens in Spain", *Plant Disease*, vol. 101, pp. 751-760, 2017. doi: 10.1094/PDIS-08-16-1177-RE

[6] B. D. L. Santos, C. Barrau, and F. Romero, Strawberry Fungal Diseases, *Journal of Food Agriculture and Environment*, vol. 1, pp.3-4, 2003.

[7] J.L. Maas, *Compendium of strawberry diseases*, 2nd Ed., American Phytopathological Society, St. Paul, Minn., pp. 98, 1998.

[8] L.R. Khot, S. Sankaran, J. M. Maja, R. Ehsani, and Schuster, "Application of nanomaterials in agricultural production and crop protection: A review", *Crop Protection*, vol. 35, pp. 64-70, 2012. doi: 10.1016/j.cropro.2012.01.007

[9] L. G. Austin, R.R. Klimpel, and P.T. Luckie, "Process Engineering of size Reduction: Ball Milling", AIME-SME, New York, USA, 1984.

[10] H. Chen, R. Yada, "Nanotechnologies in agriculture: New tools for sustainable development", *Trends in Food Science & Technology*, vol. 22, pp. 585-594, 2011.

[11] J.S. Duhan, R. Kumar, P. Kaur, K. Nehra, S. Duhan, "Nanotechnology: The new perspective in precision agriculture", *Biotechnology Reports*, vol. 15, pp. 11-23, 2017. doi: 10.1016/j.btre.2017.03.002
[12] T. Nann, J. Riegeler, G.A. Urban, "Bio-Nano-Systems", *BioMEMS*, vol. 16, pp. 351-373, 2006.
[13] A. Ditta, M. Arshad, M. Ibrahim, "Nanoparticles in Sustainable Agricultural Crop Production: Application and Perspectives", Springer International Publishing Switzerland 2015.
[14] P. Balaz, "Mechanochemistry in Nanoscience and Minerals Engineering", Springer-Verlag Berlin Heidelberg, 2008.
[15] S. Soni, A. Salhotra, and M. Suar, Handbook of Research on Diverse Applications of Nanotechnology in Biomedicine, Chemistry, and Engineering, Engineering Science Reference, 2014
[16] N. Nelson, "A Photometric Adaptation of The Somogy Method For The Determination of Glucose", *J. Biol. Chem,* pp. 375-380, 1944.
[17] O. H. Lowry, N.J. Rosebrough, A.F. Farr, and R.J. Randall, "Protein Measurement With The Folin Phenol Reagent", *J. Biol. Chem,* pp. 193-265, 1951.
[18] V. Singh, C. K. M. Tripathi, and V. Bihari, "Production, optimization and purification of an antifungal compound from *Streptomyces capoamus", Med Chem Res.,*vol. 17, pp. 94-102, 2008.
[19] T.P. Yadav, R.M. Yadav, and D.P. Singh, "Mechanical Milling: a Top Down Approach for the Synthesis of Nanomaterials and Nanocomposites", *Nanoscience and Nanotechnology*, vol. 2, pp. 22-48, 2012. doi: 10.12691/ajn-3-1-4
[20] Z.H. Loh, A.K. Samanta, and P.W.S. Heng, "Overview of milling techniques for improving the solubility of poorly water-soluble drugs", *Asian Journal of Pharmaceutical Sciences*, vol. 10, pp/ 255-274, 2015. doi: 10.1016/j.ajps.2014.12.006
[21] M. Pentimalli, M. Bellusci, and F. Padella, "High-Energy Ball Milling as a general Tool for Nanomaterials Synthesis and Processing", *Handbook of Mechanical Nanostructuring*, Chapter 28, 2015.
[22] W. Cao, "Synthesis of Nanomaterials by High Energy Ball Milling", Skyspring Nanomaterials, Inc., UnderstandingNano.com, 2007-2018, Hawk's Perch Technical Writing, LLC
[23] The British Society for Antimicrobial Chemotherapy (BSAC) Guide to Sensitivity Testing
[24] T. Fukudato , A. Matsumoto, T. Takahashi, H. Tomoda, and S. Omura, "Phenatic acids A and B, new Potentiators of antifungal miconazole activity produced by *Streptomyces sp.* K03–0132", *J Antibitot,* vol. 58, pp. 252–259, 2005.
[25] T. Amini, Z. Agapoor, and M. Ashengriph, "Evaluation of *Streptomyces spp.* against *Fusarium oxysporum* f. sp. ciceris for the management of chickpea wilt", *Journal of plant protection research,* vol. 56, 2016.
[26] C. Suryanarayana, "Progress in Materials Science", vol. 46, pp. 1-184, 2001.
[27] B.W. Jo, S. Chakraborty, H.H. Kim, Y.S. Lee, "Effectiveness of the Top-Down Nanotechnology in the Production of Ultrafine Cement (~220 nm)", *Journal of Nano materials* , pp. 1-9, 2014.
[28] L. Claudio, De Castro, S.M. Brian, "Synthesis, Functionalization and Surface Treatment of Nanoparticles", *Nanoparticles from Mechanical Attrition*, American Scientific Publishers, 2002.
[29] J. Koruza, B. Malic, O. Noshchenco, and K. Marija, "Top-Down Processing of NaNbO3 Nanopowder", *Journal of Nanomaterials*, pp. 1-7, 2012.
[30] R. Thiruvengadathan, V. Korampally, A. Ghosh, N. Chandra, Gamgopadhyay K., and S. Gamgopadhyay, "Nanomaterial processing using self-assembly", *Reports on Progress in Physics*, vol.76, IOP Publishing 2013.

IOP Conf. Series: Materials Science and Engineering **553** (2019) 012040 doi:10.1088/1757-899X/553/1/012040

Characteristics of Uranium Dioxide (UO_2) Kernel Produced by Sintering Process using Modified Sintering Reactor

Moch Setyadji[1*], Fitrotun Aliyah[2]
[1]Center for Science and Accelerator Technology, National Nuclear Energy Agency, Jl. Babarsari kotak pos 6101 ykbb, Yogyakarta 55281, Indonesia
[2]Nuclear Engineering and Engineering Physics Department, Faculty of Engineering, Universitas Gadjah Mada, Jl. Grafika No. 2, Yogyakarta 55281, Indonesia

mstyadji@batan.go.id

Abstract. Sintering process is the final stage of fuel kernel manufacturing prior to the coating process. This stage is very important part of the whole process, because it will determine the feasibility of UO_2 kernel to comply with the specifications of HTR reactor fuel. The objective of this research was to obtain UO_2 kernel with the density of $\geq$ 95% TD. The results showed that the highest density reached 92.56% TD or about 10.1441 g / cm3. This condition of sintering was gained at the temperature of 1400 °C with sintering time of 2 hours. The sintering product diameter gained was around 919 µm, the specific surface area 4.4213 m^2 / g, and a total pore volume 4,751 $x10^{-3}$ cm^3 / g. The density of UO_2 kernel produced from this research is the best compared to previous finding because of its density already approaches the HTR fuel specification requirements

Keywords: density, sintering, UO_2 kernel, reactor fuel, HTR

I. Introduction

The utilization of nuclear energy in Indonesia has become one of the highly prospective alternative energy besides fossil and renewable energy. Various studies have been done in an effort to prepare for the construction of Nuclear Power Plant (NPP) [1-2]. Generating prospective development is the type of high-temperature reactor. High-Temperature Reactor (HTR) is a privelege type of reactor having a reliably secure and highly thermal efficient. In terms of safety, HTR uses helium gas as the cooling material which is one phase and an inert fluid, fuel in the form of coated particles capable confining fission reactions at high temperatures, graphite with a high heat capacity as the moderator, as well as the reactor core made of ceramic material capable of surviving the high temperature rise. HTR also has a negative reactivity on overload the reactor would shut down automatically when the temperature rises excessively [3-5]. The use of graphite as a moderator and helium gas as the cooling material, is able to produce a higher heat so that it can be utilized a heat source for industries such as chemical process industry, the production of hydrogen, coal gasification, and desalination [6-8].

The status of HTR fuel research development at the moment has reached the stage of preparation of fuel elements which includes solution preparation sol, gelation process, aging, drying, calcination, reduction, sintering, coating and molding kernel. The series of studies aimed at getting HTR fuel in order to comply with the defined specifications [2,3,9].Various studies related to HTR fuel preparation in Indonesia have been carried out until sintering phase. However, the kernel produced has not met the required specification. The produced kernel density remains below the required specification of less than 95% theoretical density (UO_2 Theoretical Density = 10.96 g / cm^3).

Therefore, furthermore research needs to be conducted to get the kernel product which comply with the required quality of HTR fuel. Some researchers [10] conducted research previously on the effect of temperature and sintering time on the quality of UO_2 fuel kernel by using the fluidized bed furnace type in an argon gas atmosphere, the produced kernel density of approximately 91.136% of the theoretical density. Other researchers [11] conducted further research on sintering of UO_2 kernel by using fluidized bed furnace in the argon g medium as at a temperature of 1000 °C-1150 °C. The density of UO_2 kernel product obtained was 9.99 g / cm^3 (91.15% of the theoretical density). Those above kernel products did not comply with the HTR fuel standards. China has also conducted research on the preparation of UO_2 fuel kernel HTR-10. The sintering process was carried out in the research by using hydrogen gas atmosphere at temperatures above 1500 °C. The produced kernel density reaches 98% theoretical density [12].

A few researcher conducted research focused on increasing UO_2 kernel density by using the inert gas atmosphere without mathematical modeling analyzing of sintering process. Therefore, this research intends to formulate mathematical modeling of sintering process in order to obtain optimum condition. The objectives of this research are: to get UO_2 fuel kernel with a high density which is $\geq$ 95% TD, to know the optimum conditions of UO_2 fuel kernel sintering process, to know the temperature and time effect on diameter, specific surface area, and total pore volume of UO_2 fuel kernel. This research is expected to yield UO_2 fuel kernel complying with a good quality HTR fuel spesification.

Density and quality of kernels are largely influenced by the sintering temperature. Theoretically, the higher sintering temperature means the higher sintering process rate would leads the increasing of atoms mobility followed by the decreasing of particles pores so that the density of sintering product would be highly increases [13]. Research conducted by the Division of Nuclear Science and Technology at the Oak Ridge National Laboratory (ORNL), The United States. The higher the sintering temperature, the higher the rate of sintering process (solid sintered/time). Results of research conducted by the Division of Nuclear Science and Technology at the ORNL, the United States, showed that the sintering UO_2 for 5 hours at a temperature of 1550 °C in an atmosphere of hydrogen and argon (4% H2 and 96% Ar) produces a quite high density kernel at around 98% of the theoretical density (10.82 ± 0.16 g / cm^3) [14].

Sintering time affect the quality of the kernel. Sintering time too quickly can allow the imperfections at sintering process product so that the density obtained has a lower value than that of the sintering product processed in quite a long time at the same temperature [15]. Kernel size affects the specific surface area of the final products of sintering. The smaller the size of the kernel the better level of densification on the material will be. Atmospheric sintering may affect the densification and the formation of microstructure in materials. If there was gaswhich is easily absorbed, it wolud disrupt the process of densification. Heating rate of sintering is the rising of sintering temperature which changes every time. Heating rate should be regulated in order to avoid thermal shock which can cause a breakdown of the kernel. Heating done periodically by raising the temperature from low temperature to high temperature so that the particles in the kernel can adjust the temperature changes that occur in a sintering furnace. The influence of U_3O_8 powder, formed by the various oxidation temperatures, on the sintered density of CANada Deuterium Uranium (CANDU)-type UO_2 pellet was investigated. Increasing of the oxidation temperature of U_3O_8 powder causes decreasing of the sintered density of UO_2 pellet [16-18].

2. Materials and Methods

2.1. Materials

The raw material used in this research was UO_2 fuel kernel. Fuel kernel of $UO_2$was generated from the previous process of dissolution, sol-gel, soaking, drying, calcination and reduction. Specifications UO_2 kernel before sintering were a diameter of 1.015 mm, density of 8.1307 g/cm^3, the specific surface area 5.512 m^2/g, total pore volume of 0.0067 cm^3/g and the ratio of O/U 2.03. Argon gas is used as a medium or atmospheric sintering. Specifications argon gas is Argon UHP> 99.9995%, O_2 <2 ppm, H_2O <3 ppm.

IOP Conf. Series: Materials Science and Engineering **553** (2019) 012040 doi:10.1088/1757-899X/553/1/012040

Density of UO_2 fuel kernel analysis use ABM (demineralized water) of PSTA (Center for Science and Accelerator Technology) production and CCl_4 that can be obtained from local market.

2.2. Experimental Procedure

Kernel $UO_2$was put into the furnace (which has been previously vacuumed and then filled with argon gas as the medium) at a certain temperature (hereinafter the temperature was varied in accordance with the experimental design). The heating process of UO_2 kernel within the furnace was done in several stages of preparation. The process of heating covers heating rate in particular, certain fixed temperature (the sintering temperature was varied at 1300 °C, 1400 °C and 1500 °C), for a certain time (the sintering time was varied according to experimental design namely time variables were 1, 2, 2.5 and 3 hours). Meanwhile, in the cooling stage: lower heating temperature approaching the ambient temperature as well as the final termination argon gas flow and the flow of cooling water. Further analysis of kernel includes the products of the UO_2 kernel sintering process. Diameter optical microscope dinolite capture 2.0, UO_2 kernel density with a pycnometer and advance comprehensive analysis of the specific and total pore volume of kernel using NOVA-1000.

2.3. Experimental Apparatus

A series of equipments used in this research are shown in Figure 1

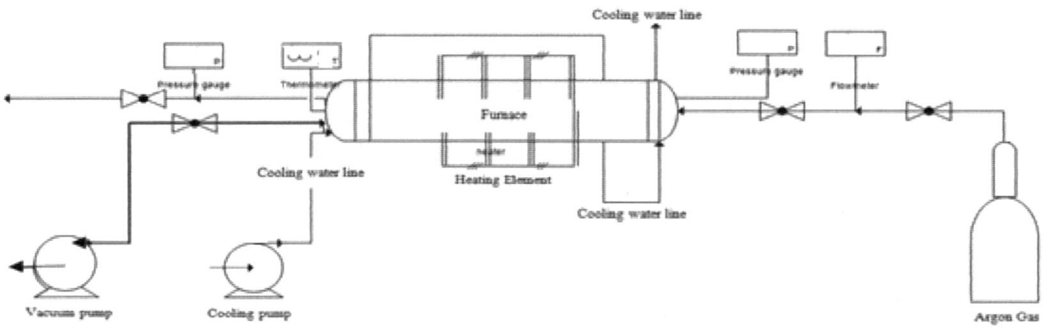

Figure 1. Sintering process equipment scheme.

Figure 2. Experimental apparatus.

IOP Conf. Series: Materials Science and Engineering **553** (2019) 012040 doi:10.1088/1757-899X/553/1/012040

3. Results and Discussion

3.1. Temperature Profile

The overall research investigating the effect of sintering temperature and time of changing the density, diameter shrinkage, changes in the specific surface area and total pore volume. Variables of the investigated temperature was 1300 °C, 1400 °C and 1500 °C. Variable of time was 1; 2; 2.5, and 3 hours. The ratio O/U of UO_2 kernel was considered fixed because it was used an inert argon gas atmosphere so that there was no oxidation. The kernel of UO_2 used was the reduction product of U_3O_8 at a temperature of 800 °C - 850 °C with a density of 8.1307 g/cm3. The rate of heating could heavily influence the sintering process, in addition to the temperature and time of sintering. Research temperature profile was shown in Figure 2 and Figure 3. Compare with the temperature profile of ORNL research [14], the heating rate of which was different. Heating rate of UO_2 kernel sintering product in this investigation was not constant. Average heating rate ranged from 8.04 to 9.02 °C/min. The temperature profile of other research conducted by Oak Ridge National Laboratory (ORNL) was shown in Figure 5.

In Figure 5, The ORNL researchers showed their experimental data of retention time at a certain temperature before reaching the maximum temperature while in this experiment showed no retention time data. This gives a different effect on the final sintering products. Retention time before the maximum temperature allow a sufficient time for the atomic particles to restructure properly, by which the pores would be reduced significantly so that the kernel density sintering products may approach the theoretical density. These results indicate that the density values obtained lower than the density of the results of sintering at ORNL. Although the temperature profile factor is not the only factor that causes the kernel density difference sintering results, but the temperature profile factor is an important factor in the success of the sintering process.

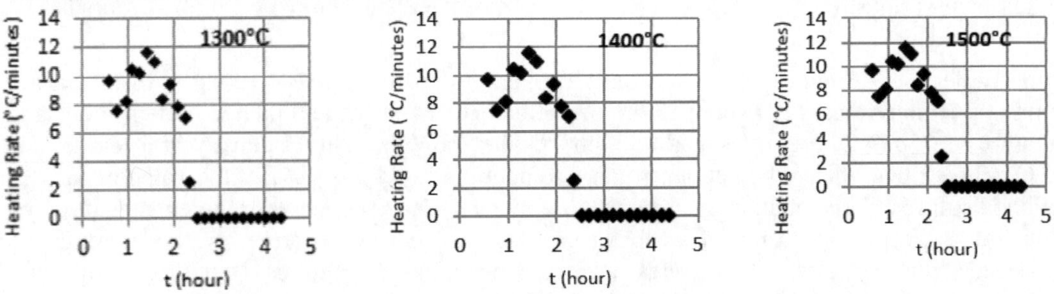

Figure 3. Heating rate of 1300 °C, 1400 °C and 1500 °C.

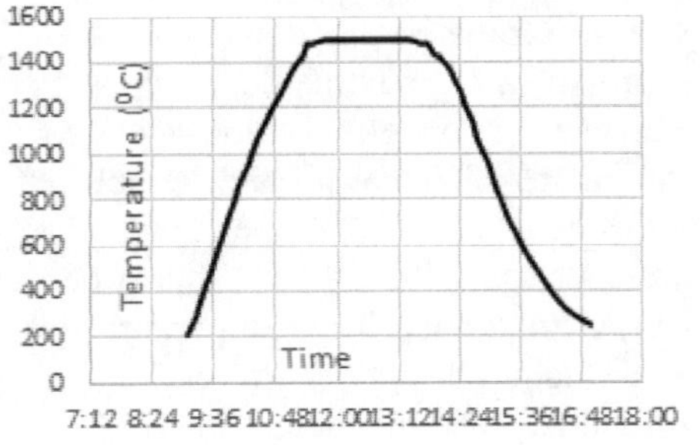

Figure 4. Profile temperatures at various times on UO_2 kernel sintering

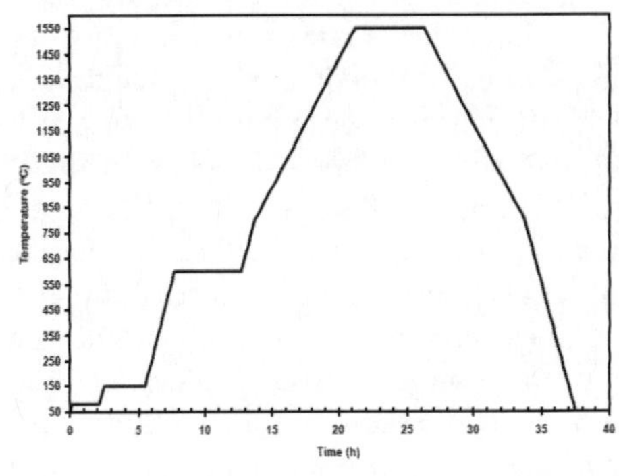

Figure 5. Profile temperatures at various times on UO_2 kernel sintering at ORNL [14]

3.2. Effect of Temperature on Density

Data effect of temperature on UO_2 kernel density changes are shown in Table 1 and Figure 5.

Table 1. Effect of temperature on the density of UO_2 kernel sintering product.

Temperature (°C)	Density UO_2 achieved (g/cm^3)	*Relative Density* (%.TD)
0	8.1307	74.19
1200	8.7998	80.29
1300	9.3378	85.20
1400	10.1441	92.56
1500	10.0341	91.55

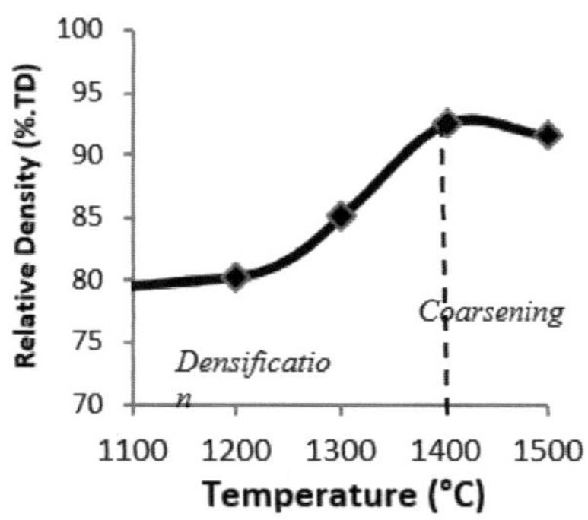

Figure 6. Correlation of relative density of UO_2 kernel at various temperature.

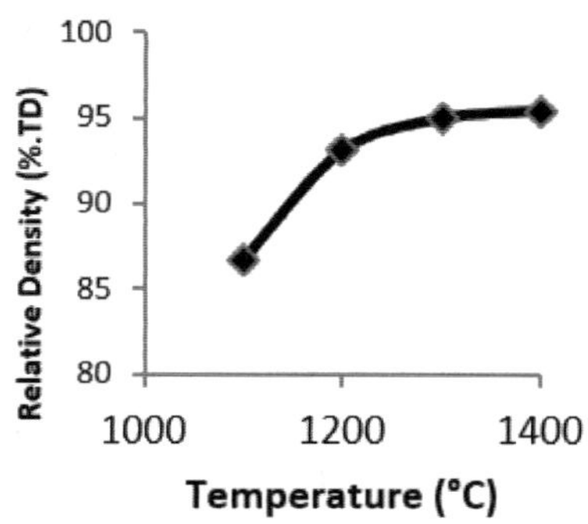

Figure 7. Correlation of relative density of UO_2 kernel at various temperature for 2 hours [19].

From Table 1 and Figure 6 show that the UO_2 kernel density increases along with increasing the sintering temperature that is at a temperature of 1300 °C to 1400 °C, and then it remains constant at a temperature of 1500 °C. From the measurement results show that kernel density at a temperature of 1500 °C is lower than that of kernel density at a temperature of 1400 °C. But the value of the density reduction is still relatively small at 0.11 g/cm^3, or about 1.084%. The increase in density caused by the densification during the occuring of sintering process. At high temperature, partial melting process occurs inside the particles. Open pores which is initially contain air started coming out and interconnected to form a closed pore so that the solid's structure becomes more dense; the pore volume to be reduced in line with the increase in sintering temperature.

The high density of the kernel sintering products due to the densification predominate compared with coarsening processes. Conversely, if the coarsening process dominates the density of UO_2 kernel sintering products will decrease. It is due to an equilibrium pressure of gas in the pores (closed pore) so that the densification process stops. Gas trapped in the pores affect the pore elimination rate. Gas trapped in the closed pores undergoes expansion due to high temperature. This causes the kernel density to decrease.

This study resulted in a lower density than that of the previous study [12,14,19]. That is because the operating condition of sintering and samples of different raw materials. The temperature profile at different times and different heating rates have an impact on the achievement of different final densities. Sintering temperature profile in the study increases continuously and no detention time at certain temperature because there is no detention facility at the temperature of the equipment used.

Detentions can only be obtained when the temperature reaches the maximum temperature. Differences in temperature versus time profile can be seen in Figure 4 and Figure 5. The detention period at a certain temperature is intended to give an opportunity to the structure of particles in performing the reduction or elimination of the pores. Samples of different raw materials also provide a different final outcome of sintering. Diameter of kernel in this study approximately 1,0 mm (10^3µm) is greater than the diameter of the kernel on the experimental data of Collins et al which is 500 ± 20 µm [14]. Theoretically that the larger the particle size so that the lower relative density would be achieved. This is consistent with the results of this investigation namely with larger diameter than that of diameter investigated on the previous investigation [14],hence the density yielded in this investigation was lower which is only about 92% TD. Research on the sintering UO_2 kernel in the medium NH_3 was also conducted by [18]. The results showed the increase density of sintering kernel along with increasing sintering temperature and reaches a maximum at a temperature of 1400 °C with the results of the density of around 95% TD. Graph of the correlation between the density and temperature of sintering UO_2 kernel research results [19] shown in Figure 7.

3.3. Effect of Time on Density

Data the effect of time on the density of UO_2 kernel sintering products are shown in Table 2 and Figure 7.

Tabel 2. Effect of time on the density of UO_2 kernel of sintering product.

Time (hour)	**UO_2 Density achieved (g/cm^3)**	***Relative Density* (%.TD)**
1	9.6830	88.35
2	10.1441	92.56
2.5	9.8066	89.48
3	9.4154	85.91
4	8.9400	81.57

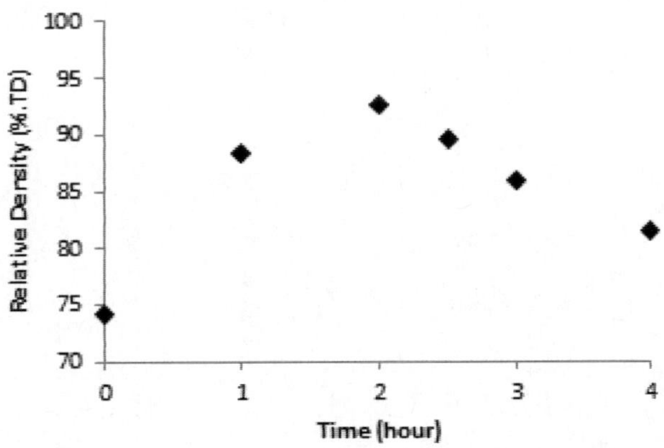

Figure 8. Correlation of relative density of UO_2 kernel at various time.

From Table 2 and Figure 8 shows that the kernel density of UO_2 increases with increasing sintering time up to 2 hours, the increase of UO_2 kernel density up to 92.56% TD and sintering time after 2 hours decreases the density up to 81.57% TD. Based on the graph relation between relative density and sintering time (sintering stage theory), the time between 0 to 2 hours is intermediate sintering, where the pores in the particles are connected so that the solid structure becomes denser and the solid density becomes higher. Sintering time provides enough opportunity for atoms to restructure atoms. Conversely, if the sintering time is more than 2 hours, the pores become isolated and when the sintering time increases, the isolated pores become larger. This causes the kernel density to decrease.

In this investigation, early kernel density was 8.1307 g/cm^3, the target density was ≥10,4 g / cm^3 (95% TD), density reached was 10.1441 g/cm^3 (92.56% TD). Achievement of UO_2 kernel density sintering results in this study is good because the minimum limit value of the required density of 92.56% TD was gained. That is, it takes only about 2.46% to comply with any kernel density qualification with HTR fuel specifications. Some reasons that the achievement of a kernel density sintering products is still slightly below specification HTR due to the heating rate is so fast and is unstable. Additionally, in this investigation also there was not any detention time at a certain temperature. The temperature increased linearly and detention time was carried out at a maximum temperature. The ratio O/U targets are $1.99 \leq x \leq 2.02$, and the reached ratio O/U was 2.03. The ratio O/U related to the initial manufacturing process of UO_2 kernel from preparation urania sol solution, gelation, soaking, to heating up the kernel. Change in the ratio O/U could be caused by the occuring of oxidation processes during the preparation process prior to the sintering. Therefore, it would be much better if the reduction and sintering processes were integrated. The completed UO_2 kernel should be reduced in the reduction furnace followed by sintering stage at the same furnace. An atmospheric gas used during sintering should be the mixture of Ar and H_2 with a composition of 4% H_2 and 96% Ar in which the residual oxygen could be reduced in order to completely reduction process.

3.4. Effect of Temperature and Time on Change Diameter

Data effect of temperature and time on the change diameter of UO_2 kernel sintering products is shown in Table 3 and Figure 9.

Table 3. Effect of temperature and time on the change diameter of UO_2 kernel sintering products.

Time (hour)	UO_2 Kernel Diameter (mm)		
	1300°C	1400°C	1500°C
0	1.015	1.015	1.015
1	1.013	1.003	0.940
2	0.988	0.919	0.920
2.5	0.986	0.931	0.954
3	0.991	0.940	0.945

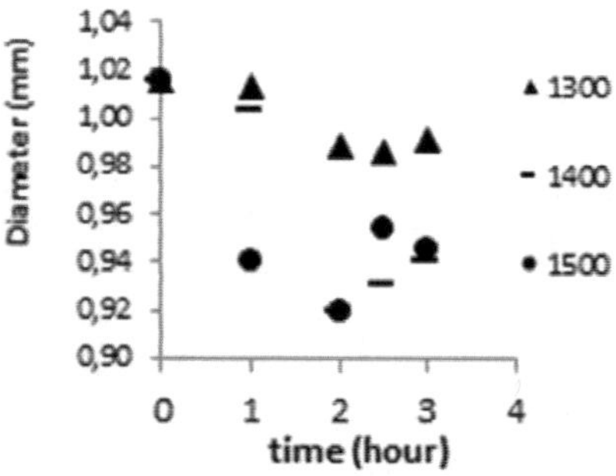

Figure 9. Correlation of Relative Diameter of UO_2 kernel at various temperature and time.

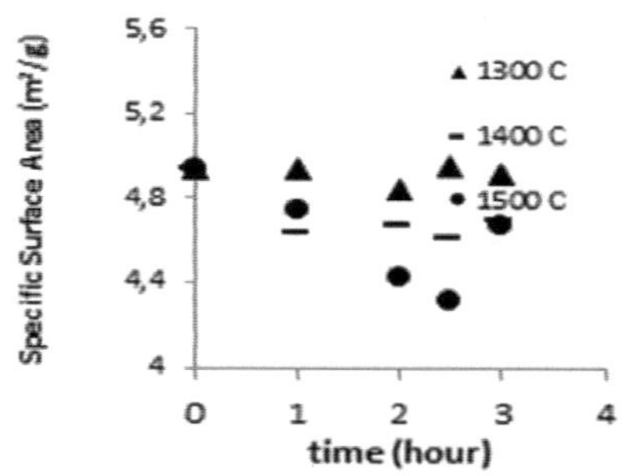

Figure 10. Correlation of SSA of UO_2 kernel at various temperatureand time.

From Table 3 and Figure 9, it was written that kernel diameter decreases very sharply at a time under 2 hours, reaching 0.919 cm at the heating temperature of 1400 °C. This happened almost at the same temperature of 1300 °C, 1400 °C and 1500 °C. After the kernel diameter has increased over 2 hours. The reduction of Kernel diameter is due to a reduction in volume and pore size while an increase in diameter indicating the existence of grain growth in the particles. In addition to the occurring of grain growth, the increase of the diameter size of the kernel could also be caused by the expansion of gas trapped in the closed poreas the result of increasing sintering temperature and time.

3.5. Effect of Temperature and Time on the Specific Surface Area of UO_2 kernel

Data influence of sintering temperature and time of the UO_2 kernel specific surface area is shown in Table 4 and Figure 10.

Table 4. Effect of temperature and time on specific surface area of UO_2 kernel.

Time (hour)	**Specific Surface Area (m^2/g)**		
	1300°C	1400°C	1500°C
0	4.9424	4.9424	4.9424
1	4.9445	4.6356	4.7389
2	4.8431	4.6761	4.4213
2.5	4.9475	4.6097	4.6798
3	4.9187	4.6893	4.6708

Table 4 and Figure 10 shows that the higher the sintering temperature, also the higher the decreaseof specific surface area. Decline of specific surface area was due to the increasing of contact point between the crystal structure inside the particles and the reduction of pores distribution inside the particles. The decrease of specific surface area greatly contributes to the sintering process due to the change in the particles free energy that encourage the occuring of densification.

In the beginning of sintering at the time of less than 2 hours the surface area decline happened, but at the sintering time over 2 hours the specific surface area increases. The increase in specifik surface area indicates that the kernel is getting more porous. This is also a proven by the decrease value of kernel density after sintering time of 2 hours.

In this investigation the initial surface area was 5.512 m^2/g, target surface area was 3.11 m^2 /g, and the reached surface area was 4.421 m^2/g. Total pore volume initially was 0.0067 cm^3/g, total pore volume target of 0,001 cm^3/g, and a pore volume achieved was 0.0044 cm^3/g. Achievement of surface area reduction and total pore volume in this investigation reached 0.45% and 0.40%. This achievement is quite good, but still needs to be improved. Surface area is closely related to the distribution and pore size inside the kernel.

Achievement of surface area and total pore volume which are still below the prerequisite specifications due to the existence pores inside the kernel. These pores formation was commenced from the gel formation process. The existence of air trapped in the gelation process causes the formation of closed pores inside the kernel. The sintering process which is not optimum could also be the cause of the reduction in distribution and pore volume less than the maximum. Data the influence of temperature and sintering time on the total pore volume are presented in Table 5 and Figure 11.

Table 5. Effect of temperature and time on total pore volume of UO_2 kernel sintering results.

Time (hour)	**Total Pore Volume (cm^3/g) x 10^3**		
	1300°C	1400°C	1500°C
0	6.7316	6.7316	6.7316
1	5.5627	5.1708	4.9203
2	5.3153	4.7515	4.8275
2.5	5.8522	5.4634	5.3422
3	6.0834	5.4829	4.8846

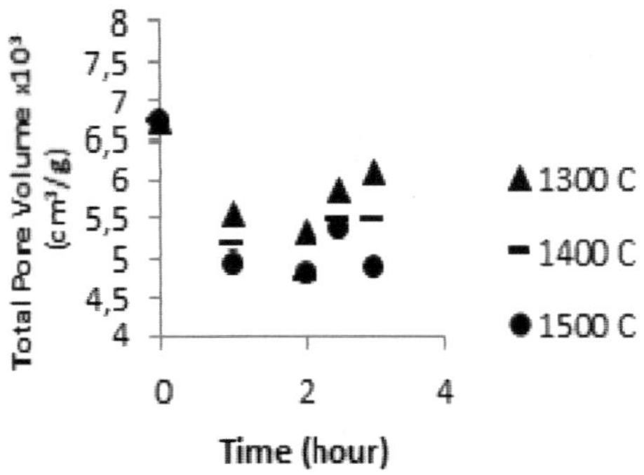

Figure 11. Correlation of relative diameter of UO_2 kernel at various temperature and time relationship between the total pore volume and the time at various temperatures.

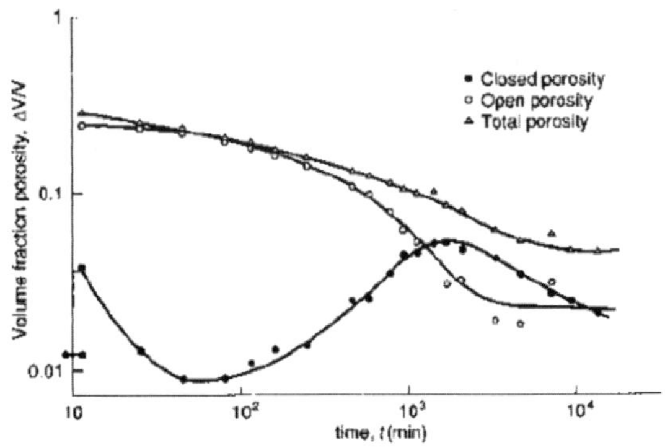

Figure 12. Correlation of porosity of UO_2 kernel at various temperatureand time Changes UO_2 porosity during sintering at a temperature of 1400 °C [13].

Initially, all the pores inside the particles were separated one another then when the sintering happened those pores would be interconnected. During sintering, the open pores turns into a closed pore. Effect of sintering time on the porosity of uo_2 sintering are shown in the graph in Figure 10. At a temperature of 1400 °C, total pore volume of uo_2 kernel decreases from 6.7316 $x10^{-3}$ cm^3/ g to 4.7515 $x10^{-3}$ cm^3/g. then the total pore volume increases after sintering time up to 2 hours. research results in figure 10 show the similar phenomenon to that results of research conducted by [12] shown in figure 11. From Figure 11 and Figure 12 it is shown that the total porosity of uo_2 decreases at initial time of sintering caused by the decrease of the open pore volume which is turns into a closed pore. while the closed pore increases continuously and decreases again after reaching its peak.

The phenomenon of open pore decrease in the beginning of sintering occurs due to the release of gases in the pores and the occurring of material diffusion and atomic restructuring led to the more densely structure. while, the continuously improvement in closed pores caused by the expansion of gas trapped in the pores due to the increase in the sintering temperature so that the pore size also increases. the pore size does not increase and tends to fall when it reaches the optimum time and temperature because of the occuring of the equilibrium between the gas pressure inside the pores with the pressure outside the pore. The ongoing material diffusion would gradually reduce the pore distribution so that the distribution and pore size decrease.

4. Conclusions

From the research and discussion above, it could be concluded that the highest density of UO_2 kernel sintering products obtained was 10.1441 g/cm^3 (92.56% TD). Achievement of UO_2 kernel density of sintering products was good because it was close to the minimum required density values of HTR fuel specification (HTR fuel kernel density is $\geq$ 95% TD). The optimum condition was achieved at a temperature of 1400 °C with an isothermal sintering time for 2 hours. The resulting kernel diameter was about 919 μm, the specific surface area was 4.421 m^2/g, and a total pore volume 4,751 $x10^{-3}$ cm^3/g.

Acknowledgement

In this occasion the writer thanks the Center for Accelerator Science and Technology BATAN for funding support. Thanks also go to Triyono who have given assistance in this research.

References

[1] Setyadji, M. (2013). Model migration of zirconium in the anion resin Dowex-1X8 in the framework of the separation of hafnium, Senior Researcher of Scientific Presentation, PTAPB BATAN Yogyakarta.

[2] Setyadji, M., Aliyah, F. (2015). Densification process models of UO_2 kernel for High temperature reactor fuel.*In*: Proceedings of tekMIRA Colloquium 2015 Ministry of Energy and Mineral Resources, The Agency of Research and Development for Energy and mineral Resources, Indonesia.
[3] Setyadji, M., Atmojo, S., Triyono (2010). Design analysis and manufacturing of kernel uo_2 sintered reactor, Jurnal Iptek Nuklir GANENDRA, 13(1).
[4] Pradhipta, A. W.(2011). Process external gelation on UO_2kernel development. Department of Chemical Engineering, Universitas Gadjah Mada:Yogyakarta.
[5] Beck, J. M., Pincock, L. F. (2011). High temperature gas-cooled reactors lessons learned applicable to the next generation nuclear plant, The Idaho National Laboratory Next Generation Nuclear Plant Project, Idaho Falls, Idaho 83415.
[6] Verfondern, K., Nabielek, H., Kendall, J. M. (2007). Coated particle fuel for high temperature gas cooled reactors. Nuclear Engineering and Technology, 39 (5), 39.
[7] Locatelli, G., Mancici, M., Todeschini, N. (2013). Generation IV nuclear reactors: Current status and future prospects. *Journal Energy Policy*. 61, 1503-1520. https://doi.org/10.1016/j.enpol.2013.06.101
[8] Hampe, J and Madlener, R. (2012). Economics of high – temperature nuclear reactors for industrial cofeneration. *FCN working paper* No.10/2012.
[9] Dewi, A.K., Sediawan, W.B., Sudiyo, R., Setyadji, M. (2015). Experimental study and mathematical modeling of reaction kinetic and transport mechanism of U_3O_8 using H_2gas reduction, Journal of Enginering science And Technology Special Issue on SOMCHE & Conference 2014, January 36-47 School of Enginering, Taylor's University.
[10] Herhady, R., and Sukarsono, R. (2007). Effect of temperature and time on the quality of fuel sintering kernel UO_2 in a fluidized bed furnace type. 13th National Seminar on Technology and Safety of Nuclear Power Plants and Nuclear Facilities, pp. 67-74.
[11] Damunir and Setyadji, M. (2011). Characterization index sintering, microstructure and morphology of UO_2 kernel sintering results in fluidization, Journal of Nuclear Materials Technology, 8 (1), 14-26.
[12] Xiaoming, F., Liang, T., Tang, Y., Xu, Z., Tang, C. (2004). Preparation of UO_2kernel for HTR-10 fuel element. 41(9), 943-948. https://doi.org/10.1080/18811248.2004.9715568
[13] Rahaman, M. N. (2003). Ceramic processing and sintering (2nd Edition ed.). New York: Marcel Dekker, Inc. https://doi.org/10.1002/maco.19960470511
[14] Collins, J. L., Hunt, R. D., Del Cul, G. D., Williams, D. F. (2004). Production of depleted UO_2kernels for the advanced gas-cooled reactor program for use in TRISO coating development. Oak Ridge National Laboratory, Nuclear Science and Technology Division. U.S .: Oak Ridge National Laboratory.
[15] Kuncoro, B. (2006). Effects of temperature and time sintering on UO_2 kernel quality in type fluidized bed furnace. Gadjah Mada University. Yogyakarta: Department of Engineering Physics.
[16] Olevsky, E. A. (2011). Sintering theory. California, USA: San Diego State University.
[17] Yang, C.M., Jung, G.D., Yoo, M.J., Lee, J.R. and Na., S.H. (2012). The effect of U_3O_8 powder on the sintered density of UO_2 pellet, Transactions of the Korean Nuclear Society Spring Meeting, Jeju, Korea, May 17-18.
[18] Sang-HoNa et al. (2013). Porous UO_2sintered pellets and method for fabricating porous sintered UO_2pellets and electrolytic reduction using same. Korean Patent. Korea Atomic Energy Research Institute, Korea Hydro and Nuclear Power Co., Ltd. Patent Number 20130148774.
[19] Halim, A. A., Alili, Y., El-Adham, K. (1987). Fabrication of UO_2 microspheres by the hydrolysis process. *Powder Metallurgy International*, 19(5).

Preparation of Polyvinyl Alcohol/ Cellulose Nano fiber Nanocomposite Isolated from Empty Oil Palm Fruit Bunches

Saharman Gea*, Indra, Yugia Muis, Nami Panindia, Yasir Arafat Hutapea
[1]Department of Chemistry, Faculty of Mathematics and Natural Sciences, University of Sumatera Utara. Jl. Bioteknologi No. 1 Medan 20155, Indonesia

s.gea@usu.ac.id

Abstract. This research was aimed to manufacture polyvinyl alcohol (PVA)/ cellulose nanofiber (CNF) nanocomposite which was isolated from oil palm empty fruit bunches (OPEFB) through steam explosion method combined with a hydrolysis process involving HCl 10%. The isolation process was performed in 2 steps; α-cellulose isolation from empty fruit bunches fibers and then cellulose nanofiber isolation from α-cellulose which was hydrolyzed with acid using ultrasonicator and homogenized in rotational speed of 8000 rpm. The functional group and morphology of nanofibres cellulose were characterized. FT-IR spectrum resulted an absorptions for C-O-C group at wavelength of 1059.99 cm-1 which indicated a glycoside bond. Then, C-H group at 2900.94 cm-1 and O-H group at 3348.62 cm-1 were revealed to indicate the cellulose nanofiber. Moreover, TEM was used for morphological analysis and showing a decrease in fibres diameters after experiencing a chemical-mechanical treatment which lead to nanofibres with final diameter size of 20-30 nm. PVA/CNF nanocomposite with the percentage ratio of 80:20 gave a tensile strength of 17.41 Mpa and Modulus Young of 0.9 Gpa. Finally, the product was also morphologically analyzed with SEM and indicating an even distribution on the PVA surface.

Keywords: Steam Explosion, HCl 10% Hydrolysis, Cellulose Nanofiber, Empty Oil Palm Fruit Bunches, Nanocomposite, Polyvinyl Alcohol

1. Introduction

Oil palm is one of the commodities traded plantation crops, both for domestic and export industry. Indonesia has known to be the largest oil palm producer around the globe, with a total production of 130-176 million tonnes of *Crude Palm Oil* (CPO) annually by 17-26 million hectares of land area. According to Afriyanti (2015), this number already represents of 39-60% the world's palm. However, the production of palm oil in Indonesia is often associated with environmental issues that adversely affected the palm oil industry itself [1]. Based on Plantation Director General of the Ministry of Agriculture, Indonesia produced 29.3 million tons of CPO in 2014. Surely, a significant amount of waste is potential to be a threat to the sustainable use of CPO in Indonesia.

The OPEFB fibers are found to be multi cellular and ligno-cellulosic compound. Ligno-cellulosic biomass is generally consisted of cellulose (glucan), hemicelluloses (xylan and mainly arabinan) and lignin. However, a pre-treatment is needed to dissolve the non-cellulosic polysaccharides and improve the proportion of mass from cellulose in the biomass [2].

Steam explosion combined with acid treatment can effectively separate the cellulose nanofibers from non-cellulosic polysaccharides of lignocellulosic biomass), including hardwoods, softwoods, herbaceous residues, sugarcane bagasse and wheat straw, are becoming the most largely employed physic-chemical pretreatment for lignocellulosic biomass [3-5].

According to Roohani *et al.* (2008), PVA has good compatibility when added with filler in the form of nanocellulose so as to produce products (nanocomposite) that are environmentally friendly [6]. Thus, by adding nanofibre cellulose, the PVA-based film is expected to increase the mechanical properties of the resulting PVA film. Some research on edible film with the addition of nanocellulose from the fibers, it can improve the mechanical properties, the nature of the barrier, and the ability to aggregate with other particles [7, 8].

Mandal and Chakrabarty (2013), stated that the addition of nanofibriled cellulose for about 5% in the manufacture of PVA/nanaocellulose nanocomposite is derived from sugarcane bagasse that showed a better thermal stability [9]. Bhatnagar and Sain (2005) in the manufacture of reinforced nanofibres composites stated that a polymer matrix/nanofibre cellulose was prepared from 10% of nanofibre cellulose showed an excellent reinforcing properties. The nanocomposite can be used in medical devices as conductor systems (drug delivery) which is biocompatible, where blood storage bags, and valves as reinforcement biomaterials. Due to the high strength and light weight of the resulting composite, it is potential to be used in the automotive sector [10].

In accordance with the development of nano technology, oil palm empty fruit bunches can be used as materials for nanofibre cellulose. Bibin et al. (2010) produced nanofibre cellulose from pineapple leaf fibers with explosive vapor method (steam explosion) [11]. Cherian et al. (2008) described that the explosive vapor method is effective in separating nanofibre from biomass [12]. A further explanation from Bibin et al. (2010), nanofiber cellulose has a wide uses in biomedicine and biotechnology, such as drug delivery, wound dressings, tissue engineering and medical implants [11].

Based on those descriptions, researchers are interested in utilizing such materials as the development of science, by isolating nanofibre cellulose from oil palm empty fruit bunches (OPEFB) through explosive vapor (steam explosion) method and were hydrolyzed with HCl 10%.

2. Experimental method

2.1 Materials

Wheat straws were taken from fields. They were washed in order to remove any impurities and dried. Sodium hydroxide (NaOH), hydrochloric acid (HCl), acetic acid (CH_3COOH) Hydrolysed polyvinyl alcohol (PVA) (MW = 84,000-124,000 g/mol; 98-99%) and hydrogen peroxide (H_2O_2) were purchased from Sigma Aldrich and were pure without any further treatment.

2.2 Isolation of nanofibriled cellulose from OPEFB fibers

OPEFB fibers were minced into random size of around 1-2 cm. The OPEFB fibers were soaked in 2% solution of NaOH overnight and then treated with 12% NaOH solution (with ratio of 1:10) in an autoclave at pressure around 168 kPa for 4 h at 130 °C. This treatment reduced the amount of lignin from the fibers. The fibers were taken from the autoclave and the fibers were washed several times in distilled water until it was free of alkaline materials

Steam exploded fibers were bleached using with a mixture of 17.5 % solution of NaOH and 7.4% acetic acid solution (ratio 1:1) and a mixture of (1:3) NaOCl solutions. Also, the steam explodes fibers were soaked in 8% of H_2O_2 solution. The fibers were washed in distilled water after the bleaching treatment and then dried.

The steam exploded bleached fibers were treated with hydrochloric acid with the concentration of 10% and putted into sonicator for 3 h. Then, the fibers were washed in distilled water until the acid was removed. Finally, the acid treated fibers were suspended in water and stirred with a WiseTis HG-15D homogenizer for about 15 minutes until the fibers were agglomerated.

IOP Conf. Series: Materials Science and Engineering **553** (2019) 012041 doi:10.1088/1757-899X/553/1/012041

Then, as many as 9 g of PVA was dissolved into 100 mL of distilled water and heated at a temperature of 75 °C for 1 hour and stirred using a magnetic stirrer. Nanofibre cellulose was added with a variation of 0.5; 1.0; 1.5; 2.0; 2.5 by weight (w/w). Mixing was done by using a homogenizer with a rotation speed of 8000 rpm for 30 minutes. After that, it was mould using acrylic glass mold and dried in an oven at a temperature of 50 °C.

2.3 Characterization

2.3.1. Transmission Electron Microscopy (TEM). Transmission electron micrographs of nanofibre cellulose was recorded with a Philips CM 30 transmission electron microscope with a voltage acceleration of 75 kV. The nanofibrils were deposited from an aqueous dilute dispersion on a micro grid covered with a thin carbon film (200 nm). The deposited fibres were subsequently stained with a 2% uranyl acetate solution to enhance the microscopic image.

2.3.2. Fourier Transform Infrared (FTIR) spectroscopy. FTIR spectra of acid hydrolysis untreated fibers and nanofibrilated cellulose were measured using ABB 3000 FTIR spectrometer equipped with a deuterated triglycine sulphate (DTGS) detector and KBr/Germanium as beam splitter. FTIR spectra were collected at the mid-infrared region 4000-450 cm^{-1} with a resolution of 8 cm-1, using 32 scans for each sample.

2.3.3. Tensile strength. Tensile strength of samples was carried out by using a tensile test Gotech AL 7000 M with tensile speed of 10 mm/ min and a load of 2000 kgf. The specimen was clamped using griff on the device, then set up stress, strain and its unit. The data load (voltage) and stroke (strain) obtained can be used to calculate the tensile strength and elongation of each specimen.

2.3.4. Thermal characterization. The DSC analysis by using a DSC-60 plus. With a sample size of 10.40 mg by using a metal cell Al and the reference sample and set the configuration on the sample and reference samples were placed on a heating with good thermal interaction, then heated slowly while increased thermal resolution.

2.3.5. Scanning Electron Microscopy (SEM) analysis. Microscopic observation using the SEM process was done on the fracture surface of the sample. Then, after the sample was cleaned with a blower, the sample was covered with gold and palladium in dionspater pressurized machine 1492 x 10^{-2} atm, the sample was then inserted into a space (vacuum evaporator) pressure of 0.2 Torr using JSM-35C engine Shumandzu. Furthermore, the sample was irradiated with an electron beam-powered 20 kV in a special room so the samples emit secondary electrons and electrons that bounce can be detected by a detector reinforced with an electrical circuit that produced the image CRT(Cathode Ray Tube) for 4 minutes. Then, it was coated with a layer thickness of 400 Armstrong and putted in the specimen chamber to do the shooting. The photographs can be adjusted to the desired magnification.

3. Results and discussion

3.1 Morphology and chemical properties

The alkaline steam explosion treatment removed the materials like hemicelluloses, pectin, lignin and etc. The fiber surfaces changed after chemical treatment. These show the fibers surface roughness and individual fibers after the removal of the cementing material. In addition, acid hydrolysis treatment helps to eliminate excessive amount of lignin from the fibers. The particle size of fibers has reduced after chemical treatment and the average size of fibers are about 15-20 μm. Reduction of fibers size because of the dissolution of the non-cellulosic polysaccharide and it is clearly related with FT-IR data. Nanofibrilated cellulose after the final treatments are showed in Figure 1 (a-b). Agglomeration treatment using a homogenizer caused in defibrillation of the nanofibrilated cellulose from the cell wall and TEM images show separation of these nanofibers from the fiber bundles. The diameters of the

IOP Conf. Series: Materials Science and Engineering **553** (2019) 012041 doi:10.1088/1757-899X/553/1/012041

fibers were observed from the figure and the average size was determined using image processing software image J. Most of the fibers sizes were found in the diameter range of 20-30 nm. Figure exhibits the distribution of nano-fiber diameter after final treatment. 20% of fibers have diameter between 20-30 nm. Over 15% of fibers has diameter between 10-20 nm and also the diameter of 40-50 nm. Around 14% fibers have diameter of 30-40 nm, but under 12% of fibers have diameter >50 nm.

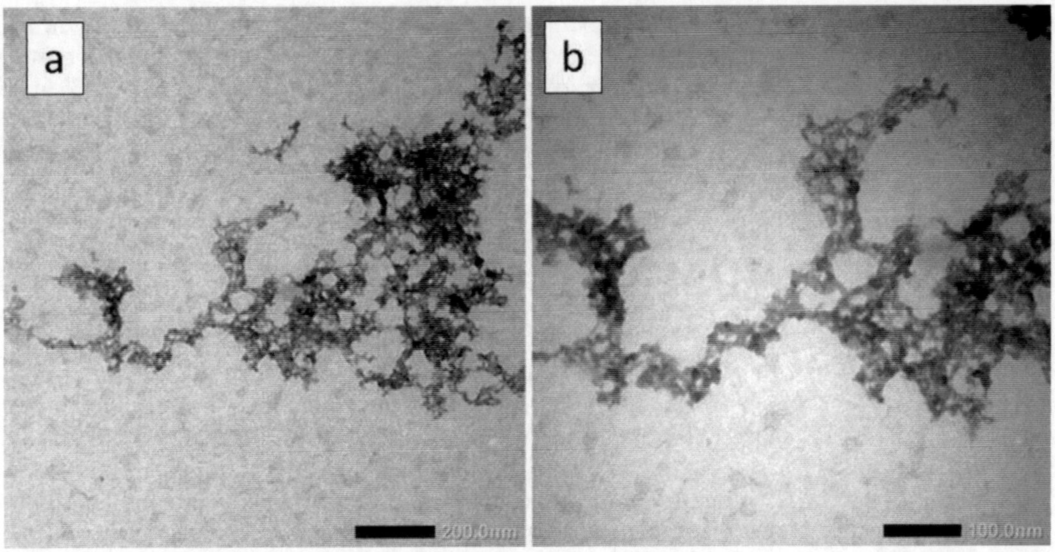

Figure 1. TEM image of the cellulose nanofibrls at different at different magnifications: (a) 20000x and (b) 40000x.

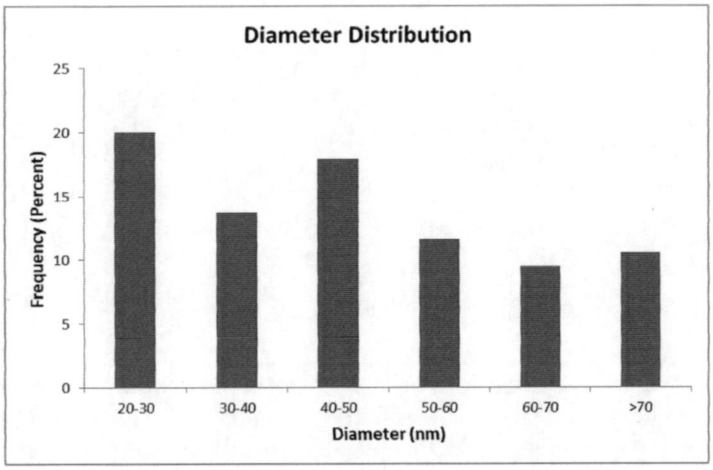

Figure 2. Diameter distribution of cellulose nanofibrils after the chemical and agglomeration treatments.

Figure 2 shows the size distribution of nano-dimensional fiber diameters after passing the final treatment. A total of 20% of the fibers with a diameter of 20-30 nm. Meanwhile, 17.89 and 16.84% respectively were 40-50 and a fiber diameter of 10-20 nm. For a fiber having a diameter of between 30-40, 50-60 and 60-70 respectively nm diameter size distribution was 13.68, 11.57 and 9.84%. Fiber having a diameter of> 70 nm, nanofiber distribution is 10.53%.

IOP Conf. Series: Materials Science and Engineering **553** (2019) 012041 doi:10.1088/1757-899X/553/1/012041

3.2 Fourier Transform Infrared (FTIR) spectroscopy

The FT-IR spectra are presented in Figure 3. Chemical constituent of fibers like cellulose, hemicelluloses and lignin has changed after the acid hydrolysis and agglomeration treatment. The absorbance peak is in area of 3348 cm^{-1} corresponds to the stretching vibration of OH and the peaks at 2916 cm^{-1} was assigned to the asymmetric stretching vibration of C-H lignin, cellulose and hemicelluloses.

The spectrum of nanofibrilated cellulose exhibited a trend difference from the acid hydrolysis untreated. The peak at 1715 cm^{-1} in the untreated fibers was attributed to the ester linkage of the carboxylic group of ferulic, the acetyl and uronic ester groups of hemicelluloses or and *p*-coumaric acid of hemicelluloses. However, the peak of nanofibrilated cellulose could not be found at 1715 cm^{-1}. The peak at 1634 cm^{-1} was assigned to the bending mode of absorbed water or quite a few contributions from carboxylate groups. The peak at 1542 cm^{-1} was attributed to the stretching mode of the aromatic C=C came from aromatic ring of lignin and also the absorbance band at 1428 cm^{-1}, observed in untreated fibers spectrum.

The absorbance bands at 1542 and 1428 cm^{-1} has significantly changed before the chemical treatment. The peak at 1542 cm^{-1} is very small and intensity of the absorbance band at 1428 cm-1 has significantly declined in nanofibrilated cellulose assigning to partial removal of lignin. The intensity of the peak at 1251 cm^{-1} has dramatically decreased after treated with the mechanical and chemical treatments attributing the removal of hemicelluloses. Some of non–cellulosic polysaccharides like hemicelluloses and lignin have removed from the fibers because of the alkaline treatment and bleaching process. The characteristic absorbance bands for all of lignin were not found in the spectrum of the nanofibrilated cellulose. The removal of lignin was occurred after the chemical treatment like alkaline treatment. In addition, the non-cellulosic polysaccharides can be also removed by acid treatment.

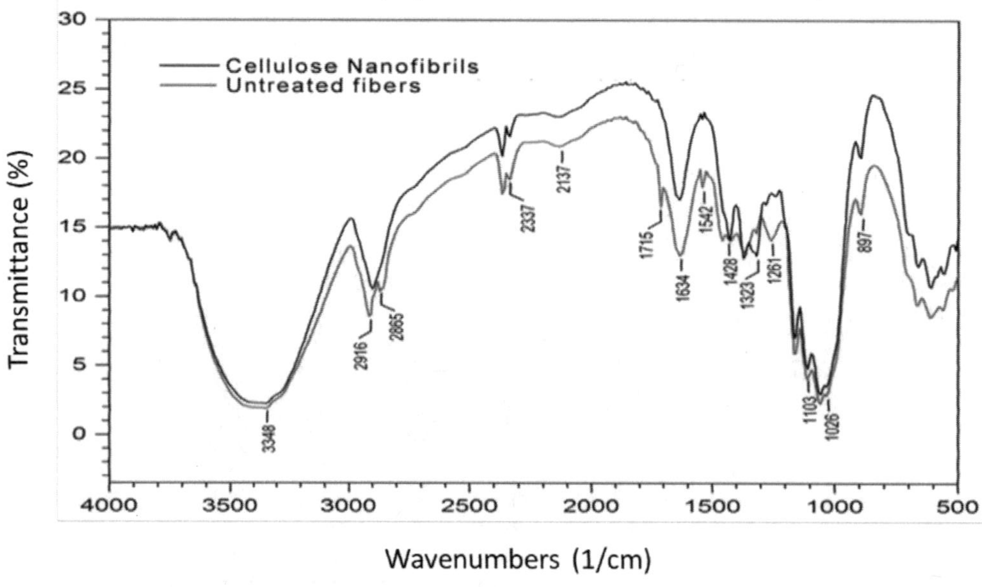

Figure 3. FT-IR spectra of nanofibrilated cellulose and untreated fibres.

3.3 Analysis of Mechanical Properties of PVA nanocomposite / NFC

The test for mechanical properties of nanocomposite polyvinyl alcohol/ Nanofiber cellulose with a variety of tested through analysis of the tensile test at room temperature using a load of 1 kN at a speed of 10 mm/ min. The average thickness of the resulting nanocomposite films is 0.1-0.2 mm. Where also the length of the nanocomposite PVA/ NSS of 80 mm and a width of 20 mm. The tensile strength of nanocomposite PVA/ NFC is displayed in Table 1.

Table 1 and figure 4 shows that the addition of the nanocomposite filler affects the results of its strength. Minimum tensile strength occured in variations of PVA/ NFC (10: 0.0) with the value of its strength was 5.68 MPa and a Young's modulus of 0.4 GPa. The maximum tensile strength variation occured in PVA/ NFC (8.0: 2.0) with 17.41 MPa and a Young's modulus of 0.9 GPa. However, the tensile strength began to decrease with the addition of cellulose nanoserat weight by 25%.

From the results of tensile strength, it can be seen that weight variation of PVA/ NFC (8.0: 2.0) (w/w) is a variation with the most severe kind because it can increase the tensile strength and Young's modulus of the nanocomposite produced. The increase in tensile strength of the resulting nanocomposite may be caused by the rigidity of the chain attached to nanofiber cellulose (due to the strength of hydrogen bonds and the inter molecular force on cellulose itself) and homogeneous distribution of the polymer and nanofiber cellulose high compatibility between the fiber and the matrix more aided by a high interfacial surface area. Nanofiber hydrogen bonding between cellulose and PVA matrix causes an increase in the tensile strength of the resulting nanocomposite. However, the increasing amount in weight gain with 2.5% cellulose nanofiber effect tensile test for impairment and Young's modulus. This is probably due to the dilution effect or tendency of agglomeration of nanometer-sized particles are very active [9].

Table 1. Mechanical Properties of PVA Film and nanocomposite PVA /NFC.

PVA nanocomposite composition/ Nanofiber Cellulose (w / w)%	Tensile Strength (MPa)	Young's Modulus (GPa)
PVA + NFC (10: 0,0)	5.68	0.4
PVA + NFC (9.5: 0.5)	8.09	0.6
PVA + NFC(9.0: 1.0)	10.88	0.7
PVA + NFC (8.5: 1.5)	14.51	0.8
PVA + NFC(8.0: 2.0)	17.41	0.9
PVA + NFC (7.5: 2.5)	15.83	0.8

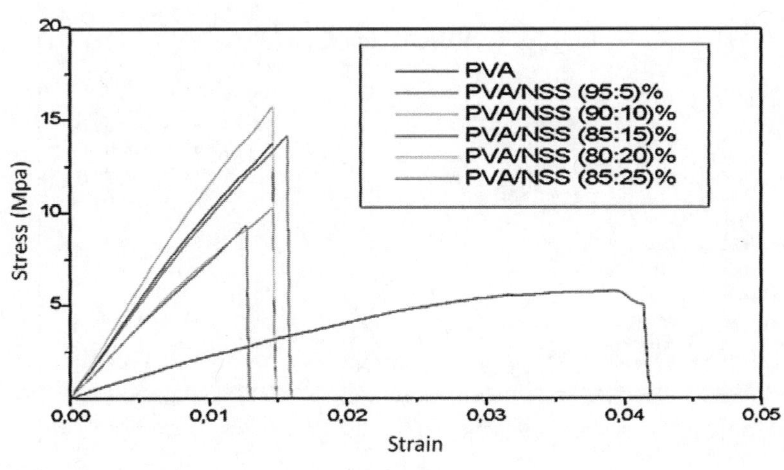

Figure 4. Stress-Strain Graph of PVA and PVA nanocomposite / NFC.

3.4 Thermal degradation analysis by DSC

Thermal analysis for PVA/ NFC nanocomposite is passed using instrumentation tool Differential Scanning Calorimetry (DSC). Variations used for thermal analysis was a PVA/ NFC nanocomposite with minimum variation in the tensile test, namely PVA/ NSS (10: 0.0) and PVA/ NFC nanocomposite with optimum variation in the tensile test, namely PVA / NSS (9.5: 0.5) using the initial temperature of room temperature at 28 °C to 500 °C.

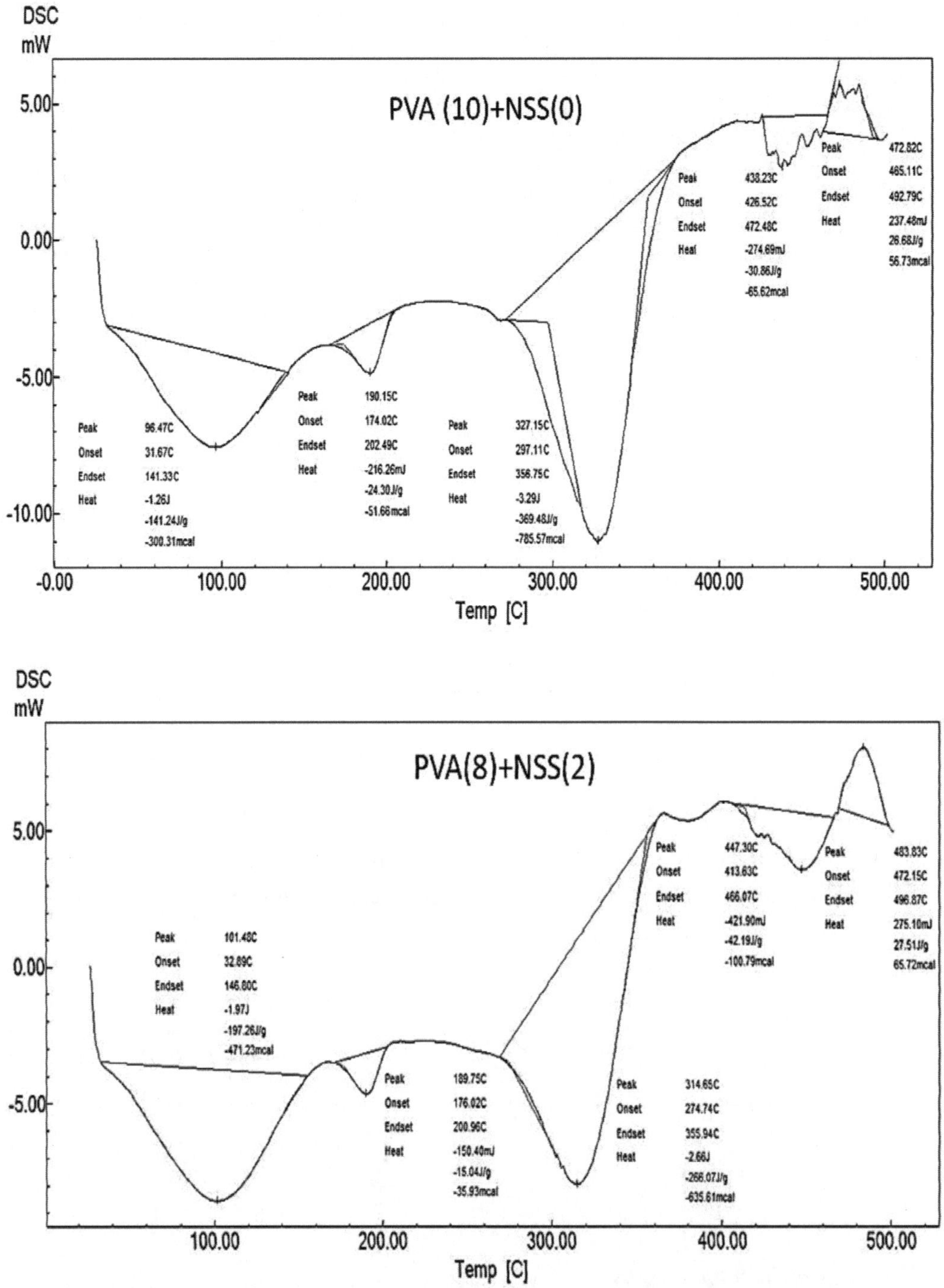

Figure 5. Spectra DSC nanocomposite PVA / NFC variation (10: 0.0)% and 8.0: 2.0%.

IOP Conf. Series: Materials Science and Engineering **553** (2019) 012041 doi:10.1088/1757-899X/553/1/012041

In a variation of PVA/ NFC (10: 0.0) DSC thermal test conducted in laboratory tests on PTKI. The result was obtained in the form of a peak that indicates an element that occurs endothermic and exothermic for the initial peak that appears to happen is marked peak endothermic reaction occurs at temperatures 96.47 °C and 300.31 mcal require heat stability which is a transition temperature (T_g), the second peak appears still state at a temperature endothermic 190.15 °C and heat stability 51.66 mcal require a crystalinity point (T_c),the third peak is still a state of endothermic and occurs at a temperature of 327.15 °C and 785.57 mcal require heat stability, the fourth peak is still in a state where there is endothermic at a temperature of 428.23°C and 65.62 mcal require heat stability is the melting point (T_m) and peak last appeared in an exothermic temperature 472.82°C and a heat of 56.73 mcal issued a decomposition temperature (T_d), the nanocomposite material starts to break down.

In the variation of (8.0: 2.0)% for the DSC test obtained similar results with minimum variation, in the form of five peaks which appeared after treatment. Peak began to appear in an endothermic area at a temperature of 101.48 °C and requires a heat stability at 471.23 mcal which is a transition temperature (T_g) the second peak appears in an endothermic at a temperature of 189.75 °C and requires a heat stability of 35.93 mcal is crystallinity point (T_c), the third peak appears in an endothermic at a temperature of 314.65 °C and requires a heat stability at 635.61 mcal, fourth peak appearing in an endothermic at a temperature of 447.30 °C and requires stability amounted to 100.79 mcal heat is melting point (T_m), and a fifth peak appears in an exothermic temperature of 483.83 °C and removing the heat of 65.72 mcal a decomposition temperature (Td).

3.5 Scanning electron microscopy (SEM) analysis

Morphology analysis by SEM using ZEISS tool, with a magnification of 500 and 1000 times. Based on morphology analysis demonstrates that the pure PVA nanocomposite has a smooth and homogeneous surface, this is caused by the PVA nanocomposite filler was not filled by nanofiber cellulose or other material that may affect the morphology of the nanocomposite surface so the surface looks more even and smooth.

In the PVA/ Nanofibre cellulose nanocomposite with a ratio of (8.0: 2.0) after the tensile test showed that nanofibre cellulose was evenly distributed on the surface of the PVA, homogeneous and smooth, so that the surface morphology is dominated by the morphology of cellulose fibers bond together between the fiber. This occurs because the cellulose fibers of nanometer-size physically interacting with PVA resulting nanocomposite surface is flat, homogeneous and produce the optimum value of the tensile test.

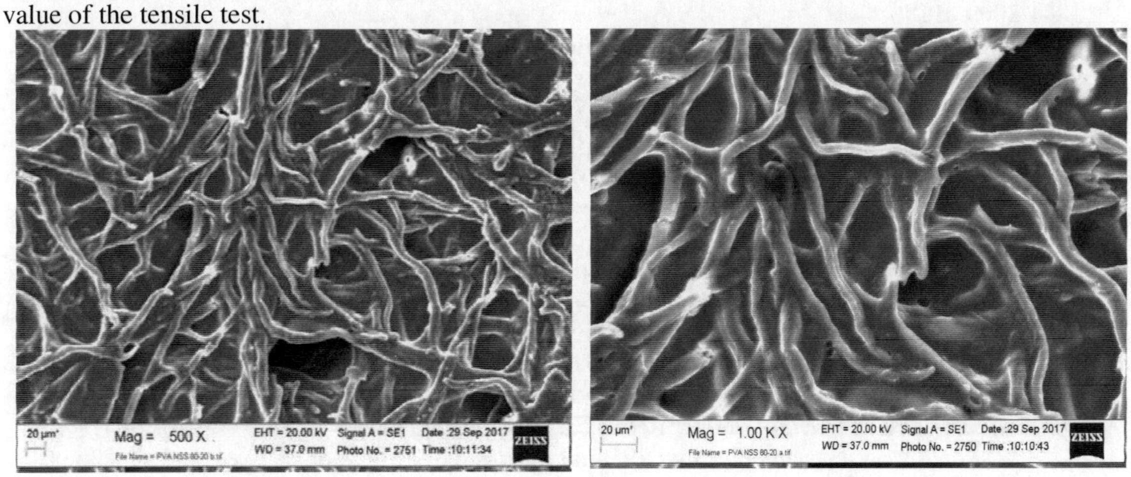

(a) (b)

Figure 6. SEM morphological analysis results PVA / NFC (8.0: 2.0)%: (a)500x (b)1000x.

4. Conclusion

Based on the research conducted, it can be concluded that Polyvinyl Alcohol/ Cellulose Nano fiber Nanocomposite was successfully isolated from the Oil Palm Empty Fruit Bunch (OPEFB). The FTIR analysis confirmed that the cellulose was completely joined the PVA, indicated by the presence of broad O-H group. Moreover, it is obvious that the composite resulted was in a nanometer scale with the average diameter of 20-30 nm. In term of mechanical properties, the nanocomposite with the ratio PVA: CNF of 80:20 promotes the optimum tensile strength and young modulus.

Acknowledgment

The authors would like to thank the Rector of University of Sumatera Utara, Medan, Indonesia, for providing the financial support via "Basic Research of University, TALENTA Project 2016" scheme.

References

[1] D. Afriyanti, C. Kroeze, and A. Saad, "Indonesia palm oil production without deforestation and peat conversion by 2050," *Science of The Total Environment*, vol. 557, pp. 562–570, 2016. https://doi.org/10.1016/j.scitotenv.2016.03.032

[2] L. Saikku, S. Soimakallio, and K. Pingoud, "Attributing land-use change carbon emissions to exported biomass," *Environment Impact Assessment Revision*, vol. 37, pp. 47–54, 2012. https://doi.org/10.1016/j.eiar.2012.03.006

[3] E. Abraham, et al., "Environmental friendly method for the extraction of coir fibre and isolation of nanofiber," *Carbohydrate Polymers*, vol. 92, pp. 1477–1483, 2013. https://doi.org/10.1016/j.carbpol.2012.10.056

[4] M. C. Fernandes, et al., "Enzymatic sacharification and bioethanol production from Cynaracardunculus pretreated by steam explosion," *Bioresources Technology*, vol. 186, pp. 309–315, 2015. https://doi.org/10.1016/j.biortech.2015.03.037

[5] M. Roohani, Y. Habibi, N. M. Belgacem, G. Ebrahim, A. N. Karimi, and A. Dufresne, "Cellulose whiskers reinforced polyvinyl alcohol copolymers nanocomposites," *European Polymer Journal* vol. 44(8), 2489-2498, 2008. https://doi.org/10.1016/j.eurpolymj.2008.05.024

[6] H. M. C. Azeredo, L. H. C. Mattoso, D. Wood, T. G. Williams, R. J. Avena-Bustillos, and T. H. McHugh, "Nanocomposite edible films from mango puree reinforced with cellulose nanofibers," *Journal of Food Science*, vol. 74(5), pp. N31-N35, 2009. https://doi.org/10.1111/j.1750-3841.2009.01186.x

[7] C. Tang, and H. Liu, "Cellulose nanofiber reinforced poly(vinyl alcohol) composites film with visible light transmittance," *Composites A*, vol. 39(10), pp. 1638-1643, 2008. https://doi.org/10.1016/j.compositesa.2008.07.005

[8] A. Mandal, and D. Chakrabarty, "Studies on the mechanical, thermal, morphological and barrier properties of nanocomposites based on poly(vinyl alcohol) and nanocellulose from sugarcane bagasse," *Journal of Industrial and Engineering Chemistry*, vol. 20(2), pp. 462-473, 2014. https://doi.org/10.1016/j.jiec.2013.05.003

[9] A. Bhatnagar, and M. Sain, "Processing of cellulose nanofiber-reinforced composites," *Journal of Reinforced Plastics and Composites*, vol. 24(12), pp. 1259-1268, 2005. https://doi.org/10.1177%2F0731684405049864

[10] M. C. Bibin, L. L. Alcides, F. Sivoney, T. Sabu, A. P. Laly, and M. Kottaisamy, "Isolation of nanocellulose from pineapple leaf fibres by steam explosion," Carbohydrate Polymer, vol. 81(3), pp. 729-725, 2010. https://doi.org/10.1016/j.carbpol.2010.03.046

[11] B. M. Cherian, L. A. Pothan, T. Nguyen-Chung, G. Mennig, M. Kottaisamy, and S. J. Thomas, "A novel method for the synthesis of cellulose nanofibril whiskers from banana fibers and characterization," *Agriculural Food Chemistry*, vol. 56, pp. 5617-5627, 2008. https://doi.org/10.1021/jf8003674

[12] S. Gea, Z. Zulfahmi, D. Yunus, A. Andriayani, and Y. A. Hutapea, "The isolation of nanofibre cellulose from oil palm empty fruit bunch via steam explosion and hydrolysis with HCl 10%,"

Journal of Physics: Conference Series, 979 012063, 2018 https://doi.org/10.1088/1742-6596/979/1/012063

Increased Electrical Conductivity of Polypyrrole Through Emulsion Polymerization Assisted Emulsifier Synthesis

Prestika Gema Sari[1], Mas Ayu Elita Hafizah[1], Andreas[2], Azwar Manaf [1*]
[1] Postgraduate Programe of Materials Science Study, Department of Physics Faculty of Mathematics and Natural Sciences, Universitas Indonesia, Depok 16424, Indonesia
[2] Indonesian Institute of Science, LIPI, Komplek PUSPITEK, Serpong, Banten. 15314, Indonesia

azwar@ui.ac.id

Abstract. A conductive Polypyrrole (PPy) has been successfully synthesized and characterized. The PPy was prepared through oxidative chemical polymerization using ammonium peroxydisulfate (APS) as an oxidizing agent in aqueous media. The reaction is prepared without heating. Anionic and non-ionic emulsifiers, sodium dodecyl sulfate (SDS), nonylphenol (NP) ethoxylate (EO) with 10 mole EO were respectively used during polymerization to increase the electrical conductivity. The FTIR spectrum of PPy indicates a change in the absorbance peaks of pyrrole monomer at 1530 cm^{-1} and those in the range region of 1100 - 1000 cm^{-1}. With the SDS addition, the temperature of solution increased with reaction time reached the highest 35 °C within 2 minutes reaction. The pH value of an SDS added solution was 3.3 higher than that of SDS free solution (1.8). The addition of SDS was also to change the color changing during the polymerization reaction. The final characteristic of PPy by measuring the electrical conductivity (σ) value concluded that the presence of emulsifier anionic in a polymerization reaction increased the electrical conductivity of PPy 150 times higher compared with that obtained of emulsifier free solution.

Keywords : Polypyrrole, emulsifier, emulsion polymerization, conductive polymer

1. Introduction

Polymeric materials have been interesting materials to be explored until today. As a macromolecule, the polymer is made of repeating huge small molecules which have long been used for insulation in various devices. However, in the 1970s, a new class of polymers possessing high electronic conductivity was introduced [1]. The so-called conductive polymers (CPs) have shown the capability of being doped or de-doped to gain their electrical properties. The conductive polymers were considered potential for various applications and able to meet current needs that require environmentally friendly because the CPs can be naturally degradable, and yet still superior against their inorganic predecessors like Li, Ni and Ni-Cd [2]. The CPs potentially can be low cost products having good environmental, more stable compared with other type of conductive materials like silver, gold and graphite based compounds by mixing with polymer material in electronics industry [3]. A few of CPs in it is series, among them are Polypyrrole (PPy), Polythiophene (PTh) and Polyaniline

(PANI) have been receiving great attention and being explored intensively by various researchers [3-7]. Polypyrrole (PPy) and its derivatives become one of the material that gain more attention due to its exceptionally high electrical conductivity, high flexibility in preparation, physically stable and have good mechanical property [4]. The PPy is suitable for electronic and Electrochromic devices and gas sensor applications like super-capacitors [5], and other important applications. In addition, the pyrrole monomer can easily be dissolved and oxidized in water [6]. The simplicity of the synthesis procedures and availability of the initial monomers is also attractive features of PPy synthesis [7].

Polypyrrole can be synthesized chemically or electrochemically through the oxidative polymerization process [8]. Chemical polymerization is a simple method that has been developed intensively involving some oxidant agents. In a chemical oxidative polymerization of pyrrole monomer, many oxidants have been used like ferric chloride ($FeCl_3$), ferric perchlorate, ammonium peroxydisulfate (APS), etc. APS was found to be the best inferior stability of electrical conductivity [9] and water is the best solvent for chemical polymerization with respect to desirable conductivity characteristic [3]. However, PPy is also having a deficiency which is rigid and brittle in physical properties. Its insolubility in some common organic solvent limits its wide range of applications [10]. Therefore, a solution to resolve the deficiency is urgently needed. It was shown that emulsion polymerization is one of the most promising methods in improving the process ability of PPy [10,11]. The mechanical properties and surface smoothness of PPy films prepared from aqueous solutions with the addition of the non-ionic surfactants were greatly improved in comparison with those of PPy prepared from the aqueous solutions in the absence of surfactants [11]. Chemical preparation of PPy from aqueous solutions containing ferric sulfate as an oxidant and a surfactant, was reported by Omastova [11]. The addition of surfactants accelerated the polymerization.

In conducting this research, polypyrrole monomer was polymerized using ammonium persulfate (APS) as an oxidation agent through emulsion polymerization assisted by anionic and nonionic surfactant under batch polymerization process. The best technique and polymerization method is required to obtain a high yield and purity of polypyrrole with higher conductivity value. The main final parameters were characterized by Particle Size Analyzer (PSA) Malvern ZS Nano series, Fourier Transform Infra-Red (FTIR) Spectrophotometer Thermo Scientific Nicolet iS5, Four Point Probe (FPP) meter, and other supporting instrument to determine the polypyrrole was synthesized.

2. Experimental method

Polypyrrole (PPy) was synthesized using pyrrole monomers and ammonium persulfate (APS) through an emulsion polymerization process assisted by anionic and non-ionic surfactants. This was compared with a sedimentation surfactant free polymerization. Both polymerization processes were conducted through a batch polymerization at a room temperature under the agitation speed of 900 rpm for 4 hours during the polymerization reaction. Analytical grade Pyrrole monomers and ammonium persulfate (APS) were obtained from Sigma Aldrich with the purity level 99.9%. Dodecyl Sulfate (SDS) which has white flake form with 99,5% purity level was used as anionic-surfactant. A selected non-ionic surfactant of Nonyphenol (NP) with 10 moles ethoxylate as a white powder with 99.7% of purity was used during polymerization reaction.

In the initial stage, the pyrrole monomers of 5% is mixture with a distillated water in a beaker glass under continuous magnetic stirring at room temperature. Meanwhile, the initiator solution of 5% of ammonium persulfate was water dissolved in another beaker glass at a room temperature. In 1 liter capacity of a reactor flask consisted of four necks the 1% of SDS solution in water was prepared under a heating treatment at 40°C for 30 minutes. A continuous agitation with a 500 rpm agitation speed was employed until all flakes of SDS dissolved completely. The SDS containing solution was then cooled naturally until the temperature reached back to a room temperature. Polymerization reaction was started by putting altogether of pyrrole monomers mixture and initiator solution into reactor flask immediately. Observation during the polymerization reaction was carried out periodically by taking up samples while the polymerization reaction in progress. Changes in the temperature of the solution, the particle sizes, viscosity and pH values, color degradation were all brought into objects of observation and evaluations. During the 4 hours of reaction, the black polymer particles were formed and collected using a Whattman filter paper no 40 and then successively washed 3 times with 100 ml distillate water. The Polypyrrole particles were then dried in a vacuum desiccator for 10 hours. The dried polypyrrole subject to further characterizations, including chemical molecule functional groups by Fourier Transform Infra-Red (FTIR) spectroscopy, particle size evaluation by Malvern Zetasizer Nano ZS Particle Size Analyzer (PSA) and electrical conductivity by Four Point Probe (FPP) method. The similar preparation polymerization experiment was repeated with different concentration of surfactants (1% and 2%), different type of surfactant (anionic and non-ionic surfactant) that represented by SDS and NP10EO respectively. These are to be compared with surfactant free PPy obtained through a similar polymerization reaction.

3. Results and discussion

During the polymerization reaction of pyrrole monomer, various main parameters are put as the object indicators for PPy formation. One of such parameters is the color changing which associated with the particle formation and changing in bonds of chemical molecules [12,13]. In the current research work, two different synthesis processes were set to be studied. There are the PPy obtained from the so-called emulsion polymerization with the presence of surfactant [14] and the sedimentation polymerization [15] with no surfactant added. The two polymerization processes have resulted a different pathway in color changing during the formation of PPy particles. The color changing of the two processes is compared in Figure 1.

(b)

Figure 1. Visual appearance of color degradation during polymerization reaction of PPy (a) surfactant free and (b) 1 % surfactant added solution.

A color changing as shown in Figure 1 (a) was started from a yellowish transparent liquid of pyrrole monomer to form a black particle of PPy through four steps. Whereas, Figure 1 (b) clearly indicates that with the presence of surfactant, a color changing occured at the initiation reaction started with a brownish hazy liquid and then gradually changing to a strong dark brown color prior to have similar color changing to that of surfactant free solution (Figure 1 (a)). A long pathway in color changing of surfactant added solution indicated that the polymerization reaction progresses in a slower reaction rate when compared with that of surfactant free.

A chemical reaction likes polymerization where the products in the form of particles, mostly accompanied by a releasing or absorbing energy. It is common that the temperature of the solution is set also as the object of observation to find out whether a chemical reaction is in progress [16-18]. The increase of solution temperature during polymerization is due to releasing of thermal energy accompanied by a breakdown of chemical bonds [19]. Figure 2 compares the temperature profile between SDS added and SDS free solutions for 240 minutes polymerization reaction. The two profiles have a somewhat similar pattern consisted of initiation, polymerization and post polymerization periods. The two profiles differ only in a temperature degree change from which the SDS containing solution has a temperature higher than that of SDS free solution during the whole period of reaction. The highest temperature of the SDS containing solution achieved 35 °C, 2 °C higher than that of surfactant free solution. This happened due to any movement of surfactant molecules released energy into the system. Such temperature profile was also obtained in the previous research works reported elsewhere [16-18]. In can be concluded that the polymerization reaction of pyrrole monomer by ammonium persulfate as the oxidation agent has no different with that of general polymerization reactions in which there is always a temperature increased due to chemical oxidation reactions. It is once again confirmed that the temperature of solution can be an important key indicator for polymerization reaction [20].

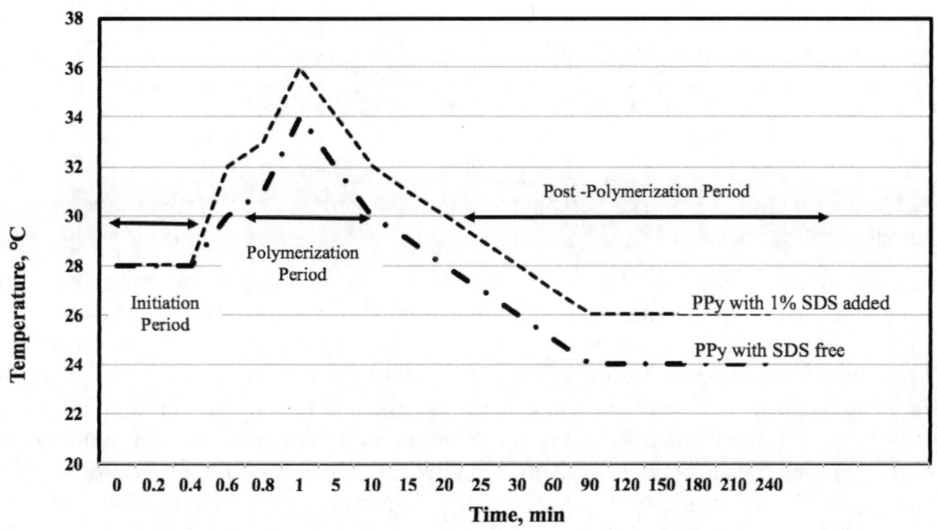

Figure 2. Comparison of the temperature profile between SDS added and SDS free solutions for 240 minutes PPy polymerization reaction.

Acidity level during polymerization reaction was also observed completely by means of real time measurement using pH meter series Metrohm 827 with double digit precision. Figure 3 compares the pH profile between SDS added and SDS free solutions for 240 minutes polymerization reaction. The two profiles differ in pH values from which the SDS containing solution has pH values higher than that of SDS free solution during the whole period of reaction. The pH of SDS added solution was 6.5 at the initial stage of the reaction which is much higher when compared with that of the SDS free solution. Obviously, the high value of SDS added solution must be due anion sulfate based (SO_4^{2-}) and cation sodium based (Na^+) of SDS molecule. Hence the pH meter measured the acidity value based on the ionic movement of SDS. The pyrrole monomer is clear and transparent liquid, slightly yellowish and have the pH value 3.9. During the polymerization reaction in SDS added solution, the increased pyrrole monomer amount of the solution along with the ionization reaction of SDS affected the pH of the solution. The pH decreased towards 3.3 after 15 minutes of reaction time and then stable by the reaction time of polymerization finished completely. On the other hand, the polymerization reaction of an SDS free solution had no significant changing on pH value which approximately 1.8 – 2.0 during the reaction time. The pH value of SDS free solutions originally came from the polymerization reaction only without any additional impact from side products. Hwang and Pyo have shown that the pH value of PPy in a basic solution will depend on the OH^- exchange against deprotonation of acidic proton of pyrrole units which described by the following chemical reaction:

$$PPyH^+X^- \leftrightarrow PPy + H^+ + X^- \text{ (in basic pH) [21].}$$

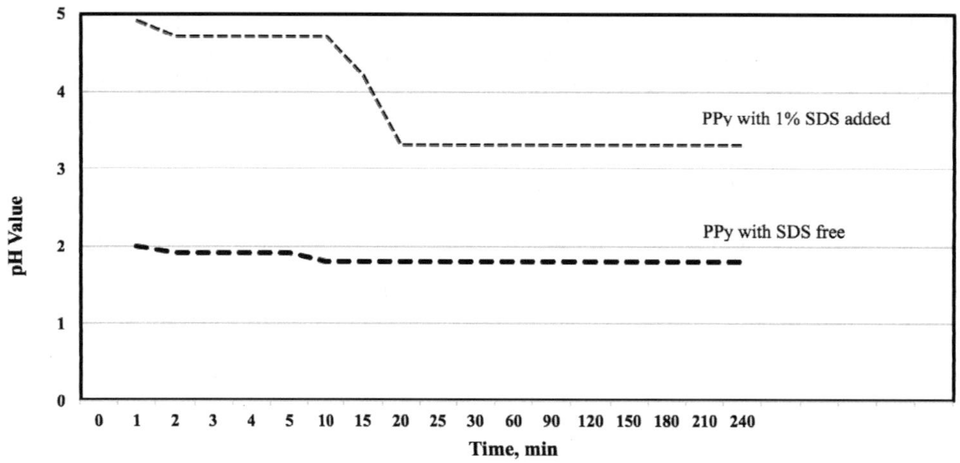

Figure 3. Comparison of pH profile between 1 % SDS added and SDS free solutions for 240 minutes polymerization PPy.

The most important parameter to find out if the polymerization reaction has occurred in the right direction can be indicated by the chemical structure of forming polymeric products. The FTIR spectrum of the products could be used for the identification. A series of study have also been carried out by several researchers pertaining the chemical molecule structure properties of polypyrrole with its derivative by FTIR spectrophotometer [11,22]. Figure 4 presents the complete FTIR spectrum for each pure single substance and the formed PPy. The spectrum of PPy in aqueous media with no surfactant added featured by the absorption band at 1433 cm^{-1} indicating a C –N stretching vibration in the pyrrole ring and a shifting of C –N vibration was observed to occur in spectrum PPy-SDS to 1435 cm^{-1}. The shifting is affected by doping and different degree of doping in the two samples [11]. Refer to spectrums plotted in Figure 4, it can be seen that a shifting in absorption peaks of pyrrole monomers occurred in both PPy with SDS and no SDS additions. The peaks shifting of PPy for the the two types occurred in the region ranging from 1250 to 1000 cm^{-1} associated with the breathing vibration of pyrrole ring.

The region of C –H and N –H vibrations in the range 1130 -1000 cm^{-1} is probably influenced for PPy –SO_4 sample by C–O vibration of hydroxypyrrole units formed by nucleophilic attack of water during the preparation [11]. It is then confirmed that the PPy has formed with a specific fingerprint of the FTIR spectrum due to a change in pyrrole monomer as a single substance to PPy. The absorbance at 689 cm^{-1} in pyrrole spectrum disappeared in the PPy spectrum. However, considering that the absorption peaks were found broadened in PPy spectrum, the absorption peak of payroll still exists. The broadening might be caused by overlapping bands of sulfate anion situated in the same position in the spectrum [11]. Polypyrrole which reacted with ammonium persulfate produced shorter conjugation lengths of polymer molecule chain which could be confirmed by FTIR in which the highest numbers of carbonyl groups occurred at formed polymer. Meanwhile the PPy with SDS addition have a specific transmittance at certain wavenumber due to the sulfate groups of oxidant agent reacted with the polymer chains to form sulfonic functional groups caused the polymer with the shorter delocalization chain length [11,23].

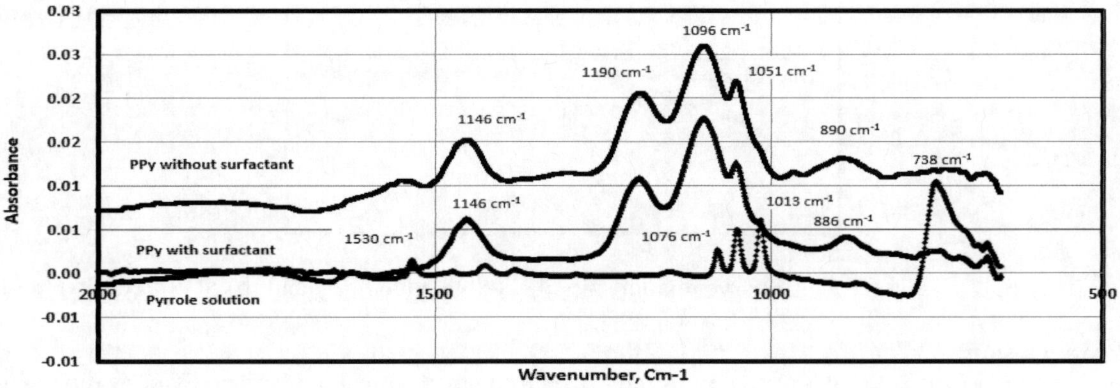

Figure 4. Spectrum FTIR of PPy With and Without Presence of Surfactant.

Electrical conductivity was evaluated through measurement of resistivity employing four point probe equipment. The voltage varied from -0.06 to 0.06 resulted in the V–I curve as shown in Figure 5. A V-I curve of the two samples showing an ohmic relationship indicating conducting samples. The resistivity values for the two PPy samples were calculated from each ohmic curve and subsequently determined its electrical conductivity. The electrical conductivity of the two samples is listed in Table 1. It is confirmed now that the addition of SDS during the polymerization reaction has increased significantly the electrical conductivity of PPy. The 1 % SDS addition PPy has increased the conductivity value more than 150 times higher than that of SDS free. The electrical conductivity value obtained from the current study is 0.16 S/cm when SDS free and increased to 24.46 S/cm when 1% of SDS added. This result is higher than values reported by Sirilak on the effect of SDS as anionic surfactant to a conductivity value of PPy through emulsion polymerization which was obtained in the range of 1.74×10^{-5} to 2.21×10^{-3} S/cm [24].

Tabel 1. Comparison of electrical conductivity value of SDS added and SDS free PPy.

Sample	Conductivity Value, σ
PPy with surfactant free	0.16 S/m
PPy with 1% of SDS	24.46 S/m

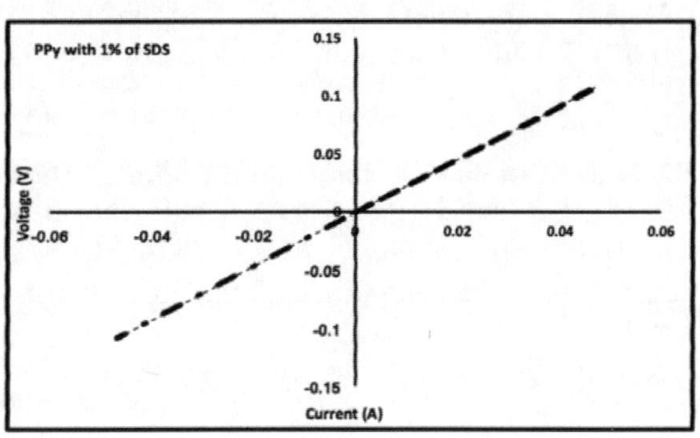

(a)

IOP Conf. Series: Materials Science and Engineering **553** (2019) 012042 doi:10.1088/1757-899X/553/1/012042

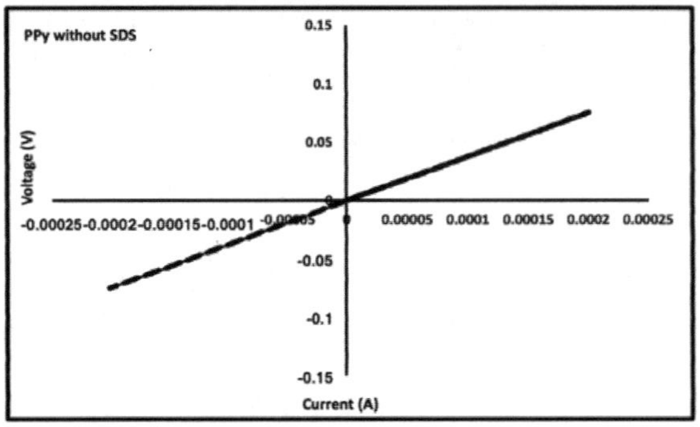

(b)

Figure 5. V-I curve relationship of (a) SDS added PPy and (b) SDS free PPy.

4. Conclusion

Conductive polypyrrole has successfully synthesized through chemical oxidative polymerization with the addition of SDS surfactant. Key parameters for PPy formation are real time of temperature and color changing together with a pH profile for 240 minutes reaction time. The reaction rate of polymerization of pyrrole with surfactant added is slower than that of surfactant free. The synthesis of polypyrrole is stable at acidic condition. The role of SDS anionic surfactant acted as a doping agent during polymerization and increased the electrical conductivity value significantly. The addition of 1 % SDS increased the electrical conductivity value to 24.46 S/m from 0.16 S/m in SDS free PPy.

Acknowledgment

The authors gratefully acknowledge the support of Postgraduate Program of Material Science Universitas Indonesia for the research facilities. We are also thankful for the financial support provided by the Directorate of Research and Community Service Universitas Indonesia under program Grants of International Publication Indexed for Final Project student 2018 (PITTA 2018) under contract no. 2237/UN2.R3.1/HKP.05.00/20018.

References

[1] G. Inzelt, *Conducting polymers, A new era in Electrochemistry*, no. 1. 2014.
[2] R. A. William, B. Sitorus, P. S. Kimia, U. Tanjungpura, P. S. Fisika, and U. Tanjungpura, "Sintesis Polianilina Pada Matriks Selulosa Sebagai Elektrolit Padat," vol. 3, no. 4, 2014
[3] R. Ansari, "Polypyrrole Conducting Electroactive Polymers: Synthesis and Stability Studies," *E-Journal Chem.*, vol. 3, no. 4, pp. 186–201, 2006.
DOI : 10.1155/2006/860413
[4] H. K. Chitte, N. V Bhat, V. E. Walunj, and G. N. Shinde, "Synthesis of Polypyrrole Using Ferric Chloride (FeCl 3) as Oxidant Together with Some Dopants for Use in Gas Sensors," vol. 2011, no. June, pp. 47–56, 2011.
DOI : 10.4236/jst.2011.12007
[5] Patil, B., H., Bulakhe, R., N dan Lokhande, C., D., *J. Mater Sci.: Master Electron*, vol. 25, pp. 2188-2198, 2014
[6] Qomariyah dan Susi, Nur, 2011, *Pengembangan Sensor Alkohol dari Bahan Polipirol Konduktif dengan Variasi Konsentrasi Dopan Flouroborat*, Fakultas Matematika dan Ilmu Pengetahuan Alam, Universitas Jember, Jember, 2011.
[7] T. V Vernitskaya and O. N. Efimov, "Polypyrrole: a conducting polymer; its synthesis, properties and applications," *Russ. Chem. Rev.*, vol. 66, no. 5, pp. 443–457, 2002.
[8] P. Saville, "Polypyrrole Formation and Use," no. January, 2005.
[9] J. Tabačiarová, M. Mičušík, P. Fedorko, and M. Omastová, "Study of polypyrrole aging by XPS, FTIR and conductivity measurements," *Polym. Degrad. Stab.*, vol. 120, pp. 392–401, 2015.

[10] S. Bilal, F. Perveen, and A. A. Shah, "Chemical synthesis of polypyrrole doped with dodecyl benzene sulfonic acid," *J. Sci. Innov. Res. 2015;*, vol. 4, no. 1, p. 33, 2015.
[11] M. Omastová, M. Trchová, J. Kovářová, and J. Stejskal, "Synthesis and structural study of polypyrroles prepared in the presence of surfactants," *Synth. Met.*, vol. 138, no. 3, pp. 447–455, 2003.
[12] A. M. Gueli, G. Bonfiglio, S. Pasquale, and S. O. Troja, "Effect of particle size on pigments colour," *Color Res. Appl.*, vol. 42, no. 2, pp. 236–243, 2017.
[13] L. M. Liz-Marzán, "Nanometals: Formation and color," *Mater. Today*, vol. 7, no. 2, pp. 26–31, 2004.
[14] T. F. Tadros, *Emulsion Science and Technology*. 2009.
[15] H. Zhang and A. I. Cooper, "Compressed Fluid Sedimentation Polymerization," *Macromolecules*, vol. 36, no. 14, pp. 5061–5064, 2003.
[16] Hafizah M.A.E, Bimantoro A, Andreas and Manaf A, *Procedia Chemistry* 19 162 – 165, 2016.
[17] Manaf A, Bimantoro A, Hafizah M A E and Andreas, *IOP Conf. Ser. Mater. Sciece Eng.* 223, 2017 .
[18] Manaf A, Prasutiyo Y J, Hafizah M A E, Andreas, *Int. J. of mechanical and Production Engineering* 5 2320-2092, 2017.
[19] Stejskal J and Gilbert R G, *Pure Appl. Chem.* 74 857867, 2002.
[20] G. J. Salomons, Y. S. Park, K. B. McAuley, and L. J. Schreiner, "Temperature increases associated with polymerization of irradiated PAG dosimeters," *Phys. Med. Biol.*, vol. 47, no. 9, pp. 1435–1448, 2002.
[21] Hwang, Ji-Hyun, dan Myoungho Pyo., *Bull. Korean Chem. Soc.*,vol. 27 (12), pp. 2067, 2006.
[22] Y. Kudoh, Synth. Met. 79 : 17, 1996**.**
[23] Hossein E., *World Journal of Chemistry*, 2 (2): 67-74, 2007.
[24] S. K. y L. S. Arunsawad Sirilak, "Effect of surfactant on conductivity of Poly(pyrrole-co-formyl pyrrole) via Emulsion Polymerization," *J. Met. Mater. Miner.*, vol. 24, no. 2, pp. 29–34, 2014.

Aerogel Based High Performance Thermal Insulation Materials

Mohanapriya Venkataraman[1*], Rajesh Mishra[1], Jiri Militky[1], Dana Kremenakova[1], Michal Petru[2]
[1*]Department of Material Engineering, Faculty of Textile Engineering, Technical University of Liberec, Czech Republic
[2] Department of Machinery Construction, Institute for Nanomaterials, Advanced Technology and Innovation, Liberec, Czech Republic

mohanapriya.venkataraman@tul.cz

Abstract. In this paper, the thermal properties of polyester/polyethylene nonwoven thermal wraps of varying thicknesses treated with aerogel were analyzed and compared. The microscopic images were taken by confocal microscope and image analysis to compare the physical structure of the aerogel treated nonwoven fabrics. Thermal conductivity and thermal resistance were measured using FOX heat flow meter instrument. The water vapor resistance and thermal resistance were measured using sweating guarded hot plate (SGHP). These tests were conducted to understand and to compare the thermal properties and water vapor resistance of aerogel treated nonwoven fabrics. The results of the experiments were statistically analyzed and showed that the fabric density and the aerogel present in the fabric have a significant effect on thermal properties and water vapor resistance of the aerogel treated nonwoven fabrics.

Keywords: Aerogel, Thermal Insulation, Thermal Conductivity, Nonwoven

1. Introduction

Silica aerogels consist of more than 96 percent air and remaining is a matrix of silicon dioxide. It is popular due to its properties such as low thermal conductivity (0.01 W/m.K), high porosity (99%), high optical transmission (99%) in the visible region, high specific surface area (1000m^2/g), low dielectric constant (1.0–2.0), low refractive index (1.05), and low sound velocity (100 m/s) [1]. Due to these features, they are used in super thermal insulation, acoustic insulation, radiation detectors, low dielectric constant aerogel films in ultra-large-scale integrated circuits, super hydrophobic aerogels for oil spill cleanup, window applications, catalysts, Internal Confinement Fusion (ICF) targets in thermonuclear fusion reactions, cosmic dust capture and waste management (gas absorption, radioactive waste confinement) etc.,[2]. In Textiles and clothing, it is used to preserve body heat loss in the cold weather where thermal transmittance properties of fabric are important. Thermal transmittance is the rate of unidirectional heat transfer per unit area in the steady state, between parallel planes, per unit difference of temperature of the plates (W/m^{2}K). Factors affecting the thermal properties are thermal conductivity of the fibre, air contained within the fabric, thickness of the fabric, bulk density of the fabric (includes the number, size and distribution of the air spaces within the fabric), heat loss by conduction/convention/radiation from skin to fabric, heat loss by evaporation of water from skin or fabric, heat gain due to water absorption by fabric, external environment likes temperature, relative humidity and movement of surrounding air. Clothing comfort includes three main considerations like thermo-physiological, sensorial and psychological comfort [3, 4]. Thermo-physiological comfort is an important parameter of clothes. The factors affect that thermo-physiological comfort are heat exchange within clothing, air permeability, transfer and evaporation of moisture. Due to metabolic processes, human body produces heat continuously. Heat generated by body is directly proportion to greater physical exertion. Because of sweat, the body begins to feel cool through the evaporation of the sweat on the skin [5]. Hence, the clothing should ensure high moisture transmission to facilitate cooling due to evaporation of sweat. Water vapour permeability is the ability to transmit vapour from the body. The sweat should be removed from the skin surface to the surface of fabric of the next-to-skin clothing for greater comfort. After the body has stopped sweating, the textile fabric should release the vapour held to the atmosphere in order to reduce the humidity on the surface of the skin [6]. The literature review reveals a lacunae in study of aerogel's application to thermal clothing. This study discusses the influence of aerogel on the thermal conductivity, thermal resistance and thermal transmittance depending on varying fabric densities of nonwoven fabrics treated with aerogel. The water vapour resistance and thermal resistance were also measured to study the effect of nonwoven layer densities on heat transfer characteristics.

2. Experimental method

2.1 Materials

In this study, polyester and polyethylene nonwoven thermal wraps, treated with amorphous silica aerogel, were used. The thermal wraps were chosen in three different thicknesses shown in Table 1. The properties of amorphous silica aerogel is shown in Table 2.

Table 1. Sample description

Sample Description	Thickness (mm)	Weight (g/m^2)	Density (kg/m^3)
Silica aerogel treated	3.424	272.56	79.66
nonwoven fabrics	6.212	499.46	80.42
(Polyester +Polyethylene)	6.608	440.7	66.73

Table 2. Properties of Silica Aerogel [7]

Properties	Value range
Particle size range	0.1–0.7mm
Pore diameter	2 ~20 nm
Particle density	120–140 kg/m^3
Surface chemistry	Fully hydrophobic
Thermal conductivity	0.012W/mK at 25°C

2.2 Methods

The thickness of the samples was measured using UNI-Thickness meter. Sweating guarded Hot Plate (SGHP) and Permetest was used to measure the water vapor resistance and thermal resistance of the samples. The fabrics were cut to dimensions of 10x10cms and the weight of the fabric was measured. All experiments were carried out in standard atmospheric conditions of about 20°±2°C and 65±2% relative humidity.

2.2.1 Microscopic analysis. The nonwoven fabric samples of three different thickness treated with aerogel were characterised using Image analysis ((with macro zoom objective and CCD camera) and confocal microscope (OLYMUPUS Confocal Scanning IR Laser Microscope, LEXT LS3000-IR).

Image analysis was done to compare the physical structures of the fabrics on microscopic scale to determine if any difference were noticeable that could explain test results.

2.2.2 Thermal transmittance measuring device. FOX heat flow meter instruments were designed for accurate thermal conductivity and thermal resistance measurements of thermal insulation materials according to the ASTM C518 and ISO 8301 standards. The FOX heat flow meter instruments also can be successfully used for testing, thermal transmittance of textile materials at conditions very much the same as in design suggested in the ASTM D1518 – "Standard test method for thermal transmittance of textile materials" – except sizes of the chamber and plate, and type of the heat flow source. After turning on the FOX instrument and re-zeroing its plates a 12"-wide (~304 mm) piece of the textile material was placed on the lower plate. Temperature of the set point was set at 10°C for upper plate and 35°C (~ human skin temperature) for the lower plate. Thermal equilibrium criteria was set as 12 and 3 blocks of all tests.

2.2.3 Water Vapour and Thermal Resistance measuring instrument. Thermal resistance Rct (m^2 K/W) is a quantity specific to textile materials or composites, which determines the dry heat flux across a given area in response to a steady applied temperature gradient, while water vapour resistance Ret (m^2 K/W) is a quantity specific for textile materials or composites, which determines the 'latent' evaporative heat flux across a given area in response to a steady applied water vapor gradient [8]. These features were measured by SDL M259B Sweating guarded hot plate apparatus following standard ISO 11092 [9], consisting of a measuring unit with temperature and water supply control, thermal guard with temperature control and the test cabinet into which is built the measuring unit that is thermal guarded, and in which the ambient air temperature and humidity are controlled according to ISO 11092. Besides, a humidity control device automatically controls the relative humidity within the cabinet and a cooling device automatically cooled the air temperature when needed [9]. The specimen to be tested was placed on an electrically heated plate with conditioned air ducted to flow across and parallel to its upper surface [9,10]. For determination of thermal resistance Rct, the heat flux through the test specimen was measured after steady-state conditions have been reached. Rct for a material was determined by subtracting the thermal resistance of the once the apparatus reached the steady condition, test time was 30 minutes.

Water vapor permeability of materials was measured with a guarded hot plate by saturating the plate surface with water. The power required to maintain the surface at a given temperature was related to the rate at which water evaporates from the surface of the plate and diffuses through the material. The thermal resistance of the material to convective heat transfer must be known before it was possible to extract the vapor permeability coefficient. Woodcock [11] developed a moisture vapor permeability index, known as i_m, which serves as a very convenient relative measure of the moisture vapor permeability of materials.

$$i_m = \frac{\left\{\frac{(Q)(R_{total})}{A}\right\} - (T_{plate} - T_{air})}{S(p_s - \phi p_a)} \tag{1}$$

Where i_m = Moisture vapor permeability index; R_{total}= Thermal resistivity of the fabric plus the boundary air layer, clo; A = Surface area of guarded plate measurement area, m^2; T_{plate} = Temperature of the saturated plate surface, °C; T_{air} = Temperature of the ambient air, °C; Q = Power required to maintain a constant saturated plate surface temperature, watts; S = Lewis relation between evaporative mass transfer coefficient and convective heat transfer coefficient, 2.2 °C/mmHg; p_s = Saturated water vapor pressure at the plate surface, mmHg; p_a = Saturated water vapor pressure of the ambient air, mmHg and ϕ = Relative humidity of ambient air, fractional relative humidity (not %).

The i_m value is a relative measure of the permeability of the material to the passage of water vapor. The i_m index should vary between 0 (for completely impermeable materials), and 1 (for completely permeable materials). In practice, the value of 1 as an upper limit is not approached until the wind speed over the plate becomes great enough to minimize the contribution of radiative heat transfer [12]. The moisture vapor permeability index, i_m, may be combined with the total dry thermal resistance, R_{total}, to yield a quantity which takes into account both convective and evaporative heat transfer. In this report R_{total}, is given in clo units, so the term becomes i_m/clo. The term i_m/clo provides a good ranking measure between materials if one is interested in materials which minimize the potential for heat stress [12].

3. Results and discussion

The results were evaluated and studied for thermal resistance, thermal conductivity, thermal effusivity, water vapor resistance and dry thermal resistance of the aerogel treated nonwoven fabrics. It was examined by one-way analysis of variance (ANOVA) with 95% confidence level. A significant difference ($p<0.05$) has been observed in the thermal resistance, thermal conductivity, thermal effusivity, water vapour permeability and dry thermal resistance properties for the three different thickness of fabrics treated with silica aerogel. It was found that there was an insignificant difference between the temperatures on thermal properties and between the instruments (SGHP & permetest) for water vapour resistance.

3.1 Microscopic analysis

(a)

(b)

Figure 1. Microscopic images taken using (a) Image analysis and (b) confocal microscope for aerogel treated nonwoven fabric.

Image analysis was done on microscopic scale for the cross-section area of the three fabrics with different magnifications. The physical structure confirmed to be different for three fabrics due to different thickness shown in Figure 1. It was observed that sample 2 has higher fabric density of 80.4 kg/m^3 as compared to other samples. The aerogel deposition in the fabric between the fibers was also observed.

3.2 Fabric density

Fabric density is the factor of weight and thickness. To obtain an indication of the effect of fabric density on thermal properties, nonwoven fabrics with comparable densities in different thicknesses and their corresponding weights were measured for aerogel treated nonwoven fabrics shown in Table 1 (a, b and c). The density difference of samples may be attributed to the fabric structure, thickness and also the percentage of aerogel particles present in the fiber.

3.3 Influence of Water Vapour Resistance

The Water Vapor Permeability (WVP) depends on the water vapor resistance which indicates the amount of resistance against the transport of water through the fabric structure. To maintain the degree of comfort of the user, the amount of water present in a fabric must be minimum. The analysis of variance (ANOVA) results show that the effect of fabric density on the water vapor resistance measured using SGHP had significant difference (p=0.001). Figure 2 shows that the water vapor resistance, measured using Sweating guarded hot plate, and it can be seen that the fabric which has the densities 80.4 and 79.6 kg/m^3 is higher than the 66.7 kg/m^{3}density. The water vapor permeability of the fabric may be attributed to the structure (open or closed cell structure) of the fabric and also the percentage of aerogel particles present in the fiber. This behavior can be explained by the moisture vapor transmission mechanism where vapor transmits through a textile layer by two processes, namely, diffusion and sorption-desorption. Water vapor diffuses through a textile structure in two ways, simple diffusion through the air spaces between the fibers and along the fiber itself.

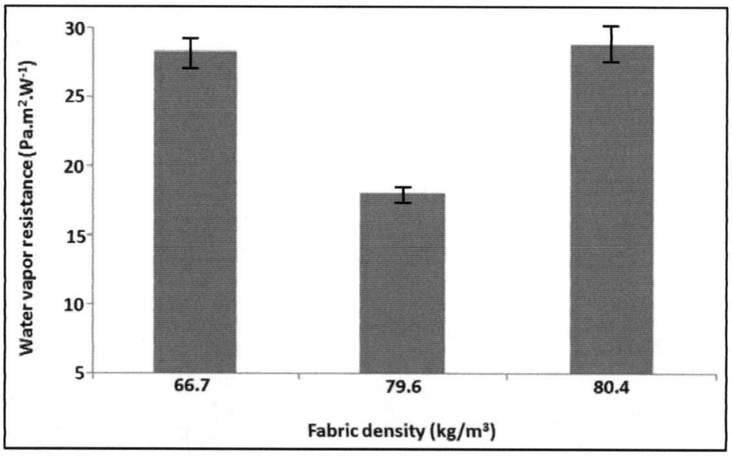

Figure 2. Water vapor resistance measured in Sweating guarded hotplate (SGHP)

3.4 Thermal Resistance (R_{ct}) measured from SGHP

Thermal resistance depends on the thickness and thermal conductivity of the fabrics. The analysis of variance (ANOVA) of dry resistance show that the effect of fabric density on the thermal resistance is significant (p=0.001). The aerogel treated nonwoven fabrics have different R_{ct} values depending on the fabric densities (Figure 3). The difference in R_{ct} values was due to the thermal conductivity of the fiber substance, which was about 0.14 (Wm^{-1}K^{-1}) for polyester and 0.34(Wm^{-1}K^{-1}) for polyethylene [16]. The thermal conductivity of aerogel particles present in the fibers was about 0.02 (Wm^{-1}K^{-1}). In any case, as the presence of fibers in these kinds of materials is very small compared to the air entrapped in the nonwoven fabrics, the fiber substance and aerogel particles used has influence on Rct and Ret.

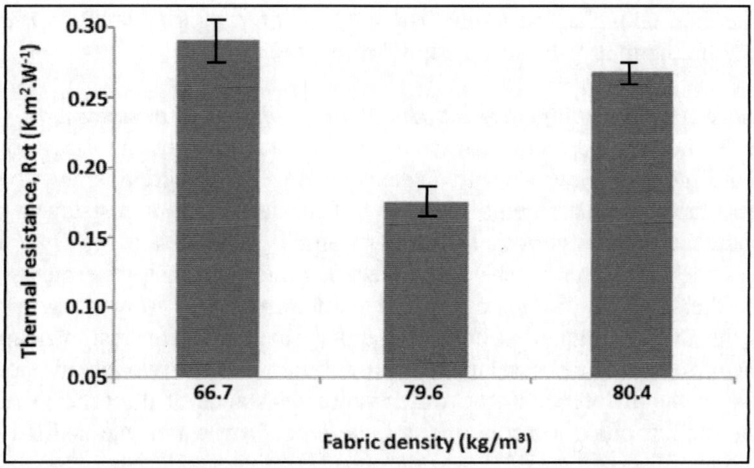

Figure 3. Dry thermal resistance, Rct from SGHP

Dry thermal resistance is calculated by measuring the temperature difference between the surface of the heated measurement area of the guarded hot plate and the temperature of the "ambient air away from the plate. It is this temperature difference which drives heat transfer through the fabric. The equation used for calculating the thermal resistance [12]:

$$R_{total} = \frac{A(T_{plate} - T_{air})}{Q} \tag{2}$$

Where R_{total} =Thermal resistivity of material plus the boundary air layer, clo (clo is a unit of thermal resistance and is equal to 0.155 °C-m2/watt); A = Surface area of guarded plate measurement area, $m^{2;}$ T_{plate} = Temperature of the plate surface, °C; T_{air} = Temperature of the ambient air, °C and Q = Power required to maintain a constant plate surface temperature, watts.

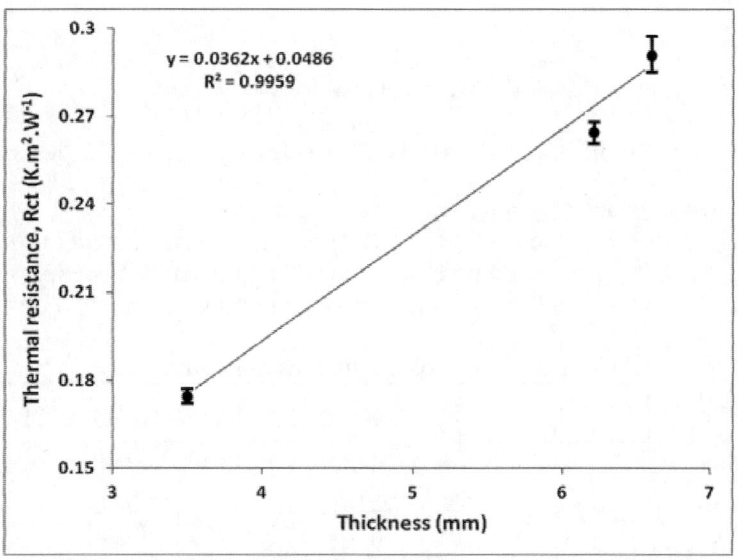

Figure 4. Thermal resistance Vs Thickness from SGHP

Figure 4 also shows the direct proportionality between thickness and thermal resistance. As the thickness of the fabric increases the thermal resistance also increases. The correlation of thermal resistance and thickness also shows higher value of about R^2= 0.99. Even the slight difference in

IOP Conf. Series: Materials Science and Engineering **553** (2019) 012043 doi:10.1088/1757-899X/553/1/012043

thickness changes the thermal resistance value. The analysis of variance (ANOVA) results show that effect of thickness on the thermal resistance is significant (p=0.005).

3.5 Comparative Analysis of Sweating Guarded Hot Plate (SGHP) and Permetest

The water vapour resistance and permetest for the aerogel treated nonwoven fabrics for three different densities are shown in Figure 5. The water vapour resistance were measured using two devices and the mean value of five separate measurements was used. The water vapour resistance measured using SGHP showed an increase for the fabric densities 80.4 and 79.6 kg/m^3 and shows a decrease for the fabric density 66.7 kg/m^3 and in the water vapor resistance measured using permetest shows that the fabric densities have the same trend for the fabric densities from 66.7 to 80.4 kg/m^3. The statistical analysis shows that the Sweating guarded hot plate and Permetest has an insignificant difference on water vapor resistance (p=0.22). The correlation (R^2=0.83) between the two test methods for materials are good. This is due to the difference in the water vapor resistance of the fabric densities 79.4 and 80.4 kg/m^3. This may be attributed to properties which depend on water vapor diffusion through air passages in the structure.

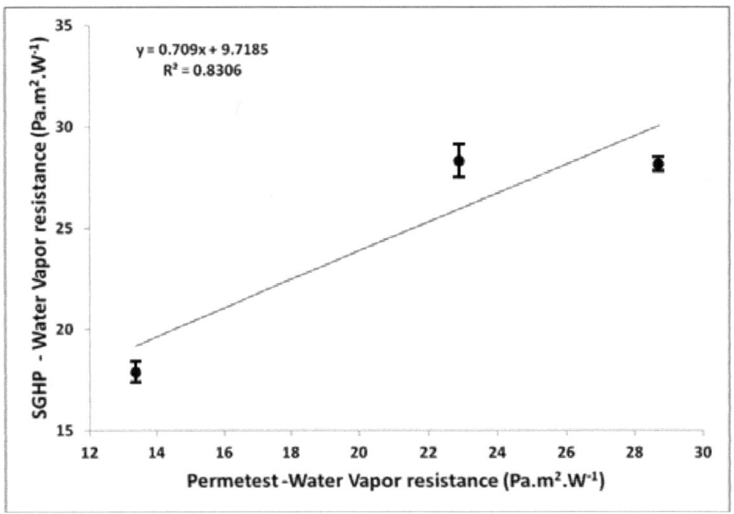

Figure 5. Correlation between SGHP and Permetest for water vapor resistance

3.6 Thermal transmittance of aerogel embedded nonwoven fabrics

Thermal transmittance (U-value) is the rate of transfer of heat (in watts) through one square meter of a structure divided by the difference in temperature across the structure. It is expressed in watts per meters squared Kelvin, or W/m²K. Insulation is inversely proportional to thermal transmittance.

Table 3. Results of thermal transmittance

Thickness (mm)	Fabric density (kg/m^3)	U1 (W/m^{2}K)	U2 (W/m^{2}K)	λ (W/mK)	R (m^{2}K/W)	R' (mK/W)	Intrinsic clo	Specific Clo
3.424	79.66	2.898	5.126	0.018	0.195	55.743	0.222	63.380
6.212	80.42	2.119	3.106	0.019	0.322	51.923	0.366	59.036
6.608	66.73	2.151	3.174	0.021	0.315	47.734	0.358	54.274

*U1 – Combined thermal transmittance, U2- Intrinsic thermal transmittance, λ – Thermal conductivity, R – Intrinsic thermal resistance, R' – Intrinsic thermal resistivity

The combined thermal transmittance or combined thermal conductance of the specimen plus the air (Eq. 3 of the ASTM D1518) can be calculated from the lower plates result (lower plate thermal conductivity) multiplied by 10

$$U_1 = \lambda_L \times 10 \ [\mathrm{W/(m^2 K)}] \tag{3}$$

i.e. divided by the 0.1m (100mm) thickness. It equals the heat flux [W/m²] through the specimen, measured by the lower heat flow meter, divided by the temperature difference ΔT of 25°C. The bare plate transmittance U_{bp} was calculated the same way after running the test without sample. Example for FOX 314 Ubp ≅ 6.67 W/(m²K). The intrinsic transmittance U_2 of the fabric alone can be calculated using Eq.4 of the ASTM D1518:

$$U_2 = (\mathrm{U}_{bp} \times \mathrm{U}_1) / (\mathrm{U}_{bp} - \mathrm{U}_1) \ [\mathrm{W/(m^2 / K)}] \tag{4}$$

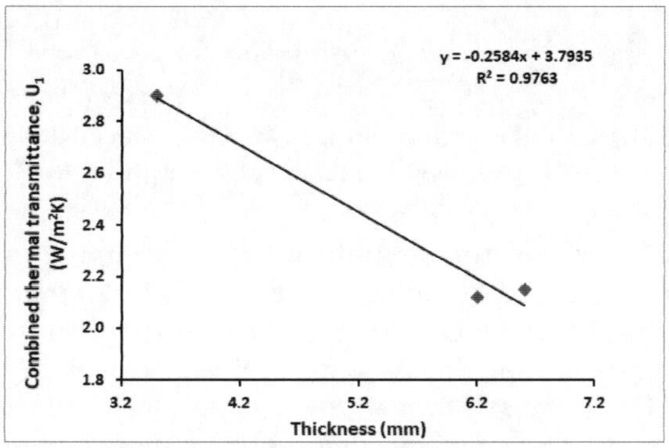

(a)

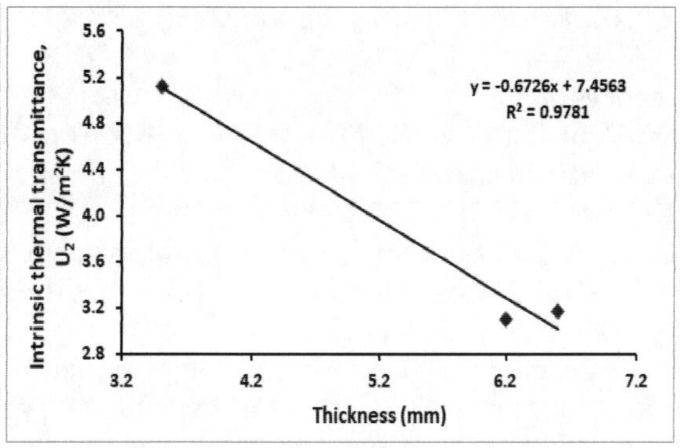

(b)

Figure 6. (a) Combined thermal transmittance and (b) Intrinsic thermal transmittance

The thermal transmittance of a fabric or batting is of considerable importance in determining its suitability for use in fabricating cold weather protective gear and clothing. The thermal interchange between man and his environment is, however, an extremely complicated subject which involves many factors in addition to the equilibrium insulation values of fabrics and battings. Therefore, measured thermal transmittance coefficients can only indicate relative merit of a material. In order to calculate U-Values it is important to first find thermal conductivity, thermal resistance and thermal transmittance which is shown in Table 3. Thermal transmittance, commonly known as the U-value, is a measure of the rate of heat loss of a component. It is expressed as watts per square metre, per degree Kelvin, W/m²K. With the increase in thickness, the combined thermal resistance decreases. The U-value is calculated from the reciprocal of the combined thermal resistances of the materials which is attributed to the element, air spaces and surfaces, which also considered as the effect of thermal bridges and air gaps of the aerogel embedded nonwoven structure. The combined thermal transmittance is shown in Figure 6 (a) and (b).

4. Conclusion

The Polyester/Polyethylene nonwoven thermal wraps treated with silica aerogel of three different thickness were measured. The thermal conductivity, resistance and transmittance of the fabric was directly proportional to fabric density. This may be attributed to decrease in heat losses due to space insulated and the structure of the fabric. Water vapor permeability is influenced by fabric structure, number of gaps and their orientation within the fabric. Water vapor resistance showed the influence of thermal resistance (R_{ct}) of the fabric which depends on fabric boundary air layer which in turn increased with the increase in fabric thickness. Also, the thermal resistance (R_{ct}) of the fabric showed a different trend when plotted with the fabric densities which was due to the weight and thickness of the fabric. The comparative analysis showed a significant difference on the thermal properties. Also, the comparative analysis was done for the SGHP and permetest instruments which showed an insignificant difference. There was a good correlation between the values of water vapor resistance from both the instruments. So, it can be concluded that the fabric density and the aerogel present in the fabric have a significant effect on thermal properties and water vapor resistance of the aerogel treated nonwoven fabrics.

Acknowledgment
This work was supported by by the Ministry of Education, Youth and Sports of the Czech Republic and the European Union – European Structural and Investment Funds in the frames of Operational Programme Research, Development and Education – project Hybrid Materials for Hierarchical Structures (HyHi, Reg. No. CZ.02.1.01/0.0/0.0/16_019/0000843) and by project "Modular platform for autonomous chassis of specialized electric vehicles for freight and equipment transportation", Reg. No. CZ.02.1.01/0.0/0.0/16_025/0007293.

References
[1] M. Venkataraman, R. Mishra, and J. Militky, "Aerogel based nanoporous fibrous materials for thermal insulation," *Fibers and Polymers*, vol. 15, pp. 1444-1449, July 2015, DOI:10.1007/s12221-014-1444-9
[2] P. R. Aravind, P. Shajesh, G. D. Soraru and K. G. K. Warrier, "Ambient pressure drying: a successful approach for the preparation of silica and silica based mixed oxide aerogels," *Journal of Sol-Gel Science and Technology*, vol. 54, pp. 105-117, 2010, DOI:10.1007/s10971-010-2164-2
[3] N. Oglakcioglu and A. Marmarali, "Thermal Comfort Properties of some Knitted Structures," *Fibres & Textiles in Eastern Europe*, vol. 64, pp. 94 – 96, 2007
[4] H. Y. Wu, W. Zhang and J. Li, "Study on Improving the Thermal-Wet Comfort of Clothing during Exercise with an Assembly of Fabrics," *Fibres & Textiles in Eastern Europe*, vol. 75, pp. 46 – 51, 2009
[5] K.H. Umbach, "Aspects of Clothing Physiology in the Development of Sportswear," *Knitting Technique*, vol. 15, pp. 165 – 169, 1993
[6] A. Bivainytė and D. Mikučionienė, "Influence of Shrinkage on Air and Water Vapour Permeability of Double-Layered Weft Knitted Fabrics," Materials Science (Medžiagotyra), ISSN 1392–1320 Vol. 18, 2012, DOI: 10.5755/j01.ms.18.3.2438.
[7] M. Venkataraman, R. Mishra, J. Wiener, J.Militky, T.M. Kotresh and M. Vaclavik, "Novel techniques to analyse thermal performance of Aerogel-treated blankets under extreme temperatures," *The Journal Of The Textile Institute*, vol.106, pp.736-747, 2015. DOI: 10.1080/00405000.2014.939808
[8] G. Mazzuchetti, G. Lopardo and R. Demichelis, "Influence of Nonwoven Fabrics' Physical Parameters on Thermal and Water Vapor Resistance," *Journal of Industrial Textiles*, vol. 36, pp. 253-264, 2007, DOI: 10.1177/1528083707072366.
[9] ISO 11902, Textiles, "Physiological Effects," *Measurement of Thermal and Water vapour Resistance under Steady Conditions* (Sweating Guarded – Hotplate Test), pp. 15, September 2014
[10] M. Venkataraman, R. Mishra, D. Jasikova, T. Kotresh and J. Militky, (2015). "Thermodynamics of aerogel-treated nonwoven fabrics at subzero temperatures," *Journal of Industrial Textiles*, vol. 45, pp. 387–404, November 2015, DOI: 10.1177/1528083714534711.
[11] A.H. Woodcock, "Moisture Transfer in Textile Systems," *Textile Research Journal*, vol. 32, pp. 622-633, 1962, DOI: 10.1177/004051756203200802.
[12] P.W. Gibson, "Factors Influencing Steady-State Heat and Water Vapor Transfer Measurements for Clothing Materials," *Textile Research Journal*, vol. 63, pp. 749-764, 1993, DOI:10.1177/004051759306301208
[13] X. Tao, "Smart Fibers, Fabrics and Clothing," *Woodhead Publishing Limited in association with the Textile Institute*, Cambridge, England, p. 60, 2003, eBook ISBN: 9781855737600.

The Roles of Process Parameters on Structures and Mechanical Properties of Polypropylene/Clay Nanocomposites

Annisa Rifathin[1*], Onny Ujianto[1], Nurul Jamilah[1], Bambang Afrinaldi[1]
[1]Center for Polymer Technology, Agency for Assessment and Application of Technology, Building 460, Puspiptek Area, Setu, South Tangerang, Banten, Indonesia 15314

annisa.rifathin@bppt.go.id

Abstract. Process parameters are crucial to produce targeted qualities in polypropylene (PP)/clay nanocomposites, due to their roles on the generation of shear and diffusion. Thus, this research aims to observe their effects on structures and properties of PP/clay nanocomposites. Samples were produced by mixing PP, PP grafting maleic anhydride (PP-g-MA), and Cloisite 20A at fixed compositions, 88/9/3 wt%, respectively, in an internal mixer with variations on temperatures (210, 220, 230 °C) and speeds (60, 80, 100 rpm). Effect of mixing parameters on nanocomposite structures and properties were investigated from XRD, SEM and flexural properties. The results showed that all samples had intercalated as well as agglomerated structures. Further analysis on XRD and SEM showed that samples produced at high conditions (230 °C or 100 rpm) had similar structures. In contrast, low setting sample (210 °C and 60 rpm), despite its similarity on dispersion level, had longer agglomerates than that of mixed at high settings. Correlated both increase of d-spacing and agglomerates length to flexural properties suggested that modulus was more influenced by dispersion level, while strength was affected by agglomerates. However, it was worth to note that improvement on d-spacing, with availability of long agglomerates might not guarantee modulus and strength improvement due to low interfacial bonding.

Keywords : Polypropylene, clay, internal mixer, XRD, SEM, flexural

1. Introduction

Polypropylene (PP) is widely applied to produce various applications as automotive, packaging, building component, etc. Due to its large applications, studies on enhancement properties of PP are still interesting field for many researchers. One possible technique that growth rapidly in the last two decades to improve PP properties as physical, mechanical, fire retardancy, and barrier is the addition of nanofiller as clay to PP matrix [1-5], due to its low loading. In order to mix this two different polarity materials, compatibilizer is needed [6, 7], and process parameters should also be considered as their contribution to disperse the fillers [8-11].

The processing parameters might provide shear, diffusion, and degradation mechanisms that influence clay exfoliation [12]. Our previous study on PP/clay nanocomposites showed that interaction between temperature and mixing time was crucial to disperse nanoclay [13]. However, correlation between this interaction and structures and properties was still unclear.

This study aimed to explore interaction effects of process temperatures and mixing speeds on clay dispersion and mechanical properties in Polypropylene matrix. Analysis on the role of mixing conditions was done according to flexural properties. The dispersion level was investigated from X-ray diffraction (XRD), while agglomerations were observed under Scanning Electron Microscopy (SEM).

2. Experimental method

2.1 Materials

Polypropylene (PP), 5169 MAS 2158 from PT. Politama Indonesia was used as matrix. It was PP with melt flow index (MFI) 1.8 g/10 min. Epolene Wax G 3015P, polypropilene-grafted-maleic anhydride (PP-g-MA), from Eastman Chemical Company was used as a compatibilizer. A commercial organophillic montmorillonite clay, Cloisite 20A (Southern Clay Product) was added as filler.

2.2 Nanocomposite preparations

Haake Rheomix 600 was used for mixing PP, PP-g-MA and clay according to process conditions presented in Table 1. All samples were produced at identical composition with 88% of PP, 9% of PP-g-MA, and 3% of clay to produce the best tensile strength [8]. Before mixing, overnight drying (80°C) was perform on clay to reduce its moisture. Nanocomposite specimens were prepared by compression molding (Collin P300P) with setting conditions shown in Table 2. Two steps of compression molding were applied to produce smooth surface as trials done in our lab.

Table 1. Setting Conditions used in an Internal Mixer.

Samples	Temperature [°C]	Speed [rpm]	Time [min]
Control	220	80	10
1	210	60	10
2	210	100	10
3	220	80	10
4	220	80	10
5	220	80	10
6	230	60	10
7	230	100	10

Table 2. Setting Conditions used in Compression Molding.

Parameters	First Step					Second Step				
	1	2	3	4	5	1	2	3	4	5
Temperature [°C]	205	205	205	205	40	205	0	195	0	40
Pressure [Bar]	0	1	1	0	0	0	0	0	0	1
Time [Min]	10	5	5	0	15	10	0	3	0	15

2.3 Characterizations

Flexural modulus and strength testing were performed using Shimadzu AGS 10 kN, Universal Testing Machine, according to ASTM D 790. Flexural speeds were calculating by considering samples thickness as stated in standard. The support span distance was set at 25.4 mm. At least five specimens were tested, and the average values were reported.

The structures of clay were observed from agglomeration analyzed using Scanning Electron Microscope (SEM) and d-spacing increase using X-Ray Difraction (XRD). The agglomerations of clay were observed to analyze the present of agglomerations in the mixtures using JEOL JSM-6510LA SEM with accelerated voltage of 15 kV. Measurement and analysis were done in 5 surface area of flexural sample number 1, 2, 6, and 7, with at least 1 measured agglomerate for each surface. Specimens were cooled with liquid nitrogen and cut. Specimens were sputter coated with Platinum to increase surface conductivity and reduce charging. The clay dispersions were analyzed using Simadzu X-Ray Diffractometer 7000 Maxima-X with Cu kα radiation. Analysis wes performed at scan speed 2.4 deg/minute, $\lambda = 1.54$ A°, 40 kV and 20mA. The intensity peak were intergrated from 1° - 10° of 2 theta.

3. Results and discussions

3.1 Effects of processing parameters on flexural properties

Flexural modulus and strength of PP/clay nanocomposite samples are presented in Table 3. It is shown that the average modulus and strength increase about 12% and 6%, respectively, attributed to clay dispersion. However, despite the increase of flexural properties for most of samples, it is shown that there was one sample (sample 1, produced at low temperature and low rpm) experiencing decrease on both modulus and strength. These phenomena were also reported in another previous study [14]. The decrease might be produced by poor polymer - clay bonding due to availability of big agglomerates. These might be happened in case of very low generated shear applied to disperse the clay [15, 16] and very high polymer viscosity that difficult to penetrate clay galleries [17]. This suggested that setting combination between low temperature (210 °C) and low rpm (60 rpm) applied to mix sample 1 might not enough to disperse most of the clay in this nanocomposite system, and resulted in some big agglomerates.

Table 3. Flexural modulus and strength of PP/clay nanocomposite samples.

Sample	Temperatures [°C]	Speed [rpm]	Time [min]	Modulus [MPa]	Strength [MPa]
Control	220	80	10	1764 ± 86.0 (± 5%)	49.15 ± 1.66 (± 3%)
1	210	60	10	1512 ± 55.0 (± 4%)	41.10 ± 2.05 (± 5%)
2	210	100	10	1974 ± 131 (± 7%)	53.20 ± 5.65 (± 11%)
3	220	80	10	2090 ± 158 (± 8%)	55.14 ± 1.13 (± 2%)
4	220	80	10	2113 ± 205 (± 10%)	56.46 ± 2.59 (± 5%)
5	220	80	10	2084 ± 184 (± 9%)	57.33 ± 2.60 (± 5%)
6	230	60	10	2119 ± 57.0 (± 3%)	50.70 ± 6.42 (± 13%)
7	230	100	10	2169 ± 105 (± 5%)	52.90 ± 7.89 (± 15%)
Average all nanocomposites				1978 ± 123 (± 6%)	52.00 ± 3.75 (± 7%)
Average of mid setting				2096 ± 182 (± 9%)	56.31 ± 2.11 (± 4%)

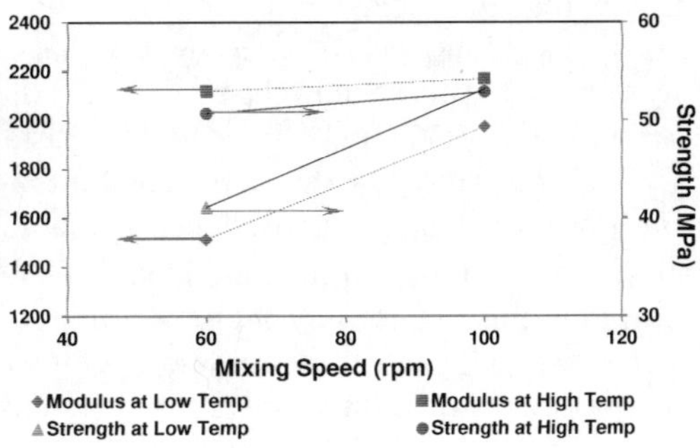

Figure 1. Effect of mixing speeds on flexural modulus and strength.

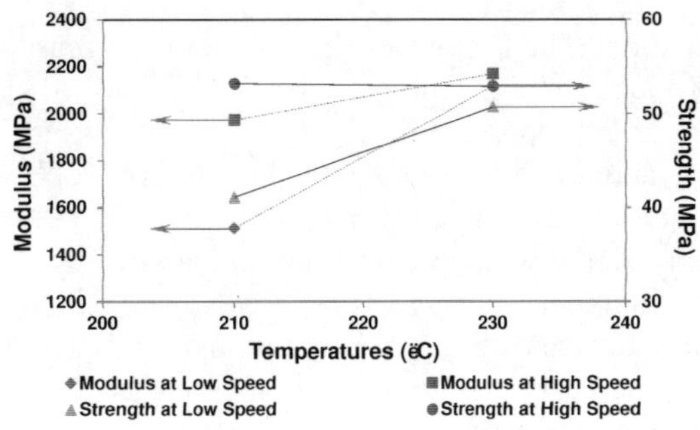

Figure 2. Effect of mixing temperatures on flexural modulus and strength.

Figure 1 and 2 show the effects of processing parameters on flexural properties. In general, both modulus and strength increased as the higher settings. These improvements were clearly obtained especially at higher setting (either mixing speed (Figure 1) or temperatures (Figure 2)) when the other parameter was set at low condition. At low temperature and low mixing speed, polymer viscosity is still high so it difficult to penetrate the interlayer galleries. At the same time, low mixing speed might unable to disperse clay agglomerates effectively. Combination these two conditions resulted in low modulus and strength. In contrast, at low temperature, higher mixing speed generated higher shear that able to disperse the clay and improve modulus and strength. However, there were slight improvements at higher mixing speed when temperature was set at high setting, and vice versa. This suggested that the effect of shear and diffusion in these boundaries might not significant change flexural properties due to competing mechanisms between shear and diffusion that work simultaneously.

3.2 Effects of processing parameters on nanocomposite structures

The structures of PP/Clay nanocomposites are presented in Figures 3 to 7. In general, all nanocomposite samples have nanostructures as well as agglomerated structures. Nanostructures were observed from peaks shifted to lower angle compared to clay peak on XRD diffractogram, while agglomerated were analyzed from the length and width of microstructures available on SEM micrographs.

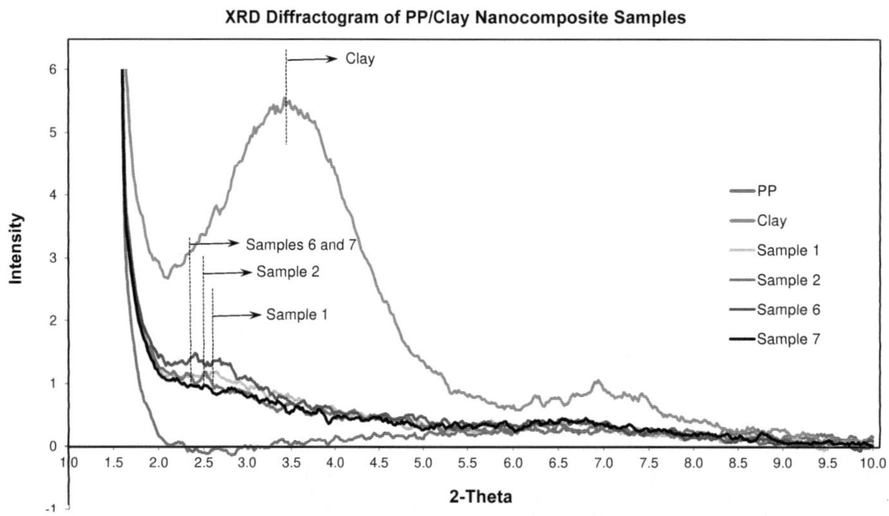

Figure 3. XRD diffractograms of PP/clay nanocomposites

Figure 3 shown that all samples have peaks at lower degree suggested intercalated of some clay structures. The best improvement on the increase of interlayer spacing was obtained by samples 6 and 7 produced at high temperature (230 °C) and either at low or high mixing speed. The peaks were shifted from 3.26° represented 2.71 nm of d-spacing to 2.34° represented 3.77 nm of inter-gallery distance. The increase of intergallery spacing from 2.71 nm to 3.77 nm suggests that the nanocomposites structure is available on sample 6 and 7. In contrast, the lowest increase was produced by sample 1 prepared at low temperature and low mixing speed (210 °C and 60 rpm) that increase from 2.71 nm to 3.39 nm, and also suggests nanocomposite structures. However, this did not support to negative results of modulus and strength. So it was hypothesized that there were both nanostructures and agglomerates available on the system and competing each other. For this reason, further analysis on the size of agglomerates were done using SEM.

IOP Conf. Series: Materials Science and Engineering **553** (2019) 012044 doi:10.1088/1757-899X/553/1/012044

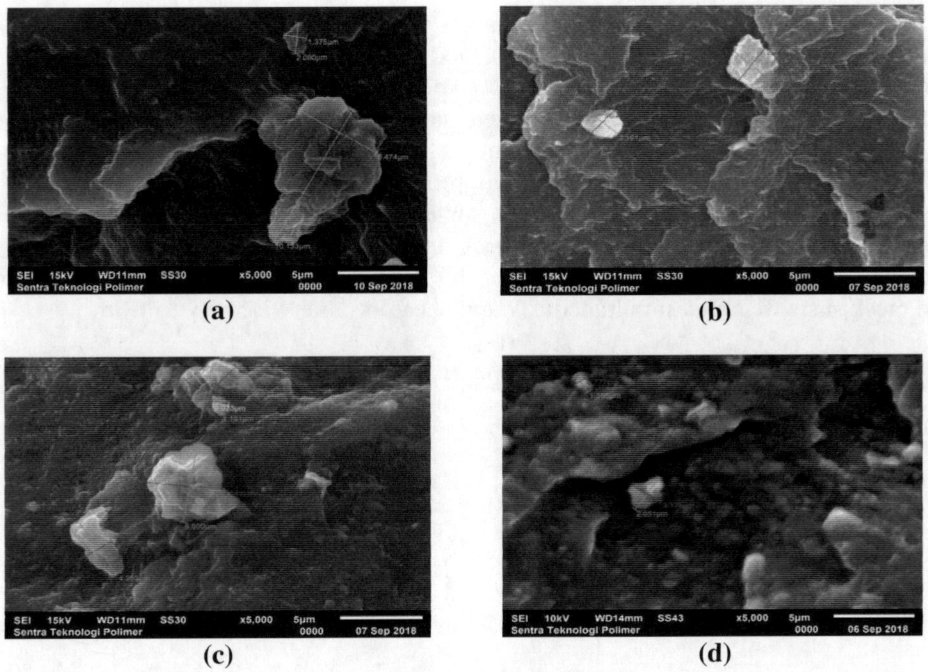

Figure 4. SEM micrograph (a) Sample 1 (210°C, 60 rpm); (b) Sample 2 (210°C, 100 rpm); (c) Sample 6 (230°C, 60 rpm); (d) Sample 7 (230°C, 100 rpm).

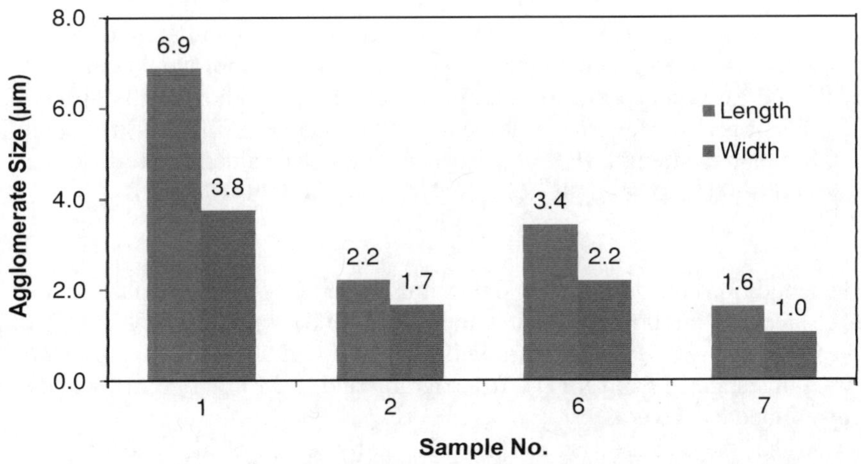

Figure 5. Size of agglomerates available on PP/clay nanocomposites.

Figures 4 and 5 show selected surface area of microstructures available on each observed samples and agglomerate size. The length of microstructures ranges from 1.6 to 6.9 μm, while the width varies from 1.0 to 3.8 μm. The smaller agglomerates were available in samples 2 and 7 those were produced at high mixing speed. On the contrary, the bigger agglomerates found at low mixing speed samples (Sample 1 and 6). The size even more than twice for sample 1 (low mixing speed, low temperature) compare to another samples, attributed to limited shear in the process. This data explains the reason behind the decrease on modulus and strength for this sample compare to control, despite the improvement on d-spacing. This finding also suggests that there is minimum ratio between level of dispersion and agglomerate size to improve both modulus and strength that is not covered in this research. For this reason, it is suggested that further experiment should be done in the future.

Combination analysis on XRD and SEM suggested that high dispersion level and small agglomerates can be achieved if there was minimum one parameter that be set at high condition (temperature or rpm). The fact that the high dispersion level was produced at high temperature or rpm, and small agglomerates were available on high rpm samples, suggested that there was synergism between shear and diffusion in order to produced desired nanostructures. The results of this research show that the shear is crucial to break the agglomerates, however this mechanism would not go further if there was no diffusion mechanism that runs simultaneously and need optimum viscosity to work.

3.3 Effects of nanocomposite structures on mechanical properties

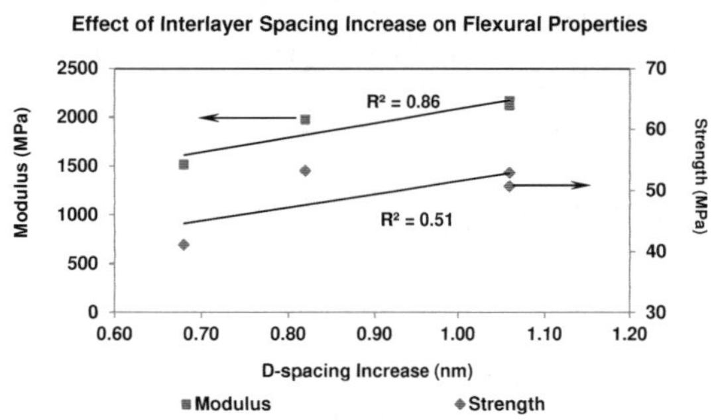

Figure 6. Effect of interlayer spacing increases on flexural modulus and strength.

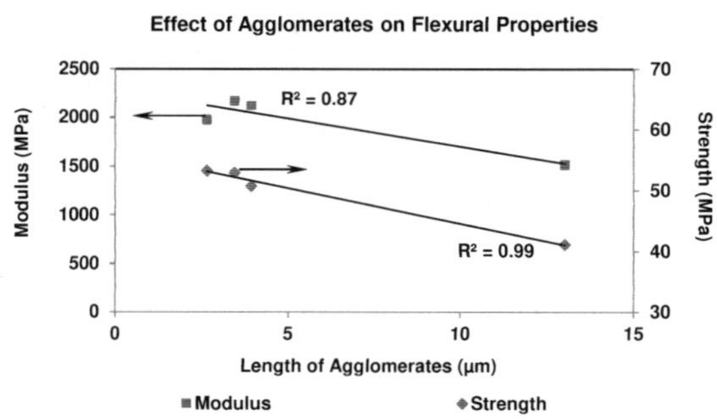

Figure 7. Effect of agglomerate sizes on flexural modulus and strength.

Structures and mechanical property of PP/clay nanocomposites are shown in Figure 6 and 7. In general, the increase on d-spacing and reduction on agglomerate length would improve modulus and strength. This is caused by the increase on aspect ratio that improves interfacial bonding between matrix and fillers. From the figures, it is also suggested that the d-spacing increase (Figure 6) has more influence to modulus ($R^2 = 0.88$) than to strength ($R^2 = 0.61$). In contrast, the reduction of agglomeration sizes (Figure 7) has more contribution to strength ($R^2 = 0.99$) than to modulus ($R^2 = 0.87$).

4. Conclusion

The roles of mixing conditions on structures and flexural properties of PP/clay nanocomposites have been explored. The samples were produced using an internal mixer with temperatures (210, 220, 230 °C) and mixing speeds (60, 80, 100 rpm) variations, and fixed composition. The structural of nanocomposites were investigated from XRD diffractograms and SEM micrograms, while mechanical properties were represented by flexural test.

The results showed that the average flexural modulus and strength of all nanocomposite samples increased by adding nanoclay on PP. These increases were influenced by the combination of intercalated structure and small agglomerate size. The samples produced at high conditions (230 °C or 100 rpm) had similar intercalated and agglomerate structures. On the other hand, bigger microstructures were found in the sample produced at low settings (210 °C and 60 rpm), despite its similarity on dispersion level. Analysis on structures and flexural property relationship shows that the intercalated structures have more contribution on modulus improvement, while agglomerates size contributes to strength. The findings of this research also suggest that the improvement on d-spacing might not successfully increase modulus and strength, if huge microstructures presents on the system.

Acknowledgment
Authors would like to thank you Center for Polymer Technology, Agency for Assessment and Application of Technology for supporting us to conduct this research.

References
[1] Delva, L., Ragaert, K., Allaer, K., Gaspar-Cunha, A., Degrieck, J., and Cardon, L., 2015 Influence of twin-screw configuration on the mechanical and morphological properties of polypropylene-clay composites *IJMPT* **52** 176-192. DOI: 10.1504/IJMPT.2016.073631
[2] Hu, Y., Wang, X., and Li, J., 2016 Regulating Effect of Exfoliated Clay on Intumescent Char Structure and Flame Retardancy of Polypropylene Composites *Ind. Eng. Chem. Res.* **55** 5892-5901. DOI: 10.1021/acs.iecr.6b00480
[3] Khalaj, M.-J., Ahmadi, H., Lesankhosh, R., and Khalaj, G., 2016 Study of physical and mechanical properties of polypropylene nanocomposites for food packaging application: Nano-clay modified with iron nanoparticles *Trends Food Schi. Technol.* **51** 41-48. DOI: 10.1016/j.tifs.2016.03.007
[4] Norouzi, M., Zare, Y., and Kiany, P., 2015 Nanoparticles as Effective Flame Retardants for Natural and Synthetic Textile Polymers: Application, Mechanism, and Optimization *Polym Rev.* **55** 531-560. DOI: 10.1080/15583724.2014.980427
[5] Pappalardo, S., Russo, P., Acierno, D., Rabe, S., and Schartel, B., 2016 The synergistic effect of organically modified sepiolite in intumescent flame retardant polypropylene *Eur. Polym. J.* **76** 196-207. DOI: 10.1016/j.eurpolymj.2016.01.041
[6] Kiliaris, P. and Papaspyrides, C.D., 2010 Polymer/layered silicate (clay) nanocomposites: An overview of flame retardancy *Prog. Polym. Sci.* **35** 902-958. DOI: 10.1016/j.progpolymsci.2010.03.001
[7] Shafiee, M., Ramazani, S.A.A., and Danaei, M., 2010 Investigation of the Gas Barrier Properties of PP/Clay Nanocomposite Films with EVA as a Compatibiliser Prepared by the Melt Intercalation Method *Polym. Plast. Technol. Eng,* **49** 991-995. DOI: 10.1080/03602559.2010.482075
[8] Akbari, B. and Bagheri, R., 2012 Influence of Compatibilizer and Processing Conditions on Morphology, Mechanical Properties, and Deformation Mechanism of PP/Clay Nano composite *J. nanomater* **2012** 1-8. DOI: 10.1155/2012/810623
[9] Lertwimolnun, W. and Vergnes, B., 2005 Influence of compatibilizer and processing conditions on the dispersion of nanoclay in a polypropylene matrix *Polym* **46** 3462-3471. DOI: 10.1016/j.polymer.2005.02.018
[10] Lertwimolnun, W. and Vergnes, B., 2006 Effect of processing conditions on the formation of polypropylene/organoclay nanocomposites in a twin screw extruder *Polym. Eng. Sci* **46** 314-323. DOI: 10.1002/pen.20458
[11] Lertwimolnun, W. and Vergnes, B., 2007 Influence of screw profile and extrusion conditions on the microstructure of polypropylene/organoclay nanocomposites *Polym. Eng. Sci.* **47** 2100-2109. DOI: 10.1002/pen.20934
[12] Ujianto, O., Jollands, M., and Kao, N., 2015 Polyethylene/Clay Nanocomposites Prepared in an Internal Mixer: Effect of Processing Variable on Mechanical Properties *Adv. Mater. Res.* **1105** 46-50. DOI: 10.4028/www.scientific.net/AMR.1105.46
[13] Zaiby, Y. A. Swasono, A. L. Juwono, and Ujianto, O. 2017 Polypropylene / Clay Nanocomposites Prepared in an Internal Mixer: Optimization of Processing Conditons to Improve Flexural Modulus *IOP Materials Science and Engineering* **432** 1-6. DOI: 10.1088/1757-899X/432/1/012014.
[14] Dong, Y. and Bhattacharyya, D., 2008 Effects of clay type, clay/compatibiliser content and matrix viscosity on the mechanical properties of polypropylene/organoclay nanocomposites *Compos. Part A Appl. Sci. Manuf.* **39** 1177-1191. DOI: 10.1016/j.compositesa.2008.03.006

[15] Domenech, T., Peuvrel-Disdier, E., and Vergnes, B., 2012 Influence of Twin-Screw Processing Conditions on Structure and Properties of Polypropylene – Organoclay Nanocomposites *Int. Polym. Proc* **27** 517-526. DOI: 10.3139/217.2591

[16] Spencer, M.W., Hunter, D.L., Knesek, B.W., and Paul, D.R., 2011 Morphology and properties of polypropylene nanocomposites based on a silanized organoclay *Polym* **52** 5369-5377. DOI: 10.1016/j.polymer.2011.09.034

[17] Moghri, M., Shahabadi, S.I.S., and Garmabi, H., 2013 Effect of nanoclay content on nanoclay dispersion mechanism: Diffusion-controlled mechanism versus shear-controlled mechanism *J. Reinf. Plast. Compos* **32** 1956-1965. DOI: 10.1177/0731684413493950

Shape Memory Thermoplastic Natural Rubber for Forensic Applications

Theerarat Sengsuk [1], Sililak Intom[1], Ladawan Songtipya [2], Jobish Johns[3], Krisadakon Chawengsaksopak[4], Ekwipoo Kalkornsurapranee[1*]
[1]Department of Materials Science and Technology, Faculty of Science, Prince of Songkla University, Hat-Yai, Thailand
[2]Department of Material Product Technology, Faculty of Agro-Industry, Prince of Songkla University, Hat-Yai, Thailand
[3]R&D Center, Department of Physics, Rajarajeswari College of Engineering, Bangalore, India
[4]Police Forensic Science Center 9, Songkhla, Thailand

ekwipoo.k@psu.ac.th

Abstract. Shape memory thermoplastic natural rubber was successfully developed in our laboratory. The main aim of this work is to obtain the suitable recipe and condition for fabrication of shape memory polymers (SMPs) based thermoplastic natural rubber (TPNR). It can be easily used as 3D imprint material for forensic application. The effect of blending ratio of thermoplastic (TP) and natural rubber (NR) on the mechanical and shape memory properties has been investigated. It was found that the mechanical properties show a decreasing trend with increasing of NR content while the shape fixity ($\geq 70\%$) and recovery efficiency ($\geq 60\%$) increased when adding NR in the blends. In addition, the prepared SMPs based TPNR in this work was also compared with commercially available forensic grade silicone. The resulting material provided better performances in terms of mechanical properties and shelf life. Low cost and ease of processing are the advantages and benefits of this system.

Keywords: Thermoplastic Natural Rubber, Forensic, Shape Memory Polymer, Blend Ratio

1. Introduction

Shape memory polymers (SMPs) have been developed very recently and become more interested in the field of materials science. SMPs can be called " smart materials" which can be programed to temporary shape until the exposure or external stimulation with temperature, light, water or pH value, which initiates to activate the shape recovery process to return its permanent or original shape [1]. A number of reports are available in literature about the various fabrication methods to fabricate SMPs with different conditions such as temperature, PH and light [2-4]. Furthermore, various applications of SMPs in the areas of medical smart devices (magnetically sensitive), sensors (electrically sensitive), self-tightening structure and self-retractable materials (thermally sensitive) [5] has been introduced. Recently, a novel technique to prepare SMPs by using thermoplastic natural rubber (TPNR) for forensic applications has been developed [6] in our laboratory. The newly introduced "Thermoplastic Natural Rubber" can be used to collect the 3D evidences (i.e., tool marks and human identification) from the locations of crime scene. It can also be used to overcome the limitations of presently existing investigation techniques. The material can be transformed at approximately 75 ºC to clay-like soft putty and pressed on the identified marks. The imprint marks of the evidences will be clearly appear on the TPNR on cool down to room temperature after 10 seconds and the investigator can easily copy the identified evidences on a mold. It can be used to trace bullet holes, cracks and damaged spots from small tools including human identification from corpse's fingerprints and teeth. The advanced features of SMPs include its ability to create 3-Dimensional imprint and mold which can be used in crime scene investigation. The advantages such as ease of handling, fast setting, high quality makes the system more attractive in the field of forensic science. Multiple copies of the marks can be taken, and its original imprint will be kept for a long time. The rubber-like mold is easy for investigators to carry and reuse for further investigation in the laboratory.

In this work, an attempt has been made to prepare SMPs based TPNR by blending technique for different blend ratios. Effect of blend ratio on the mechanical and shape memory properties was investigated. The optimized SPMs based on TPNR were also compared with the other evident collecting techniques. The concept used in the present investigation to develop material has high potential in manufacturing medical and sports items, environmental field research and maintenance work in construction industry.

2. Experimental method

2.1 Preparation of SMPs based TPNR

SMPs based TPNR were prepared using blending technique. Natural rubber (NR) was first compounded and blended with thermoplastic called "block-co-polymer based polyolefin (BCO)" by following the reported previous formulation [6] using simple blend technique. Influence of different weight ratios between NR and BCO (i.e., NR/BCO = 30/70, 40/60 and 60/40) on properties was investigated. Blending was carried out by melt mixing process using a Haake Rheocord 600 laboratory internal mixer with a rotor speed of 60 rpm at 100 ºC for 10 min. The blend product was then prepared as test specimen using Yong Fong cold feed extruder.

2.2 Mechanical Testing

Tensile test of the dumbbell-shaped specimens was performed according to ASTMD412 using universal testing machine (UTM) at a fixed extension speed of 500 mm/min^{-1}. Surface hardness was tested using a Shore A durometer (Frank GmbH, Hamburg, Germany) according to ASTM D2240.

2.3 Shape Memory Testing

The shape memory behavior of SMP was studied based on the work reported by Brostowitz et al. (2014) [7]. An adjustable clamp was used to test the shape memory behavior under tension as shown in Figure 1. The test specimens were prepared by 30 mm x 10 mm size with thickness of 2 mm and clamped on the adjustable clamps. Gauge length of the initial sample, l_i is marked and measured with calipers. Then, the sample was dipped under warm water at 80 ºC for 15 seconds and stretched to 100% strain. The stretched length, l_s was measured using calipers. The sample was then immediately quenched in cold water at 10 ºC for 15 seconds. The stretched rubber sample was further removed from the clamp and the fixed length, l_f was measured. Finally, the fixed length sample was immersed back to the warm water at 80 ºC for another 15 seconds, quench under cold water at 10 ºC for 15 seconds and the recovery length, l_r was measured with calipers. The shape fixity (F) and shape recovery (R) efficiencies of the SMP based TPNR have been calculated by using Equations (1) and (2):

$$F=\frac{l_f}{l_s}\times 100 \qquad (1)$$

$$R=\frac{l_f-l_r}{l_s-l_i}\times 100 \qquad (2)$$

where, l_f is the change in fixed length, l_s is the change in stretched length, l_r is the change in recovery length and l_i is the change in initial gauge length.

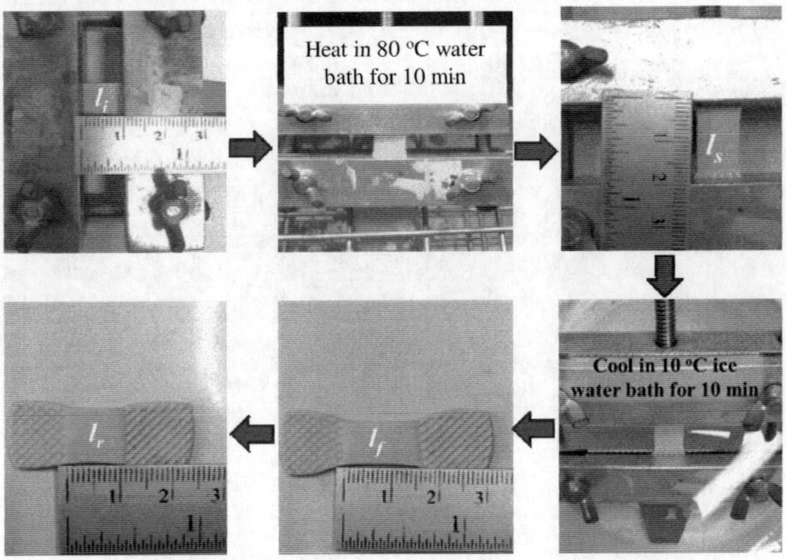

Figure 1. Shape memory cycle tested with an adjustable clamps.

3. Results and discussion

3.1 Effect of blend ratio on properties

The SMPs based TPNR have been successfully prepared with various weight ratios of NR/BCO (30/70, 40/60 and 60/40 compositions). Modulus, tensile strength, hardness and imprint images of the blends are depicted in Table 1. A decreasing trend in modulus, tensile strength and hardness is observed for the blends by increasing the NR contents in the SMPs based TPNR. This may be attributed to the increase in soft phase of NR and reduce the crystalline phases into the SMPs based TPNR. Higher NR content caused better flexibility between polymer chains. Therefore, the surface of the SMP is changed easily by applied force. Furthermore, the shape fixity efficiency (F) increased by adding NR in all the blend ratios. The NR/BCO with 40/60 provided highest shape recovery efficiency (R) (Figure 2) which is much better than SPMs prepared from previous technique [1]. The shape fixity and recovery efficiency agrees very well with the imprint of fingerprint as shown in Table 1. The blends based NR provides better 3D-imprint (fingerprint) images, while the imprint (fingerprint) without NR (NR/BCO = 0/100) has not been observed. This may be attributed by the SMPs based TPNR prepared from NR and BCO with different characters. The crystallinity of TPNR at above 80 °C which acts as pseudo-crosslink point (rigid point) in thermoplastic (BCO) has been disappeared on heating and hence under compressing force condition the material becomes soft as clay. This is due to both phases of NR and BCO in the blend (TPNR) can easily passed through (without physical or pseudo-crosslink) each other at above 80 °C. On recurring to low temperature, the soft phase (NR) which consists of higher loss modulus (at high temperature) recovered back to original shape immediately as compared to the recovery of crystalline phase in thermoplastic as shown in Figure 3. Therefore, the imprint of the blends with NR is immediately fixed while the imprint on BCO without NR is not obtained. From these studies, the blend 40/60 ratio of NR/BCO has been selected for further investigation.

Table 1. Properties of SMPs based TPNRs prepared from different NR/BCO blends.

Properties	NR/BCO			
	0/100	30/70	40/60	60/40
100% modulus (MPa)	1.83±0.14	1.43±0.05	1.23±0.09	0.89±0.02
Tensile strength (MPa)	7.61±0.53	8.05±0.72	5.99±0.45	5.75±0.30
Hardness (Shore A)	59.47±0.55	52.60±0.55	47.67±0.51	38.67±0.51
Imprint- fingerprint image			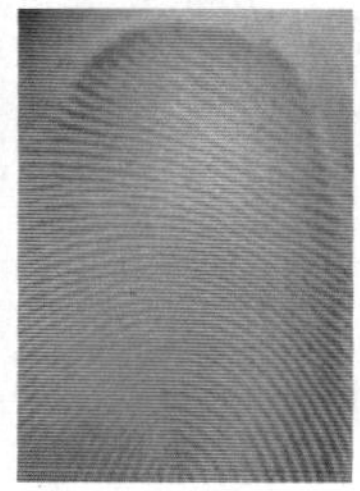	

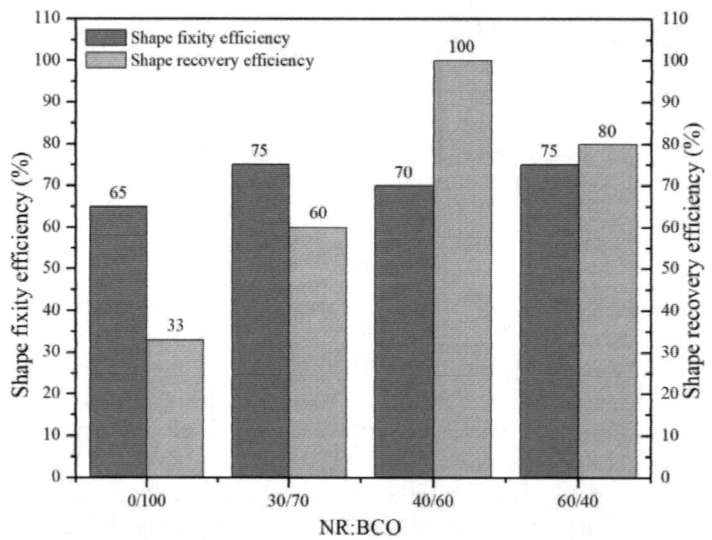

Figure 2. Shape fixity and shape recovery efficiencies of SMPs based TPNR with different NR:BCO ratios.

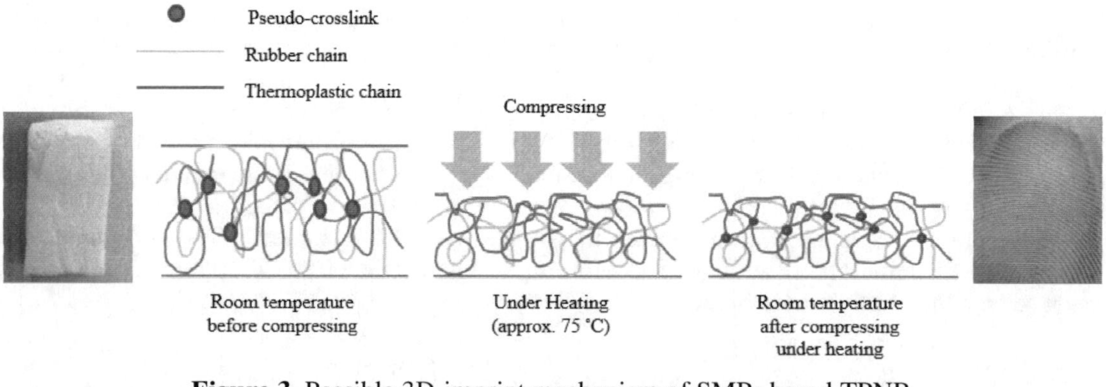

Figure 3. Possible 3D-imprint mechanism of SMPs based TPNR.

3.2 Comparison with previous techniques

The properties of newly developed SMPs based TPNR have been compared with silicone (general purpose and forensic grade) and the values are depicted in Table 2. It is found that SMPs based TPNR exhibits better performance in terms of shorter working time and higher mechanical properties on compared to both the silicones. Table 3 shows the quality of imprints collected from different materials. It is seen that the SMPs material developed from TPNR shows fine imprint with improved performance and it is easy to handle on compared to the other silicone-based materials. Furthermore, the skill of evidence collector is much important when using silicone. The ingredients must be carefully mixed with suitable proportions before applying. In contrast, the evidence collecting with SMPs based TPNR requires only a heating set up to raise the temperature above 75 °C and press on the object until it gets cool down. The risk of destroying evidence collected from the crime scene can be minimized by using this newly introduced method. This technology has already been applied in the real crime scene investigation by Thai Forensic Police and the satisfaction survey is also studied. The satisfaction level has been analyzed by collecting the data from Forensic Department of Thailand. The survey on the level of satisfaction among the methods shows a great extent of satisfaction.

Table 2. Comparison of the properties of different materials.

Properties	Silicone (General grade)	Silicone (Forensic grade)	SMPs based TPNR
Working time	6-12 h	10 min	2 min
100% Modulus (MPa)	0.44	0.41	0.90
Tensile strength (MPa)	2.60	1.95	3.12
Elongation at beak (%)	325	400	1,133
Hardness (Shore A)	25	31	29

Table 3. Comparison of imprint prepared from different techniques.

Object types	Marking object	Silicone (General grade)	Silicone (Forensic grade)	SMPs based TPNR
Cracked object	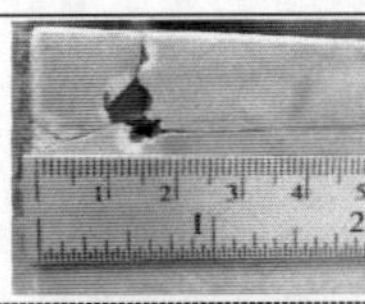	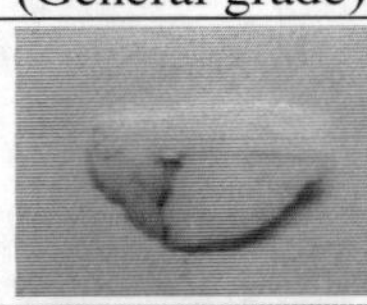		
Bullet marks	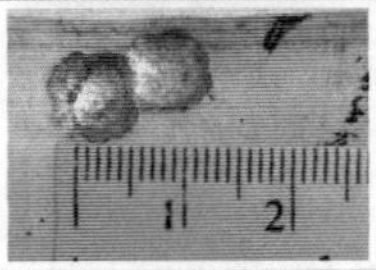	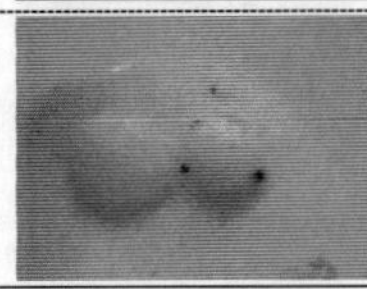		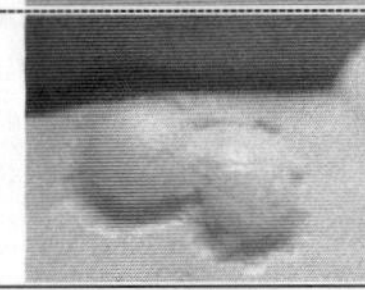

4. Conclusion

Shape memory polymer based thermoplastic natural rubber was successfully prepared by blending of NR and BCO. All blends provided high shape fixity and shape recovery efficiency than that of the one with blending. The blend of NR/BCO with 40/60 provided finest 3D-imprint The newly developed SMPs based TPNR was found to be an excellent 3D-imprint material for collecting evidences from crime scenes. Better performance and advantages that includes mechanical properties, shelf life, ease of processing and lower cost makes the newly introduced material more attractive in the field of forensic science.

Acknowledgment
The authors gratefully acknowledge the financial support by a grant from National Research Council of Thailand (NRCT), (contact number SCI6200069M), Thailand and we would like to thank the Rubber Product Development Group, Faculty of Science, Prince of Songkla University for providing the facilities.

References

[1] J. Wee, A. B. Chai and J-H. Ho, "Fabrication of shape memory natural rubber using palmitic acid", *Journal of King Saud University – Science*, Vol. 29, pp. 494–501, 2017. DOI.org/10.1016/j.jksus.2017.09.003.
[2] G. Tsukada, M. Tokuda and M. Torii, "Temperature triggered shape memory effect of trans polyisoprene-based polymer" *Journal of Endodontics*. Vol. 40, pp. 1658–1662. 2014. DOI: 10.1016/j.joen.2014.05.003.
[3] H. Feil, Y.H. Bae, J. Feijen and S.W. Kim, "Mutual influence of Ph and temperature on the swelling of ionizable and thermosensitive hydrogels". *Macromolecules*. Vol. 25, pp. 5528–5530. 1992. DOI: 10.1021/ma00046a063.
[4] K.M. Lee, H. Koerner, R.A. Vaia, T. J. Bunning, T.J. White, "Light-activated shape memory of glassy, azobenzene liquid crystalline polymer networks". *Soft Matter*. Vol. 7, pp. 4318–4324. 2011. DOI: 10.1039/C1SM00004G.
[5] F. Pilate, A. Toncheva, P. Dubois, J.-M. Raquez, "Shape-memory polymers for multiple applications in the materials world". *European Polymer Journal*. Vol. 80, pp. 268-294. 2016. DOI.org/10.1016/j.eurpolymj.2016.05.004.
[6] E. Kalkornsurapranee, T. Sengsuk, W. Kaewsakul, B. Sawatarmara, and K. Chawengsaksopak, 3D tracing thermoplastic natural rubber for forensic applications, Thailand Patent 11992 (2016)
[7] N.R., Brostowitz, R.A., Weiss and K.A., Cavicchi, "Facile fabrication of a shape memory polymer by swelling cross-linked natural rubber with stearic acid". *ACS Macro Letter*. Vol. 3, pp. 374–377. 2014. DOI: 10.1021/mz500131r.

Enhancing Properties of Cured NR/PVA Blends Using Glutaraldehyde as a Crosslinking Agent: Effect of Nano-Clay Loading

Nussana Lehman [1*], Ladawan Songtipya [2], Jobish Johns[3], Ekwipoo Kalkornsurapranee[1*]
[1] Department of Materials Science and Technology, Faculty of Science, Prince of Songkla University, Hat-Yai, Thailand
[2] Department of Material Product Technology, Faculty of Agro-Industry, Prince of Songkla University, Hat-Yai, Thailand
[3] R&D Center, Department of Physics, Rajarajeswari College of Engineering, Bangalore, India

nus_na1708@hotmail.com
ekwipoo.k@psu.ac.th

Abstract. A novel and inexpensive method to vulcanize natural rubber (NR) was successfully developed and reported in our previous work. NR can be easily vulcanized using glutaraldehyde (GA) at low temperature without adding any specific activators. However, there is some limitation (i.e., poor film formation and properties) to apply as many product items. Therefore, the main objective of this work is to improve the properties and film formation of this cured system. The effect of loading levels of nano-clay on the properties of GA cured NR/PVA blends has been investigated. Mechanical, thermal and morphological properties were characterized. From the overall characterization, cured NR/PVA blends with 1 phr of nano-clay exhibited the highest tensile strength, elongation at break and hardness of 8.02 MPa, 690% and 62.4 shore A, respectively. Enhancing in thermal stability of the sample was confirmation by shifting of degradation temperature based TGA technique. The experimental results reveal that the properties and film formation of this system can be developed with the proper additives in the preparation. From obtained knowledge, it is interesting due to it can be applied for various industrial applications with eases of producing at low temperature and cost.

Keywords: Natural rubber, Nano-clay, Glutaraldehyde, Cured NR/PVA blends, Vulcanization.

1. Introduction

Natural rubber (NR) is a natural material which is derived from typical plants. NR is normally known as *cis*-1,4-poly (isoprene) [1]. It is an unsaturated elastomer with many superior properties such as high strength, outstanding resilience, and high elongation at break [2]. In addition, these properties can be improved by different methods such as grafting, blending, addition of fillers etc. Blending is a mechanical mixture of more than two polymers and it is a convenient route for the modification of properties to meet specific need and reasonable demands of the polymers [3]. The common way is blending with other latex or water soluble polymer to obtain better performance or fabrications. Polyvinyl alcohol has good water solubility, film-forming property, adhesive force and emulsification [4]. However, the natural rubber is a non-polar polymer, whereas the PVA is a polar polymer. Therefore, NR/PVA blend could be partial compatibility, leading to low physical properties. Nowadays there is great interest in the development of rubber composites using nano-clay as reinforcing material [5]. The functional properties such as mechanical properties, barrier resistance and flame retardancy can be enhanced considerably at lower loading level of nano clay due to their high aspect ratio [6]. The reinforced rubber depends on several factors: filler particle size, concentration, aspect ratio, dispersion and morphology. All these factors lead to the concept of filler network formation and enhancing the mechanical properties of rubber matrix. It makes the final products less expensive [7]. However, the vulcanization process is very important for rubber applications. Mechanical properties of NR increase during the vulcanization process due to the cross-linking of NR chains [1]. In general, there are many potential vulcanization techniques including sulfur, peroxide, and phenolic resin curing [8]. These curing systems need high temperature together with various chemicals, i.e., activators and accelerators. Natural rubber can be easily vulcanized using glutaraldehyde (GA) at low temperature without adding any specific activators, which is highly favorable to the environment [9]. According to this study, glutaraldehyde initially reacts with ammonia present in the natural rubber latex and generates pentane-1,5-diylidenediamine. It has been shown that, this curing system provides better thermal stability than that of the sulfur cured system with greater ease of processing. However, the GA-cured system exhibits low mechanical properties [10]. The aim of this study is to reinforce nano-clay into NR/PVA blends to improve the properties and stability of the blends using glutaraldehyde as a crosslinking agent.

2. Experimental method

2.1 Materials
Commercial high ammonia latex (HA Latex) with 60% dry rubber content (DRC) was purchased from Chalong Latex Industry, Thailand. Glutaraldehyde (GA) and Polyvinyl alcohol (PVA) were purchased from Boss Official Limited Partnership, Songkhla, Thailand.

2.2 Preparation of nano-clay dispersion
The nano-clay (Cloisite 30B) used in this study was collected from Rockwood Clay Additives, GmbH, Germany. The aqueous dispersion of 30 wt% nano-clay was prepared by conventional ball milling technique for 72 h.

2.3 Preparation of Cured NR with various nano-clay contents using Glutaraldehyde
12.5% Glutaraldehyde (GA) solution was prepared by diluting in distilled water. The ammonia content present in natural rubber was estimated according ASTM D-1076-02 into the mixture of NR/PVA blends with 90/10 weight ratios. Nano-clay was added into the mixture of NR/PVA blends with various loading levels under continuously stirring at room temperature. Glutaraldehyde solution was slowly added into the reinforced latex using continuously stirring at a mole ratio 1:1 between ammonia and glutaraldehyde. The mixture was cast on a glass plate and dried in hot air oven at 50 °C for 24 hours [8]. The samples were introduced for testing and characterization.

2.4 Characterization
The overall chemical composition of the filler was determined by an EDX spectrometer attached to scanning electron microscope (SEM). The specimens were mounted on aluminum stubs using carbon tape and then coated with a thin layer of gold to prevent charging before the observation. Swelling measurement was

IOP Conf. Series: Materials Science and Engineering **553** (2019) 012046 doi:10.1088/1757-899X/553/1/012046

carried out in engine oil as a test liquid at room temperature for 24 h. The degree of swelling was calculated by using Equation (1).

$$\text{Change in weight } (\%)=\frac{W_s\text{-}W_o}{W_o}\times 100 \quad (1)$$

where W_o and W_s are the weight of the specimen before and after immersion in the test liquid, respectively.

The tensile strength was performed according to the ASTM D412 method using dumbbell-shaped test specimens at a crosshead speed of 500 mm/min using a universal testing machine. Indentation hardness was tested using a Shore A durometer according to ASTM D2240. The thermal properties were characterized using thermogravimetric analysis (TGA). A small amount (1–10 mg) of sample was taken in a crucible and placed inside the machine from 40°C to 800°C at a rate of 10°C/min in oxygen atmosphere.

3. Results and discussion

3.1 Scanning Electron Microscopy

The overall chemical composition of the cured NR/PVA blends with various nano-clay contents is identified and qualified by EDX spectrometry and SEM. The SEM image of the cured NR (Figure 1a) reveals a very smooth surface. Figure 1b shows that the nano-clay is dispersed uniformly in the NR matrix. Figure 1c shows the nano-clay particles are poorly dispersed and agglomerated due to the higher content of nano-clay. The presence of overall chemical composition of nano-clay in the NR/PVA blends matrix has been confirmed by EDX analysis (Table 1). EDX profile concludes that skeleton of nano-clay is composed of elements: mainly silicon, second most oxygen, third most aluminum and finally carbon. The amount of these elements is increased with increase in nano-clay content. Especially, the percentage of silicon and aluminum atom increases with increasing the amount of nano-clay into NR/PVA blends (Figure 2). In addition, the distribution of silicon and aluminum atoms further support the observation of aggregates of nano-clay in the matrix. Better dispersion can be seen in the cured NR with 1 phr nano-clay compared to the composite with 2.5 phr nano-clay. However, the analysis shows an increase in the percentage of carbon atom for all samples. This may be due to the presence of long chains molecules of natural rubber (polyisoprene) in the sample.

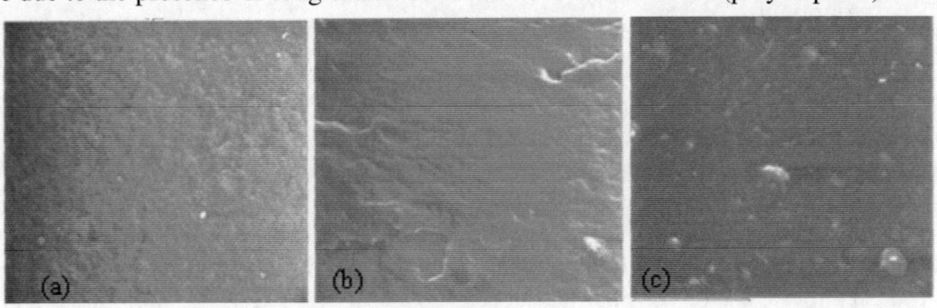

Figure 1. SEM images of cured NR/PVA blends with various nano-clay contents: (a) 0 phr (b) 1 phr and (c) 2.5 phr.

IOP Conf. Series: Materials Science and Engineering **553** (2019) 012046 doi:10.1088/1757-899X/553/1/012046

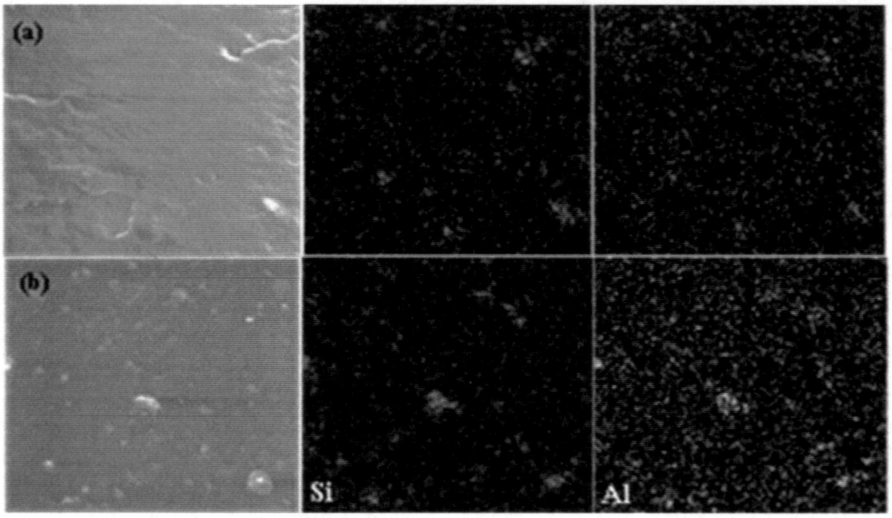

Figure 2. Mapping of the elements found on the image of cured NR/PVA blends with various nano-clay content using GA as crosslinking agent; Si: silicon and Al:aluminum: (a) 1 phr and (b) 2.5 phr.

Table 1. Overall chemical composition obtained by EDX spectrometry.

Elements	0 phr (%)	1 phr (%)	2.5 phr (%)
Carbon (C)	94.0	95.4	94.3
Oxygen (O)	6.0	4.4	5.4
Silicon (Si)	0	0.1	0.3
Aluminum (Al)	0	0.1	0.1

3.2 Swelling Behaviors

Figure 3 shows the degree of swelling of the cured NR/PVA blends with nano-clay for 0, 1 and 2.5 phr in engine oil. It is seen that the degree of swelling of cured NR with 1 phr nano-clay are found to be much less when compared to the NR with 0 phr and 2.5 phr. This can be explained in terms of the composite materials containing layered silicate with a high aspect ratio, it strongly limits the deformation of macromolecules due to a highly efficient stress transfer [11]. In addition, the reduction of swelling upon the addition of layered silicates is due to the enhanced rubber/filler interaction. The reduction in solvent transport can be explained in terms of tortuosity of the path and decreased transport area in the membrane. However, the degree of swelling for cured NR with 2.5 phr nano-clay shows higher values because of the weak interface and poor clay dispersion. As a result, the available free voids within the matrix increase. Therefore, diffusion of solvent molecules through rubbery polymers is easy. This result confirms the fine dispersion of nano-clay in the NR matrix. However, crosslinking reduces the rubber chain mobility and thus reduces the transport of solvent through the rubber. In our previously published article [9], it was reported that the cured NR prepared using glutaraldehyde exhibited better oil resistance properties than of the pure NR. It is seen that glutaraldehyde vulcanizes rubber phase and does not allow the chains to rearrange easily under solvent stress.

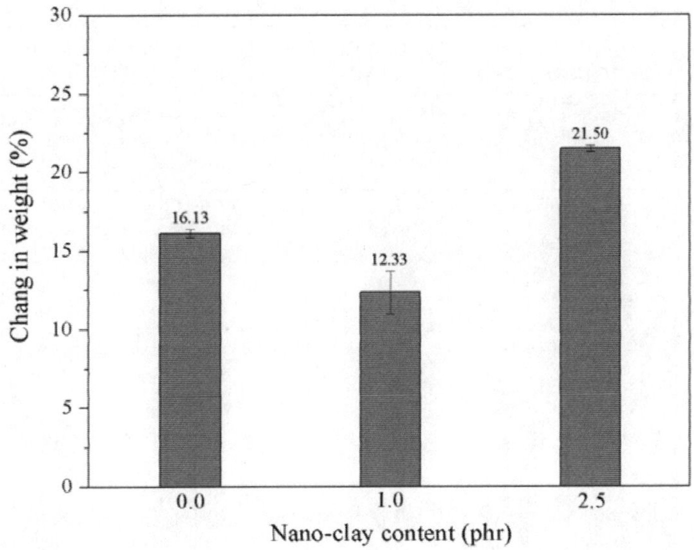

Figure 3. Swelling behavior of cured NR/PVA blends using GA as crosslinking agent with different nano-clay content.

3.3 Mechanical properties

Mechanical properties of the cured rubber with nano-clay varying from 0 to 2.5 phr are listed in Table 2. Improvement in the the properties of NR/PVA blends can be seen by adding nano-clay. The highest tensile strength and elongation at break are observed a at clay loading of 1 phr (Figure 4). This is directly related to the dispersion and distribution of nano-clay as shown in previous part. For higher loading level (2.5 phr), the failure of sample may be due to the agglomerations and the formation of filler–filler networks in NR/PVA blends matrix. The proposed model of interaction between matrix and nano fillers are shown in Figure 7. Incorporation of filler into the NR/PVA blends matrix reduces the elasticity of rubber chains and the resulting composite becomes more rigid [12]. Rigidity of the material leads to increase in modulus and hardness with increasing the percentage of nano-clay (Figure 5). In addition, it might be attributed by the effect of PVA which consists of coarse crystalline phase. Formation of cross-links in PVA and NR phases when it is crosslinked with glutaraldehyde enhances the stability. Crosslinking of natural rubber is occured by the reaction between rubber and pentane 1,5-diylidenediamine as shown in Figure 7. It leads to the formation of three-dimensional network and restricts the movement of rubber chains. From the overall results, it is found that the sample with 1 phr nano-clay loading level exhibited higher mechanical properties. It is clearly confirmed from the fine and uniform dispersion of nano-clay in the rubber matrix by SEM image (Figure 1b and Figure 7).

Table 2. Mechanical properties of cured NR with different nano-clay content.

Amount of nano-clay (phr)	100% modulus (MPa)	Tensile strength (MPa)	Elongation at break (%)	Hardness (Shore A)
0	2.37	4.46	438	61.20
1.0	2.51	8.02	690	62.40
2.5	2.64	5.61	503	68.40

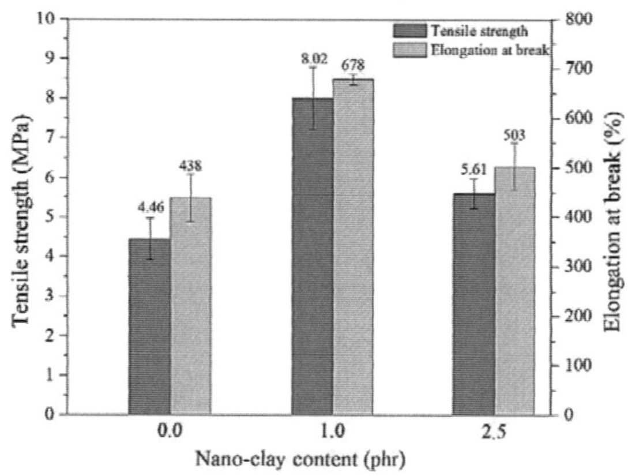

Figure 4. Tensile strength and elongation at break of cured NR/PVA blends using GA as crosslinking agent with different nano-clay content.

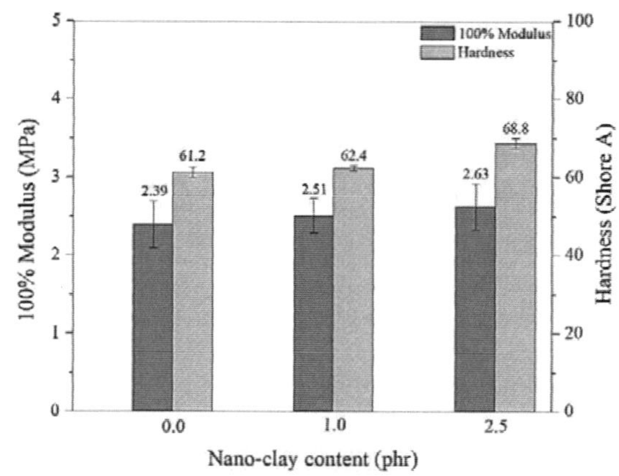

Figure 5. 100% modulus and hardness of cured NR/PVA blends using GA as crosslinking agent with different nano-clay content

3.4 Thermal properties

Results of TGA for the composites with various percentages of nano clay are shown in Figure 6. The onset degradation temperature of cured NR/PVA blends with 2.5 phr nano-clay contents is higher (245 °C) when compared to composites with 0 and 1 phr (230 and 233 °C). The results show that the thermal stability increases with the addition of nano-clay. Due to the barrier effect from nano-clay may hinder diffusion of small molecules generated during thermal decomposition. Lengthened paths of diffused low molecular gases in the polymeric matrix from the uniform distribution of the layered silicates delay the degradation of the cured NR/PVA blends [13]. In addition to this, vulcanization of rubber phase with glutaraldehyde in the presence of ammonia increases the thermal stability. Newly generated three -dimensional networks between rubber molecules is also a reason for the enhancement in the thermal stability of rubber. Reaction between rubber and pentane 1,5-diylidenediamine molecules crosslinks natural rubber (Figure 7).

IOP Conf. Series: Materials Science and Engineering **553** (2019) 012046 doi:10.1088/1757-899X/553/1/012046

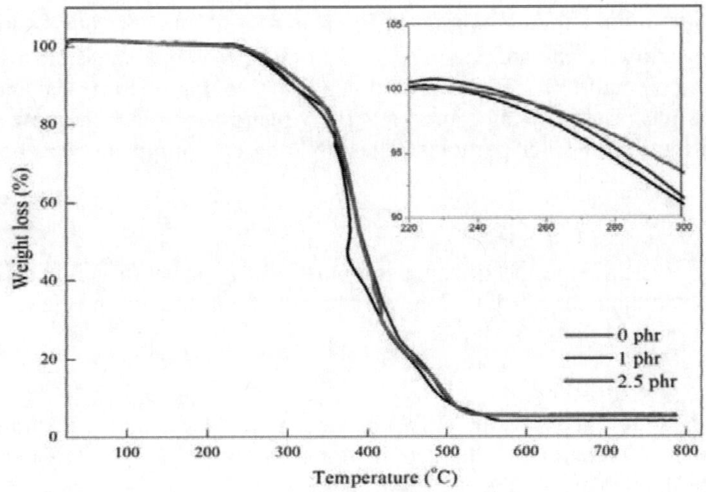

Figure 6. TGA curves for cured NR/PVA blends using GA as crosslinking agent with various amounts of nano-clay.

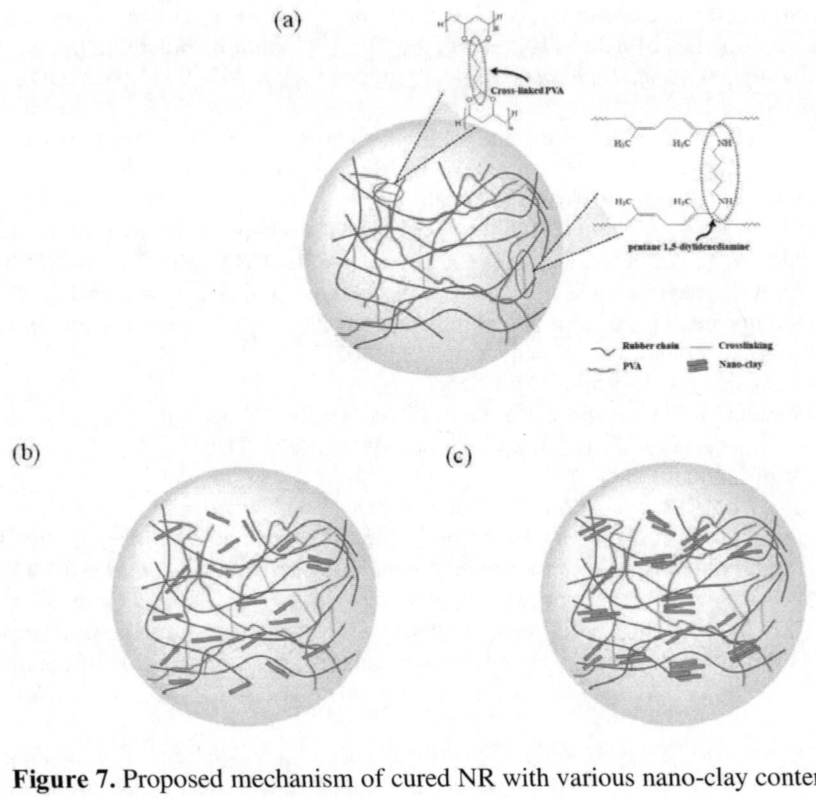

Figure 7. Proposed mechanism of cured NR with various nano-clay content using GA as crosslinking agent (a) 0 phr, (b) 1 phr and (c) 2.5 phr.

4. Conclusion

Cured NR/PVA blends with various nano-clay contents using glutaraldehyde as curing agent was successfully prepared at low temperature. Dispersion of nano-clay was studied using SEM and EDX. Addtion of nano-clay into cured NR samples provided better swelling, mechanical and thermal properties. However, loading of 1 phr nano-clay into cured NR/PVA blends exhibited the best properties due to the fine and uniform distribution of filler particles. These NR based composites can be applied for various industrial applications.

Acknowledgement

The authors gratefully acknowledge the financial support by a grant from National Research Council of Thailand (NRCT), (contact number SIT590171a), Graduate School, faculty of science and Rubber Product Development Group, Prince of Songkla University.

References

[1] O. Sivaraman, N. Ghosh, S. Gayathri, P. Sudhakara, S. K. Misra, and J. Jayaramudu, "Natural Rubber Nanoblends: Preparation, Characterization and Applications," *Polymer and Composite Materials,* 2017. DOI 10.1007/978-3-319-48720-5_2

[2] T. A. Dung *et al.*, "Modification of Vietnam natural rubber via graft copolymerization with styrene," *J. Braz. Chem. Soc.*, vol. 28, no. 4, pp. 669–675, 2017. DOI:10.21577/0103-5053.20160217

[3] M. A. Haque, M. U. Ahmad, F. Akhtar, N. C. Dafader, and M. E. Haque, "Improvement of physicochemical properties of natural rubber and polyvinyl alcohol blends by radiation vulcanization," *Polym. - Plast. Technol. Eng.*, vol. 46, no. 10, pp. 927–931, 2007. DOI: 10.1080/03602550701280323

[4] K. Chen and R. Wang, "The Mechanical and Aging Properties of the Natural Rubber/Polyvinyl Alcohol (NR/PVA) Blends," *Adv. Mater. Res.*, vol. 1073–1076, pp. 12–15, 2014. DOI:10.4028/www.scientific.net/AMR.1073-1076.12

[5] S. Mathew and S. Varghese, "Natural Rubber Latex-based Nanocomposites with Layered Silicates," *J. Rubb. Res.*, vol. 8, no. 1, pp. 1–15, 2016. DOI: 10.13140/RG.2.1.2189.2726

[6] A. Amarasiri, U. N. Ratnayake, U. K. De Silva, S. Walpalage, and S. Siriwardene, "Natural rubber latex-clay nanocomposite: Use of montmorillonite clay as an alternative for conventional $CaCO_3$," *J. Natl. Sci. Found. Sri Lanka*, vol. 41, no. 4, pp. 293–302, 2013. DOI: http://dx.doi.org/10.4038/jnsfsr.v41i4.6258

[7] T. Pojanavaraphan and R. Magaraphan, "Prevulcanized natural rubber latex/clay aerogel nanocomposites," *Eur. Polym. J.*, vol. 44, no. 7, pp. 1968–1977, 2008. DOI:10.1016/j.eurpolymj.2008.04.039

[8] E. Kalkornsurapranee, W. Yung-Aoon, B. Thongnuanchan, A. Thitithammawong, C. Nakason, and J. Johns, "Influence of grafting content on the properties of cured natural rubber grafted with PMMAs using glutaraldehyde as a cross-linking agent," *Adv. Polym. Technol.*, 2017. DOI 10.1002/adv.21806

[9] A. Johns, M. P. S. Aan, J. Johns, M. S. Bhagyashekar, C. Nakason, and E. Kalkornsurapranee, "Optimization study of ammonia and glutaraldehyde contents on vulcanization of natural rubber latex," *Iran. Polym. J. English Ed.*, vol. 24, no. 11, pp. 901–909, 2015. DOI 10.1007/s13726-015-0382-8

[10] E. Kalkornsurapranee, W. Yung-Aoon, L. Songtipya, and J. Johns, "Effect of processing parameters on the vulcanisation of natural rubber using glutaraldehyde," *Plast. Rubber Compos.*, vol. 46, no. 6, pp. 258–265, 2017. DOI: 10.1080/14658011.2017.1323610

[11] Kong, Lingxue, Chen, Ying, Peng, Zhen and Li, Puwang, "Latex-based nanocomposites," *Progress in Polymer Nanocomposite Research.*, Hauppauge, N.Y., pp.83-104, 2008. ISBN: 978-1-60456-484-6
[12] C. M. Ewulonu, C. M. Obele, I. O. Arukalam, and S. R. Odera, "Effects of local clay incorporation technique on the mechanical properties of natural rubber vulcanizates," vol. 4, no. 3, pp. 307–320, 2015. DOI: 10.6088.ijaser.04031
[13] C. Zhang and J. Wang, "Natural rubber/dendrimer modified montmorillonite nanocomposites: Mechanical and flame-retardant properties," *Materials (Basel).*, vol. 11, no. 1, 2017. DOI: 10.3390/ma11010041

Poly(lactic acid)-based Composites Incorporated with Spent Coffee Ground and Tea Leave for Food Packaging Application: A Waste to Wealth

Ladawan Songtipya[1], Thitirat Limchu[1], Suphatchara Phuttharak[1], Ponusa Songtipya[1], and Ekwipoo Kalkornsurapranee[2]

[1] Department of Material Product Technology, Faculty of Agro-Industry, Prince of Songkla University, Hat-Yai, Songkhla, 90112, Thailand
[2] Department of Material Science and Technology, Faculty of Science, Prince of Songkla University, Hat-Yai, Songkhla, 90112, Thailand

ladawan.so@psu.ac.th

Abstract. Polymer composites of polylactic acid (PLA) and polybutylene adipate terephthalate (PBAT) incorporated with spent coffee grounds (SCG) and tea leave (TL) were prepared by two-roll mill mixer. 4,4-methylene diphenyl diisocyanate (MDI), toluene 2,4-diisocyanate (TDI), and bis[3-(triethoxysilyl) propyl] tetrasulfide (TESPT) were used as coupling agents. The influences of coupling agent types, coupling agent content, and weight ratios of polymer to filler on the mechanical properties, melt flow index, and overall migration (OM) of the composites were studied. The results showed that MDI and TDI had better performance compared to TESPT for both tensile and elongation at break. The tensile strength and elongation at break of PLA+PBAT/SCG composites with weight ratio of polymer to filler = 70/30 increased from 19.6 MPa to ~23.0 to 25.0 MPa, and 6.6 to ~10.0%, respectively when using these coupling agents (MDI and TDI) of 3 g/100 g polymers. Moreover, the addition of MDI and TDI greatly increased the viscosity of the melted composites (4-fold), while TESPT made the viscosity decrease. However, the mechanical properties of the composites decreased drastically with increasing SCG proportion. Compared to PLA+PBAT/SCG, interfacial adhesion of PLA+PBAT/TL was higher confirming by tensile strength and SEM images. However, there was no significant difference between PLA+PBAT/TL and PLA+PBAT/SCG composites in terms of elongation at break, impact strength and melt flow index. The OM of PLA+PBAT/SCG and PLA+PBAT/TL composites with coupling agents were in the range of ~0.03–0.28 mg/dm^2 when using 3% acetic acid and 10% ethanol as food simulants, which not excess the migration limit (10 mg/dm^2) according to Food Contact Materials EU No. 10/2011 legislation. It means that they might be safe for use as food contact materials for packaging and containers.

Keywords: Polylactic acid, Polybutylene adipate terephthalate, Coupling agent, Spent coffee ground, Tea Leave, Migration

1. Introduction

Plastics are used in many types of food packaging and containers in order to protect foods from damage, provide food safety and extend the shelf-life of foods. However, majority of plastics used to pack food normally dumped as a waste and it leads to serious environmental pollution. In order to reduce the pollution from plastic waste, government has been promoted to implement on the use biodegradable polymers such as polylactic acid, cellulose esters, polyhydric butyl ether polyhydroxyl butyrates and starch etc. to replace the synthetic polymers. This is due to their ability to break down into non-toxic gas, water, biomass and inorganic chemicals after the consumption of food. However, plastics manufactured from petroleum-based derivatives is not a suitable option for the preparation of biodegradable plastics due to its expense. Normally, a reduction of material cost can be achieved by the addition of low-cost extension fillers such as clay, calcium carbonate and talc [6]. These systems normally provide a higher weight of composites compared to natural plant fillers.

Recently the properties of micro- and nanocomposites prepared from several polymers (e.g., polypropylene, polyvinyl alcohol, polypropylene carbonate, polylactic acid, and poly(butylene adipate-co-terephthalate) [1, 4–5, 8–9, 11] and natural fillers (spent coffee grounds (SCG) [1, 4–5, 8–9] and tea leave (TL)) [11] have been reported. From these works, the effect of filler content (filler content is in a range of 5–30 wt%) [1, 4] particle size distribution of coffee particles [8], and hydrophobic treatment on SCG powder [5] on the thermal, rheological, physical and mechanical properties of the composites were investigated. Moreover, some researchers also reported that one possible application of the composite incorporated with SCG is a food container or food packaging [3]. However, the safety of food contact materials has not been evaluated and reported in the examination documents. But in fact, the chemical migration of composite materials should be concerned due to some small size chemicals like coupling agents that have been used to increase the compatibility of immiscible polymer blends such as 4,4-methylene diphenyl diisocyanate (MDI) [1], PP-g-MA [10], silane, and styrene-ethylene- butene-styrene-graft-maleic anhydride [4] might be migrated from the materials into foods.

From these reasons, in the present work, spent coffee ground (SCG) and tea leaves (TL) have been employed as fillers to prepare the composite of polylactic acid (PLA)+polybutylene adipate terephthalate (PBAT) (PLA/PBAT = 70/30) to make use of large quantities of waste from coffee shops or coffee industrials. Overall migration (OM) of the composites with various coupling agents including 4,4-methylene diphenyl diisocyanate (MDI), toluene 2,4-diisocyanate (TDI) and bis[3-(triethoxysilyl) propyl] tetrasulfide (TESPT) and several content (0, 3, 5, and 7 g/ 100 g polymers) for the weight ratios of polymer to filler (80/20, 70/30, and 60/40) in food simulants have been tested according to Food Contact Materials EU No. 10/2011 legislation in order to determine and compare the quantity of the chemical substances that migrate from the composite samples to the food simulants. Moreover, the effect of blend composition and coupling agent on the mechanical properties and melt flow index have been studied in detail.

2. Experimental methods

2.1 Materials

Polylactic acid (PLA) and Polybutylene adipate terephthalate (PBAT) were purchased from Nature-Works LLC, USA. 4,4-methylene diphenyl diisocyanate (MDI), toluene 2,4-diisocyanate (TDI) and bis[3-(triethoxysilyl) propyl] tetrasulfide (TESPT) were used as coupling agent were supplied by Sigma-Aldrich. Spent coffee ground (SCG) and tea leaves (TL) were obtained from the local coffee shops in Songkhla province, Thailand.

2.2 Methods

The SCG and TL were first sieved through a 35 mesh screen and dried in a hot-air oven at 50 ºC for 48 h. Dried fillers were mixed along with PLA/PBAT (70/30 by weight) and coupling agent using a two-roll mill (150 ºC for 5 min, nip = 0.5 mm.). The resulting composites were then suddenly compressed by compression molding at 120 ºC for 10 sec. The test specimens were stored under room temperature with 50±5 % RH for 5 days before test. The samples were characterized according to the international standards.

2.3 Morphology
The surface morphology of the composites was studied using scanning electron microscope (SEM; Quanta 400, FEI Company, Hillsboro, OR). The samples were first cracked under liquid nitrogen and then coated with gold. The SEM was performed at an accelerating voltage of 20 kV to observe the fractured surface.

2.4 Melt flow index
The melt flow index machine (Charoen TUT, MFR-2, Thailand) was used to measure the melt flow index according to ISO 1133 under the temperature of 160 °C for 10 min. Experiment was performed in triplicate.

2.5 Mechanical properties
Tensile strength was performed using the Hounsfield Tensometer model H 10 KS (Hounsfield Test Equipment Co. UK) at a crosshead speed of 1 mm/min. Impact strength was tested by impact tester (Charoen TUT, IMP-25J, Thailand) according to ISO 8256. Five specimens were exploited and the average value of $p<0.05$ was considered as significant.

2.6 Overall Migration
The composite sheets (thickness = ~ 1 mm) were cut into small pieces of 1×1 cm for overall migration studies. The experiment was performed with 200 mL of food simulants (3% acetic acid and ethanol) by total immersion method under 70 °C for 2 h according to Regulation (EU) No 10/2011 (surface area of test sample = 100 cm^2). After incubation, food simulants were poured into an evaporated dish and evaporated by using water bath at 90–95 °C. The evaporated dishes were placed again in a hot air oven at 150 °C for 2 h to ensure the food simulants dry completely. The mass of the residue was determined by subtracting the original weight of the dish from the weight of dish and residue by using Equation 1.

$$OM\ (mg/dm^2) = \frac{(w_b - w_s)}{S} \times 1000 \quad (1)$$

Where, OM is the overall migration into the food simulant. W_s is the weight of the residue (g) from the test specimen after evaporation. W_b is the weight of residue (g) from the food simulant (blank). S is the surface area of the test specimen in contact with foodstuffs.

3. Results and Discussion

3.1 Effect of coupling agent types
The mechanical properties of PLA+PBAT/SCG composites (weight ratios of polymer to filler = 70/30) with different coupling agents (TDI, MDI, and TESPT, 3 g/100 g polymers) are measured and compared the composites without coupling agent as shown in Table 1. It is seen that the composite with MDI and TDI gave better performance in term of tensile strength and elongation at break while TESPT provides lower tensile strength. The addition of MDI and TDI enhances the tensile strength from 18.7 MPa to 23.5 MPa and 25.2 MPa, respectively owing to the addition of coupling agent induced an interaction between isocyanate group of coupling agents and hydroxyl group of the filler which have cellulose, hemicellulose, and lignin as a composition [1]. Therefore, the interaction between polymer and SCG phases are improved and showed smoother surface as shown in SEM image (Figure 1. (a-d)). MFI value of the composite of PLA+PBAT/SCG + MDI is found to be 1.1 g/10 min as same as the composite of PLA+PBAT/SCG+TDI which is lower than the composite of PLA+PBAT/SCG (4.4 g/10 min) as shown Table 1. It means that the viscosity of composites was increased when using MDI and TDI as the coupling agent which is caused by the increase of intermolecular interaction forces and limitation of chain mobility [4]. Whereas, the composite of PLA+PBAT/SCG+TESPT had higher MFI up to 13.4 g/10 min (Table 1.). This might be attributed to the plasticization effect caused by TESPT in the polymer molecules. Interfacial adhesion between the

phase of polymers and hydrophilic fillers cannot be improved by using TESPT. However, the impact strength shows no significant change for all composite samples.

Table 1. Mechanical properties and melt flow index of PLA+PBAT/SCG composites using different coupling agents (3 g/100 g polymers) compared to the composites without coupling agents.

Samples	Mechanical Properties			Melt Flow Index (g/10 min)
	Tensile Strength (MPa)	Elongation at break (%)	Impact Strength (J/m²)	
PLA+PBAT/SCG	19.6 ± 0.2[b]	6.6 ± 0.8[b]	81.2 ± 0.2[b]	4.4 ± 0.3[b]
PLA+PBAT/SCG + TESPT	18.7 ± 0.2[b]	4.6 ± 0.7[c]	81.7 ± 0.3[b]	13.4 ± 0.2[a]
PLA+PBAT/SCG + MDI	23.5 ± 0.7[a]	10.9 ± 0.4[a]	82.7 ± 0.2[ab]	1.1 ± 0.1[c]
PLA+PBAT/SCG + TDI	25.2 ± 0.5[a]	10.6 ± 0.5[a]	81.3 ± 0.1[b]	1.1 ± 0.1[c]

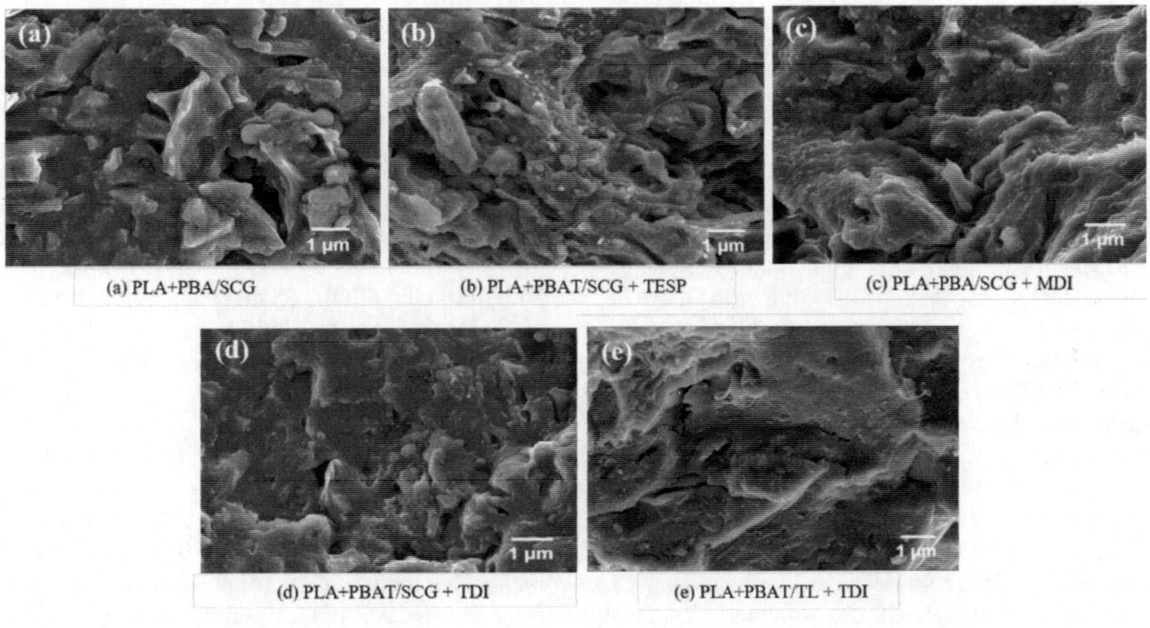

Figure 1. SEM micrographs of PLA+PBAT/SCG composites with and without coupling agents: (a) PLA+PBAT/SCG, (b) PLA+PBAT/SCG + TESPT, (c) PLA+PBAT/SCG +MDI, (d) PLA+PBAT/SCG + TDI compared to (e) PLA+PBAT/TL + TDI composite.

OM of PLA+PBAT/SCG composites with and without coupling agents are shown in Figure 2. The results showed that OM values from all composite samples in all food simulants (3% acetic and 10% ethanol) are lower than the maximum limit (10 mg/dm²), which followed according to the EU regulation. OM of PLA+PBAT/SCG + TESPT composites was higher than that of other composites containing TDI and MDI. This confirms that the chemical bonds or chemical interaction has no exist between TESPT and other molecules. On the other hand, the addition of MDI and TDI leads to reduce the OM might be due to the favourable interaction between coupling agent and SCG changes the chemical nature of small molecules in polymer and also filler components. Therefore, the chemicals cannot be diffused easily through the polymer composites into food simulants. The OM of almost all the composites immersed in ethanol shows significantly higher than water, which can be explained by the affinity of the chemical substances in the SCG and coupling agents [2].

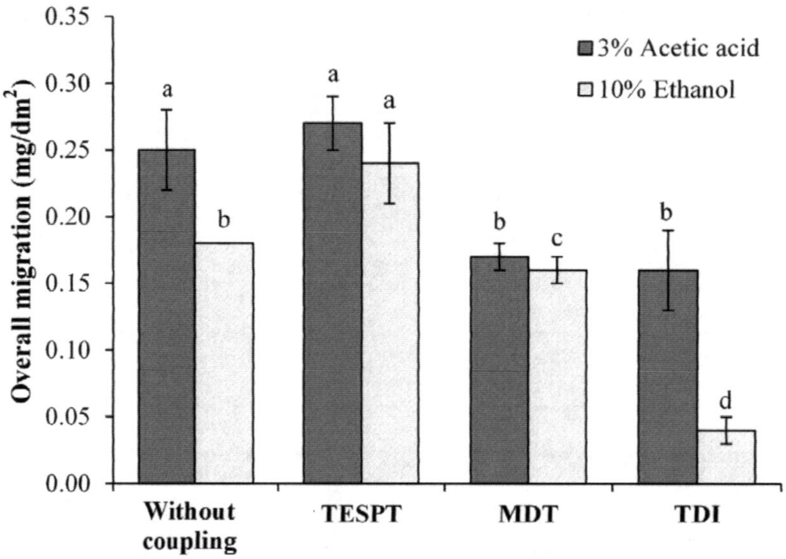

Figure 2. OM of PLA+PBAT/SCG composites with and without coupling agents. 3% acetic and 10% ethanol were used as food simulants.

3.2 Effect of weight ratios of polymer to filler

The effect of weight ratios of polymer to filler (i.e., 80/20, 70/30, and 60/40) on mechanical properties and MFI are depicted in Table 2. TDI is used as a coupling agent with 3 g/ 100 g polymers for all formulation of composites because of its better performance. Tensile strength, elongation at break and impact strength values are decreased with increasing filler (SCG) content from 20 to 40 wt%. Reduction in the tensile properties on increasing the percentage of filler can be explained based on the higher phase separation and weak compatibility between the hydrophobic polymer and the hydrophilic SCG particles [11]. This results correspond well with the reported data. Baek et al. reported decreasing tensile strength with increasing SCG filler loading from 10 to 40 wt% [1]. Essabir et al. also found that tensile strength slightly decreases with increasing SCG from 10 to 20 wt% [4]. Furthermore, flow behavior (MFI) is also decreased on increasing SCG content. This reduction is probably because chain mobility interception of the polymer was occurred by SCG particles during melt state [4] and solid/solid friction between filler particles [7]. The OM of composites in this part has not been investigated due to the TDI content present in the same level of all samples.

Table 2. Mechanical properties and melt flow index of PLA+PBAT/SCG composites with different weight ratio of polymer to filler (80/20, 70/30, and 60/40).

Samples	Weight ratio (g)	Mechanical Properties			Melt Flow Index (g/10 min)
		Tensile Strength (MPa)	Elongation at break (%)	Impact Strength (J/m^2)	
PLA+PBAT/SCG + TDI	80/20	27.6 ± 0.1^a	12.2 ± 1.6^a	82.8 ± 0.9^a	2.3 ± 0.2^a
	70/30	25.2 ± 0.5^a	10.6 ± 0.5^a	81.3 ± 0.1^{ab}	1.1 ± 0.1^b
	60/40	20.8 ± 1.2^b	5.7 ± 0.1^b	80.6 ± 0.2^b	1.2 ± 0.1^b

3.3 Effect of coupling agent content

Table 3. shows the mechanical properties and melt flow index (MFI) of the PLA+PBAT/SCG composites with different TDI contents i.e., 0, 3, 5, and 7 g/100 g polymers. In this part, the composite with weight ratio of polymer/filler of 70/30 has been chosen due to its similarity in mechanical properties with the weight ratio of polymer/filler of 80/20 but a higher cost reduction can be achieved. Based on the tensile strength of the composites, compatibility improvement between phases of polymer and SCG can be confirmed. Increases in tensile strength on increasing the content of TDI up to 5 g/100 g polymers. Upon further addition, there is no significant change in tensile strength. On the other hand, the elongation at break and MFI are slightly decreased due to the higher interaction which lowers mobility of polymer chains. However, there is no significant change in the impact strength of all samples. OM values of the composites in both food simulants slightly increased when increasing TDI content, but they were still lower than the standard (see Figure 3.). This is clearly reflected in the amounts of TDI that not reacted with the filler and polymer.

Table 3. Mechanical properties and melt flow index of PLA+PBAT/SCG composites (weight ratios of polymer to filler = 70/30) with different TDI content (3, 5, and 7 g/100 g polymers).

Samples	Coupling agent amounts (g/100 g polymers)	Mechanical Properties			Melt Flow Index (g/10 min)
		Tensile Strength (MPa)	Elongation at break (%)	Impact Strength (J/m^2)	
PLA+PBAT/SCG + TDI	0	19.6 ± 0.2^c	6.6 ± 0.8^b	81.2 ± 0.2^a	4.4 ± 0.3^a
	3	25.2 ± 0.5^{ab}	10.6 ± 0.5^a	81.3 ± 0.1^a	1.1 ± 0.1^b
	5	25.3 ± 0.3^{ab}	7.5 ± 1.0^b	81.2 ± 0.1^a	0.6 ± 0.0^c
	7	26.8 ± 0.6^a	5.8 ± 0.5^c	81.1 ± 0.1^a	0.4 ± 0.1^c

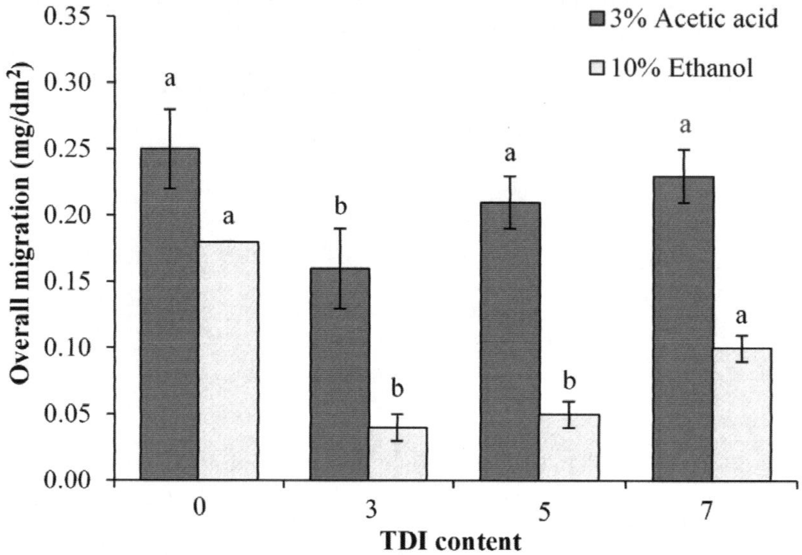

Figure 3. Overall migration of PLA+PBAT/SCG composites with different TDI content varying from 0–7 g/ 100 g polymers.

3.4 Comparison of the composites based on SCG and TL

Table 4. shows the mechanical properties and melt flow index (MFI) of the PLA+PBAT/SCG and PLA+PBAT/TL composites with the weight ratios of polymer to filler of 70/30 and TDI content of 3 g/ 100 g polymers. It was found that the composite based on TL provided better tensile strength than that of the composite of SCG. The result gave good agreement with SEM image as shown in Figure 1(e). This might be attributed by the differences in shape and chemical structure of both materials. TL may consists of higher polyphenol group, which develops the interaction and improvement in the properties of the composites can be observed. The composite of PLA+PBAT/TL shows smoother cracked surface, implying that the better compatibility of TL compared to the composite of SCG. However, elongation at break, impact strength and melt flow index (MFI) showed no significant changes. The OM values of both composites are lower than the migration limit. Higher OM values observed in the composite of PLA+PBAT/TL indicates the chemical substances present in the TL have more affinity with food simulants compared to SCG as shown in Figure 4.

Table 4. Mechanical properties and melt flow index of PLA+PBAT/SCG composite compared to PLA+PBAT/TL composite with constant TDC content (3 g/100 g polymers).

Samples	Mechanical Properties			Melt Flow Index (g/10 min)
	Tensile Strength (MPa)	Elongation at break (%)	Impact Strength (J/m^2)	
PLA+PBAT/SCG + TDI	25.2 ± 0.5^a	10.6 ± 0.5^a	81.3 ± 0.1^a	1.1 ± 0.1^a
PLA+PBAT/TL + TDI	37.4 ±1.7^b	10.5 ± 1.7^a	82.1 ± 0.4^a	1.1 ± 0.1^a

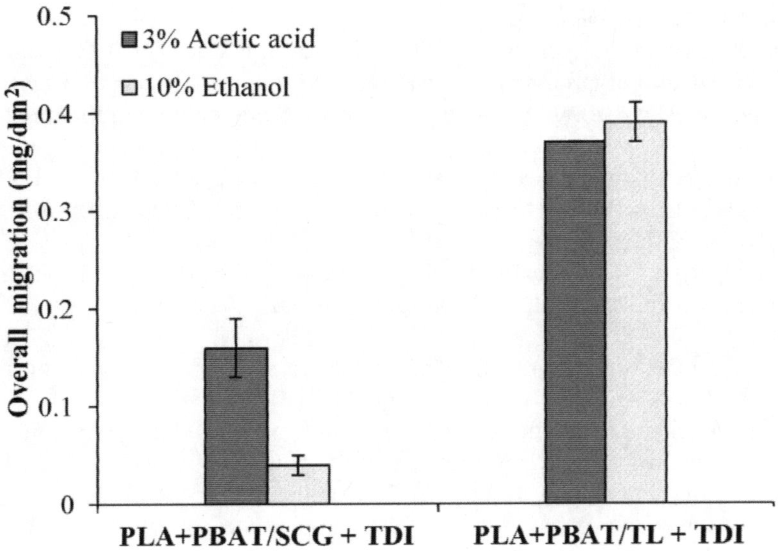

Figure 4. Overall migration of PLA+PBAT/SCG composites compared to PLA+PBAT/TL using TDI (3 g/100 g polymers) as a coupling agent.

4. Conclusions

Phase compatibility of poly (lactic acid)-based composites incorporated with the spent coffee ground (SCG) and tea leave (TL) have been improved using coupling agents. Among the three different coupling agents, TDI exhibited the highest mechanical properties followed by MDI, without coupling agent and TESPT, respectively. The properties of the composites increased with increasing TDI content. However, adding TDI 3 g/100 g polymers exhibited lowest migration value. Furthermore, the composite based on TL showed better tensile strength, while elongation at break, impact strength and melt flow index (MFI) showed no significant change. The results gained from this study exhibited the beneficial advantages (i.e., cost and waste reduction) for applying as a food contact material. However, specific migration of coupling agent should be studied in the future. Finally, it can be concluded that the method involved in this investigation is a waste to wealth approach.

Acknowledgement

The authors gratefully acknowledge financial support by Faculty of Agro-Industry, Prince of Songkla University, Hat-Yai, Songkhla, Thailand.

References

[1] Baek BS, Park JW and Lee BH 2013 Development and application of green composites: using coffee ground and bamboo flour *J. Polym. Environ.* **21** 702–09 doi: 10.1007/s10924-013-0581-3

[2] Coltro L, Pitta, BP, Costa PA, Perez MÂF, Araújo VA and Rodrigues R 2014 Migration of conventional and new plasticizers from PVC films into food simulants: a comparative study *Food Control* **44** 118–29 doi: 10.1016/j.foodcont.2014.03.058

[3] Canavarro V, Alves JL and Rangel B 2017 *Coffee powder reused as a composite material* (Advanced Structured Materials vol 65). ed FM Lucas da Silva (Switzerland: Springer) pp 114–2

[4] Essabir H, Raji M, Laaziz SA, Rodrique D, Bouhfid R and Qaiss AEK 2018 Thermo-mechanical performances of polypropylene biocomposites based on untreated, treated and compatibilized spent coffee grounds *Compos. Part B-Eng.* **149**, 01–11 doi:

10.1016/j.compositesb.2018.05.020
[5] García-García D, Carbonell A, Samper MD, García-Sanoguera D and Balart R 2015 Green composites based on polypropylene matrix and hydrophobized spend coffee ground (SCG) powder. *Compos. Part B-Eng.* **78** 256–65 doi: 10.1016/j.compositesb.2015.03.080
[6] Khan I and Bhat AH 2014 *Micro and nano calcium carbonate filled natural rubber composites and nanocomposites* (Composites and Nanocomposites vol 2) ed S Thomas, HJ Maria, JP Joy, CH Chan and LA Pothenr (The Royal Society of Chemistry) pp 467–87 doi: 10.1039/9781849737654-00467
[7] Kiehl J, Huser J, Bistac S and Delaite C 2012 Influence of fillers content on the viscosity of unsaturated polyester resin/calcium carbonate blends *J. Compos. Mater.* **46** 1–6 doi: 10.1177/0021998311427780
[8] Lee SH and Wang S 2006 Biodegradable polymers/bamboo fiber biocomposite with bio-based coupling agent *J. compos. Part A.* **37** 80–91 doi: 10.1016/j.compositesa.2005.04.015
[9] Moustafa H, Guizani C and Dufresne A 2017 Sustainable biodegradable coffee grounds filler and its effect on the hydrophobicity, mechanical and thermal properties of biodegradable PBAT composites. *J. Appl. Polym. Sci.* **134** 01–11 doi: 10.1002/APP.44498
[10] Wu CS 2015 Renewable resource-based green composites of surface-treated spent coffee grounds and polylactide: Characterisation and biodegradability *Polym. Degrad. Stabil.* **121** 51–59 doi: 10.1016/j.polymdegradstab.2015.08.011
[11] Xia G, Reddy KO, Maheswari CU, Jayaramudu J, Zhang J, Zhang J and Rajulu AV 2015 Preparation and properties of biodegradable spent tea leaf powder/poly (propylene carbonate) composite films *Int. J. Polym. Anal. Ch.* **20** 377–87 doi: 10.1080/1023666X.2015.1019257

IOP Conf. Series: Materials Science and Engineering **553** (2019) 012048 doi:10.1088/1757-899X/553/1/012048

Wood Plastic Composites from Waste Plastic Bags: Formula Optimization Using Extreme Vertices Mixture Design

Y A Swasono[1*], L Adhyati[2], O Ujianto[1], B Ramadhoni[1], W Wuryanti[2]

[1]Center for Polymer Technology, Agency for Assessment and Application of Technology, Building 460, Puspiptek Area, Setu, South Tangerang, Banten, Indonesia 15314

[2]Research Institute for Housing and Human Settlements (RIHHS), Ministry of Public Works and Housing, Cileunyi Wetan, Bandung, Indonesia 40393

yogi.angga@bppt.go.id

Abstract. Waste plastic bags (WPB) from high density polyethylene (HDPE) have been reprocessed for many applications. However, it is limited to be used as wood plastic composites (WPC) for building applications as the decrease of mechanical properties. This research examined the optimal composition of WPB that can be added to WPC and complied with Indonesian Standard (SNI) 8154 – 2015. Samples were produced according to extreme vertices mixture design using twin screw extruder (TSE). The materials were WPB bounded in the range of 0 to 20 wt%, sawdust at 50 to 70 wt%, neat HDPE at 30 to 50 wt%, and HDPE grafting maleic anhydride (PE-g-MA) at 0 to 10 wt%. The compositions were optimized to obtain desired flexural modulus and strength as required on SNI standard. The optimized formula was then reproduced and characterized to analyze its water content, density, swelling, and formaldehyde content. The results showed that the optimum formula was predicted at 3 wt% of WPB, 50 wt% of sawdust, 37 wt% of neat HDPE, and 10 wt% of PE-g-MA, giving 2111 MPa and 27.3 MPa of predicted values that above the minimum requirement for WPC material (2000 MPa for modulus, 20 MPa for strength). Meanwhile, the modulus and strength of verification sample were 2173 MPa and 25.7 MPa. Further analysis on the other properties showed 0.6 ± 0.1% of water content, 1.7 ± 1.4% of swelling, and 0.015 ± 0.000 mg/L of formaldehyde that lower than maximum specifications, and 1.13 ± 0.01 g/cm^3 of density that above minimum the requirement.

1. Introduction

Wood plastic composites (WPC) are composed of wood and plastic, thus reduce wood consumptions. It has capability to replace traditional material such as metal and wood due to mechanical properties improvement [1]. It has been applied in building, such as panels, roofing, fencing and outdoor structure [2]. However, despite its benefit, the use of virgin plastics as matrix should also be reduced as it is non-renewable character. Therefore, utilization of waste plastics is an alternative that can be done.

In asia, Indonesia become one of the second largest producer of waste plastic after China [3]. It was predicted that one of contributor to the huge amount of waste plastic in Indonesia was waste plastic bag (WPB) which was produced almost more than 10 million annually [4] from high density polyethylene (HDPE). This is a big problem to environment and waste management. On the other hand, this is also potential as a material resource for WPC production because of its slight changes on mechanical properties compare to the virgin HDPE [2].

The use of recycled HDPE from post-consumer goods, including WPB for WPC production has been explored by many studies [2, 5, 6]. However, there is lack information on how material composition is optimized in order to comply with standard. Therefore, further study is needed to formulate the optimum compositions of WPC consisted of mostly wasted material including waste wood and WPB.

A technique that could be applied to formulate WPC is mixture design of experiment. It is a special technique of response surface analysis in which investigated product is made from several ingredients that have 100% in total [7]. It was reported that this technique has been applied to formulate many material compositions [8-13]. The formulation could be done using simplex method (lattice or centroid) or extreme vertice. However, due to the nature of materials and mixture boundaries that need to be done in very small portion, extreme vertice is more reasonable to be performed.

In this work, WPC from WPB was produced and optimized using extreme vertice mixture design. The formulation was performed to obtain minimum flexural modulus and strength that comply to Indonesian Standard (SNI). Other properties of optimized and verified sample as water content, density, swelling and formaldehyde content were then tested and analyzed compared to standard requirement.

2. Experimental method

2.1 Materials

Neat high-density polyethylene (HDPE) Titanvene 6070EA (flexural modulus 1700 MPa) from Lotte Chemical Titan Nusantara was utilized as matrix in WPC along with shredded waste plastic bags WPB as co-matrix. 80-mesh size sawdust from various hard woods was used as filler for reinforcement. Sawdust was pre-dried at 80 °C for 24 hours in advance of usage to reduce water content. Polyethylene-graft-maleic anhydride (PE-g-MA) used as compatibilizer was produced based on the route developed in our laboratory [14] using HDPE Titanvene 6070EA, maleic anhydride (MA) and benzoyl peroxide (BPO). MA and BPO were purchased from Merck.

2.2 Formulations and manufacturing of WPC

WPC were compounded according to formulas listed in Table 1. These formulas were generated in Minitab 16 based on extreme vertices mixture design. Sawdust content was chosen within the range of 50% to 70% due to expectation to prediction optimum amount that existed in this boundary, and processing issue. Neat HDPE contents were ranged from 0% to 50%, WPB from 0% to 30%, PE-g-MA from 0% to 20% to give polymer bi-matrix concentration of 50% in each formula and economical reason.

Table 1. Formula of WPC using HDPE, Shredded WPB, Sawdust and PE-g-MA.

Formula	HDPE [wt%]	WPB [wt%]	Sawdust [wt%]	PE-g-MA [wt%]
1	50.0	0.00	50.0	0.00
2	30.0	0.00	70.0	0.00
3	30.0	0.00	60.0	10.0
4	40.0	0.00	50.0	10.0
5	30.0	20.0	50.0	0.00
6	30.0	10.0	50.0	10.0
7	35.0	5.00	55.0	5.00

WPC compounding was done using Collins ZK 25 twin screw extruder. All ingredients were hand-mixed prior to compounding. The resulted mixture was poured into hopper at once and processed according to conditions listed in Table 2. The resulted pellets were air dried and transferred to

IOP Conf. Series: Materials Science and Engineering **553** (2019) 012048 doi:10.1088/1757-899X/553/1/012048

compression moulding machine Collins P300P for specimen preparations. Conditions for compression moulding are listed in Table 3.

Table 2. Twinscrew Conditions for Manufacturing WPC.

Temperature (°C)								Screw Speed
T_1	T_2	T_3	T_4	T_5	T_6	T_7	T_8	(rpm)
40	170	175	180	185	190	190	190	60

Table 3. Compression Molding Conditions.

Parameter	Steps				
	1	2	3	4	5
Temperature [°C]	190	190	190	190	40
Pressure [Bar]	5	5	5	5	5
Time [Minute]	5	5	5	5	15

2.3 Characterizations

2.3.1. Flexural modulus and strength. In advanced of testing, the molded samples were cut to 50 mm x 150 mm and pre-conditioned at humidity 50 ± 5 % and temperature 23 ± 2 °C for more than 40 hours. Flexural characteristics of WPC were obtained according to ISO 16978 standard using Universal Testing Machine (UTM) Shimadzu AG-X Plus 50 kN. The load speed was set at 0.3 MPa/sec. Flexural modulus was calculated from the load between 10 % and 40 % of maximum load. The ultimate flexural strength was recorded automatically. The results were compared to required value in SNI 8154 – 2015.

2.3.2. Water content. Water content was measured following the procedure in SNI ISO 16979 standard. 20 miligrams of samples were weighted, followed by drying at 103 °C in an oven until mass equilibrium was reached. The dried sample was weighted. Water content of sample was calculated using equation (1) and averaged for 5 test specimens, where M_0 represents initial mass and M_1 is dried mass.

$$\text{Water content (\%)} = \frac{M_0 - M_1}{M_1} \text{ x } 100\% \tag{1}$$

2.3.3. Density. Density of WPC was measured according to SNI ISO 9427 standard. The sample was cut to a size of 50 x 50 mm. Thickness was measured at the position of the diagonal intersection of the sample. Length and width were measured in the center of the sample, which passes through the midpoint of the intersection between diagonals. The mass of the sample was measured using a density balance with an accuracy of 0.0001 grams. Sample density was calculated according to equation (2) and averaged for 6 test specimens.

$$\text{Density } (\rho)\left(\frac{kg}{m^3}\right) = \frac{m}{b_1 \text{ x } b_2 \text{ x } t} \text{ x } 10^6 \tag{2}$$

Where m = initial mass
b_1 = length
b_2 = width
t = thickness

2.3.4. Swelling. Percentage of swelling of sample WPC was carried out according to SNI ISO 16983 standard. The sample was cut to 50 x 50 mm specimen, and soaked in water of pH 7.0, at 20 ± 1 °C, for 24 hours. The specimen was vertically placed in thermal humidity chamber. The thickness of the specimen before and after soaking was measured using micrometer at the point of diagonal intersection of the specimen. Percentage of swelling was calculated based on equation (3) and averaged for 8 test specimens.

$$\text{Percentage of swelling (\%)} = \frac{t_2 - t_1}{t_1} \times 100\% \qquad (3)$$

Where t_1 = thickness before soaked
t_2 = thickness after soaked

2.3.5. Formaldehyde content. Formaldehyde content was measured according to JIS A 1460 – 2001 standard. Prior to measurement, standard solution containing formaldehyde i.e 0, 0.2, 0.4, 0.8, 2.0 and 4.0 mg/L were made and the absorbance of solution was measured using UV-Vis. Calibration curve was obtained. To measure formaldehyde content of WPC, 10 WPC specimens were placed above a cup containing 300 mL distilled water. The specimens and the cup were isolated in a dessicator for 24 hours. The test environment was controlled at temperature 20 ± 2 °C, and humidity 65 ± 5%. After 24 hours, the DI water is reacted with acetylacetone ammonium acetate (AAAA). The absorbance of formaldehyde contained distilled water was measured. Formaldehyde content was calculated using equation (4).

$$\text{Formaldehyde content (mg/L)} = \frac{F \times (A_s - A_b) \times 1800}{S} + \text{Intercept} \qquad (4)$$

Where F = Slope of calibration curve (mg/L)
A_s = Absorbance of formaldehyde contained distilled water
A_b = Absorbance at standard curve
S = Total area of samples (cm^2)

3. Results and discussions

3.1 Flexural modulus and strength

The resulted flexural modulus and strength is listed in Table 4. It shows that flexural modulus of WPC is within the range of 1630 MPa (formula 5) to 2173 MPa (formula 4), while strength of WPC is spread between 11.2 MPa (formula 2) to 29.1 MPa (formula 4). The average modulus of all samples is higher than neat HDPE.

Table 4. Flexural Strength and Modulus of WPC.

Formula	HDPE [wt%]	WPB [wt%]	Sawdust [wt%]	PE-g-MA [wt%]	Sawdust : HDPE	Modulus [MPa]		Strength [MPa]	
						Actual	Target	Actual	Target
1	50.0	0.00	50.0	0.00	1.00	1748	2000	21.5	18
2	30.0	0.00	70.0	0.00	2.33	1799	2000	11.2	18
3	30.0	0.00	60.0	10.0	2.00	2093	2000	21.1	18
4	40.0	0.00	50.0	10.0	1.25	2173	2000	29.1	18
5	30.0	20.0	50.0	0.00	1.66	1630	2000	17.2	18
6	30.0	10.0	50.0	10.0	1.66	1924	2000	23.0	18
7	35.0	5.00	55.0	5.00	1.57	1922	2000	20.5	18
Average						1898		20.5	
Standard Deviation						191		5.5	

However, comparison on the obtained flexural properties and SNI 8154 – 2015 show that there are only two samples (formula 3 and 4) comply to standard, namely those with limited amount of WPB and high PE-g-MA level success. In contrast, most of samples (except formula 2 and 5) have strength that higher than that of stated in SNI. The highest strength is produced by sample with low content of sawdust and high level of PE-g-MA, while the lowest is obtained by that of with high percentage of sawdust and limited amount of PE-g-MA. Unexpected low strength for sample 2 and 5 might be influenced by weak interfacial bonding by the absent of compatibilizer (PE-g-MA).

Meanwhile, Table 4 also presents the effect of sawdust, WPB, and PE-g-MA to neat HDPE. The Table shows that mixing sawdust with neat HDPE (formula 1) successfully increase the modulus of the latter. The modulus increases from 1700 MPa to 1748 MPa. This means that sawdust has positive impact on modulus of WPC. However, further addition of sawdust, increase modulus only slightly (51 MPa) by increasing amount of sawdust to 70% as in formula 2 (ratio of sawdust to HDPE is 2.33). In contrary, unexpected decrease of strength is observed by the increase of sawdust. The strength is down from 21.5 MPa in formula 1 to 11.2 MPa in formula 2. This is probably due to insufficient amount of HDPE to wrap the surface of sawdust that tend to be agglomerated, as well as the absent of PE-g-MA, leaving poor interconnection of WPC constituent in plastic region.

In formula 3, the incorporation of PE-g-MA while keeping the neat HDPE constant, reduces the amount of sawdust, hence the ratio is down to 2.00. The strength of WPC then arises to 21.1 MPa. Further lowering the ratio to 1.25 by raising the amount of neat HDPE in formula 4, results in the highest strength of WPC (29.1 MPa). These results indicate that the lower the ratio of sawdust to HDPE, the more HDPE is available to wrap the surface of sawdust, the higher the strength. Other than that, the incorporation of PE-g-MA into WPC promotes good bonding between sawdust and neat HDPE that produce better modulus and strength simultaneously.

Moreover, introducing WPB into HDPE and sawdust as in formula 5, reduced modulus of WPC compared to that of neat HDPE. However, modulus and strength of formula 5 (1630 MPa and 17.2 MPa) are lower than those of formula 1 (1748 MPa/21.5 MPa). This result indicates that WPB has inferior effect compare with neat HDPE probably due to truncated polymer chain of the former as result of reprocessing as well as the existence of other ingredients in WPB such as colorant, extender and etc. Improved results are achieved by adding PE-g-MA to the formulation (formula 6). The resulted WPC has higher properties than that of without PE-g-MA due to good bonding among materials.

The result in Table 4 was then analyzed using Minitab 16 and the models are generated. Equation (5) and equation (6) predict the flexural modulus and strength of WPC according to its composition. Equation (5) shows that neat HDPE, WPB and sawdust independently have positive effect on flexural modulus of WPC. In contrary, sole PE-g-MA has the opposite influence. This might be caused by can be explained by the fact of its low intrinsic mechanical properties. The entanglement and interconnection among HDPE chains are loosen by grafting maleic anhydrate pendant groups on polymer backbones. This results in separation of polymer molecules that deprive the mechanical properties thereof. However, interaction between neat HDPE and PE-g-MA has the highest positive influence on this property. The interaction of both materials effects flexural modulus 100 times higher than that of PE-g-MA alone. This is because PE-g-MA and neat HDPE have identical polymer backbone which results in good interaction between both polymers as well as between sawdust. This is also true on flexural strength of WPC as describe on equation (6). Interaction between neat HDPE and PE-g-MA as well as neat HDPE and WPB alone are beneficial to flexural strength of WPC. In contrary, sawdust and PE-g-MA independently have detrimental effect on the property.

$$\text{Modulus elasticity} = 1577*\text{HDPE} + 907*\text{WPB} + 1921*\text{Sawdust} - 146.5*\text{PE-g-MA} + 14976*\text{HDPE}*\text{PE-g-MA} \quad (5)$$

$$\text{Strength} = 46.7*\text{HDPE} + 24.0*\text{WPB} - 3.8*\text{Sawdust} - 12.0*\text{PE-g-MA} + 336.0*\text{HDPE}*\text{PE-g-MA} \quad (6)$$

Predictions of optimized compositions are shown in Figure 1 to 3. Figure 1 depicts WPC formulations with maximum flexural properties. It is shown that the maximum modulus would be obtained at composite with no additional WPB. However, this was not the desired output of this research, as the WPB should be maximized. For this reason, the composition was predicted by manipulating the vertical red line on the generated response optimizer to left or right side. By doing this at particular line, it would change another composition. The result shows that the maximum WPB that can be added to mixture and satisfy the SNI 8154 – 2015 requirement was 8%, resulted 2002 MPa for modulus and 24.5 MPa for strength (Figure 2). However, due to safety factor and standard deviation, the formula was predicted to produce modulus higher than 2100 MPa (Figure 3).

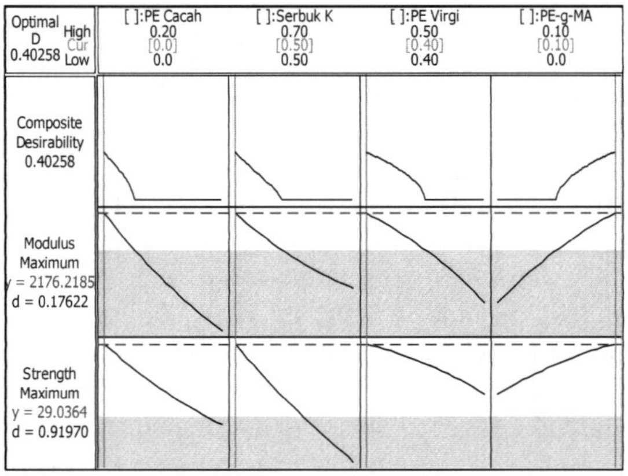

Figure 1. WPC formulation for maximum flexural properties.

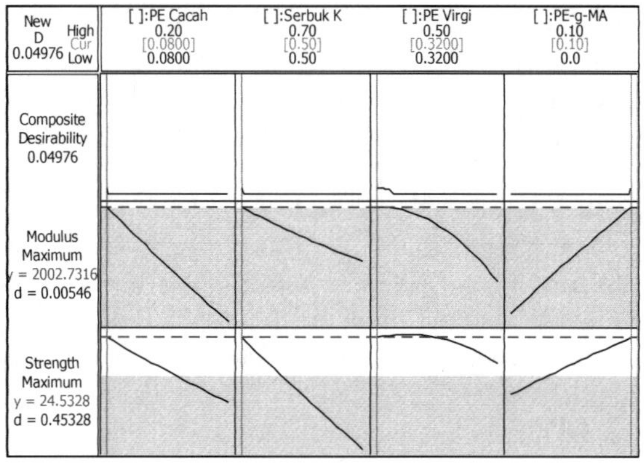

Figure 2. WPC formulation at maximum WPB level that satisfy SNI 8154 – 2015 for flexural.

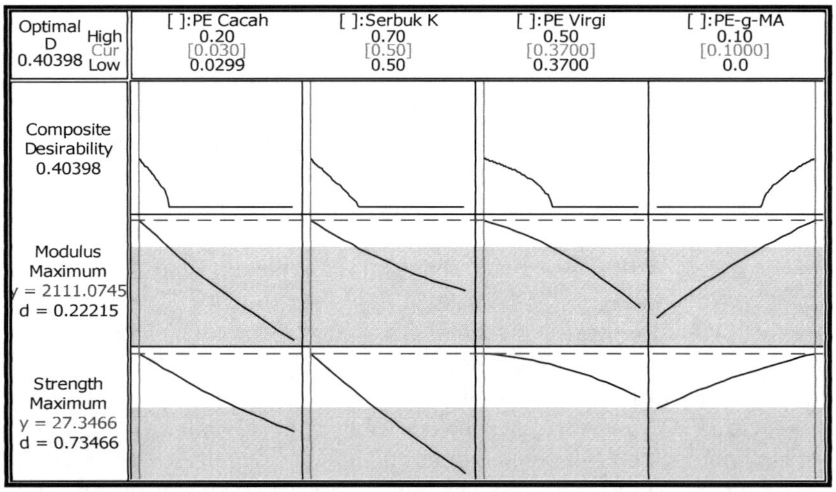

Figure 3. WPC formulation that satisfy SNI 8154 – 2015 for flexural.

Table 5. Flexural Strength and Modulus of Verified WPC formula.

WPB [wt%]	Sawdust [wt%]	HDPE [wt%]	PE-g-MA [wt%]	Modulus [MPa]		Error [%]	Strength [MPa]		Error [%]
				Prediction	Actual		Prediction	Actual	
3.00	50.0	37.0	10.0	2111	2173	3%	27.3	25.7	6%

To verify the desired composition, a verification sample was prepared and tested. Results of verification experiment were summarized in Table 5. It is shown that there is good agreement between models and verification data. 3% and 6% of errors are achieved for modulus and strength, respectively. Slightly higher error for strength can be explained by the fact that the composition was calculated based on the model for modulus. However, these relatively low errors indicate that good manufacturing of WPC from HDPE, WPB, sawdust and PE-g-MA is possible.

3.2 Characterization of optimized sample

The verified WPC sample is tested according to SNI 8154 – 2015 and JIS A 5741. All the testing results is compiled in Table 6. In general, the verified WPC sample is able to satisfy all the criteria specified by the standard.

Table 6. Test Results of Verified WPC formula.

No	Test Item	Result	Specification
1	Water content [%]	0.6 ± 0.1	< 12%
2	Density [g/cm^3]	1.13 ± 0.01	≥ 0.60
3	Swelling [%]	1.7 ± 1.4	$\leq 4\%$
4	Formaldehyde content [mg/L]	0.015 ± 0.000	≤ 0.4

The water content of the sample is 0.5687 which is much lower than specified by the standard. This value illustrates the good bonding between polymer matrix and sawdust, which manage to prevent the return of moisture into sawdust after manufacturing.

The sample density of verified WPC is 1.13 g/cm^3. This value is almost double the specifications determined by SNI (> 0.6 g / cm3). This indicates the good interaction between matrix and filler as well as less void that occurs in the sample, hence increases the flexural modulus of elasticity and strength.

The swelling of verified WPC sample that occurred after 24 hours is 1.71% is listed in Table 6. This value is below the maximum limit allowed by SNI. As explained in density test section, the compactness and the reduce void in the sample as well as good bonding between matrix and filler prevent the re-entry of water into the sample, hence hindering the swelling to be occur.

(a) (b) (c)

Figure 4. (a) Specimens set up before test, (b) specimens position in dessicator, (c) distilled water containing formaldehyde in desiccator reacted with AAAA solution

Test of formaldehyde content was depicted in Figure 4. The test results show that formaldehyde content in the sample is 0.015 mg/L. This formaldehyde content can be originated from recycle material used in making verified WPC. However, the amount of formaldehyde released can be minimized if there is good interaction and strong bonds between matrix and filler. Polymer matrices that wrap sawdust well, not only prevent the return of moisture to sawdust, but also prevent the release of formaldehyde from sawdust.

4. Conclusion

Formula optimization of wood plastic composites (WPC) from waste plastic bags (WPB) high density polyethylene (HDPE) has been done using extreme vertice mixture design in order to comply to Indonesian Standard (SNI) 8154 – 2015. Formulation was conducted using twin screw extruder with material boundaries were set in the range 0 to 20 wt% of WPB, 50 to 70 wt% of sawdust, 30 to 50 wt% of neat HDPE, and 0 to 10 wt% of HDPE grafting maleic anhydride (PE-g-MA). The optimized formula was then verified and observed its water content, density, swelling, and formaldehyde content.

IOP Conf. Series: Materials Science and Engineering **553** (2019) 012048 doi:10.1088/1757-899X/553/1/012048

The results showed that 3 wt% of WPB, 50 wt% of sawdust, 37 wt% of neat HDPE, and 10 wt% of PE-g-MA would give optimum values of flexural modulus and strength (2111 MPa and 27.3 MPa), that higher than minimum specification stated in SNI 8154 – 2015 (2000 MPa for modulus, 20 MPa for strength). Further analysis on verification samples resulted 2173 MPa and 25.7 MPa in modulus and strength, respectively. It also had $0.6 \pm 0.1\%$ of water content, $1.7 \pm 1.4\%$ of swelling, and 0.015 ± 0.000 mg/L of formaldehyde content that lower than maximum specifications, and 1.13 ± 0.01 g/cm^3 of density that above the minimum requirement. Good results of verification sample show that WPC from WPB is possible to be processed and complied to Indonesian Standard (SNI).

Acknowledgment
Authors would like to thank you Center for Polymer Technology, Agency for Assessment and Application of Technology and Research Institute for Housing and Human Settlements (RIHHS), Ministry of Public Works and Housing for supporting us to conduct this project.

References
[1] Turku I, Karki T, and Puurtinen A 2018 *Heliyon* **4** 1 doi : 10.1016/j.heliyon.2018.e00559
[2] Najafi S K, Kiaefar A, Hamidina E, and Mehdi T 2016 *J. Reinf. Plast. Compos.* **26** 341 doi : 10.1177/0731684407072519
[3] Jambeck J R, Geyer R, Wilcox C, Siegler T R, Perryman M, Andrady A, Narayan R, and Law K L 2015 *Sci.* **347** 6223 doi : 10.1126/science.1260879
[4] T.H. Mintarsih."*Kebijakan Kantong Belanja Plastik Tidak Gratis"*. 2016 [cited 2018September30];Availablefrom:http://www.menlhk.go.id/tinymcpuk/gambar/file/Kebijakan%20Kantong%20Belanja%20Plastik%20Tidak%20Gratis.pdf.
[5] Islam M N and Islam M S 2013 *J. Therm. Compos. Mat.* **28** 1135 doi : 10.1177/0892705713503671
[6] Rajendran S, Scelsi L, Hodzic A, Soutis C, and Al-Maadeed M A 2012 *Resour., Conserv. Recycl.* **60** 131 doi:10.1016/j.resconrec.2011.11.006
[7] Lundstedt T, Seifert E, Abramo L, Thelin B, Nyström A, Pettersen J, and Bergman R 1998 *Chemom. Intell. Lab. Syst.* **42** 3
[8] Asiaban S and Moradian S 2012 *Dyes Pigm.* **92** 1 doi:10.1016/j.dyepig.2011.05.019
[9] Bilal A, Lin R J T, and Jayaraman K 2014 *J. Appl. Polym. Sci.* **131** doi: 10.1002/app.40647
[10] da Costa H M, Ramos V D, da Silva W S, and Sirqueira A S 2010 *Polym. Test.* **29** 572 doi:10.1016/j.polymertesting.2010.04.003
[11] Fehri M K, Mugoni C, Cinelli P, Anguillesi I, Coltelli M B, Fiori S, Montorsi M, and Lazzeri A 2016 *eXPRESS Polym. Lett.* **10** 274 doi : 10.3144/expresspolymlett.2016.26
[12] Mali S, Debiagi F, Grossmann M V E, and Yamashita F 2010 *Ind. Crops Prod.* **32** 353 doi : 10.1016/j.indcrop.2010.05.014
[13] Raja R S, Manisekar K, and Manikandan V 2014 *Mater. Des.* **55** 499 doi : 10.1016/j.matdes.2013.10.026
[14] Ujianto O, Marzuki, and Radini F A 2016 *Mater. Sci. Forum* **864** 13 doi : doi.org/10.4028/www.scientific.net/MSF.864.13

Relaxation Behavior of Natural Rubber Composites Through Recovery Measurement after Tension and Compression Set

I N Indrajati*, I Setyorini
Center for Leather, Rubber and Plastics, Ministry of Industry Republic of Indonesia, Sokonandi St. Numb. 9, Yogyakarta, Indonesia, 55166

i-novia@kemenperin.go.id

Abstract. Relaxation behavior of natural rubber (NR) composites in term of the recovery after tension and compression set had been successfully studied. The effect of crosslink density, either physical or chemical, on the recovery percentage was evaluated. Filler carbon black N-330 and sulfur crosslinking agent were varied for this purpose. N-330 loading ranging from 30, 35, 40 and 45 phr with baseline of 2 phr sulfur. Meanwhile, sulfur loading ranging from 1, 2, 3 and 4 phr with the baseline of 30 phr N-330. Tension set was performed on constant elongation with 25, 50 and 100% strain. The measurements were performed at 100ºC for 72 h. While compression set was carried out in constant deflection 25% of initial thickness, carried out both at room temperature and 100ºC for 72 h. The results findings showed that the tension set was independent to crosslink densities, but greatly depended on the strain level. Low strain level provided higher recovery percentage. The filler increment markedly increased the compression set, but sulfur tended to improve it. Compression set was higher at elevated temperature since the deterioration not only generated by physical process but also by chemical oxidative thermal process.

Keywords: relaxation, natural rubber, tension set, compression set, crosslink

1. Introduction

Rubber is versatile engineering materials due to its elasticity and durability [1]. It can be formulated to obtain wide range of mechanical properties for many applications. Designing rubber product should consider many factors. It is not enough to characterized rubber product by its mechanical properties because they only describe the short term behavior. Consideration should have to be taken involving the long term behavior such as creep, stress relaxation and dynamic properties.

In the application, rubber is exposed to various forces whether as a stress, strain or shear either in static or dynamic loading. If definite forces are worked on the system, then all parts of the system are involved in the effort resisting the force. When these forces are removed out from the system, the molecule chains relax until achieves new equilibrium. Thus, relaxation is rate related to the how fast the transition of the system from one stage of equilibrium to another [2]. The relaxation process can be physical or chemical. Physical relaxation is occurred mostly by rearrangement of rubber molecule chains and the filler when subjected to deformation. Chemical relaxation is more prominent at higher temperature and longer time exposure involving the oxidative thermal aging which results in chain scission, crosslink breaking or even crosslink formation [3]. Chains scission and crosslink breaking

IOP Conf. Series: Materials Science and Engineering **553** (2019) 012049 doi:10.1088/1757-899X/553/1/012049

evidently slower the rate of relaxation. These could be a permanent deformation known as set properties. Set is correlated to the measurement of the recovery after removal of an applied forces. In practice, set is a convenient way to describe the relaxation phenomenon of a material. Moreover, it is quite simple method and relatively fast measurement, that makes it very useful to adopt in industry. Set properties can be measured under either tension or compression force.

Many efforts have been attempted by researchers in studying relaxation behavior of rubber to dates. However, to date only few of them discussed relaxation in term of recovery. Ayoub et al. [4] modeled permanent set in terms of a visco-hyperelastic damage. Maiti et al. [5] applied Tobolsky model to interpret the radiation effect of silicone rubber on the permanent set. Rosszainily et al. [6] measured permanent set of unvulcanized natural rubber filled by carbon black through tensile stress experiment. High percentage recovery was given by low carbon black loading as a result of strain induced crystallization. Mohaved et al. [7] studied the effect of efficient vulcanization (EV) curing system on compression set of nitrile rubber filled by various carbon black. Kömmling et al. [8] revealed the effect of chain scission of EPDM rubber during aging through relaxation and recovery experiments under compression force. Dai et al. [9] studied recovery after compression of γ-ray irradiation vulcanized HNBR. Compression set was decreased with increasing adsorbed γ-ray dose, which means improved crosslink density. Vishvanthperumal and Gopalakannan [10] investigated the effect of nano-clay loading and curing systems on compression set of EPDM nanocomposite. Compression set was found to increase incrementally with increasing nano clay loading, while sulfur vulcanization also showed the highest compression set among mixed- and peroxide curing.

In this present work, relaxation behavior of NR composite is investigated in term of the recovery after deformations, either in tension or compression. For these purposes, filler (carbon black N-330) or sulfur loading is varied to evaluate physical or chemical crosslink formed within the matrix. Bound rubber (BdR) has measured to identify the physical crosslinking, whilst total crosslink density is determined by equilibrium swelling. Tension set is carried out in constant elongation with various strain level, while compression set is done with constant deflection both in room and elevated temperature.

2. Experimental method

2.1 Materials

Natural rubber grades Standard Indonesian Rubber (SIR) 10 was supplied by local estate PTPN IX. Carbon black N-330 and N-774 (Ex. OCI, Korea), zinc oxide active (Indoxide), Aflux 42 processing promotor. Paraffinic oil was purchased from CV. Indrasari, Semarang. Antidegradant system consisted of 2,2,4-Trimethyl-1,2-Dihydroquinoline (TMQ) (Kemai) antioxidant, N-(1,3-Dimethylbutyl)-N'-phenyl-p-phenylenediamine (6PPD) antiozonant (Northeast) and paraffin wax Antilux 654 A (Rhein Chemie). Accelerators N-Cyclohexylbenzothiazole-2-Sulfenamide (CBS) (Northeast) and diphenyl guanidine (DPG) (Shandong Shianxian). Sulfur Midas SP 325 (Miwon) was used as vulcanizing agent. All chemicals were industrial grade.

2.2. Rubber compound preparation

The rubber compound was prepared according to the formulation as seen on Table 1. Carbon black filler (N-330) and sulfur loading were varied to evaluate the effect of crosslink formed onto the recovery percentage after tension or compression set, while another ingredient was incorporated in a fixed amount. The detail of the variable taken in this study was listed on Table 2.

Mixing process were performed in a two roll mill laboratory scale. Rubber was first masticated and then followed by additive incorporation in a sequence. After completed, the rubber compound was soaked in the cold water for 5 minutes to stop the pre-vulcanization which might be proceeded. The rubber compound was then allowed to rest for 24 h at 25°C before testing.

Table 1. Rubber Compound formulation

Ingredients	Concentration (phr)
SIR 10	100.0
ZnO	5.0
Aflux 42	1.5
N-330	Varied
N-774	30.0
Paraffinic oil	10.0
TMQ	2.0
6PPD	1.0
Paraffin wax	0.5
CBS	1.9
DPG	0.2
Sulphur	Varied

Table 2. Variables of the experiment

Variable	Baseline	Range
N-330	30 phr	30, 35, 40, 45 phr
Sulphur	2 phr	1, 2, 3, 4 phr

2.3. *Testing samples preparation*
The rubber samples for both tests were prepared by compression molding. Uncured rubber compounds were cured in a hydraulic press with hot platen in both sides (upper and lower) according to the dimension requirements for each test. The pressure applied was 150 kg/cm^2 and at temperature of 150 °C. The curing time was determined from curing test that performed in a Moving Die Rheometer (MDR) Gotech M3000A at the same temperature. Curing was performed corresponding to t_{90} obtained from MDR.

2.4. *Bound rubber (BdR) and crosslink density measurement*
Bound rubber (BdR) was essentially measured by swelling method of uncured rubber compound. The sample was cut in small pieces and placed in the wire basket, then immersed in toluene for 72 h at room temperature. After completion immersion period, the sample was taken out from the solvent. Rubber sample was then carefully removed from the wire basket and dried in air-circulated oven until achieving constant weight. The weight of the sample before and after swelling were taken, and BdR was calculated using Equation (1),

$$BdR\ (\%) = 100\left(\frac{w_i-\left(\frac{CPD}{100}\right)w_f}{w_i}\right) \tag{1}$$

where w_i is the initial weight of the sample before immersion (in grams), w_f is the final weight after immersion and dried (in grams), and *CPD* is the total filler in the formulation (in phr).

Crosslink density was determined by equilibrium swelling. The swelling experiment was carried out by immersing the rubber sample in toluene at ambient temperature for 72 h. Swelling test was carried out for both samples, before and after compression of sample was taken before immersion. After completion swelling period, the sample was taken out and blotted with tissue paper to remove excess solvent. The mass of the sample, initial (w_0) and final (w_f), were weighed using analytical balance with 0.0001 g accuracy. The crosslink density was determined according to Flory-Rehner as expressed on Equation (2),

$$M_c = \frac{-\rho_p\, V_s\, V_r^{\frac{1}{3}}}{\left[\ln(1-V_r)+V_r+XV_r^2\right]} \tag{2}$$

IOP Conf. Series: Materials Science and Engineering **553** (2019) 012049 doi:10.1088/1757-899X/553/1/012049

where M_c is the number average molecular weight of the rubber chains between crosslink; ρ_p is density of the polymer; V_s is molar volume of the solvent (for toluene V_s=106.2 cm^3 mol^{-1}), V_r is molar volume of rubber, X is polymer-solvent interaction parameter (for NR-toluene=0.42). The term V_r was calculated by using the relation as expressed on Equation (3).

$$V_r = \frac{\frac{w_1}{\rho_1}}{\left(\frac{w_1}{\rho_1}+\frac{w_2}{\rho_2}\right)} \quad (3)$$

w_1, ρ_1 are the weight and density of rubber, while w_2, ρ_2 are the weight and density of solvent. The swelling experiment was performed on original and compressed samples. Flory-Rehner equation calculates total crosslink density of the composites, included physical and chemical crosslinks. Thus, chemical crosslink is then obtained by subtracting the total and physical crosslink. In this calculation, physical crosslink is determined from the intercept of the extrapolated curve (N-330 or sulfur loading vs crosslink density) as illustrated on Figure 1.

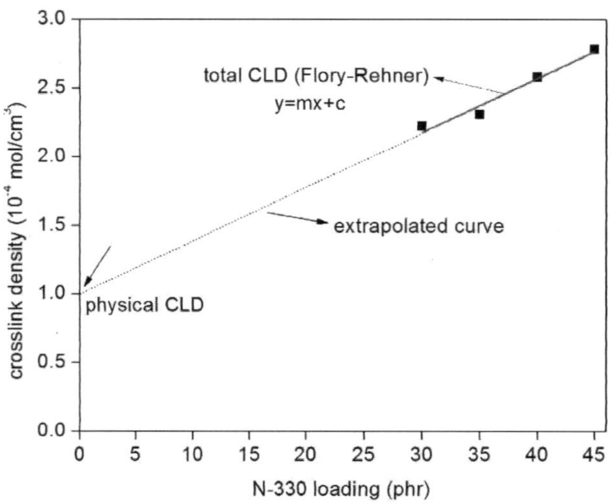

Figure 1. illustration of physical crosslink determination based on Flory-Rehner Equation.

2.5. Tension and compression set tests

Tension set measurements were performed at constant load according to ISO 2285:2013. A bar-type sample with 2-mm thickness was used. A fixed strain was applied to each sample and aged for 72 h at 100 °C in dry-air circulated oven. Three different strains were employed, i.e. 25, 50 and 100% of initial length. The detail of tension set testing is presented on Figure 2. The final length was (*Lf*) measured 30 minutes after the sample was removed from the test device using electronic digital Mitutoyo calliper with 0.01 mm accuracy. Tension set is expressed in percentage of recovery after strained, calculated according to Equation (4), where *L0* denotes for initial length (before strained), while *Lf* is the final length after being strained.

$$tension\ set = \frac{L0-Lf}{L0} \times 100\% \quad (4)$$

Figure 2. Tension set experiment.

Figure 3. Compression set testing.

Compression set testing was carried out in accordance to ISO 815-1:2014. The test was performed both on ambient and elevated temperature using assembled device as depicted on Figure 3. A cylindrical disc standard sample was employed for each test. The samples were placed between the space bars. The distance of the samples was adjusted to accommodate sufficient clearance for bulging when the rubber samples were compressed. Each bolt shall be tightened in order to get the uniform drawn throughout the compression area. The compressive stress was taken 25% of initial thickness (*h0*). For elevated temperature testing, the assembled device was aged in the dry-air circulated oven for 24 h at 100 °C. After completion of compression, the samples were removed from the device and placed in non-heat conductive material, such as wooden bar. The samples were allowed to recover for 30 minutes, then the final thickness was measured using electronic digital Mitutoyo calliper with 0.01 mm accuracy. Compression set is expressed as the recovery percentage of rubber sample respect to compress thickness. It is calculated using Equation (5),

$$compression\ set = \frac{h0-hf}{h0-hs} \times 100\% \tag{5}$$

where h_0 is the initial thickness, h_f is the final thickness after testing and h_s is the height of the spacer. Compression set tests at ambient temperature were conducted in similar way to the elevated temperature test, but without aging. The assembled device was placed in dry room with controlled-ambient temperature of 25 °C. The effect of compression on rubber samples were monitored by crosslink densities measurement of the original and compressed samples.

3. Results and discussion

3.1. Crosslink densities of NR composites

Rubber composites are usually composed by two types of crosslinking, e.g. physical and chemical crosslink. Physical crosslink is a result of filler-rubber chain interaction or chain entanglement. Carbon black has surface activities resulted from active functional groups that can provide strong interaction with rubber chains. A thin layer of rubber covers single filler particle or agglomerates which is stabilized by physical bonds such as van der Waals forces as well as chemical bond. This phenomenon is well-known as bound rubber (BdR). It is solvent-resistant interphase that often used to characterized the physical crosslink [11]. The BdR is reduced by increasing N-330 amount (Figure 4), indicating filler agglomeration occurred within the rubber matrix. It should be noticed because it will bring negative effect to the composite when certain deformation forces work on it. Meanwhile, BdR tends to unchanged by increasing sulfur loading (Figure 5) attributed to constant filler loading employed in these composites.

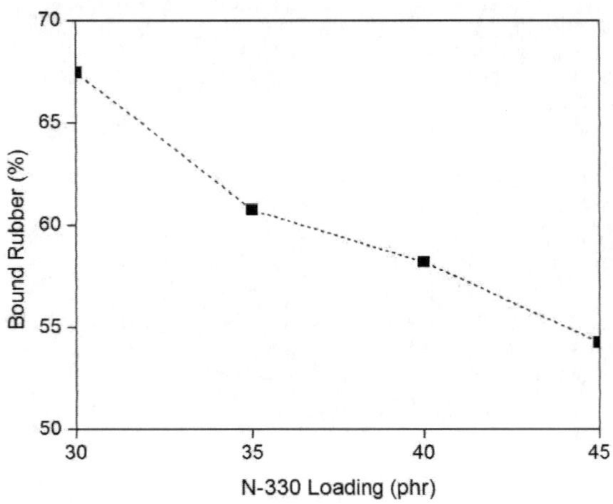

Figure 4. Bound rubber of NR composites on various N-330 loading.

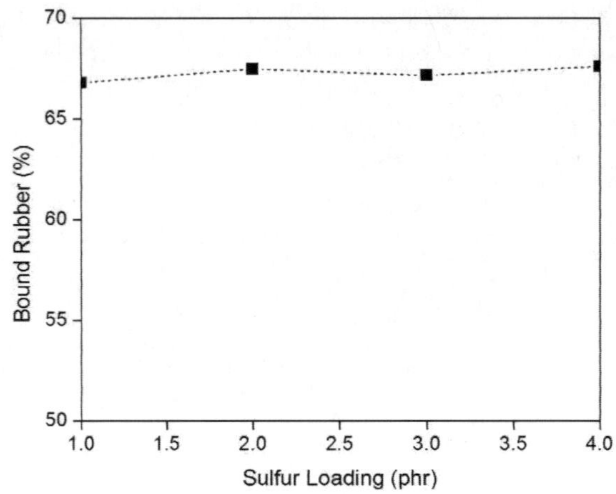

Figure 5. Bound rubber of NR composites on various sulfur loading.

Chemical crosslinks are formed during curing as a result of reaction between sulfur and rubber chain. Their formation can be determined by monitoring the torque changes over the time during curing test, or by equilibrium swelling method of vulcanized rubber. Increasing N-330 loading does not change the chemical crosslink density significantly since the sulfur used in the formulation is in the same level. In the other side, chemical crosslink rises markedly by increasing sulfur amount as expected. Sulfur reacts with unsaturation sites of NR backbone resulting in crosslinked chains through sulfur bridge. Thus, the more sulfur amount theoretically will result in the higher crosslink density.

Table 3.Crosslink density determination using Flory-Rehner equation.

Loading	Crosslink Density (x10^{-4} mole/cm^3)		
	Total*	Physical**	Chemical
N-330			
30 phr	2.222	1.004	1.218
35 phr	2.306	1.004	1.302
40 phr	2.580	1.004	1.576
45 phr	2.783	1.004	1.780
Sulfur			
1 phr	1.482	0.831	0.652
2 phr	2.222	0.831	1.391
3 phr	3.234	0.831	2.403
4 phr	3.525	0.831	2.694

*Flory-Rehner
**intercept of extrapolated curve

3.2. *Effect of N-330 and sulfur loading on tension set*

Tension set characterizes the ability of rubber to recover after being stretched. it is dealing with the residual strain that is generated as a result of new equilibrium state of molecular network after stress is removed. Upon stretching small chains reach their extensibility limit and brake, and free chains are in their relaxed configuration. Thus, the molecular network is more organized and causes reduction of entropy compared to virgin network. Tension set is depended on strain level and the composition of rubber, i.e. filler [3-5].

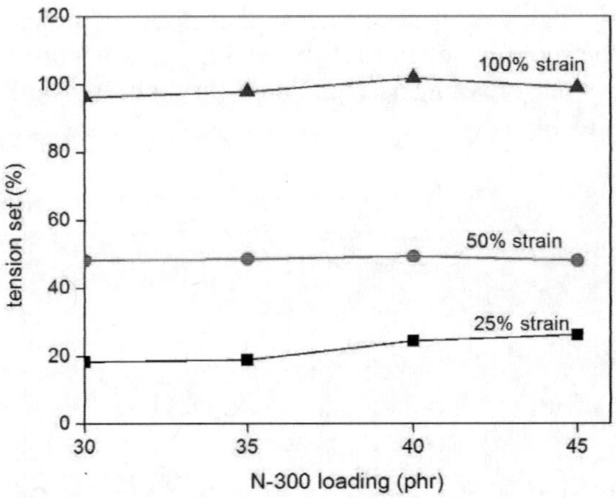

Figure 6. The tension set of NR composites on various N-330 loading.

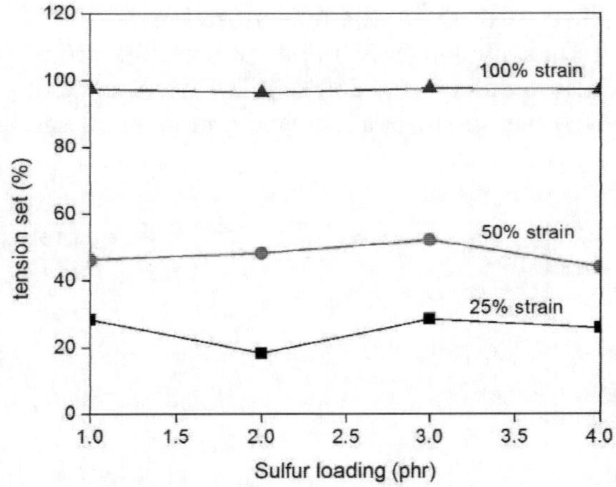

Figure 7. The tension set of NR composites on various sulfur loading.

Figure 6 and Figure 7 clearly depict that tension set of NR composites is remain unchanged upon increasing either N-330 or sulfur loading. But, the level of strains markedly increases the tension set. Many crosslinked chains are probably broken during stretching resulting more free chains that is in relaxed configuration. Since the experiments were performed at elevated temperature, the chain breaking of crosslinked network could be due to stretching and heat aging. Heat aging may be caused chain scission that produces short-free chain. This lowers the entropy and hence increases the residual strain. Therefore, the recovery ability of NR composites after stretched is reduced.

3.3. Effect of N-300 and sulfur loading on compression set

Similar to tension set, compression set also measures the ability of rubber to recover after deformation force is removed. The difference is relied on the deformation force acted on the rubber. In compression set, the deformation force is compressive. In general, low compression set is better. In this study, compression set is carried out at ambient and elevated temperature. Ambient temperature compression set only describes the set phenomena caused by physical changes, whilst testing on elevated temperature will provide set description attributed by physical and chemical changes. Figure 8 depicts the compression set of NR composites on various N-330. Compression set tends to increase gradually with the increment of N-330. It is probably attributed by poor interaction between rubber and filler within the matrix, or agglomerates of filler are presented in large amount at higher loading of filler. Consequently, the physical crosslinking is decreased indicated by lower BdR content (Figure 4). Interaction between filler particles in agglomerate is weaker than of filler-rubber chain. It makes the agglomerates of filler is easier to break down by applied stress. Thus, better filler particle dispersion will result in a low compression set. Meanwhile breaking agglomeration of filler under compressive stress tends to increase compression set [13]. High filler loading also induces an increase of stiffness so that increases the compression set [10].

In the other sides, increasing sulfur amount rises compression set as seen on Figure 9. During compression, enormous crosslinks are involved in resisting the forces indicated by an increase of stress in the rubber network. This resistance may cause some crosslinks to break down. Then, the number of crosslinks responsible for the strain recovery are less than the number of crosslinks responsible to resist the compression [14]. These breakdown crosslinks may be categorized as permanent set that cause reduction of recovery percentage (lower than 100%). At low sulfur loading (1 phr), the total crosslink density is somewhat lower (Table 3). Of course the number of crosslinks involved in resisting compressive force are much lower than those of composite with high loading sulfur. Surprisingly, compression set deviates when 4 phr sulfur is employed. In this formulation, sulfur amount is much higher than of accelerator (high sulfur to accelerator ratio) that possible to form polysulfide crosslink

(C-S_x-C, with x>2), with longer S-S bond. The S-S bond has low bonding energy that is easily broken out. Thus, the longer S-S bond in crosslink probably the more chance of chain breaking that lead to a reduction on compression set. The occurrence of crosslink breaking is confirmed through crosslink density measurement as showed on Figure 10 and Figure 11.

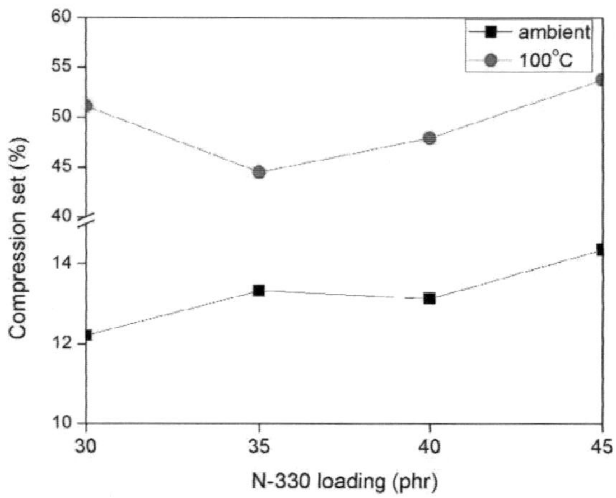

Figure 8. The compression set of NR composites on various N-330 loading.

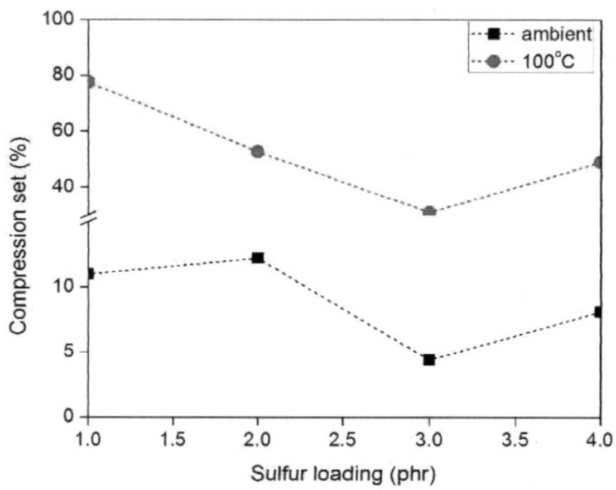

Figure 9. The compression set of NR composites on various sulfur loading.

In general, compression set results in lower value at ambient temperature compared to those at elevated temperature (in this case 100ºC). At low temperature, the chain breaking is mainly due to the physical deterioration, such as resisting the compressive force. The effect becomes more severe at elevated temperature because thermal degradation is taken account. Degradation at elevated temperature is mainly oxidative aging that may cause chain scission, crosslinking or even modify the filler-rubber interface [5]. Chain scission may dominate resulting in lower crosslinking. Thus, the number of broken crosslink is the sum of physical deterioration and thermal degradation. This explains the reduction of compression set at elevated temperature. A marked reduction on crosslink density is observed at 4 phr sulfur. This confirms the existence polysulfidic linkage at higher sulfur loading. This linkage is unstable at elevated temperature.

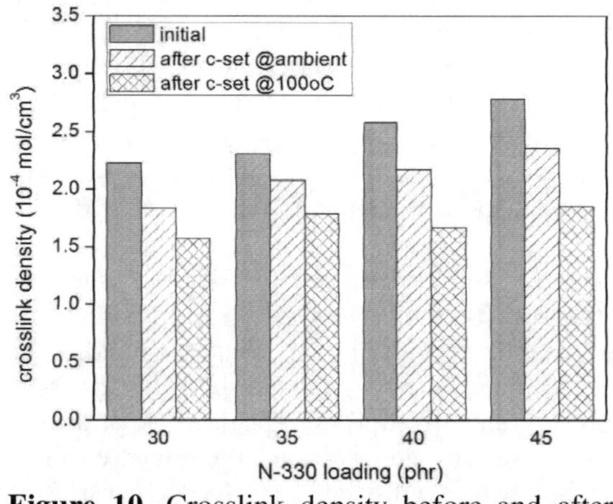

Figure 10. Crosslink density before and after compression set on various N-330 loading

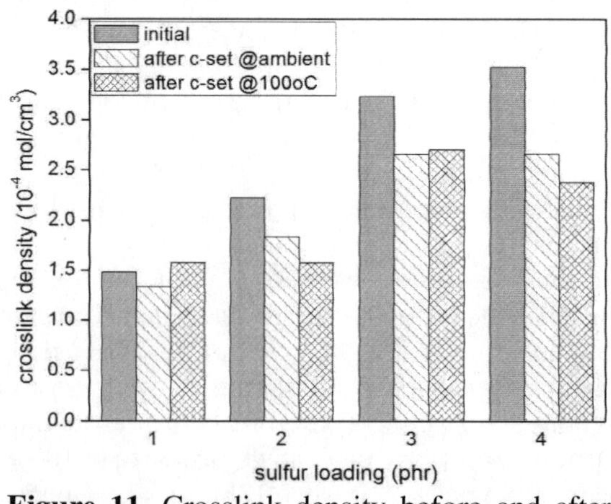

Figure 11. Crosslink density before and after compression set on various sulfur loading

4. Conclusion

Studies on relaxation behavior of NR composites has been successfully done through comprehensive studies on the tension and compression set. Tension and compression set represented the recovery of the rubber after certain forces worked on it. In this study, Carbon black N-330 and sulfur loading were varied to evaluate their effect on recovery of the composites. Tension set was carried out at constant elongation, i.e. 25, 50 and 100% strain of initial length, whilst compression set was conducted at constant deflection of 25% of initial height. Physical crosslinks were observed to reduce by increment of carbon black, but it stayed constantly on various sulfur loading. Meanwhile, the total crosslink density was found to increase upon sulfur loading, but it tended to steady on N-330 loading. Tension set was greatly affected by the level of strain instead of crosslink density. The set properties rose by increasing strain, either for N-330 or sulfur loading. Consequently, the recovery was decreased. Compression set was conducted both at room temperature and elevated temperature (100 °C). Compression set increased with the increment of N-330 that meant decreasing recovery percentage. Bound rubber measurement has confirmed this finding. Increasing sulfur loading improved the compression set. But deviation was found at higher sulfur loading (4 phr). High sulfur to accelerator ratio resulted in polysulfidic linkage that unstable to either stress or heat. High temperature imparted negative impact on compression set correlated to heat aging effect on rubber network. Chain scission could be majorly occured that increase the chain breaking possibility during compression.

Acknowledgment

Authors would like to express their gratitude for Sutarto, ST, MT (head of Research and Standards Division of the Center for Leather, Rubber and Plastics) for facilitating this in-house research. The authors also would like to thank to Dr. Muhammad Chalid for intensive discussions.

References

[1] E. B. D. da Rocha, F. N. Linhares, C. F. S. Gabriel, A. M. F. de Sousa, and C. R. G. Furtado, "Stress relaxation of nitrile rubber composites filled with a hybrid metakaolin/carbon black filler under tensile and compressive forces," *Appl. Clay Sci.*, vol. 151, pp. 181–188, Jan. 2018. DOI. 10.1016/j.clay.2017.10.008

[2] H. J. Maria, N. Lyczko, A. Nzihou, K. Joseph, C. Mathew, and S. Thomas, "Stress relaxation behavior of organically modified montmorillonite filled natural rubber/nitrile rubber nanocomposites," *Appl. Clay Sci.*, vol. 87, pp. 120–128, Jan. 2014. DOI. 10.1016/j.clay.2013.10.019

[3] M. Abu-Abdeen, "Single and double-step stress relaxation and constitutive modeling of viscoelastic behavior of swelled and un-swelled natural rubber loaded with carbon black," *Mater. Des.*, vol. 31, no. 4, pp. 2078–2084, Apr. 2010. DOI. 10.1016/j.matdes.2009.10.006

[4] G. Ayoub, F. Zaïri, M. Naït-Abdelaziz, J. M. Gloaguen, and G. Kridli, "A visco-hyperelastic damage model for cyclic stress-softening, hysteresis and permanent set in rubber using the network alteration theory," *Int. J. Plast.*, vol. 54, pp. 19–33, Mar. 2014. DOI. 10.1016/j.ijplas.2013.08.001

[5] A. Maiti, R. H. Gee, T. Weisgraber, S. Chinn, and R. S. Maxwell, "Constitutive modeling of radiation effects on the permanent set in a silicone elastomer," *Polym. Degrad. Stab.*, vol. 93, no. 12, pp. 2226–2229, 2008. DOI. 10.1016/j.polymdegradstab.2008.05.023

[6] I. R. A. Rosszainily, M. A. Salim, M. R. Mansor, M. Z. Akop, A. Putra, M. T. Musthafah, M. Z. Hassan, M. N. Abdul Rahman, and M. N. Sudin, "Effect of carbon black fillers on tensile stress of unvulcanized natural rubber compound," *J. Mech. Eng. Sci.*, vol. 10, no. 2, pp. 2043–2052, 2016. DOI. 10.15282/jmes.10.2.2016.9.0193

[7] S. O. Movahed, A. Ansarifar, and F. Mirzaie, "Effect of Various Efficient Vulcanization Cure Systems on the Compression Set of a Nitrile Rubber Filled with Different Fillers," *J. Appl. Polym. Sci.*, vol. 41512, pp. 1–10, 2015. DOI. 10.1002/app.41512

[8] A. Kömmling, M. Jaunich, and D. Wolff, "Revealing effects of chain scission during ageing of

EPDM rubber using relaxation and recovery experiment," *Polym. Test.*, vol. 56, pp. 261–268, Dec. 2016. DOI. 10.1016/j.polymertesting.2016.10.026

[9] P. Dai, A. Cheng, X. Zhao, Y. Jiao, Y. Lu, L. Wang, and X. Zeng, "Fabrication and mechanical properties of γ-ray radiation vulcanized hnbr with high recovery after compression," *Radiat. Phys. Chem.*, vol. 151, pp. 271–275, Oct. 2018. 10.1016/j.radphyschem.2018.03.012

[10] S. Vishvanathperumal and S. Gopalakannan, "Effects of the Nanoclay and Crosslinking Systems on the Mechanical Properties of Ethylene-propylene-diene Monomer/styrene Butadiene Rubber Blends Nanocomposite," *Silicon*, Apr. 2018. DOI. 10.1007/s12633-018-9836-x

[11] D. Gabriel, A. Karbach, D. Drechsler, J. Gutmann, K. Graf, and S. Kheirandish, "Bound rubber morphology and loss tangent properties of carbon-black-filled rubber compounds," *Colloid Polym. Sci.*, vol. 294, no. 3, pp. 501–511, Mar. 2016. DOI. 10.1007/s00396-015-3802-6

[12] M. Rebouah and G. Chagnon, "Permanent set and stress-softening constitutive equation applied to rubber-like materials and soft tissues," *Acta Mech.*, vol. 225, no. 6, pp. 1685–1698, 2014. DOI. 10.1007/s00707-013-1023-y

[13] L. Shen, L. Xia, T. Han, H. Wu, and S. Guo, "Improvement of hardness and compression set properties of EPDM seals with alternating multilayered structure for PEM fuel cells," *Int. J. Hydrogen Energy*, vol. 41, no. 48, pp. 23164–23172, 2016. DOI. 10.1016/j.ijhydene.2016.11.006

[14] Y. Lu, J. Zhang, P. Chang, Y. Quan, and Q. Chen, "Effect of filler on the compression set, compression stress-strain behavior, and mechanical properties of polysulfide sealants," *J. Appl. Polym. Sci.*, vol. 120, no. 4, pp. 2001–2007, May 2011. DOI. 10.1002/app.33298

Shape Memory Behavior Of Shape Memory Natural Rubber

N M Setyadewi* and I N Indrajati
Center for Leather, Rubber, and Plastics - Ministry of Industry
9 Sokonandi Road, Yogyakarta, 55166, Indonesia

nmsetyadewi@gmail.com

Abstract. Natural rubber compound with different blend ratios of sulfur was prepared. Fabrication of shape memory polymer by swelling natural rubber compound with palmitic acid at 75 ^{0}C (100 wt% palmitic acid loading). Shape recovery experiments run by stretching to 100% (1 hour) at permanent set after heating treatment in a water bath (75°C for 30 s) and then freezing in 10°C water for 30 s. The samples were subsequently reheated in a water bath to recover the permanent shape (three stretching cycles). Shape memory compound swollen palmitic acid exhibited shape recovery of 96-100% and shape fixity 57-98%. Repeated shape recovery cycles had little effect on its shape memory property and mechanical property.

Keywords: natural rubber, shape memory polymer, palmitic acid, shape fixity, shape recovery

1. Introduction

Shape Memory Polymers are materials that can be programmed become a temporary shape so that an external stimulus as light [1], temperature [2] is initiated to activate the shape recovery process to return to their original shape [3]. At least, there are two different phases in shape memory polymers that are stable network phase and a crystalline phase. Stable phase assists to stabilize and resist the original shape. The shape recovery at this phase driven by deformation. The stable phase can be made by the introduction of cross-linking chemicals or interpenetrating network [4]. Furthermore, the crystalline phase is liable in impeding shape recovery until temperature reaches above melting temperature or glass transition temperature [5].

There are two types of SMP's, namely SMP's based on melting transition (T_M) and SMP's based on glass transition (T_g). Natural rubber (NR) is a common example of T_M based SMP's. NR crystallized when it's strained, which imparts outstanding green strength and tack, and gives natural rubber high resistance to crack growth at severe deformation [3]. Benefits of shape memory natural rubber (SMNR) reside in mechanical energy, its ability to store large strain [6], tunable trigger temperature during and after programming [7,8].

There were two strategies that have been developed to induce shape memory effect on natural rubber, firstly, strain-induced crystallization in weakly cross-linked NR was reported by Katzenberg et al [9]. The second method is based upon blending of natural rubber with additives such as fatty acids [10]. Brostowitz et al. proposed fabrication of SMP using sulfur cross-linked filled rubber band by swelling it in molten stearic acid to create shape fixing the temporary network.

IOP Conf. Series: Materials Science and Engineering **553** (2019) 012050 doi:10.1088/1757-899X/553/1/012050

Wee et al., have been exploring the possibility of applying palmitic acid as the swelling agent in the fabrication process and researched the thermomechanical response of SMNR. Basically, palmitic acid could serve as a retard the onset of curing and vulcanization of rubber [11]. Palmitic acid could be one of the factors supported to increase of cross-linking density rubber, which leads to higher tensile strength, but after optimum level, softening the effect of acid expropriate the cross-linking effect and decrease tensile strength of SMNR [3].

In this research, SMNR was made from a compound formula with variations of sulfur addition and swelling palmitic acid method (100 wt%) to explore mechanical properties and repeatability of SMNR.

2. Experimental method

2.1. Material

SMNR compound was made from natural rubber (NR) Ribbed Smoke Sheet-1, Zinc oxide (ZnO, Indoxide) and stearic acid (Aflux 42M Rhein Chemie) was used as activator, Carbon black N660 was used as filler, antidegradant 2,2,4-Trimethyl-1,2-Dihydroquinoline (TMQ) and N-(1,3-Dimethylbutyl)-N'-phenyl-p-phenylenediamine (6PPD, Starchem), paraffinic oil (Indrasari), CBS, vulcanizing agent sulfur (Miwon) and palmitic acid from Graha Jaya Pratama Kinerja.

2.2. Method

The compounds were prepared by mixing the ingredients for SMNR compounds with a laboratory size two-roll mill as illustrated in Table 1. SMNR compounding was started by mastification of NR approximately 5 min, then followed adding with ZnO and stearic acid, paraffinic oil and carbon black, then 6PPD, TMQ, CBS, and the last is sulfur. After obtained a homogeneous compound, approximately 21 min, the compound was stored for 24 h in room temperature. Curing time of the compound was tested by Moving Die Rheometer Gotech M-3000A at 150°C (Gotech 3000 A). SMNR compounds were vulcanized in a hydraulic press (Toyoseiki A-652) at 150 °C with a pressure of 150 kg/cm2 with time-based on the rheometer data.

Table 1. SMNR compound formulation.

Material	an amount, phr (per hundred rubber)		
	S7	S12	S15
NR	100	100	100
ZnO	0.2	0.2	0.2
Stearic acid	1.4	1.4	1.4
CB N660	3	3	3
TMQ	1	1	1
6 PPD	0.5	0.5	0.5
Paraffinic Oil	3	3	3
CBS	1	1	1
Sulfur	0.75	1.25	1.5

2.3. Mechanical properties

The mechanical properties of SMNR vulcanizates were observed using Universal Materials Testing Machine (UTM) Tinius Olsen. The parameters tested were tensile strength and elongation for 3 cycles.

2.4. Shape memory properties

The shape memory properties of the palmitic acid swollen SMNR vulcanizates (code PA100) were made by modified method refers to Wee et al. [3], Brostowitz et al. [10] and Sun et al. [12]. SMNR vulcanizates were swollen in molten palmitic acid (100 % w/w) at 75°C for 1 h then quenching in a beaker of a 10 °C water.

Sun et al., proposed a method to measure shape memory behavior and it's adopted in this research. SMNR vulcanizates with length L_0 were immersed in a 75 °C water bath for 30 s, stretched to 100% strain with a permanent set, and then immersed in a 10 °C water for 30 s. Then the stretched samples were left at permanent set approximately 1 h at temperature room and the length was measured as L_1. After 1 h, SMNR vulcanizate unclamped from a permanent set and the length were measured again as L_2. Shape recovery was initiated by immersing the unclamped SMNR vulcanizate in the 75 °C water bath for 30 s, and then immersed in a 10 °C water for 30 s, the final length was measured and taken as L_3. Gauge length was marked on the sample with a pen and measured with calipers. The shape memory phenomena were evaluated by calculating the shape fixity was defined as the strain fixed by crystallin (SF) using equation (1) and shape recovery (SR) is defined as Equation (2).

$$SF = \frac{L_2}{L_1} x100\% \quad (1)$$

$$SR = \frac{(L_2 - L_3)}{(L_2 - L_0)} x100\% \quad (2)$$

3. Results and discussion

3.1. Mechanical properties

Mechanical properties of SMNR is depicted by tensile strength and total elongation as shown in Figure 1 and Figure 2. The lowest tensile strength has resulted from the blank sample at cycle 1. Generally, tensile strength increases with the increasing of a sulfur amount.

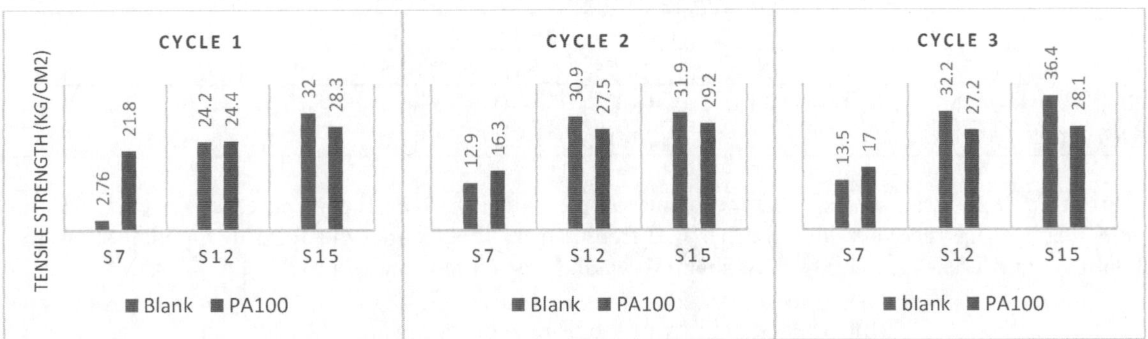

Figure 1. Effect of a sulfur amount on tensile strength for 3 cycle

From Figure 1. it can be seen that tensile strength for Blank sample increases with the increment of a sulfur amount as well as PA100 sample. But the value of tensile strength PA100 sample decrease when compared with the blank sample. This phenomenon is similar to elongation value (as shown in Figure 2.)

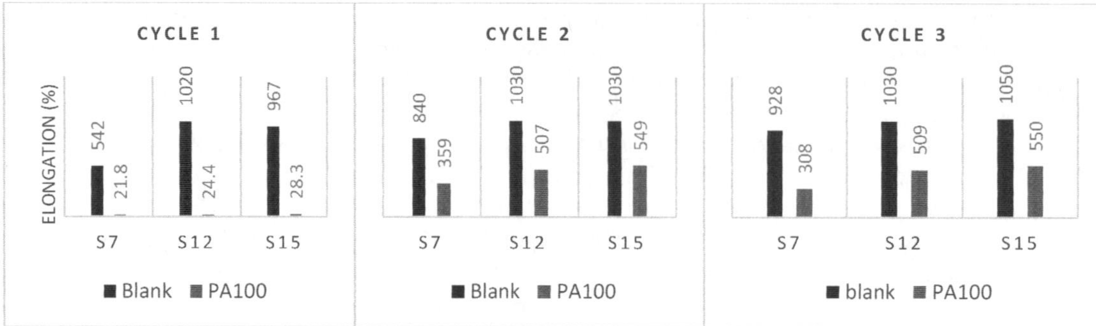

Figure 2. Effect of a sulfur amount on elongation for 3 cycle

Tensile strength values and elongation of SMNR Blank and PA100 samples for 3 cycles did not change much. But when compared between the Blank and PA100 sample, a decrease occurs. This showed that swelling treatments using palmitic acid was very influential on vulcanizate crosslinking. SMNR vulcanisate that swollen in palmitic acid becomes softer and brittle fracture which decreased the mechanical properties.

3.2. Shape memory behavior

Immersion treatment with palmitic acid makes microscopic crystalline is able to form a temporary network with a permanent crosslinking of the rubber compound. The key to the shape memory effect is the melting crystallization of the palmitic acid network and the entropic recovery force of the cross-linked NR [10]. The shape recovery of SMNR vulcanizates comes from the elastic strain and elastic recovery of the SMNR network.

Table 2. Repeatability of the SMNR properties for the blank sample.

BLANK	cycle 1			cycle 2			cycle 3		
	S7	S12	S15	S7	S12	S15	S7	S12	S15
SF (%)	26.4	10.1	11.0	18.1	10.0	9.3	15.4	9.9	8.3
SR (%)	58.7	74.6	76.2	77.0	65.8	77.0	87.8	67.4	75.8

From Table 2, it can be seen that the blank sample value of shape fixity decreases for each code in cycle 1 and shape recovery increases. It is different from the second and third cycles that show the phenomenon decreasing trend both of shape fixity and shape recovery value.

Table 3. Repeatability of the SMNR properties for PA100

PA100	cycle 1			cycle 2			cycle 3		
	S7	S12	S15	S7	S12	S15	S7	S12	S15
SF (%)	98.5	76.6	84.8	94.9	81.8	88.9	93.7	58.0	86.9
SR (%)	96.9	99.7	99.5	100.0	99.2	100.4	99.7	99.6	99.7

When compared with the PA100 sample, the value of shape fixity and shape recovery SMNR vulcanizate rises dramatically. This happened because more crystals are formed in PA100, which the crystal in SMNR vulcanizate straiten the mobility of the molecular chains and hence fixes the strain temporarily.

4. Conclusion

The sulfur amount had an influence on mechanical and shape-memory properties. Generally, tensile strength and degree of shape recovery increase with the increasing of a sulfur amount. Immersion treatment with palmitic acid makes microscopic crystalline is able to form a temporary network with a permanent crosslinking of the rubber compound. The key to the shape memory effect is the melting crystallization of the palmitic acid network and the entropic recovery force of the cross-linked NR. This research needs to be studied further to obtain the SMNR compound formula with better mechanical properties.

Acknowledgment

The authors would like to acknowledge the Center of Leather, Rubber, and Plastics - Ministry of Industry Indonesia for financial support and Dr. rer. nat. Noviyan Darmawan, MSc (Department of Chemistry IPB) for helpful discussion.

References

[1] Wu Y P, Lin Y H, Zhou F, Zheng Z H and Ding X B 2012 Light-induced shape memory polymer materials *Prog. Chem.* **24** 2004-10

[2] Luo H, Li Z, Yi G B, Wang Y, Zu X, Wang H, Huang H and Liang Z 2015 Temperature sensing of conductive shape memory polymer composites *Materials Letters.* **140** 71-4 http://dx.doi.org/10.1016/j.matlet.2014.11.010

[3] Wee, J S H, Chai A B and Ho J H 2017 Fabrication of shape memory natural rubber using palmitic acid *Journal of King Saud University - Science.* **29** 494-501 https://doi.org/10.1016/j.jksus.2017.09.003

[4] Hu J, Zhu Y, Huang H and Lu J 2012 Recent advances in shape-memory polymers: structure, mechanism, functionality, modeling, and applications. *Progress in Polymer Science.* **37** 1720-63 https://doi.org/10.1016/j.progpolymsci.2012.06.001

[5] Lendlein A and Kelch S 2002 Shape-memory polymers. *Angew. Chem. Int. Ed.* **41** 2034-57

[6] Huewers B, Beckel A, Krieger A, Katzenberg F and Tiller J C 2013 Shape-memory natural rubber; an exceptional material for strain and energy storage *Macromolecular Chemistry and Physics* **214** 912-23 https://doi.org/10.1002/macp.201200649

[7] Quitmann D, Reinders F M, Heuwers B, Katzenberg F and Tiller J C 2015 Programming of shape memory natural rubber for near discrete shape transitions. ACS Applied Materials and Interfaces **7** 1486-90 https://doi.org/10.1021/am507184c

[8] Heuwers B, Quitmann, D, Katzenberg F and Tiller J C 2012 Stress-Induced Melting of Crystals in Natural Rubber : a New Way to Tailor the Transition Temperature of Shape Memory Polymers. *Macromolecular Rapid Communications* https://doi.org/10.1002/marc.201200313

[9] Katzenberg F and Tiller J C 2016 Shape memory natural rubber *Journal of Polymer Science Part B: Polymer Physics* **54** 1381–88 https://doi.org/10.1002/polb.24040

[10] Brostowitz N R, Weiss R A and Cavicchi K A 2014 Facile fabrication of a shape memory polymer by swelling cross-linked natural rubber with stearic acid ACS Macro Letters 3(4) 374–7 http://dx.doi.org/10.1021/mz500131r

[11] Kirk-Othmer 1997 *Kirk Othmer Encyclopedia of Chemical-Technology* 24 pp 233-250

[12] Sun H, Zhao F, Peng H and Zhao S 2011 Influence of amount of sulfur on Mechanical, Dynamic, and Shape Memory Properties of Trans-1,4-Polyisoprene *Journal of Macromolecular Science Part B: Physics* **50** 871-9 http://dx.doi.org/10.1080/00222348.2010.497034

IOP Conf. Series: Materials Science and Engineering **553** (2019) 012051 doi:10.1088/1757-899X/553/1/012051

Biodegradability of Cassava Starch/High Density Polyethylene Reactive Blend During Compost Burial

Dwi Wahini Nurhajati*, Noor Maryam Setyadewi, Ihda Novia Indrajati
Center for Leather, Rubber and Plastics, Jl.Sokonandi 9, Yogyakarta 55166, Indonesia

dwiwahini@gmail.com

Abstract. Starch is promising material to answer the environmental hot issues related to plastic wastes. Plasticized and gelatinized starch can behave as thermoplastic that completely biodegrade in nature. This study was aimed to investigate the biodegradability of cassava starch (CS) blended with high density polyethylene (HDPE). Simulated soil burial was applied using the compost. The blends were prepared by reactive mixing using Haake Rheomix internal mixer. The ratio of starch and HDPE was ranging from 30/70, 40/60, 50/50, 60/40, and 70/30. Dual hydrophobization was employed, i.e. addition acetic acid and polyethylene-grated maleic anhydride (PE-g-MA). Compost burial was performed indoor for 56 days. The surface appearance and morphology were investigated. Weight loss and change in tensile strength and elongation at break after burial also determined. Results findings showed that the increasing of cassava starch content and burial time decrease the tensile strength and elongation at break, however increased the weight loss of CS/HDPE reactive blend. In addition, evidence of biodegradation is shown by the appearance of some microorganism colonies on the surface of reactive blend and morphological changes in CS/HDPE reactive blend.

Keywords : Cassava starch, HDPE, reactive blend, biodegradability, compost burial

1. Introduction

A material is categorized as an ideal when stables during its service life but biodegrade rapidly in the environment [1]. By this definition, plastic is not an ideal even though it is relatively stable during service life but it hardly biodegrades when discarded into the environment. Conventional plastics, originate from petroleum such as polyolefin groups, degrade within decades or even hundreds year. Nowadays, large volumes application of plastics in most aspect of daily life has led to serious environmental problems. Therefore, there is a growing demand for environmentally friendly plastic that can compete with conventional plastic for both mechanical properties and cost-performance basis. Developing a fully biodegradable plastic, such as poly-lactic acid (PLA), from renewable resources could be one good solution. Unfortunately, this development needs high investment and is complicated production process involving large volumes of chemicals. Starch and starch-based polymer are promising for inducing biodegradability property. They are inexpensive, renewable resources and naturally biodegradable [2].

Plasticized and gelatinized starch (often called by thermoplastic starch/TPS) can behave similar to thermoplastic [2,3]. The application of TPS is limited by its hydrophilic nature [3] and low thermal stability [5]. TPS fails to fulfill several essential requirements in general application because of its sensitivity to water that lead to brittleness [6]. The main point taken to improve these limitations is hydrophobization of starch, either through chemical modification or blending with other hydrophobic materials. One chemical modification is the addition the carboxylic acid into starch during processing. This addition has proven to increase the mechanical properties, thermal properties, aging and moisture resistivity of the resultant TPS [6]. TPS often blends with synthetic polymers such as polypropylene, polyethylene, poly-lactic acid and poly-glycolic acid [4] or polyesters to attain higher mechanical properties [7]. Many studies have been successfully blended starch with PE through reactive blending. Various types of starch such as corn starch [1,3,6,8], wheat starch [3] and cassava starch [7–9] were employed within these studies. High-density polyethylene (HDPE) is one of the polyethylene family and used in a variety of applications where excellent impact resistance, high tensile strength, low moisture absorption and chemical as well as corrosion resistance properties are required. However, the blend of starch and HDPE is immiscible with high interfacial tension. HDPE is highly hydrophobic, while on the other hand starch is hydrophilic. Thus, the blend of these materials is non-compatible that lead to inferior properties. The addition of compatibilizer could overcome this problem. The most common compatibilizer used for this purpose is PE grafted with maleic anhydride (PE-g-MA) [1–3]. PE-g-MA improve mechanical properties to some degree by improving the interfacial adhesion and reduces the dispersed phase [8].

The biodegradability term is related to how a material can be degrade biologically in nature. The key parameter of biodegradable material is the sensitivity to water and moisture. It becomes important because most microorganisms are likely to be active in high moisture atmosphere. On the other hand, material also must have sufficient strength to withstand the load during service life. Thus, comprehensive study should be taken in designing phase to obtain balanced properties. The study of biodegradation of a material usually describes in the term of low wettability, high water absorption and high weight loss after soil burial [2]. Soil burial, either natural or compost simulation, is probably the simplest way to investigate the biodegradability characteristic. Several studies have been attempted using this method [7,10–13]. They reported that weight loss dan reduction in mechanical properties of the starch/PE blends were observed after soil burial test.

In present work, the biodegradability of cassava starch/HDPE blends was investigated through simulated soil burial. Acetic acid was employed to improve starch hydrophobization and compatibilizer PE-g-MA was applied to enhance the interfacial interaction between starch and HDPE. By this dual modification, it was expected that the mechanical properties of starch/HDPE would increase, balanced with the biodegradation characteristic.

2. Experimental method

2.1 Materials

HDPE extrussion grade of Asrene SF 5007 were supplied from PT. Chandra Asri (density 0.950 g/cm^3 and melt flow index 0.05 g/10 min (190 °C/2.16 kg)). Local tapioca starch with approximately 17.87% wt of amylose and 82.13% wt of amylopectin. Polyethylene-grafted-maleic anhydride (PE-g-MA) compatibilizer was purchased at Sigma-Aldrich with melting point and density of 156°C and 0.934g/cm^3 respectively. Antioxidant Songnox 1010 was supplied by Songwon International. The commercial glycerol and aquadest as a plasticizer. Stearic acid and acetic acid were purchased from Bratachem. The compost was purchased from a flower nursery.

2.2 Preparation of CS/HDPE reactive blend Materials

Cassava starch (CS) was premixed with acetic acid, glycerol and water with a ratio of 70% of cassava starch, 1.3% of acetic acid, 20% glycerol, and 8.7% water. These materials were mixed thoroughly untill attaining a homogeneous powdered for 15 min. The mixture then keep 2 days to allow the starch granules to swell. The CS/HDPE were prepared by reactive blending in a Haake Rheomix internal mixer at a temperature of 130°C and a rotor speed of 40 rpm for 15 minutes. The ratio CS/HDPE were

IOP Conf. Series: Materials Science and Engineering **553** (2019) 012051 doi:10.1088/1757-899X/553/1/012051

varied from 30/70; 40/60; 50/50; 60/40; and 70/30. PE-g-MA was employed 5% wt based on polymer blend. The amount of other additives (antioxidant, and stearic acid) were kept constant for all formulation.

2.3 Compost Burial Test

Compost burial test was carried out on a laboratory scale using the compost in accordance to Obasi et al. [14] with some modifications. The compost were purchased from local market that composed of manure, charcoal husk, sand, soil, and cocopeat. Therefore, the compost certainly contain microorganisms such as fungi and bacteria. Dumbbell shaped specimens of definite sizes were cutted from each of the blends and placed into plastic pot containing moist compost. The specimens were buried in the compost at a depth of 10 cm from the surface in order to be the subject of attacking microorganisms in compost. The test was carried outside the room and lasted for 56 days. After the test, the specimens were removed, gently cleaned to remove impurities, and condisioned at 23±2°C for 24 hours for further testing. The weight loss of the samples was calculated according to Equation (1),

$$weight\ loss = \frac{W_0 - W_1}{W_0} \times 100\% \quad (1)$$

where W_0 is the initial weight of the samples before test, and W_1 is the residual weight of the samples after test. Surface appearance also observed to investigate the changes after soil burial test.

2.4 Tensile Properties

Tensile tests for CS/HDPE blends were carried out with a universal testing machine Tinius Olsen-H25K according to ASTM D 882. A dumbbell shaped type was employed. Tensile strength and elongation at break were determined as the average of 5 replications. The tensile test was performed on the sample before and after soil burial.

2.5 Morphology Properties

The morphology of the CS/HDPE blends before and after soil burial was evaluated using a Scanning Electrone Microsscope SEC-SNE 3200M with 500x magnification.

3. Results and discussion

3.1. Surface appearance and morphology during compost burial test

HDPE is synthetic polymer with high hydrophobic level and high molecular weight, but is naturally non-biodegradable. Improvement of biodegradability of PE can be achieved by increasing its hydrophilicity level and/or reducing the polymer chain length by oxidation to be accessible for microbial degradation [9].

Figure 1 shows CS/HDPE reactive blend photos after compost burial test. It is visually observed that several black to gray and brown stains appear on the surface of the blends. These stains are assumed as the microorganism colonies. Water and moisture present in the compost during burial test facilitate the growth of microorganism. Further, these microorganisms utilize starch as sole carbon source to support their life, resulting in partial degradation of the starch/HDPE blends [11]. Similar result is reported by Heydari et al. [15], in which black stains are observed on the surface of corn-starch nanocomposite film after soil burial associated with colonies of microorganisms. The higher starch content and the longest burial period, the more microorganisms present on the blend surface. It is related to the higher moisture absorption on higher starch content and longest burial period. This condition will accelerate the microorganism's growth. These result show that the microorganisms initially attack starch in the blend, resulting in an increase in the porosity and consequent enhancement of its biodegradability as seen in Fig. 2b (i) to (iii).

IOP Conf. Series: Materials Science and Engineering **553** (2019) 012051 doi:10.1088/1757-899X/553/1/012051

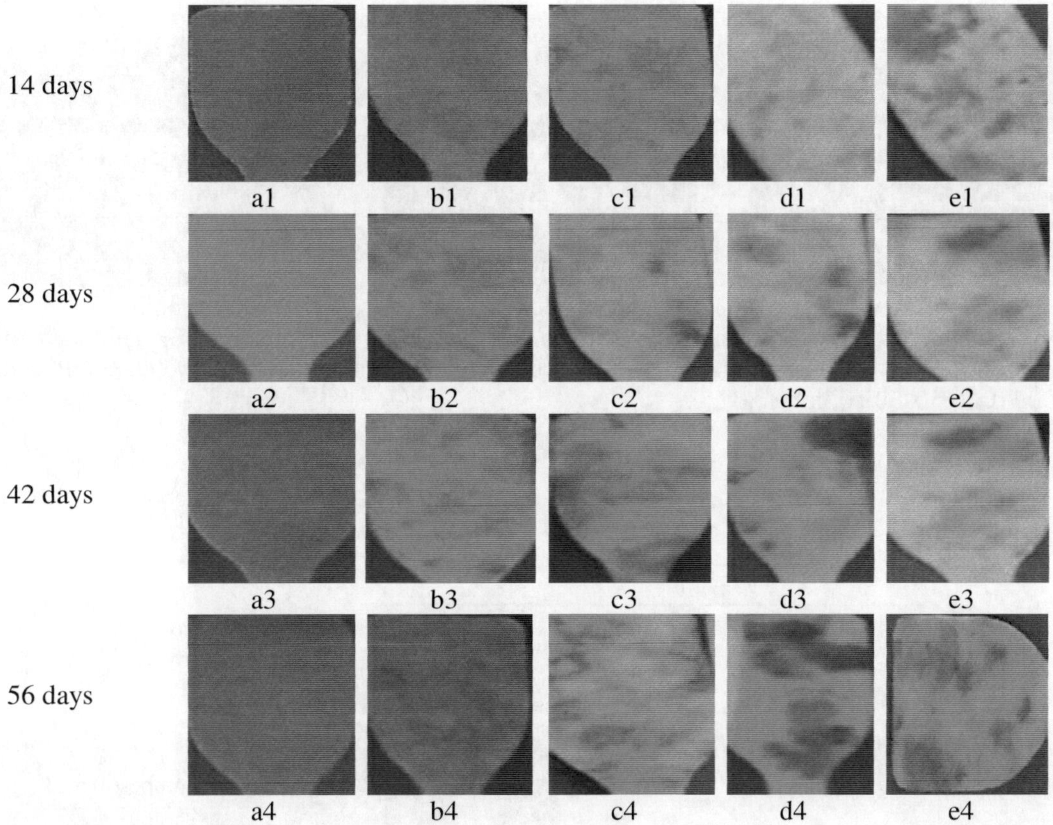

Figure 1. Surface appearance during compost burial of various CS/HDPE ratio a) 30/70, b) 40/60, c) 50/50; d) 60/400 and e) 70/30

Compost burial affect the morphology of CS/HDPE blend surface as shown in Figure 2. Initially, the CS/HDPE blends (2ai-iii) exhibit homogeneous and relatively smooth surface morphology. But they are slightly changed after 14 days burial (2bi-iii). The surface becomes rough and small holes and cracks on the surface are observed. The numbers of the hole and crack are more upon increasing starch content. Abd El-Rehim et al. [9] and Roy et al. [16] reported similar findings. It is presumably that the holes and crack are form as a result of microorganism activity during burial time. As they use starch as the food for their life, the blend starts to degrade. The holes and cracks formation also a sign that the CS/HDPE blend was degraded and decayed in the compost environment, but needed a time for a significant degradation.

Before burial

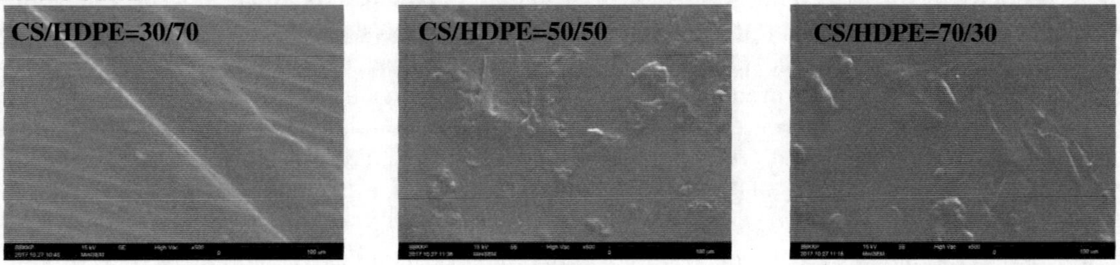

2ai 2aii 2aiii

IOP Conf. Series: Materials Science and Engineering **553** (2019) 012051 doi:10.1088/1757-899X/553/1/012051

After 14 days burial

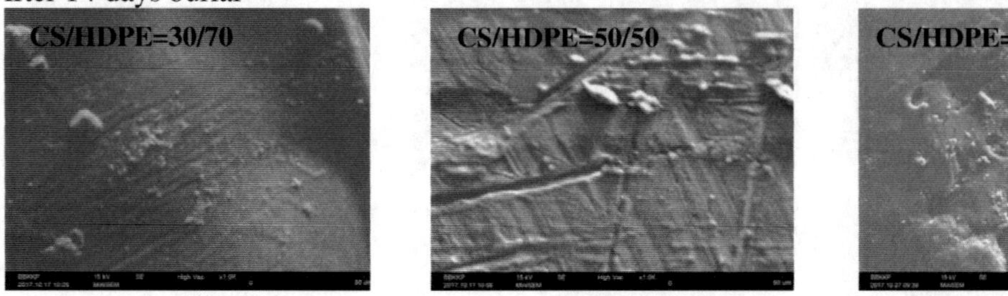

2bi 2bii 2biii

Figure 2. Scanning electron micrograph of the blend containing various CS/HDPE ratio a) before burial test b) after 14 days burial test. Note: nomenclature i, ii and iii associated to CS/HDPE ratio of 30/70, 50/50, 70/30, respectively.

3.2. Weight loss

The weight loss of CS/HDPE reactive blends at different CS/HDPE ratio after compost burial test shown in Figure 5.

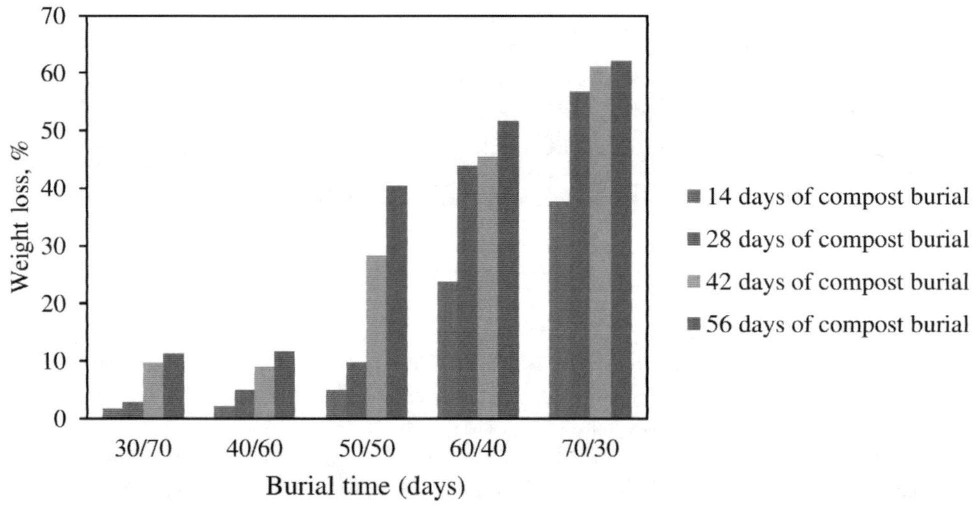

Figure 3. Weight loss of CS/HDPE blends during burial time.

Figure 3 shows the weight lost of CS/HDPE reactive blend at different CS/HDPE ratio content generally increase with increase in CS content and burial time. This weight loss may be due to the invasion of microorganisms into the reactive blends sample due to reactive blends absorpt moisture from the compost during the burial. Reactive blends with higher moisture absorption is usually more prone to microorganism attack. According to Maran et al. [4], Sahari et al. [16] increase in hygroscopic characteristics of material will promote the growth of microorganisms during degradation and increases the weight loss of material. The reactive blend samples with highest CS content in compost degraded rapidly in the first 14 days. Over the next 56 days a gradual decrease of starch occurred. In the 14 days burial test period, weight reductions of reactive blend containing CS 30 phr, 40 phr, 50 phr, 60 phr and 70 phr were 1.75%, 2.19%, 4.97%, 23.77% and 37.75% respectively. In the longest burial test period (56 days) weight reduction of reactive blend containing CS 30 phr, 40 phr, 50 phr, 60 phr and 70 phr were 11.2%. 11.68%, 40.42%, 51.64% and 62.08% respectively.

3.3. *The change of tensile properties*

Structural degradation as a result of microorganism activity in compost burial consequently affect the mechanical properties. Effect of compost burial test on the tensile strength and elongation at break are shown in Figures 4 and Figure 5 respectively.

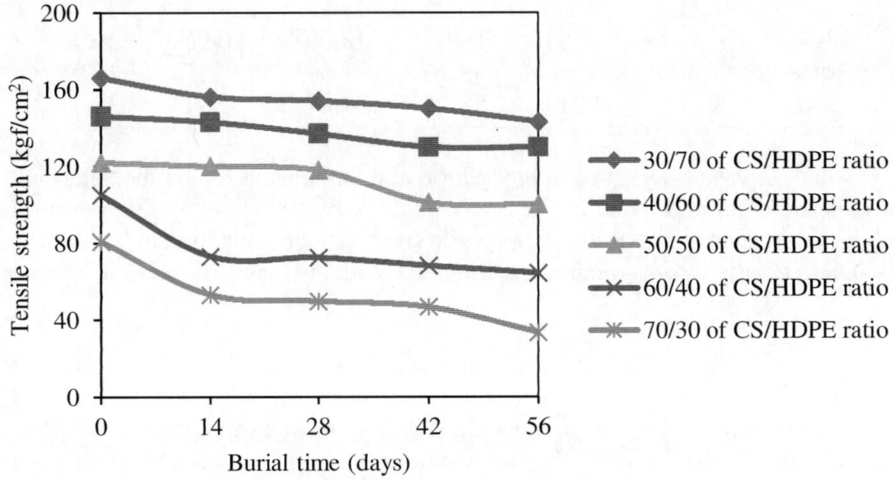

Figure 4. Tensile strength change as a function of number of burial days.

Tensile strength tends to decrease by increasing burial time. It confirms the structural changes occurred during burial. Figure 4 clearly depicts that the reduction of tensile strength runs little bit slow up to 50% starch in the blend. In the other hand, fast reduction is observed when starch content is 60 and 70% the first 14 days of burial, then the reduction rate is slower. Similar to tensile strength, elongation at break of the blends also reduces upon increasing burial times. This reduction probably related to water absorption capability of the starch. As mentioned earlier, high water absorption of starch will stiffen the starch. Thus, the blend becomes more brittle and will fails at low elongation. The severe effect is found on higher starch content (70%). Details data related to loss tensile strength and elongation at break is listed on Table 1.

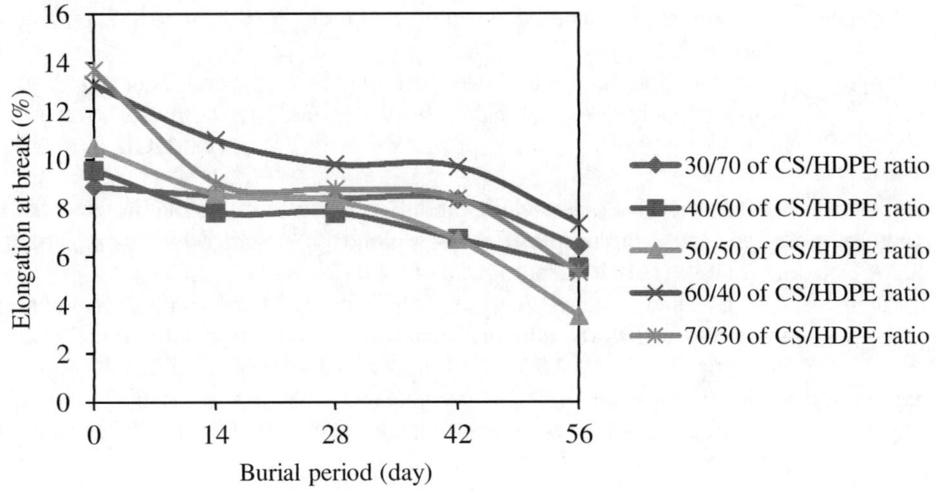

Figure 5. Elongation at break as a function of number of burial times.

Table 1. Percent decrease in tensile strength and elongation at break.

CS/HDPE ratio	% loss in tensile strength				% loss in elongation at break			
	burial time				burial time			
	14 days	28 days	42 days	56 days	14 days	28 days	42 days	56 days
30/70	6.02	7.23	9.64	13.85	4.17	5.74	5.97	27.93
40/60	2.05	6.16	10.96	10.96	18.11	18.32	29.74	42.20
50/50	1.64	3,28	17.21	18.03	18.09	20.95	20.95	66.38
60/40	30.48	31.05	35.14	38.76	17,56	25.27	25.95	43.97
70/30	34.2	38.41	42.13	58.74	34.01	35.91	38.69	61.09

The reduced value of tensile strength and elongation at break are caused by the deterioration of the sample that occurs more and more due to the breaking of polymer bonds by microorganisms in the compost which ultimately leads to a decrease in tensile strength and elongation at break of the reactive blend samples. These results are in agreement with Obasi et al. [14], and Ali et al. [17].

4. Conclusion

Compost burial test result showed that the increasing of cassava starch content and burial time reduced the tensile strength and the elongation at break, however increased weight reduction of CS/HDPE reactive blends. At the 56 days of burial test, the highest drop of tensile strength (58.74%) was indicated by a reactive blend with 70/30 of CS/HDPE ratio, whereas the highest percent decrease of elongation at break (61.09%) indicated by reactive blend with 50/50 of CS/HDPE ratio. The physical appearance of the reactive blend surface shows that some black and red stains are formed. The morphology test result showed that voids cracks, and damages on the surface are formed after burial in compost.

Acknowledgment

This research was funded by the Indonesian government under DIPA 2018. The authors are grateful to the Head of Center for Leather, Rubber and Plastics for facilitating this research. The authors also want to thank to the Metal Industries Development Center (MIDC) for providing internal mixer machine facilities for reactive blend preparation. Great thankful also honor to PT. Inter Aneka Lestari for providing the antioxidant.

References

[1] A. P. Abott, T. Z. Abolida, W. Qu, W. R. Wise, and L. A. Wright, "Thermoplastic starch-polyethylene blends homogenised using deep eutetic solvent," *RSC Adv.*, vol. 7, pp. 7268–7273, 2017. DOI. 10.1039/c7ra00135e

[2] S. Pichaiyut, C. Nakason, and S. Wisunthorn, "Biodegradability and Thermal Properties of Novel Natural Rubber/Linear Low Density Polyethylene/Thermoplastic Starch Ternary Blends," *J. Polym. Environ.*, vol. 26, no. 7, pp. 2855–2866, Jul. 2018. DOI. 10.1007/s10924-017-1174-3

[3] A. Taguet, M. N. Bureau, M. A. Huneault, and B. D. Favis, "Toughening mechanisms in interfacially modified HDPE/thermoplastic starch blends," *Carbohydr. Polym.*, vol. 114, pp. 222–229, Dec. 2014. DOI. 10.1016/j.carbpol.2014.07.073

[4] H. A. Pushpadass, P. Bhandari, and M. A. Hanna, "Effects of LDPE and glycerol contents and compounding on the microstructure and properties of starch composite films," *Carbohydr. Polym.*, vol. 82, no. 4, pp. 1082–1089, Nov. 2010. DOI. 10.1016/j.carbpol.2010.06.032

[5] K. Jantanasakulwong, N. Leksawasdi, P. Seesuriyachan, S. Wongsuriyasak, C. Techapun, and T. Ougizawa, "Reactive blending of thermoplastic starch and polyethylene-graft-maleic anhydride with chitosan as compatibilizer," *Carbohydr. Polym.*, vol. 153, pp. 89–95, Nov. 2016. DOI. 10.1016/j.carbpol.2016.07.091

[6] A. H. M. Zain, M. K. Ab Wahab, and H. Ismail, "Biodegradation Behaviour of Thermoplastic Starch: The Roles of Carboxylic Acids on Cassava Starch," *J. Polym. Environ.*, vol. 26, no. 2,

pp. 691–700, Feb. 2018. DOI. 10.1007/s10924-017-0978-5

[7] H. Vieyra Ruiz, E. S. M. Martínez, and M. Á. A. Méndez, "Biodegradability of polyethylene-starch blends prepared by extrusion and molded by injection: Evaluated by response surface methodology," *Starch - Stärke*, vol. 63, no. 1, pp. 42–51, Jan. 2011. DOI. 10.1002/star.201000075

[8] M. Sabetzadeh, R. Bagheri, and M. Masoomi, "Study on ternary low density polyethylene/linear low density polyethylene/thermoplastic starch blend films," *Carbohydr. Polym.*, vol. 119, pp. 126–133, Mar. 2015. DOI. 10.1016/j.carbpol.2014.11.038

[9] H. . Abd El-Rehim, E.-S. A. Hegazy, A. . Ali, and A. . Rabie, "Synergistic effect of combining UV-sunlight–soil burial treatment on the biodegradation rate of LDPE/starch blends," *J. Photochem. Photobiol. A Chem.*, vol. 163, no. 3, pp. 547–556, May 2004. DOI. 10.1016/j.jphotochem.2004.02.003

[10] D. M. Nguyen, T. V. V. Do, A. C. Grillet, H. Ha Thuc, and C. N. Ha Thuc, "Biodegradability of polymer film based on low density polyethylene and cassava starch," *Int. Biodeterior. Biodegrad.*, vol. 115, no. September, pp. 257–265, 2016. DOI. 10.1016/j.ibiod.2016.09.004

[11] J. P. Maran, V. Sivakumar, K. Thirugnanasambandham, and R. Sridhar, "Degradation behavior of biocomposites based on cassava starch buried under indoor soil conditions," *Carbohydr. Polym.*, vol. 101, pp. 20–28, Jan. 2014. DOI. 10.1016/j.carbpol.2013.08.080

[12] M. D. Samper-Madrigal, O. Fenollar, F. Dominici, R. Balart, and J. M. Kenny, "The effect of sepiolite on the compatibilization of polyethylene–thermoplastic starch blends for environmentally friendly films," *J. Mater. Sci.*, vol. 50, no. 2, pp. 863–872, Jan. 2015. DOI. 10.1007/s10853-014-8647-8

[13] Y. Tanetrungroj and J. Prachayawarakorn, "Effect of starch types on properties of biodegradable polymer based on thermoplastic starch process by injection molding technique," *Songklanakarin J. Sci. Technol.*, vol. 37, no. 2, pp. 193–199, 2015.

[14] C. Henry and O. Isaac, "Cassava Starch-Mixed Polypropylene Biodegradable Polymer : Preparation , Characterization and Effects of Biodegradation Products on Growth of Plants," *Int. J. Sci. Res.*, vol. 3, no. 7, pp. 802–807, 2014.

[15] A. Heydari, I. Alemzadeh, and M. Vossoughi, "Influence of glycerol and clay contents on biodegradability of corn starch nanocomposites," *Int. J. Eng. Trans. B Appl.*, vol. 27, no. 2, pp. 203–214, 2014. DOI. 10.5829/idosi.ije.2014.27.02b.05

[16] S. B. Roy, S. C. Shit, R. A. Senguptaξ, and P. R. Shukla, "Studies on Biodegradability, Morphology, Physicomechanical and Thermal Properties of Pp/Potato Starch Bio-Composite," *Int. J. Curr. Eng. Technol.*, vol. 5, no. 2, pp. 1074–1084, 2015.

[17] R. R. Ali, W. A. W. A. Rahman, R. M. Kasmani, N. Ibrahim, S. N. H. Mustapha, and H. Hasbullah, "Tapioca starch biocomposite for disposable packaging ware," *Chem. Eng. Trans.*, vol. 32, pp. 1711–1716, 2013. DOI. 10.3303/cet332286

Fabrication and Characterization Of Poly Lactic Acid (PLA)-Starch Based Bioplastic Composites

Akbar Hanif Dawam Abdullah*, Anti Khoerul Fikriyyah, Oceu Dwi Putri, Putri Putih Puspa Asri
Research Unit for Clean Technology, Indonesian Institute of Sciences (LIPI), Jalan Sangkuriang Komplek LIPI Gedung 50 Bandung, Indonesia 40135

akba001@lipi.go.id

Abstract. In this study, to make a good bioplastic composite, starch-based bioplastic is produced by adding polylactic acid (PLA) to improve its properties. PLA was added into starch-based bioplastic with various concentrations of 0, 3, and 10 wt.%. The extrusion was performed at 90-150 °C and compression moulding process was conducted at 150 °C and pressured at 50 kgf/cm^2. Bioplastic composites have been characterized to know its properties. FTIR analysis indicated shifting and increasing spectra of interaction between PLA and starch-based bioplastic. Contact angle and solubility analysis revealed that adding PLA can increase the stability of hydrophobic characteristic and insoluble properties. The combination of PLA and starch-based bioplastic can improve the mechanical properties. In addition, thermal properties of bioplastic composites have a better thermal stability and produce a lower melting point thus the energy needed to melt for bioplastic composites becomes milt as raising PLA composition. The density of bioplastics was in the range of 1.2 - 1.3 g/cm^3 that would be good for light bioplastic. The results of this study showed that the combination of starch-based bioplastics and PLA at low concentration (10wt.%) potentially could enhance the properties of bioplastic composites for food packaging.

Keywords: Bioplastics, Starch, Poly Lactic Acid, Composites

1. Introduction

The use of conventional plastics (non-biodegradable polymer) as food packaging produced waste and threatens the environment because the decomposition process is so slow for decades. The accumulation of plastic waste can damage natural ecosystems thus needs an alternative to reduce the plastic waste used primarily for food packaging. Plastic also has health security deficiencies since plastic compounds are not friendly to body tissues. This has become one of the world's concerns in handling plastic waste in the environment and the influence of using plastic as a food packaging for health.

One alternative that can be used to deal with this problem is fabricating plastic from natural materials (bioplastic). Bioplastic research and manufacture has been completed by researchers with multiple material resources and techniques. In general, bioplastic can be fabricated from various types of natural polymers such as cellulose [1], protein [2], starch [3] etc. [4]. Starch is one of the basic ingredients that have been widely used to make bioplastics. Starch is a biopolymer that is abundant in nature, biocompatible, biodegradable and easily renewed [5]. Starch generally contains amylose and amylopectin in different percentage of each source that would be affected on properties of bioplastics.

Research of starch-based bioplastics has been studied recently with various starch sources such as corn starch [6], potato starch [7], cassava starch, rice starch [8], canna starch [9], etc. However, the use of starch has disadvantages that its hydrophilic characteristic will decrease its properties as bioplastic [5]. Therefore, the addition of other ingredients is needed to improve typical starch-based bioplastic.

Poly Lactic Acid (PLA) is one of a synthetic polymer that can be degraded, renewable, eco-friendly, biocompatible and has hydrophobicity characteristic [10]. PLA has composed from lactic acid that synthesized from polysaccharide fermentation [3]. Blends of starch-based bioplastic and PLA have been investigated recently [11]. Several studies also add compatibilizers [12], modifying starch to be high crystallinity [13] and made layers to improve the hydrophobic properties of starch-based bioplastics with the addition of other materials such as chitosan [14], etc. Sanyang et al. (2016) conducted research of sugar palm starch-based bioplastics and PLA as layers, thus, the results showed that the addition of PLA improved the mechanical and hydrophobic properties of bioplastic [15]. Muller et al. (2016) also produced bioplastic from corn starch and PLA to identify interaction and miscibility of PLA and starch-based bioplastic with adding glycerol on PLA [16]. In this case, the addition of PLA in blending with starch-based bioplastic is above 10% which can affect the production cost of bioplastics for industrial scale. Besides, some solvents such as chloroform in the process of PLA dissolution is also dangerous since it ruins the health.

Researches of synthesis starch-based bioplastic and PLA addition at low concentration weren't much performed, so we aim to produce PLA - starch-based bioplastic blend (bioplastic composites) at low percentages of PLA (0, 3, 10%) using extrusion method to know the effectiveness of the addition of PLA with small concentration. Bioplastic composites were characterized by physical, chemical, thermal, mechanical, hydrophobicity and morphological observations of bioplastic composites.

2. Experimental method

2.1 Materials and chemicals
The native corn starch was obtained from PT. TEREOS (Indonesia) and moisture analyzer showed that water content of starch was around 15% and density measurement using pycnometer (PYREX) was around 1.45 g/cm^3; PLA was purchased from NatureWorks (U.S.A), and glycerol was obtained from PT Wilmar Nabati Indonesia.

2.2 Preparation of Starch-based Bioplastic
Starch-based bioplastic was prepared by dry-blending of corn starch and glycerol with ratios 3:1 (w/w based on starch) by using Philips Blender HR2116 through 600 W power for 3 minutes. Then, forming pellets of starch-based bioplastics were extruded by using a single screw extruder from Labo Plastomill, (Toyoseki, Japan). The heating sections were 80-130 ^oC and the screw speed was 35-50 rpm. Starch-based bioplastic also will be called as J-PLA 0%.

2.3 Fabrication of Bioplastic Composites
Pellet of starch-based bioplastics and PLA were combined with various concentrations of PLA at 3wt.% and 10 wt.% (J-PLA 3% and J-PLA 10%). Extrusion process was performed at heating temperature 90-150 ^oC. and samples were formed to be plates by compression moulding machine (Toyoseki, Japan) conducted in temperature 150 ^oC and pressured at 50 kgf.cm^{-2} for 5 min for J-PLA 0%, J-PLA 3% and J-PLA 10%.

2.4 FTIR Analysis
To identify the chemical interaction of bioplastic composites, samples were analyzed by Fourier Transform Infrared Spectroscopy in Attenuated Total Reflectance mode (FTIR-ATR) using Thermo

IOP Conf. Series: Materials Science and Engineering **553** (2019) 012052 doi:10.1088/1757-899X/553/1/012052

Scientific Nicolet iS5. Samples were determined by using air as background, recording 64 scans and resolution from 4000 – 400 cm^{-1}.

2.5 Density Analysis

The density of composites was measured by calibrated pycnometer (10 ml volume, PYREX) and distilled water was used as the immersing liquid with three repetitions. The sample was weighed before putting into pycnometer that contains liquid. After that, the sample in the pycnometer was weighed (m) and the density (ρ) was calculated by the equation:

$$\rho = \frac{m}{V} \tag{1}$$

2.6 Contact Angle Analysis

The contact angle of samples was captured triplicate for each sample manually and imageJ software was used to analyze hydrophobicity degree of composites.

2.7 Solubility Analysis

The samples were dried at 105 °C and weighed as Wi. Then, samples were immersed at 50 mL of distilled water for 24 h at room temperature. After that, samples were dried at 105 °C and weighed as Wo. Water solubility of samples was calculated with the equation:

$$Solubility\ (\%) = \left[\frac{Wi - Wo}{Wi}\right] \times 100 \tag{2}$$

2.8 Mechanical Analysis

Mechanical properties of samples were measured by Universal Tensile-meter (Tensilon, Japan) with three repetitions of each sample. It was controlled with room temperature and (23 ± 2 °C) and 60 ± 1% RH. The samples were analyzed using standard method JIS 7113-2-1/2 and load cell 40% from 100 kgf. The samples was pulled with crosshead speed 5 mm/min.

2.9 Morphological Analysis

Morphology image of composites was obtained by SEM (JEOL JSM-IT300) under the operating voltage of 20 kV at low vacuum. Samples were coated with gold and strip by adhesive tape (carbon tape).

2.10 Thermal Analysis

Thermal properties of the sample were carried out using TGA-DSC thermogram (Netzsch 214 Polyma) to analyze decomposition stability and melting point of the samples. The TGA test was measured at temperature 40 – 450 °C and scan rate 10 °C/min. then, DSC test was obtained at temperature 40 – 200 °C with scan rate 10 °C/min.

3. Results and discussion

3.1 FTIR analysis

Chemical interaction of bioplastic composites can be identified using FTIR spectrum analysis as shown in Figure 1. The addition of PLA into starch-based bioplastic produces functional groups formed in J-PLA 3% and J-PLA 10%. The functional group of –C=O stretching produced shifting peak from 1721 cm^{-1} (PLA) to 1750 cm^{-1} (J-PLA 3% and J-PLA 10%). The functional group of -C-O stretching also generates a shifting area from 1198 cm^{-1} to 1180 cm^{-1}. Then, the functional group of –CH_3 stretching was also developed at wavenumbers 1450 and 1370 cm^{-1}.

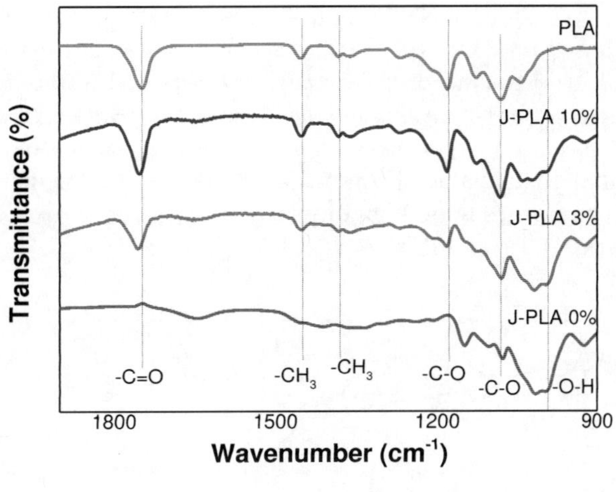

Figure 1. FTIR Spectra of PLA and bioplastic composites.

-C-O stretching was also grown from PLA and starch-based bioplastics in the wavenumber area of 1080 cm^{-1} that created a constructive peak, thus the absorbance values of J-PLA 3% and J-PLA 10% is higher than J-PLA 0%. The -OH bending in the wavenumber area of 1020 cm^{-1} is developed from starch-based bioplastic. Shifting and constructing peaks indicate the interaction between PLA and starch-based bioplastic developed on J-PLA 3% and J-PLA 10%. However, molecular bonds on FTIR spectra is not sufficient to justify the miscibility of PLA and starch-based bioplastic. Some functional groups were not strong enough to make changes in typical vibration in composite of PLA and starch-based bioplastic. Sanyang et al. (2016) confirmed that PLA and bioplastic from sugar palm starch produced the physical interaction only between PLA and bioplastic. In addition, Gonzales at al. (2013) verified that PLA and bioplastic from soy protein could not mix each other because the chemical carriage of them is appeared and not established a new chemical bond [2]. Muller et al. (2016) recognized the interaction between PLA and starch-based bioplastic by thermomechanical analysis and thermodynamic modeling indicate that interaction between PLA and thermoplastic starch (bioplastic) give poor miscibility and heterogeneous structures [16].

3.2 Density Analysis

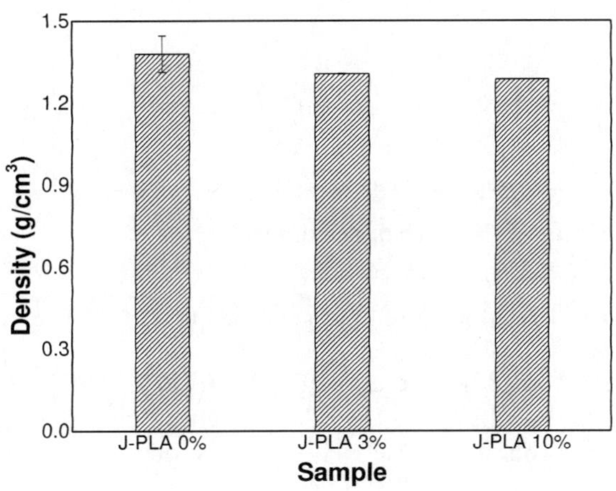

Figure 2. The density of bioplastic composites.

IOP Conf. Series: Materials Science and Engineering **553** (2019) 012052 doi:10.1088/1757-899X/553/1/012052

Physical properties of PLA – starch-based bioplastic composites were exhibited from density measurement in Figure 2. Based on that, J-PLA 0% had the highest density (1.3 g/cm^3) than others. Then, J-PLA 3% and J-PLA 10% produced density derivation related to the raising of PLA content. Starch and glycerol were included in J-PLA 0% in which the density values of starch and glycerol are 1.45 g/cm^3 and 1.21 g/cm^3 (based on experiment). While PLA density is around 1.24 g/cm^3[17] that mean PLA would decrease the density value of composites. It designates that increasing of PLA content can improve the density of composites to be light bioplastics.

3.3 Thermal Analysis

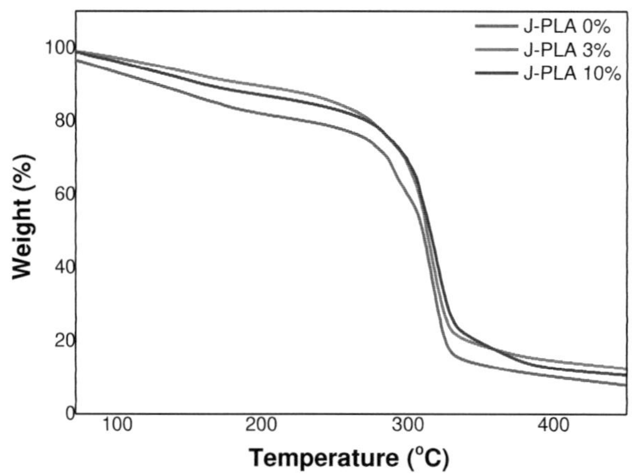

Figure 3. TGA thermogram of bioplastic composites.

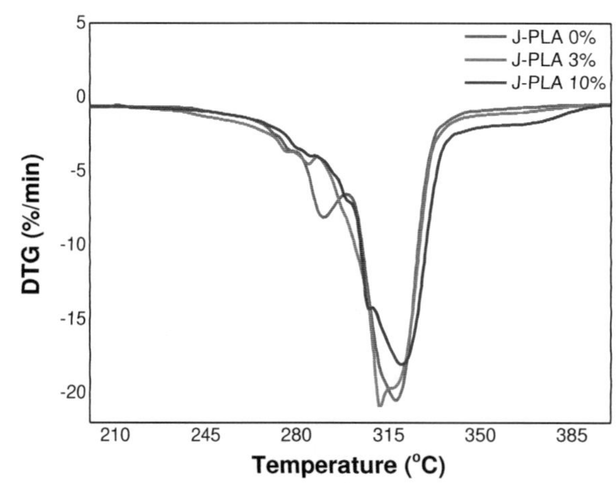

Figure 4. DTG thermogram of bioplastic composites.

Thermal properties of bioplastic composite were analyzed by TGA-DTG Thermogram as shown in Figure 3 and Figure 4. The thermal degradation of PLA – starch-based bioplastic was observed in the TGA curve (Figure 3). The samples has losing the weight at a temperature under 200 °C which is water molecules according to previous literature [2]. Furthermore, the DTG curve (Figure 4) shows decomposition at 290,6 °C with 20.86% weight loss in J-PLA 0% that indicated the evaporation of glycerol, which concurs with the literature [15]. J-PLA 3% and J-PLA 10% didn't present the peak of glycerol loss directly since the processes were heated twice. Then, decomposition process significantly established at temperature 317.9 °C with 52.4 % weight loss for J-PLA 0% indicating where the starch was degraded. And sharp peaks of J-PLA 3% and J-PLA 10% were established at temperature 311.9 °C and 319.9 °C related with 62.64% and 76.28% weight loss. Also, thermal degradation process records a residual mass until temperature 450 °C, where J-PLA 0% produced the lowest residual mass (7.96 %) than J-PLA 3% and J-PLA 10% which leaves 12.51% and 10.77% residue. The addition of PLA wasn't changed the thermal stability of bioplastic composites significantly and by adding PLA content can maintain the temperature so that the material mass is not easily decreased. This concurs with the literature that mass loss of PLA/starch blend increased as raising PLA content [15].

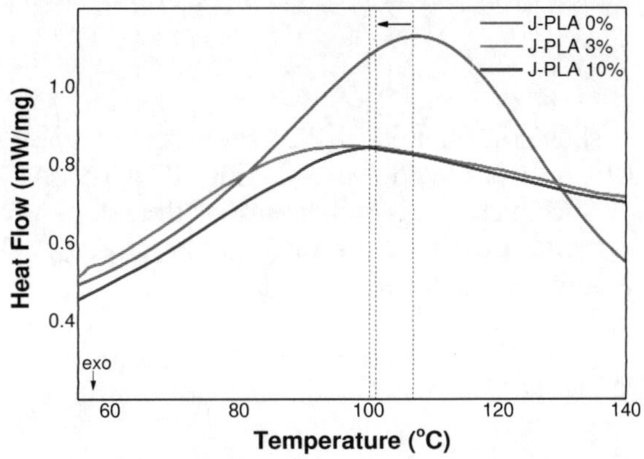

Figure 5. DSC thermogram of bioplastic composites.

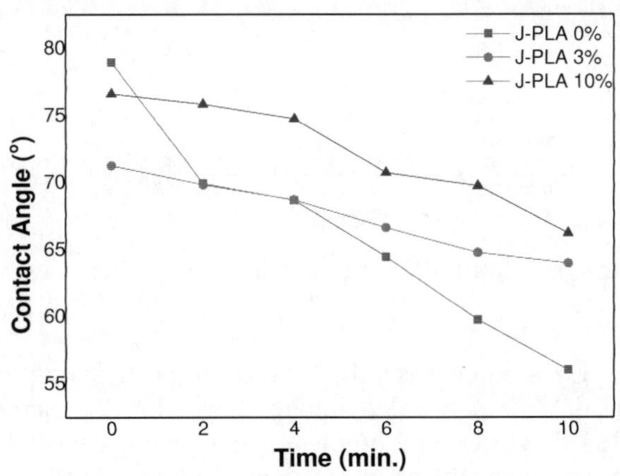

Figure 6. Contact angle measurements of bioplastic composites.

Melting point temperature of bioplastic composites was carried out by DSC thermogram analysis as shown in Figure 5. The results showed that the addition of PLA decreased melting temperature and energy requirements. J-PLA 0% had a melting temperature 107.6 °C and ΔH at 182 J/g. Thus, J-PLA 3% held melting temperature 96.7 °C with ΔH of 53.28 J / g and J-PLA 10% develop a melting point at 94.3 °C with ΔH of 54.71 J / g. Combination of PLA and starch-based bioplastic reduced melting point temperature and energy. This indicated that PLA addition increasing its free volume[18]. It is comparable with the density measurements that PLA addition can lessen the density value. It is good for the industry to minimalize the use of energy in the processing materials.

IOP Conf. Series: Materials Science and Engineering **553** (2019) 012052 doi:10.1088/1757-899X/553/1/012052

3.4 Contact Angle Analysis

The wettability and hydrophobicity properties of composites were analyzed by contact angle measurement as seen in Figure 6. The results indicated that J-PLA 0% had the highest degree, but also decreased significantly after a few minutes. J-PLA 0% contains materials that can absorb water such as starch. It has –OH bond so that bioplastics can bind H_2O molecules. The –OH bending also appears in FTIR spectra that is parallel with the literature [19].

Furthermore, J-PLA 3% and J-PLA 10% have the stability degree after a few minutes. It shows that the addition of PLA can be improving the hydrophobicity of composites. Nonetheless, appending PLA doesn't influence hydrophobicity degree at the beginning of measurement that is similar to the statement of literature [2].

3.5 Solubility Analysis

The solubility of composites is shown in Figure 7. J-PLA 0% has the highest solubility percentage at 48%. Then, J-PLA 3% and J-PLA 10% have lower solubility than J-PLA 0% where the solubility percentage is 27% and 25% respectively. This indicates that the addition of PLA can reduce the solubility of composite bioplastic and perform better water barrier properties. This is matched with the results of the contact angle measurement and literature [2].

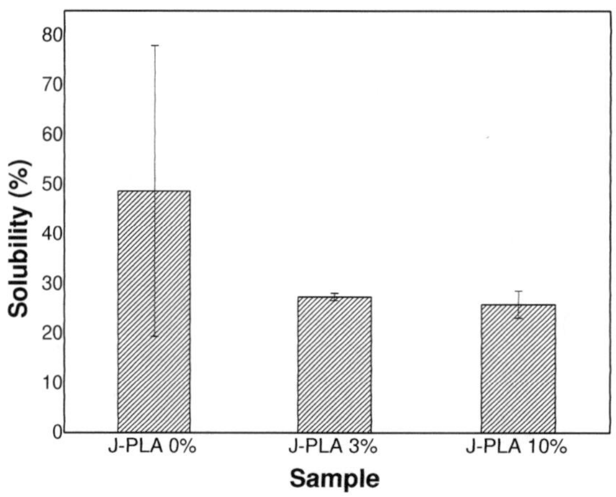

Figure 7. Solubility of bioplastic composites.

3.6 Mechanical Analysis

The mechanical properties of PLA - starch-based bioplastic composites are presented in Figure 9. It can be seen that J-PLA 0% has the lowest tensile strength, thus, with PLA addition in J-PLA 3% and J-PLA 10% increase the tensile strength values completely. On the other hand, the elongation at break of J-PLA 3% and J-PLA 10% are lower in the proportion of increasing PLA. It proved that adding PLA can raise tensile strength and decrease elongation at break. Mechanical properties of PLA is brittle and have high tensile strength values (55.4 MPa [11]) capable of enhancing tensile strength properties from bioplastic composites. This consists with previous studies that starch-based bioplastics have higher bending properties than PLA because starch-based bioplastic has amylose/amylopectin bonds which produce flexural properties and the addition of glycerol to starch-based bioplastic also improve the flexibility of bioplastic. Meanwhile, PLA has a very low elongation at break (<10% [11]) since PLA has strong bonds between lactic acid monomers which make PLA more rigid and have no flexibility like starch [5].

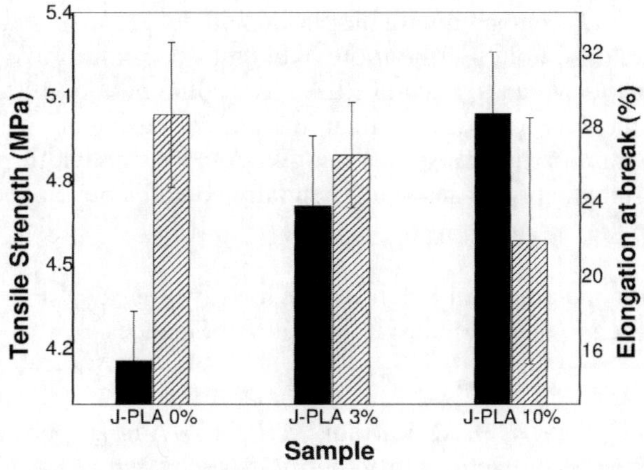

Figure 8. Mechanical properties of bioplastic composites.

3.7 Morphological Analysis

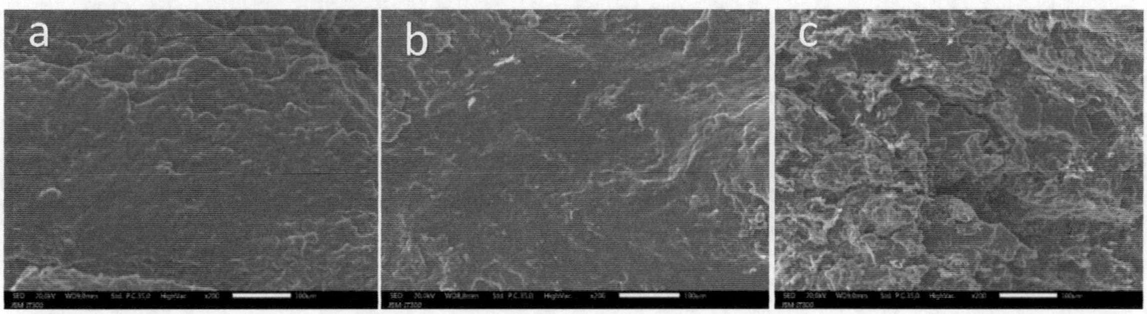

Figure 9. SEM image cross section of bioplastic composites J-PLA 0 % (a), J-PLA 3 % (b), and J-PLA 10 % (c).

Figure 10 shows a cross-sectional image of a morphological PLA – starch-based bioplastic composites. It can be considered that the fracture of J-PLA 0% (Figure 10a) has emerged. It means that J-PLA 0% arranges ductility characteristic on mechanical properties. However, J-PLA 3% and J-PLA 10% have a mixed fracture that requires the transition properties between ductile and brittle characteristics to have formed. J-PLA 3% exhibits PLA existence on wrinkle sheet structure that sticks on starch-based bioplastic. Furthermore, J-PLA 10% presents agglomeration structures and PLA content at wrinkle sheet structures, it appears much more than on J-PLA 3%. Besides, J-PLA 10% has a crack considering tensile strength between J-PLA 3% and J-PLA increase insignificantly. On the other hand, the solubility of J-PLA 10% and J-PLA 3% consume more than 20%. It means that crack on the samples influences water barrier solubility and mechanical properties. Also, the crack affected higher free volume in the samples that cause decreasing melting point temperature on thermal properties.

4. Conclusion

Bioplastic composites were produced successfully by extrusion and compression molding of starch-based bioplastic and PLA at low matter to determine its properties. The FTIR analysis indicated shifting wavenumbers and increasing spectra that means the interaction between starch-based bioplastic and PLA were happened in bioplastic composites. Nevertheless, raising PLA concentration can decrease bioplastic composite density to be light bioplastic that would be a good point for bioplastic. Afterwards, the contact angle measurement revealed hydrophobic properties of bioplastic composites by increasing PLA contents. The Solubility analysis also observed that adding PLA can resist water molecules up to 20% compared with bioplastic without PLA, so that bioplastic composites have a good resistance of water soluble. Furthermore, adding PLA can improve mechanical properties of bioplastic composites up to 5 MPa (10wt.% of PLA). The combination of PLA 10wt.% and starch-based bioplastic showed the best possession than the others based on its properties especially mechanical and water barrier properties. Despite adding PLA in low matter, these results affirmed that bioplastic composites had suitable properties and potentially could be developed as food packaging.

Acknowledgment
The authors are grateful for the financial support provided by Insinas Kemenristekdikti 2018 and Research Units for Clean Technology, Indonesian Institute of Sciences.

References
[1] Akrami, M., Ghasemi, I., Azizi, H., & Karrabi, M. (2016). A new approach in compatibilization of the poly (lactic acid)/ thermoplastic starch (PLA / TPS) blends. *Carbohydrate Polymers*, *144*, 254–262. http://doi.org//10.1016/j.carbpol.2016.02.035
[2] Bledzki, A. K., & Gassan, J. (1999). Composites reinforced with cellulose based fibres. *Progress in Polymer Science (Oxford)*. http://doi.org/10.1016/S0079-6700(98)00018-5
[3] González, A., & Alvarez Igarzabal, C. I. (2013). Soy protein - Poly (lactic acid) bilayer films as biodegradable material for active food packaging. *Food Hydrocolloids*, *33*(2), 289–296. http://doi.org/10.1016/j.foodhyd.2013.03.010
[4] Koh, J. J., Zhang, X., & He, C. (2018). Fully biodegradable Poly(lactic acid)/Starch blends: A review of toughening strategies. *International Journal of Biological Macromolecules*, *109*, 99–113. http://doi.org/10.1016/j.ijbiomac.2017.12.048
[5] López, O. V., Castillo, L. A., García, M. A., Villar, M. A., & Barbosa, S. E. (2015). Food packaging bags based on thermoplastic corn starch reinforced with talc nanoparticles. *Food Hydrocolloids*, *43*, 18–24. http://doi.org/10.1016/j.foodhyd.2014.04.021
[6] Martin, O., & Ave, L. (2001). Poly (lactic acid): plasticization and properties of biodegradable multiphase systems. *Polymer*, *42*, 6209–6219. http://doi.org/0032-3861/01/$
[7] Muller, J., Casado Quesada, A., González-Martínez, C., & Chiralt, A. (2017). Antimicrobial properties and release of cinnamaldehyde in bilayer films based on polylactic acid (PLA) and starch. *European Polymer Journal*, *96*(September), 316–325. http://doi.org/10.1016/j.eurpolymj.2017.09.009
[8] Muller, J., González-Martínez, C., & Chiralt, A. (2017). Combination of Poly(lactic) Acid and Starch for Biodegradable Food Packaging. *Materials*, 1–22. http://doi.org/10.3390/ma10080952
[9] Müller, P., Bere, J., Fekete, E., Moczo, J., Nagy, B., Kallay, M., … Pukanszky. (2016). Interactions , structure and properties in PLA / plasticized starch blends. *Polymer*, *103*, 9–18. http://doi.org/10.1016/j.polymer.2016.09.031
[10] Nafchi, A. M., Moradpour, M., Saeidi, M., & Alias, A. K. (2013). Thermoplastic starches : Properties , challenges , and prospects. *Starch/Staerke*, *65*, 61–72. http://doi.org/10.1002/star.201200201
[11] Phetwarotai, W., Potiyaraj, P., & Aht-Ong, D. (2012). Characteristics of Biodegradable Polylactide/Gelatinized Starch Films: Effects of Starch, Plasticizer, and Compatibilizer. *Journal Application Polymer Science*, *5*(4), 1295–1307. http://doi.org/10.1002/app.36736
[12] Prachayawarakorn, J., & Hwansanoet, W. (2012). Effect of silk protein fibers on properties of thermoplastic rice starch. *Fibers and Polymers*, *13*(5), 606–612. http://doi.org/10.1007/s12221-012-0606-x
[13] Sanyang, M. L., Sapuan, S. M., Jawaid, M., Ishak, M. R., & Sahari, J. (2016). Recent developments in sugar palm (Arenga pinnata) based biocomposites and their potential industrial applications: A review. *Renewable and Sustainable Energy Reviews*, *54*, 533–549.

http://doi.org/10.1016/j.rser.2015.10.037
[14] Soares, F. C., Yamashita, F., Müller, C. M. O., & Pires, A. T. N. (2013). Thermoplastic starch / poly (lactic acid) sheets coated with cross-linked chitosan. *Polymer Testing*, *32*, 94–98. http://doi.org/10.1016/j.polymertesting.2012.09.005
[15] Widyastuti, E., Sedyadi, E., & Prabawati, S. Y. (2016). Effect of Addition of Soursop Leaf Extract to Ganyong (Canna edulis Ker.) Starch Edible Film and its application in Red Grape Storage Time. *Biology, Medicine, & Natural Product Chemistry*, *5*(2), 55. http://doi.org/10.14421/biomedich.2016.52.55-59
[16] Witono, J. R., Noordergraaf, I. W., Heeres, H. J., & Janssen, L. P. B. M. (2014). Water absorption, retention and the swelling characteristics of cassava starch grafted with polyacrylic acid. *Carbohydrate Polymers*, *103*, 325–332. http://doi.org/https://doi.org/10.1016/j.carbpol.2013.12.056
[17] Yang, Y., Tang, Z., Xiong, Z., & Zhu, J. (2015). Preparation and characterization of thermoplastic starches and their blends with poly(lactic acid). *International Journal of Biological Macromolecules*, *77*, 273–279. http://doi.org/10.1016/j.ijbiomac.2015.03.053
[18] Yuan, H., Liu, Q., Hrymak, A., Thompson, M., & Ren, J. (2010). Thermoplastic potato starch blends and bioplastic films. In *Annual Technical Conference - ANTEC, Conference Proceedings* (Vol. 2, pp. 1463–1467). Retrieved from http://www.scopus.com/inward/record.url?eid=2-s2.0-77956741259&partnerID=tZOtx3y1
[19] Zhuang, C., Jiang, Y., Zhong, Y., Zhao, Y., Deng, Y., Yue, J., … Mu, H. (2018). Development and characterization of nano-bilayer films composed of polyvinyl alcohol, chitosan and alginate. *Food Control*, *86*, 191–199. http://doi.org/10.1016/j.foodcont.2017.11.024

Corrosion Resistance of 316L Biomaterial in Simulated Body Fluid by Modification of Shot Distance and Shot Angle of Shot Peening

Erich Umbu Kondi Maliwemu[1,2*], Viktor Malau[1], Priyo Tri Iswanto[1]
[1] Department of Mechanical and Industrial Engineering, Universitas Gadjah Mada, Yogyakarta, Indonesia.
[2] Department of Mechanical Engineering, Universitas Nusa Cendana, Kupang, East Nusa Tenggara, Indonesia.

erich.umbu@mail.ugm.ac.id

Abstract. Corrosion resistance of 316L biomaterial can be improved by mechanical surface treatment such as shot peening. Shot peening can create plastic deformation as a nanocrystal surface layer on the metal surface to protect the material from corrosion attack. This study aims to investigate the effect of shot peening on the corrosion resistance of 316L to reach the best rate of corrosion. The shot peening parameters in this research are shot distance (6, 8, 10, and 12 cm), shot angle (30, 60, and 90°), compressor pressure (7 kg/cm^2), shot duration (20 minutes), and steel balls S-170 (diameter of 0.6 mm). Simulated Body Fluid (SBF) was used as a corrosion media with ion concentrations nearly equal to human blood plasma. Corrosion tests were used versaSTAT, and the data were analyzed by VersaStudio. The result shows that the shot peening can decrease the corrosion rate of the 316L biomaterial. The larger the shot angle, the lower the corrosion rate and the nearer the shot distance, the lower the corrosion rate. The nearest shot distance (6 cm) and the largest shot angle (90°) give the best result to decrease the corrosion rate from 0.8617 mpy to 0.1096 mpy.

Keywords: Biomaterial, corrosion resistance, shot peening, simulated body fluid

1. Introduction

Medical science and improvement in biomedical materials have an outstanding increase in advanced year population and quality of life. Medical implants can fail due to wear, fatigue, chemical destruction, and corrosion. Metallic biomaterials are used extensively for implants fabrication. The surface of the implant must be accepting the ability for long-term usage contact to biological connection with tissue. The corrodible metals are used to minimize the corrosion of metallic biomaterials in medical device applications [1-3]. AISI 316L stainless steel is generally used as components of biomaterial due to its outstanding corrosion resistance in corrosion environments such as in the human body as an orthopedic implant, cardiovascular, and dental devices [4].

Shot-peening is a cold-working treatment that creates a layer with compressive residual stress by plastic deformation, modifying the surface and forming the ultrafine grain and nanostructure with high productivity [5-7]. Austenitic stainless steel has excellent corrosion resistance because it contains high chromium contents. Metastable austenitic stainless steels have a phase transformation to induce martensitic by cold working such as shot peening. The corrosion rate of AISI 304L with shot-peening is lower than AISI 316L without shot-peening. It means that the AISI 304 with shot-peening has better than non-treated AISI 316L [8-10]. Shot size is important to decrease the corrosion rate. The coverage and Almen intensity markedly improved the micro-hardness and compressive residual stresses on the surface layer. Shot peening increases the surface roughness by modifying the different shot peening parameters. Shot peening can increase the surface roughness and hardness by changing the shot distance and shot angle of shot peening [11, 12].

High corrosion, wear resistance and biocompatibility are good properties for biomedical materials. Corrosion was commonly deeper in the cold-drawn steels than in the solution-annealed [13, 14]. Chloride ions in simulated body fluid can cause severe pitting corrosion on the surface of a metal. The pitting potential (E_{pit}) as a critical parameter can evaluate the pitting response at a critical voltage under a corrosive media [15]. Simulated Body Fluid (SBF) and Bovine Serum Albumin (BSA) are general methods to create apatite interface layer. Hank solution can also be used as SBF. The simulated abiotic sulfide solution is not able to create a corrosive environment with the biotic sulfate-reducing bacteria culturing medium. In vivo apatite formation can be applied in SBF with ion concentrations nearly equal to human blood plasma. The in vivo bone bioactivity can be predicted by the apatite formation on the surface in SBF [16-19]. In the human body fluids, inorganic salts such as organic compounds like amino acids and proteins have an important role. The effects of the composition and buffering ability of immersion solutions are commonly used in evaluations of in-vitro corrosion [20].

This research aims to investigate the effect of shot peening as a suitable mechanical surface treatment by investigating the modification of shot distance and shot angle of shot peening to enhance the corrosion resistance of 316L biomaterial.

2. Experimental method

Stainless steel of 316L was used as the corrosion specimen with chemical composition as shown in Table 1. This chemical composition was gained by spectrometry measurement.

Table 1. The chemical composition of 316L.

Element	C	Si	P	Mn	Ni	Cr	Mo	Fe
wt.%	0,030	0,969	0,037	1,078	10,867	16,782	1,894	67,78

This research used steel balls S-170 with a diameter of 0.6 mm. Compressor pressure was kept constant at 7 kg/cm^2 for 20 minutes. Shot peening was applied by steel balls impact on the surface of 316L by modifying the shot distance at 6, 8, 10, and 12 cm, and shot angle at 30, 60, and 90°.

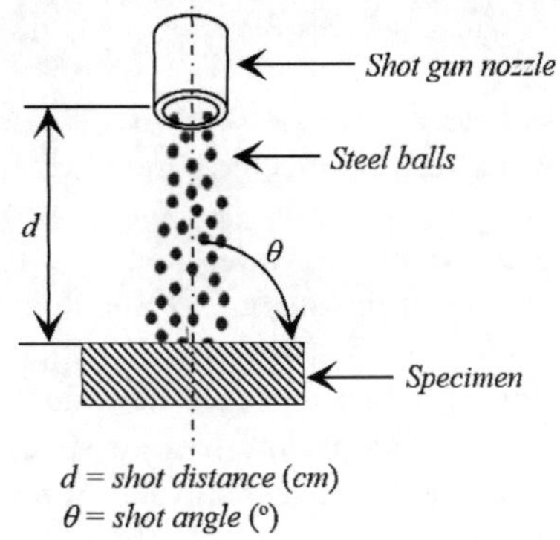

Figure 1. The scheme of shot peening [12].

IOP Conf. Series: Materials Science and Engineering **553** (2019) 012053 doi:10.1088/1757-899X/553/1/012053

SBF was used in this research because the ion concentration of SBF nearly equal to human blood plasma with composition as shown in Table 2.

Table 2. The composition of Simulated Body Fluid [19].

Reagent	Amount	Purity (%)	Formula weight
NaCl	8.035 g	99.5	58.4430
$NaHCO_3$	0.355 g	99.5	84.0068
KCl	0.225 g	99.5	74.5515
K_2HPO_4 $3H_2O$	0.231 g	99.0	228.2220
$MgCl_2$ $6H_2O$	0.311 g	98.0	203.3034
1.0M-HCl	39 ml	-	-
$CaCl_2$	0.292 g	95.0	110.9848
Na_2SO_4	0.072 g	99.0	142.0428
Tris ($HOCH_2$)3CNH_2)	6.118 g	99.0	121.1356
1.0M-HCl	0-5 ml	-	-

Corrosion tests were used versaSTAT, and the data were analyzed by VersaStudio with parameters as shown in Table 3.

Table 3. The parameters of the corrosion test.

Endpoint properties			Scan properties			Threshold properties	
Initial potential (V)	Vertex potential (V)	Final potential (V)	Step high (mV)	Step time (s)	Scan rate (mV/s)	Start level (V)	Threshold (mA)
-0.1	1.2	-0.1	2	1	2	0.2	1

3. Result and discussion

The corrosion test has a purpose to investigate the corrosion properties of 316L as a non-treatment specimen and shot peening specimen as shown in Figure 2.

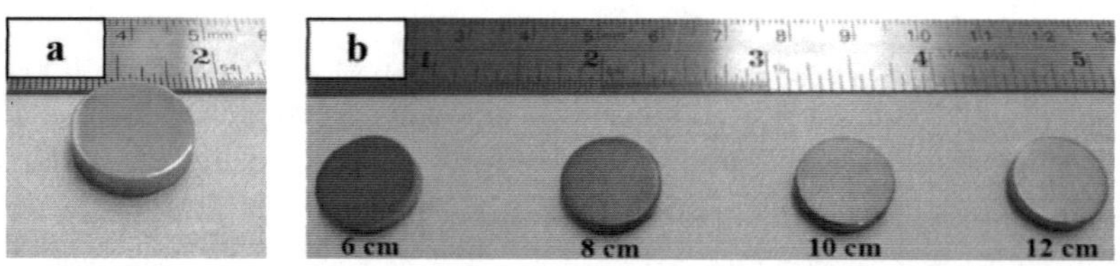

Figure 2. The corrosion test specimens of 316L; (a) Non-treatment specimen, (b) Specimens after shot-peening in different shot distance.

Figure 2 shows the different color between non-treatment specimen and specimens after shot peening. This is the effect of shot peening on the surface of the specimen that makes surface darker than the non-treatment specimen. It can happen because the shot intensity gives a significant effect of changing the morphology on the surface of the specimen. The reduction in corrosion resistance has a correlation with the morphology on the surface of the treated specimens [11]. The distance of the shot peening has a significant impact that causes surface modification which is marked by discoloration on the surface of the specimen, where the color of the specimen becomes brighter at the farthest of the shot distance, while the smaller the shot angle, the specimen becomes brighter. This is due to the greater shot intensity at nearer distances, so the surface looks darker/gray [12].

IOP Conf. Series: Materials Science and Engineering **553** (2019) 012053 doi:10.1088/1757-899X/553/1/012053

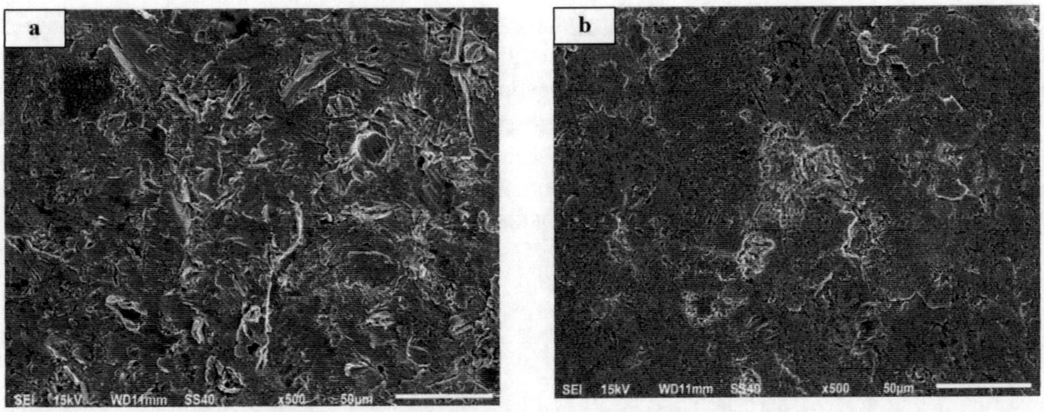

Figure 3. SEM photograph of the specimen after corrosion test; (a) Non-treatment specimen, (b) Specimen with shot peening.

Figure 3 shows that uniform corrosion attacks the surface of the specimen. A specimen without shot peening (non-treatment) has more oxide on the surface as the effect of corrosion because the specimens without shot peening are more easily oxidized. Thin layer on the surface of the specimen as the effect of shot peening can protect the substrate of material from corrosion attack. This phenomenon can happen because shot peening can produce a layer containing huge compressive residual stresses, modifying the metal surface structure, and creating the ultrafine grain and nanostructure [5].

Corrosion test can produce the polarization curve that can be used to determine the parameters of corrosion potential (E_{corr}) and corrosion current density (I_{corr}) as a result of the intersection between Tafel lines of anodic and cathodic. Both of these parameters are used to determine the corrosion rate.

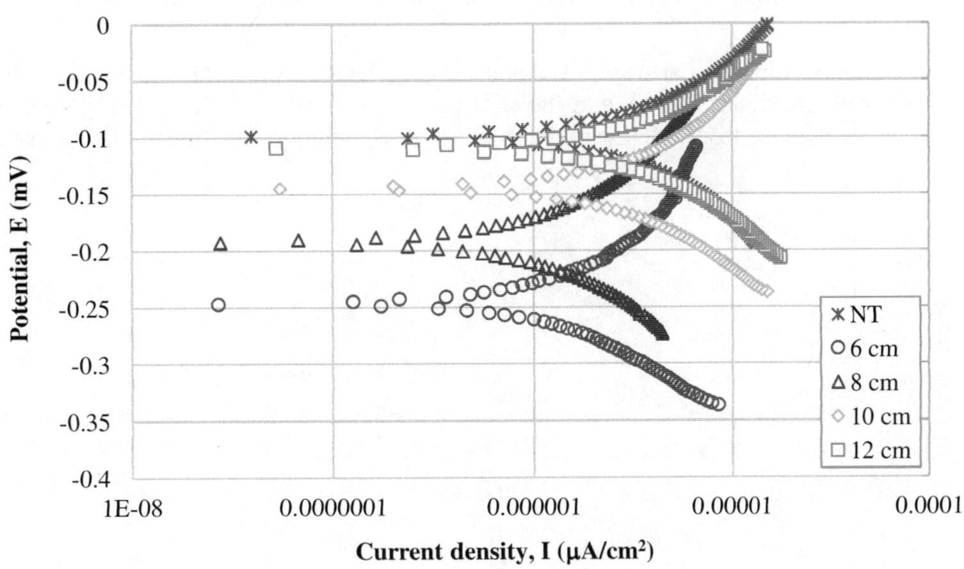

Figure 4. The polarization curve of 316L by shot distance modification of shot peening.

Figure 4 shows that the shot distance can reduce the corrosion current density of 316L from 2.059 µA/cm^2 (non-treatment) to 0.262 µA/cm^2 (shot distance of 6 cm), 0.468 µA/cm^2 (shot distance of 8 cm), 1.040 µA/cm^2 (shot distance of 10 cm), and 1.589 µA/cm^2 (shot distance of 12 cm). The nearest distance at 6 cm has a good corrosion current density. The shot angle of shot peening also affects the corrosion resistance of 316L. The shot angle can reduce the corrosion current density from 2.059 μA/cm² (non-treatment) to 1.331 μA/cm² (shot angle of 30°), 0.573 μA/cm² (shot angle of 60°), and 0.262 μA/cm² (shot angle of 90°) as shown in Figure 5. The largest shot angle gives the best result of corrosion resistance indicated by the lowest corrosion current density. The corrosion current density has a directly proportional relation with corrosion rate. The low corrosion current density indicates that the specimen has a low corrosion rate.

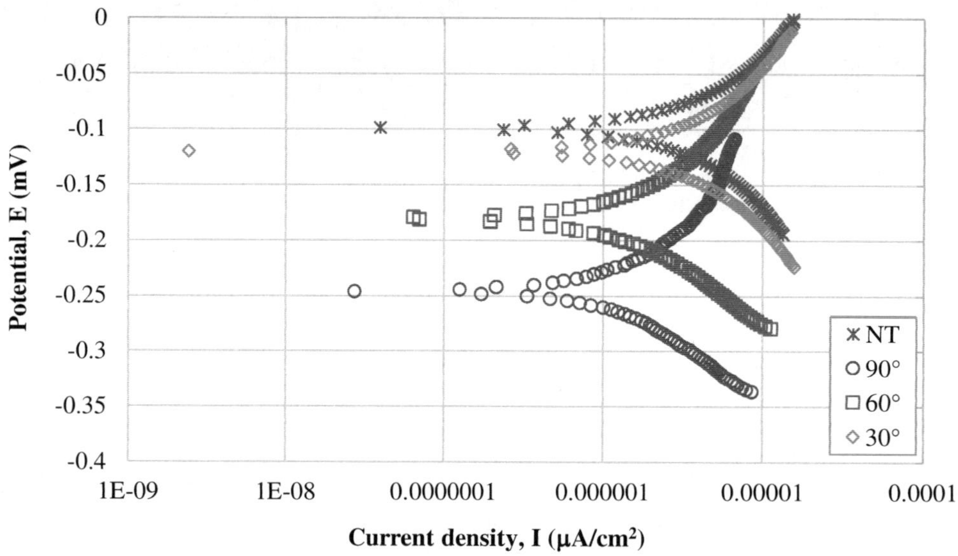

Figure 5. The polarization curve of 316L by shot angle modification of shot peening.

The effect of shot peening on corrosion resistance was indicated by a reduction in corrosion rate after shot peening. Figure 6 and 7 show that the shot peening can affect the corrosion resistance by decreasing the corrosion rate from 0.8617 mpy to 0.1096 mpy at the shot distance of 6 cm and shot angle of 90°.

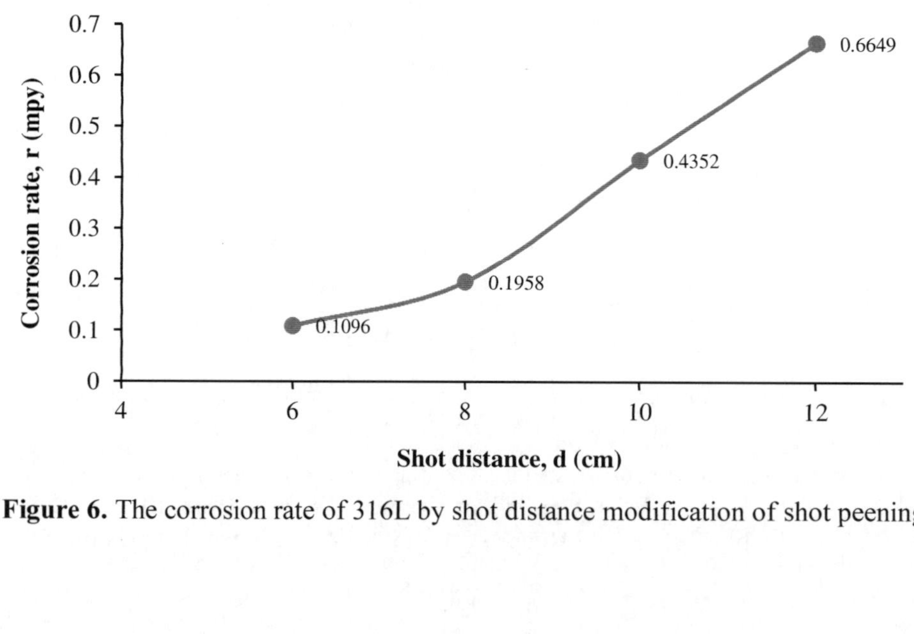

Figure 6. The corrosion rate of 316L by shot distance modification of shot peening.

The shot distance of shot peening can affect the corrosion rate as shown in Figure 6. The nearest shot distance gives the best result of corrosion rate because the shot intensity of the nearest distance is higher than the farthest distance. Multi-impact at the nearest distance is higher than the farthest distance so that the surface of the specimen will be denser than the farthest distance and can cover the area of the surface of the specimen.

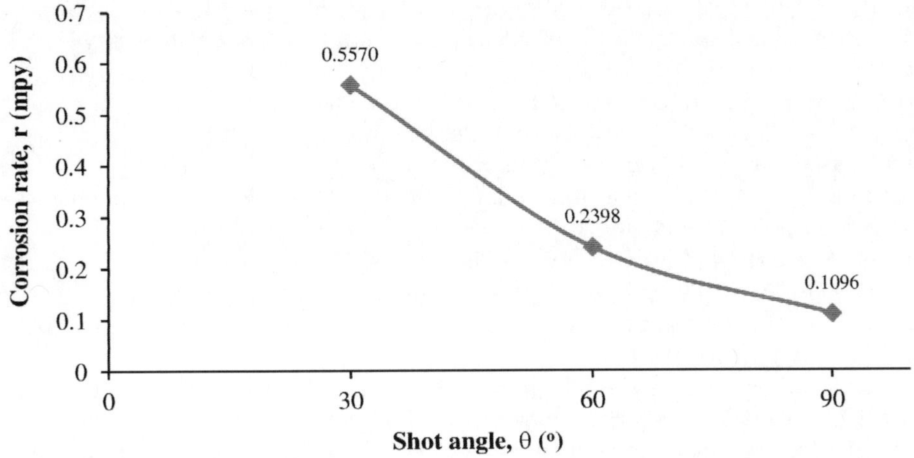

Figure 7. The corrosion rate of 316L by shot angle modification of shot peening.

Shot angle also can affect the decreasing of corrosion rate of the 316L biomaterial as shown in Figure 7. The larger the shot angle, the lower the corrosion rate. The largest shot angle is the best result of corrosion rate because the area of the contact angle between the position of shot nozzle and surface of the specimen is larger than the smallest shot angle, so the shot intensity increased too and centered on the surface of the specimen. These effects can cover the surface of the specimen from corrosion attack.

4. Conclusion

The shot peening can improve the corrosion resistance of the 316L biomaterial by decreasing the corrosion rate. The larger the shot angle, the lower the corrosion rate and the nearer the shot distance, the lower the corrosion rate. The nearest shot distance (6 cm) and the largest shot angle (90°) give the best result to decrease the corrosion rate from 0.8617 mpy to 0.1096 mpy.

Acknowledgments

This research was supported by a grant of Doctoral Dissertation Research (PDD-RISTEKDIKTI 2018) from The Ministry of Research, Technology and Higher Education of The Republic of Indonesia.

References

[1] Roy M 2012 *Surface Engineering for Enhanced Performance against Wear* Springer http://dx.doi.org/10.1007/978-3-7091-0101-8

[2] Pilliar RM Metallic 2009 *Biomedical Materials* Springer http://dx.doi.org/10.1007/978-0-387-84872-3_2

[3] Zheng YF, Gu XN and Witte F 2014 *Materials Science and Engineering R* Elsevier **77** pp 1-34 http://dx.doi.org/10.1016/j.mser.2014.01.001

[4] Escalada L, Lutz J, Brühl SP, Fazio M, Márquez A, Mändl S, Manova D and Simison SN 2013 *Surface and Coatings Technology* Elsevier **223** pp 41-46

http://dx.doi.org/10.1016/j.surfcoat.2013.02.025

[5] Umemoto M, Todaka Y and Tsuchiya K 2003 *Materials Transactions* The Japan Institute of Metals **44** pp 1488-1493 http://dx.doi.org/10.2320/matertrans.44.1488

[6] Bagherifard S, Pariente IF, Ghelichi R and Guagliano M 2010 *Procedia Engineering* Elsevier **2** pp 1683-1690 http://dx.doi.org/10.1016/j.proeng.2010.03.181

[7] Okada H, Tange A and Ando K 2006 *Fracture of Nano and Engineering Materials and Structures* Springer pp 1315-1316 http://dx.doi.org/10.1007/1-4020-4972-2_653

[8] Biehler J, Hoche H and Oechsner M 2017 *Surface & Coatings Technology* Elsevier **313** pp 40-46 http://dx.doi.org/10.1016/j.surfcoat.2017.01.050

[9] Iswanto PT, Malau V, Priyambodo BH, Wibowo TN and Amin N 2017 *Materials Science Forum* Trans Tech Publications **901** pp 91-96 http://dx.doi.org/10.4028/www.scientific.net/msf.901.91

[10] Priyambodo BH, Malau V and Iswanto PT 2018 *Advanced Science Letters* American Scientific Publishers **24** pp 9545-9547 http://dx.doi.org/10.1166/asl.2018.13071

[11] Ahmed AA, Mhaede M, Basha M, Wollmann M and Wagner L 2015 *Surface & Coatings Technology* Elsevier **280** pp 347-358 http://dx.doi.org/10.1016/j.surfcoat.2015.09.026

[12] Maliwemu EUK, Malau V and Iswanto PT 2018 *IOP Conf. Ser.: Mater. Sci. Eng.* **299** 012051 pp 1-6 http://dx.doi.org/10.1088/1757-899X/299/1/012051

[13] Singh R and Dahotre NB 2007 *Journal of Materials Science: Materials in Medicine* Springer **18** pp 725-751 http://dx.doi.org/10.1007/s10856-006-0016-y

[14] Lambrinou K, Delville R, Charalampopoulou E, Donck V and Schryvers D 2017 *Journal of Nuclear Materials* Elsevier **490** pp 9-27 http://dx.doi.org/10.1016/j.jnucmat.2017.04.004

[15] Huang CH, Lai JJ, Huang JC, Lin CH and Jang JSC 2016 *Materials Science and Engineering C* Elsevier **62** pp 368-376 http://dx.doi.org/10.1016/j.msec.2016.01.080

[16] Song G 2007 *Corrosion Science* Elsevier **49** pp 1696-1701 http://dx.doi.org/10.1016/j.corsci.2007.01.001

[17] Huang L, Zhouc B, Wuc H, Zheng L and Zhaoa J 2016 *Materials Science and Engineering: C* Elsevier **70** pp 955-961 http://dx.doi.org/10.1016/j.msec.2016.05.115

[18] Chen S, Cheng YF and Voordouw G 2017 *International Biodeterioration & Biodegradation*, Elsevier **120** pp 91-96 http://dx.doi.org/10.1016/j.ibiod.2017.02.014

[19] Kokubo T and Takadama H 2006 *Biomaterials* Elsevier **27** pp 2907-2915 http://dx.doi.org/10.1016/j.biomaterials.2006.01.017

[20] Witecka A, Bogucka A, Yamamoto A, Máthis K, Krajňák T, Jaroszewicz J and Święszkowski W 2016 *Materials Science and Engineering C* Elsevier **65** pp 59-69 http://dx.doi.org/10.1016/j.msec.2016.04.019

IOP Conf. Series: Materials Science and Engineering **553** (2019) 012054 doi:10.1088/1757-899X/553/1/012054

Comparative Studies on the Synthesis of Polypyrrole through Polymerization Reaction Between Batch and Continuous Processes

Gina Ayu Wardhani[1], Mas Ayu Elita Hafizah[1], Andreas[2], Azwar Manaf[1*]
[1] Postgraduate Program of Materials Science Study, Department of Physics, Faculty of Mathematics and Natural Sciences, Universitas Indonesia, Depok 16424, Indonesia
[2] Indonesian Institute of Science, LIPI, Komplek PUSPITEK, Serpong, Banten 15314, Indonesia

azwar@ui.ac.id

Abstract. Comparative studies on conductive polymer of Polypyrrole (Ppy) synthesized through oxidative polymerization in batch and continuous techniques is reported. The common step for the preparation of solutions consisted of mixing between monomer and ammonium peroxidisulphate (APS) solutions with volume ratio 1:1 under an agitation speed of 400 rpm for 4 hours of reaction time. The two different techniques of polymerization resulted in powder of Ppy which confirmed by the standard Fourier Transform Infra Red (FTIR) spectrum. Formation of Ppy occurred in a much faster speed in a batch technique in comparison with that of continuous one. The advantage of continuous process in obtaining the pure Ppy is that it requires a simple washing step with distillate water. Ppy that resulted from a synthesis of the two processes has electrical conductivity value, making Ppy a conducting polymer.

Keywords: Conductive polymer, Polypyrrole (PPy), Chemical oxidative polymerization, Polymerization technique

1. Introduction

Polymeric based materials in general being considered as lighter weight and insulating materials. Current development on the use of polymer materials shown that the polymer has met a variety of applications which are not limited only being used as passive materials due to its isolative properties, but also being used as active materials that hold optical, conducting, and electromagnetic properties [1]. In addition, polymer materials have also extended its potential uses in the areas of energy storage devices [2] and possibly in the area of magnetic applications since polymer with magnetic properties has been reported [3]. The study of polymers that related to the electronics and electrical properties is being carried out intensively to date [4]. One of the objectives is exploring its electrical conductivity which can be generated through ordered conjugation with extended π electrons and sufficiently large electric charge concentration. Hence, the presence of conjugated double bonds along the backbone of the polymer is a key property which must be owned by conductive polymers [5].

In the past, the conductive properties of polymer based material were obtained through embedding the highly conductive particles like graphite, gold, silver powders into a polymer matrix [6]. However, the conducting polymers that prepared through this manner have been proven ineffective due to insufficient value of electrical conductivity obtained from the polymer and deteriorations of physical properties of the polymer itself. In the last two decades, intensive studies have been addressed to generate the electrical conductivity in polyaromatic backbone polymers. Among them, polyaniline (PANI), Polypyrrole (Ppy) and polythiophene (PT) are considered potential candidates for development of conducting polymer type materials [7]. However, large attentions have been given to the Ppy because in addition to its high electrical conductivity value, Ppy is chemically stable with good mechanical properties and can be synthesized through a simple process. Not surprisingly, the conducting Ppy has met applications in various areas like light-weight batteries, electronic devices, sensors and membrane separation [8]. Another potential application areas of Ppy are drug delivery, rechargeable batteries, supercapacitors, microwave shielding and corrosion protection [9].

The preparation of conducting polymer is mostly carried out either through chemical reaction or electrochemical method. The synthesis of polymer by either of these methods depends upon the intended application of the polymer. The chemical route usually involves a polymerization reaction of monomer which carried out in the reactor [10]. It is a simple, fast and the chief synthesis method since no special instruments are required. The polymerization reaction is ended with resulting in a fine powder obtained after oxidative polymerization of the monomer by chemical oxidants in aqueous or non-aqueous solvents taken place during reaction [11]. Hence, large quantities of polypyrrole (Ppy) can be made through such preparation method. Nevertheless, a number of alternative synthesis methods of Ppy had been introduced by the researchers. Among the methods are electrochemical [12], chemical oxidative polymerization [13], a method which involving self-degradation of template [14] and micro emulsion polymerization technique [9]. Subject to the sequence of method development which are available, the chemical oxidative polymerization was found to be the most attractive and favorite method for synthesis of Ppy. This was indicated by the number of research results that reported results of synthesis of the PPy through chemical oxidative polymerization reaction [15]. When the oxidative chemical polymerization method is selected for synthesis of Ppy, an oxidation agent is required. For instance, ferric chloride, ammonium persulfate can be used as an oxidant for synthesis of pyrrole through oxidative chemical polymerization [8]. It was reported that PPy was formed due to a coupling agent reaction between radical cations from oxidant agent and monomer.

In the present paper, we report results of comparative studies on the preparation of polypyrrole through chemical oxidation method between continuous and batch processes. Ammonium Peroxidisulphate was used as the oxidant to be reacted with pyrrole monomers. Results of comparison studies include molecule formation, color degradation during washing and the electrical conductivity of Ppy.

2. Experimental method

Pyrrole (Sigma Aldrich, 99.9%) and Ammonium Peroxidisulphate (Sigma Aldrich, 99,8%) as a monomer and oxidizing agent respectively were used for Ppy preparation. The monomer solution was prepared by mixing 5 ml pyrrole with distillated water and stirred for 10 minutes to form 100 ml monomer solution. The initiator solution was found by mixing 5 grams of APS with distillated water and stirred for 10 minutes to form 100 ml solution.

In a batch process, the polymerization reaction was carried out under mixing between the monomer and initiator solutions with the volume ratio 1:1, and continuously stirred with a constant speed 400 rpm for 4 hours at room temperature. The process has resulted in powder containing solution which was then filtered out under vacuum and subsequently washed using distiled water for several times to remove any impurities present. The resulted powder was then dried in desiccator for 3 days. For a continuous process, the same ratio of monomer and initiator solutions was applied to prepare the solution. However, the mixing of the two solution was carried out under drop by drop mixing during the polymerization reaction at room temperature for 4 hours under the same constant stirring speed. Afterwards in the polymerization process finished, the resulted powder was filtered, washed and dried by the same treatment as batch technique.

The chemical bonding was analyzed using FTIR spectroscopy by FTIR spectrometer model Thermo Scientific Nicolet iS5 system using Diamond ATR during the whole reaction and washing process.

Gardner Color Scale by Lovibond Spectrophotometer was used to see any product or impurities left after washing process. The conductivity was tested by Four-Point Collinear Probe. This technique involves bringing four equally spaced probes in contact with a material of unknown resistance. The electrical conductivity was then calculated from the slope obtained from an ohmic relation between the voltage and electrical current.

3. Results and discussion

3.1. Polypyrrole Characterization

Figure 1 compares the FTIR spectra of powder obtained after 5 and 240 minutes polymerization reaction in a batch process. The two spectra is almost identical but a slight different in values of peak absorbances. The specific finger print of PPy consisted of a molecule group with peaks absorbance at the wavenumbers 1197 cm^{-1}, 1103 cm^{-1} and 1054 cm^{-1} belongs to respectively C-N stretching, C-H in-plane deformation and N-H in-plane deformation [8,16]. FTIR spectra in Figure 1 show the presence of characteristic absorption bands at 1452 cm^{-1} (C=C stretching of pyrrole ring), 1196 cm^{-1} (C-N stretching vibration in the ring), 1102 cm^{-1} (C-H in-plane deformation), 1052 (N-H in-plane deformation), 887 cm^{-1} (C-H out-of-plane deformation), 748 cm^{-1} (C-H out-of-plane ring deformation). Such set of absorption characteristic is match with that of standard Ppy [8,16]. Obviously, the formation of Ppy molecules already occurred within 5 minutes reaction but the yield is still relatively low.

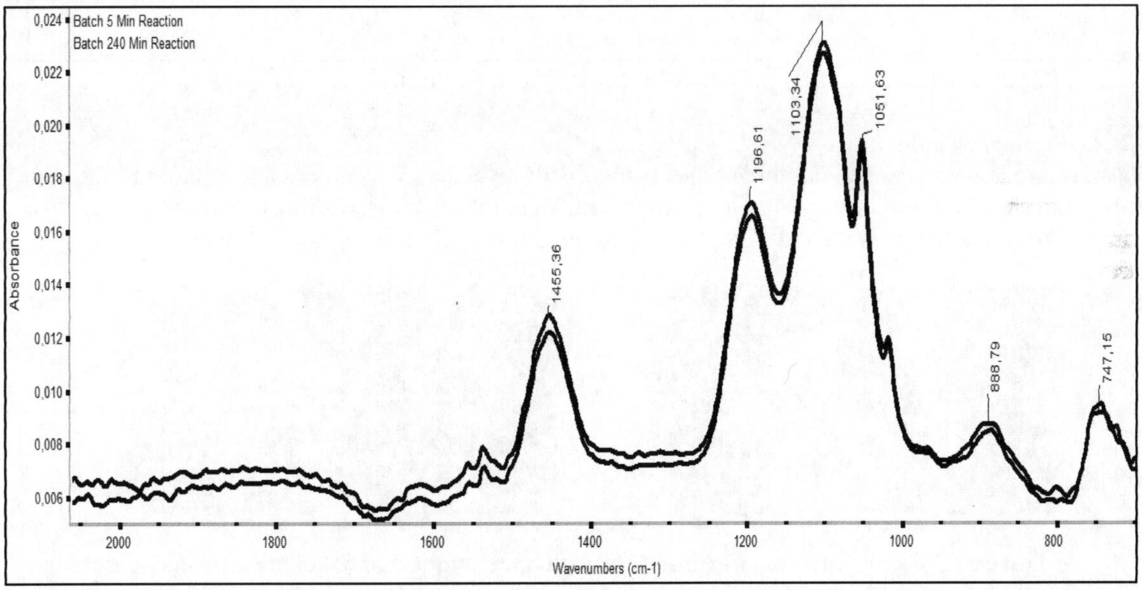

Figure 1. FTIR spectra of powder obtained in the batch technique after 5 and 240 minutes reaction time.

FTIR spectra of the powder obtained after 5, 15, 30 and 240 minutes polymerization reaction in a continuous process is compared in Figure 2. Once again, the all spectrums are almost identical, but those of shorter reaction time (5 and 15 minutes), an additional absorbance peak present in each spectrum indicating the Ppy molecules were still not formed completely. Only after 30 minutes reaction time such additional absorbance peak disappears from the spectrum. Hence, when compare results of Figure 1 and 2, it can be concluded that the two processing methods have resulted in different mechanism of Ppy formation. In the batch technique, polymerization reaction occurred through the effective reaction which indicated by the no change spectrum profile.

The formation rate of Ppy in the batch process is much faster than that of continuous processes. The Ppy molecules have formed in the batch process at only in the first 5 minutes reaction time. This was not the case in the continuous process from which the reaction time for a complete formation of Ppy molecules is required up to 30 minutes.

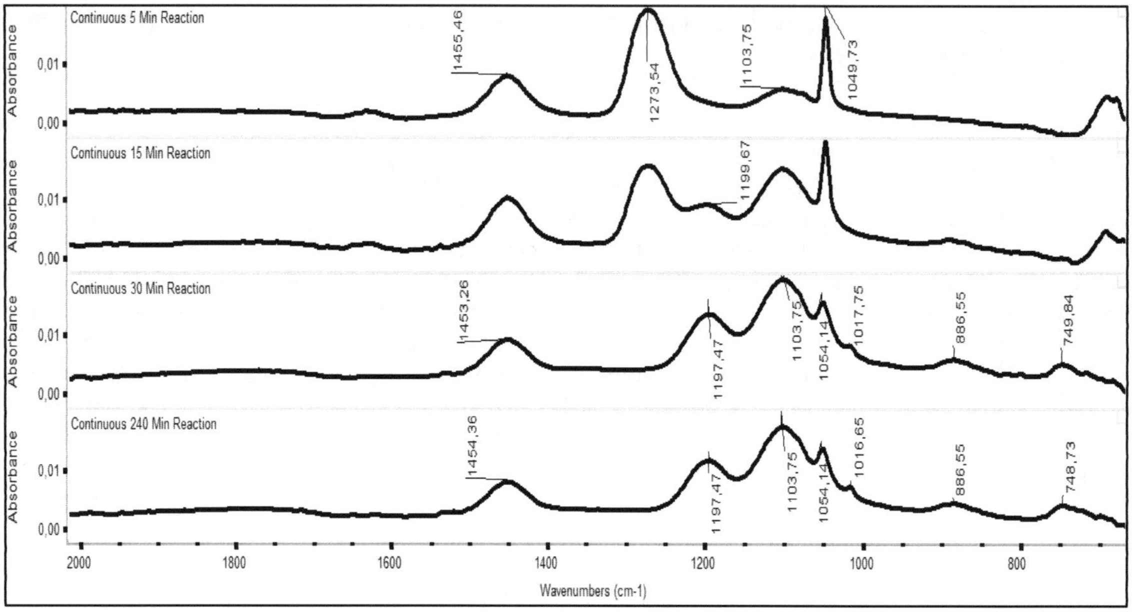

Figure 2. FTIR spectra of polypyrrole was synthesized through continuous technique.

3.2. Color degradation

Washing process was carried out three times using distillate water. This process is required to find out if any impurities of product still left after filtered with Whatman 42 and washing process. Products that resulted from this step was then evaluated by Gardner Color Scale Spectrophotometer.

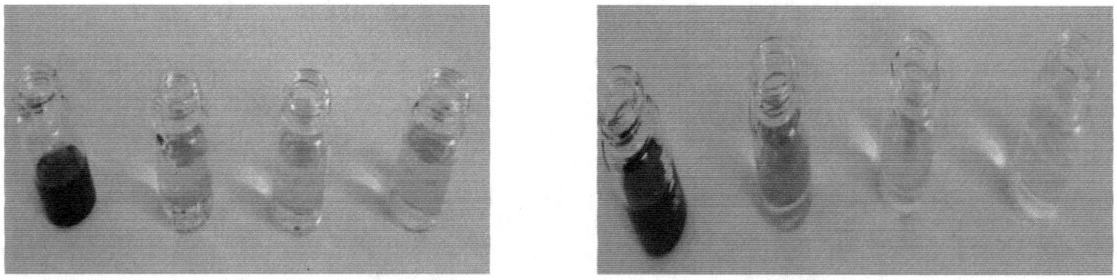

(a) (b)

Figure 3. Color degradation of filtrates (a) batch technique and (b) continuous technique.

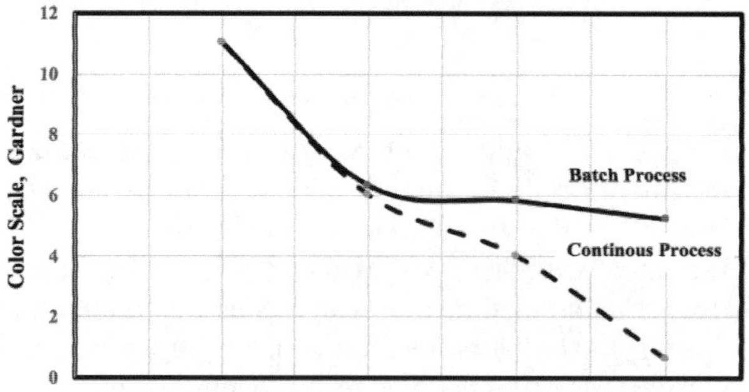

Figure 4. Color Scale Gardner value of batch and continuous techniques.

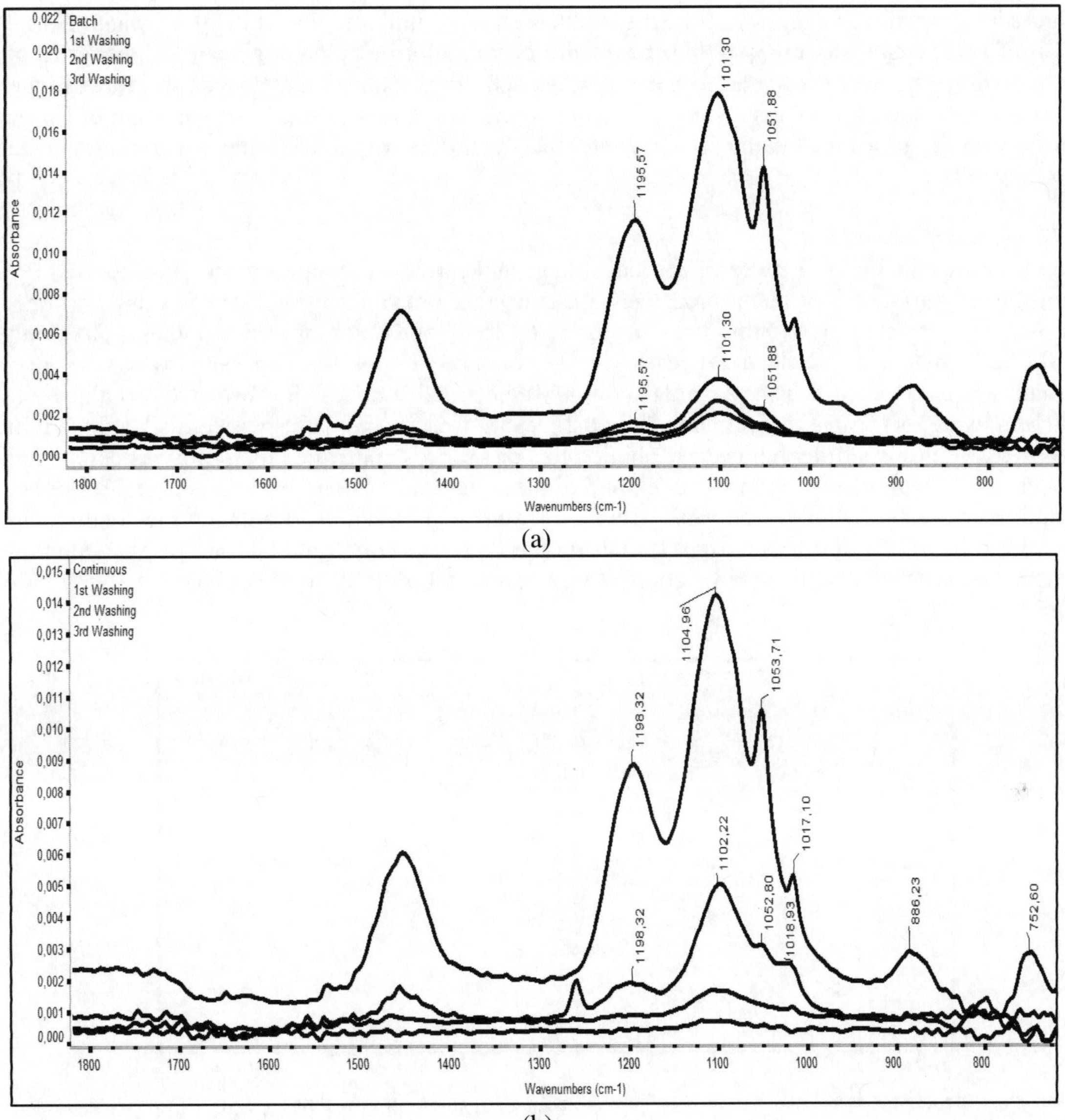

Figure 5. (a) FTIR spectra of each washing process on batch technique; (b) FTIR spectra of each washing process on continuous technique.

Figures 3(a) and 3(b) show photos of color degradation of the filtrate after each washing treatment for the Ppy powder obtained from the batch and continuous processes. The color scale for each filtrate of batch process is compared with that of continuous process in Figure 4. Referring to result in Figure 3 and 4, results show that for batch system, the color degradation did not significantly change after three times washing process by water. It is also confirmed by results of FTIR evaluation in Figure 5(a) for washing powder obtained from a batch process. Other chemical molecule probably the monomer in addition to polypyrrole still exists after washing process. Hence, washing ppy of batch process with water seems less effective. Other suitable solvent for the washing process in this batch system is required. On the other hand, the continuous system gives different result as the color degradation changed significantly and almost nothing left on the filtrate after the whole washing process. It is shown in Figure 3 (b) where at the third washing the filtrate has color similar to that of distillate water. Hence, in term of color degradation in washing treatment the continuous technique gives much better result after three times washing compared with that of the batch. It was confirmed by FTIR spectra in Figure 5(b) that there are almost no absorbance peaks of Ppy can be observed in a FTIR spectrum of filtrate obtained after third washing treatment. It indicates that the filtrate was already free from other chemical molecules.

3.3. Electrical Conductivity

Figure 6 shows ohmic plots between electrical voltage and current of samples made of compacted Ppy obtained from batch and continuous processes. The resistance (R) value of each sample was determined from the slope of respective ohmic lines in Figure 6. The slope value of each samples differs only slightly indicating that the electrical resistivity or conductivity for the two samples has no large different. It was obtained that the calculated value of electrical conductivity for batch technique was 0.160 S/m, whereas for that of continuous was 0.163 S/m. Despite the value of electrical conductivity of the two samples is still low, however, ohmic plots of samples confirmed the sample is conducting polymer. In the previous reports, we explored the electrical conductivity properties of doped polyaniline in which doping the polyaniline with a strong acid of $HClO_4$ resulted in the conductivity value of 3.6 S/cm [3]. Result of current studies on the Ppy has confirmed that the Ppy is conducting polymer. Hence, the enhancement of conductivity value in Ppy will be the subjects of our future studies.

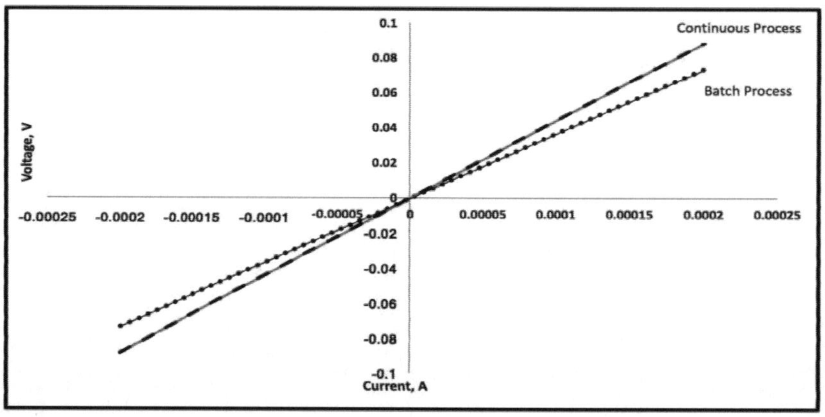

Figure 6. Electrical conductivity of Ppy samples obtained from batch and continuous processes

4. Conclusion

Results of current works have confirmed that Polypyrrole (PPy) can be obtained through the oxidative polymerization reaction in batch and continuous process. The two processes differ only in the rate of Ppy formation where the batch process is much faster than the continuous one. The advantage of the continuous process in comparison with that of batch is that the continuous process requires only a simple washing treatment in which within three steps washing with distillate water, the Ppy has freed from other molecules. PPy obtained from the two processes has property of electrical conductivity.

Acknowledgment

The authors gratefully acknowledge the support of Postgraduate Program of Material Science Universitas Indonesia for the research facilities. We are also thankful for the financial support provided by the Directorate of Research and Community Service Universitas Indonesia under program Grants of International Publication Indexed for Final Project student 2018 (PITTA 2018) under contract no. 2237/UN2.R3.1/HKP.05.00/20018.

References

[1] Chao D, Chen J, Lu X, Chen L, Zhang W and Wei Y 2005 Synth. Met. 150 47–51.
[2] R. Kumar, S. Singh and B.C. Yadav, Conducting Polymers: Synthesis, Properties and Applications (International Advanced Research Journal in Science, Engineering and Technology, India, 2015), Vol. 2, ISSN 2393-8021.
[3] M.A. Elita Hafizah, A. Bimantoro, Andreas and A. Manaf, "Synthesized of Conductive Polyaniline by Solution Polymerization Technique", Procedia Chemistry, Vol. 19, pp. 162 – 165, 2016.
[4] Inzelt, Gyorgy, 2012, *Conducting Polymer : New Era in Electrochemistry*, 2nd Edition, Berlin : Spinger-Verlag.
[5] Ansari, Reza, 2006, *E-Journal of Chemistry*, vol. 3, pp. 186-201.
[6] Chitte, H.K., Narendra V. B, Vasant E.W, and Ganesh N. S, 2011, *Journal of Sensor Technology*, pp. 47-56.
[7] Patil, B., H., Bulakhe, R., N dan Lokhande, C., D., 2014, *J. Mater Sci.: Master Electron*, vol. 25, pp. 2188-2198.
[8] Hermant K.C, Narendra V.B, Vasant E.W, Ganesh N.S, Journal of Sensor Technology, 2011, vol. 1, p 47-56.
[9] Vernitskaya, T.,V. dan Efimov, O., N, 1997, *Russian Chemical Reviews*, pp. 443-457.
[10] Manaf A, Bimantoro A, Hafizah M A E and Andreas 2017 *IOP Conf. Ser. Mater. Science Eng.* 223.
[11] Saville, Paul, 2005, "Polypyrrole Formation and Use", Technical Memorandum, Canada: Defence Research and Development Canada.
[12] Tabaciarova, Jana, Matej Micusik, Pavol Fedorko, and Maria Omastova,2015, pp. 392-401.
[13] Bilal, Salma, Fouzia Perveen, and Ali A. Shah, 2015, *Journal of Scientific & Innovative Research*, pp. 33-42.
[14] Omastova, Maria, Miroslava Trchova, Jana Kovarova, and Jaroslav Stejskal, 2003, *Synthetic Metals*, pp. 447-455.
[15] Marzan, Luis, M. Liz, 2004, *Materialstoday,* pp 1369.
[16] Natalia V.B, J. Stejskal, M. Trchova, J.Prokes and M. Omastova, European Polymer Journal, 2007, vol 43, pp. 2331-2341.

IOP Conf. Series: Materials Science and Engineering **553** (2019) 012055 doi:10.1088/1757-899X/553/1/012055

Tensile Properties of Glass Roving and Hybrid Tapes

Mohanapriya Venkataraman[1], Jiří Militký[1], Rajesh Mishra[1], Dana Kremenakova[1], Michal Petru[2]
[1] Technical University of Liberec, Faculty of Textile Engineering, Department of Material Engineering, Liberec 46117, Czech Republic
[2] Technical University of Liberec, Faculty of Mechanical Engineering, Department of the Design of Machine Elements and Mechanisms, Czech Republic.

mohanapriya.venkataraman@tul.cz

Abstract. Among many other materials for the reinforcement of composites, technical fabrics are increasingly being used, especially from glass tows, which have good mechanical properties. In this paper, glass fiber based tapes containing special types of resins, in pre-condensate form with application in technology of composites preparation by precise winding is prepared. Also, the selected suitable resin was loaded by suitable types of nanoparticles for enhancing mechanical properties. The tensile properties of the tapes made from glass were investigated. Measurements of resultant stabilized hybrid composite tapes (SHCT) were focused on the breaking force, deformation at break and initial modulus. It was found that the durability of the tapes was sufficient for long term storing. The main benefit was avoiding of resin penetration into winded structure which is due to tight arrangements of fibrous phase which is very complicated. The results show that SHCT can be stored and used after a longer period of time.

***Keywords*:** hybrid tapes, tensile properties, precise winding, composites, epoxy resin

1. Introduction

Glass fibres are known for their nonflammability, strength, abrasion resistance and cost effective processing in textile industry applications. Multifilament yarns are used for the production of glass fabrics, where filaments are glued together (roving yarn) into tape shapes (tows) to ensure compactness and improved properties (strength, evenness) [1,2]. The starting material for composites is multifilaments (wires), i.e., the strands that are formed by a combination of filaments (filaments) of low fineness (diameter below 10 μm) required to ensure low bending stiffness. The number of cross-sectional fibers ranges from 1k to 24k (number of filaments in roving in thousands). Twisted yarns, i.e. yarns, have an approximately circular cross-section and the density of the arrangement increases with increasing twist. However, the curl affects the initial modulus and yarn strength. Also, penetration of the resin or its components is not easy. Rovings (tow) are with slight (protective) twist (less than 50 curbs / m) and are suitable for the manufacture of prepregs. They usually have an oval cross section and their porosity decreases toward the center of the body. Spread Tow Tape (STT) is characterized by the fact that their width is multiplied by more than 3 times the diameter of the original multifilament. The individual filaments are much more closely arranged and behave like a bunch of identical long parallel threads. This leads to a better utilization of the strength of individual filaments and to a reduction in the thickness (by 20 to 30%) and hence the density of the composites [3]. The problem is to ensure cohesion and then resin saturation. The resin must ensure the storage life of the tapes. To improve mechanical properties of resins, it is preferred to use suitable fillers (particulate systems). The resulting stabilized hybrid composite tapes (SHCT) comprising an optimally arranged resin-bonded fibrous phase and surface nanoparticles with functional particles will allow the production of composites in particular by the technique of precision robotic winding. Among the end-use properties of glass fabrics, one should note generally high strength along with a relatively low density, high thermal resistance and nonflammability, good insulative characteristics, resistance to microorganisms, as well as good chemical resistance to oils, fats and organic solvents, acids and alkalis in the pH-range from 3 to 9. Major disadvantages are sensitivity to strong alkalis, surface corrosion, fragility and relatively hard processability [4-9]. This paper deals with the preparation and mechanical properties (breaking force, deformation at break, initial modulus) of stabilized hybrid composite tapes (SHCT). The objective was to study the durability of the tapes for long term storing. Also to understand avoiding of resin penetration into the winded structure which is very complicated due to tight arrangements of fibrous phase.

2. Materials and Method

The stabilized hybrid composite tape SHCT (Figure 1) consists of an expanded tape of a glass uncoated multifilament cable (E-glass 1310 fibers) parallelized to the direction of the belt axis. The parameters are given in Table 1. The sample details are given in Table 2.

Table 1. E-glass fiber description.

Parameter	Confidence Interval (C.I) *	Lower Limit (LL)*	Upper Limit (UL)*
Fineness (dtex)	3.16	2.60	3.72
Strength (cN)	30.64	26.29	35.0
Deformation at break(%)	6.72	6.07	7.35
Initial modulus (GPa)	26.12	17.30	34.95

(*) 95% confidence interval

Table 2. Sample description.

Sample no.	Sample description
Sample 1	E-glass without epoxy without modification
Sample 2	E-glass 1310/1 epoxy coated
Sample 3	E-glass 1310/1 epoxy dissolved in acetone (fast), washed in water and air-dried.

Since the strength of the epoxy-free tape (sample 1) was 285.6 N, and the strength of the fibers were 30.65 cN fiber, the number of fibers in the tape were estimated as 922 fibers. According to the manufacturer, the fibers in the tape are 1310 fibers. This difference indicates the measurement uncertainty. Epoxy type resins with a special catalyst ensuring durability (in the order of months) under storage conditions and subsequent heat curing. As a textile reinforcement, glass fabrics of filaments of 600 Tex, thickness of 2.15 mm and of a fiber content of 0.34 were used. LH 288 epoxy resins and hardener HY 951 were purchased from Havel Composites Czech Republic. The epoxy resin was mechanically mixed with the hardener at room temperature until a homogeneous mixture was obtained. The composite was cured in an oven at 120 ° C and 50 kPa for 30 minutes.

2.1. Preparation of Tapes

Preparation of SHCT pre-pregnated tapes (prepregs) was carried out on pilot plant, which was built by company Večerník s.r.o. The pilot plant is a model which however fulfilled the purpose of producing a reproducible prepreg made of glass roving in the form of thin endless tapes. At present, the development of a drive unit is being carried out at the department of textile and single-purpose machines. The aim of this device is to automatically control the speed of initial unwinding of roving and final winding of finished prepreg. The drive unit will also include an element that will provide a constant strain of roving.

The current form of a semi-operating device for preparing a prepreg is shown in Figure 1. The overall path of roving before the wrap of the prepreg is about 15m.

Figure 1. Semi-operating device for preparing a prepreg.

In the first part roving passes through the smooth steel rod system. In this part, the effort is to mechanically separate the individual fibres and pull on the roving to ensure a constant disconnection of the individual fibres. For industrial glass roving, the major problem is due to very high speeds in the production itself and the subsequent spinning of the cross-winding fibres, uneven disconnection of the individual fibres occurs. If composite material from such a roving is prepared, optimal mechanical properties cannot be achieved. It is due to the reason that when the composite is loaded, only the percentage of fibres can be completely turned off during the preparation of the prepreg. The roving is made from a mixture of acrylate polymers and epoxysilane, suitable for epoxy matrices. In Czech Republic, glass roving prepared in the industries are unsuitable for epoxy mats due to starch sizing fibre blending. The possibility of modifying the sizing, which would be compatible with the epoxy matrix, but at the same time would not cause the fibres to be interconnected. It would be possible to achieve a homogenous shutdown of the individual fibres to ensure their perfectly parallel arrangement which would at least ensure the approximation of the theoretical value of the glass fibre strength. This is the main objective of the study that will have a major impact on improving the mechanical properties of composites made from these prepregs. In order to ensure sufficient epoxy matrix content, three epoxy dispersion-based coating stations are used in the line and at each of the stations the impregnated roving is dried at elevated temperatures. In terms of sifting the fibrous structure, the first deposit is most important, further deposits are mainly due to the fact that the layers are interconnected with each other in the formation of the composite. After the last coating of the epoxy dispersion with the catalyst, a cross-winding is performed on a paper roll, which is currently performed manually and awaits completion of the drive unit development. The tensile properties were measured by 1000N (Instron instrument) head with pneumatic jaws and rubber, a clamping length of 200mm, a jaw velocity of 50mm / min, i.e. a strain rate of 0.25 / min, a tape thickness of 0.16mm (differences without epoxy) and a width of 4mm. Fiber properties were measured in Vibrodyn 400 instrument. The weathering test was done in Atlas weathering instrument at 80 deg, C 65% RH for 24 hrs in VUTS, Czech Republic.

3. Results and Discussion

The microscopic images of hybrid composite tapes are shown in Figure 2. It is clearly visible that there is a physical bonding between the glass filaments and the epoxy resin matrix. SEM was used to determine the microscopic bonding characteristics of the prepregnated glass filament tapes.

IOP Conf. Series: Materials Science and Engineering **553** (2019) 012055 doi:10.1088/1757-899X/553/1/012055

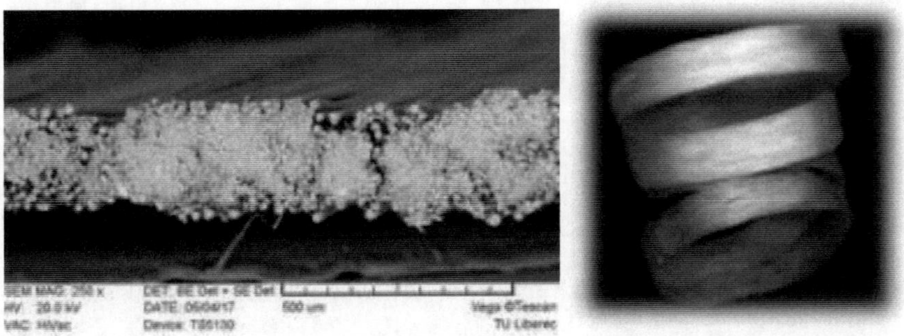

Figure 2. Cross-section and view of a stabilized hybrid composite tape.

The mechanical characteristics of the glass tape without modification (black), SHCT (red) and SHCT after removal of resin in acetone (blue) was measured. The initial modulus in MPa applies to all samples on a constant surface of 4x0.16 = 0.64mm^2. The highest values of strength, deformation at break and initial modulus are achieved with glass with epoxy. The values of these properties for glass with epoxy dissolved in acetone between the epoxy and non-epoxy glass values increase the variability of the values and indicate the imperfect removal of the epoxy. The results showed a significant difference between the samples. Values in brackets are 95% confidence intervals of mean values of breaking force, deformation at break and initial modulus. The upper and lower limits of the mechanical properties was charted in the graphs.

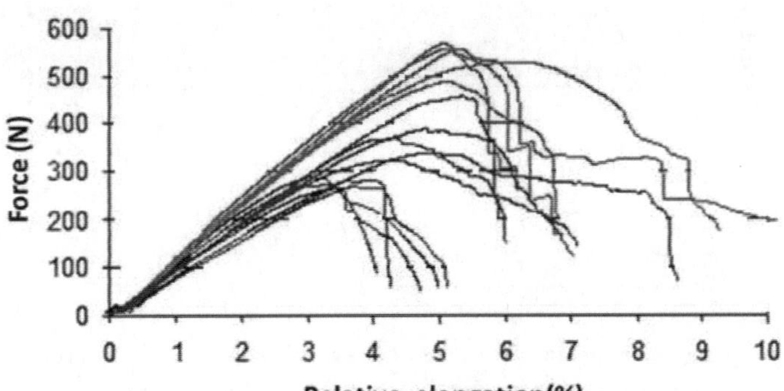

Figure 3. Relative elongation, red – SHCT (sample 2), blue- glass tape after resin removed with acetone (sample 3), black – glass tape without modification (sample 1).

Figures 3, 4, 5 and 6 shows the results of relative elongation, breaking force, deformation at break and initial modulus of the samples given in Table 1. The sophisticated hybrid composite tapes exhibited the greatest improvement in relative elongation with respect to force (the relative elongation increased from 4.5% to 10%). A maximum increase of the initial modulus from 12.9 MPa to 18.22 MPa was achieved. With the addition of epoxy by weight, the initial modulus changed by 40% and breaking strength increased by 75% relative to samples without resin. The sample with resin had a breaking force increase of 14.30%. However, the sample without resin and samples washed with acetone was only 5.64% and 10.2%, increase in impact strength compared to samples washed with acetone. Due to the binding effect of epoxy in the glass multifilament, the breaking force was higher. For glass with epoxy resin, an improvement in tensile characteristics was observed. The initial modulus of sample coated with resin improves over the entire range compared to the samples without resin coating. The maximum improvement was observed for the epoxy coated glass sample where the deformation at break was increased from 4.1% to 5.1%.

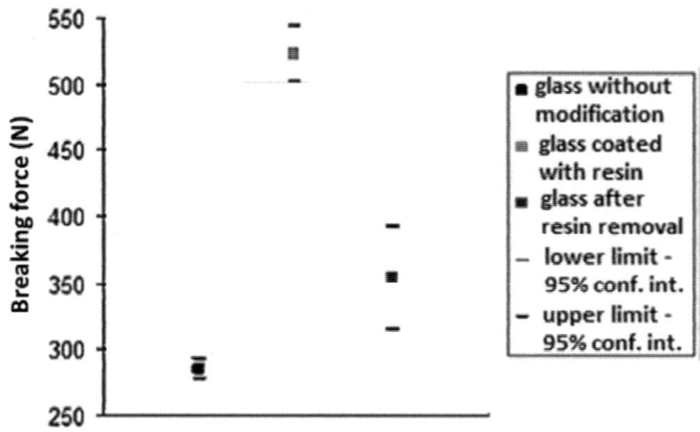

Figure 4. Breaking force of glass tape without modification (black), SHCT (red) and SHCT removed by acetone (blue).

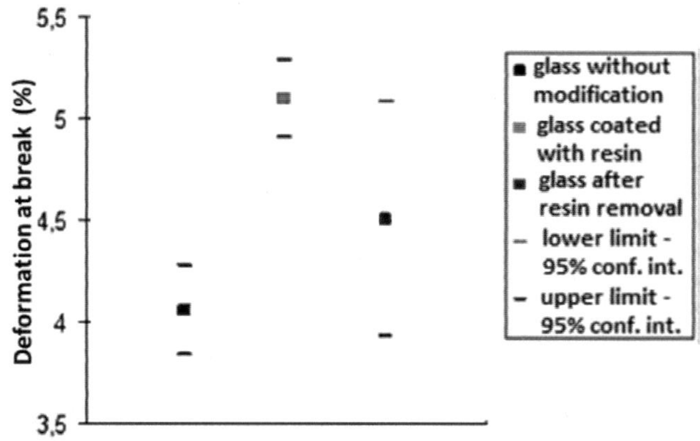

Figure 5. Deformation at break of glass tape without modification (black), SHCT (red) and SHCT removed by acetone (blue).

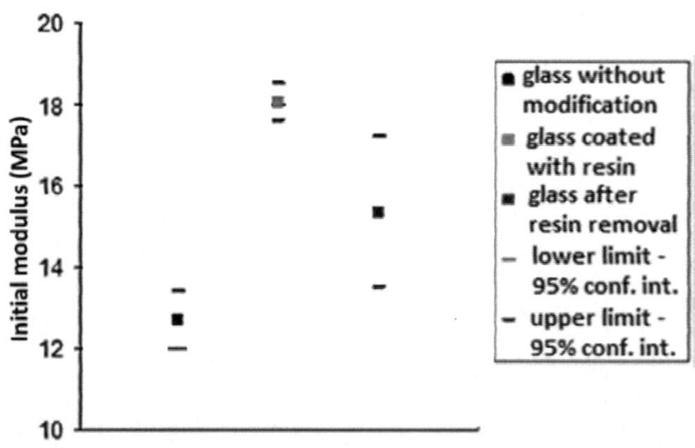

Figure 6. Initial modulus of glass tape without modification (black), SHCT (red) SHCT removed by acetone (blue).

4. Conclusion

Advanced glass fibre based tapes containing special types of resins in pre-condensate form for use in technology of composites preparation by precise winding was prepared. Main benefit is avoiding of resin penetration into winded structure, optimization of tape geometry and composition according to the requirement of precise winding, maximum use of tow filaments strength and enhanced durability for long term storage of tapes. Mechanical properties (breaking force, deformation at break and initial modulus) were enhanced by adding suitable type of resin which showed a significant different between the samples. The glass tape with epoxy resin increased the strength and modulus significantly over the spread of the glass tape. Based on durability tests, it has been shown that SHCT can be stored and used after a longer period of time.

Acknowledgement

This work was supported by the Ministry of Education, Youth and Sports of the Czech Republic and the European Union – European Structural and Investment Funds in the frames of Operational Programme Research, Development and Education – project Hybrid Materials for Hierarchical Structures (HyHi, Reg. No. CZ.02.1.01/0.0/0.0/16_019/0000843), project "Modular platform for autonomous chassis of specialized electric vehicles for freight and equipment transportation", Reg. No. CZ.02.1.01/0.0/0.0/16_025/0007293 and project "Sophisticated hybrid tapes for fabrication of composites by precise winding" [Project no: TJ01000292, 14014/136].

References

[1] Skoko M. Investigation of the Properties with Multiaxial Strengths and Deformations of Coated Fabrics. Tekstil 1998; 7: p. 339-344.
[2] Kovačević S. Brnada S. Dubrovski P.D. Analysis of the Mechanical Properties of Woven Fabrics from Glass and Basalt Yarns: Fibres Text East Eur. 2015; 6: p. 83-91; doi.org/10.5604/12303666.1167424
[3] EL-Dessouky H.M. Ultra-light weight thermoplastic composites: tow spreading technology. ECCM15 – 15th European Conference on Composite Materials, Venice, Italy, 24-28 June 2012.
[4] Venkataraman M. Mishra R. Subramaniam V. Fibers Polym. 2016; 17: p. 925-931; doi.org/10.1007/s12221-016-5882-4
[5] Artemenko SE, Kadykova Yu A. Polymer composite materials based on carbon, basalt, and glass fibres. Fibre Chem: 2008; 1: 37-39. doi.org/10.1023/A:1026170209171
[6] Singha KA. Short Review on Basalt Fibre. Int J Text Sci: 2012; 4: p. 19-28. doi.org/10.5923/j.textile.20120104.02
[7] Cerny M, Glogar P, Golias V, Hruska J, Jakes P, Sucharda Z, Vavrova I. Comparison of mechanical properties and structural changes of continuous basalt and glass fibres at elevated temperatures. Ceramics – Silikáty: 2007; 2: p. 82-88.
[8] Kovačević S, Ujević D, Brnada S, Brlobašić Šajatović B. Structural Multi-layered Composite Textiles Materials: Textile Science and Economy III. Ed. Pavlović M, Milan, Zrenjanin, University of Novi Sad., Technical faculty: 2011; p. 50-67.
[9] Šomođi Ž, Kovačević S, Dimitrovski K. Fabric Distortion After Weaving – An Approximate Theoretical Model. 5th International Textile, Clothing and Design Conference 2010 - Magic World of Textile. Ed. Dragčević Z, Hursa Šajatović A, Vujasinović E, Dubrovnik, Faculty of Textile Technology. Croatia, 3-6 October 2010, p. 729-734.

Growth of ZnO Nanorods Synthesized via Chemical Bath Deposition at Different Reaction Times and Precursor Concentrations

Nofrijon Sofyan*, Sadra Partogi Panjaitan, Akhmad Herman Yuwono

Department of Metallurgical and Materials Engineering, Faculty of Engineering Universitas Indonesia, Depok 16424, Indonesia

nofrijon.sofyan@ui.ac.id

Abstract. The characteristics of ZnO grown via chemical bath deposition on an FTO glass substrate at different reaction time and precursors concentration have been examined. The ZnO seed was firstly spin coated at 500 rpm for 5 seconds onto an FTO glass substrate and then at 3000 rpm for another 30 seconds. The coated substrate was heated at 130 °C to remove the solvent. The growth of ZnO nanorods was performed via chemical bath deposition (CBD) at various precursor concentrations and reaction times. The morphology of the obtained ZnO nanorods were characterized using a field-emission electron microscope (FE-SEM) equipped with energy dispersive X-ray spectroscopy (EDX) to reveal the morphology and elemental composition of the nanorods, whereas X-ray diffraction (XRD) was used to examine the crystal structure. The results showed that the ZnO products have nanorod structure and sizes for each concentration. The results of this morphology were supported by the results from XRD. XRD patterns revealed that the formation of nanostructure of ZnO has been obtained at the reaction time of 2 hours, however, the optimum formation time was proven to be at 3 hours. The results also revealed that ZnO nanorods with a precursor concentration of 0.0375 M has the optimum formation of ZnO nanorod in terms of size and distribution of the nanorods.

***Keywords*:** Chemical Bath Deposition, Nanorods, Zinc Oxide

1. Introduction

Zinc Oxide (ZnO) is one of important semiconductors and has been one of well-researched, especially in the forms of nanomaterials, because of its unique properties and advantages. Its unique properties include direct wide band gap of 3.37 eV, large exciton binding energy of 60 meV, and low cost of production [1-5]. This material has also been used for many applications in various fields such as gas sensor [6], dye-sensitized [7], perovskite solar cells [8], light-emitting diodes [9], photocatalyst [10], and piezoelectric device [11]. Many varieties of ZnO nanostructures such as nanorods [12], nanotube [13], snow-flake [11], nanobelts [14], nanosheets [15], nanowires [16], nanospheres [17], lotus like [18], brush-pen-like and flower-like [11] have been reported by many investigators.

ZnO can be synthesized via several methods such as hydrothermal process [19], thermal evaporation process [11], solvothermal [20], sol-gel [17], solid state reaction [18], electrodeposition and wet chemical method [21]. Electrodeposition, hydrothermal process, and chemical bath deposition (CBD) are chemical based methods that can be used for one dimensional ZnO nanostructure. In terms of growth process, CBD is one of the simple methods because it can be grown at low temperature and cheap equipment. The chemical bath deposition involves 2 step processes, i.e. nucleation on a substrate followed by the growth to form a solid phase from a solution [22]. During the formation, however, the parameters of this two-step process such as precursor concentrations and time [23], seeding layer [24], and temperature [25] may affect the structure and morphology of ZnO nanorods and thus its properties.

Despite its wide use and important application, there are not many references discusses the effect of different reaction times and precursors concentration on the growth of ZnO nanorods. The present work aiming at evaluating the effect of different reaction times and precursor concentrations on the characteristics of ZnO nanorods synthesized and grown on a glass substrate via chemical bath deposition method. The results of the grown ZnO nanorods could be applied as a semiconductor in the renewable energy devices such as dye-sensitized and or perovskite solar cells.

2. Experimental method

2.1. ZnO Seeding

The seed was synthesized through a sol-gel process using precursors of NaOH 0.03 M and zinc acetate dihydrate (Zn(CH3COOH)2·2H2O) 0.02 M solved in methanol under magnetic stirring at 60 °C for 2 hours. The ZnO seed in the forms of solution was cooled to room temperature and was firstly spin-coated at 500 rpm for 5 seconds onto an FTO glass substrate followed at 3000 rpm for another 30 seconds until the ZnO seeds spread equally on the FTO surface. The coated substrate was heated at 130 °C to remove the remaining solvent. Thin white layer of precipitates was observed on the surface of the FTO glass.

2.2. Growth ZnO Nanorods

The growth of ZnO nanoparticles was performed via chemical bath deposition at various precursor concentrations and reaction times. For the precursor concentration, the ZnO seed was soaked in the precursors with a position ZnO seed layer facing the bottom of the beaker glass containing hexamethylenetetramine (HMTA) solution and zinc nitrate hexahydrate ($Zn(NO_3)_2 \cdot 6H_2O$) in water with concentration variations of 0.025, 0.0375, 0.05, and 0.075 M at 130 °C for 3 hours. For the variation in reaction times, the precursor concentration was fixed at 0.025 M with different reaction times of 2, 4, and 6 hours. After CBD reaction was completed, the glass substrate was washed using deionized water and was heated at 130 °C for 5 minutes before being ready for the characterization.

2.3. Characterization

The characteristics of the ZnO nanorods were investigated through surface morphology and crystal structure. The morphology of ZnO nanorods was investigated using a field-emission electron microscope (FE-SEM, FEI Inspect F50) to reveal the morphology of the nanorods, whereas X-ray diffraction (XRD, PANalytical X'Pert PRO) operated at 40 kV and 30 mA with Cu Kα radiation (λ = 1.5406 Å) was used to examine the crystal structure.

3. Results and Discussion

3.1. Structure and Surface Morphology

3.1.1. Effect of precursors concentration on ZnO nanostructure. In this study we set deposition time at its optimum value of 3 hours and the precursor concentration variations of 0.025 M, 0.375 M, 0.05 M, and 0.075 M at 130 °C. Figure 1(a) shows secondary electron image of the obtained ZnO nanorods with 0.025 M precursor concentration. As can be seen from the figure, the ZnO nanorods have the

IOP Conf. Series: Materials Science and Engineering **553** (2019) 012056 doi:10.1088/1757-899X/553/1/012056

smallest diameter compared to the others, but with random ZnO nanorod orientation. Figure 1(b) shows secondary electron image of the obtained ZnO nanorods with 0.035 M precursors concentration. The image reveals that the nanorod diameter is larger with the direction of nanorod is more upright and straighter than that of the nanorods from the 0.025 M. Figure 1(c) shows secondary electron image of the obtained ZnO nanorods with 0.05 M precursors concentration. The ZnO nanorods have larger diameters as compared to the previous two lower precursor concentrations. As can be seen from the figure, the ZnO nanorods structure are arranged tightly with each other and show a distinct hexagonal structure. Figure 1(d) shows secondary electron image of 0.075 M precursors concentration. As can be seen, ZnO nanorods show the largest diameter compared to the others with hexagonal structure but almost clump together. It is expected that this is because at 0.075 M concentration, the precursor has been saturated in the solution. The average diameters of the obtained nanorods at these different concentrations are 68 nm, 86 nm, 235 nm, and 375 nm for 0.025 M, 0.035 M, 0.05 M, and 0.075 M, respectively. These results are in agreement with the results found by others [12, 26].

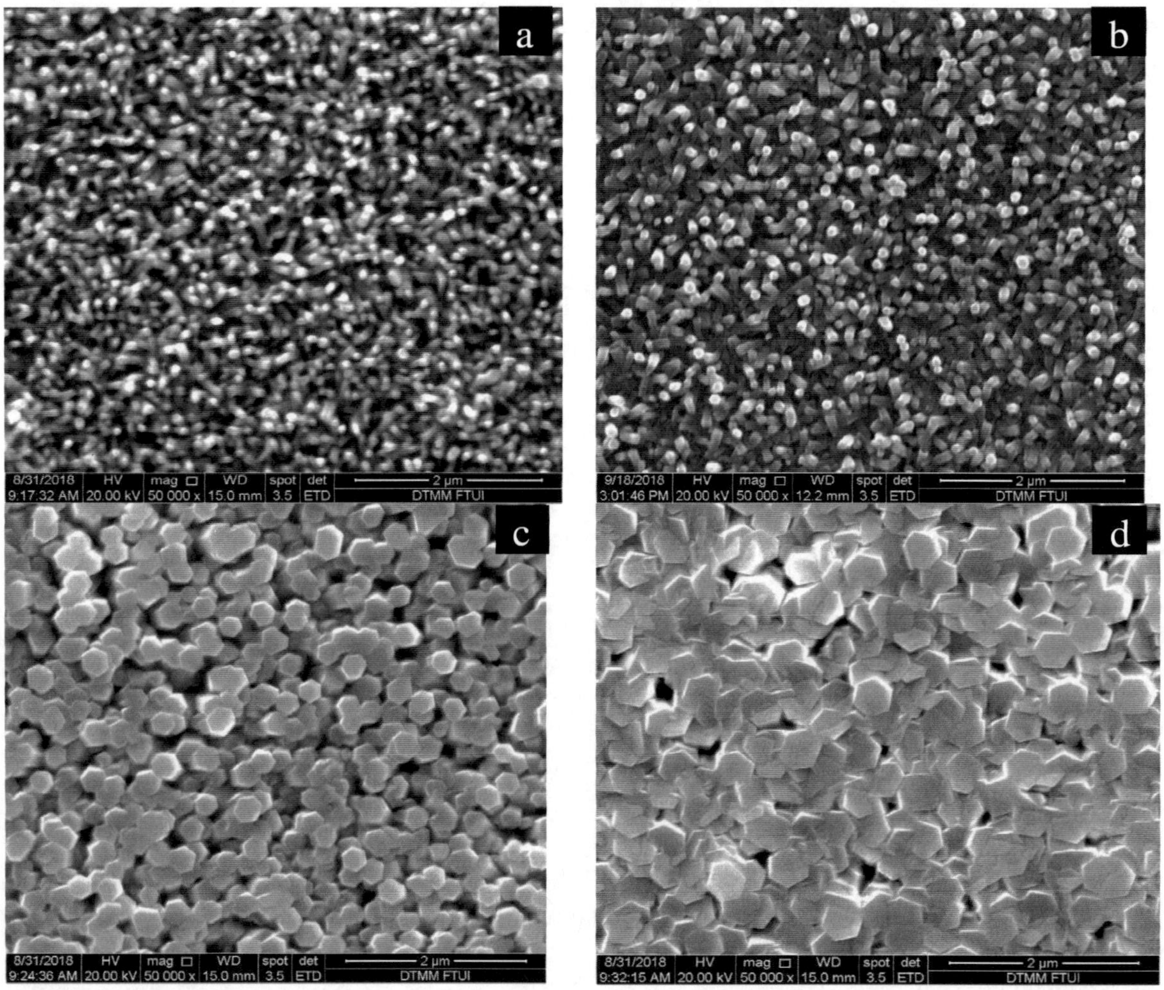

Figure 1. Secondary electron images of ZnO nanorods synthesized with different precursor concentrations: (a) 0.025 M, (b) 0.0375 M, (c) 0.05 M, and (d) 0.075 M.

3.1.2. Effect of reaction time. Figure 2 shows all secondary electron images of the microstructure ZnO nanorods synthesized at 130 °C with precursors concentration of 0.025 M but different reaction

IOP Conf. Series: Materials Science and Engineering **553** (2019) 012056 doi:10.1088/1757-899X/553/1/012056

times. Figure 2(a) shows the secondary electron images of ZnO nanorods synthesized for 2 hours. The image reveals an early establishment of the nanorods growth formation with random direction and small nanorod diameters. Figure 2(b) shows ZnO nanorods morphology with reaction time of 4 hours. The product is showing a denser morphology with larger diameter with more upright and straight direction compared to the previous reaction time of 2 hours. Figure 2(c) is ZnO nanorods morphology with the reaction time of 6 hours. Interestingly, as can be seen in the figure, the morphology of the ZnO nanorods seem to grow with a tendency to form flower-like nanorod structure. This tendency could be as a result of the prolonged reaction time in which instead of growing in an upright position, it tends to form branches resulting in a flower-like morphology.

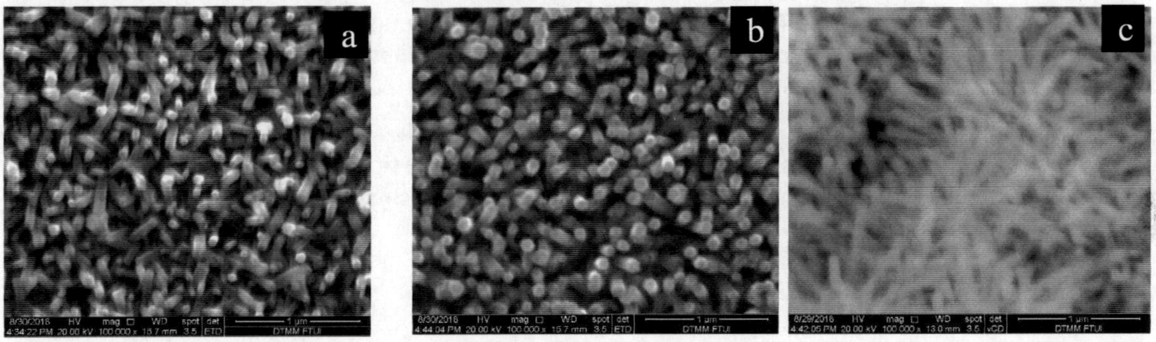

Figure 2. SEM images of ZnO Nanorod Synthesized with different time reaction. (a) 2 hours. (b) 4 hours. (c) 6 hours.

3.2. Phase Structure Analysis

3.2.1. Influence of precursor concentration. In this study, to examine the effect of precursor concentration on the phase formed, the deposition time was fixed for 3 hours at different precursor concentrations. The X-ray diffraction patterns are given in Figure 3.

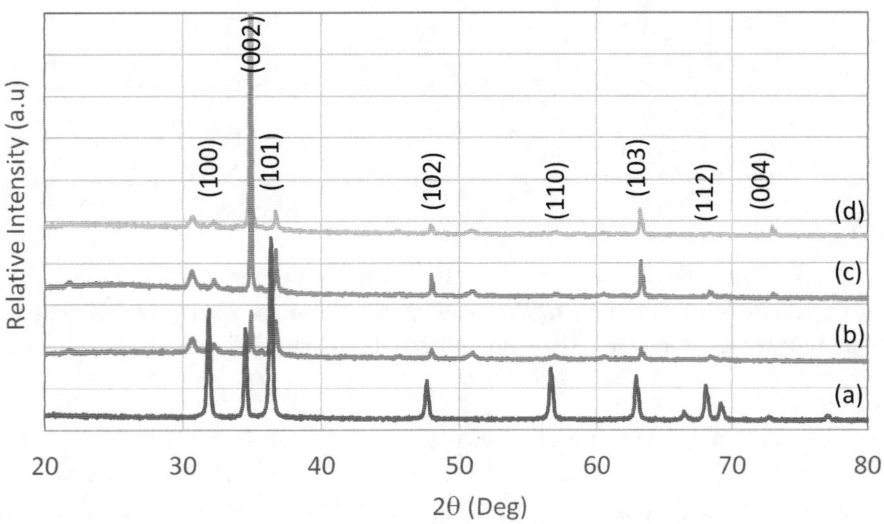

Figure 3. X-ray diffraction patterns of ZnO nanorods synthesized with different precursor concentrations. (a) Pure ZnO, (b) 0.025 M, (c) 0.05 M, and (d) 0.075 M.

IOP Conf. Series: Materials Science and Engineering **553** (2019) 012056 doi:10.1088/1757-899X/553/1/012056

As can be seen in Figure 3, most of the diffraction patterns can be indexed to the ZnO nanorods with the planes (100) (002) (101) (102) (110) (103) (200) (112). Some other diffraction patterns of impurities peaks, however, are also still detected. It seems that during the synthesis, not all of the precursors form the end product. This result is in accordance with what have been found by other investigators on the diffraction patterns of ZnO nanorods [20].

The strong intensity peak indicates that ZnO nanorods product have good crystallinity. In this result, however, there is a tendency that at precursor concentrations of 0.05 M and 0.075 M, the crystal grows to form (002) preferred orientation. It is expected that the growing of the crystal at these precursor concentrations is at the plane (002) as other investigators have been also found [16]. The results from the X-ray diffraction patterns reveal that the ZnO nanorods layer on the glass substrate have been formed characterized by three dominant peaks corresponding to (100), (002), and (101) orientations. The (002) reflection peak has very strong intensity in all precursor concentration, with the highest intensity is the one from 0.075 M, followed by 0.05 M and 0.025 M.

3.2.2. Influence of reaction time. X-ray diffraction patterns of the different reaction time at a fix precursor concentration are shown in Figure 4. As can be seen in the figure, most diffraction peaks can be indexed to the phase of pure ZnO similar to that of different precursors concentration with several peaks from the impurities. The diffraction patterns show that the peaks intensity from the obtained ZnO nanorods at 2 and 4 hours of reaction time are still are still low as an indication of low crystallinity due to the formation of amorphous phase and or nanosized materials. On the contrary, at 6 hours of reaction time, the intensity increases. This can be understood since the longer the reaction time the more the time for self-arrangement of the atoms to form a crystalline phase.

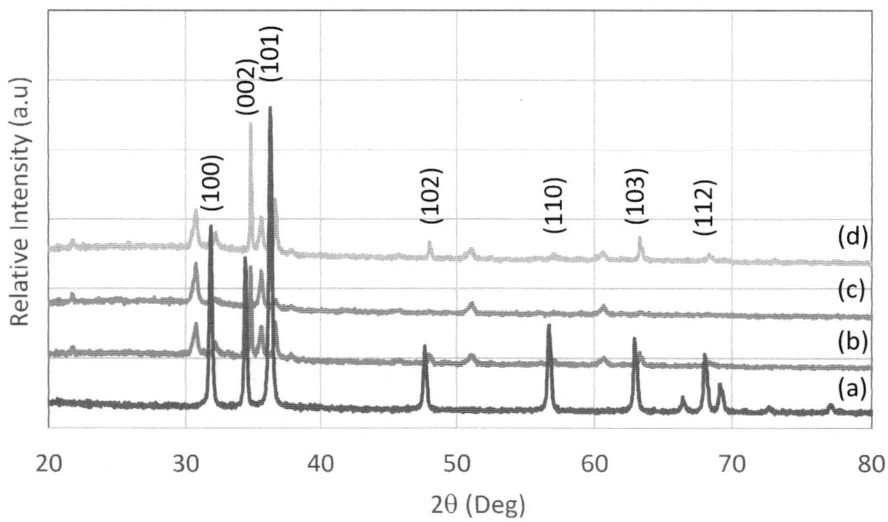

Figure 4. X-ray diffraction patterns of ZnO nanorod synthesized with different reaction times (a) Pure ZnO, (b) 2 hours, (c) 4 hours, and (d) 6 hours.

4. Conclusion

A simple synthesis of ZnO nanorods on a glass substrate via chemical bath deposition with different precursor concentrations and reaction times has been successfully performed. In this work, precursor concentration and reaction time both play crucial roles in determining the morphology and structure of the obtained ZnO nanorods as revealed by the results from secondary electron images and X-ray diffraction patterns. It has also been shown that ZnO nanorods with a precursor concentration of 0.0375 M has the optimum formation of ZnO nanorod in terms of size and distribution of the nanorods.

Acknowledgment
The funding from Directorate of Research and Community Services (DRPM) Universitas Indonesia under grant Hibah PITTA No. 2504/UN2.R3.1/HKP.05.00/2018 is acknowledged.

References

[1] X. S. Chen, J. Y. Liu, X.Y. Jing, J. Wang, D. L. Song, L. H. Liu, Self-assembly of ZnO nanosheets into flower-like architectures and their gas sensing properties, *Materials Letter*, vol. 112, pp. 23-25, 2013. https://doi.org/10.1016/j.matlet.2013.08.118

[2] S. Bai, W. W. Wu, Y. Qin, N. Y. Cui, D. J. Bayerl, X. D. Wang, High-performance integrated ZnO nanowire UV sensors on rigid and flexible substrates, *Advanced Functional Materials*, vol. 21, pp. 4464-4469, 2011. https://doi.org/10.1002/adfm.201101319

[3] B. L. Zhu, C. S. Xie, A. H. Wang, D. W. Zeng, W. L. Song, X. Z. Zhao, The gas-sensing properties of thick film based on tetrapod-shaped ZnO nanopowders, *Materials Letter*, vol. 59, pp. 1004-1007, 2005. https://doi.org/ 10.1016/j.matlet.2004.10.075

[4] M. Law, L. E. Greene, J. C. Johnson, R. Saykally, P. Yang, Nanowire dye-sensitized solar cells, *Nature Materials*, vol. 4, pp. 455-459, 2005. https://doi.org/10.1038/nmat1387

[5] X. H. Han, G. Z. Wang, J. S. Jie, W. C. H. Choy, Y. T. I. Yuk, J. G. Hou, Controllable Synthesis and Optical Properties of Novel ZnO Cone Arrays via Vapor Transport at Low Temperature, *Journal of Physical Chemistry* B, vol. 109, pp. 2733-2738, 2005. https://doi.org/10.1021/jp0475943

[6] L. Zhu, W. Zeng, Room-temperature gas sensing of ZnO-based gas sensor: A review, *Sensors and Actuators A: Physical*, vol. 267, pp. 242-261, 2017. https://doi.org/10.1016/j.sna.2017.10.021

[7] S. B. Zhu, X. Tian, J. J. Chen, L. M. Shan, X. L. Xu, Z. W. Zhou, A facile approach to construct multiple structured ZnO crystals by trisodium citrate-assisted hydrothermal growth toward performance enhancement of dye-sensitized solar cells, *Journal of Physical Chemistry* C, vol. 118, pp. 16401–16407, 2014. https://doi.org/10.1021/jp410325p

[8] J. Luo, Y. Wang, Q. Zhang, Progress in perovskite solar cells based on ZnO nanostructures, *Solar Energy*, vol. 163, pp. 289-306, 2018. https://doi.org/10.1016/j.solener.2018.01.035

[9] S. J. Pearton, F. Ren, Advances in ZnO-based materials for light emitting diodes, *Current Opinion in Chemical Engineering*, vol. 3, pp. 51-55, 2014. https://doi.org/10.1016/j.coche.2013.11.002

[10] S. C. Li, K. Yu, Y. Wang, Z. L. Zhang, C. Q. Song, H. H. Yin, Q. Ren, Z. Q. Zhu, Cu_2S@ZnO hetero-nanostructures: facile synthesis, morphology-evolution and enhanced photocatalysis and field emission properties, *CrystEngComm*, vol. 15, pp. 1753-1761, 2013. https://doi.org/10.1039/C2CE26692J

[11] C. Li, G. J. Fang, N. S. Liu, Y. Y. Ren, H. M. Huang, X. Z. Zhao, Snowflake-like ZnO structures: Self-assembled growth and characterization, *Materials Letter*, vol. 62, pp. 1761-1764, 2008. https://doi.org/10.1016/j.matlet.2007.10.009

[12] Y. Zhou, W. B. Wu, G. Hu, H. T. Wu, S. G. Cui, Hydrothermal synthesis of ZnO nanorod arrays with the addition of polyethyleneimine, *Materials Research Bulletin*, vol. 43, pp. 2113-2118, 2008. https://doi.org/10.1016/j.materresbull.2007.09.024

[13] N. Wang, X. Y. Li, Y. X. Wang, Y. Hou, X. J. Zou, G. H. Chen, Synthesis of ZnO/TiO2 nanotube composite film by a two-step route, *Materials Letter*, vol. 62, pp. 3691-3693, 2008. https://doi.org/10.1016/j.matlet.2008.04.052

[14] T. J. Sun, J. S. Qiu, C. H. Liang, Controllable fabrication and photocatalytic activity of ZnO nanobelt arrays, *Journal of Physical Chemistry* C, vol. 112, pp. 715-721, 2008. https://doi.org/10.1021/jp710071f

[15] H. Algarni, M. M. El-Gomati, M. S. Al-Assiri, Low-temperature growth and properties of nanocrystalline thin ZnO nanosheet interconnects on zinc foil, *Science of Advanced Materials*, vol. 4, pp. 961-968, 2012. https://doi.org/10.1166/sam.2012.1398

[16] Y. C. Chang, Low temperature growth of ZnO nanowire arrays with enhanced high-performance photocatalytic activity and reusability, *Catalysis Communications*, vol. 56, pp. 45-49, 2014. https://doi.org/10.1016/j.catcom.2014.07.009

[17] J. Xu, C. H. Liu, Z. F. Wu, Direct electrochemistry and enhanced electrocatalytic activity of hemoglobin entrapped in graphene and ZnO nanosphere composite film, *Microchimica Acta*, vol. 172, pp. 425-430, 2011. https://doi.org/10.1007/s00604-010-0515-x

[18] J. Liu, X. Huang, J. Duan, H. Ai, P. Tu, A low-temperature synthesis of multiwhisker-based zinc oxide micron crystals, *Materials Letter*, vol. 59, pp. 3710-3714, 2005. https://doi.org/10.1016/j.matlet.2005.06.043

[19] M. Podlogar, J. J. Richardson, D. Vengust, N. Daneu, Z. Samardzija, S. Bernik, A. Recnik, Growth of transparent and conductive polycrystalline (0001)-ZnO films on glass substrates under low-temperature hydrothermal conditions, *Advanced Functional Materials*, vol. 22, pp. 3136-3145, 2012. https://doi.org/10.1002/adfm.201200214

[20] B. Pal, S. Dhara, P. K. Giri, D. Sarkar, Room temperature ferromagnetism with high magnetic moment and optical properties of Co doped ZnO nanorods synthesized by a solvothermal route, *Journal of Alloys and Compounds*, vol. 615, pp. 378-385, 2014. https://doi.org/10.1016/j.jallcom.2014.06.087

[21] J. Yang, X. F. Wang, T. T. Jiang, Y. C. Li, Q. Ma, J. Han, J. Y. Chen, J. X. Wang, Y. Q. Wang, Controllable preparation, growth mechanism and the properties research of ZnO nanocrystal, *Superlattices Microstructure*, vol. 72, pp. 91-101, 2014. https://doi.org/10.1016/j.spmi.2014.04.006

[22] K. A. Wahid, W. Y. Lee, H. W. Lee, A. S. Teh, D. C. S. Bien, I. A. Azid, Effect of seed annealing temperature and growth duration on hydro-thermal ZnO nanorod structures and their electrical characteristics, *Applied Surface Science*, vol. 283, pp. 629-635, 2013. https://doi.org/10.1016/j.apsusc.2013.06.159

[23] F. Xu, Z. Y. Yuan, G. H. Du, T. Z. Ren, C. Bouvy, M. Halasa, B. L. Su, Simple approach to highly oriented ZnO nanowire arrays: large-scale growth, photoluminescence and photocatalytic properties, *Nanotechnology*, vol. 17, pp. 588-594, 2006. https://doi.org/10.1088/0957-4484/17/2/041

[24] L. E. Greene, M. Law, J. Goldberger, F. Kim, J. C. Johnson, Y. Zhang, R. J. Saykally, P. Yang, Low-temperature wafer-scale production of ZnO nanowire arrays, *Angewandte Chemie International Edition*, vol. 42, 3031-3034, 2003. https://doi.org/10.1002/anie.200351461

[25] J. H. Tian, J. Hu, S. S. Li, F. Zhang, J. Liu, J. Shi, X. Li, Z. Q. Tian, Y. Cheng, Improved seed less hydrothermal synthesis of dense and ultra-long ZnO nanowires, *Nanotechnology*, vol. 22, pp. 245601, 2011. https://doi.org/10.1088/0957-4484/22/24/245601

[26] A. H. Yuwono, L. Suhaimi, N. Sofyan, D. Dhaneswara, G. Ramahdita, A. Sholehah, C. Hudaya, Nanostructural Growth Investigation of ZnO Nanorods Derived from Chemical Bath Deposition for Transparent Heater Application, *International Journal of Technology*, vol. 9(6), pp. 1216-1224, 2018. https://doi.org/10.14716/ijtech.v9i6.2452

IOP Conf. Series: Materials Science and Engineering **553** (2019) 012057 doi:10.1088/1757-899X/553/1/012057

Effect of Activated Carbon Addition on Electrochemical Performance of $Li_4Ti_5O_{12}$/Nano Si Composite as Anode Material for LiB

N Annisa[1], I Orlando[1], Faizah[1], A Z Syahrial[1]

[1]Departement of Metallurgical and Materials Engineering, Universitas Indonesia, Depok, 16424, Indonesia

anne@metal.ui.ac.id

Abstract. Lithium titanate ($Li_4Ti_5O_{12}$/LTO) is one of the materials being developed as an anode in Li-ion battery. LTO has zero-strain properties that do not volume change during charge and discharge, does not cause SEI, and can be used for the high rate. However, LTO has a weakness such as low electrical conductivity of 10^{-13} S/cm and low capacity of 175 mAh/g. Therefore, it needs to be combined with high-capacity materials such as silicon and materials that have high electrical conductivity such as carbon. In this study, the LTO-C/nano-Si composite was made to obtain an anode with high capacity and electrical conductivity. Carbon is added with a variation of 1, 3, and 5 wt% during the sol-gel process, while nano-Si is added by 10 wt% of the total active material ingredient in the slurry making. The carbon added is activated carbon which has previously been activated by using NaOH. Activated carbon is characterized by BET and SEM, while LTO-C/nano-Si composite was characterized by XRD and SEM. For the battery electrochemical performance, EIS, CV, and CD tests are performed. Activated carbon has a surface area of 490,007 m^2/g. In this study, the optimum result achieved by the addition of 3% activated carbon (LTO-3%C/nano-Si) with 182.87 Ω resistivity and 17.14 mAh/g discharge capacity at 20C.

Keywords : $Li_4Ti_5O_{12}$, nano Si, LTO/nano Si composites, activated carbon, Li-ion battery

1. Introduction

In the industrial and economic development, people have been heavily dependent on fossil fuels such as oil, coal, and natural gas. The use of fossil fuels in the automotive world is still a major energy source to operate many vehicles. On the other hand, the use of fossil fuels can lead to serious problems such as global warming and environmental pollution [1]. One of the handlings to the effects of fossil fuels is by developing Electric Vehicle (EV) which in its operation using electrical energy and ultimately producing less harmful emissions [2]. In the storing that electrical energy needed a component in the form of batteries, such as lithium-ion batteries that are currently being developed for applications on electric vehicles. The simplest lithium-ion battery (LIB) uses graphite as the anode. Graphite has a good capacity which is 370 mAh/g [1], but graphite has a weakness because pulverization can happen in high rate application so that it can lose its capacity [3]. The other disadvantage is the formation of a solid electrolyte interface (SEI) which affects battery performance [4]. Lithium Titanate or $Li_4Ti_5O_{12}$ (LTO) anode is material that can be substituted on graphite anodes. LTO has several advantages such as zero strain characteristic that the battery does not go through volume change or very low volume change

(<1%) during charged-discharged and does not cause SEI [5]. Beside the advantages, LTO still has its downsides such as low conductivity (10^{-8}-10^{-13} S/cm), poor lithium ion diffusion capability (10^{-8}-10^{-13} cm^2/s), and low enough capacity (175 mAh/g) [1]. The disadvantages of lithium titanate can be overcome by combining it with other elements.

One element that has a huge capacity is silicon (Si). Silicon has a theoretical specific capacity of 4200 mAh/g [6] so that when combined with lithium titanate will increase the capacity of $Li_4Ti_5O_{12}$. However, silicon also has a limitation in which the process of charge and discharge on the battery will cause volume changes in silicon. However, the weakness can be overcome because lithium titanate has zero strain properties, so the addition of silicon to $Li_4Ti_5O_{12}$ matrix does not cause significant volume change. The addition of activated carbon to the LTO also can be an alternative to improve battery performance. Coating with carbon is considered to be a suitable surface treatment method because of its low cost and efficiency. The use of activated carbon to the LTO can increase the surface area that is formed due to the porous structure. Based on the electrochemical results of the tests performed by G.J.Wang et al. [7], the carbon-coated $Li_4Ti_5O_{12}$ will influence the diffusion of lithium ions to become larger and have excellent reversibility capability.

In this study, we discussed how to improve the capacity and conductivity properties of LTO anode by adding nano Si and activated carbon. With the volume stability of LTO, the large capacity of silicon, as well as the increase of LTO surface area due to the porous structure of activated carbon, is expected to produce a battery anode that has good volume stability and has a capacity higher than LTO theoretical capacity. To find out the best performance, the nano Si is added to LTO and some variations of active carbon addition to seeing the effect of its addition to lithium battery performance.

2. Experimental

2.1 Activated Carbon

Commercial carbon was activated in advance to obtain activated carbon. Commercial carbon sources have a form of granules with non-uniform sizes, so before doing the activation, carbon needs to be smoothed. Reducing the size of carbon can be done by using high energy ball milling tools involves the use of zircon balls. The fine coal of the milling result was heated in a furnace at the rate of 20°C min^{-1} from room temperature up to 500°C and maintained at this temperature for 2 h. The heated carbon was mixed with NaOH pellets under magnetic stirring for 2 h. The ratio of NaOH pellets and carbon is 3: 1 accompanied by the addition of 10 mL of water [8]. The mixture of carbon and NaOH then heated at 130°C for 4 h and continued with heated into a furnace under N_2 flow to the final temperature of 700°C with maintained for 1.5 h. After cooling, the resulting mixture was washed with a 0.1 M solution of HCl followed by hot distilled water until pH ~6.5 and then filtered with filter paper. At the end of the process, the carbon obtained was dried at 110°C for 24 h. The activated carbon then subjected to BET testing by using NovaWin Quantachrome.

2.2 Synthesis of spinel $Li_4Ti_5O_{12}$-C powder

The synthesis of LTO-C can be done through several stages of the sol-gel process, calcination, hydrothermal, mixing with a lithium source, and sintering. The first stage is the sol-gel process performed to produce xerogel TiO_2-C. It takes 2 main solutions to get xerogel TiO_2-C that is a primary solution and secondary solution. The primary solution is a mixed solution of titanium tetra-n butoxide, pH 3 ethanol solution, and activated carbon with variations of 1, 3, 5 wt%. The secondary solution is a mixed solution of pH 3 ethanol with H_2O. The measurement for each variable is presented in Table 1. The sol-gel process can be done by slowly adding a secondary solution to a primary solution with a magnetic stirrer for 1.5 h. The result of this process is a clear blackish TiO_2-C gel. The precursor was then left in the beaker glass for 24 h. In the drying process, we crushed the precursor then subjected to an open-air drying for 120 h that there will be liquid solvent separation by an evaporation method. Xerogel TiO_2-C produced by drying process then crushed to make it into powder form. The powder calcined in a furnace at 300°C for 2 h to remove any solvents which may still be present in the powder and to promote the formation of TiO_2-C anatase phase. The O_2 gas that flows in the furnace aims to maximize the formation of TiO_2-C anatase. The next step to make a particle with a broader surface area is using the hydrothermal process which the powder heated at 135°C for 15 h. The hydrothermal process of TiO_2-C anatase can increase the surface area, reduce the particle, and increase the interaction between TiO_2 and LiOH to form $Li_4Ti_5O_{12}$ [9]. Commercial LiOH was used as lithium sources to LTO. TiO_2-C and LiOH were mixed by the mechanochemical process with high energy ball milling tools using zircon balls for 15 min. The LTO-C obtained was sintered at 750°C for 3 h to obtain the $Li_4Ti_5O_{12}$ spinel phase and increase the density of the powder.

Table 1. Mass composition of each sample

C (%)	Primary solution: pH 3 ethanol (mL)	Primary solution: Activated Carbon (gram)	Primary solution: $Ti(OR)_4$ (gram)	Secondary solution: pH 3 ethanol (mL)	Secondary solution: H_2O (gram)
1	36	0.017	6.119	9	1.133
3	36	0.051	6.119	9	1.133
5	36	0.087	6.119	9	1.133

2.3 $Li_4Ti_5O_{12}$-C/nano-Si composite anode coin cell

In this study, the mixing process of LTO-C and nano-Si was carried out in a slurry making the process to avoid the formation of silicon oxide. LTO-C and nano-Si as the active material, acetylene black as the conductive material, and PVDF as the binder was measured with a ratio of 8:1:1. Every sample's composition is in Table 2. The slurry was made by using 6 grams of DMAC as the solvent. The material was added while the DMAC was stirred using a magnetic stirrer with a temperature of 80°C to make the slurry homogeneous. The slurry then coated on Cu foil by a doctor blade and crimped with $LiPF_6$ electrolyte. The samples then subjected to XRD, SEM, and electrochemical performance studies.

Table 2. LTO-C/nano-Si composite composition

Material	Mass (gram): LTO	Mass (gram): Nano-Si	Mass (gram): Acetylene Black	Mass (gram): PVDF
LTO-1%C/nano-Si	1.8	0.2	0.25	0.25
LTO-3%C/nano-Si	1.8	0.2	0.25	0.25
LTO-5%C/nano-Si	1.8	0.2	0.25	0.25

3. Results and discussion

The activated carbon from the activation process has been done by BET testing. The purpose of this test is to find out the large surface area of activated carbon. Addition of activated carbon in the LTO is to increase the surface area of the sample because the activated carbon is a material that has a high surface area and high adsorption capability [8]. With the increased surface area of activated carbon, the place for lithiation- delithation reactions is increasing so that the addition of activated carbon is expected to increase the capacity and conductivity value of the LTO sample.

BET testing was performed with involving nitrogen gas adsorption at 77.3 K. Based on BET test results, the surface area owned by activated carbon was 490.007 m^2/g. X-Ray Diffraction (XRD) was performed to determine the phase formed and measure the crystallites in the $Li_4Ti_5O_{12}$-C/nano-Si samples. XRD result data is analyzed by X'pert HighScore Plus program. XRD pattern in figure 1 showed the result of synthesized LTO-C/nano-Si. Those phases are the $Li_4Ti_5O_{12}$ spheres (JCPDS No.49-0207), TiO_2 rutile (JCPDS No.01-1292), TiO_2 anatase (JCPDS No.21-1272), silicon (JCPDS No.27-1402), silicon oxide (JCPDS No.30-1127), Li_2TiO_3 (JCPDS No.33-0831). This indicates that the synthesis of $Li_4Ti_5O_{12}$-C/nano-Si is not entirely successful, because not only $Li_4Ti_5O_{12}$ and silicon phases, but there are phases such as TiO_2 rutile, TiO_2 anatase, Li_2TiO_3, and silicon oxide. Those TiO_2 and Li_2TiO_3 compounds are retained from the synthesis process, while the silicon oxide is indicated by Si powder. The presence of undesirable compounds can negatively affect to the quality of battery performance, as when TiO_2 rutile undergoes high degree lithiation ($x>0.5$) to Li_xTiO_2, its crystal structure will change so that can lead to loss of capacity [10]. In this study, the carbon added to the LTO did not show the diffraction peak. The undetectability of carbon in XRD results can be due to the amount of carbon added being too low and influenced by the amorphous nature of the carbon itself.

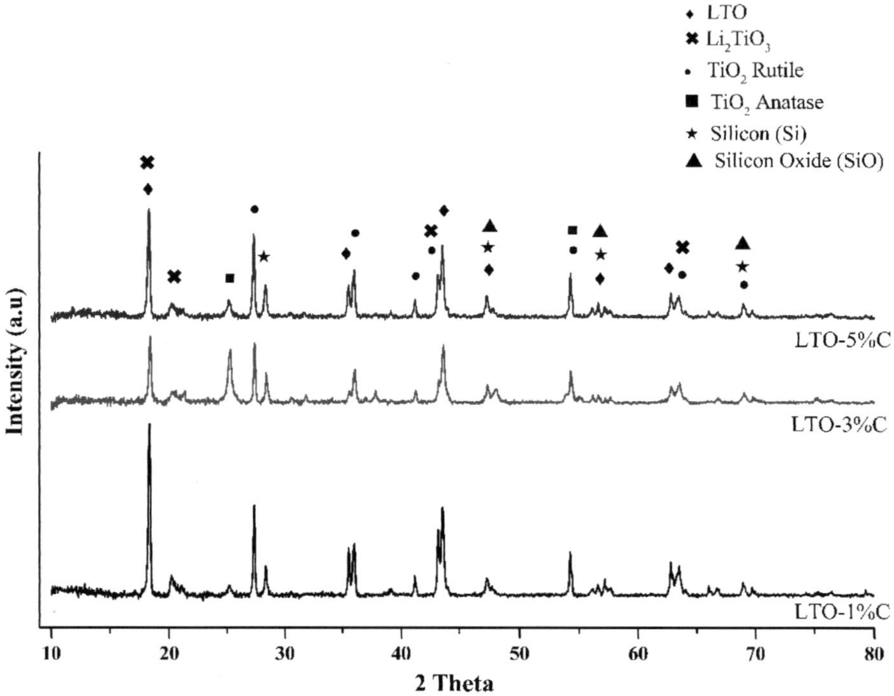

Figure 1. XRD pattern of synthesized $Li_4Ti_5O_{12}$-C/nano-Si

The observation of morphology of synthesized LTO was done by using SEM with 1000X magnification of the sample LTO-1%C/nano-Si, LTO-3%C/nano-Si, and LTO-5%C/nano-Si are shown in Figures 2(a), (b), and (c), respectively. It can be seen that the morphology of each anode sheet showed crack. The cracks are supported by massive volume expansion of silicon. The volume expansion might have happened after the slurry that coated to the Cu foil was in a drying process, where the heat has been given to the sheet. Besides that, in Fig.2 can be seen the agglomerates of the sample. The agglomeration can negatively affect to the sample because it could reduce the surface area and the expansion volume that occurred was too large in one part only, so it can initiate the pulverization at that place which ends on the capacity loss [11]. The agglomeration may occur due to less slurry agitation time, so the slurry product becomes less homogeneous.

Figure 2. SEM result of (a) LTO-1%C/nano-Si, (b) LTO-3%C/nano-Si, (c) LTO-5%C/nano-Si

From the EIS result in table 3, it is shown that the addition of carbon increases the resistivity of the samples. The conductivity value is reverse to the resistivity value, which means that the lower resistivity is, the higher the conductivity. Therefore, LTO-1%C/nano-Si has the highest conductivity followed by LTO-3%C/nano-Si and LTO-5%C/nano-Si. The highest conductivity found in the sample with the lowest carbon content is supported by XRD results that indicates the LTO-1%C/nano-Si sample has the highest number of LTO intensities and high crystallinity. Also the number of impurities in LTO-1%C/nano-Si sample tends to be less compared to the other samples. Y.J.Gu et al. [12], investigated that the presence of TiO_2 and Li_2TiO_3 may negatively affect the electrochemical properties of $Li_4Ti_5O_{12}$.

Table 3. Charge transfer resistivity of LTO-C/nano-Si composites

Sample	Rct (Ω)
LTO-1%C/nano-Si	27.01
LTO-3%C/nano-Si	182.87
LTO-5%C/nano-Si	208.57

Cyclic voltammetry (CV) test was performed to determine the working voltage and capacity of the material. From Figure 3, it can be seen the result of the CV test on the LTO-C/nano-Si samples. By calculating the average of anodic and cathodic peaks, the working voltage can be obtained. The LTO-1%C/nano-Si sample has two peaks at 1.69 V and 1.46 V which the average is 1.575 V, that relatively close to the theoretical working voltage of LTO which is 1.55 V [1]. Also the LTO-3%C/nano-Si and LTO-5%C/nano-Si samples has the working voltage around 1.55 V that represents of synthesized LTO by sol-gel, hydrothermal, and mechanochemical methods succeed to be obtained the working voltage of ~1.55 V. The highest capacity that can be obtained from cyclic voltammetry test showed by LTO-3%C/nano-Si with 168 mAh/g followed by LTO-1%C/nano-Si with 150 mAh/g and LTO-5%C/nano-Si with 131 mAh/g. Besides the LTO peak in Fig.3, there are also other observable peaks such as at 2.092 V on anodic peak and 1.67 V on a cathodic peak in the LTO-1%C/nano-Si sample. It shows peak from TiO_2 that the presence on the CV test result can be reviewed from the XRD characterization results, which detected the TiO_2 rutile and TiO_2 anatase phases in the sample. Based on figure 3 can also be observed the existence of peak-like mound at the beginning of each CV curve. It indicates the presence of silicon in each sample, such as at 0.395 V in the LTO-5%C/nano-Si sample. The existence of Si on the CV curve result was supposed to be at around 0.34 V and 0.51 V that corresponds to the formation of intermediate Li_xSi and amorphous silicon [13].

IOP Conf. Series: Materials Science and Engineering **553** (2019) 012057 doi:10.1088/1757-899X/553/1/012057

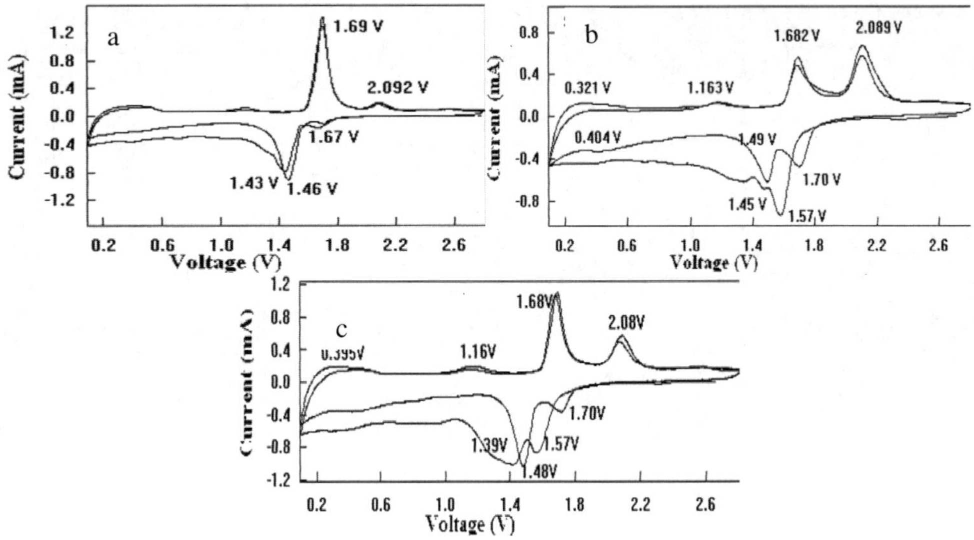

Figure 3. Cyclic voltammetry curve of (a) LTO-1%C/nano-Si, (b) LTO-3%C/nano-Si, (c) LTO-5%C/nano-Si

Figure 4 shows the charge-discharge curve and voltage on a given C. The charge-discharge (CD) test was performed to know the capacity of the battery on certain C-rate. In this study, the CD test is done at several current rates, from 0.2 C until 20 C. The result showed that sample with 5 wt% activated carbon has the highest capacity of the battery, which is shown by the plateau that is longer than the other samples, followed by LTO-1%C/nano-Si and LTO-3%C/nano-Si. In this condition, Si and carbon have successfully contributed its capacity to the mixture of the sample even though it is operated in a high rate condition. The LTO-3%C/nano-Si sample has a lower capacity than LTO-1%C/nano-Si can be due to the number of impurities from synthesis process of LTO-3%C/nano-Si is more than the LTO-1%C/nano-Si and LTO-5%C/nano-Si samples that imparted to the battery performance.

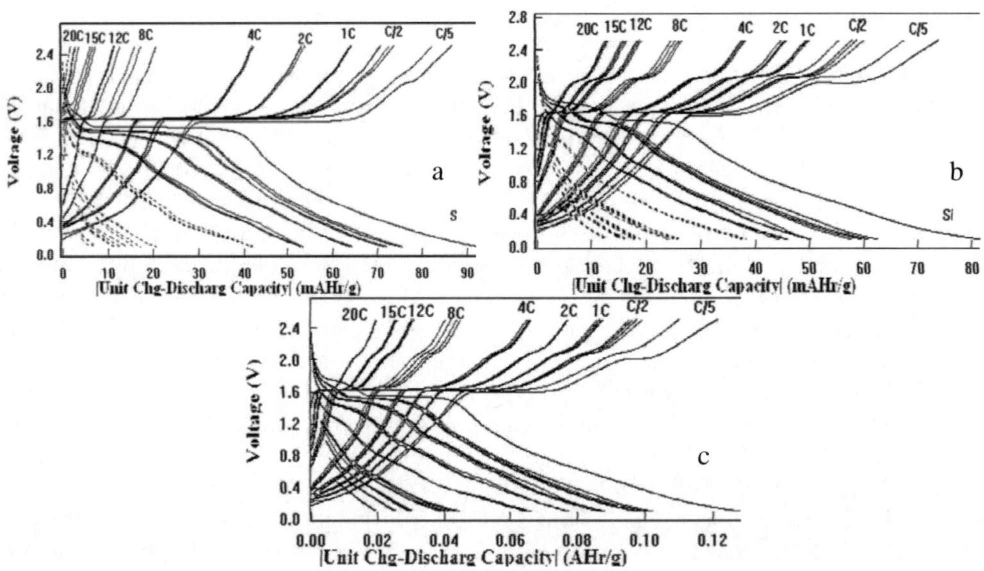

Figure 4. Charge and discharge curve of (a) LTO-1%C/nano-Si, (b) LTO-3%C/nano-Si, (c) LTO-5%C/nano-Si

Table 4 illustrates the efficiency of the coulomb obtained from the ratio of charging and discharging capacity with the fullest potential at 0.2 C. In its use, EV need C rate around 4 C. The capacity at 4 C represents almost half of the battery capacity was a loss, then continue at high C rate.

Table 4. Capacity loss from charge-discharge

C Rate	Capacity Loss (%)					
	C LTO-1%C/Nano Si	D LTO-1%C/Nano Si	C LTO-3%C/Nano Si	D LTO-3%C/Nano Si	C LTO-5%C/Nano Si	D LTO-5%C/Nano Si
0.2	-	-	-	-	-	-
4	51.51	57.39	48.21	52.48	45.42	47.87
20	95.71	97.25	82.86	84.78	83.65	84.38

4. Conclusion

$Li_4Ti_5O_{12}$-C has been successfully synthesized by a sol-gel process in which the activated carbon was added and produces xerogel TiO_2-C, calcination process, hydrothermal process, mechanochemical process with LiOH as the lithium ion source, and sintering process. The addition of nano-Si into synthesized LTO was carried out in a slurry making process and obtained LTO-C/nano-Si. The optimum result achieved by the addition of 3% activated carbon (LTO-3%C/nano-Si) with 182.87 Ω resistivity and 17.14 mAh/g discharge capacity at 20C.

Acknowledgments
The authors would like to thank the Directorate Research and Public Services Universitas Indonesia for their financial support to do this research under HIBAH PITTA 2018 with contract No: 2377/UN2.R3.1/HKP.05.00/2018.

References

[1] B. Zhao, R. Ran, M. Liu, and Z. Shao, "A comprehensive review of $Li_4Ti_5O_{12}$-based electrodes for lithium-ion batteries: The latest advancements and future perspectives," *Mater. Sci. Eng. R*, vol. 98, pp. 1–71, 2015. doi: 10.1016/j.mser.2015.10.001

[2] J. Jaguemont, L. Boulon, and Y. Dubé, "A comprehensive review of lithium-ion batteries used in hybrid and electric vehicles at cold temperatures," *Appl. Energy*, vol. 164, pp. 99–114, 2016.

[3] N. Nitta, F. Wu, J. T. Lee, and G. Yushin, "Li-ion battery materials: present and future," *Mater. Today*, vol. 18, no. 5, pp. 252–264, 2015. doi: 10.1016/j.apenergy.2015.11.034

[4] P. Verma, P. Maire, and P. Novák, "A review of the features and analyses of the solid electrolyte interphase in Li-ion batteries," *Electrochim. Acta*, vol. 55, no. 22, pp. 6332–6341, 2010.

[5] B. Scrosati and J. Garche, "Lithium batteries: Status, prospects and future," *J. Power Sources*, vol. 195, no. 9, pp. 2419–2430, 2010. doi: 10.1016/j.electacta.2010.05.072

[6] C. K. Chan, H. Peng, G. Liu, K. McIlwrath, X. F. Zhang, R. A. Huggins, and Y. Cui, "High-performance lithium battery anodes using silicon nanowires," *Nat. Nanotechnol.*, vol. 3, no. 1, pp. 31–35, 2008. doi: 10.1038/nnano.2007.411

[7] G. J. Wang, J. Gao, L. J. Fu, N. H. Zhao, Y. P. Wu, and T. Takamura, "Preparation and characteristic of carbon-coated $Li_4Ti_5O_{12}$ anode material," *J. Power Sources*, vol. 174, no. 2, pp. 1109–1112, 2007. doi: 10.1016/j.jpowsour.2007.06.107

[8] A. L. Cazetta, A. M. M. Vargas, E. M. Nogami, M. H. Kunita, M. R. Guilherme, A. C. Martins, T. L. Silva, J. C. G. Moraes, and V. C. Almeida, "NaOH-activated carbon of high surface area produced from coconut shell: Kinetics and equilibrium studies from the methylene blue adsorption," *Chem. Eng. J.*, vol. 174, pp. 117–125, 2011. doi: 10.1016/j.cej.2011.08.058

[9] K. Byrappa and M. Yoshimura "Handbook of Hydrothermal Technology - A technology for Crystal Growth and Materials Processing," *Byrappa, M. Yoshimura–Noyes Publ. Park Ridge, NJ*, pp. 62–69, 2001.

[10] R. Wen, "Nanostructured $Li_4Ti_5O_{12}$ as Anode Material for Lithium Ion Batteries," M.Sc Thesis, Faculty of Science, The University of New South Wales, 2012.

[11] S. A. Firnadya, A. Z. Syahrial, and A. Subhan, "Enhancing battery performance by nano-Si addition to $Li_4Ti_5O_{12}$ as anode material on the lithium-ion battery," *Ionics*, 2017. doi: 10.1007/s11581-017-2284-6

[12] Y.-J. Gu, Z. Guo, and H.-Q. Liu, "Structure and electrochemical properties of $Li_4Ti_5O_{12}$ with Li excess as an anode electrode material for Li-ion batteries," *Electrochim. Acta*, vol. 123, pp. 576–581, 2014. doi: 10.1016/j.electacta.2013.12.159

[13] C. Chen, R. Agrawal, and C. Wang, "High-Performance $Li_4Ti_5O_{12}$/Si Composite Anodes for Li-Ion Batteries," *Nanomaterials*, vol. 5, pp. 1469–1480, 2015.

Current State of Lithium Ion Battery Components and Their Development

Maria Kayra Saskia[1], Evvy Kartini[2]
[1] BINUS Serpong graduate, Science Stream, South Tangerang, Indonesia
[2] Science and Technology Center for Advanced Materials, National Nuclear energy Agency (BATAN), Indonesia, Puspiptek Area, Serpong, South Tangerang 15314, Indonesia

skyegenet@gmail.com

Abstract. LIB is a secondary (rechargeable) battery and have been in the spotlight of battery development due to its advantageous qualities in our era of mobile electronics. To improve the LIB for its application in day-to-day life and beyond, development of its separate components, namely the cathode, anode and electrolyte, is essential. Hence, this paper compiles past developments of Lithium Ion Battery (LIB) components, focusing on a general overall view of cathode, anode and electrolytes materials, and including the recent development of LIB in Indonesia.

Keywords: Lithium-ion battery (LIB), cathode, anode, electrolyte, solid electrolyte, neutron diffraction

1. Introduction

Batteries are so ubiquitous today that they're almost invisible to us. Yet they are a remarkable invention with a long and storied history, and an equally exciting future. The first electric battery was built by Volta in 1800, which before was called the voltaic pile. Since then numerous scientists have been undergoing research on improving and developing better materials for electric batteries, using a variety of combinations of elements. Batteries are very important in our current daily lives due to the omnipresence of electrical gadgets. From mobile phones to electric cars, everything seems to run on batteries these days. Not only are lighter and better batteries needed as technology develops, for example the invention of smartphones that has long replaced old mobile phones, we also expect batteries to become safer and less hazardous in their applications. It is even more so with the advent of Electric Vehicles (EVs) coming much closer than ever. With manufacturers aiming to sell EVs commercially, it is important to find longer-lasting batteries that are reliable and light [1]. Consumers would not invest in EVs if replacing its batteries cost the same or even more than a fossil fuel-using car.

This is why it is very important for research on batteries to proceed and bring new solutions to improve our everyday electrical needs. This review will highlight the development of batteries, more specifically Lithium Ion Batteries (LIBs), through the development of its separate components.

2. Development of Batteries

Batteries work by converting chemical potential energy into electric energy, or vice versa when it is recharged, through electrochemical, otherwise known as redox, reactions. These reactions occur between the anode and the cathode, which has a potential difference, when in contact with an electrolyte,

IOP Conf. Series: Materials Science and Engineering **553** (2019) 012058 doi:10.1088/1757-899X/553/1/012058

which acts as a medium in which ions migrate from one end to the other. So the cathode and the anode determines the voltage of the battery. A separator between the anode and the cathode allows movement of protons but not of electrons. The mechanism can be seen as shown in Figure 1. The potential difference between the 2 electrodes acts as an electromotive force that forces the electrons to move through the circuit, hence forming electricity.

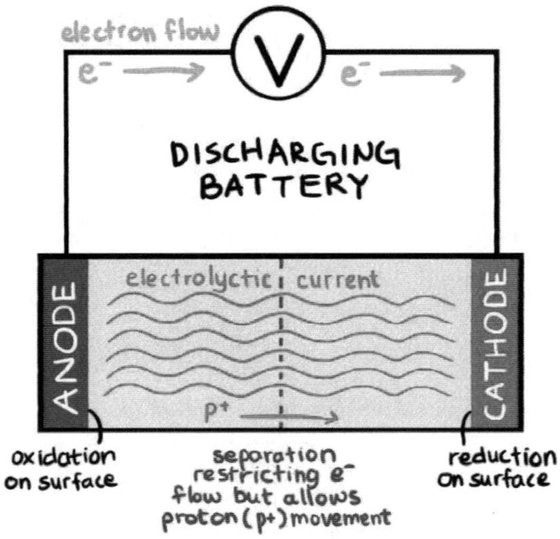

Figure 1. General diagram showing process of discharging battery.

Batteries can be divided mainly into 2 categories: primary batteries and secondary batteries. Primary batteries are also known as non-rechargeable batteries, and actually plays an important role as they have relatively high specific energy, long storage times and instant readiness. They have been used in military combat, pacemakers, smart meters, animal-tracking, light beacons, watches, and other uses. A few examples of primary batteries are as follows: Carbon Zinc, Alkaline, Silver Oxide, Zinc Air and Lithium Cells. The lithium is not to be mistaken with Lithium Ion Batteries, Lithium Batteries are not rechargeable [2].

Secondary batteries, meanwhile, are batteries that can be recharged and used over and over again. This is not just due to its recharge ability, but also because they have low internal resistance, therefore able to form high current that is required by most electrical devices in the current industry. The following are a few examples of secondary batteries, excluding LIBs: Rechargeable Alkaline, Nickel-Cadmium, Nickel-Metal Hydride and Lead-Acid. The later was the most popular rechargeable battery worldwide due to lower cost and high reliability. Not used in consumer applications due to its heavy weight, and contains lead, which is toxic and carcinogenic.

This paper focuses on Lithium Ion Battery, which is a secondary battery. They are a relatively recent breakthrough and one of the most researched types of batteries due to their potential in consumer applications [3].

3. Current State of Lithium Ion Battery

One of the best batteries of today that are commercially available, due to its lightness and high energy density, among other advantages, are Lithium Ion Batteries (LIB). The first LIB was first formulated by Armand in the late 1970s, using intercalation materials of different potentials for the two electrodes, and is often called a rocking chair battery because of the flow of lithium ions back and forward between the two electrodes. The idea was then quickly taken up by other laboratories, which soon figured out that mixing Lithium with other elements (ex. transition metal oxides) [4]. A typical LIB works through the reversible transfer of lithium ions between the cathode and the anode. The general mechanism of an LIB is shown in Figure 2.

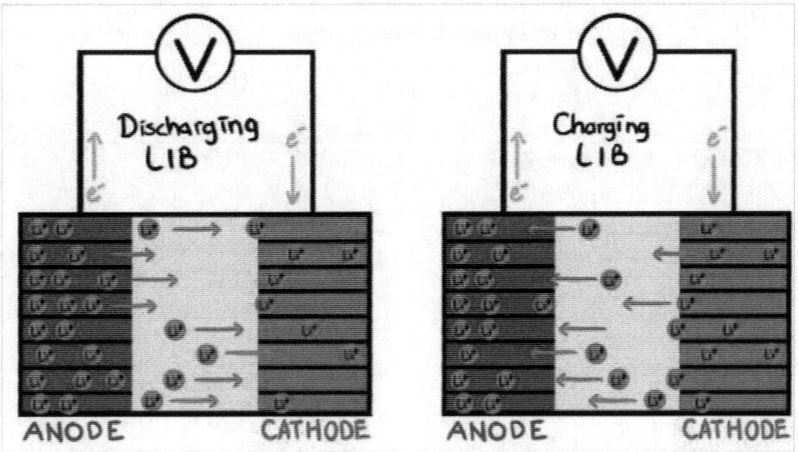

Figure 2. General mechanism of discharging (left) and charging (right) Lithium-ion Battery.

It had been debated that LIB cathodes may develop a layer called the interfacial protective film (IPF) during storage, which increases internal resistance and may lead to lithium plating. However the nature of this layer is still not fully understood. Solid electrolyte interphase (SEI) layer is formed on the electrode surface when the solvent and electrolytic salt in the electrolyte are reduced. However, the layer can show positive and/or negative effects. Good SEI formation can improve the LIB. Composition of electrolyte and electrode micro- and nanostructure may affect the type of SEI formation. SEI insulates the anode but does not block ionic diffusion, though it does lower the capacity. Furthermore, it protects the battery and increases longevity. Bad SEI layer may affect battery performance, rate capability, cycle life, safety, and cause irreversible charge loss. A good SEI layer is porous and allow ionic diffusion. SEI formation may also get destabilized. High temperatures (75-90°C) may cause the breakdown of SEI layer of most commercial batteries.

Coulombic Efficiency (CE) of a battery defines the degree of transfer of electrons during its charge and discharge. It is able to measure the lithium lost due to SEI growth on the anode and the EO at the cathode. However, it varies with temperature and charge rate. CE decreases as cycle life increases due to self-discharge, but large CE causes large cycle life as well.

4. Development of Cathode

A cathode is also known as the positive electrode, where reduction takes place, of an electrochemical cell. Hence they have a more positive electrode potential than the anode. LIB cathodes, however, usually work by storing guest ions, in this case Li^+ ions, which move reversibly from the cathode to the anode. They do this through one of 2 mechanisms: either they are made of a material with a structure that allows insertion and separation of the ions, in other words a 'host network', or, the cathode itself undergoes a reversible redox reaction. LIBs are dual intercalation systems, in which both the cathode and the anode have structures that allow reversible insertion and extraction of lithium cations. Originally, LIB cathodes use cobalt as its main active component, however, nowadays there had been many different synthesis of cathode materials so as to be researched and optimized [5-8]. LIB has been named due to Lithium Ion (Li^+) being its active material, then shortened to form the LIB acronym. Similarly, LIB have been known to be named after the other active Li^+-containing material used, shortened to form easier names when discussing them, instead of using their chemical formula. Table 1 summarizes the averaged abilities of LIBs with the 5 currently most common cathodes.

IOP Conf. Series: Materials Science and Engineering **553** (2019) 012058 doi:10.1088/1757-899X/553/1/012058

Table 1. Summary of Abilities of LIBs with Commonly Found Cathodes [9].

Name	**Lithium Cobalt Oxide (LCO)**	**Lithium Manganese Oxide (LMO)**	**Lithium Nickel Manganese Cobalt Oxide (NMC)**	**Lithium Iron Phosphate (LFP)**	**Lithium Nickel Cobalt Aluminum Oxide (NCA)**
Formula	$LiCoO_2$	$LiMn_2O_4$	$LiNiCoAlO_2$	$LiFePO_4$	$LiNiCoAlO_2$
Voltage (V)	3.0 – 4.2	3.0 – 4.2	3.0 – 4.2	2.5 – 3.65	3.0 – 4.2
Energy Density (Wh/kg)	150 – 240	100 – 150	150 – 220	90 – 120	200 – 260
Charge rate (C)	0.7 – 1.0	0.7 – 3.0	0.7 – 1.0	1.0	0.7; faster charge for some
Discharge rate (C)	1.0	1.0; 10 for some; 30 if pulse (5s)	1.0; 2.0 for some	1.0; 25 for some; 40 if pulse (2s)	1.0
Cycle Life	500 – 1000	300 – 700	1000 – 2000	1000 – 2000	500
Thermal Runaway (°C)	150	250	210	270	150
Applications	Small portable electronics	Power tools, medical devices, electric vehicles	Portable electronics, electric vehicles, industrial use	Portable and stationary appliances requiring high load current and endurance	Medical devices, industrial uses, electric vehicles
Disadvantages	Short life span, low thermal stability, low specific power, high cost	Low energy density	Lower voltage, patent issues	Low energy density, high self-discharge, lower shelf life, sensitive to moisture, require special preparation	Low thermal stability, Sensitive to moisture, higher cost

All the cathode materials in the table have ‘host networks’, and the material itself does not undergo redox reactions. Cathode materials which do can go through one of 2 reactions, both reversible. In first type reaction, metal halides react with lithium to form metal nanoparticles and lithium halides, which contains lithium ions. Second type reactions are undergone by lithium and one of the following elements: sulfur, selenium, tellurium, iodine and oxygen. However, only a few of these type of cathode material is suitable for commercial use. Design of electrodes may also affect its capabilities. For example, their placement, thickness, carbon content of cathode, their porosities, and the type of carbon used as the anode. Carbon-coated $LiFePO_4$ cathode in battery prepared by hydrothermal route also improve the electronic conductivity [10, 11].

5. Development of Anode

The anode can be said to be almost completely opposite of the cathode, in both mechanism and placement within the LIB. Whereas reduction occurs in the cathode during discharging, oxidation occurs at the anode. The anode have also been called as the negative electrode. Anode material for current LIBs require high porosity, conductivity, durability, light weight, low cost, and suitable voltage when matched with the cathode. Hence, for the most common cathodes as shown above, carbon-based anodes are used. Usually this consists of processed graphite that is baked onto copper anodes, and meet the requirements of the cathodes. However, due to safety, cost and environmental concerns of future batteries, graphite may no longer satisfy these requirements. Increasing energy demand also plays a part in possible commercial anode material change.

Silicon has been tested as an anode material due to its large capacity; over ten times that of graphite. However, it experiences drastic volume change during charging and forms an unstable SEI layer. Furthermore, it has poor mechanical stability and is chemically passive. Lithium metal has also been used as an anode material for both Lithium Battery and LIBs. It has a similar capacity to silicon, which is why it had been researched extensively in the past. However lithium metal is highly reactive and may cause safety issues. Its usage as an anode also cause bad SEI formation and dendrite growth. Alloy anodes seem to have a distinct advantage compared to purely carbon anodes, due to their potential for higher capacity and decreased corrosion. For example, addition of silicon particles in small amounts on carbon-based anodes have been shown to improve cycle life and capacity with little difference in cost. Corrosion of the anode may lead to SEI growth and degradation. This can be prevented by improving anode coating, modifying surface structure of anode, or changing the composition of the electrolyte.

Another popular choice for LIB anodes is lithium titanate (Li_4Ti_5O12, battery name LTO), which has been known since the 1980s and has shown potential due it being structurally stable, environmentally friendly, and having lower cost. However, it has relatively lower charge/discharge capability. The material have faced significant improvement when combined with carbon nanotubes to form a composite. Its use have been shown to improve rate capability, cycle life, as well as safety of LIBs. Usage of lithium titanate material as the anode minimizes SEI formation and volume change [12, 13]. It has a wide temperature range and long life, and have been applied in electric trains and street lighting. Lithium Titanate seems to be open to further improvement through different synthesis methods, as shown by various researches in recent years, such as $Li_4Ti_5O_{12}$ anode material prepared by hydrothermal process [11, 14].

6. Development of Electrolytes

Cathode and anode materials are not the only components that are being improved in order to create better LIBs. Qualities such as energy density, power density, lower resistance and safety need to be reflected in the electrolytes used as well. Furthermore, electrolytes must have good ionic conduction while preventing electron transfer, is stable and tolerant to possible environmental changes, and especially for commercial batteries, must be non-toxic, environmentally friendly and costs cheaper. Most current commercially available batteries use liquid electrolytes due to its high conductivity, relatively wide temperature range and the added potential for additives. The first electrolyte used commercially in LIBs is liquid organic electrolyte, which is non-aquatic and consists of a lithium salt dissolved in an organic solvent. They are widely used in handheld electronics and energy storage due to its stability, good voltage, and high energy density. These solvents include carbonates, esters, and ethers, while the salts may be $LiBF_4$, $LiPF_6$, Li Triflate, Li Imide, LiFAP, Li imidazolide or various others. Moreover, they are highly flammable, which has raised concerns of the safety of its application. Inappropriate insulation and heat management of the battery have caused electronics to short-circuit and catch on fire.

Aqueous electrolytes solves this problem as they are non-flammable, while also being low cost. Research on these electrolytes, which replaces organic solvents with water, has shown promising findings. However, it is not as stable as other electrolytes as water may evaporate or be electrolyzed to form gases. Water also limits the electrochemical reactions that can occur within the electrolyte, so that its voltage may be lower than other types of electrolytes. Ionic liquids (ILs), in comparison, consists of

IOP Conf. Series: Materials Science and Engineering **553** (2019) 012058 doi:10.1088/1757-899X/553/1/012058

a liquid with a cation and an anion, as well as a lithium-containing salt. Its properties are largely dependable on the cation and anion selection. They are non-flammable, have high chemical stability, ion conductivity, and solubility of both organic and inorganic compounds. However, their viscous nature may reduce ionic conductivity, and they costs more than other organic solvents [15]. Another type of electrolyte is the solid polymer electrolyte. They have advantages over liquid electrolytes as they have better structural properties, better safety, and may prevent formation of dendrites. Unfortunately, they only work at elevated temperatures, and are no longer commercially available. Gel polymer electrolytes (GPEs) have similar properties as solid polymer electrolytes, but they are more commercially suitable as they also have some similarities with liquid organic electrolytes. They are composed of a salt, a liquid organic solvent, and a polymer.

The development of solid electrolytes is one of the most significant advancements in LIB research [16-18]. It solves the problem of flammability of LIBs that operate using other types of electrolytes, will not cause leakage, mitigates dendrite formation, and have higher energy, voltage and power density than previously expected. Solid Electrolytes may replace commercially used organic electrolytes due to its better safety and toxicity. However, it does have room for improvement as it experiences several disadvantages.

7. Development of Solid Electrolyte

There are two kinds of solid electrolytes (SEs). The first is ceramics solid electrolytes, otherwise known as crystalline electrolytes, which in turn also have several sub-types. Both of these types of electrolytes must allow diffusion of lithium ions, which would require a different mechanism when compared to liquid electrolytes. The difference between liquid and solid electrolytes is shown in Figure 3.

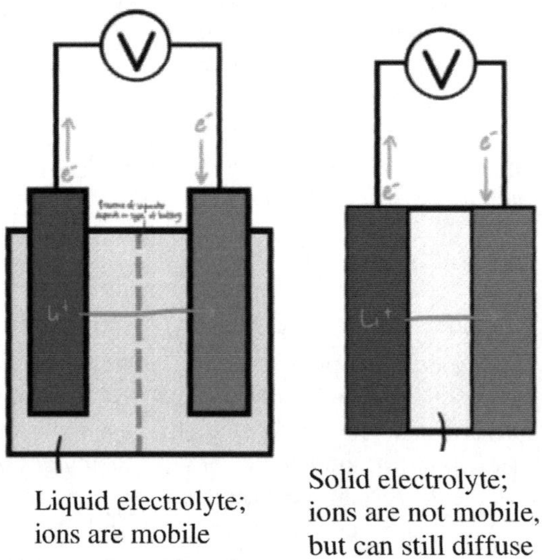

Figure 3. Simple battery mechanism with liquid electrolyte (left) and solid electrolyte (right).

Both types of solid electrolytes are single ion conductors, which means that they allow the conduction of only one ion, while other cations and anions form the framework or structure of the electrolyte. This may also prevent side reactions and decomposition of the electrolyte. Most other electrolytes cannot replicate this kind of property, and usually have downsides when they do. Furthermore, they do not cause leakage and have relatively stable temperatures.

LISICON-type SEs have low conductivity during room temperature which improves at 300°C. But improvements on the material, for example by replacing oxide by sulfur in its framework, can highly improve its ionic conductivity. Garnet-type electrolytes have high conductivity and chemical stability.

IOP Conf. Series: Materials Science and Engineering **553** (2019) 012058 doi:10.1088/1757-899X/553/1/012058

$Li_7La_3Zr_2O_{12}$ (LLZO) is one example, which exhibits very good safety and electrochemical stability, able to perform excellently when applied properly in an LIB. Perovskite-type electrolytes have high bulk conductivity. $Li_{0.29}La_{0.57}TiO_3$ (LLTO) is an example of it. It has high ionic conductivity when amorphous and is used as a thin film, thus reducing resistance as ions diffuse through a shorter distance. It is also stable even at high voltages. The above SEs are highly researched upon as a commercial electrolyte for day-to-day applications, especially for EVs. The second type of SEs, namely glass-based electrolytes, also have potential in commercial use.

Glassy electrolytes do not suffer from any grain boundary resistance, are easy to form into a thin film, which is especially useful in cylindrical battery applications, and have relatively high ionic conductivity [19]. They can also prevent dendrite formation. Glassy electrolytes are typically composed of a lithium salt and modifier, the glass which forms the structure, and if applicable, additives to improve its properties [18, 20, 21]. The glass must allow ion diffusion in order to behave as an electrolyte. Usually, the glassy electrolyte is ground into a fine powder or milled so as to produce amorphous materials, which can allow ion diffusion. Another subtype of glass-based electrolytes are glass-ceramic electrolytes, which involved crystallization of the precursor glass. They have grain boundaries filled with amorphous phases, therefore causing lower resistance.

Indonesia has a lab specifically for battery research in BATAN, Serpong, and has various research on developing materials for battery components, also for contributing battery production and testing at laboratory scale. Recent development at our laboratory on glassy electrolytes based on phosphate glasses, such as $LiPO_3$, Li_3PO_4, Li_2O-P_2O_5 and its corresponding glasses with variation of doping salts to increase the ionic conductivity [16, 18, 22]. Several methods has been developed such as Li_3PO_4 powder as solid electrolyte material prepared by wet chemical reaction and solid-state reaction Li3PO4-MMT-PVDF composites as solid electrolyte material is also being developed in this laboratory [18, 19].

8. Conclusion

It is concluded that the research and development of on LIB moving forward. This paper has summarized the early development of batteries and the current state of LIB. Most specific, development of materials cathode, anode and electrolyte as battery components. The research on solid electrolyte is growing due to safety reason. The future of solid state battery will be replacing the current commercial lithium ion battery which based on liquid electrolyte. Until now, both private companies and the Indonesian government has further developed LIB research for various reasons. Profit and an improved economy seems to be their main driving force, as well as the pride of being able to keep up and overcome other, more developed countries in battery research. However, much more time is needed before Indonesia's citizens would be able to take the lead in battery research, as their attitude is unfortunately currently more apathetic than not towards these kinds of developments within their own country. But with the help of researchers, the push of the government, and financial aid of interested sponsors, battery research in Indonesia has made leaps and bounds from its starting point as almost negligible, until now.

Acknowledgment

The author would like to thank Battery team at BATAN, for their guidance and expertise throughout the writing of this paper, and K parents for their support and understanding. Materials Research Society Indonesia (MRS-INA) for giving opportunity to join the International Conference in Asia of the International Union of Materials Research Society (ICA-IUMRS), Bali, October 30-November 2, 2018.

References

[1] B. Scrosati and J. Garche, "Lithium batteries: Status, prospects and future," *J. Power Sources*, vol. 195, pp. 2419–2430, 2010.
[2] J. Hwang, S. Myung, and Y. Sun, "Chem Soc Rev Sodium-ion batteries : present and future," *Chem. Soc. Rev.*, 2017.
[3] E. Kartini, "Emerging battery research in Indonesia: The role of nuclear applications," in *AIP*

Proceeding "5th International Conference on Women in Physics, Waterloo, Canada, 5-8 August 2015"., 2015, vol. 1697, pp. 100001–2.

[4] L. Zhao, D. Zhou, W. Huang, X. Kang, Q. Shi, Z. Deng, and X. Yan, "Electrochemical Performances of Maricite NaFePO 4 / C as cathode material for Sodium-Ion and Lithium-Ion Batteries," vol. 12, pp. 3153–3165, 2017.

[5] S. Oh, S. Myung, J. Hassoun, B. Scrosati, and Y. Sun, "Electrochemistry Communications Reversible NaFePO 4 electrode for sodium secondary batteries," *Electrochem. commun.*, vol. 22, pp. 149–152, 2012.

[6] Y. Long, Y. Shu, X. Ma, and M. Ye, "In-situ synthesizing superior high-rate LiFePO4/C nanorods embedded in graphene matrix," *Electrochim. Acta*, vol. 117, pp. 105–112, 2014.

[7] W. Honggowiran and E. Kartini, "Characterization of LiFePO 4 Cathode by Addition of Graphene for Lithium Ion Batteries," in *6th Nanoscience and Nanotechnology Symposium (NNS2015)*, 2016, vol. 030045, pp. 1–6.

[8] W. Honggowiranto, Sudaryanto, E. Kartini, and A. Purwanto, "Electrochemical Performance of LiFePO 4 Cylinder Cell Battery," in *AIP Proceeding "6th Nanoscience and Nanotechnology Symposium (NNS2015),"* 2016, pp. 030042–1–7.

[9] M. M. Doeff, "Batteries: Overview of Battery Cathodes," in *Springer Encyclopedia of Sustainability Science and Technology*, pp. 709-739, 2012.

[10] X. Li, T. Li, Y. Zhang, X. Zhang, H. Li, and J. Huang, "Graphene nanoribbon-wrapping LiFePO4 by electrostatic absorbing with improved electrochemical performance for rechargeable lithium batteries," *Electrochim. Acta*, vol. 139, pp. 69–75, 2014.

[11] W. Honggowiran and E. Kartini, "Characterization of LiFePO 4 Cathode by Addition of Graphene for Lithium Ion Batteries," in *AIP Proceeding*, 2016, vol. 030045.

[12] B. Priyono, A. Z. Syahrial, A. H. Yuwono, E. Kartini, M. Marfelly, and W. R. F. Muhamad, "Synthesis of Lithium Titanate (Li4Ti5O12) through Hydrothermal Process by using Lithium Hydroxide (LiOh) and Titanium Dioxide (TiO2) Xerogel," *Int. J. Technol.*, vol. 4, pp. 555–564, 2015.

[13] A. Z. Syahrial, B. Priyono, A. H. Yuwono, E. Kartini, H. Jodi, and Johansyah, "Synthesis of Lithium Titanate (Li4Ti5O12) through Hydrothermal Process by using Lithium Hydroxide (LiOh) and Titanium Dioxide (TiO2) Xerogel," *Int. J. Technol.*, vol. 3, pp. 392–400, 2016.

[14] N. Sofyan, G. T. Setiadanu, A. Zulfia, and E. Kartini, "Effect of Different Calcination Temperatures and Carbon Coating on the Characteristics of LiFePO4 Prepared by Hydrothermal Route," *IJET*, vol. 9, no. 4, pp. 3310–3317, 2017.

[15] A. Reni, S. Evvy, K. Wagiyo, H. Sudaryanto, and R. Hidayat, "Perfomance of Lithium Ion Battery Based on LiTFSI Lithium Salt as Electrolyte Compared to LiPF 6," vol. 1, no. 1, pp. 1–5.

[16] N. I. P. Ayu, E. Kartini, L. D. Prayogi, M. Faisal, and Supardi, "Crystal structure analysis of Li3PO4 powder prepared by wet chemical reaction and solid-state reaction by using X-ray diffraction (XRD)," *Ionics (Kiel).*, vol. 353, pp. 1–7, 2016.

[17] E. Kartini and M. Manawan, "Solid Electrolyte For Solid-State Batteries: Have lithium-ion batteries reached their technical limit?," in *AIP Proceeding (in press)*, 2015.

[18] E. Kartini, M. Nakamura, M. Arai, Y. Inamura, K. Nakajima, T. Maksum, W. Honggowiranto, and T. Y. S. P. Putra, "Structure and dynamics of solid electrolyte (LiI)0.3(LiPO3)0.7," *Solid State Ionics*, vol. 262, pp. 833–836, 2014.

[19] E. Kartini, M. Manawan, M. F. Collins, and M. Avdeev, "Physica B : Condensed Matter Neutron diffraction study on Li 3 PO 4 solid electrolyte for lithium ion battery," *Phys. B Phys. Condens. Matter*, vol. 551, no. August 2017, pp. 320–326, 2018.

[20] E. Kartini, T. Sakuma, K. Basar, and M. Ihsan, "Mixed cation effect on silver-lithium solid electrolyte (AgI) 0.5(LiPO3)0.5," *Solid State Ionics*, vol. 179, pp. 706–711, 2008.

[21] H. Jodi, A. Z. Syahrial, Sudaryanto, and E. Kartini, "Synthesis and Electrochemical Characterization of New Li2O-P2O5 Compounds For Solid Electrolytes," *IJTech 2017*, vol. 6,

no. 8, pp. 328–336, 2017.
[22] E. Kartini, T. Y. S. P. Putra, I. Kuntoro, and T. Sakuma, "Recent Studies on Lithium Solid Electrolytes (LiI) x (LiPO 3) 1-x for Secondary Battery," *J. Phys. Soc. Jpn*, vol. 79, no. Suppl.A, pp. 54–58, 2010.

IOP Conf. Series: Materials Science and Engineering **553** (2019) 012059 doi:10.1088/1757-899X/553/1/012059

Synthesis and Characterization of $LiFePO_4$ By Variation Addition Of $(Li_3PO_4)_{0.5}(AgI)_{0.25}(LiI)_{0.25}$ as Cathode Material For All-Solid-State Li-ion Battery

Lisda Ainiya[1*], Evvy Kartini[2], Wagiyo Honggowiranto[2]
[1]Department of Physics, Kalimantan Institute of Technology, Kampus ITK, Karang Joang, Balikpapan 76127, Indonesia
[2]Center for Science and Technology of Advanced Materials, National Nuclear Energy Agency, Kawasan Puspiptek Serpong, Tangerang Selatan15314, Banten, Indonesia

lisdaainiya.21@gmail.com

Abstract. The safety of lihium-ion batteries have always been a concerned for industrial application. Therefore, there is a strong interest to produce an all-solid-state battery. An all-solid-state battery has several advantages, including easy fabrication, non-flammable, low leakage, but mainly because it is safer. In this study, $(Li_3PO_4)_{0.5}(AgI)_{0.5}(LiI)_{0.25}$ was synthesized as solid electrolyte. Meanwhile, as cathode, $LiFePO_4$ was added by variations of: 0%, 10%, and 30% of $(Li_3PO_4)_{0.5}(AgI)_{0.5}(LiI)_{0.25}$ to minimize the presence of solid state interface between solid electrolyte and cathode. The samples were characterized using Scanning Electron Microscope (SEM), X-Ray Diffraction (XRD), and conductivity measurement. For SEM images, the particle distribution was not thoroughly homogeneous because it was prepared manually by only grinding the samples. The X-ray data showed that $(Li_3PO_4)_{0.5}(AgI)_{0.5}(LiI)_{0.25}$ has halo pattern and the crystal structure of synthesized cathode belongs to *P m n a* space group (Olivine Structure). The range of measured AC conductivity was around 10^{-4} S/cm - 10^{-8} S/cm.

Keywords : Lithium- ion battery, $LiFePO_4$, $(Li_3PO_4)_{0.5}(AgI)_{0.25}$ $(LiI)_{0.25}$, Solid Electrolyte

1. Introduction

A rechargeable battery plays a significat role to provide an efficient storage of electrical energy as chemical energy [1]. There are numerous types of rechargeable batteries, including rechargeable Li-ion batteries (LIBs). LIBs have several advatages, such as, long life cycle, high power density, low self-discharge property, high gravimetric, and volumetric energy [2-8]. Currently, the major cathode materials being explored are $LiMn_2O_4$, $LiNiO_2$, $LiCoO_2$, and their composites [1, 10-12]. Unfortunately, $LiMn_2O_4$ is struggling with cycling and thermal stability and pure $LiNiO_2$ is isothermal and cannot commercialize due to safety concerns. Even though, $LiCoO_2$ has been widely commercialized, there are safety concerns considering $LiCoO_2$ is toxic and very expensive [13].

On the other side, $LiFePO_4$ is a good choice to be a cathode material for Li-ion batteries, safer charging/discharging, good theoritical capacity at 170mAh/g and a flat voltage of 3.4 V. But the major drawback of $LiFePO_4$ is its much lower electronic conductivity (~10^{-9} S.cm^{-1}) compares to other cathode, such as, $LiCoO_2$ (10^{-3} S.cm^{-1}) [14]. However, the increasing number of fire incidents in electrical vehicles and in airplanes is a huge concerned for Li-ion batteries. One way to improve the safety and improving the lifecycle of the batteries is to replace the liquid electrolytes by solid electrolytes to avoid the grow of dendrite formation [15-17]. Even though, liquid electrolytes have a high ionic conductiviy (12 mS/cm for 1 M $LiPF_6$ in EC-DMC at 27°C) but it is risky because its being flammable [18].

A previous work on solid electrolyte was performed by E.Kartini et al. [19-20] on Li_3PO_4 and $LiPO_3$. Furthermore, T.Utomo [21] using $(Li_3PO_4)_x(AgI)_y(LiI)_z$ as percursor. The reason why Li_3PO_4, AgI, LiI, are chosen to be solid electrolyte because of several reasons. Li_3PO_4 has been proved to be a good candidate for solid electrolyte, due to its easy in preparation, high melting temperature, low cost, and good compatibility with the electrode materials [22]. While AgI and LiI as dopants can increase the electrical conductivity of Li_3PO_4 [23]. The focus on the previous work was on synthesis and characterization of molar ratio between x:y:z to make the perfect solid electrolyte. The molar ratio which used was vary from 1:0:0 to 0.34:0.33:0.33. However, the optimum conductivity was achieved for the composition of 0.5:0.25:0.25 or $(Li_3PO_4)_{0.5}(AgI)_{0.25}(LiI)_{0.25}$ namely 8.3349x10^{-5} S/cm and the rest were around 10^{-9} S/cm and 10^{-6} S/cm. The X-ray diffraction data shows that all samples have similar crystal structure with different intensity due different compositions.

The main purpose of this study is to synthesis solid electrolyte from $(Li_3PO_4)_{0.5}(AgI)_{0.25}(LiI)_{0.25}$ and cathode material using $LiFePO_4$ by the addition of 0%, 10%, and 30% of $(Li_3PO_4)_{0.5}(AgI)_{0.25}(LiI)_{0.25}$ in order to obtain cathode and solid electrolyte that is applicable for solid state battery. The addition of solid electrolyte to the cathode is to prevent the solid state interface between solid electrolyte and cathode. Further application, all-solid-state lithium-ion battery will be made from these cathode and solid electrolye.

2. Experimental method

All of activies of synthesis and characterizations were carried out at the Integrated Battery Laboratory, Center for Science and Technology of Advanced Materials, National Nuclear Energy Agency, Indonesia.

2.1 Sample preparation of $(Li_3PO_4)_{0.5}(AgI)_{0.25}(LiI)_{0.25}$

In this study, $(Li_3PO_4)_{0.5}(AgI)_{0.25}(LiI)_{0.25}$ has been prepared manually. The appropriate amounts of lithium phosphate, silver iodide, and lithium iodide with the molar ratio 0.5:0.25:0.25 were mixed inside of glove box. It was mixed manually by grinding on the agate mortar for approximately 45 minutes.

2. 2 Sample preparation of $LiFePO_4$-$(Li_3PO_4)_{0.5}(AgI)_{0.25}(LiI)_{0.25}$

The solid electrolyte of $(Li_3PO_4)_{0.5}(AgI)_{0.25}(LiI)_{0.25}$ that has been made was added to $LiFePO_4$ by mixing and grinding them together using agate mortar inside of glove box for approximately 45 minutes. The appropriate amounts of lithium iron phosphate and solid electrolyte with the molar ratio 10:0, 9:1, and 7:3.

2. 3 Characterization of $(Li_3PO_4)_{0.5}(AgI)_{0.25}(LiI)_{0.25}$ and $LiFePO_4$-$(Li_3PO_4)_{0.5}(AgI)_{0.25}(LiI)_{0.25}$

The characterization of solid electrolyte of $(Li_3PO_4)_{0.5}(AgI)_{0.25}(LiI)_{0.25}$ and cathode of $LiFePO_4$-$(Li_3PO_4)_{0.5}(AgI)_{0.25}(LiI)_{0.25}$ were measured by XRD and SEM measurements. The purpose of XRD measurement is to see if solid electrolyte was successfully added to cathode and to find out the crystal structure of $LiFePO_4$-$(Li_3PO_4)_{0.5}(AgI)_{0.25}(LiI)_{0.25}$. The meausrements were operated at 45 kV and 40mA, a can rang from 10° to 80° [2θ] that is conducted at room temperature by PANalytical Empyrean diffractometer equipped with PIXcel 1D detector SEM measurements were operated at 20kV, a chamber pressure of 20 Pa, wide distance of 10 mm and Spot Size of 65 using JEOL JSM-6510LA. The measurements were done to see the morphology of samples. AC Impedance

IOP Conf. Series: Materials Science and Engineering **553** (2019) 012059 doi:10.1088/1757-899X/553/1/012059

Measurement of samples were carried out using HIOKI LCR HiTester 3532-50 to see the electronic conductivity of each samples.

3. Results and discussion

The reason why all the preparations were done inside the glove box because lithium iodide is easily oxidized when it's exposed to air. The observation of morphology analysis was obtained by scanning electro microscopy (SEM). All the entire images of SEM show the morphology that was obtained by backscattered electron. Backscattered electron determine the density difference of atom in samples. Atom which has bigger density will produce more bright than the atom which has a smaller density.

Figure 1. The SEM micrograph of (a) Solid Electrolyte $(Li_3PO_4)_{0.5}(AgI)_{0.25}(LiI)_{0.25}$ (b) $LiFePO_4$-0%$(Li_3PO_4)_{0.5}(AgI)_{0.25}(LiI)_{0.25}$ (c) $LiFePO_4$-10%$(Li_3PO_4)_{0.5}(AgI)_{0.25}(LiI)_{0.25}$ (d) $LiFePO_4$-30%$(Li_3PO_4)_{0.5}(AgI)_{0.25}(LiI)_{0.25}$

Figure 1 shows micrrostructures of $(Li_3PO_4)_{0.5}(AgI)_{0.25}(LiI)_{0.25}$ and $LiFePO_4$ by variation addition of 0%, 10%, and 30% of $(Li_3PO_4)_{0.5}(AgI)_{0.25}(LiI)_{0.25}$ are shown in Figure 1(a), 1(b), 1(c), and 1(d), respectively. All samples produced partially crystalline samples which consist of porous agglomerates. The agglomerates indicate the compound has particles smaller than the size shown in the micrograph. As shown in the Figure 1 that the samples were not thoroughly homogeneous throughout the sample due to the preparation by only manually grading using agate mortar estimately 45 minutes. Further increasing amount of $(Li_3PO_4)_{0.5}(AgI)_{0.25}(LiI)_{0.25}$, the sample crystalized with similar crystal structure since the addition of of $(Li_3PO_4)_{0.5}(AgI)_{0.25}(LiI)_{0.25}$ was not significant.

The XRD characterization is used to identify if solid electrolyte (SE) of $(Li_3PO_4)_{0.5}(LiI)_{0.25}(AgI)_{0.25}$

is added successfully into the cathode of $LiFePO_4$. The wavelength for x-ray diffraction is $\lambda_x \sim 1.54$ Å (from CuKα target) and for neutron $\lambda n \sim 1.82$ Å (from Ge (331) monochromator).

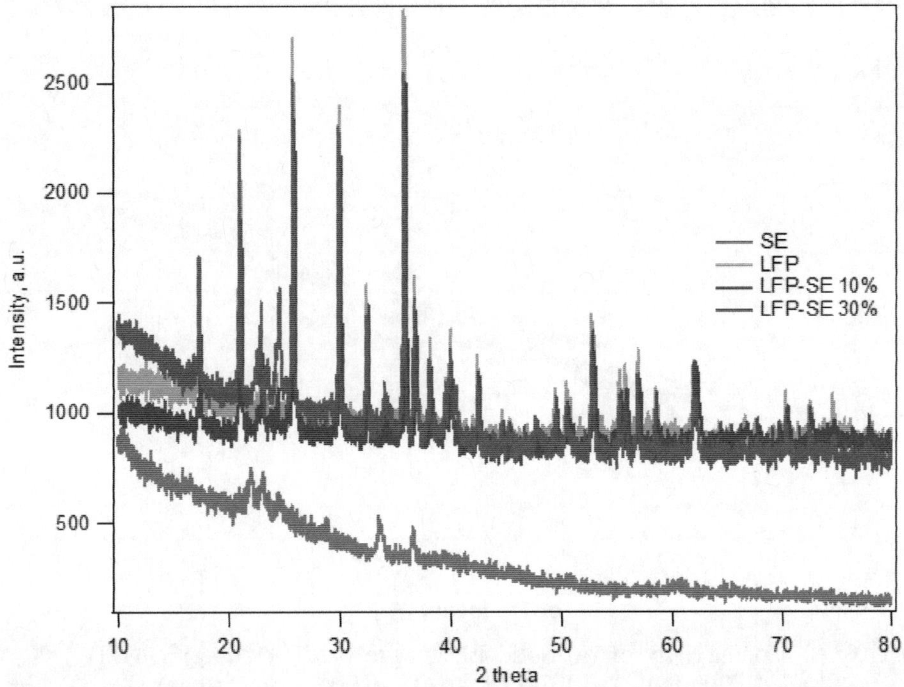

Figure 2. XRD Characterization Data Individually: (a) Solid Electrolyte $(Li_3PO_4)_{0.5}(LiI)_{0.25}(AgI)_{0.25}$ (Black) (b) $LiFePO_4$ + 0% $(Li_3PO_4)_{0.5}(LiI)_{0.25}(AgI)_{0.25}$ (Red) (c) $LiFePO_4$ + 10% $(Li_3PO_4)_{0.5}(LiI)_{0.25}(AgI)_{0.25}$ (Violet) (d) $LiFePO_4$ + 30%$(Li_3PO_4)_{0.5}(LiI)_{0.25}(AgI)_{0.25}$ (Green).

Figure 2 shows the data characterization of samples of $(Li_3PO_4)_{0.5}(LiI)_{0.25}(AgI)_{0.25}$ and $LiFePO_4$-0% $(Li_3PO_4)_{0.5}(LiI)_{0.25}(AgI)_{0.25}$ before and after the addition of $(Li_3PO_4)_{0.5}(LiI)_{0.25}(AgI)_{0.25}$ with variations of 10% and 30%. As the figure shows can be concluded that the solid electrolyte was successfully added to $LiFePO_4$. From Figure 2(a), it shows that $(Li_3PO_4)_{0.5}(LiI)_{0.25}(AgI)_{0.25}$ has an amorphous phase. The X-ray diffraction data is showing only a halo pattern. The broad peak position at $2\Theta \sim 23°$ shows that the solid electrolyte mostly consists of Li_3PO_4 crystal. From Figure 2(b), it shows that the whole peaks of diffraction were proper to the standard $LiFePO_4$ indexed to olivine structure, *orthorhombic* space group *P n m a* (JCPDS No.083-2092). The strong and sharp peaks show that the $LiFePO_4$ sample was very crystalline. Figure 2(c) and Figure 2(d) show the crystal structure after the addition of 10% and 30% of $(Li_3PO_4)_{0.5}(LiI)_{0.25}(AgI)_{0.25}$. The broad peak position at $2\Theta \sim 23°$ shows that the solid electrolyte was successfully added to $LiFePO_4$. The figures also show different peaks and different intensity. Different peaks show that peaks caused by the addition of solid electrolyte and different intensity of peaks caused by the amount of the addition. But, more detail is still needed.

One of the disadvantages of Li_3PO_4 as a solid electrolyte is the low conductivity which was the reason why AgI and LiI were added as dopants. AgI and LiI compounds are believed to increase the conductivity because they have crystal defects which play an important role in ion transport. Figure 3 shows the AC conductivity of samples, it can be seen that those with the highest AC range conductivity was $LiFePO_4$-10%$(Li_3PO_4)_{0.5}(LiI)_{0.25}(AgI)_{0.25}$. That was the reason why $LiFePO_4$-10%$(Li_3PO_4)_{0.5}(LiI)_{0.25}(AgI)_{0.25}$ will be chosen to be coated onto the alumunium foil.

IOP Conf. Series: Materials Science and Engineering **553** (2019) 012059 doi:10.1088/1757-899X/553/1/012059

Because it had higher range of AC conductivity compared to $LiFePO_4$-0%$(Li_3PO_4)_{0.5}(LiI)_{0.25}(AgI)_{0.25}$ and $LiFePO_4$-30%$(Li_3PO_4)_{0.5}(LiI)_{0.25}(AgI)_{0.25}$. But the challenge was that the conductivity was still in AC, which means it was frequency dependence. There were several reason why it was still in AC range conductivity, including, the voltage used for the experiment was low.

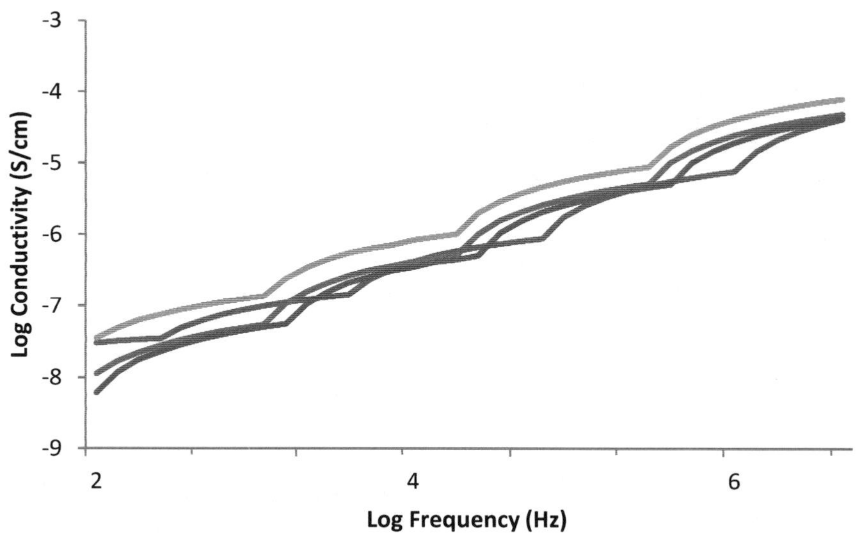

Figure 3. AC Conductivity of (a) Solid Electrolyte $(Li_3PO_4)_{0.5}(AgI)_{0.25}(LiI)_{0.25}$ (red) (b) $LiFePO_4$-0.%$(Li_3PO_4)_{0.5}(AgI)_{0.25}(LiI)_{0.25}$ (blue) (c) $LiFePO_4$-10%$(Li_3PO_4)_{0.5}(AgI)_{0.25}(LiI)_{0.25}$ (green) (d) $LiFePO_4$-30%$(Li_3PO_4)_{0.5}(AgI)_{0.25}(LiI)_{0.25}$ (violet).

The values of AC conductivity were vary from 10^{-8} S/cm to 10^{-4} S/cm in the frequency from 1 Hz to 10^6 Hz. There was an increase for the $LiFePO_4$–10%$(Li_3PO_4)_{0.5}(LiI)_{0.25}(AgI)_{0.25}$ shows that the optimum composition for cathode was by the addition of 10%$(Li_3PO_4)_{0.5}(LiI)_{0.25}(AgI)_{0.25}$. This cathode is to be used for further application in all-solid-state battery.

4. Conclusion

The sythesis of solid electrolyte $(Li_3PO_4)_{0.5}(LiI)_{0.25}(AgI)_{0.25}$ by molar ratio of 0.5:0.25:0.25 and cathode $LiFePO_4$ with variation of 0%, 10%, 30% of solid electrolyte $(Li_3PO_4)_{0.5}(LiI)_{0.25}(AgI)_{0.25}$ by molar ratio of 10:0, 9:1, and 7:3 were successfully prepared manually by grinding each of the $LiFePO_4$, Li_3PO_4, AgI, and LiI materials inside of the glove box. SEM images of samples show that the samples were not homogeneous thoroughly throughout the sample, possibly because it was prepared manually. XRD measurement data is used to see the success of adding solid electrolyte to the cathode $LiFePO_4$, this has been proved by the presence of peaks after the addition of solid electrolyte mostly peaks of Li_3PO_4. The intensity of the peaks are changed with the increasing amount of solid electrolyte on the cathode $LiFePO_4$. The impedance measurement was used to calculate the conductivity of each material, and it was found that the highest conductivity was obtained $LiFePO_4$-10%$(Li_3PO_4)_{0.5}(LiI)_{0.25}(AgI)_{0.25}$. This study gives the idea of all-solid-battery preparation and how to reduce the presence of solid electrolyte interface problem.

Acknowledgment

Thanks to Agus Rifani S.Si., M.Si. and Atut Reni Septiana S.Pd., M.Si., for their guidance on the synthesis and analysis. This study supported by PSTBM, BATAN, Indonesia.

References

[1] L.-X. Yuan, Z.-H. Wang, W.-X. Zhang, X.-L. Hu, E. P. Wigner, J. –T. Chen, et al., 2011, *Energy Environ. Sci.*, 4, 269–284, DOI: 10.1039/C0EE00029A.

[2] B. Scrosati, J. Garche, J., 2010, *Power Sources,* 195(9):2419-2430, DOI: 10.1016/j.jpowsour.2009.11.048.

[3] G. Girishkumar, B. McCloskey, A.C. Luntz, S. Swanson, W. Wilcke, J., 2010, *Phys. Chem. Lett. 1*, 2193-2203, DOI: 10.1021/jz1005384.

[4] J.B. Goodenough, K.-S. Park, 2013, *J. Am. Chem. Soc.*, 135 (4), pp 1167-1176, DOI: 10.1021/ja3091438.

[5] M. Armand and J.M. Tarascon, 2008, *Nature*, 451(7179):652-7, DOI: 10.1038/451652a.

[6] S.F. Amalraj, R. Marom, N. Leifer, D. Jacob, D. Aurbach, 2011, *J. Mater. Chem.*, 2, 9938-9954, DOI: 10.1039/C0JM04225K.

[7] T.-H. Kim, J.-S. Park, S.K. Chang, S. Choi, J.H. Ryu, H.-K. Song, 2012, *Adv. Energy Mater.*, 2, 860-872, DOI: 10.1002/aenm.201200028.

[8] V. Etacheri, R. Marom, R. Elazari, G. Salitra, D. Aurbach, 2011, *Energy Environ. Sci.*, 4, 3243-3262, DOI: 10.1039/C1EE01598B.

[9] C.H. Jiang, S.X. Dou, H.K. Liu, M. Ichihara, H.S. Zhou, 2007, *Journal of Power Sources*, 172(1):410-415, DOI: 10.1016/j.jpowsour.2007.07.039.

[10] J.-M. Chen, Y.-D. Cho, C.-L. Hsiao, G.T.-K. Fey, 2009, *Journal of Power Sources*, 189(1):279-287, DOI: 10.1016/j.jpowsour.2008.09.049.

[11] J. Lee and A.S. Teja, 2005, *Journal of The Supercritical Fluids* 35(1):83-90, DOI: 10.1016/j.supflu.2004.12.005.

[12] J.F. Whitacre, K. Zaghib, W.C. West, B.V. Ratnakumar, 2008, *Journal of Power Sources*, 177(2):528-536, DOI: 10.1016/j.jpowsour.2007.11.076.

[13] W. Honggowiranto, Sudaryanto, E. Kartini, AIP (2016) *Jurnal Sains Materi Indonesia*, Vol. 17, No. 2, Januari 2016, hal. 49-55, DOI: 10.17146/jsmi.2016.17.2.4202.

[14] J.-M. Tarascon and M. Armand., 2001, *Nature,* 414(6861)::359–67, 2001, DOI: 10.1038/35104644.

[15] V. Borgel, E. Markevich, D. Aurbach, G. Semrau, M. Schmidt, 2009, *Journal of Power Sources*, 189(1):331-336, DOI: 10.1016/j.jpowsour.2008.08.099.

[16] John B. Goodenough and Y. Kim, 2010, *Chem. Mater.*, 22(3), pp 587–603, DOI: 10.1021/cm901452z.

[17] P. Knauth, 2009, *Solid State Ionics*, 180(14-16):911-916, DOI: 10.1016/j.ssi.2009.03.022.

[18] Sveinbjörnsson, D. Þ., Vegge, T., Norby, P., & Mogensen, M. B. 2014. Design and Characterisation of Solid Electrolytes for All-Solid-State Lithium Batteries. Department of Energy Conversion and Storage, Technical University of Denmark.

[19] E. Kartini, M. Manawan , M. F. Collins, M. Avdeev, 2017, *Physica B: Condensed Matter*, DOI: 10.1016/j.physb.2017.11.079.

[20] E. Kartini, M. Nakamura, M. Arai, Y. Inamura, K. Nakajim, et al., 2013, *Solid State Ionics*, 262. DOI: 10.1016/j.ssi.2013.12.041.

[21] T. Utomo, "Sintesis dan Karakterisasi $(Li_3PO_4)_X(LiI)_y(AgI)_Z$ Sebagai Elektrolit Padat Untuk Baterai Li-ion," undergraduate thesis, Sekolah Tinggi Teknologi Nuklir, Yogyakarta, Indonesia, 2015.

[22] Makhsun, T. Sakuma, and E. Kartini, 2015, *International Journal of Innovation in Science and Mathematics*, 2, 2347-9051.

[23] N.I.P. Ayu, E. Kartini, L. Prayodi, M. Faisal, Supardi, 2016, *Ionics*, 22(7), DOI: 10.1007/s11581-016-1643-z.

Improving Performance of $LiFePO_4$ by addition of Carbon Nano Tube for Lithium Ion Battery

W. Honggowiranto[1*], E. Kartini[1], Sudaryanto[1], R. N. S. Rofika[2], E. Hutamaningtyas[1], A. Subhan[3], A. Sudjatno[1]

[1]Center for Science and Technology of Advanced Material, National Nuclear Energy Agency, Serpong, South Tangerang, Banten, 15314, Indonesia.
[2]Department of Physic, Faculty of Mathematics and Sciences, Bandung Institute of Technology, Ganesha 10, Bandung, 40132, Indonesia.
[3]Reseach Center of Physic, Indonesia Indonesian Institute of Sciences, Serpong, South Tangerang, Banten, 15314, Indonesia.

wagiyo@batan.go.id

Abstract. $LiFePO_4$ (LFP) cathode material has been synthesized with hydrothermal method. The reaction was done by reacting a mixture of FeSO4.7H2O, H_3PO_4, LiOH and CNT. In order to improve performance of LFP, the carbon nano tube (CNT) was added with the variation of 5, 10 and 15 mmol, before hydrothermal process. The material was stirred using a magnetic stirrer for 30 minutes, and then autoclave was heated at 180°C for 6 hours then sintered at 700°C for 6 hours. The results were characterized by X-ray diffraction (XRD), and Scanning Electron Microscope (SEM), and Impedance Spectroscopy (EIS). The X-ray data shows that the crystal structure of synthesized $LiFePO_4$ has a group of Pmn with a space (olivine structure) which is in agreement with the LFP standard material. The addition of CNT does not change the crystal structure. This shows in SEM images that the crystallite size of $LiFePO_4$ particles does not have much effect on the composite. The battery cell performance was measured by Impedance Spectroscopy and charge / discharge Battery Analyzer BST-8. The EIS data, showed the decreasing of battery impedance total from $LiFePO_4$ material without CNT to addition of 5, 10 and 15 mmol CNT namely 214; 128.1; 88.6 and 70.1 Ω, and the specific capacity 0.1C are 38.78; 51.53; 106.84; 92.79 mAh/g, respectively. It is shown that the maximum specific capacity was obtained for LFP composite with the addition of 10mmol CNT. It can be concluded that the addition of CNT increases the conductivity and specific capacity, thus improving performance of lithium ion battery.

Keywords *: $LiFePO_4$, Hydrothermal, CNT, Lithium ion battery*

1. Introduction

The lithium ion battery has been considered to be an attractive power source not only for small electronic devices, such laptop, cell phones, flash light, but also for large scale applications like energy storage for photo voltage, electric vehicles (EV), hybrid electric vehicles (HEVs). The lithium ion battery cell has been regarded as one of promising power source candidates for electric vehicles in recent years [1,2]. $LiFePO_4$ batteries have become a main of research related to the fact that they are relatively safe and low manufacturing cost. In theories, LFP have a high capacity of 170 mAh/g, good cycling ability and good stability during charge-discharge [3-5]. The limitation of this materials are the inherent low electronic conductivity (10^{-9}-10^{-10} S cm^{-1} at room temperature). However the wide commercialization are limited by its low electronic conductivity and low lithium ion diffusivity [3].

Many researches had been done to improve the performance of LFP[3]. Metal doping and carbon coating were conducted to increase the electronic conductivity [6]. As for the low ionic conductivity, the reduction of particle size may be improved it. LFP can be synthesized by solid state method, sol-gel, precipitation and also hydrothermal synthesis [7,8]. One of the most relevant methods to reduce particle size was reported to be hydrothermal method. In order to reduce the energy usage and costs of synthesizing LFP cathode materials (at high temperature), the hydrothermal synthesis can indeed be used at low temperature and it is favourable method [9,10]. Carbon coating is used to improve the weak performance and electrical conductivity of LFP [11]. In order to get the maximum electrochemical performance, carbon as active material must be in nano particle size with high crystallinity and uniform size distribution [12,13]. The nano size carbon related to support the ionic transport and electronic conduction of LFP [14-16]. Currently, carbon nanotube (CNT) are the strongest and offer superior electrical and thermal conductivities. The addition of CNT on LFP has no effect on the structure of pure LFP, and also exhibits high discharge capacity[9].

Aim of this work is to synthesize LFP by hydrothermal method with addition of CNT during the process in order to improve performance of lithium ion battery. The crystal structure and microstructure of the LFP will be characterized, including the electrochemical and battery performance.

2. Experimental method

Cathode materials of $LiFePO_4$ were prepared by hydrothermal process. The main precusors were $FeSO_4.7H_2O$ (Sigma-Aldrich, 99.0%), H_3PO_4(Aldrich, 85.0%), LiOH (Merck, 98.0%). In addition, variations of carbon nano tube (CNT) with 5 mmol, 10 mmol, 15 mmol were added to the precursors before hydrothermal.The LiOH was dissolved in 60 mL of deionized water, then was added subsequently into $FeSO_4{\cdot}7H_2O$ suspension under vigorous stirring, then, the H_3PO_4 were dropped into it and CNT with stirring to get suspension for 30 minutes. After sterringthe mixed solution was transferred into a 90 mL Teflon lined stainless steel autoclave. It was sealed off tightly followed by heating in an electric muffle oven at temperatures of 180°C for 6 hours, then cooled down to room temperature. Details of the hydrothermal method has been described elsewhere [15]. The solid grey precipitates was filtered and washed with demineral water several times with ethanol and distilled water. It was dried at vacuum oven at 120°C for 10 hours, then annealed in tubular furnace at 700°C for 6 hours in atmospheric argon.

All the samples are marked as LFP1, LFP2, LFP3 and LFP4 (respectively, 0 mmol, 5 mmol, 10 mmol, and 15 mmol CNT). The crystal structure of those $LiFePO_4$ were characterized by X-ray diffraction (XRD) using a with Cu Kα target, scan range from 10° to 80°. The morphology of the samples was observed by using scanning electron microscopy.

The cathode was prepared by mixing the active materials with acetylne black and then was grounding the mixture for 30 minutes and heated in vacuum oven at 120°C for 2 hours. Polyvinylidene fluoride (PVDF) was mixed with active materials for 15 minute, then N-methyl-2-pyrrolidinone (NMP) was added to make slurry while mixing for 30 minutes. The weight ratio of active materials with acetylne black and PVDF was 85:3:5. The slurry was pasted onto an aluminum current collector using doctor blade. It was dried on oven vacuum at 120°C for 10 hrs. A half- cell battery was prepared under the glovebox. The electrochemical performance and charge-discharge of the lithium ion battery was done using an impedance spectroscopy and a battery analyzer.

IOP Conf. Series: Materials Science and Engineering **553** (2019) 012060 doi:10.1088/1757-899X/553/1/012060

3. Results and discussion

The SEM image and XRD pattern of multiwall carbon nano tube (CNT) commercial are shown in Figure 1(a) and (b), respectively. The microstructure of CNT shows circular porous and some aglomerations. The diffraction peaks at 2θ~ 25.944, 72.703 and 43.095 corresponds to CNT standard.

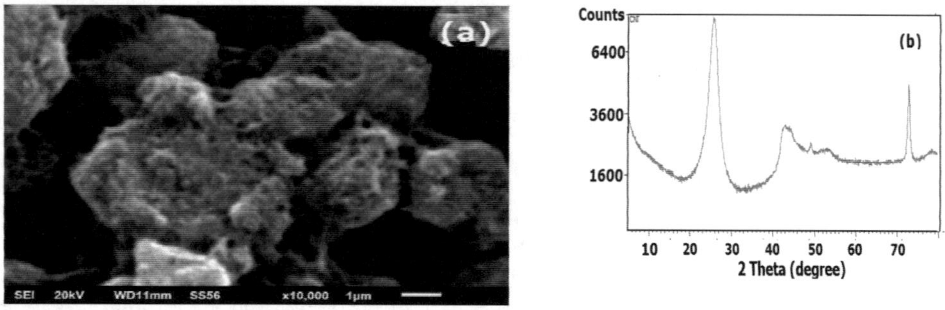

Figure 1. (a) SEM image and 1(b) X-ray diffraction of Carbon Nano Tube commercial.

Figure 2(a) shows image of the $LiFePO_4$ (LFP) prepared by hydrothermal method without any additional CNT. Figures 2(b), 2(c) and 2(d) are the images of LFP prepared by hydrothermal with the addition of CNT 5, 10 and 15 mmol, respectively.

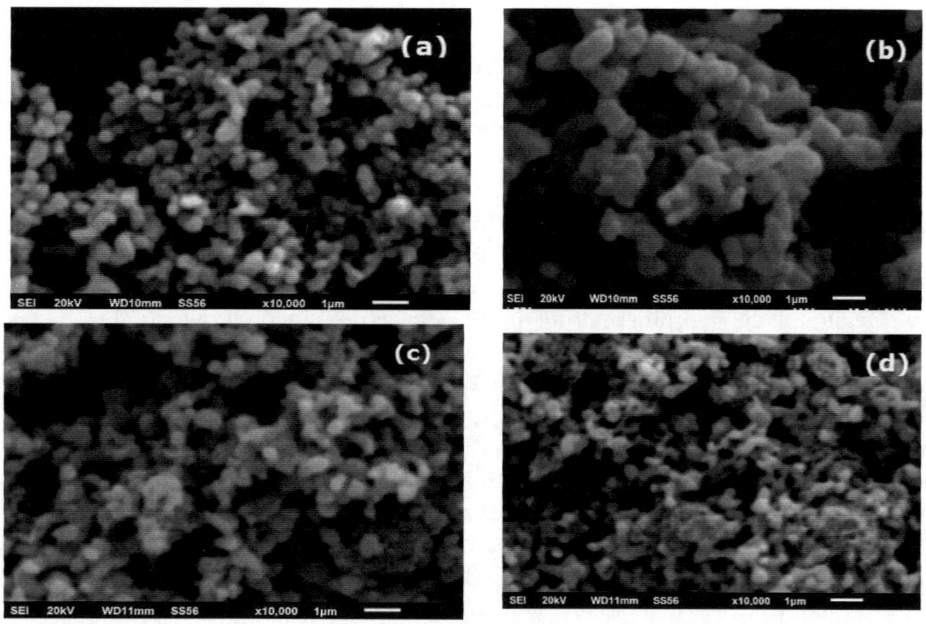

Figure 2. The SEM images of (a) LFP1, (b) LFP2, (c) LFP3 and (d) LFP4.

The distribution of LFP particles without CNT are dispersed homogenously, with the particle size between ~ 0.3-1.3 μm.The addition of CNT made agglomeration, and the sizes increased to ~0.4-1.7 μm and ~0.4- 1.5 μm, for LFP2 and LFP3, respectively. Further increasing CNT the size decreased again to 0.2-1 μm for LFP4.

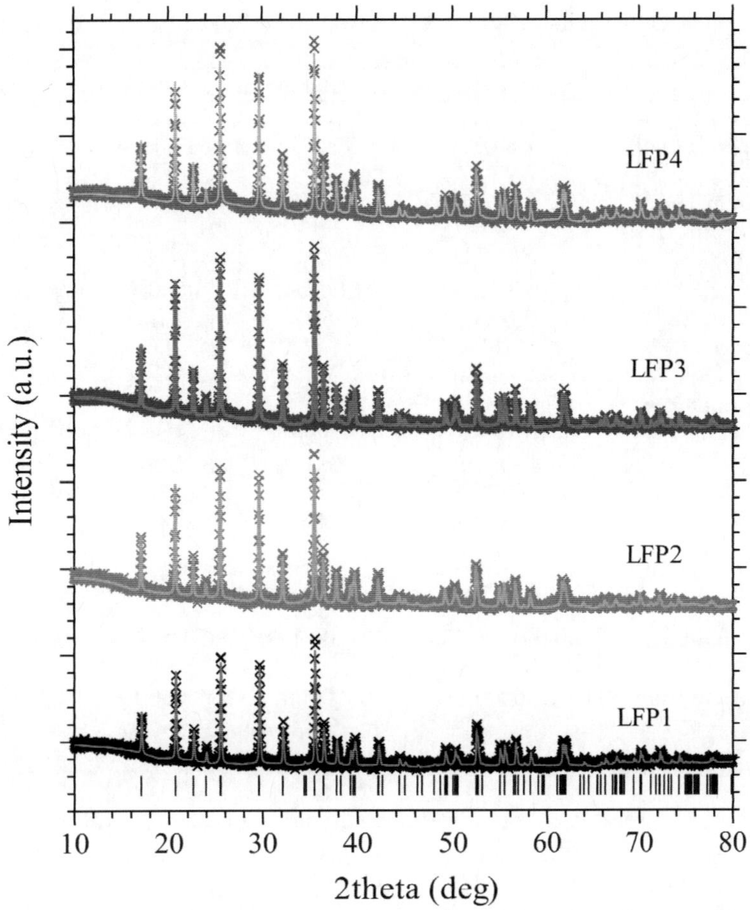

Figure 3. X-ray diffraction patterns of LFP1, LFP2, LFP3 and LFP4.

Further examination of the LFP shape, they did not change with addition of CNT. These implied that the addition of CNT did not give the impact to the physical and structural of LFP as observed by x-ray diffraction, as shown in Figure 3. The refinement results of diffraction data showed that all LFP can be indexed as the orthorhombic phase with space group *Pnma 62* which belongs to $LiFePO_4$. This means that LFP prepared by the hydrothermal process has been done successfully. There is no change in the crystal structures of LFP2, LFP3 and LFP4 after the addition of CNT with 5, 10 and 15 mmol. Similar result by Feng et al. identified that no characteristic peaks from CNT are observed, due to the low content of CNT, which show that the additions have no obvious impact on the structure of LFP [13]. There is a slightly change in the crystallite sizes namely 1000.2Å, 1413.8Å, 1506.7Å and 952.3Å, for LFP1, LFP2, LFP3 and LFP4, respectively. The refinement results of the diffraction patterns are listed in Table 1.

The impedance spectroscopy (EIS) of the half-cell was measured at SOC (0%) before charge and disharge process. The EIS results of LFP1, LFP2, LFP3 and LFP4 are shown in Figure 4. The intercept on x-axis in the high frequency region indicates an internal resistance of the cell, and the diameter of the semicircle represents the effect caused by charge transfer and electrochemical double layer. As shown in Figure 4, the equivalent diagram reflecting the process internal process. The total impedance could be regarded as the electrolyte resistance R1 and the charge transfer resistance R2, for LFP1, LFP2, LFP3 and LFP4 were 214;128.1; 88.6 and 70.1Ω (at SOC 0%). This result indica-

IOP Conf. Series: Materials Science and Engineering **553** (2019) 012060 doi:10.1088/1757-899X/553/1/012060

Table 1. Refinement Results of X-ray diffraction from LFP1, LFP2, LFP3 and LFP4.

	LFP1	LFP2	LFP3	LFP4
Formula /Unit Crystal	O16P4 Fe4Li4	O16P4Fe4Li4	O16P4Fe4Li4	O16P4Fe4Li4
Formula mass (g/mol)	631.0376	631.0376	631.0376	631.0376
Calculate Density (g/cm3)	3.5942	3.5942	3.5991	3.6025
R (expected)/%	2.87967	2.56783	2.63627	2.62593
GOF	1.11669	1.06554	1.05008	1.17499
R(weighted profile)/%	3.04305	2.65064	2.70147	2.84642
Weight Fraction (%)	100	100	100	100
Space group (No.)	P n m a (62)	P n m a (62)	P n m a (62)	P n m a (62)
Lattice Parameters :				
a (Å)	10.3299(5)	10.3323(6)	10.3286(4)	10.3248(4)
b (Å)	6.0065(3)	6.0081(3)	6.0061(2)	6.0035(3)
c (Å)	4.6929(3)	4.6957(3)	4.6926(2)	4.6919(2)
alpha (°)	90	90	90	90
beta (°)	90	90	90	90
gamma (°)	90	90	90	90
Preferred orientation direction /hkl	0.00 0.00 1.00	0.00 0.00 1.00	0.00 0.00 1.00	0.00 0.00 1.00
Preferred orientation parameters	1.000000	1.000000	1.000000	1.000000
Crystallite Size (Å)	1000.2	1413.8	1506.7	952.3

ting addition of CNT decreased the charge transfer polarization resistance which related to the unique CNT network allowed easy electronic transfer, thus maximizing the efficiency of current collection and increased the mobility of lithium ions [9,17]. As for the double-layer capacitance in form of CPE and Ws is the Warburg impedance that reflects the diffusion of lithium-ion in the solid. The simulation results are tabulated in Table 2.

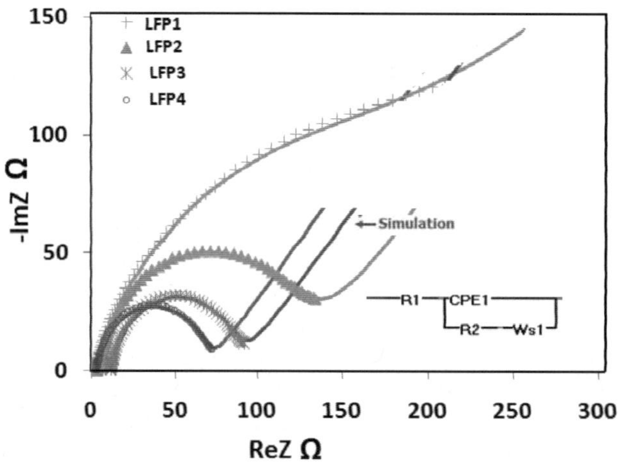

Figure 4. Impedance plots of half cells of LFP1,LFP2,LFP3 and LFP4.

IOP Conf. Series: Materials Science and Engineering **553** (2019) 012060 doi:10.1088/1757-899X/553/1/012060

Table 2. Electrochemical performance of LFP batteries.

Sample	R1 (Ω)	R2 (Ω)	Ws (Ω)	Cyl.1 Specific Capacity (mAh/g)	Cyl.10 Specific Capacity (mAh/g)	Retention after 10th cycle
LFP1	4.0	210.0	1284.7	38.78	27.12	69.881
LFP2	3.8	124.3	274.9	51.53	39.50	76.654
LFP3	12.4	76.2	122.7	106.84	104.50	97.809
LFP4	3.4	66.7	85.9	92.79	90.49	97.521

Figure 5 shows the charge-discharge analysis of all LFP batteries at 0.1C rate measured at room temperature. The spesific capacity for both cycle 1 and 10 increased with the addition of CNT as for LFP1, LFP2 and LFP3, then decreased at LFP4. The discharge capacity of LFP3 is observed at 106.84 mAh/g, which exhibited the highest initial discharge capacity, with the capacity retention of 97.8% after 10th cycles. Similar results was done for mesoporous LFP microspheres embedded homogeneously with 3D CNTs conductive networks, where the initial discharge capacity is 150 mAh/g [17]. The lower discharge capacity of LFP4 relative to LFP3 show that the current density is large and reducing the discharge capacity in charge-discharge process as the result of the excess CNT as active material at LFP4 [13]. The first discharge capacity of LFP4 is observed at 92.79 mAh/g with retention is 97,5% after 10th cycle.

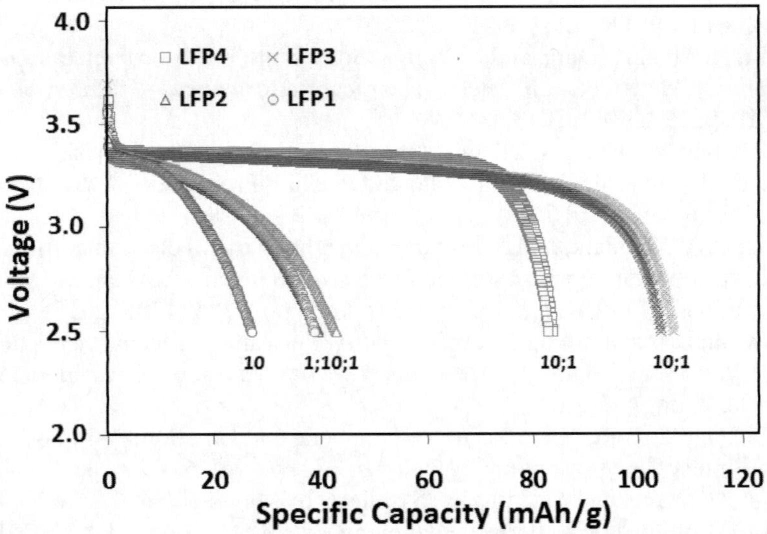

Figure 5. (a). Specific capacity of (a) LFP1, (b) LFP2, (c) LFP3 and (d) LFP4.

4. Conclusion

LiFePO4 as cathode material for Li-ion battery was successfully synthesized by hydrothermal method. The addition of CNT before the hydrothermal has reduced the internal resistance, thus increased the mobility of lithium ions due to the existing of CNT. The charge–discharge results show good performance of the lithium ion battery, and the highest specific capacity was 106.84 mAh/g obtained for LFP3 (CNT 10mmol)

Acknowledgment
This work is financially supported by the Ministry Research Technology and Higher Education through the Research Grant of National Innovation System Consortium (INSINAS) with the contract no. IRPK-042-2018.

References

[1] B. Scrosati and J. Garche, "Lithium batteries: Status, prospects and future," *J. Power Sources*, vol. 195, pp. 2419–2430, 2010.
[2] E. Kartini, A. Purwanto, S. Sudaryanto, H. Honggowiranto, and R. N. S. Rofika, "Indonesian consortium of lithium ion battery for solar street lamp," in *IOP Conf. Series: Materials Science and Engineering*, 2018, vol. 432, pp. 012063–1.
[3] A. K. Padhi, K. S. Nanjundaswamy, and J. B. Goodenough, "Phospho-olivines as Positive-Electrode Materials for Rechargeable Lithium Batteries," *J. Electrochem. Soc.,* vol. 144, no. 4, pp. 1188–1194, 1997.
[4] Y. Xu, S. Chung, J. T. Bloking, and Y. Chiang, "Electronic Structure and Electrical Conductivity of Undoped $LiFePO_4$," *Electrochem. Solid-State Lett.*, vol. 7, no. 6, pp. 131–134, 2004.
[5] U. K. Muhsin Mazman, Omur Cuhadar, Davut Uzun, Ercan Avci, Emre Bicer Tevhit Cem Kaypmaz, "Optimization of $LiFePO_4$ synthesis by hydrothermal method," *Turkish J. Chem.*, vol. 38, pp. 297–308, 2014.
[6] Y. Dai, L. Ma, H. Zhang, Y. Chen, Y. Chen, and W. Wang, "Hydrothermal Synthesis of Spindle-Like Nanostructured LiFePO4 Powders Mediated by Organic Acid," *Int. J. Electrochem. Sci.*, vol. 11, pp. 1730–1737, 2016.
[7] Y. Yin, X. Li, X. Mao, X. Ding, and S. Yang, "Solid State Reaction Preparation of $LiFePO_4$ /(C D Cu) Cathode Material and Its Electrochemical Performance," *J. Mater. Sci. Technol.*, vol. 29, no. 10, pp. 937–942, 2013.
[8] Q. Jiang, X. Wang, and H. A. N. Zhang, "One-Pot Hydrothermal Synthesis of $LiMn_2O_4$ Cathode Material with Excellent High-Rate and Cycling Properties," *J. Electron. Mater.*, vol. 45, no. 8, pp. 4350–4351, 2016.
[9] S. R. Rao and U. V Varadaraju, "Hydrothermal synthesis of $LiFePO_4$ nanorods composed of nanoparticles from vivianite precursor and its electrochemical performance for lithium ion battery applications," *Bull. Mater. Sci.*, vol. 38, no. 5, pp. 1385–1388, 2015.
[10] W. Honggowiranto, Sudaryanto, E. Kartini, and A. Purwanto, "Electrochemical Performance of $LiFePO_4$ Cylinder Cell Battery," *AIP Conf.Proc. 6th Nanosci. Nanotechnol. Symp.*, vol. 1710, no. 030042, pp. 1–7, 2016.
[11] H. Wu, Q. Liu, and S. Guo, "Composites of Graphene and $LiFePO_4$ as Cathode Materials for Lithium-Ion Battery : A Mini-review," *Nano-Micro Lett.*, vol. 6, no. 4, pp. 316–326, 2014.
[12] L. Yu, D. Cai, H. Wang, and M. Titirici, "Synthesis of Microspherical $LiFePO_4$-Carbon Composites for Lithium-Ion Batteries," *Nanomaterials*, vol. 3, pp. 443–452, 2013.
[13] W. Feng, Y. Cao, X. Zhao, J. Gang, and W. Su, "Effect of Carbon Nanotubes on the Electrochemical Performance of $LiFePO_4$ Particles in Lithium Ion Batteries," *Int. J. Electrochem. Sci.*, vol. 12, pp. 5199–5207, 2017.
[14] K. P. Abhilash, P. C. Selvin, B. Nalini, H. Xia, and S. Adams, "Electrochemical Analysis of the Carbon-Encapsulated Lithium Iron Phosphate Nanochains and Their High-Temperature Conductivity Profiles," *ACS Omega 2018*, vol. 3, p. 6446–6455, 2018.
[15] C. P. Supriadi, E. Kartini, W. Honggowiranto, and K. T. Basuki, "Synthesis and Characterization of Carbon Material Obtained From Coconut Coir Dust by Hydrothermal and Pyrolytic Processes," *IJTech 2017*, vol. 8, pp. 1470–1478, 2017.
[16] W. Honggowiran and E. Kartini, "Characterization of $LiFePO_4$ Cathode by Addition of Graphene for Lithium Ion Batteries," *AIP Conf.Proc. 6th Nanosci. Nanotechnol. Symp.*, vol. 1710, no. 030045, pp. 1–6, 2016.

[17] Z. S. Guan Wu, Ran Ran, Bote Zhau, Yujing Sha, Chao Su, Yingke Zhou, "3D-amorphous-carbon-and-graphene-co-modified-$LiFePO_4$-composite-derived-from-polyol-process-as-electrode-for-high-power-lithium-ion-batteries_2014_Jour.pdf," *J. Energy Chem.*, vol. 23, pp. 363–375, 2014.

Interfacial Properties of Solid Polymer Electrolyte for Lithium-Ion Battery

Sudaryanto,[1*], Evi Yulianti[1], Deswita[1], Wahyudianingsih[1], Desy N. P. Anggreyni[2], and Evvy Kartini[1]
[1]Center for Science and Technology of Advanced Materials, National Nuclear Energy Agency (BATAN), PUSPIPTEK Area, South Tangerang 15314, Indonesia
[2] Faculty of Natural Science and Mathematics, Lambung Mangkurat University, Jend. Ahmad Yani Street, Km 35, Banjarbaru 70714, Indonesia

dryanto@batan.go.id

Abstract. The problem in developing solid electrolyte is the interface between solid electrolyte and electrode in an all solid lithium-ion batteries (LIB). Due to this reason, the interfacial properties of solid polymer electrolyte (SPE) for LIB have been studied using scanning electron microscopy (SEM), and electrochemical impedance spectrometry (EIS). The aim of this study is to elucidate the effect of interfacial structure onto the properties of SPE for LIB. Two kind different electrolyte / cathode interfacial arrangements were constructed by coating and stacking Chitosan based SPE onto lithium permanganate based cathode. The effect of interfacial structure onto the impedance has been studied. It was found that in case the SPE stacked on the electrode, micro-gap appeared at the interface and resulted in the existence of Warburg impedance or interfacial resistance. However, when the SPE was coated using solvent casting onto the electrode the micro-gap did not appear and the interfacial resistance could be neglected.

Keywords: Solid Polymer Electrolyte, Interfacial Properties, Lithium-ion Battery

1. Introduction

Lithium-ion batteries [LIB], having a high voltage and a high energy density, are widely used in various mobile devices such as smart phone and tablet PC. However, due to using a flammable liquid electrolyte, there is a risk of explosion or fire accidents. To reduce the risk, the liquid electrolytes are substituted by solid ones, making all-solid-state lithium batteries are proposed as the best solution [1]. An all-solid-state battery is also safer and more easily packaged [2].

Solid-state electrolyte (SSE) can be divided into inorganic solid electrolytes (ISE) and solid polymer electrolytes (SPE). The ISE are mostly based on ceramic or glassy material such as NASICON type phosphates, lithium phosphorus oxynitride (LiPON), garnet, and so on [3]. In this last decade, we have also studied the synthesis and characterization of solid-state electrolyte based on silver-lithium phosphate [4]. While SPE are formed by the incorporation of inorganic such as lithium salt into polymer matrices which are having functional groups such as polyether, polyacrylonitrile, polyester, polysiloxane and soon. Compared with ISEs, SPE have several advantages: They are lightweight and flexible, have interfacial compatibility, and are very processable. Previously, several efforts to obtain Chitosan based SPE have been done [5-7].

One of the problems in developing SSE is the interface between solid electrolyte and electrode [8]. Recent computational studies have shown that various interfaces are not always stable chemically or electrochemically leading to loss of capacity and increase in impedance [9, 10]. Other study showed that proper engineering of the electrode/ electrolyte interface was essential for long term cell performance [11]. Several approaches had been applied to reduce interfacial impedance between SSE and cathodes, including surface coating [12], interface softening [13] and so on. It is known that the interfacial properties are affected by the physical as well as chemical structure at the interface.

In this study, we will elucidate the relationship between interfacial structure and properties of SPE for LIB. Two kind different electrolyte - cathode interfacial arrangement were constructed by coating and stacking Chitosan based SPE onto lithium permanganate based cathode. Scanning electron microscopy (SEM) and electrochemical impedance spectrometry (EIS) were used to study the structure and properties, respectively.

2. Experimental method

2.1. Materials

The chitosan (low molecular weight, 86% deacetilated degree) was purchased from Bogor Agricultural Institute Laboratory. Analytical grade of acetic acid as solvents was purchased from Merck, and used without further purification. Deionized water was used to make 1 wt.% acetic acid solution. Lithium perchlorate ($LiClO_4$) as lithium salt and lithium permanganate (LMO, $LiMn_2O_4$) as cathode material were purchased form Sigma-Aldrich and used as-received. The LMO however, was dried using vacuum oven at 80°C for 12 hours priory.

2.2. Sample preparation

The Chitosan powder was dissolved into the acetic acid to make a 4 wt. % polymer solution. A certain amount of $LiClO_4$ was then added to get a 30 wt.% polymer electrolyte solution. The SPE film was prepared as in previous method, i.e. by casting the electrolyte solution in a cast mold and kept for two days to evaporate the solvent [5-7]. The obtained SPE film, (a) sample, was then stored under vacuum condition at room temperature to complete the solvent evaporation. The cathode was prepared by mixing 1.5 g. LMO powder into 28 g. polymer electrolyte solutions to get 15 : 85 solid to solid proportion. The mixing solution was then casted onto 17 μm Aluminum foil and kept for one hour under heating temperature at 100°C to remove the solvent. The obtained LMO cathode film, (b) sample, was stored under vacuum. SPE coated LMO cathode, (c) sample, was prepared by coating polymer electrolyte solution on the surface of LMO cathode and followed by drying process under heating temperature at 100°C for one hour. The stacking sample, (d) sample, was prepared by stacking and pressing SPE film on the surface of LMO cathode. The samples thickness was controlled to be about 100μm and the structure are illustrated as in Table 1. The cross section was observed by fracturing them after immersed in liquid nitrogen.

Table 1. Samples name and their illustration.

Samples	Illustration	Remarks
(a) SPE		Chitosan based SPE film; (Chitosan + $LiClO_4$)
(b) LMO		LMO cathode, composed of LMO : SPE = 85:15 wt%. ; (LMO+ SPE)
(c) SPE coated LMO		SPE coated LMO cathode, prepared by casting technique; ((LMO+SPE)/SPE)
(d) SPE stacked LMO		SPE stacked LMO cathode, prepared by pressing method ; (LMO+SPE)&SPE

2.3. Sample characterization

A Scanning Electron Microscope (SEM, JEOL JSM 6510 LA) equipped with Energy Dispersive X-ray Spectrometer (EDS) were used to study the surface as well as the cross section structures. While the impedance characteristic was studied by using a computer controlled impedance analyzer (HIOKI LCR Hi-Tester Model 3532, Japan) within frequency range of 42 Hz to 5 MHz. Data fitting were performed using EIS Spectrum Analyzer (EISSA) software.

3. Results and discussion

SEM photograph were observed to confirm the surface as well as the cross section morphologies of each sample. Figure 1 shows the surface micrograph of SPE film, LMO cathode, SPE coated on LMO, and SPE stacked on LMO samples with 200X magnification except SPE stacked on LMO (500X). As in previous study, the Chitosan based SPE film (Figure 1(a)), shows relative smooth surface exhibits the domination of Chitosan polymer on the surface [5-7]. LMO cathode (Figure 1(b)) shows the existence of LMO particle on the surface. The SPE stacked on LMO sample (Figure 1(d)), was completely same with SPE film (a), except a small effect of pressing in the stacking process. In contrast with SPE film sample, the surface of SPE coated on LMO cathode (Figure 1(c)), was not smooth even if dominated by Chitosan film and some agglomeration of the lithium salt, as confirmed by EDS (Figure 2). This difference can be due to the difference of drying method that is the sample of SPE film was dried slower than SPE coated on LMO.

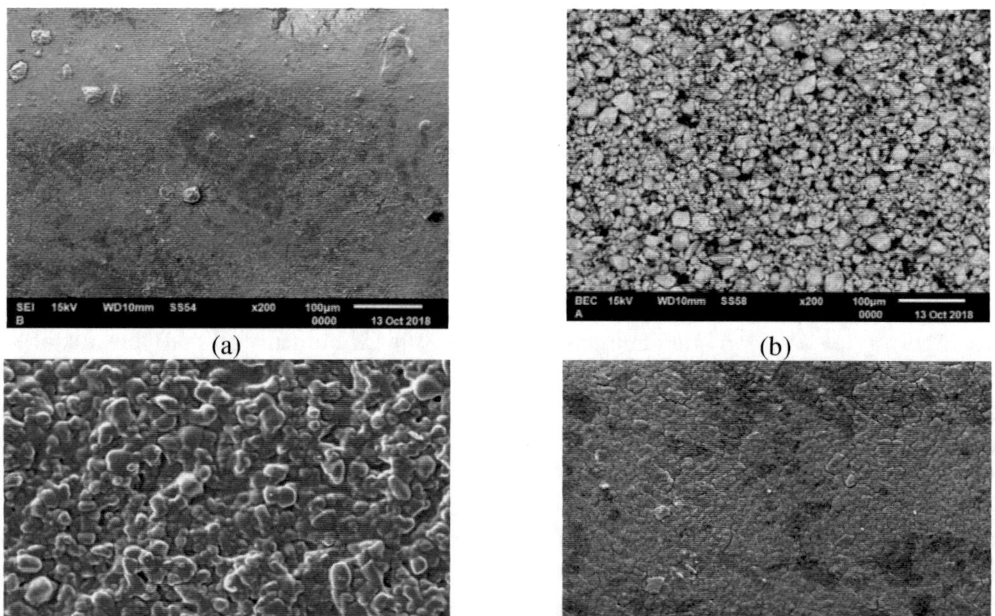

(a) (b) (c) (d)

Figure 1. SEM micrograph from surface of (a) SPE, (b) LMO, (c) SPE coated on LMO, and (d) SPE stacked on LMO samples.

Figure 2 shows the EDS results of the sample SPE coated on LMO. It was found that the surface was dominated by carbon (60.83 atom%) which is originated from Chitosan. However, an interesting phenomena was found in which Manganese (the main component of LMO) was observed on the surface (3.76 atom %). It means that some part of cathode material dissolved into the SPE layer. Of course, it is unexpected since it will contact with the anode and cause short contact.

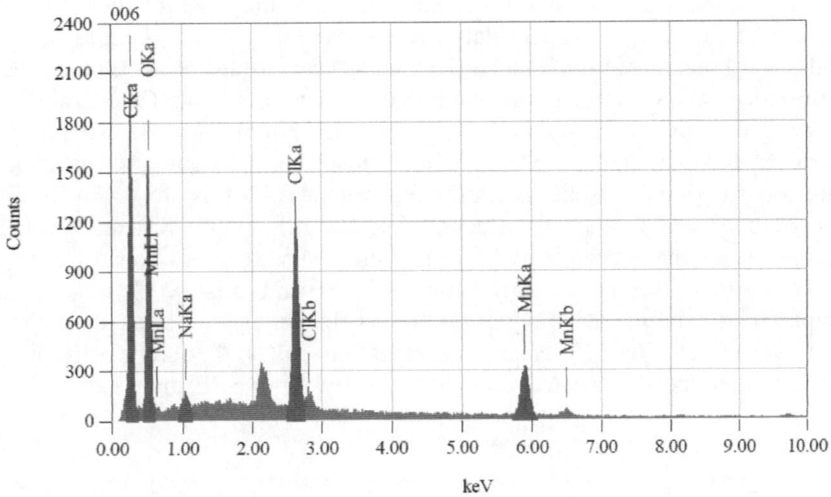

Figure 2. EDS spectrum from surface of Chitosan SPE coated on LMO cathode ((c) sample).

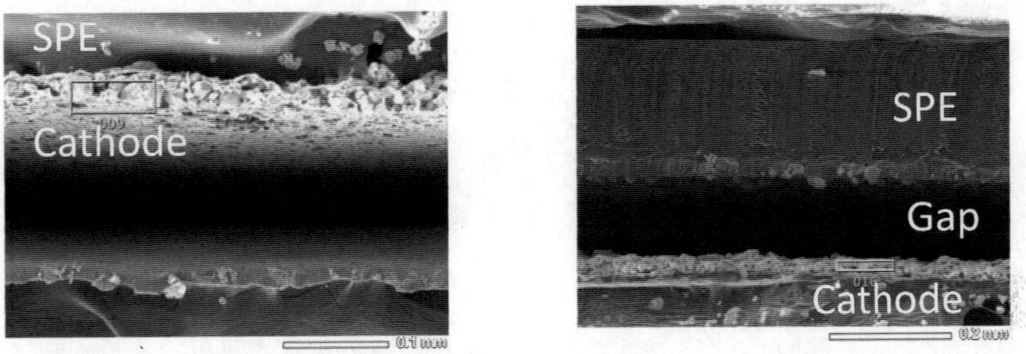

Figure 3. SEM micrograph from cross section of (c) SPE coated on LMO and (d) SPE stacked on LMO.

The different between coated and stacked sample was observed from the cross sectional micrograph as shown in Figure 3. When SPE was coated on the cathode, the SPE was completely contacted with cathode and some part of cathode dissolved in the SPE layer as shown in sample of SPE coated on LMO. In other hands, when SPE was just stacked on the cathode surface micro-gap or void was found at the SPE cathode interface as indicated on sample of SPE stacked on LMO. In other word, the coated sample has better SPE cathode interface than that of stacked one.

Electrochemical Impedance Spectroscopy (EIS) is a technique currently used to study the electrical properties of the bulk materials and their interfaces (i.e., electrode electrolyte interfaces) over a wide range of frequency and temperature. The bulk and interface contributions can be separated by using this technique [14]. Therefore, effect of the interface onto the electrical properties was studied by impedance measurement as shown in Figure 4. The Nyquist plots of the impedance spectra, shown in Figure 4, consist of semicircle alike in appearance for all samples. The SPE film shows the appearance of semicircle at high frequency and smallest size. The LMO cathode shows two semicircles at high and medium frequency. The semicircle of coated sample seems to be similar in appearance but smaller in size compared to the LMO cathode only. In contrast, the stacked sample shows the appearance of a big semicircle. However, all samples show the existence of capacitance and resistance.

In order to investigate the difference in the EIS spectra, we employed equivalent circuits and fitted to the data as displayed in Table 2. The EIS data was interpreted by fitting them to equivalent circuit models. The model was constructed by using resistor (R) and constant phase element (CPE). CPE was commonly introduced to the circuit when using pure capacitor could not fit in simulation. CPE has no physical meaning. It is just a calculated parameter when circuit element behaving in between capacitor and resistor [15]. Therefore, we consider mostly on the role of resistance (R). The experimental data was fitted and matched with the equivalent circuit as shown in the Table 2. The model was taken account based on the experimental data and the sample system. The EIS spectra of SPE film for example, show the appearance of 2 semicircles. Therefore, we constructed an equivalent circuits composed of a single CPE and a parallel circuit of CPE and resistor. We assumed that the SPE sample was composed of lithium salt ($LiClO_4$) dissolved in polymer (Chitosan) matrix. Thus, the impedance of SPE was affected by the distribution of lithium salt in the matrix. Similarly, the LMO cathode also showed 2 semicircles spectra, can be originated from SPE matrix and dissolved LMO particles.

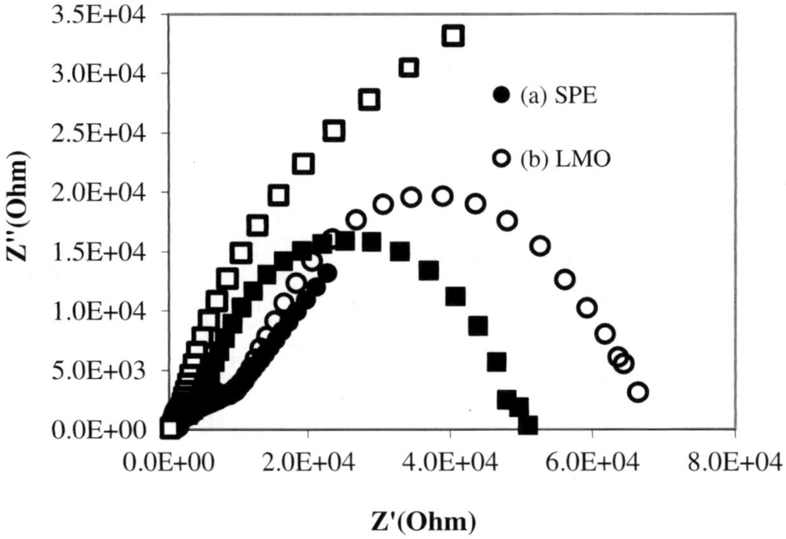

Figure 4. Nyquist Plot of (a) SPE film, (b) LMO cathode, (c) SPE coated LMO, and (d) SPE stacked LMO.

Table 2 clearly shows that the equivalent circuit model can be fitted on the experimental data. Having initial resistance, the LMO cathode seems to be different with SPE film, it may due to the existence of $LiMn_2O_4$ powder as cathode active material. The SPE coated LMO ((c) sample), however, appear to be same with the LMO cathode rather than SPE film only. While the SPE stacked LMO matched with different circuit, in which Warburg impedance must be involved. Comparing the equivalent circuit between (c) and (d) sample, it was clear the source of difference between coating (c) and stacking (d) samples. The SPE coated LMO shows a similar circuit with the uncoated LMO. It means that there was no effect of solid electrolyte cathode interface. While, at SPE stacked LMO (d sample) the effect of interface seems to appear in form of Warburg impedance ($1,5x10^3 k\Omega$) result in high resistance. The stacked sample also have a higher initial resistance ($R1 = 0.6\ k\Omega$) than the coated one which is mostly zero ($R1 = 1.1 \times 10^{-15} k\Omega$). It can also be due to the existence of micro-gap at the interface of the stacked sample. In other word, the problem in the interface between SPE and electrode in an all solid lithium-ion batteries can be reduced by casting the electrolyte solution on the cathode surface.

IOP Conf. Series: Materials Science and Engineering **553** (2019) 012061 doi:10.1088/1757-899X/553/1/012061

Table 2. Equivalent circuits, Nyquist Plot and fitting results of EIS spectra.

S.N.	Circuits	Fitting result	Sample remarks
(a)	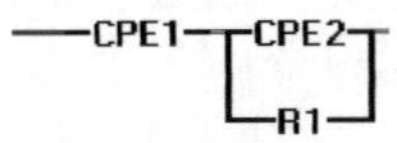 R1 = 19.7 kΩ	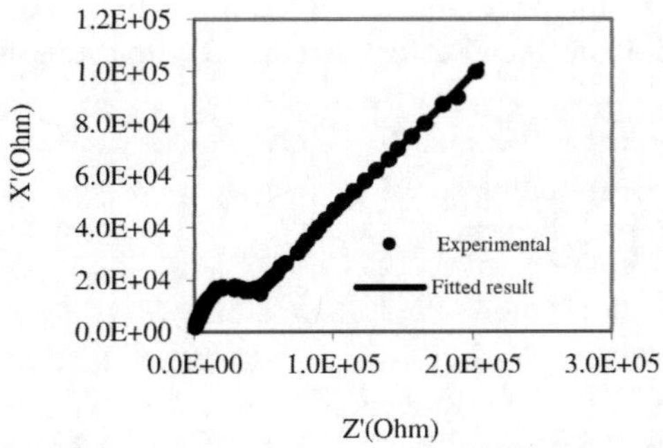	 SPE (Chitosan + $LiClO_4$)
(b)	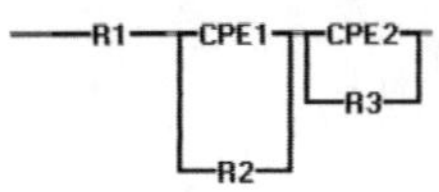 R1 = 3.8 x 10^{-16}kΩ R2 = 57.3 kΩ R3 = 8.9 kΩ	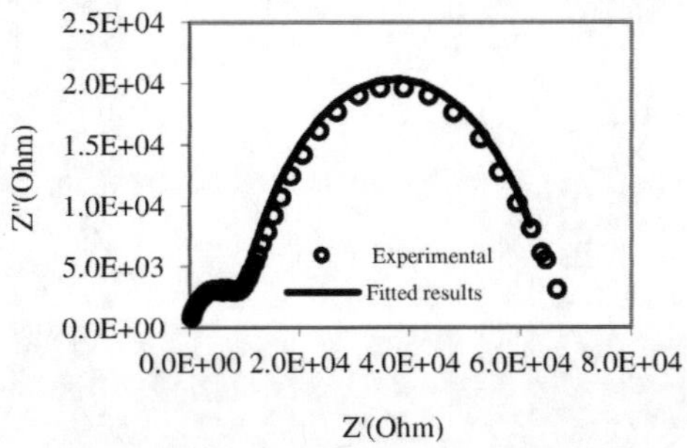	 LMO (SPE + $LiMn_2O_4$)
(c)	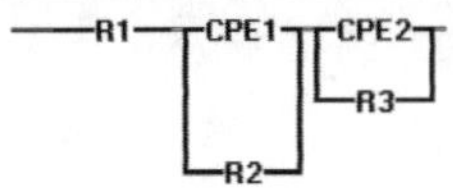 R1 = 1.1 x 10^{-15}kΩ R2 = 45.6 kΩ R3 = 3.9 kΩ	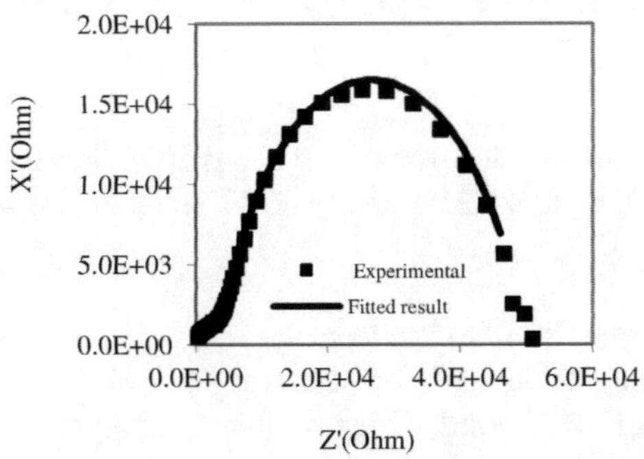	 Coated (SPE coated on LMO)
(d)	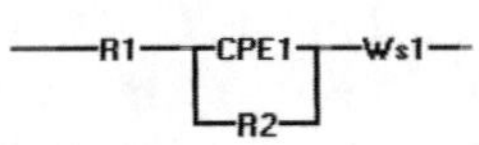 R1 = 0.6 kΩ R2 = 56.9 kΩ	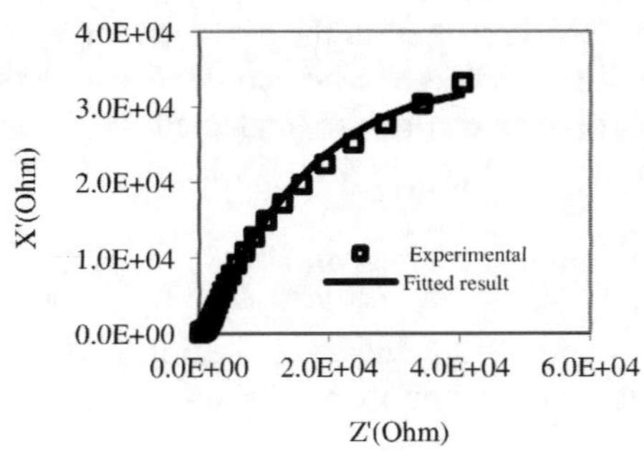	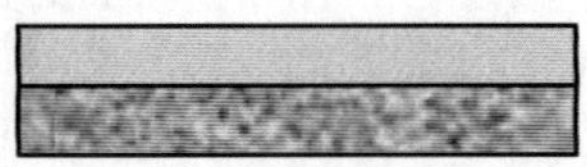 Stacked (SPE stacked on LMO)

The effect of interfacial structure to the properties can be illustrated as in Figure 5. Micro-gap between electrolyte and electrode was observed when the Chitosan based SPE just stacked on the cathode surface result in the appearance of Warburg impedance. The results of this study recommend how to control the interfacial structure and the properties of SPE in all solid battery system. That is, we must eliminate the gap in the interface and make the SPE stick to the cathode. Surface coating is one of the methods to reduce the interfacial effect and agree with previous study [12].

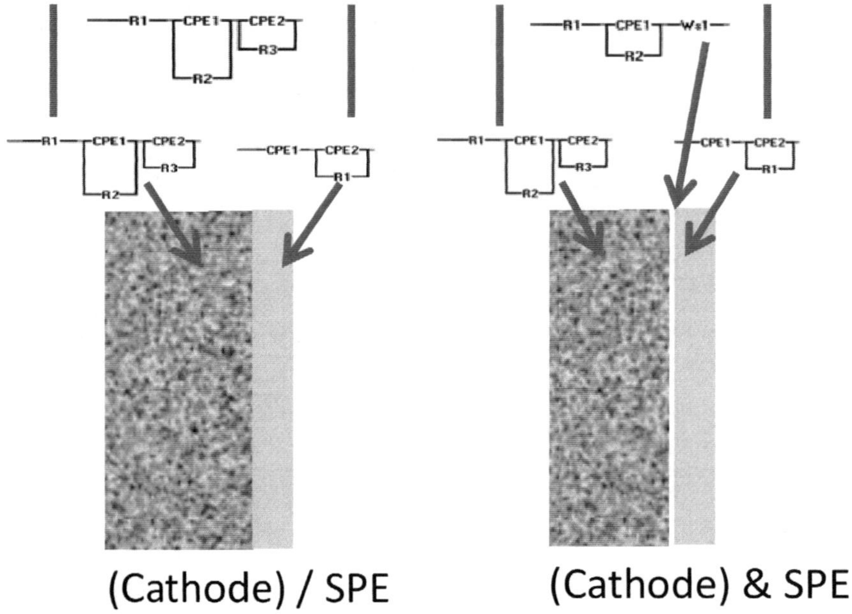

(a) (b)

Figure 5. Schematic illustration for two types difference SPE cathode interfacial arrangements (a) SPE coated LMO and (b) SPE stacked LMO.

4. Conclusion

The interfacial structure and properties of solid electrolyte and the cathode for battery application have been studied. $LiClO_4$ containing Chitosan were used as the solid electrolyte and $LiMn_2O_4$ (LMO) was the cathode active material. Two kind different interfacial arrangements were constructed by coating and stacking method. SPE coated and stacked cathode was used as the sample to study the effect of interfacial structure to the resistance. When the Chitosan SPE stacked on the LMO cathode, micro-gap appeared at the interface and resulted in the existence of Warburg impedance or interfacial resistance. However, when the SPE was coated using solvent casting onto the cathode the micro-gap did not appear and the interfacial resistance could be neglected.

Acknowledgment

This study was carried out under Center for Saints and Technology of Advanced Materials – BATAN Budget Implementation List (DIPA) number 3446.007.001. This work was also supported by Lithium Ion Battery consortium project under the Ministry Research, Technology and Higher Education (SINAS 2018). Discussing with the consortium members always bring up to a new idea.

References

[1] E. Kartini and M. Manawan, 2016, *AIP Conference Proceedings* 1710, 020001
[2] M. Tatsumisago, M. Nagao, A. Hayashi, 2013, *J. As. Ceram. Soc.* 1, 17–25
[3] Q. Zhang, K. Liu, F. Ding, and X. Liu, 2017, *Nano Research* 10(12): 4139–4174
[4] E.Kartini, T.Sakuma, K.Basar and M.Ikhsan, 2008, *Solid State Ionics* 179, 706-711
[5] Sudaryanto, E Yulianti, and H. Jodi, 2015, *Polym. Plast. Technol. Eng.*, 54, 290–295.
[6] Sudaryanto, E. Yulianti, and Patimatuzzohrah, 2016, AIP Conference Proceedings 1710, 020003 (2016); doi: 10.1063/1.4941464
[7] E .Yulianti, A K. Karo, L. Susita, and Sudaryanto, 2012, *Procedia Chem.*,4, 202–207.
[8] A.C. Luntz, J. Voss, K. Reuter, 2015, *J. Phys. Chem. Lett.* 6, 4599-4604.
[9] W.D. Richards, L.J. Miara, Y. Wang, J.C. Kim, G. Ceder, 2016, *Chem. Mater.* 28 266-273.
[10] L.J. Miara, W.D. Richards, Y.E. Wang, G. Ceder, 2015, *Chem. Mater.* 27, 4040-4047.
[11] Z. Wang, J. Z. Lee, H.L.Xin, L. Han, N. Grillon, D. Guy-Bouyssou, E. Bouyssou, M. Proust, Y.S. Meng, 2016, *J. Power Sources* 324, 342-348
[12] Yada, C.,2014, *Adv. Energy Mater.* 4, 1301416-1301420.
[13] Y. Seino, T. Ota, K.Takada, 2011, *J. Power Sources* 196, 6488-6492. doi:10.1016/j.jpowsour.2011.03.090
[14] S. B. Aziz, T. J. Woo, M.F.Z. Kadir, H. M. Ahmed., 2018, *J. Sci. Adv. Mat. Dev.*, 3, 1-17
[15] I.D. Raistrick, D.R. Franceschetti, J.R. Macdonald, "Theory" in *Impedance Spectroscopy Theory, Experiment, and Applications*, 2nd Ed. New Jersey: John Wiley & Sons, Inc., 2005, pp. 27-117

Synthesis of $Li_4Ti_5O_{12}$ (LTO) by Sol-Gel Method for Lithium Ion Battery Anode

Yustinus Purwamargapratala[1*], Agus Sujatno[1], Yohanes Lugas Sabayu[2], Evvy Kartini[1]

[1] Centre of Advanced Materials Science and Technology, National Nuclear Energy Agency,South Tangerang13310, Indonesia
[2] Physical Engineering, Faculty Of Engineering, Surya University Tangerang, Indonesia

pratala_yustinus@yahoo.com

Abstract. $Li_4Ti_5O_{12}$ (LTO) is one of the material that has the potential as anode in lithium ion battery that is safe to use and has a long life cycle. In this research, LTO was synthesed by sol-gel method using oxalic acid as chelating material, it is expected to produce materials with smaller particles and homogeneous distribution. The LTO sol-gel compared to commercial LTO. Characterization was carried out by X-ray diffraction (XRD), scanning electron microscopy (SEM), electrochemical impedance spectroscopy (EIS). The measurement of LTO battery performance is carried out by charge discharge and electron impedance spectroscopy. The sol-gel LTO has a purity of 99%, which is higher than 57% commercial LTO. The average particle size is 1 μm, and the conductivity is $7.6x10^{-7}$ $S.cm^{-1}$. While commercial LTO has a purity of 57%, the average particle size is 30 μm and conductivity is $7.4x10^{-7}$ $S.cm^{-1}$. Batteries with synthesized LTO anodes have a capacity of 25.9% higher in the first cycle and 110% higher in fifth cycles incomparison with batteries used commercial LTO anodes.

Keywords : $Li_4Ti_5O_{12}$, LTO, oxalic acid, sol-gel

1. Introduction

The development of technology has created a lot of equipment requiring electrical energy, especially batteries. A battery is a device that can convert the chemical energy it has into electricity. Batteries can be divided into two, namely primary and secondary batteries. Primary batteries are batteries that cannot be recharged while batteries that can be recharged and can be used again are called secondary batteries [1].

Lithium ion batteries are batteries that have great potential for the future. The existence of lithium ion batteries in the sales market began with Sony's company announcing its new product called lithium ion batteries in 1991 [2]. The demand for lithium ion batteries continues to increase. The price of lithium ion batteries in 2016 has decreased by 73% since 2010. Although the price has decreased, the growth in demand for lithium ion has increased throughout the year [3]. Lithium ion batteries have energy density, efficiency, and a high life cycle, can store a lot of energy in the smallest volume, and can work well for a long time [4].

Anode is a negative electrode that functions as an electron release which then moves to the external circuit and undergoes oxidation when the electrochemical reaction occurs. Anode material affects the nature of the battery [1]. The ideal properties of anodes are low atomic weights, cheap, high ionic and

IOP Conf. Series: Materials Science and Engineering **553** (2019) 012062 doi:10.1088/1757-899X/553/1/012062

electrical conductivity [5]. With safety considerations, lithium ion batteries use lithium metal as an anode, hence the type of carbon anode, mixed metal, metal oxide and sulfide is developed [5-6]. Graphite is an anode material that is most widely used in the manufacture of lithium ion batteries because it has a high energy density, however it has the disadvantage of having a short life cycle and short circuit prone characteristics, an alternative anode is $Li_4Ti_5O_{12}$. The use of $Li_4Ti_5O_{12}$ has the advantage of safe batteries and a long life cycle [4,7-10].

$Li_4Ti_5O_{12}$ can be synthesized by solid state, sol-gel, hydrothermal, and so on. The synthesis method that is often used in the manufacture of $Li_4Ti_5O_{12}$ is solid state synthesis because it is easier, and cost-effective, but it is difficult to adjust particle size, particle size distribution, and lack of stoichiometric control [7,11-13]. Particle size can affect the performance of the anode, therefore alternative synthesis is needed that can overcome the problem. Sol-gel synthesis is a synthesis that can make particles smaller and spread more homogeneously [7]. In the sol-gel synthesis the number of chelating agents can affect the end result of the [11-12]. Chelating agent plays a major role in the formation of $Li_4Ti_5O_{12}$ because it affects particle size. Oxalic acid is one of the chelating agents that can be used in the synthesis of $Li_4Ti_5O_{12}$ [11].

The aim of this study to synthesis LTO with sol-gel method by titanium butoxide and compared the results with commercial $Li_4Ti_5O_{12}$ both in the form of anode material. The performance of the lithium batteries will be characterized.

2. Experimental method

Sol-gel synthesis was carried out according to the steps of Qiu [14] but using titanium butoxide as a source of titanium and lithium acetate dihydrate as sources of lithium. The lithium dihydrate acetate is dissolved in ethanol and deionized air with the ratio of ethanol and deionized air to 5:1. Oxalic acid is added to the solution with the ratio of oxalic acid and ethanol to 1. The dissolved titanium butoxide is added to the solution and stirred for 4 hours, then left for 12 hours. The material that has reached maturity is dried at 80 °C for 6 hours. Then it is crushed and put into a container and heated at 200 °C for 30 minutes, at a temperature of 500 °C for 30 minutes and calcined at a temperature of 800 °C for 1 hour. The product from $Li_4Ti_5O_{12}$ synthesis by the sol-gel method is called solgel LTO and from commercial materials called commercial LTO [13].

Observation of the morphology and particle size of LTO was carried out by scanning electron microscopy (SEM), while the crystalline phase was measure by X-ray diffraction (XRD). LTO sheet was produc by LTO mixing with PVDF and NMP using a vacuum mixer. The LTO mixture is coated on a copper sheet. The electrochemical profiles was measured by LCR and a battery tester.

3. Results and discussion

Observation of the crystal structure was carried out using X-ray diffraction and analyzed using the High Score Plus method on the sol-gel LTO and commercial LTO, as shown in Figure 1.

The results of X-ray diffraction with the main peaks at 2.3^0; 18.3^0; 35.5^0; 43.2^0; 47.2^0; 57.1^0; 62.7^0; 65.9^0; 74.2^0; and 79.2^0 shown in Figure 1. Data were analyzed using the Hight Score Plus method. In accordance with the database, high scores plus used numbers 98-008-4973 and number 98-026-1239, sol-gel LTO and commercial LTO contain $Li_4Ti_5O_{12}$ and Li_2TiO_3.

Table 1 shows the refinement results of XRD data by HighScore Plus (HSP). It is shown that sol-gel LTO and commercial LTO have 2 phases with different compositions. LTO sol-gel resulted about 99% $Li_4Ti_5O_{12}$ and 1% Li_2TiO_3, whereas commercial LTO contained 57.7% $Li_4Ti_5O_{12}$ and 42.3% Li_2TiO_3. The analysis also shows that $Li_4Ti_5O_{12}$ corresponds to a space group F-3 m with a cubic while Li_2TiO_3 belongs to a space group C 1 2/c 1 with a monoclinic.

IOP Conf. Series: Materials Science and Engineering **553** (2019) 012062 doi:10.1088/1757-899X/553/1/012062

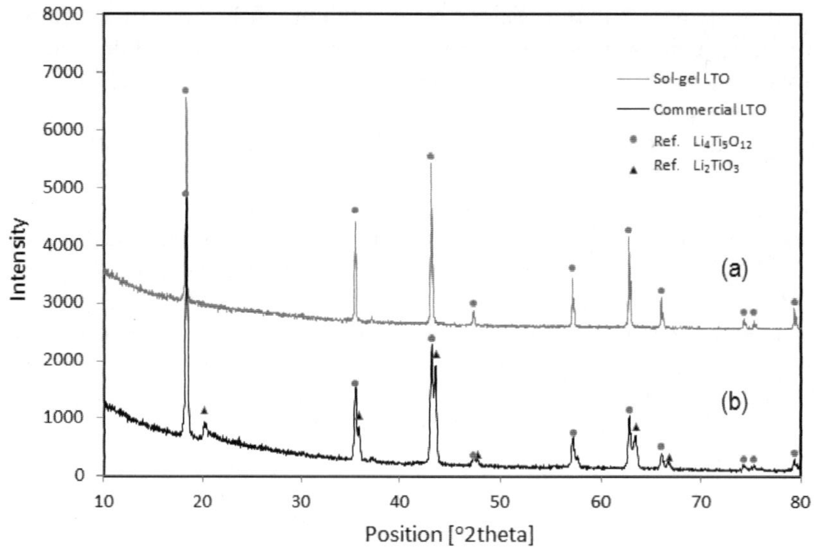

Figure 1. X-ray diffraction pattern from LTO (a) sol-gel LTO (b) commercial LTO.

Table 1. X-ray Refinement results of sol-gel LTO and commercial LTO.

Sample	Chemical Formula	Space group	QPA wt%	Crystallite Size (nm)	Strain (%)	Lattice Parameters a (Å)	b (Å)	c (Å)	R_{factor} R_{wp} (%)	R_{exp} (%)	X
Sol-gel LTO	$Li_4Ti_5O_{12}$	F d -3 m	99.0	561.50	0.000403	8.3561	8.3561	8.3561	7.61	5.05	1.43
	Li_2TiO_3	C 1 2/c 1	1.0	855.03	0.000416	5.0700	8.9000	9.5900			
Commercial LTO	$Li_4Ti_5O_{12}$	F d -3 m	57.7	396.42	0.000143	8.3590	8.3590	8.3590	7.48	5.08	1.47
	Li_2TiO_3	C 1 2/c 1	42.3	383.30	0.000136	5.0650	8.7910	9.7670			

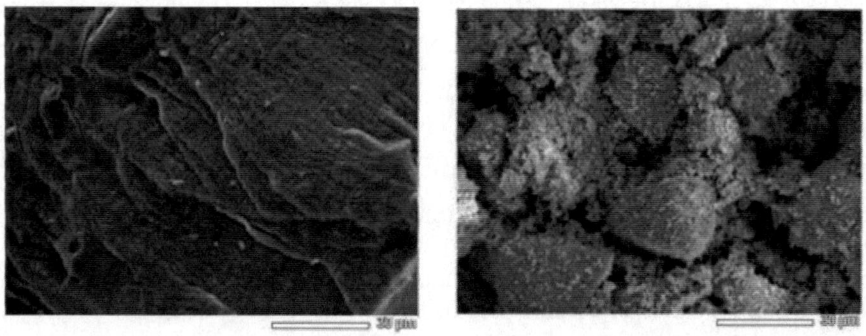

(a) (b)

Figure 2. SEM Observation Results 1000x magnification from LTO (a) Sol-gel LTO (b) Commercial LTO.

Analysis uses a scanning electron microscope (SEM) to study sample morphology, both in shape and size. In Figure 2 shows the SEM photo with a magnification of 1000 times, from LTO synthesized by the commercial sol-gel and LTO method. Figure 2(a) shows that the particles of sol-gel LTO appear to be in the form of sheets with an average particle size of 1 μm. The sheet form will improve to material conductivity, because it will facilitate the distribution of lithium ions. Figure 2(b) shows granular commercial LTO particles with an average size of 30 μm.

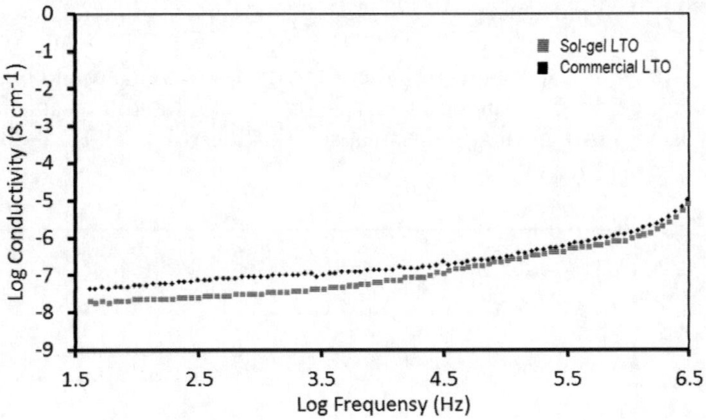

Figure 3. The Conductivity of sol-gel LTO and commercial LTO.

Table 2. Conductivity of sol-gel LTO and commercial LTO.

Sample	Conductivity
Sol-gel LTO	7.6×10^{-7} S.Cm^{-1}
Commercial LTO	7.4×10^{-7} S.Cm^{-1}

The conductance measure by EIS with a voltage of 1V in the frequency range 42 Hz - 5M Hz as shown in figure 3 and Table 2. Table 2 shows conductivity sol-gel LTO is 7.6×10^{-7} S.cm^{-1} and the commercial LTO is 7.4×10^{-7} S.cm^{-1}. This conductivity value is not quite different in the same order. This shows that the sol-gel LTO conductivity is similar to the commercial LTO.

The charge discharge of battery measurement with a rate of 0.05C and a voltage range of 0.2 V to 3.0 V for battery cell systems sol-gel LTO/$LiFePF_6$/$LiFePO_4$ and commercial LTO/$LiFePF_6$/$LiFePO_4$ were shown in Figure 4.

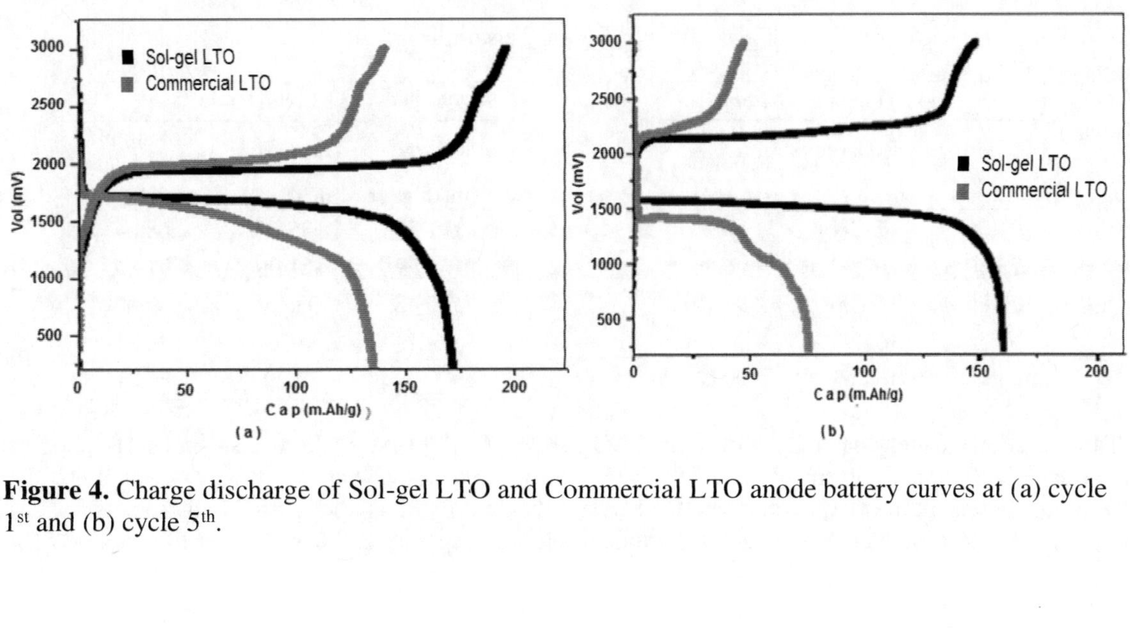

Figure 4. Charge discharge of Sol-gel LTO and Commercial LTO anode battery curves at (a) cycle 1st and (b) cycle 5th.

Figure 4 shows the result of first cycle charge discharge battery with the sol-gel LTO as anode material and using a commercial LTO. In cycle 1, the battery of sol-gel LTO as anode has the first output voltage of around 1.75 V and capacity is 170 $mAh.g^{-1}$and capacity of commercial LTO as battery is 135 $mAh.g^{-1}$, this means capacity of the sol-gel LTO as battery was 25.9% higher than commercial LTO as battery. Figure 4(b) shows the status of batteries in the fifth cycle around high voltage 1.4 V - 1.6 V. The capacity of sol-gel LTO as anode battery is 160 $mAh.g^{-1}$ and the capacity of commercial LTO as anode battery is 75 $mAh.g^{-1}$, this means capacity of the sol-gel LTO as battery was 110% higher than commercial LTO as battery.

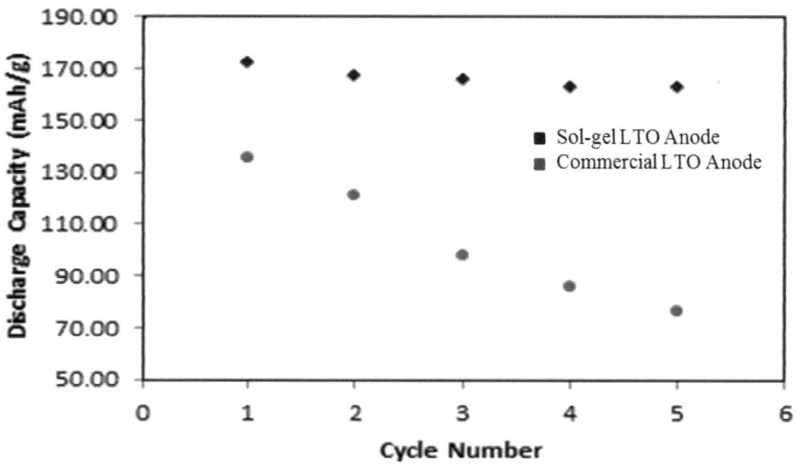

Figure 5. Battery performance of sol-gel LTO anode and commercial LTO anode.

Figure 5 shows the 1st to 5th cycle of the battery performance. In the 1st cycle it can be seen that sol-gel LTO anode batteries have a higher capacity than commercial LTO anode batteries. As cycles increase, sol-gel LTO anode batteries have a more stable capacity when compared to commercial LTO anode batteries. The sol-gel LTO anode battery has a slope of -2.3 and a commercial LTO anode battery of around -15.4. The more flat the battery performance cycle, the longer the battery life cycle [11].

Table 3. Capacity of the sample.

Ciclus	Sol-gel LTO anode battery			Ciclus	Comercial LTO anode battery		
	Charge ($mAh.g^{-1}$)	Discharge ($mAh.g^{-1}$)	Coloumbic Efficiency ($mAh.g^{-1}$)		Charge ($mAh.g^{-1}$)	Discharge ($mAh.g^{-1}$)	Coloumbic Efficiency ($mAh.g^{-1}$)
1	204.05	171.94	84.27	1	197.59	135.62	68.64
2	173.30	167.30	96.54	2	131.69	121.00	91.89
3	170.86	166.06	97.19	3	115.48	98.15	85.00
4	167.55	162.92	97.24	4	96.76	85.93	88.81
5	164.15	162.63	99.07	5	87.78	76.20	86.81

Table 3 shows battery capacity with both LTO anodes. In the first cycle, the sol-gel LTO anode battery had Coulumbic Efficiency (CE) of 84.27% and for the next cycle above 96%. The commercial LTO anode battery in the first cycle has CE 68% and the next cycle is above 85%. The results show that the bottery performance of sol-gel LTO anode has higher capacity and has a longer life cycle [15].

4. Conclusion

The synthesis of LTO with the sol-gel method has been carried out and the results was compared with commercial LTO. LTO synthesized using the sol-gel method has a purity of 99%. The sol-gel LTO has average particle size 1 μm, and conductivity 7.6x10^{-7} S.cm^{-1}. While commercial LTO has a purity of 57%, the average particle size is 30 μm and conductivity is 7.4x10^{-7} S.cm^{-1}. Batteries with a sol-gel LTO anode has a capacity of 25.9 % higher in the first cycle and 110 % higher in fifth cycles using with commercial LTO anodes batteries.

Acknowledgment

This work was supported by the 2018 DIPA PSTBM BATAN and partially funded by INSINAS grant (No. IRPK-2017-076) Ministry of Research, Technology and Higher Education of the Republic of Indonesia (KEMENRISTEK DIKTI).

References

[1] Morris Melissa and Tosunoglu Sabri 2012, Comparison of rechargeable battry technology, *ASME Early Career Technical Conference J., November 2012*

[2] Blomgren, G.E., 2017. The Development and Future of Lithium Ion Batteries. *The Electrochemical Society*, pp.A5019-25

[3] Jung Youn Mo and Wooyoung Jeon 2018, *Sustainability J.* 10, 2870, doi:10.3390/su10082870

[4] Nitta, N., Wu, F., Lee, J.T. &Yushin, g., 2015. Li-ion battery materials: present and future. *Materials today*, pp.252-64.

[5] Fu, R. et al., 2017. Comparison of Lithium-Ion Anode Materials Using an Experimentally Verified Physics- Based Electrochemical Model.*Energy Storage and Application*, p.2174.

[6] Vinodkumar Etacheri, Rotem Marom, Ran Elazari, Gregory Salitra and Doron Aurbach 2011, Energy & Environmental Science, DOI : 10.1039/c1ee01598b

[7] Sandhya, C.P., John, B. &Gouri, C., 2014. Lithium titanate as anode material for lithium-ion cells: a review. *Ionics*, pp.601-20.

[8] Sun, X., Radovanovic, P.V. & Cui, B., 2015. Advances in spinel LTO anode materials for lithium-ion batteries.*New Journal of Chemistry*, pp.38-63.

[9] Wang Y. et al., 2014. Solid-state Synthesis of Submicron-sized LTO/Li_2TiO_3 Composites with rich grain boundaries for Lithium ion batteries. *Journal of Power Sources*, pp.114-120.

[10] Yang L. et al, 2015. $Li_4Ti_5O_{12}$ nanosheet as high-rate and long-life anode materials for sodium ion batteries, Journal of Materials Chemistry A 3(48), DOI: 10.1039/C5TA07403G.

[11] Hao Y.J. et al., 2006.Synthesis and characterization of spinel LTO anode material by oxalic acid-assisted sol–gel method.*Power Sources*, p.1358–1364.

[12] Hao, Y. et al., 2005.Synthesis by TEA sol–gel method and electrochemical properties of LTO anode material for lithium-ion battery.*Solid State Ionics*, p.1201–1206.

[13] Purwamargapratala Y. *et al* 2018. Synthesis of $Li_4Ti_5O_{12}$-Sn by ultrasonic method as anode materials for lithium ion battery, *IOP Conf. Ser.: Mater. Sci. Eng.* **432** 012062 doi:10.1088/1757-899X/432/1/012062.

[14] Qiu, C. et.al., 2013, Sol-gel preparation and electrochemical properties of La-doped LTO anode material for lithium ion battery, J.Solid State, Electrochem,841-847

[15] Buchmann I., 2016. *Batteries in a Portable World: A Handbook on Rechargeable Batteries for Non-Engineers, Fourth Edition*. British Columbia: Cadex.

IOP Conf. Series: Materials Science and Engineering **553** (2019) 012063 doi:10.1088/1757-899X/553/1/012063

Study of Al-Zn-Cu-Sm Efficiency for Low-Voltage Sacrificial Anode Candidate

D Ferdian[1*], J A Sibuea[1], Y R Dharmawan[2], Y Pratesa[1]
[1] Department of Metallurgy and Materials Engineering, Faculty of Engineering Universitas Indonesia, Kampus Baru UI Depok 16424, Indonesia
[2] Materials and Metallurgical Engineering Department, Kalimantan Institute of Technology, Kampus ITK Karang Joang, Balikpapan, 76127, Indonesia

deni@metal.ui.ac.id

Abstract. Sacrificial anode is one of the most effective methods for preventing corrosion in seawater environment. Aluminum (Al) anode works very well in seawater environment with high electrochemical capacity and high driving voltage. Commonly used aluminum alloy sacrificial anodes have the possibilities to trigger the occurrence of Stress Corrosion Cracking and Hydrogen Embrittlement due to overprotection on high-strength steel. The low voltage sacrificial anode was developed to overcome this problem. Electrochemical behavior of Al-5Zn-0.5Cu alloy with varied samarium (Sm) addition of 0.02 wt% and 0.1 wt% was investigated in this study. The commonly used Al-5Zn-0.02In alloy is also tested as a comparison. Anode efficiency test with weight loss method according to DNVGL-RP-B401 standard was conducted to determine the performance of Al-5Zn-0.5Cu-xSm alloy. The addition of 0.1 wt% Samarium shows the best anode performance with electrochemical capacity 2809.80 Ah/kg and 98% efficiency. Al-5Zn-0.5Cu-0.1Sm alloy is the most potential alloy to be developed as a low potential sacrificial anode to replace commonly used aluminum sacrificial anode with indium addition.

1. Introduction

Aluminum-based sacrificial anodes are the most common type of sacrificial anode used in seawater environments due to its high electrochemical capacity and high efficiency. The most commonly used sacrificial anode is Al-Zn-In. With the addition of indium to Al-Zn alloys, it will activate aluminum that will prevent the formation of a passive layer resulting increase of the efficiency of the sacrificial anode. This activation process will even more easily occur when in environments with high chloride ions such as in seawater environments[1]. However, the disadvantage of this sacrificial anode is that Al-Zn-In can initiate the occurrence of stress corrosion cracking and hydrogen embrittlement in high strength steels[2]. The low voltage sacrificial anode was developed to overcome the aforementioned problems. Low voltage sacrificial anode would have to supply a working potential around -0.8V vs SCE, because at this range of potentials the amount of hydrogen liberated is highly reduced and steel is still protected[2]. In its development, rare earth metals were used as additional alloying elements to modify the nature of the sacrificial anode. In previous studies, gallium[2], cadmium[2], and samarium[3] were used as alloying elements. The previous research suggests that samarium has promising results. This study aims to compare the efficiency of Al-Zn-In anodes with Al-Zn-Cu-Sm to develop low voltage sacrificial anode candidate. Electrochemical behavior of Al-5Zn-0.5Cu alloy with varied samarium (Sm) addition of 0.02 wt% and 0.1 wt% compared to Al-5Zn-0.02In was investigated in this study.

Adding zinc (Zn) to aluminum-based anodes has the purpose of destabilizing the passive layer of the aluminum. The addition of zinc up to 5 wt% can optimize the working efficiency of aluminum anodes by increasing their electrochemical properties. The presence of undissolved β-phase in the aluminum matrix causes damage to the passive alumina layer on aluminum anodes which will increase the outgoing current, so that the efficiency of the sacrificial anode increases[4]. Although the addition of copper (Cu) in aluminum will increases strength, hardness, and machinability, it will reduce corrosion resistance of the aluminum, as well as increasing susceptibility to stress corrosion[5]. The addition of copper in Al-Zn alloy also effects the microstructure. Based on previous research, the initiation of pitting corrosion occurs due to the formation of Al_2Cu precipitates which tend to segregate at the grain boundary. Based on immersion testing, the area first attacked by pitting starts from the precipitate around the grain boundary. The lowest value of corrosion resistance is found at alloy Al-5Zn-1Cu[6].

Addition of samarium to aluminum alloy plays a role in changing the microstructure. The addition of the samarium element can change the morphology of α-Al dendrites to be finer. Moreover, the addition also the phase rich in eutectic iron and silicon. Iron-rich phases will change morphology from irregular to thin morphologies. The volume fraction will also decrease[7]. The addition of the samarium also effected the silicon eutectic phase morphology. With the addition of 0.3 wt% samarium, the silicon eutectic becomes smoother and its sharp edges begin to disappear. The optimum morphology was achieved by adding 0.6 wt% samarium. However, excessive addition of samarium will cause the eutectic silicon to return to its original shape[8]. Pratesa et al. showed that the addition of samarium to Al-5Zn-0.5Cu alloy results in $SmCu_4Al_{11}$ precipitates after the solidification process[3].

2. Experimental method

2.1 Material Preparation

This research begins with casting the alloy from high purity aluminum, zinc, copper, and samarium ingot with 99.99% grade. The materials were cut into pieces, weighed and melted in graphite crucible in electric resistance furnace under atmospheric condition at 700°C. The mixture was gently stirred using a carbon bar to obtain a homogeneous mixture, as well as heat dissipation. The molten alloy was poured in a preheated cast iron die and allowed to cool in air. The process started with casting the master alloy of Al-5Zn-0.5Cu and Al-10Sm. The casting process continued with cutting master alloy then drying and weighing in accordance to obtain the alloying target. There are two types of alloy with samarium addition that will be made Al-5Zn-0.5Cu-0.02Sm and Al-5Zn-0.5Cu-0.1Sm. The next casting carried out is Al-5Zn-0.02In alloy which begins with cutting aluminum, zinc, and indium ingots according to their material balance and repeated the same casting procedure. Actual chemical composition of casted alloy was determined by Optical Emission Spectrometer and is given in Table 1.

2.2 Anode Current Efficiency

Anode efficiency testing has the purpose of determining the work efficiency of the sacrificial anode. The size of the samples was Φ 5 mm × 10 mm. Mild carbon steel and a silver/silver chloride were used as cathode and reference electrode, respectively. The minimum surface area ratio of test anode to cathode was 1:20. Anode and cathode were coupled and immersed together in aerated 10 liters of natural seawater. The temperature for this test must be maintained at 20°C ± 3°C. This test is carried out based on the DNVGL-RP-B401 standard[9]. The basic principle of this test is to discharge current with a certain predetermined current density for 96 hours of testing on the electrochemical cell circuit. Different current densities were employed ca. 1.5, 0.4, 4.0, 1.5 mA/cm^2. At the end of the test, the sample is cleaned with running water and soaked in a solution of 20 grams of CrO_3 and 30 ml of H_3PO_4 per liter of water for 10 minutes at 80°C. The sample is then washed with water followed by alcohol

IOP Conf. Series: Materials Science and Engineering **553** (2019) 012063 doi:10.1088/1757-899X/553/1/012063

and then dried. Final weight calculation is done after sample cleaning. The lost anode weight was calculated to determine the electrochemical capacity of the anode during the test. The electrochemical capacity has a unit of Ah / kg. This study was conduct at Department of Metallurgical and Materials Engineering Universitas Indonesia, Depok, Jawa Barat, Indonesia.

2.3 Metallographic Examination

Microstructure observation is carried out on samples after casting. This observation is carried out using an optical microscope. Each sample grinded with #80 - #1500 silica paper and polished with 1μm diamond paste. In the sample preparation stage, no etching process done to prevent corrosion on the grain boundaries to be observed because precipitates tend to form at grain boundaries. This observation aims to determine the effect of the addition of alloying elements on grain shape, grain boundaries and precipitates formed. Microstructure examination also done after anode efficiency test using optical microscope. Samples that have been tested for efficiency are cut in the middle with 2.5 cm from the top with a cross section. Samples were prepared by grinding with silica paper up to 1200 grit and observed with optical microscopy. The whole samples are also observed by digital camera before cut. The purpose of this observation is to determine the form of corrosion that occurs on the surface of the sacrificial anode during efficiency testing

3. Results and discussion

3.1 Chemical Composition

The chemical composition of the aluminum alloy sacrificial anodes after casting is obtained through Optical Emission Spectroscopy (OES) testing. Table 1 shows the results of chemical composition test for Al-5Zn-0.5Cu (Alloy A), Al-5Zn-0.5Cu-0.02Sm (Alloy B), Al-5Zn-0.5Cu-0.1Sm (Alloy C), and Al-5Zn-0.02In (Alloy D). The results of the composition testing in Table 1 show a deviation of chemical composition from the target. The desired zinc content in alloy A, B, and C samples is 5.0 wt%, however the actual composition has a deviation of 0.54, 0.07, and 0.04 respectively. The compositions of copper in alloy A, B, and C samples has a deviation of 0.06, 0.009, and 0.082 respectively of the desired targets of 0.5 wt%. The composition test results for alloy D sample have a deviation of 0.04 from the target of 5 wt% zinc and 0.0063 from the target of 0.02 wt% indium. The deviation of the main constituent elements of the alloy may be ignored because the deviation value is insignificant, so that the aluminum alloy from casting can be considered to have reached the target.

Table 1. Chemical composition of the alloys (wt%).

Alloys	Zn		Cu		In		Al	
	Actual	Deviation	Actual	Deviation	Actual	Deviation	Actual	Deviation
A	5.54	0.54	0.56	0.06	< 0.01	-	93.8	0.20
B	4.93	0.07	0.509	0.009	< 0.01	-	94.3	0.18
C	5.04	0.04	0.582	0.082	< 0.01	-	94.1	0.30
D	5.04	0.04	< 0.001	-	0.0263	0.0063	94.8	0.18

3.2 Electrochemical Performance

Based on the data obtained in Table 2, the electrochemical capacity obtained in Al-5Zn-0.5Cu alloy was 1852.83 Ah/kg with 65.08% efficiency. The electrochemical capacity and current efficiency increased with addition of samarium. This electrochemical capacity value indicates the ability of an alloy to work as a sacrificial anode in protecting the structure. The higher the value of the electrochemical capacity of an alloy, the better the protective ability[10]. Based on the results obtained on the two samples of the sacrificial anode with variations of samarium content, Al-5Zn-0.5Cu-0.1Sm alloy has the best protection capability based on its electrochemical capacity value with 2809.80

A.hr/kg and 98 % current efficiency. Based on previous research, the addition of samarium to Al-5Zn-0.5Cu alloy results in $SmCu_4Al_{11}$ precipitates after the solidification process[3]. The presence of these precipitates results in a difference in potential values between the aluminum matrix and $SmCu_4Al_{11}$ precipitates, which increases galvanic corrosion occurrence in alloys.

The addition of samarium by 0.02 wt% in the sacrificial anode alloy results in a greater electrochemical capacity value with a value of 2604.01 Ah/kg compared to the sacrificial anode of aluminum alloy with same amount of indium addition with a value of 2398.98 Ah/kg. The larger electrochemical capacity value indicates that the anode protection capability of the samarium alloy sacrifice is superior to the indium alloy. Based on the standards used, the minimum electrochemical capacity of the aluminum alloy sacrificial anode is 2500 Ah/kg[9]. After electrochemical test performed, only alloy with the addition of 0.02 wt% and 0.1 wt% samarium meets this prerequisite.

Table 2. Electrochemical performances of the alloys.

Alloys	OCP (V vs SCE)	CCP (V vs SCE) at Different Current Densities (mA/cm^2)				Capacity (A.hr/kg)	Efficiency (%)
		1.5	0.4	4	1.5		
A	-0.913	-0.735	-0.715	-1.033	-1.047	1852.83	65.08
B	-0.911	-0.890	-0.599	-0.649	-0.610	2604.01	91.00
C	-0.884	-0.834	-0.755	-0.897	-0.914	2809.80	98.00
D	-0.942	-0.779	-0.748	-0.681	-0.553	2398.98	84.00

Anode potential measurement during efficiency test can be seen in Table 2. The open circuit potential (OCP) and the close circuit potential (CCP) of aluminum alloy with samarium addition showed more positive potential compared to the other alloys. However, the potential value obtained from the measurement shows decreasing and increasing at each measurement, this is due to the current from the external power source during the testing of the sacrificial anode efficiency which affects the work and potential value of the sacrificial anode. Wen et al. showed that lower fluctuation of CCP value implies the uniform corrosion, corresponding to the high anode efficiency[11]. According to Pautasso et al., low voltage aluminum sacrificial anode would have to supply a working potential around -0.8V vs SCE, because at this range of potentials the amount of hydrogen liberated is highly reduced and steel is still protected[2]. In this research, the most suitable candidate for low voltage aluminum anode is Alloy C with 0.1 wt% samarium addition, because it has OCP and CCP around -0.8V vs SCE. Furthermore, compared to aluminum anode with indium doping, alloy with samarium addition has better electrochemical performance for low voltage anode.

3.3 Macrostructure Analysis

The macro structure photographs in Figure 1 will present the physical differences produced by the sample of aluminum sacrificial anodes given the addition of samarium and indium alloys. The following are photos taken after weight loss testing and cleaned using a solution of 20 grams of CrO_3 and 30 ml of H_3PO_4 per liter of water for 10 minutes at 80°C. Based on four samples tested for efficiency, the most striking difference is that the macro structure changes to be rougher and more perforated. Samples with addition of samarium have a blackish appearance after testing, while samples with the addition of indium have silver appearance as tested. This is due to differences in the alloying elements used in casting the sacrificial anode. Based on the observations, the Al-5Zn-0.5Cu-0.1Sm alloy has a more even black color appearance compared to other samples with the addition of different samarium concentrations. This indicates that the Al-5Zn-0.5Cu-0.1Sm alloy has more uniform corrosion on its surface. On the other hand, three other samples are still visible in silver. This silver colored area shows no uniform corrosion throughout the surface of the sacrificial anode. The addition of samarium to the aluminum alloy anode sample causes the corrosion on the sacrificial anode to worsen up to the concentration of 0.1 wt% samarium. This occurs due to the segregation of samarium at the grain boundary on the surface of the sacrificial anode which results in the surface concentration of Cl^- ions from seawater to rise on the oxide or passive layer of the sacrificial anode sample so that the passive layer becomes unstable and the surface of the sample becomes active[12].

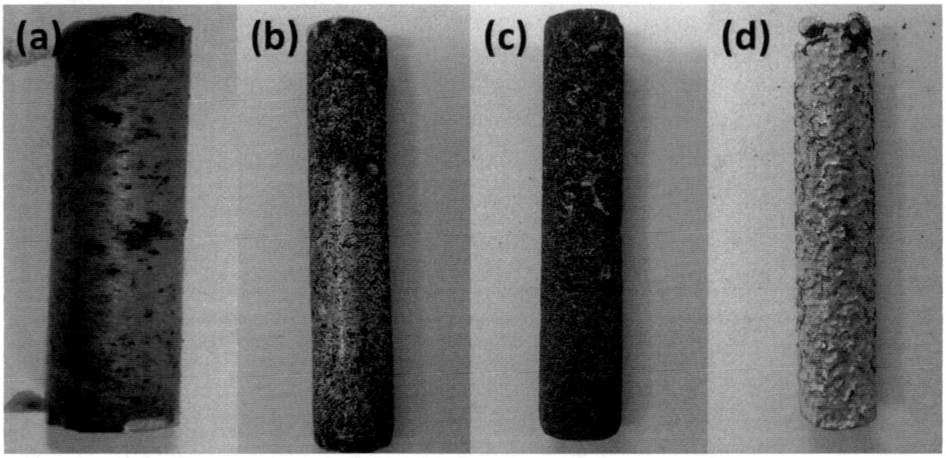

Figure 1. Macro Corrosion Photograph of the Alloys After Efficiency Test: (a) Al-5Zn-0.5Cu alloy; (b) Al-5Zn-0.5Cu-0.02Sm alloy; (c) Al-5Zn-0.5Cu-0.1Sm alloy; (d) Al-5Zn-0.02In alloy.

On the other hand, the use of indium in the aluminum alloy sacrificial anode resulted in a uniform corrosion appearance on the sacrificial anode. Although the sacrificial anode macro structure is silver in color, the surface of this anode is corroded evenly. The form of corrosion that occurs on the surface of the sacrificial anode with the addition of indium is also different from the sacrificial anode with the addition of samarium. The sacrificial anode with indium has widely shaped corrosion throughout the surface, while with the addition of samarium, shaped as pitting corrosion.

4. Conclusions

Electrochemical behavior of Al-5Zn-0.5Cu alloy with varied samarium (Sm) addition of 0.02 wt% and 0.1 wt% compared to Al-5Zn-0.02In was investigated in this study. The highest value of electrochemical capacity and efficiency of sacrificial anode is found at Al-5Zn-0.5Cu-0.1Sm alloy with a value of 2809.80 Ah/kg with an efficiency of 98%. The use of samarium produces sacrificial anodes with better work efficiency compared to the addition of indium. Samarium increased the pitting resistance of aluminum, hence caused local corrosion. Al-5Zn-0.5Cu-0.1Sm alloy is the most potential alloy to be developed as a low potential sacrificial anode to replace commonly used aluminum sacrificial anode with indium addition.

Acknowledgement

The authors would like to give our thanks and highest appreciation to Indexed International Publication Grants for Student Final Project (PITTA) 2018 Universitas Indonesia (Grant No. 2413/UN2.R3.1/HKP.05.00/2018) for supporting the funding in this research.

References

[1] N. Idusuyi, O.O. Oluwole, 2012 *Int. J. Sci. Technol.*, **2** 8 561–566
[2] J. Pautasso, B. Victor, H. L. E. Guyader, V. Debout, 1998 *Nace Int* **725**
[3] Y. Pratesa, D. Ferdian, F. Y. Ramadhan, B. Maulana 2018 *AIP Conference Proceedings* **1964,** 020013
[4] A. Muazu, S. A. Yaro 2011 *J. Miner. Mater. Charact. Eng.*, **10** 2 185–198.
[5] R. S. Rana, R. Purohit, S. Das, 2012 *IJSRP*, **2** 6 1–7

[6] D. Ferdian, Y. Pratesa, I. Togina, I. Adelia, 2017 *Procedia Eng.,* 184, 418–422, 2017
[7] Z. Hu, H. Yan, Y. Rao 2013 *Trans. Nonferrous Met. Soc. China* **23** 11 3228–3234
[8] Q. Hongxu, Y. Hong, H. Zhi, 2013 *J. Alloys Compd.*, **567** 77–81
[9] M. A. Talavera, S. Valdez, J. A. Juarez-Islas, B. Mena, J. Genesca, 2002 *J. Appl. Electrochem.*, **32** 8 897-903
[10] J. Genescà, J. Juárez, 2000 *Contribution to Science* **1** 3 331–343
[11] J. Wen, J. He, X. Lu *2011 Corros. Sci.,* **53** 11 3861–3865
[12] T. M. Umarova, I. N. Ganiev 2008 *Russ. J. Appl. Chem.*, **81** 5 796–799

IOP Conf. Series: Materials Science and Engineering **553** (2019) 012064 doi:10.1088/1757-899X/553/1/012064

Infinite Energy Creates Better Life

Ambar Pertiwi[1], Rudy Harjanto[1], Evvy Kartini[2]
[1]University of Prof. Dr. Moestopo (Beragama), Jl. Hang Lekir I/8, Central Jakarta, 10270, Indonesia
[2]Center for Science and Technology of Advanced Materials, National Nuclear Energy Agency, Puspiptek Serpong, Banten 15314, Indonesia

ambar.pertiwi@gmail.com
abrudyharjanto@gmail.com
evvy.kartini@gmail.com

Abstract. As the world moves towards sustainability, renewable energy development is driven in most countries as an alternative offer to the use of non-renewable and polluting fossil fuels. One such alternative is solar energy, which is the versatile technology used to harness the sun's energy and make it usable. It refers also to capturing the energy from the sun, collected elsewhere (i.e. the Earth) and subsequently converting it into electricity. Due to the non-constant nature of solar energy, collector and storage unit components are required to have a functional solar energy generator. The collector simply collects the radiation that falls on it and converts a fraction of it to other forms of energy (electricity or heat). The storage unit is required to store excess energy during the periods of maximum productivity and release it when the productivity drops. The proposed study is aimed to explore the feasibility of solar energy as an infinite energy to be restored and used as it needs. The merit of this study lies in the fact that solar energy creates better life (such as cost reduction, resource availability, eco-friendly. and sustainability) should it is harvested, stored and utilized considerably in a holistic way. Moreover, it may enhance independency of electricity supply, particularly in remote areas where access to power supply are limited. It is anticipated that the study will not only create a further interdisciplinary research platform for fundamental studies on solar energy harvesting, storage, and utilization, but also a research that leads to technology transfer to industry as it is lauded as a renewable and an inexhaustible fuel source, environmentally friendly due to its nature of pollution – and often noise – free.

Keywords: solar energy, solar energy storage, green economic development

1. Introduction

Low carbon economic growth approach as a model of green economic development has become the main focus in many states' policies, particularly in anticipating the depletion of non-renewable natural resources. As non-renewable natural resources in particular fossil based oil is nearing its limit, alternative extraction method such as fracking might cause environmental damage and negative impacts of climate change.

Climate change studies as caused by global warming has been peer reviewed and agreed upon will negatively affect the world at large. Although the present state of knowledge regarding regional and

IOP Conf. Series: Materials Science and Engineering **553** (2019) 012064 doi:10.1088/1757-899X/553/1/012064

local climate change does not allow scientists to identify precisely the most vulnerable areas, they are able to indicate the sort of area which will be most affected [1].

As fossil fuel dominates the world current energy requirement, how about energy for the future? If we continue to generate most of our energy from coal, oil and gas, do we have enough to keep us going? Current knowledge of proven recoverable reserves indicates that known reserves of fossil fuel will meet demand for the period up to 2020 [1].

Despite the development of modern solar energy over the past forty or fifty years, the technology still needs a higher profile and more involvement from scientists, engineers, environmentalists, entrepreneurs, financial experts, publishers, architects, politicians and civil servants. A new generation of solar energy pioneers has to be nurtured [2,3].

The formulation of the problem from the above-mentioned data are (1) how is the future of renewable energy, i.e. solar energy?; (2) what is the importance of solar harvesting and storage?; and (3) what is green economy's contribution to the world?. The purpose of current study intends to determine the usage of solar energy as it found to be infinite, environmentally friendly, cost effective, and reachable to remote areas is under discussion in present study. Qualitative approach has been taken in study through a collection and analysis of data and information, including a causal analysis of innovation information contained in various journals and reference books and a series of focus group discussion among scholars in the Universitas Prof. Dr. Moestopo (Bergama) and the Materials Research Society Indonesia (MRS- INA), conducted in Jakarta during period January – March 2018.

2. Literature Review

Many countries today strategically pursue development and economic natural resources diversification to overcome the current circumstances. One of the focus is on solar energy. Solar energy is a renewable energy source that can generate electricity, provide hot water, heat and cool a house and provide lighting for buildings. The sun has produced energy for billions of years. Sunshine is an infinite resource and investment in the technology that harnesses its energy as a general-purpose technology for development and export could be a priority of a progressive government [4].

Solar energy is the solar radiation that reaches the earth and it has several major advantages when compared with other sources, as it is distributed, though unequally, to every location on the globe. The resource is abundant, to the extent that many countries already harvested it [5].

Solar energy offers solution to the problem of scarcity resources, as it still has 6.5 billion years of life according to NASA. Solar energy can potentially play a very important role in providing most of the heating, cooling and electricity needs of the world [6]. Indeed, in a rather less time, solar technology in several countries has evolved to compete with conventional power generation sources.

Today solar sources provide around 10% of the energy used worldwide, but in the developing countries their share is still of the order of 40%. This contribution could start growing again, thanks to progress in solar technology and the pressure of recurrent energy and environmental crises related to fossil fuels and nuclear power [2]. In just a few decades, this will be a major part of a sustainable energy system for the world. In a matter of speak, solar energy will never stop shining.

The current study entails the exploration of energy alternative solutions, specifically solar energy to benefit most people in the world. Solar storage and harvesting have potentially been making contribution to alternative solution in providing energy for remote areas, anticipating the problem of non-renewable energy, pollution and highly cost. In addition, it is also a sustainable energy and can lead to a green economy.

3. Methodology

This study is a collection and analysis of data and information about solar energy harvesting, storage and utilization run by Universitas Prof. Dr. Moestopo (Beragama) in collaboration with MRS-INA parsed selected local and international literature. The study includes a causal analysis of innovation information contained in various journal and reference books.

IOP Conf. Series: Materials Science and Engineering **553** (2019) 012064 doi:10.1088/1757-899X/553/1/012064

The secondary data and a series of focus group discussion among scholars in the Universitas of Prof. Dr. Moestopo (Beragama) have also been extracted and implemented. The research was conducted in January – March 2019. Limited scholars and short time period of research is limitation of study; covers only among Universitas Prof. Dr. Moestopo (Beragama) and MRS-INA with limits the generalized findings.

4. Results and discussion

The New York Times on April 4, 1931 made the headline "Use of solar energy is near a solution", quoted the German solar energy scientist, Dr. Bruno Lange, as precursor that 80 years later electricity was supplied to millions of people in the world from renewable energy. Today, the whole world is aware of the limitations of fossil fuels and their effects on the environment as a cause of global warming. For this reason, the world states its readiness to accelerate the transition to a low-carbon economy.

The Government of Indonesia itself through the website of the Directorate General of New and Renewable Energy and Energy Conservation (EBTKE) launched the Low Carbon Economic Development Strategy on the occasion of the third annual meeting of the Low Carbon Emission Development Strategies (LEDS Forum) in Yogyakarta, attended by more than 250 forum participants consisting of government officials, experts, and representatives from international institutions, NGOs, and businesses identifying policies and activities that can increase economic growth, create jobs and other priorities in various countries in Asia, through a low carbon economic approach to leading green economic development.

Solar power as an energy is radiation from the sun that is converted into heat or electricity. It is also an energy that is always renewable and does not require fuel, is clean and environmentally friendly, and suitable for tropical regions such as Indonesia.

Solar energy creates clean, renewable power from the sun and benefits the environment. Generating electricity with solar power instead of fossil fuels can dramatically reduce greenhouse gas emissions, particularly carbon dioxide (CO_2). Greenhouse gases, which are produced when fossil fuels are burned, lead to rising global temperatures and climate change. Climate change already contributes to serious environmental and public health issues in some parts of the world, including extreme weather events rising sea levels. By going solar, we can reduce demand for fossil fuels, limit greenhouse gas emissions, and shrink our carbon footprint.

Moreover, solar energy results in very few air pollutants, reduce nitrous oxides, sulphur dioxide, and particulate matter emissions, all of which can cause health problems. In another word, solar power results in fewer cases of chronic bronchitis, respiratory and cardiovascular problems, and lost workdays related to health issues.

Individual off-grid Photovoltaics, solar application as a stand-alone power source for various applications such as CCTV and communication stations can also be implemented in public sector.
For areas like flat lands, forests and islands, which do not receive power, owing to their remote location, solar power is certainly a blessing in disguise. Remote and rural areas can now take advantage of power to initiate different development projects in their areas. Consequently, education and medicinal facilities can also be increased in these areas by the introduction of solar power.

Street lighting, general lighting systems for roads, parks, red lights, can be upgraded using more advanced technology that use energy efficient LED lights with independent electricity sources from solar power as well as solar panel roads. The U.S. Department of Energy predicted in 2009 that by the year 2030, solar power would produce almost 0.5 percent of the grid electricity generated by renewable sources. In comparison, the department said that, in 2007, solar power provided only about 0.3 percent of the electricity from renewable sources. Overall, renewable sources—including solar power—were expected to account for about 15 percent of the electricity produced for the grid in 2030 [7].

Overall, the sun gives off far more energy than we will ever need. The only limitation is our ability to convert it to electricity in a cost-effective way and environmentally friendly for the benefit of human's welfare.

IOP Conf. Series: Materials Science and Engineering **553** (2019) 012064 doi:10.1088/1757-899X/553/1/012064

Solar power is also scalable. This means that it can be deployed from industrial scale or it can be used to power a single household. When it is used on a small scale, extra electricity can be stored in a battery or fed back into the electricity grid.

Storage on batteries will reduce electricity costs, increase efficiency, and support the green economy and can also be considered as a business growth opportunity.

The advantages of batteries as storage is due to their ability to store large amounts of solar energy in a cheap and efficient manner. A study in the Journal of Sustainable Finance & Investment predicts that a combination of battery storage and renewable energy can make fossil fuels obsolete.

The breakthrough in battery technology came from Professor John Goodenough of the Cockrell School of Engineering at the University of Texas at Austin, who helped create lithium-ion batteries. Goodenough's lithium battery technology is said to have energy density three times higher than the current battery market. That means: higher energy density, extra long-lasting power, longer life, and reduced overall costs [8].

Lithium ion battery has become the basis of the huge market for cellular phones and laptop computers, and these mobile communication market continues to grow at a rapid rate, supported by the demand all over the world [9]. Even so, intensive efforts are still under way to further improve battery technology. The main target of the effort is not only the automobile industry by achieving higher energy density, but also the energy storage market supplementing environmentally friendly power source such as solar energy and wind turbine [10].

At present batteries are produced to fulfil the increasing demand as more and more mobile electronic and electric devices ranging from mobile phones to electric vehicles are entering into our life [11]. A national consortium of lithium ion battery has been started since 2016 to develop the lithium ion battery modules for solar street lamp [12,13].

To simplify our daily life, there are several things we can benefit with solar energy. Solar battery charger, calculator and watch with natural sunlight to power itself. Indoor lighting through LEDs (light emitting diodes) which are highly energy efficient, can be connected to a battery charged with solar generation system. Outdoor lighting and heating water for domestic, commercial or industrial use, water pumps, air pumps, water heaters, generators, refrigerators, air conditioners, fans, gas stoves, and many other appliances can benefit from solar energy produced power.

Consumers are not interested in intricacies of battery technology. The proliferation of battery technologies comes from widespread usage of electricity consuming devices. Past battery technology such as nickel cadmium make way for lithium because of the higher energy density it offers compared to nickel based battery. As such, due to higher energy density thus electricity capacity from lithium battery, technology vendors are now able to create handheld computers that have telephony ability we now call smartphones. A bright future in science and technology relating to advanced batteries can only be expected through continuing basic and applied research on lithium material its uses [10].

Not just consumers, purchasers of battery are also international agencies, governments and their agencies, and companies. This market will continue to expand as decision makers in these institutions become convinced of the reliability and cost-effectiveness of photovoltaics compared to the alternatives in an increasing number of applications [14].

The main applications include systems for lighting, refrigeration, radio and TV, water pumping and general village electrification, includes all applications where the individual is both purchaser and user, e.g. in a remote home or farm, caravans, boats or the leisure industry.

In June 2001, the European Council of Gothenburg (Sweden) added an environmental dimension to the Lisbon Agenda, constituting the European plan for sustainable development. Under this new framework, public policies in Europe should adopt a long-term vision to deal with issues such as ratification of the Kyoto Protocol and promotion of renewable energies. In this regard, the European Union (EU) has set a binding target for renewable energy at 20% of the EU's total energy needs by 2020 [15].

IOP Conf. Series: Materials Science and Engineering **553** (2019) 012064 doi:10.1088/1757-899X/553/1/012064

5. Conclusions

The study found very interesting findings that majority advanced countries have considered solar energy usage as part of their domestic policy towards sustainable and green economy due to its less polluted nature. The result of the current study also verify that through a proper system of solar storage and harvesting, one can reach a better life in terms of the availability, easy access, cost effective, abundant, sustainable and green.

Further, solar energy can be regarded as an abundant near infinite energy resource and is on the verge of a massive boom. Together with wind energy, it directly challenges the incumbent dominant forms of traded energy, fossil and nuclear. We are already seeing the beginnings of divestment of fossil fuelled energy by influential investors, as fears grow about stranded assets. Humanity needed alternative energy resources which the solar energy makes it possible. We need battery as energy storage and conversion of solar energy as an infinite energy to make our live better, reduce electricity costs, increase efficiency, and support the clean environment.

Researchers has come up with interesting findings of current study which includes 1) impact of solar energy usage for a better life, 2) strong positive impact of cost effectiveness has been found in study, 3) as an infinite energy, solar storage can be used to remote areas, 4) environmentally friendly as it is non-polluted energy. As for practical implication, the innovative and better storage system of solar energy is expected by many countries to cover widely energy needs. The right storage system is necessary as it can bring a better life.

References

[1] J. Jung, S. Han, and B. Kim. Digital numerical map-oriented estimation of solar energy potential for site selection of photovoltaic solar panels on national highway slopes. 2019. *Appl. Energy*, vol. 242, no. February, pp. 57–68.

[2] H. Kong, X. Jiajia, L. Liping, B. Zhenmao, and X. Jiajia. Smart Grid Privacy Protection Using Solar Energy and District Solar Heating Energy Grid Privacy Protection Using and Rechargeable Battery Rechargeable Battery Assessing the feasibility of using the heat demand-outdoor temperature function for a Liu district heat demand forecast. 2019. *Energy Procedia*, vol. 158, pp. 6170–6175

[3] G. Raina and S. Sinha. Outlook on the Indian scenario of solar energy strategies : Policies and challenges. 2019. *Energy Strateg. Rev.*, vol. 24, no. August 2018, pp. 331–341.

[4] A. Hai. Energy Research & Social Science Rethinking the social acceptance of solar energy : Exploring ' states of willingness ' in Finland. 2019. *Energy Res. Soc. Sci.*, vol. 51, no. December 2018, pp. 96–106.

[5] A. Kasaeian, F. Rajaee, and W. Yan. Osmotic Desalination by Solar Energy : A Critical Review. 2018. *Renew. Energy.*

[6] E. Kartini, M. Nakamura, M. Arai, Y. Inamura, K. Nakajima, T. Maksum, W. Honggowiranto, and T. Y. S. P. Putra. Structure and dynamics of solid electrolyte (LiI)0.3(LiPO3)0.7. 2014. *Solid State Ionics*, vol. 262, pp. 833–836.

[7] Balat, Havva, 2005. Solar Energy Potential in Turkey. *Energy Exploration & Exploitation.* Volume 23 · Number 1 · 2005 pp.61–69.

[8] C.Masquelier, A.K.Padhi, K.S.Nanjundaswamy. J.B.Goodenough : New Cathode Materials for Rechargeable Lithium Batteries: The 3-D Framework Structures $Li_3Fe_2(XO_4)_3$(X=P, As). 1998. *Journal of Solid State Chemistry*, 228-234 https://doi.org/10.1006/jssc.1997.7629

[9] Scrosati, B., & Garche, J. 2010. Lithium batteries: Status, prospects and future. *Journal of Power Sources*, *195*, 2419–2430. http://doi.org/10.1016/j.jpowsour.2009.11.048

[10] Ozawa, Kazunori, 2009. Lithium Ion Rechargeable Batteries. Weinheim: WILEY-VCH Verlag GmbH & Co.

[11] Besenhard, Jurgen O.1999**.** Handbook of battery materials. Weinheim: Wiley-VCH

[12] Kartini, E. 2015. Emerging battery research in Indonesia: The role of nuclear applications. In B. A. Cunningham, O. Catherine, & G. Shohini (Eds.), *AIP Proceeding "5th International Conference on Women in Physics, Waterloo, Canada, 5-8 August 2015"*. (Vol. 1697, pp.

100001–1). AIP Publishing LLC. http://doi.org/10.1063/1.4937703
[13] Kartini, E., Purwanto, A., Sudaryanto, S., Honggowiranto, H., & Rofika, R. N. S. 2018. Indonesian consortium of lithium ion battery for solar street lamp. In *IOP Conf. Series: Materials Science and Engineering* (Vol. 432, pp. 012063–1). http://doi.org/10.1088/1757-899X/432/1/012063
[14] Laughton, Michael A. 2003. Renewable Energy Sources. Watt Committee Report Number 22. London: *Elsevier Applied Science*, Taylor & Francis Books, Inc.
[15] Vincenzi, Marco and Deniz Ozabaci, 2017. The Effect of Public Policies on Inducing Technological Change in Solar Energy. *Agricultural and Resource Economics Review* 46/1 44–72 http://creativecommons.org/licenses/by/4.0/

Grading of Lithium Ion Battery Module for Public street lighting

E Kartini[1], S Sudaryanto[1], W Honggowiranto[1], A Sudjatno[1], A Fadjri[2], I Marto[2], R N S Rofika[3]
[1]Science and Technology Center for Advanced Materials, National Nuclear Energy Agency, Puspiptek Serpong, South Tangerang 15314, Banten, Indonesia
[2]Engineering and Development Center, ASTRA Autopart, Cikarang, Indonesia
[3]Physics Department, Faculty of Mathematics and Natural Sciences, Bandung Institute of Technology, Bandung, Jawa Barat, Indonesia

kartini@batan.go.id

Abstract. A module of lithium ion battery has been constructed to replace ion battery for public street lighting. The module was designed to deliver a power of minimum~120 Wh for running 10 Watt solar street llighting, with a solar panel of 80 Wp. The module consisted of 40 cylinder cells 18650 of $LiFePO_4$. Before designing the module, all the cells have to be formatted, graded and sorted out, to obtain an optimum results. The charge-discharge testing and internal resistance were measured to every single cell using a battery analyzer. The cells grading,sorting and grouping were consecutively done to obtain an optimum LIB module. The results showed each cylinder cell delivering discharge capacity, voltage and internal resistance of ~1.2-1.4 Ah, ~ 3.2-3.3 V, and 50-70 Ω, respectively. The cells were arranged into 4 serial and 10 parallel, to produce LIB module with the power of ~130-140 Wh, which is higher than expected.The LIB module made in Indonesia, with high local content can run the public street lighting and replace the conventional Lead Acid battery.

Keywords: *grading, module battery, lithium ion battery, solar panel, public street lighting*

1. Introduction

The Indonesian government has targeted the use of 30 % mixed energy in 2030 that mostly from the renewable energy resources, such as geothermal, wind, solar, hydro, biomass and nuclear power plant [1]. In order to store the energy form the renewable sources, the energy storages are required and very important aspects [2,3]. The combination of a pollution-free, battery energy storage system and a'clean' photovoltaic energy supply results in an efficient peak-power supply without harmful environmental effects. At the same tune, the economics are greatly unproved as the expensive solar energy is used to supply electricity at peak-power rates. In certain cases, a primary battery, e g, aluminum/a~, may be used as an emergency power supply. Reliability and flexibility are intrinsic characteristics of electrochemical energy storage. The electrical features of a solar generator have to be taken into account when charging the battery. The particular requirements for both the charge and the operation of the lead/acid battery must be considered in order to obtain the most efficient, i.e, the most economic, use of solar energy [4]. The great interest in load-levelling systems, using lead/acid batteries for energy storage, has already resulted in several very large storage units in storing energy to meet the peak-power requirements. Solar energy is certainly very attractive for load-levelling applications. However, the used of conventional lead acid batteries are limited due to short life cycles, non-rechargeable, and not friendly to the environment, due to toxic lead materials. Therefore, the replacement of lead acid with the lithium ion battery is important and one of the scientific breakthroughs. The use of LIB module will have more advantageous in comparison with the conventional lead acid battery [5]. The LIB module is much lighter, smaller size, long life cycles, and higher energy density.Thus requires an extensive investigation for each application [1,2,6,7].

Solar energy has to be consumed locally in order to avoid conversion and distribution losses. The same advantages are present in small, stand-alone installations. The use of battery storage enables the use of smaller photovoltaic generators as the consumption peaks are levelled out by the battery. The larger the batteries, the less generators will have to be operated [8,9]. The installation of the solar systems feeding into the street lights is one of the most interest, due to the large applications in isolated area, housing or along the highway. It is also simple and no electrical power supply needed. The solar energy itself can be run unlimited and of course renewable. Thus the research on application of solar panel for the public street lights and household has becoming a significant issue.

The aim of this study is to design a module of lithium ion batteries to turn on the public street lighting. The LIB module will be designed and constructed from the LIB cells that previously investigated [1,10]. The process of grading of lithium ion cells will be discussed in order to optimise the system configuration that will present the closest match between generator, storage, and user. Finally performance of the module will be characterized using a charge discharge instruments.

2. Experimental Detail

There are several steps on obtaining the module of lithium ion battery (LIB), namely grading, sorting, grouping, welding and battery module testing. These series of experimental procedures have to be conducted to obtain an optimum module for Lithium Ion Battery. The size and capacity of the LIB module was first determined from the targeted output, such as the minimum power to run the solar street lamp. Design of the modules is decided by arranging numbers of cylinder cells 18650. The detail of cylindrical battery 18650 for this research has been described previously [1,10,11]. The cell consists of $LiFePO_4$, polymer, $LiPf_6$ and C, as cathode, separator, electrolyte and anode materials, respectively. The cathode was coated on the Al-foil as positive current collector, while the anode was coated on the Cu-foil as the negative current collector. The electrolyte was pour into the cylindrical cell in the glove-box. Detail of cylindrical cell assembly has been described elsewhere [10]. The cell formation was done for each cell using a battery analyzer, in order to obtain the discharge capacity, voltage, and internal resistance. The process is called grading of cells. All experiments were done at Integrated Battery Laboratory – BATAN Indonesia. The measurement of internal resistance, capacity, and voltage were done by using NEWARE battery testing system-5V6A.

IOP Conf. Series: Materials Science and Engineering **553** (2019) 012065 doi:10.1088/1757-899X/553/1/012065

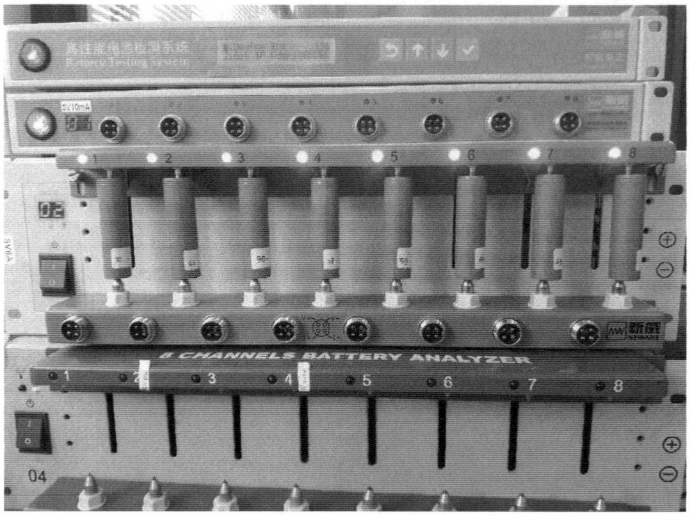

Figure 1. Grading of cylindrical batteries.

2.1 Grading

2.1.1 Measurement of battery capacity and voltage.

Figure 1 shows process of testing battery from cylindrical cell. The charge discharge measurements were performed using NEWARE battery testing system-5V6A. As shows in Table 1, the charging process was performed with two steps, these are the constant current and constant volatge (CCCV_Chg) of 0.28 A, which is equal to 0.2 C-rate until reaching Vmax= 3.65 V. For discharge process, the battery was drained by current of 0.28 A until battery voltage drops to 2.6 V. The voltage measurement was done before and after cycling. For the batteries using LFP cathode, the voltage is around 3.2 volt. Using the charge-discharge battery analyzer the data is more accurate.

Table 1. Set of charge discharge for batteries grading.

Step ID	Step Name	Time (h:min:s.ms)	Voltage (V)	Current (A)	End Current (A)
1	Rest	0:05:00.000			
2	CC_DChg		2.6000	0.2800000	
3	Rest	0:05:00.000			
4	CCCV_Chg		3.6500	0.2800000	0.0280
5	Cycle	Start from step ID: 1	Cycle Count: 5		
6	End				

2.1.2 Measurement of internal resistance.

Internal resistance (iR) is very important variable to be measured before arranging a module. Without testing, selecting and sorting this item, the module will not properly function. The iR was determined by dischraging a battery in 1 second. Current and voltage limit are the same with the charge discharge process as explained above. The difference during 1 second discharge step, the current used was 1.4 A which was equal to 1 C-rate. The set of the process is shown in Table 2.

Table 2. Set of determination internal resistance (iR).

Step ID	Step Name	Time (h:min:s.ms)	Voltage (V)	Current (A)	End Current (A)
1	Rest	0:02:10.000			
2	CC_DChg		2.6000	0.2800000	
3	CC_DChg	0:00:01.000		1.4000000	
4	Rest	0:05:00.000			
5	CCCV_Chg		3.6500	0.2800000	0.0280
6	End				

iR can be calculated by equation (1):

$$iR = \left|\frac{\Delta V}{\Delta I}\right| = \left|\frac{V_t - V_0}{I_t - I_0}\right| \quad (1)$$

where V_0 and V_t are start and end voltage when the battery was discharging at in a second. I_0 and I_t are start and end current (see Table 2).

2.2. Selection, Sorting and Grouping

The selection process was done by choosing batteries with internal resistance value of less than 100 mΩ and sorting them from the smallest values. The batteries should have capacity at least 1 Ah, and for the batteries with capacity less than 1 Ah, will be rejected. Similar to the batteries with the voltage are less than 3.2 Volt, will also be rejected. The battery module can be customized and used for energy storage on Public street lighting, electric vehicle, and electric bike, etc. Public street lighting need 12 volt battery. So that we need 4 seri connection to generate it.

3. Results and Discussions

3.1 Charge discharge Module battery

During formation, the batteries were charged and discharged for 5 cycles. Battery capacity data was taken from the fifth cycle is shown in Table 3. The result showed that the capacity of battery was 1.4797 Ah, at 5th cycle, identifying that the battery performance was good. The discharge capacity is depicted in Figure 2.

Table 3. Charge-discharge Capacity of cylinder cell 18650 at 5th cycle.

Cycle ID	Step	Time (h:min:s.ms)	Cap_Chg(Ah)
5	CCCV_Chg	5:22:44.700	1.4797
	Rest	0:05:00.000	0.0000
	CC_DChg	5:17:19.800	1.4750
	Rest	0:05:00.000	0.0000

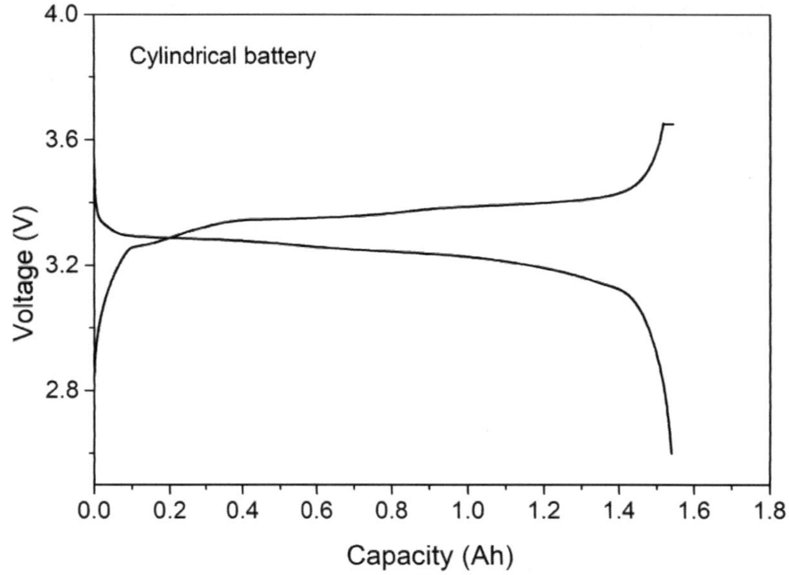

Figure 2. Charge Discharge curve one of cylindrical battery 18650 at 5th cycle.

All the cells were charged and discharge for 5 cycles, then were selected that fulfill the requirements. Furthermore, the internal resistance were measured and calculated for each cell with the measurement data as listed in Table 4. By calculating the data of start and end voltage, the iR was determined using equation (1). In this case, for example, hence I_0 and I_t are 0.28 A and 1.4 A, respectively, so ΔI is 1.12 A. The voltage different is marked in Table 4, and resulted ΔV ~0.0543 V [12]. By using equation (1) the internal resistance for all the battery can be calculated, and some of the results are shown in Table 5.

Table 4. Measurement data iR of battery.

Step ID	Step Name	Time (h:min:s.ms)	Start Voltage (V)	End Voltage (V)
1	Rest	0:05:00.000	3.3293	3.3293
2	CC_DChg	4:56:20.700	3.3175	2.5998
3	Rest	0:02:02.900	2.6140	2.7625
4	CC_DChg	0:00:01.000	2.6866	2.6323
5	Rest	0:05:00.000	2.6931	2.8134
6	CCCV_Chg	5:36:21.900	2.8270	3.6495

Table 5. Grading data of the batteries.

Cell	iR a (mΩ)	Voltage a (V)	DCHG Cap (Ah)	RCHG Cap (Ah)	Voltage t (V)	iR t (mΩ)
1	21.6	3.33	1.5632	1.5725	3.34	21.7
2	22.6	3.33	1.3665	1.3981	3.34	23.3
3	22.1	3.33	1.5122	1.5613	3.33	24.4
4	23.0	3.28	1.3842	1.4602	3.29	24.8
5	25.0	3.32	1.4571	1.4728	3.32	25.7

Table 5 shown how the cells have been selected, sorted and grouped into a parallel or serial. The ten paralel and four seri connection has been chosen. The battery module design is shown in Figure 5. Batteries are arranged in parallel in order of iR values, as well the seri connection listed in Table 5. All the four series are grouped into four parallel to increase the capacity. From the calculation the expected module output 14.48 Ah and 13.33 V. The monitoring system will be designed and constructed in parallel in order to evaluate and analyze the results during the operation. An integrated system will be made to support this, and called as a smart public solar lighting [12].

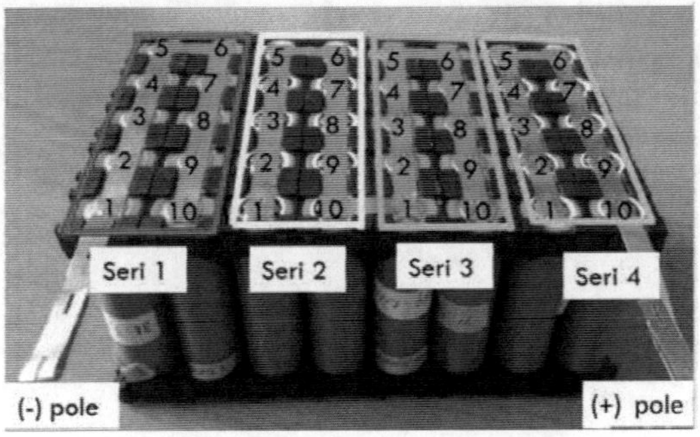

Figure 3. Battery module arrangement and design of welding in top view.

4. Conclusion

This study has focused on developing the module of lithium ion battery (LIB) for public solar street lighting. The interest of using LIB module is due to more advantageous in comparison with the conventional lead acid batter. The "battery module" is usually only used associated with high-power batteries, consisting of an assembly of cell packages, including safety features, as well as a battery management system (BMS), and a base plate or housing. The cells are connected in parallel or in series, depending on the capacity or voltage needs for an application. Parallel configuration adds capacity, leaving the voltage constant, serial configuration adds voltage, leaving the capacity constant, respectively.

It has been shown here, that the battery grading are very important process before designing a module. Through grading will determine the discharge capacity, the voltage and the internal resistance of each cell. Then the sorting of those only cells that fulfill the requirements another important aspect before combining into parallel and serial. The module of LIB that consists of 40 cells parallel and serial has been successfully constructed, and exhibited the power of 140 Wh, with good ability and efficiency. This result has fulfilled the requirement to turn on the public solar street light, and open the challenge of replacing the lead acid battery overall. Finally, it is concluded that the Indonesia national battery consortium has successfully proven to produce a module lithium ion battery with high local content and made by Indonesian researchers.

Acknowledgment

This work is financially supported by the Ministry Research Technology and Higher Education through the Research Grant of National Innovation System Consortium (INSINAS) with the contract no. 01/INS-1/PPK/E2/2018 and IRPK-2018-148.

References

[1] E. Kartini, A. Purwanto, S. Sudaryanto, H. Honggowiranto, R.N.S. Rofika, in:, IOP Conf. Ser. Mater. Sci. Eng., 2018, pp. 012063–1. doi:10.1088/1757-899X/432/1/012063

[2] E. Kartini, in:, B.A. Cunningham, O. Catherine, G. Shohini (Eds.), AIP Proceeding "5th Int. Conf. Women Physics, Waterloo, Canada, 5-8 August 2015"., AIP Publishing LLC, 2015, pp. 100001–1. doi: 10.1063/1.4937703

[3] D. Henkensmeier, D.K. Kim, S. Kim, 15 (2018) 1–21.
[4] X. Guo, Y. Zhang, D. Qin, Y. Luo, D. Li, Y. Pang, Q. Meng, J. Power Sources 195 (2010) 7684–7690.
[5] M. Brand, S. Gläser, J. Geder, S. Menacher, S. Obpacher, A. Jossen, D. Quinger, (2013) 1–9.
[6] M. Lowe, S. Tokuoka, T. Trigg, G. Gereffi, (2010).
[7] E. Kartini, M. Manawan, M.F. Collins, M. Avdeev, Phys. B Phys. Condens. Matter 551 (2018) 320–326.
[8] F. Steffens, J. Power Sources 35 (2000) 1–20.
[9] J. Ordoñez, E.J. Gago, A. Girard, Renew. Sustain. Energy Rev. 60 (2016) 195–205.
[10] W. Honggowiranto, Sudaryanto, E. Kartini, A. Purwanto, AIP Conf.Proc. 6th Nanosci. Nanotechnol. Symp. 1710 (2016) 1–7. doi: 10.1063/1.4941511
[11] J. Liu, Z. Wang, G. Zhang, Y. Liu, A. Yu, Int. J. Electrochem. Sci., 8 (2013) 2378–2387.
[12] A. Mulyono, T. Mustika, H. Sulaikan, E. Kartini, in:, IOP Conf. Ser. Mater. Sci. Eng., 2018, pp. 1–8. doi:10.1088/1757-899X/432/1/012057

IOP Conf. Series: Materials Science and Engineering **553** (2019) 012066 doi:10.1088/1757-899X/553/1/012066

Remaining Life and Corrosion Rate Prediction of Well Casing Material in the CO_2 Environment

Muhamad Yudi Masduky Sholihin[1]*, Hendry Saputra Kusyanto[2], Roni Hadi Wijaya[2], Wahyu Haryadi[3]

[1] Department of Mechanical Engineering, Universitas Pancasila, Jl. Srengseng Sawah, Lenteng Agung, Jakarta 12640, Indonesia

[2] Material Science Study, Department of Physic, Universitas Indonesia, Depok 16424, West Java, Indonesia

y.masduky@ui.ac.id

Abstract. Prediction of CO_2 corrosion is required for predicting remaining life and determination of the required material selection. This paper give prediction of corrosion rate and remaining life of casing in the North Aceh well which is operated almost 14 years. Simulation method application is appropriate with the condition of North Aceh field since the well fluids contain high CO_2 gas (35 – 37%) and small amount of H_2S gas (0.016%). Based on historical data North Aceh well's after production almost 14 years, it is possible that production packer leaks and degradation of cement could cause formation gas migrate through micro channelling inside cement. The CO_2 gas which is accumulate in casing could cause CO_2 corrosion. The corrosion rate calculation based on temperature 141°C, pressure 103.4 Barg and CO_2 contain 37% is 0.871 mm/year. After calculating the thickness loss, we found thickness loss of casing K55 is 12.741 mm. meanwhile thickness of K55 casing 10.92 mm, it means the casing already leak before year 14th. Based on simulation and calculation the North Aceh well is assumed to have tubing and packer leaks in annulus #0 and channeling on annulus #1 so that the well is not recommended for re-production because of the communication between the annulus and the loss of the well barrier that can cause wellhead leakage. This condition is very dangerous because the well fluid is flammable and can cause pollution in the sea.

Keywords : Remaining life, prediction, corrosion

1. Introduction

Since the 40s, oil and gas industry has always faced corrosion problems. Corrosion can affect the main important equipment which is design in metal. Because of this deterioration, the carbon steel corrosion mechanism has been studied to reduce the contamination in the oil industry. Many documents associated to carbon steel corrosion have been published, to relate the carbon steel contamination mechanisms and different interpretation to resolve the problem of corrosion have been reported. Several research and studies has shown the cause of corrosion in petroleum industry to be the existence of hydrogen sulphide, carbon dioxide, oxygen and water in the reservoir. Another material that is affected by corrosion is the surface facility and wells material. Corrosion damage can be internal or external. Factors that can influence internal corrosion in pipelines and wells can be the surface of the steel, temperature, flow velocity, water chemistry, oil or water wetting, hydrogen sulphide (H2S) and carbon dioxide (CO2) content.

Well structure consists of conductor, surface casing, tubing, casing shore and liner head with varying diameters according to the design as shown in Figure 1. The well is drilled with differently sized drill bits that get smaller as the well gets deeper. Each point when a casing diameter changes, the joint need to be sealed, cements is pushed between casings and bedrock. It protects the metal wall from gas pressure and from gas leaking up the outside of well pipe.

In this study, the well that located in the offshore of the North Sumatra Basin in the Straits of Malacca, covering an area of 77 Km^2 and 6287 ft depth with directional well type. The well has been shut in since 2004 due to production facility problem.

Based on historical data the well after production almost 14 years, it produced gas and oil that contains 37% CO_2 and 0.016% H_2S. In order to reproduce, the integrity of the well need to be assess because high amount of CO_2 in fluid can cause CO_2 corrosion in tubing and casing of the well [1-3]. It is also possible that the production packer leaks and degradation of cement caused the formation gas to migrate through micro channeling inside cement that can cause cracking [4,5].
In this paper we study the corrosion that occurs in old wells in the North Sumatra Basin using Norshok method which will be used as recommendation if it will be operated again for production.

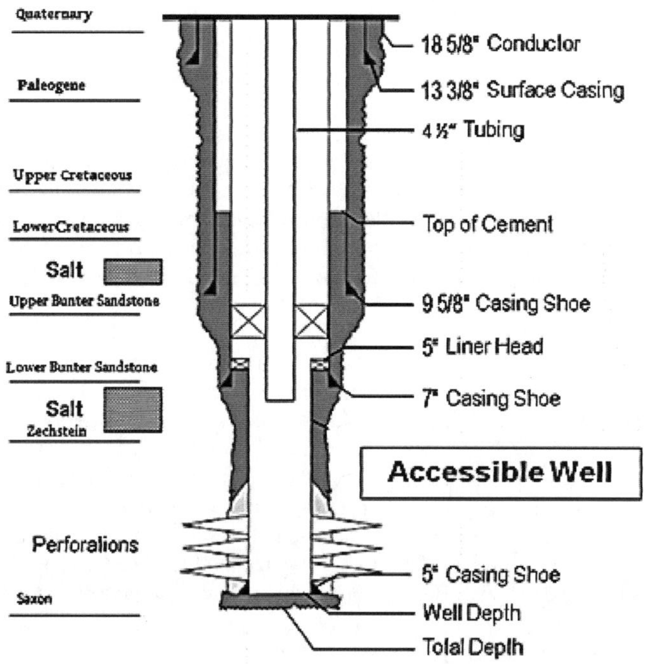

Figure 1. Operational Well Structure [6].

2. Experimental Method

Studied the different corrosion prediction models to evaluate field data by running different corrosion prediction models for a set of cases taken from the corrosion field database. It varied considerably from case to case which models were most successful in their prediction, and it was not possible to declare one or two models as better than the others. An evaluation of any model against field data can be strongly dependent on the selection of field data used for the evaluation or validation and the validity and accuracy of the field data. It is important to understand the limitations and uncertainties in the corrosion prediction models [6]. The Norsok M-506 model is an empirical model developed by the Norwegian oil companies Statoil, Norsok Hydro and Saga Petroleum for prediction of CO_2 corrosion. The model is fitted to a large amount of laboratory data. The model takes larger account for the effect of protective. In order to be able to run the different corrosion prediction models the following data

IOP Conf. Series: Materials Science and Engineering **553** (2019) 012066 doi:10.1088/1757-899X/553/1/012066

from the field cases were specified as a minimum: Temperature inlet and outlet, Pressure inlet and outlet, CO_2 mole %, Bicarbonate, acetate and calcium content in the water (preferably full ionic water composition), production rates (Gas, water, and oil), Pipe inner diameter, Bubble point pressure or pressure in last separator stage for full liquid systems Corrosion films at higher temperature and higher pH than several of the other models, as it is fitted to high temperature data up to 150 °C [7].

Provide the main limitations for Norsok M-506 model are the following: The model cannot be used for ratio between CO_2 and H_2S partial pressure less than 20 and when the H_2S partial pressure is above 0.5 bars. The model cannot be used for total organic acid content above 100 ppm when the partial pressure of CO_2 is less than 0.5 bars. The model cannot be used for prediction of corrosion rates for pH-stabilised systems. The model cannot be used for prediction of top-of-line corrosion. The model is a point model and change in pH along a pipeline due to corrosion has to be determined separately

Corrosion in pipeline, casing, and tubing can always be modelled, but the corrosion model should consider the limitation when modelling corrosion. The model should be part of an overall corrosion risk assessment methodology, consider existing corrosion rate data and experience with similar production fluids when it is available to backup corrosion modelling. When there is uncertainty, consider accelerated corrosion testing to support corrosion assessment [8]. This simulation is focusing to predict corrosion rate and remaining life of well tubing and casing using Norsok M-506 method as shown in Figure 2 and the basic input parameter in Table 1.

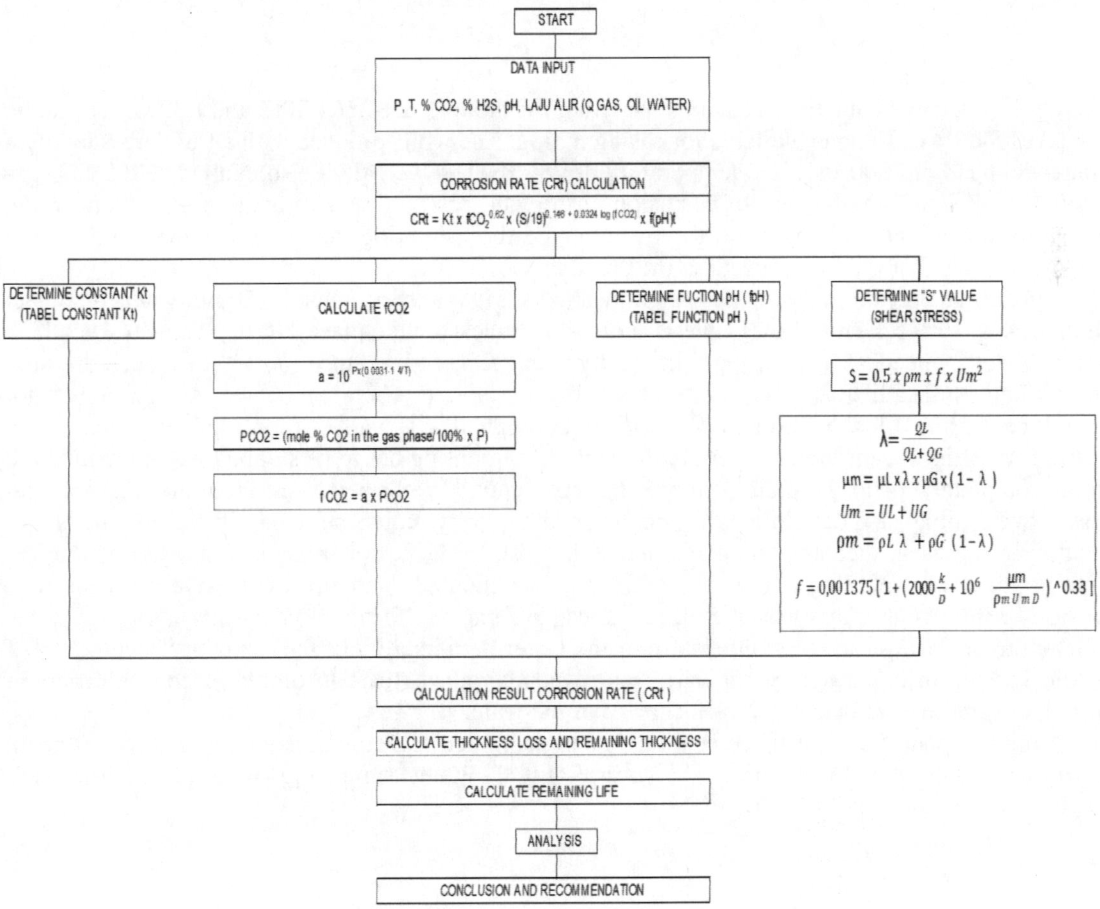

Figure 2. Experimental And Prediction Method Using Norsok M-506.

Table 1. Basic Input Parameter Norsok M-506.

Parameter	Units	Range	Comments
Temperature	°C	20 to 150	
	°F	68 to 302	
Total pressure	bar	1 to 1000	
	psi	14.5 to 14500	
Total mass flow	kmole/h	10^{-3} to 10^{6}	Only relevant when CO_2 is given in kmole/h
CO_2 Fugacity in the gas phase	Bar	0,1 to 10	The CO_2 partial pressure shall be
	Psi	1,45 to 145	≤ the total pressure. The allowed
	Mole %	Variable	ranges of mole % and kmole/h
	Kmole/h	Variable	CO_2 are dependent on the total pressure.
Wall shear stress	Pa	1 to 150	Can be calculated by use of other input parameters.
pH		3,5 to 6,5	Can be calculated by use of other input parameters.
Glycol concentration	Weight %	0 to 100	
Inhibitor efficiency	%	0 to 100	See NORSOK M-001

3. Result and discussion

Based on the available data, the casing material used for the well are 25Cr, K55 and L80 which can be seen in API 5 CT as a standard suitable for casing tubing. The wells produce with a flow pressure at an average wellhead of 1500 Psig and temperature of 286 °F. The well fluid also contains 37% CO_2 gas and 0.016% H_2S gas which cause of corrosion in the well casing. The degradation of cement in the annulus allows the gas coming from the reservoir of the migrating well to enter the annulus and accumulate so that corrosion can occur in the annulus.

The most fundamental in casing corrosion mechanism is that there must be liquid water in contact with the casing surface. This allows aqueous corrosion attack to take place. If the cement sheath is intact and well-bonded to the casing surface then there is little chance for water to reach the steel surface. Cement sheath quality is rarely so reliably perfect. In some cases well designs have not required cementing of the casing over critical geological intervals. More typically, cementing operations face significant losses in certain intervals so cementing has to be stopped without filling the annulus completely, or to the planned top of cement depth. Well operations, including thermal and pressure cycles, may also deteriorate the cement sheath quality with time. Under these circumstances ground water or subsurface aquifers may contact the casing. If the water has a composition which is aggressive to steel, including one or more constituent which can support a corrosive reaction, then corrosion of the casing may then take place. These are varying factor of corrosion at casing: water contact with the casing surface, temperature of the water, temperature of the casing, concentration of cathodic species in the water (particularly dissolved oxygen or carbon dioxide), concentration of chloride ions in the water, ease of replenishment of the water [9].

Poor mud displacement during cementing was high-lighted as being a major cause of poor sheath integrity and loss of zonal isolation as shown in Figure 3. Water channeling along a bad cement job with potential for corrosion at the casing surface.

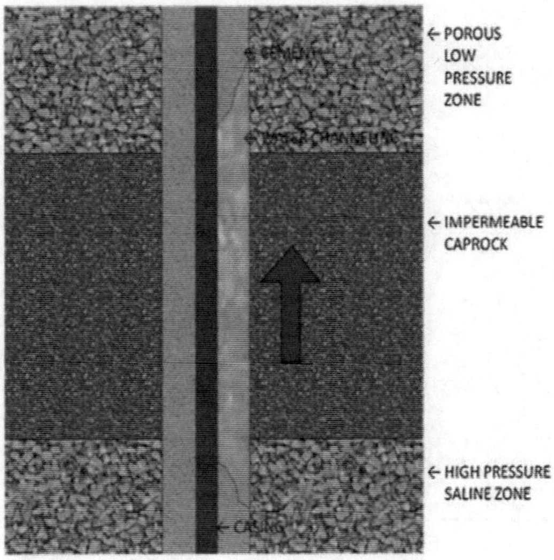

Figure 3. Water Channelling Along The Casing [10].

When CO_2 is dissolved in water, it forms carbonic acid (HCO-3) which reacts with compounds in hydrated. Portland cement matrix such as calcium silicate gel (C–S–H) and calcium hydroxyl ($Ca(OH)_2$) [6]. The major reaction products are calcium carbonate and amorphous silica gel. Leaching of the resulting $CaCO_3$ and Ca $(HCO_3)_2$, leads to rapid reduction in strength, varied permeability, and corrosion on the casing [11]. The process can be described with the following reactions:

$$CO_2 + H_2O \leftrightarrow H_2CO_3 \tag{1}$$

$$H_2CO_3 + Ca(OH) \longrightarrow CaCO_3 + 2H_2O \tag{2}$$

$$C - S - H + H_2CO_3 \longrightarrow CaCO_3 + \text{amorphous silica} \tag{3}$$

$$CaCO_3 + H_2CO_3 \longrightarrow Ca(HCO_3)_2 \tag{4}$$

$$Ca(HCO_3)_2 + Ca(OH)_2 \longrightarrow 2CaCO_3 + 2H_2O \tag{5}$$

In the initial reaction, carbonic acid is built out of dissolved carbon dioxide and water. Then the acid reacts with calcium hydroxide and the calcium–silicate–hydrate gels and forms calcium carbonate. The calcium carbonate continues to react with the carbonic acid and builds water-soluble calcium bicarbonate. The last reaction is very critical, because the formation of water allows dissolution of more CO_2. Micro annulus further indicated that CO_2 migrated along the cement/cap rock and cement/casing interfaces, in the latter case speculating that CO_2 infiltration may have occurred due to migration up the casing wall, through casing threads or at particular points of corrosion. Figure 4 shows CO_2 corrosion caused by micro annulus and channeling. Microannulus further indicated that CO_2 migrated along the cement/cap rock and cement/casing interfaces, in the latter case speculating that CO_2 infiltration may have occurred due to migration up the casing wall, through casing threads or at particular points of corrosion.

Based on data well parameter in Table 2, Kt factor in Table 3 and f(pH) could be determined. A constant Kt has been generated, is found by a linear extrapolation between the calculated corrosion rate at the temperature above and below the desired temperature. Kt factor based on temperature table is shown below in Table 4. Kt is a temperature dependant constant. It increases corrosion rate from 5^0 to 60^0 C and then decreased up to 150 °C.

IOP Conf. Series: Materials Science and Engineering **553** (2019) 012066 doi:10.1088/1757-899X/553/1/012066

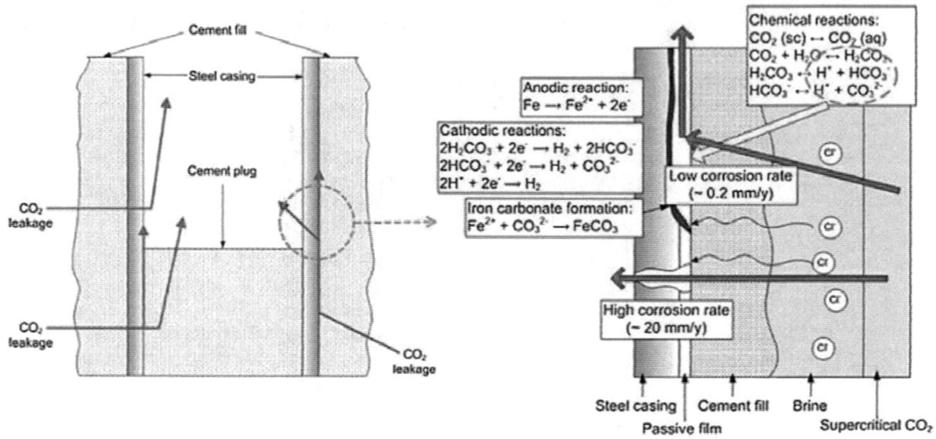

Figure 4. CO_2 Corrosion Process By Microannulus and Channelling [9].

Table 2. Well Parameter for Calculating Corrosion Rate.

Well	Parameter	Value	Units	Range	Comments
	Temperature	286	°F	68 to 302	
		141.1	°C	20 to 150	
	Pressure	1500	Psi	14.5 to 14500	
L3		103.4	Bar	1 to 1000	
	% mole CO_2	37	%		
	pH	5		3.5 to 6.5	
	Glycol Concentration		%	0 – 100	Not available
	Inhibitor efficiency		%	0 – 100	Not available

Table 3. Kt factor Based on Temperature.

Temperature °C	Kt
5	0.42
15	1.59
20	4.762
40	8.927
60	10.695
80	9.949
90	6.250
120	7.770
150	5.203

Temperature of the well is 1410 °C, Kt factor based on extrapolated between 1200 °C and 1500 °C is 5.973. Determine f(pH) based on is f(pH) 0.037. The effect of temperature, which is well and 1500 °C is 5.973. Determine f(pH) based on Table 4 is f(pH) 0.037. The effect of temperature, which is well known to accelerate all processes involved in CO_2 corrosion, has been clarified. At low pH when protective scales do not form, temperature accelerates corrosion and vice versa when conditions are favorable for iron carbonate to form, temperature leads to faster precipitation and often to lower corrosion rates. The effect of pH on CO_2 corrosion is well established and successfully modelled, at low pH typical for condensed water will accelerate corrosion rate. In almost all cases higher pH leads to lower corrosion rates.

Table 4. f(pH) based on temperature and pH range.

Temperature (°C)	pH	F(pH)
5	$3.5 \le pH \le 4.6$	f(pH) = 2.0676 – (0.2309 x pH)
5	$4.6 \le pH \le 6.5$	f(pH) = 4.342 – (1.051 x pH) + (0.0706 x pH^2)
15	$3.5 \le pH \le 4.6$	f(pH) = 2.0676 – (0.2309 x pH)
15	$4.6 \le pH \le 6.5$	f(pH) = 4.986 – (1.191 x pH) + (0.0706 x pH^2)
20	$3.5 \le pH \le 4.6$	f(pH) = 2.0676 – (0.2309 x pH)
20	$4.6 \le pH \le 6.5$	f(pH) = 5.1885 – (1.2353 x pH) + (0.0706 x pH^2)
40	$3.5 \le pH \le 4.6$	f(pH) = 2.0676 – (0.2309 x pH)
40	$4.6 \le pH \le 6.5$	f(pH) = 5.1885 – (1.2353 x pH) + (0.0706 x pH^2)
60	$3.5 \le pH \le 4.6$	f(pH) = 1.836 – (0.1818 x pH)
60	$4.6 \le pH \le 6.5$	f(pH) = 15.444 – (6.1291 x pH) + (0.8204 x pH^2) - (0.0371 x pH^2)
80	$3.5 \le pH \le 4.6$	f(pH) = 2.6727 – (0.3636 x pH)
80	$4.6 \le pH \le 6.5$	f(pH) = 331.68 x $e^{(-1.2618 \text{ x pH})}$
90	$3.5 \le pH \le 4.57$	f(pH) = 3.1355 – (0.4673 x pH)
90	$4.57 \le pH \le 5.62$	f(pH) = 21254 x $e^{(-2.1811 \text{ x pH})}$
90	$5.62 \le pH \le 6.5$	f(pH) = 0.4014 – (0.0538 x pH)
120	$3.5 \le pH \le 4.3$	f(pH) = 1.5375 – (0.125 x pH)
120	$4.3 \le pH \le 5$	f(pH) = 5.9757 – (1.157 x pH)
120	$5 \le pH \le 6.5$	f(pH) = 0.546125 – (0.071225 x pH)
150	$3.5 \le pH \le 3.8$	f(pH) = 1
150	$3.8 \le pH \le 5$	f(pH) = 17.634 – (7.0945 x pH) + (0.715 x pH^2)
150	$5 \le pH \le 6.5$	f(pH) = 0.037

The value of f(CO_2) based on formula:

$$a = 10^{\,Px(0.0031-1.4/T)} \quad P \le 250 \text{ Bar} \tag{6}$$

where; a is fugacity coefficient, P = Pressure (Bar), T = Temperature (°C),
The CO_2 partial pressure is found by

$$PCO_2 = (\text{mole \% } CO_2 \text{ in the gas phase}/100\% \text{ x P}) \tag{7}$$

Where mole % CO_2 is 37% and P = 103.4 B, the result is 38.26
The real CO_2 pressure can then be expressed as:

$$fCO_2 = a \text{ x } PCO_2 \tag{8}$$

where fCO_2 is CO_2 pressure, a is fugacity coefficient and PCO_2 is CO_2 partial pressure
With a = 0.936 and PCO_2 = 38.26, the result of fCO_2 is 35.802

Fugacity of CO_2 decreases with increasing temperature. It decrease with distance due to stronger dependency on system pressure (P) which decreases along the pipeline with assuming partial pressure (PCO_2) is constant. In the case of scale-free CO_2 corrosion, an increase of CO_2 partial pressure PCO_2

typically leads to an increase in the corrosion rate. At a high pH, higher PCO_2 leads to an increase in bicarbonate and carbonate ion concentration and a higher supersaturation, which accelerates precipitation and scale formation [12]. Mean wall shear stress in straight pipe sections is used. Shear stress formula is:

$$S = 0.5 \text{ x } \rho m \text{ x } f \text{ x } Um^2 \quad (9)$$

$$\lambda = QL / (QL + QG) \quad (10)$$

$$\mu m = \mu L x \; \lambda \text{ x } \mu G \text{ x } (1- \lambda) \quad (11)$$

$$Um = UL + UG \quad (12)$$

$$\rho m = \rho L \, . \, \lambda + \rho G \; (1- \lambda) \quad (13)$$

$$f = 0.001375 \; [1 + (2000 \; (k/D) + 10^6 \text{ x } (\mu m / (\rho m.\mu m.D)) \text{ ^}0.33] \quad (14)$$

The result of shear stress is S = 0.55 with Q_{gas}: 4.2906 MMSCFD and Q_{oil} : 4227 BBLD. Wall shear stress increases with increasing of viscosity and density which in turn decrease with increasing of temperature [13]. Wall shear stress can be calculated from production rates and pipe diameter. The effect of flow on CO_2 corrosion still remains a contentious area in the predictive processes. The flow parameter currently favored for determining the effect of velocity on corrosion rate and scale and inhibitor film formation/stability is liquid shear stress at the pipe wall [14]. The tangential force produced by the higher velocity flow of the liquid may hinder the formation of the protective film on the surface of the metal, in that case, the corrosion rate intensified [15]. The corrosion rate based on temperature 141 °C, pressure 103.4 Barg and CO_2 contain 37% by Norsok M-506 is 0.871 mm/year.

$$CRt = Kt \text{ x } fCO_2^{\,(0.62)} \text{ x } (S/19)^{0.146 + 0.0324 \log (fCO_2)} \text{ x } f(pH)t \quad (15)$$

Well has production casing K55 type. The thickness is 10.92 mm. The well had been produced since 8 November 2004. Based on calculation we found that corrosion rate is 0.871 mm/year so the thickness loss after 14 years is 12.741 mm, meanwhile thickness of K55 casing 10.92 mm, it means he casing already leak before year 14th.

The complex interaction of multiple factors that may cause production casing leaks is most likely: cement sealing performance, age of well, production casing thickness, CO_2 content in the indigenous aquifer. Assessment of well integrity for both operational and abandoned wells can be performed in different ways. For operational wells, different tests can be implemented with varyin accuracy. However, no logging tool is available which can explicitly detect the presence and extent of corrosion in cement. A commonly used one is CBL complimented with Variable Density Log (VDL) to identify different defects such as channeling, micro-annulus and deboning.

4. Conclusion

Norsok M-506 method application is appropriate with the condition of the well since the well fluids contain high CO_2 gas (35 – 37%) and small amount of H_2S gas (0.016%). Based on historical data well after Shut in almost 14 years it is possible that production packer leaks and degradation of cement could cause formation gas to migrate through micro channeling inside cement. The CO_2 gas which is accumulate in casing could cause CO_2 corrosion. The corrosion rate based on temperature 141°C, pressure 103.4 Barg and CO_2 contain 37% by Norsok M-506 is 0.871 mm/year. After calculating the thickness loss, we found thickness loss of casing K55 is 12.741 mm. meanwhile thickness of K55 casing 10.92 mm, it means the casing already leak before year 14th. Based on simulation and calculation the North Aceh well is assumed to have tubing and packer leaks in annulus #0 and channeling on annulus #1 so that the well is not recommended for re-production because of the communication between the annulus and the loss of the well barrier that can cause wellhead leakage.

This condition is very dangerous because the well fluid is flammable and can cause pollution in the sea. Assessment of well integrity for both operational and abandoned wells can be performed in different ways. For operational wells, different tests can be implemented with varying accuracy. However, no logging tool is available which can explicitly detect the presence and extent of corrosion in cement. A commonly used one is CBL complimented with Variable Density Log (VDL) to identify different defects such as channeling, micro-annulus and deboning.

References

[1] S. Azuma, H. Kato, Y. Yamashita, K. Miyashiro, S. Saito, "The Long-term Corrosion Behaviour of Abandoned Wells Under CO2 Geological Storage Conditions: (2) Experimental Results for Corrosion of Casing Steel", *Energy Procedia* 37, pp. 5793-5803, 2013.

[2] H. M. A. El-Lateef, V. M. Abbasov, L. I. Aliyeva, T. A. Ismayilov. "Corrosion Protection of Steel Pipelines against CO2 Corrosion-A Review", *Chemistry Journal*, 2(02), pp. 52-63, 2012.

[3] D. Li, D. Han, L. Zhang, M. Lu, L. wang, W. Ma. "Effects of Temperature on CO2 Corrosion of Tubing and Casing Steel", *NACE Int.* NACE-2013-2426, 2013.

[4] B. Li, H. Li, F. Zhou, B. Guo, X. Chang, "Effect of cement sheath induced stress on well integrity assessment in carbon sequestration fields" *J. Nat. Gas Sci. and Eng*. 46, pp 132-142, 2017.

[5] A. S da Silva, D. Schebath, T. Z. Phyoe, S. Kapoor, E. H. Albuja, M. Sajid, " Engineered Flexible Expandable High-Density Cement System in Loss Prone High-Pressure Salt Section in Northern UAE" *Society of Petroleum Engineers* SPE-193292-MS, 2018.

[6] M. Bai, J. Sun, K. Song, L. Li, and Z. Qiao, "Well completion and integrity evaluation for CO 2 injection wells," *Renew. Sustain. Energy Rev.*, vol. 45, pp. 556–564, 2015.

[7] R. Nyborg, "Field Data Collection, Evaluation and Use for Corrosivity Prediction and Validation of Models," *NACE Int. Corros.* NACE-06118, 2006.

[8] E. Allen and A. Sturgeon, "Pipeline Corrosion – Can it always be modelled with confidence ?," *SUT/Hysoc Pipeline Design & Integrity Seminar* - 17th June, 2015.

[9] L. Smith, D. Milanovic, C. H. Lee, M. Billingham, "Modeling and Prediction of the Corrosion of Onshore Well Casings," *NACE Intl. Corros.* NACE-2012-1378, 2012.

[10] Y. S. Choi, D. Young, S. Nesic, L. G. S. Gray, "Wellbore integrity and corrosion of carbon steel in CO_2 geologic storage environments : A literature review", *Int. J. Greenh. Gas Con.* vol.16, pp. S70-S77, 2013.

[11] M. Bai, Z. Zhang, and X. Fu, "A review on well integrity issues for CO 2 geological storage and enhanced gas recovery," *Renew. Sustain. Energy Rev.*, vol. 59, pp. 920–926, 2016.

[12] S. Nešić, "Key issues related to modelling of internal corrosion of oil and gas pipelines - A review," *Corros. Sci.*, vol. 49, pp. 4308–4338, 2007.

[13] M. E. Mohyaldin and N. Elkhatib, "Coupling Norsok CO2 Corrosion Prediction Model With Pipelines Thermal / Hydraulic Models To Simulate CO2 Corrosion Along Pipelines," *Eng. Sci. Technol.* vol. 6, pp. 709–719, 2011.

[14] M. B. Kermani and A. Morshed, "Carbon Dioxide Corrosion in Oil and Gas Production—A Compendium," *Corrosion*, vol. 59, no. 8, pp. 659–683, 2003.

[15] Z. Zhi, L. Jing, Z. Yushan, Z. Cheng, Z. Qingsheng, and Z. Dezhi, "Journal of Petroleum Science and Engineering Finite service life evaluation method of production casing for sour-gas wells," *J. Pet. Sci. Eng.*, vol. 165, pp. 171–180, 2018.

AUTHOR INDEX

Institute of Physics
Dirac House, Temple Back
Bristol BS1 6BE UK

ISSN: 1757-8981
ISBN 978-1-7138-0895-4

9 781713 808954

CURRAN ASSOCIATES INC.
proceedings.com